PROCEEDINGS OF SYMPOSIA
IN PURE MATHEMATICS
Volume XIII, Part I

AXIOMATIC SET THEORY

AMERICAN MATHEMATICAL SOCIETY
Providence, Rhode Island
1971

Proceedings of the Symposium in Pure Mathematics
of the American Mathematical Society

Held at the University of California
Los Angeles, California
July 10–August 5, 1967

Prepared by the American Mathematical Society
under National Science Foundation Grant GP-6698

Edited by

DANA S. SCOTT

AMS 1970 *Subject Classifications*
Primary 02-02, 04-02, 02K15
Secondary 02B15, 02B25, 02C15, 02F27, 02F29, 02F35, 02H13, 02J05, 02K05, 02K10, 02K20.
02K25, 02K30, 02K35, 04A10, 04A15, 04A20, 04A25, 04A30, 18A15, 28A05, 28A10 28A60

International Standard Book Number 0-8218-0245-3
Library of Congress Catalog Number 78-125172

Printed in the United States of America

TABLE OF CONTENTS

FOREWORD

The Fourteenth Annual Summer Research Institute, sponsored by the American Mathematical Society and the Association for Symbolic Logic, was devoted to Axiomatic Set Theory. Financial support was provided by a grant from the National Science Foundation. The institute was held at the University of California, Los Angeles, from July 10 to August 5, 1967, and was attended by more than 125 participants. The Organizing Committee consisted of Paul J. Cohen, Abraham Robinson (chairman), and Dana S. Scott (editor). Special thanks are due to the Department of Mathematics of UCLA for providing facilities and assistance which contributed in large measure to the excellent success of the meeting.

The program for the four weeks of the institute was organized into two ten-lecture series, given by Dana S. Scott and Joseph R. Shoenfield, plus individual contributions generally in one-hour sessions at the rate of four lectures per day. By the last week this was reduced to three per day, as the strength of the participants had noticeably weakened. Nevertheless, most of the success of the institute was due to the fact that nearly everyone attended all of the sessions.

The papers in this volume of the proceedings represent revised and generally more detailed versions of the lectures presented at the institute. In view of the large number of papers, which resulted in delaying the receipt of papers from some authors, it was felt advisable to divide the proceedings into two volumes so as not to delay the publication of these papers any longer.

DANA S. SCOTT

SETS CONSTRUCTIBLE USING $L_{\kappa\kappa}$

C. C. CHANG[1]

Introduction. At the end of 1966, I started thinking about using the infinitary languages $L_{\kappa\kappa}$, κ a regular infinite cardinal, to form a hierarchy much in the same way Gödel [5] formed the ramified heirarchy of sets from the usual first-order language $L_{\omega\omega}$. At that time I was interested in locating more precisely the Hanf number of $L_{\omega_1\omega_1}$, and I had just proved some results about the behavior of the ramified hierarchy with respect to the infinitary languages $L_{\lambda\kappa}$ under the assumption of existence of large cardinals. (See Chang [1] and Theorem IX of this paper.) Therefore, I thought it was significant to look more closely at the hierarchy of κ-constructible sets. At the very beginning, the concept and notion of κ-constructibility seemed difficult to handle, as one was always afraid of saying something nonsensical like, for example, that an infinitary formula is absolute. Soon after, however, I realized that practically all of the Gödel results on the ramified hierarchy have natural translations to the hierarchy of κ-constructible sets. The most important of these is, of course, the isomorphism theorem which has an analog given here in Theorem V. From this one can deduce a version of the GCH, Theorem VI, assuming that all sets are κ-constructible. Furthermore, it turns out that the construction of Scott [9] using measurable cardinals also generalizes, thus yielding the present Theorem VIII. The solution of these easy problems led to other problems about κ-constructible sets. Many of these have been solved by other people since the Set Theory conference at UCLA.

In this paper, I shall first give the basic notions and definitions, then state the list of fairly standard results taken from the ramified hierarchy but given here for

[1] The reserach and preparation of this paper was partially supported by NSF Grants #5600, #8827, and partially by a Fulbright Grant.

κ-constructible sets. In the remarks at the end, I shall outline some more recent results about this notion and raise some further problems.

Basic notions and definitions. Unless specifically stated otherwise, our theorems are proved in ZFC (Zermelo-Fraenkel with choice). Notation: κ, λ, μ range over infinite regular cardinals, ν ranges over infinite cardinals, and α, ξ, η range over ordinals. The power set of a set x is Sx, the cardinal of a set X is $|X|$, the cardinal successor of ν is ν^+, and 2^ν is the cardinal of $S\nu$. $L_{\kappa\kappa}$ is the usual infinitary language formalized within ZFC with:

$$\begin{array}{ll} \text{predicates:} & \in, =; \\ \text{variables:} & v_\xi,\ \xi < \kappa; \\ \text{connectives:} & \neg,\ \bigwedge_{\xi<\nu<\kappa},\ \bigvee_{\xi<\nu<\kappa}; \\ \text{quantifiers:} & \forall_{\xi<\nu<\kappa},\ \exists_{\xi<\nu<\kappa}. \end{array}$$

Atomic formulas are sequences $v_\xi \in v_\eta$ and $v_\xi = v_\eta$, and formulas of $L_{\kappa\kappa}$ are obtained in the usual way from the atomic formulas by closure under the connectives and quantifiers. Because κ is regular, each formula of $L_{\kappa\kappa}$ has fewer than κ free variables. Let Form^κ be the set of all formulas of $L_{\kappa\kappa}$. We require that Form^κ be not too complicated and that if $\lambda < \kappa$ then $\mathrm{Form}^\lambda \subset \mathrm{Form}^\kappa$. $L_{\omega\omega}$ is the usual language of set theory and Form^ω is the usual set of finitary formulas. We assume that the reader is familiar with the usual interpretations of the connectives and quantifiers in any model $\mathfrak{A} = \langle A, E\rangle$. We use the notation $\mathfrak{A} \equiv_\kappa \mathfrak{B}$ and $\mathfrak{A} \prec_\kappa \mathfrak{B}$ for $\mathfrak{A}$ and $\mathfrak{B}$ are $L_{\kappa\kappa}$-equivalent and $\mathfrak{A}$ is an $L_{\kappa\kappa}$-elementary submodel of $\mathfrak{B}$, respectively.

The predicate $\mathrm{Sat}^\kappa(x, \phi, A)$ shall mean: the sequence $x \in {}^\kappa A$ which is eventually zero satisfies the formula $\phi \in \mathrm{Form}^\kappa$ under the interpretation of x_ξ for v_ξ in the model $\langle A, \epsilon_A\rangle$. (For our purpose 0 shall always belong to A.) Given a set A, we let

$$D^\kappa A = \{\{t \in A : \mathrm{Sat}^\kappa(t ^\frown x, \phi, A)\} : x \in {}^\kappa A \text{ eventually zero and } \phi \in \mathrm{Form}^\kappa\}.$$

$D^\kappa A$ is read as the set of all κ-definable subsets of A. Note that Sat^κ and D^κ are, respectively, a predicate and a term in the original (finite) language of ZFC. Specializing κ to ω we get the usual definitions of satisfaction and the set of all definable subsets. By transfinite induction we define the term C^κ_α, depending on α and carrying κ as a parameter as follows:

$$\begin{aligned} C^\kappa_0 &= 0; \\ C^\kappa_{\alpha+1} &= D^\kappa C^\kappa_\alpha; \\ C^\kappa_\eta &= \textstyle\bigcup_{\alpha<\eta} C^\kappa_\alpha \quad \text{if } \eta \text{ is a limit ordinal.} \end{aligned}$$

We let C^κ denote the class of κ-constructible sets, namely

$$C^\kappa(x) \leftrightarrow \exists\alpha(x \in C^\kappa_\alpha).$$

The axiom of κ-constructibility is the formula $\forall x C^\kappa(x)$. Note again that this formula has κ as a parameter. If κ is definable in ZFC then the axiom can be written as a sentence.

We list some simple properties of C^κ. (R_α denotes the sets of rank less than α: Ord is the class of all ordinals.)

$$\text{Form}^\kappa \subset C^\kappa_\kappa;$$
$$C^\kappa_\alpha = R_\alpha \text{ if } \alpha \leq \omega; \quad C^\kappa_\alpha \subset R_\alpha;$$
$$C^\kappa_\alpha \text{ is transitive; if } \alpha < \beta \text{ then } C^\kappa_\alpha \subset C^\kappa_\beta; \quad C^\kappa \text{ is transitive;}$$
$$\alpha \subset C^\kappa_\alpha; \quad \alpha \in C^\kappa_{\alpha+1} - C^\kappa_\alpha; \quad \text{Ord} \subset C^\kappa;$$
$$C^\omega = L,\ C^\omega_\alpha = L_\alpha, \text{ the Gödel ramified hierarchy.}$$

Results. After each result is stated we shall indicate the proof and give credit where it is due.

THEOREM I. $(\text{ZF})^{(C^\kappa)}$.

The proof here is quite standard. One uses the fact that C^κ has certain closure properties.

THEOREM II. (i) $[\forall x C^\kappa(x)]^{(C^\kappa)}$.

(ii) *C^κ is the least transitive model M of* ZF *containing all the ordinals and closed under less than κ termed sequences.*

For II(i) we show that Sat^κ, D^κ, and, finally, C^κ are absolute in C^κ. Showing that these notions are also absolute in M proves II(ii). I am indebted to Lévy for pointing out II(ii); II(ii) was also noticed by several people independently. In certain cases, for example $\kappa = \omega$, we may substitute for κ the definition of κ inside the square brackets. The fact that II(ii) is true shows that the transfinite hierarchy defined by

$$M^\kappa_0 = 0,$$
$$M^\kappa_{\alpha+1} = D^\omega M_\alpha \cup \{x \in {}^\nu M_\alpha : \nu < \kappa\},$$
$$M^\kappa_\eta = \textstyle\bigcup_{\alpha<\eta} M^\kappa_\alpha,$$
$$M^\kappa = \textstyle\bigcup_\alpha M^\kappa_\alpha,$$

is such that $M^\kappa = C^\kappa$. One can show that there are arbitrarily large ordinals α such that $M^\kappa_\alpha = C^\kappa_\alpha$; however, it is not true that $M^\kappa_\alpha = C^\kappa_\alpha$ for all ordinals α. In general most of our results have analogs in this new hierarchy.

THEOREM III. (i) $\lambda < \kappa \to C^\lambda \subset C^\kappa$;

(ii) $\lambda < \kappa \to (\forall x C^\lambda(x) \to \forall x C^\kappa(x))$;

(iii) $\forall x \exists \kappa C^\kappa(x)$.

III(i) is proved by showing that Sat^λ, D^λ, C^λ are absolute in C^κ. III(iii) follows from the axiom of regularity.

THEOREM IV. *Let M be a countable transitive model of* ZFC *and let κ be any infinite regular cardinal of M. Then there are Cohen extensions N of M such that* $\text{card}^N = \text{card}^M$ *which satisfy the following:*

(i) $\text{ZF} + \text{GCH} + \forall\lambda, \mu(\lambda < \mu \to C^\lambda \neq C^\mu)$;

(ii) $\text{ZF} + \text{GCH} + \forall x C^{\kappa^+}(x) + \neg\forall x C^\kappa(x)$;

(iii) $\text{ZF} + (\neg\text{AXCH}) + \forall x C^\kappa(x) + \forall\lambda < \kappa\, \neg\forall x C^\lambda(x)$.

These results follow easily from the works of Cohen [**2**] and Easton [**3**], as was pointed out to me by several people including Lévy, Silver, and Solovay. A question that has not yet been settled is whether IV(iii) can be improved to: there is an extension N of M which satisfies:

(iv) $\mathrm{ZFC} + \forall x C^{\kappa}(x) + \forall \lambda < \kappa \neg \forall x C^{\lambda}(x)$.

THEOREM V. *Suppose that $\kappa < \eta$ and η is a limit ordinal.*
(i) *If $\kappa = \omega$, $\mathfrak{A} = \langle A, E\rangle$ is well founded, and $\mathfrak{A} \equiv_{\kappa} C^{\kappa}_{\eta}$, then $\exists \alpha(\mathfrak{A} \cong C^{\kappa}_{\alpha})$.*
(ii) *If $\kappa = \nu^{+}$ and $\mathfrak{A} \equiv_{\kappa} C^{\kappa}_{\eta}$, then $\exists \alpha(\mathfrak{A} \cong C^{\kappa}_{\alpha})$.*
(iii) *If $\mathfrak{A} \prec_{\kappa} C^{\kappa}_{\eta}$ and $\kappa \in A$, then $\exists \alpha(\mathfrak{A} \cong C^{\kappa}_{\alpha})$.*

The proof employs the downward Löwenheim-Skolem theorem for $L_{\kappa\kappa}$ and the absoluteness of Sat^{κ}, D^{κ}, C^{κ} in certain transitive models. The use of the downward Löwenheim-Skolem theorem is the first place where we require the axiom of choice. One does not in fact need the full axiom of choice. For example, if $\kappa = \omega_1$, then it is sufficient to assume the axiom of dependent choice.

THEOREM VI. *Let $|\xi|^{\underset{\smile}{\kappa}}$ denote the cardinal sum $\sum_{\nu<\kappa}|\xi|^{\nu}$.*
(i) $\kappa \leq \xi \rightarrow |C^{\kappa}_{\xi}| \leq |\xi|^{\underset{\smile}{\kappa}}$;
(ii) $x \subset C^{\kappa}_{\xi}$ and $C^{\kappa}(x) \rightarrow \exists \eta < [|\xi|^{\underset{\smile}{\kappa}}]^{+} (x \in C^{\kappa}_{\eta})$;
(iii) $\forall x C^{\kappa}(x) \rightarrow \forall \nu \geq \kappa\, (2^{\nu} \leq [\nu^{\underset{\smile}{\kappa}}]^{+})$;
(iv) $\forall x C^{\kappa}(x) \rightarrow 2^{[\nu^{\underset{\smile}{\kappa}}]} = [\nu^{\underset{\smile}{\kappa}}]^{+}$;
(v) $\forall x C^{\omega}(x) \rightarrow \mathrm{GCH}$.

The proof of VI(ii) requires Theorem V. All other parts are straightforward. VI(iv) shows that $\forall x C^{\kappa}(x)$ implies that the GCH holds on a cofinal class of cardinals. VI(v) is of course the famous result of Gödel; it can be proved without the axiom of choice because there is a natural definable well-ordering of C^{ω}. There appears to be no way in which C^{κ}, for $\kappa \neq \omega$, can inherit a 'nice' well-ordering from the universe.

Exploiting an idea of Keisler-Rowbottom [**6**], we have

THEOREM VII. *Let λ be a cardinal such that no cardinal $\mu \leq \lambda$ is measurable, and let κ be such that $\mu^{\underset{\smile}{\lambda}} < \kappa$ for all $\mu < \kappa$. Suppose that $\forall x C^{\lambda}(x)$. Then the model $\mathfrak{A} = \langle C^{\lambda}_{\kappa}, \in, \lambda\rangle$ is a model of power κ which has no proper $L_{\lambda\lambda}$-elementary submodels of power κ.*

The proof of Theorem VII leans heavily on Theorems V and VI. Notice that in case $\lambda = \omega$, then assuming that, $\forall x C^{\omega}(x)$ (or, in other words, $V = L$), there is a Jónsson algebra in every infinite power.

We write κ is m.c. for κ is a (two-valued) measurable cardinal. The next theorem is motivated by the construction and result of Scott [**9**].

THEOREM VIII. *Suppose that κ is m.c. Then $\neg \forall x C^{\kappa^{+}}(x)$; in fact, if D is any κ-complete ultrafilter over κ, then $\neg C^{\kappa^{+}}(D)$.*

This is an improvement of a result of Scott. I first only proved the weaker result where κ^{+} above is replaced by any $\lambda < \kappa$. The truth of the present stronger form of

Theorem VIII was pointed out to me (separately and independently) by Lévy, Silver and Solovay. Using the known result (Silver, Jensen; see their papers in this volume) that it is consistent to assume GCH together with measurable cardinals, it is easy to show that Theorem VIII now is the best possible result, as obviously $C^{\kappa^{++}}(D)$. The proof uses the Scott ultrapower construction and the facts that

(i) the ultrapower of the universe using κ and D is isomorphic to a transitive model M of ZF containing all ordinals and closed under κ-termed sequences, and

(ii) $D \notin M$.

The result now follows from II(ii).

A consequence of Theorem VIII is that if the class of m.c. is cofinal in the class of ordinals, then the universe is not λ-constructible from any set. Compare this result with one of Vopěnka which says that if there exists a strongly compact cardinal then the universe is not ω-constructible from any set. Kunen has pointed out that the method of Vopěnka and Hrabáček also establishes that: if κ is strongly compact, then the universe is not κ^+-constructible from any set.

Finally we mention a very special result concerning the hierarchy C^ω_α. We assume the reader is familiar with the partition relations $(\lambda \to \kappa^{<\omega})$.

THEOREM IX. *Suppose there exists a cardinal ν such that $(\nu \to \omega_1^{<\omega})$. Let $\kappa < \lambda < \mu$. Then C^ω_λ is an $L_{\lambda\kappa}$-elementary submodel of C^ω_μ.*

Theorem IX improves results due to Rowbottom [8], Gaifman [4], and Silver [10]. We point out that the formulas of $L_{\lambda\kappa}$ are closed under fewer than λ conjunctions and disjunctions and fewer than κ quantifications, and are assumed to have fewer than κ free variables. By putting in various values for κ, λ, μ in Theorem IX we see that in particular we have

$$C^\omega_{\omega_1} \text{ is an } L_{\omega_1\omega}\text{-elementary submodel of } C^\omega$$

and

$$C^\omega_{\omega_2} \text{ is an } L_{\omega_2\omega_1}\text{-elementary submodel of } C^\omega.$$

It follows that any ordinal α in C^ω which is definable by a formula of $L_{\omega_1\omega_1}$ must be such that $\alpha < \omega_2$.

REMARKS. We use these remarks to state some problems and to communicate some latest (unpublished) results. I am indebted to Silver and, more recently, to Kunen for clarifying many points in remarks (b) and (c).

REMARK (a). Assume the hypothesis of Theorem IX. Solovay [11] has shown that

$$S\omega \cap C^\omega \notin \Delta^1_3.$$

We ask here two questions. Under the stronger conclusions of IX can one prove something more about the set of Gödel numbers of formulas satisfied by increasing sequences of indiscernibles? Secondly, are there any "interesting" ordinals which are definable in C^ω by means of a (real) $L_{\omega_1\omega_1}$ formula? We feel that such ordinals are hard to find because there is a difference between $L_{\omega_1\omega_1}$ and $(L_{\omega_1\omega_1})^{C^\omega}$, whereas $L_{\omega\omega} = (L_{\omega\omega})^{C^\omega}$.

REMARK (b). Again, under the hypothesis of IX one can show that

$$|S\omega \cap C^{\omega}| = \omega \text{ and, in fact, } C^{\omega}_{\omega_1} \prec_{\omega} C^{\omega}.$$

Suppose we assume the existence of a m.c. Then can we prove that

$$C^{\omega_1}_{\omega_2} \prec_{\omega} C^{\omega_1}?$$

This was first answered in the negative by Silver using a lemma of Solovay. More precisely Silver showed that

(1) if (ZFC + $\exists$ m.c.) is consistent then so is

(i) (ZFC + $\exists$ m.c. + $\neg(C^{\omega_1}_{\omega} \prec_{\omega} C^{\omega_1})$ + $|S\omega_1 \cap C^{\omega_1}| > \omega_1$).

This result shows that it is impossible to extend Gaifman's method and result for C^{ω} to C^{ω_1}. More recently Kunen [7] has shown that (1) can be improved so that the conclusion (i) now reads as

(ii) (ZF + GCH + $\exists$m.c. κ + $\forall \lambda \leq \kappa^{+} \neg(C^{\omega_1}_{\lambda} \prec_{\omega} C^{\omega_1})$ + $S\kappa \cap C^{\omega_1} = S\kappa$).

Kunen accomplishes this by using the model $L[U]$ where U is a normal ultrafilter over the measurable cardinal κ. Thus the existence of m.c. κ and the GCH do not cut down the size of $S\kappa$ in C^{ω_1} and do not allow us to conclude that some small initial segment of C^{ω_1} is an elementary submodel of C^{ω_1}. This is quite interesting as compared with Theorem IX.

REMARK (c). It was proved in Silver's thesis **[10]** that

(2) if (ZF + $\exists \nu(\nu \to \omega^{<\omega})$) is consistent then so is

$$(\text{ZF} + \forall x C^{\omega}(x) + \exists \nu(\nu \to \omega^{<\omega})).$$

Of course, it already follows from the work of Rowbottom **[8]** that

$$(\text{ZF} + \forall x C^{\omega}(x) + \exists \nu(\nu \to \omega_1^{<\omega}))$$

is inconsistent. It is implicit in Silver's thesis (pointed out by Silver) that

(3) if (ZF + $\exists$ m.c.) is consistent, then so is

$$(\text{ZF} + \forall x C^{\omega_1}(x) + \exists \nu(\nu \to \nu^{<\omega})).$$

In [7], Kunen has improved the conclusion of (3) to:

$$(\text{ZF} + \text{GCH} + \forall x C^{\omega_1}(x) + \exists \nu(\nu \to \nu^{<\omega})).$$

He uses the model $L[U] \cap R_{\kappa}$, where U is a normal ultrafilter over the measurable cardinal κ. This result shows that the Rowbottom technique for cutting down the size of C^{ω} using Ramsey cardinals fails for C^{ω_1} even in the presence of GCH. Thus Remarks (b) and (c) show that the original Scott method is (so far) the only one that is applicable to show that C^{ω_1} is not the universe. This will have to stand until (possibly) a new class of large cardinals is discovered which lies between $\kappa \to \kappa^{<\omega}$ and κ m.c.; unless of course one can prove that κ = 1st m.c. is in some sense the smallest cardinal whose existence reduces C^{ω_1}. One technical problem remains:

Can the hypothesis of (3) be weakened to the consistency of (ZF + $\exists \nu(\nu \to \nu^{<\omega})$)?

REMARK (d). Let M be a countable transitive model of ZF + $V = L$. Let α be the set of all ordinals in M. Then, as is well known, we can find a generic subset G of ω over M and a model $N = L_\alpha[G]$ (the sets constructible from G up to rank α) such that

(i) N is an extension of M;
(ii) N is a countable transitive model of ZF;
(iii) M and N have the same ordinals and cardinals;
(iv) M is the constructible universe in N;
(v) $G \in N$ and G is not constructible in N.

It turns out that with very little modification of the original argument of Cohen **[2]**, one can establish results of the following kind. Let $L_\alpha^{\omega_1}[G]$ denote the set of all sets ω_1-constructible from G, up to rank α. Assume $2^\omega = \omega_1$. Let M be a transitive model of ZF of power ω_1 such that

$$M \text{ is closed under countable sequences}$$

and

$$M \text{ satisfies } \forall x C^{\omega_1}(x).$$

Let α be the set of all ordinals in M. Then there is a generic (in the sense of $L_{\omega_1\omega_1}$) subset G of ω_1 over M and a model $N = L_\alpha^{\omega_1}[G]$ such that

(i′) N is an extension of M closed under countable sequences;
(ii′) N is a transitive model of ZF of power ω_1;
(iii′) M and N have the same ordinals and cardinals;
(iv′) M is the class of all ω_1-constructible sets in N;
(v′) $G \in N$ and G is not ω_1-constructible in N.

The proof is an exact analog of Cohen's proof **[2]**, except that ω is replaced by ω_1, $L_{\omega\omega}$ by $L_{\omega_1\omega_1}$, and finite conditions by countable conditions.

This simple remark allows us to infer that Cohen's method (together with $2^\omega = \omega_1$) will give us new transitive models of power ω_1 closed under countable sequences from old transitive models of power ω_1 closed under countable sequences. I do not know if this observation will help in establishing new independence results.

REMARK (e). Recalling the notions of OD, HOD, and second order definability (see the Myhill-Scott paper in this volume), we see that OD and HOD have natural extensions to OD^κ and HOD^κ where κ is a parameter and formulas of $L_{\kappa\kappa}$ are used. Are there any theorems relating these new notions to C^κ?

REMARK (f). Kripke's paper on admissible ordinals (Lecture Notes, Summer Institute on Axiomatic Set Theory, University of California, 1967) shows that there are many results of the form $\Phi(\alpha) \leftrightarrow \Psi(C_\alpha^\omega)$ where $\Phi(\alpha)$ is a property of α stated in terms of Kripke's equational calculus, and $\Psi(C_\alpha^\omega)$ is a model-theoretical property of C_α^ω. Suppose that Kripke's equational calculus can be extended in a natural manner to an (even more) infinitary setting so that the notion of κ-admissibility makes sense. Then can one prove similar theorems of the sort $\Phi(\alpha) \leftrightarrow \Psi(C_\alpha^\kappa)$?

REMARK (g). Finally, on a more philosophical level, on my visit to Warsaw in May, 1967, Mostowski and some of his students asked if it is not more natural to study κ-constructibility using a system of ZF which is itself infinitary. Suppose we call such a system ZF^{κ} (the precise form of ZF^{κ} is difficult to guess at) then is it possible that better theorems about C^{κ} can be obtained in ZF^{κ}? In particular, can one show that the AXCH holds in C^{ω_1} using ZF^{ω_1}?

References

1. C. C. Chang, *Infinitary properties of models generated from indiscernibles*, to appear in Proc. of 1967 Internat. Congr. Logic, Methodology, and Philos. Sci., Amsterdam.

2. P. J. Cohen, *Set theory and the continuum hypothesis*, Benjamin, New York, 1966.

3. W. Easton, *Powers of regular cardinals*, Ph.D. thesis, Princeton Univ., Princeton, N.J., 1964.

4. H. Gaifman, *Measurable cardinals and constructible sets*, Amer. Math. Soc. Notices **11** (1964), 770.

5. K. Gödel, *The consistency of the axiom of choice and of the generalized continuum hypothesis with the axioms of set theory*, Ann. of Math. Studies No. 3, Princeton Univ. Press, Princeton, N.J., 1940.

6. H. J. Keisler and F. Rowbottom, *Constructible sets and weakly compact cardinals*, Amer. Math. Soc. Notices **12** (1965), 373.

7. K. Kunen, *Inaccessibility properties of cardinals*, Ph.D. thesis, Stanford Univ., Stanford, Calif., 1968.

8. F. Rowbottom, *Large cardinals and small constructible sets*, Ph.D. thesis, Univ. of Wisconsin, 1964.

9. D. Scott, *Measurable cardinals and constructible sets*, Bull. Acad. Polon. Sci., Sér. Sci-Math., Astronom. Phys. **7** (1961), 145–149.

10. J. Silver, *Some applications of model theory in set theory*, Ph.D. thesis, Univ. of California, Berkeley, 1966.

11. R. Solovay, *A Δ^1_3 nonconstructible set of integers*, Trans. Amer. Math. Soc. **127** (1967), 50–75.

University of California, Los Angeles

COMMENTS ON THE FOUNDATIONS OF SET THEORY

PAUL J. COHEN[1]

It is, of course, obvious that to give a talk on the philosophical problems of set theory is not quite the same as to give one on set theory itself. For me, at least, it is a rather strange and uncomfortable position. I am struck at once by the futility of trying to state opinions which will be universally or even widely accepted, and also by the inconsistencies and difficulties of my own point of view. Certainly the perilous transition from mathematics to philosophy has customarily been made at a somewhat later stage in one's career. Finally, to compound the problem, there is the enormous difficulty of contributing something new to this ancient discourse. For, on such fundamental problems, it is my feeling that technical achievements shed little light, although they may certainly affect prevailing opinion. Despite all these objections, I find that I do have a certain enthusiasm to state my opinions, not too dogmatically I trust, and to point out some things which I think should be said. We are still close enough to the time of the fundamental discoveries in logic to share the thrill of groping with these problems and perhaps the upsurge in activity in set theory, as exemplified by this meeting, contributes further to our sense of enthusiasm. There does, however, seem to be some change in the tone of philosophical discussions nowadays. Perhaps it is due to the fact that mathematicians have already spent themselves in the violent polemics of the past, or maybe it is merely that their audience has tired of the debate, but in any event one is more accustomed to having opinions stated without an attempt to convert the listener on the spot. It is in this spirit that I too wish to proceed and I earnestly assure my listeners of my tolerance of other points of view.

[1] I wish to acknowledge the support of the Air Force Office of Scientific Research through their grant AF-AFOSR-1071-66 and of the Office of Naval Research through contract NR-043-317.

Although I cannot conceive of what might be called "true" progress in foundational questions, it is quite interesting from a historical point of view, to see how various generations have expressed themselves and to conjecture, if we can, how their opinions have been colored by the prevailing spirit of the times. My own prejudice is to see mathematics as a highly personal endeavor and not as an impersonal advance of knowledge immune from all human weaknesses. Thus, the attitudes that people profess towards the foundations seem to be greatly influenced by their training and their environment. It seems to me that the desire to accept principles which lead to beautiful and interesting mathematics in the past has certainly overcome many serious criticisms. In this talk I should like to point out similar tendencies which exist today.

In the past, much discussion has centered about points which I shall quite arbitrarily choose to ignore, for example, the use of the law of the excluded middle. Although the law of the excluded middle is related to problems in set theory such as the use of impredicative definitions, it is not a question in set theory itself and we shall ignore it. We shall also ignore all other questions about the use of the predicate calculus, as well as questions about the nature of the formalization of mathematics, and all questions of a purely philosophical nature having little to do with the special nature of mathematics. For me the essential point is the existence of infinite totalities. The attitude towards infinite sets has traditionally been the great dividing line between mathematicians. The famous antinomies in logic never played a role in mathematics simply because they were totally alien to the type of reasoning one normally uses. One never considers all possible objects in the universe nor the length of descriptions, etc. These difficulties belong more properly to the development of the notion of a formal system. Similarly, Zeno's paradoxes do not really strike us as presenting serious difficulties in the way that they were intended. In general, I would say that many of these problems were historically associated with the transition period from classical philosophy to mathematics as we know it today.

There certainly are some cases in which the use of infinite sets presents no essential difficulties. For example, to say that a property holds for all integers or that it holds for all members of the set of integers, is clearly equivalent. Similarly, to say n belongs to the set of even integers is equivalent to saying that n is even. Thus the use of some sets can be avoided by referring back to a suitable property. If this could always be done we would have little problem. In number theory, if we do not allow the notion of an arbitrary set of integers, we must formulate the induction principle by stating it for every property we are concerned with. Nevertheless, because of the great complexity of set theory, especially its impredicative nature, we cannot simply think of sets as a kind of shorthand for properties. Nevertheless, the most powerful and characteristic axioms of set theory, the Power Set and Replacement Axioms, describe sets through properties and, of course, Gödel's constructible sets show that only sets which in some sense can be defined by properties need to be used to obtain a model for set theory. The fact that the Replacement Axiom is an infinite schema in some ways is a defect because it seems that we are too specifically allowing only certain properties, whereas what we might hope for is some basic description of how sets are formed. Of course, behind this is Gödel's Incompleteness Theorem by which no finitely axiomatizable system is complete. This theorem still is the greatest barrier to any attempt to totally understand the nature of infinite sets. At the same time, since it shows that higher infinities have repercussions in number theory, allowing us to prove otherwise unprovable statements, it makes it extremely difficult to maintain that higher infinities can merely be dismissed. Our familiarity with the Incompleteness Theorem should not make us lose sight of the great deficiency of all formal systems, a fact which is more far-reaching than the independence of particular statements such as the Continuum Hypothesis. It underlies my basic pessimism that no purely technical achievement will ever shed much light on the basic philosophical problems.

To the average mathematician who merely wants to know that his work is securely based, the most appealing choice is to avoid difficulties by means of Hilbert's program. Here one regards mathematics as a formal game and one is only concerned with the question of consistency. This point of view has perhaps become even more appealing with the passage of time, as the operational approach has spread to other fields, such as physics. That is, one is concerned only with the objects with which one has direct contact, which in mathematics is the formal language rather than infinite sets. Indeed, this Formalist Program of Hilbert still remains as the one totally precise (as distinct from correct) point of view in these matters. This is certainly an example of how time has done little to present new and original points of view on the Foundations. Of course, Formalism has its difficulties and before returning to it we shall examine its main alternative, a view which may be called Platonism, or as we prefer, Realism.

In the Realist philosophy, one wholeheartedly accepts traditional mathematics at face value. All questions such as the Continuum Hypothesis are either true or false in the real world despite their independence from various axiom systems. The Realist position is probably the one which most mathematicians would prefer to take. It is not until he becomes aware of some of the difficulties in set theory that he would even begin to question it. If these difficulties particularly upset him, he will rush to the shelter of Formalism, while his normal position will be somewhere between the two, trying to enjoy the best of two worlds. The chief advantage of Realism is that there is no necessity to justify the axioms of set theory. There is no need to show their consistency, and what is equally important to me, no need to explain why these particular axioms are so successful and worthy of special attention. Correspondingly, the greatest weakness of the Formalist position is to explain why the axioms of set theory, presumably reflecting no underlying reality, are able to prove arithmetical statements unprovable by more finitistic means. A weakness which I believe any realist would have to accept is his inability to explain the source of the never-ending sequence of higher axioms, such as the higher axioms of infinity. Certainly even the staunchest realist must flinch when contemplating cardinals of a sufficiently inaccessible type. Also there are axioms such as that of the Measurable Cardinal which are more powerful than the most general Axiom of Infinity yet considered, but for which there seems absolutely no intuitively convincing evidence for either rejection or acceptance. The recent independence results are another challenge to the Realist position. As regards the last, there are some people who feel some intuitively acceptable axiom may eventually settle the Continuum Hypothesis and similar problems. On the other hand, the Axiom of the Measurable Cardinal does not hold even the dimmest prospect for such a resolution and seems destined to be adopted by zealous set theoreticians as an axiom. Yet, even in this regard the Realist position is more enviable than that of the Formalist, since for the latter there are propositions in number theory itself, such as Consis (ZF), which are beyond resolution. An optimistic Realist position would seem to be that somehow Consis (ZF + Measurable Cardinal) will be reduced to the consistency of suitably powerful statements which will be regarded as Axioms of Infinity. The most optimistic position would be that every question of number theory can be decided by a suitable axiom of infinity.

Historically, mathematics does not seem to enjoy tolerating undecidable propositions. It may elevate such a proposition to the status of an axiom, and through repeated exposure it may become quite widely accepted. This is more or less the case with the Axiom of Choice. I would characterize this tendency quite simply as a form of opportunism. It is, of course, an impersonal and quite constructive opportunism. Nevertheless, the feeling that mathematics is a worthwhile and relevant activity should not completely erase in our minds an honest appreciation of the problems that beset us. In the case of the Continuum Hypothesis (CH), this tendency may possibly, though unlikely, lead to a splitting of set theory into branches depending on how one evaluates the power of the continuum. Taking a cynical view, one might say that Opportunism will judge philosophical questions from the point of view that mathematics should develop so as to give employment to the greatest number of mathematicians. Recently much work has been done on independence questions in set theory. A curious result has been that the greater facility for handling these questions has given people greater motivation to believe that set theory refers to "real" mathematical objects. It would be truly sad if this wave of success should succeed in totally dismissing all philosophical concern over CH and similar questions as inconsequential. Of course, good mathematics is beautiful, while most philosophical discussion is barren and certainly not beautiful.

From the Realist point of view one can speculate as to the fate of CH. It would seem that only axioms such as the Axiom of Constructibility which limit the nature of sets can possibly decide it. On the other hand, there seems little hope that such an axiom will come to be accepted as intuitively obvious. A more likely development is that its negation will be accepted as an axiom. The justification might be that the continuum, which is given to us by the power set construction, is not accessible by any process which attempts to build up cardinals from below by means of a construction based on the Replacement Axiom. Thus C would be considered greater than $\aleph_1$, $\aleph_n$, $\aleph_\omega$, etc. Of course, this is sheer speculation. Some attention has already been given to the technical consequences of various axioms which relate to CH. Although such work may be of great esthetic value, it seems highly unlikely that it can lead to basic philosophical clarification.

By now it may have become obvious that I have chosen the Formalist position. This is hardly a courageous choice since probably most of the famous mathematicians who have expressed themselves on the question have in one form or another rejected the Realist position. In my own case, I can say that it was the address by Abraham Robinson at Jerusalem in 1964 which caused me to make my choice quite explicit. It is a choice which carries with it certain heavy weights. Certainly one of the heaviest is the admission that CH, perhaps the first significant question about uncountable sets which can be asked, has no intrinsic meaning. If the Hilbert program had not been shattered by Gödel, life would be much more comfortable. I firmly believe that the Hilbert program can in no sense be resuscitated. There is always a great sense of frustration about consistency proofs and they seem to retain a circular character which makes them most unsatisfying.

As mentioned, the great weakness in the Formalist position is of course the fact that it must explain how the purely formal axioms which constitute set theory can be successful. My opinion, often expressed before, is that the axioms are an extrapolation from the language of more finitistic mathematics. The tendency to such an extrapolation is strong. To illustrate this, let me first remind you of a situation which all logicians eventually encounter. When speaking to competent mathematicians who are ignorant of logic, one finds that communication begins to break down whenever one begins speaking about formal systems and the analysis of how formulas are constructed. The mathematician would much rather speak about models of an axiom system than about the set of all formulas provable from those axioms. Of course, the Completeness Theorem says that the two points of view are equivalent. Yet there is a natural tendency to replace discussions of methods and statements by discussions of suitable abstractions which are considered as "objects". For example, the development of real analysis in the nineteenth century was marked by a change in attitude towards what constituted a function. A function had previously been thought of as an explicit rule which assigns numbers to numbers. Eventually a function was thought of as an independent entity with no explicit concern for how it is to be calculated. The continuous nondifferentiable function of Weierstrass had as legitimate an existence as $\sin x$. When Cantor first discussed set theory, probably much of the opposition to it was simply that only sets that have been explicitly defined can be legitimately discussed. As we all know, Cantor's point of view triumphed completely. The essential reason was probably ultimately one of convenience. Namely, it is simpler to speak about abstract sets than constantly to be concerned with specific constructions. A more recent example of the same tendency is category theory. Here one speaks of the category of groups, for example. One might ask, what is the advantage of saying "G belongs to the category of groups", rather than "G is a group". The answer is simply that it may be quite suggestive to transfer the methods from one category to another and even to prove general theorems about categories. Yet, unless I am not sufficiently aware of current trends, the set-theoretical difficulties in handling categories have not inspired many set theorists and it has had little impact on logic as a whole. Thus, although we have thoroughly accepted highly impredicative set theory because we understand its internal cogency, we, as logicians, are less likely to accept category theory whose roots lie in algebraic topology and algebraic geometry. It could be retorted that the existing axioms of infinity are ample to cover formalizations of category theory, yet an obstinate categorist could say that categories themselves should be accepted as primitive objects. In a sense, categories resemble the classes of Gödel-Bernays set theory. Here again, although classes merely axiomatize the infinite axiom scheme of Zermelo-Fraenkel set theory, they have gained wide acceptance as objects in themselves. Another example of how repeated exposure dulls the critical faculties is the Axiom of Inaccessible Cardinals. The usual justification given for assuming it is a negative one, namely, that it is unreasonable to think that every set is accessible. The analogy is drawn with the passage from finite to infinite sets. If one performs by induction a transfinite sequence of suitable closure operations, it is argued, we can still go further and presumably find an inaccessible cardinal beyond. However, I feel that this is a specious argument, since it tends rather to justify the existence of a standard model for set theory, which is an incomparably weaker assertion. A more honest reason for accepting inaccessible cardinals is that experience has shown that they do not lead to contradictions and we have developed some kind of intuition that no such contradiction exists.

Since I feel it incumbent on me, having chosen Formalism, to explain why I do not advocate abolishing all infinitistic mathematics, I should like to put forward the view that we do set theory because we feel we have an informal consistency proof for it. This feeling is based on the fact that in any given case we are only speaking about specific sets defined by properties and by tracing back a contradiction we could eventually reduce it to the integers. The use of impredicative definitions complicates our intuition since unrestricted impredicativity certainly leads to the well-known paradoxes. The usual Replacement Axiom still gives us some handle on a set to start our reduction, since every set in the set to be defined must be paired to some element of a previously given set. Since I have already stated that technical developments do not lead to basic clarification, I do not want to attempt to give a rigorous consistency proof based on some abstruse higher principle equivalent to ZF set theory. However, I shall give only an outline of the general framework in which this intuitive thinking is done.

One way to think of consistency proofs is as follows. One starts with finitely many axioms, say S_1. One invents a symbol for each set which is asserted to exist and replaces it in the corresponding statement to form a system of statements S_2. At S_k, we invent new symbols for all the sets asserted to exist and substitute them, and for each statement of the form $\forall x A(x)$ and for each symbol c already used we add $A(c)$. For some n we assume there is a contradiction among the quantifier free statements. For the sake of convenience we may, at certain stages, wish to split the proof into two branches; in one we adjoin some statement A and in another $\sim A$. We assume then that a contradiction is reached in both branches. Also we may simplify the contradiction by not adjoining all the statements in the above process. Our object is to sketch a reduction in the complexity of contradictions. We start with the symbols ϕ and ω. Suppose at some stage we encounter a set x_1 defined by the Replacement Axiom for some property. If another set x_2 eventually appears in the formula $x_2 \in x_1$, one can attempt to eliminate the use of x_1 by substituting the corresponding property of x_2. Then we split our proof according to the two cases that x_2 does or does not satisfy this property. If the set x_1 occurs later we attempt to replace the use of x_1 by the finite set of those sets occurring in the proof which are members of it. Of course to make this precise would require an analysis of the impredicative definitions used and an ordering of degrees of impredicativity. Since we know that the Incompleteness Theorem makes this quest essentially hopeless we do not pursue it. The key point is that an element of our new set must be tied to some element in a previous set so that the reduction can be continued. In the Russell Paradox there is a circularity which prevents this. As is well known, Gentzen carried out such a proof for number theory and was led to the ordinal ϵ_0. In the case of ZF, it is not clear if one can define an analogous ordinal. If one can, it would be interesting to study its relation to the known invariant of ZF, namely, the countable ordinal of the minimal model. The latter is the least ordinal such that M_α, the set given by Gödel's construction at stage α, is a model for ZF. The ordinal of proof theory should be smaller since it "constructs" the smallest nonstandard model of ZF.

Even at best the intuition I have tried to sketch can only handle the problems created by the Replacement Axiom. For inaccessible or measurable cardinals our intuition is probably not yet sufficiently developed or at least one cannot communicate it. Nevertheless I feel that this is a useful task, to develop our mystical feeling for which axioms should be accepted. Here, of course, we must abandon the scientific program entirely and return to an almost instinctual level, somewhat akin to the spirit with which man first began to think about mathematics. I, for one, cannot simply dismiss these questions of set theory for the simple reason of their reflections in number theory. I am aware that there would be few operational distinctions between my view and the Realist position. Nevertheless, I feel impelled to resist the great esthetic temptation to avoid all circumlocutions and to accept set theory as an existing reality.

The reader will undoubtedly sense the heavy note of pessimism which pervades these attidues. Yet mathematics may be likened to a Promethean labor, full of life, energy and great wonder, yet containing the seed of an overwhelming self-doubt. It is good that only rarely do we pause to review the situation and to set down our thoughts on these deepest questions. During the rest of our mathematical lives we watch and perhaps partake in the glorious procession. Great questions of set theory that seemed untouchable eventually yield. New axioms are investigated and larger and larger cardinals are somehow brought closer to our intuition. Through all this, number theory stands as a shining beacon. If, as I hope does not happen too often, our doubts begin to overpower us, we retreat back into the safe confines of number theory until refreshed, we venture out again into the unsafe ground of set theory. This is our fate, to live with doubts, to pursue a subject whose absoluteness we are not certain of, in short to realize that the only "true" science is itself of the same mortal, perhaps empirical, nature as all other human undertakings.

UNSOLVED PROBLEMS IN SET THEORY

P. ERDÖS AND A. HAJNAL

1. **Introduction.** Since 1958 we have published a number of joint papers on set theory [**1**] . . . [**12**] and some triple papers with R. Rado, E. C. Milner, J. Czipszer, G. Födor [**13**] . . . [**17**]. During this period we collected a fair amount of problems we could not solve. Some of them are stated in the papers we have published, some are connected with unpublished results of ours. We were both enthusiastic when we learned that the organizing committee of this symposium was willing to give us an opportunity to publish a paper on these problems.

After having started the work, we immediately realized that the task we have undertaken is not quite as easy and pleasant as we thought it was. First of all, we have problems of very different types.

(A) There are some which seem to be unsolvable, or connected with problems whose independence has already been proved.

(B) There are some which we tried very hard to solve and failed and that is why we feel they are difficult.

(C) There are some which seem to be difficult but we suspect that the difficulty is only technical.

(D) There are some which we only know of, we find them interesting but we simply did not have the time to look at them properly.

(E) There are some which would seem uninteresting to anyone who did not think about them and we would like to publish them all the same, since for one reason or other we are interested in the answer.

On the other hand, there are many cross connections between the different problems and, for lack of time and space, we will be able to give very few of them. In many cases, it will be difficult to tell whose problem it is we are going to state. There are many other problems which arose, but, to make a long story short, we have decided to accept the following principles.

We will restate here quite a lot of the published problems for two reasons. They seem to be more important than some of the unpublished ones, and it is quite difficult to understand the latter without knowing the former

We are going to collect all our problems of type (A), the time and the place being appropriate to put them in the hands of logicians.

We will make comments only on some carefully selected problems, and we will state a lot of others, giving only references (if any) and leave the reader to find out for himself what they are worth. The only help we can give is to indicate to which of the categories (A) . . . (E) we believe the problem stated belongs.

We will try our best to make the references and historical remarks as complete and as fair as possible. If nothing else is stated, we think that the problem in question is due to the two of us, except in §3 where all the problems, if not indicated differently, are due to P. Erdös and R. Rado.

If an open problem depends on several parameters, we usually formulate the instance of it which seems to be the simplest for us.

The order in which the problems are stated does not express any opinion on their importance. We will try to give them in some logical order and to avoid as much new notation as possible.

During our work in set theory, whenever we could not solve a problem, we tried to solve it assuming the generalized continuum hypothesis (G.C.H in what follows). If we still could not solve it, we said that even assuming G.C.H. we do not know the answer. This will be done in this paper too. The word "even" used here is not intended to express any considered opinions or preference. It just describes the way we have been thinking about these problems. If a problem is stated in §§1–7 under the assumption of G.C.H., it means that for various reasons the problem as stated there does not make sense if we do not assume G.C.H.

In the end of our paper we do not bother to state problems without assuming G.C.H., since this would complicate the notations, or the formulation. We think it will be clear for the reader in many cases how to formulate a corresponding problem without assuming G.C.H.

Though many of the difficulties mentioned before seem insurmountable, we still hope that this survey of our problems will be useful.

2. **Notations.** We are going to use the usual classical notations of set theory. It is not appropriate for our present purposes to identify cardinals with initial ordinals. However, finite cardinals and ordinals will not be distinguished.

We point out one difficulty. In many of the different papers on the subject, different kinds of $\rightarrow$ (arrow) relations are introduced and the same notation has been used for different purposes in different papers. Whenever an $\rightarrow$ relation occurs in the text, we will give its definition.

$a, b, c, \ldots, m, n, \ldots$	denote cardinals,
i, j, l	denote integers,
$\alpha, \beta, \gamma, \ldots, \xi, \zeta, \mu, \nu, \rho, \ldots$	denote ordinals.

If a is a cardinal, a^+ is the smallest cardinal greater than a. $\sum$ and $+$ are used to denote both cardinal and ordinal addition. If a is a cardinal, $\Omega(a)$ is its initial number.

By a graph $\mathcal{G} = \langle g, G\rangle$, we mean an ordered pair where g is the set of vertices, G is a set of subsets of two elements of g. The elements of G are the edges of $\mathcal{G}$. For a detailed explanation of the terminology we refer to our paper **[10]**.

3. **Problems formalized with the ordinary partition symbol.** If S is a set, r a cardinal, we put

$$[S]^r = \{X \subset S : |X| = r\}; \qquad [S]^{<r} = \{X \subset S : |X| < r\}.$$

If $[S]^r = \bigcup_{\nu<\varphi} \mathcal{T}_\nu$ the sequence $(\mathcal{T}_\nu)_{\nu<\varphi}$ is said to be an *r-partition of S of type* $|\varphi|$.

DEFINITION OF SYMBOL-I. Let a, b_ν, $\nu < \Omega(c)$ be cardinals or order types, and let c, r be cardinals. Assume further that each b_ν is a cardinal if a is a cardinal. We write $a \to (b_\nu)^r_c$ if the following statement is true.

Let S be a set if a is a cardinal and let S, $\prec$ be a (simply) ordered set if a is an order type, such that $|S| = a$ or typ $S(\prec) = a$ respectively. Let $(\mathcal{T}_\nu)_{\nu<\Omega(c)}$ be an r-partition of type c of S.

Then there exist a $\nu < \Omega(c)$ and a subset $S' \subset S$ such that $[S']^r \subset \mathcal{T}_\nu$ and $|S'| = b_\nu$ or typ $S'(\prec) = b_\nu$ if a is a cardinal or an order type, respectively.

We write $a \nrightarrow (b_\nu)^r_c$ (and in the case of all other symbols to be defined) if this statement is false.

If all the b_ν's equal b, we write $a \to (b)^r_c$. If $c = c_0 + \cdots + c_{n-1}$ and c_i of the b_ν's equal b_i for $i < n < \aleph_0$ we sometimes write

$$a \to ((b_0)_{c_0}, \ldots, (b_{n-1})_{c_{n-1}})^r.$$

If $c_i = 1$, we omit it. Using this terminology, Ramsey's classical theorem **[33]** can be expressed as follows:

$$\aleph_0 \to (\aleph_0)^r_k \quad \text{if} \quad k, r < \aleph_0.$$

P. Erdös and R. Rado were the first who started to investigate consciously and methodically the possible transfinite generalizations of this theorem, though several other people, e.g. D. Kurepa, have published results which can be expressed using the ordinary partition symbol. A survey of the history of the problem is given in [13]. Erdös and Rado have published with other authors a series of papers on this subject. The symbol in this generality was actually defined in [**21**]. In their paper [**21**] they gave a survey of the results and problems known at that time. It is fair to say that their work started all the investigations we are now talking about and though the different problems crystalized by theorems proved by different people it seems to be logical to attribute all the problems concerning Symbol-I (except those involving inaccessible cardinals) to P. Erdös and R. Rado.

3.1. *The ordinary partition symbol in the case of cardinals.* Note that by an early result of P. Erdös and R. Rado, we have

$$a \nrightarrow (\aleph_0, \aleph_0)^{\aleph_0} \quad \text{for every } a;$$

hence we can always assume that $r < \aleph_0$ and the case $r = 1$ is trivial in case of cardinals.

It was realized by P. Erdös and A. Tarski in 1942 **[23]** that while a direct generalization of Ramsey's theorem fails for cardinals not strongly inaccessible there might be cardinals for which $a \rightarrow (a, a)^2$ holds.

The history of this problem is well known and in this chapter we avoid mention of any problems for Symbol-I in which strongly inaccessible cardinals are involved.

In **[13]** with Rado we gave a discussion of Symbol-I for cardinals. Using G.C.H. our discussion is almost complete. See main Theorems I and II of **[13]** on pp. 130 and 138 respectively. The only unsolved problem not involving strongly inaccessible cardinals is highly technical.

Problem 1. *Assume G.C.H.*

$$\aleph_{\omega_{\omega+1}+1} \nrightarrow (\aleph_{\omega_{\omega+1}}, (4)_{\aleph_0})^3?$$

(See **[13**, Problem 2].)

Recently we have been investigating how far our results and methods cover the problems if we do not assume G.C.H. In case $r = 2$, $b_\nu \geq \aleph_0$ one can obtain a rather complete discussion. These results will be published in detail in a forthcoming book by the three of us. We would like to mention some of the open problems.

In **[13]** for obtaining negative partition relations our major tool was the negative stepping up lemma.

Problem 2 (Erdös, Hajnal, Rado). *Assume* $2 \leq r < \aleph_0$, $a \geq \aleph_0$, b_ν, $(\nu < \Omega(c))$ *are cardinals and* $a \nrightarrow (b_\nu)^r_c$ *holds. Does then* $2^a \nrightarrow (b_\nu + 1)^{r+1}_c$ *hold*? (*Here* + *denotes cardinal additions, i.e.,* $b_\nu + 1 = b_\nu$ *if* $b_\nu \geq \aleph_0$.)

Lemmas 5A, 5F of **[13]** give this result under different additional assumptions, e.g. if two of the b_ν's are infinite and one is regular. One might guess that this is a problem of type (C) and the answer is affirmative. The most difficult case is when $r = 2$, one b_ν is singular and the others are finite. In this case, we cannot prove the statement even assuming G.C.H. (see Problem 1).

Problem 3 (Erdös, Hajnal, Rado). *Assume that there is an increasing sequence of integers* $(n_k)_{k<\omega}$ *such that the sequence of cardinals* $2^{\aleph_{n_k}}$ *is strictly increasing and* $2^{\aleph_{n_0}} > \aleph_\omega$. *Does then*

$$a = \sum_{k<\omega} 2^{\aleph_{n_k}} \rightarrow (\aleph_\omega, \aleph_\omega)^2?$$

Note if $\aleph_\omega$ is replaced by $\aleph_\alpha$ we have $a \rightarrow \cdots$ or $a \nrightarrow \cdots$ if $\alpha < \omega$ or $\alpha > \omega$, respectively.

Problem 3 might be important, since (without speaking too precisely) it might give an opportunity to show that the truth value of $a \rightarrow (b, c)^2$, $a, b, c \geq \aleph_0$, cannot be computed from the function $\aleph_\alpha^{\aleph_\beta}$.

The only other typical instance in which we cannot tell the truth value of the $\overset{\mathrm{I}}{\rightarrow}$ in case $r = 2$ with infinite cardinal entries is the following:

Problem 4 (Erdös, Hajnal, Rado). *Put* $\varphi = \Omega((2^{\aleph_0})^+)$ *and assume that the sequence* $\aleph_\alpha^{\aleph_0}$, $\alpha < \varphi$, *is not eventually constant. Put* $a = \sum_{\alpha<\varphi} \aleph_\alpha^{\aleph_0}$. *Does then*

$$a \rightarrow (\aleph_\varphi, \aleph_1)^2$$

hold?

(The answer is affirmative if we assume G.C.H. or some other additional assumptions on the function $\aleph_\alpha^{\aleph_\beta}$.)

In case some of the b_ν's are finite many of our results given in **[13]** make use of G.C.H. heavily. It follows easily from the results of **[13]** that

$$\aleph_\omega^{\aleph_0} \nrightarrow (\aleph_{\omega+1}, (\aleph_0)_{\aleph_0})^2$$

and using G.C.H. we proved **[13**, Theorem 10]

$$\aleph_{\omega+1} \nrightarrow (\aleph_{\omega+1}, (3)_{\aleph_0})^2.$$

We cannot fill up the gap between these results if we do not assume G.C.H.

Problem 5 (Erdös, Hajnal, Rado). *Can one prove without assuming G.C.H. that*

$$\aleph_{\omega+1} \nrightarrow (\aleph_{\omega+1}, (3)_{\aleph_0})^2$$

holds?

The general problem of Symbol-I for order types seems to be very ramified. There are only scattered partial results even in case of ordinal numbers.

3.2. *Symbol*-I *in case of denumerable ordinals*. Note that α^β always denotes ordinal power. It seems to be reasonable to consider powers of ω in the first entry. It is easy to see that $\eta \nrightarrow (\omega + 1, 4)^3$ (where η denotes the type of rational numbers); hence we consider only the case $r = 2$. $\omega \rightarrow (\omega, \omega)^2$ follows from Ramsey's theorem, $\alpha \nrightarrow (\omega + 1, \omega)^2$ is trivial for every $\alpha < \omega_1$.

E. Specker proved in **[37]**

(1)
$$\omega^2 \rightarrow (\omega^2, k)^2 \quad \text{for} \quad k < \omega,$$
$$\omega^n \nrightarrow (\omega^n, 3)^2 \quad \text{for} \quad 3 \leq n < \omega.$$

E. C. Milner proved

(2)
$$\omega^{\alpha\cdot 3} \nrightarrow (\omega^{\alpha\cdot 2+1}, 3)^2 \quad \text{for} \quad \alpha < \omega_1.$$
$$\omega^4 \rightarrow (\omega^3, 3)^2$$
$$\omega^3 \rightarrow (\omega^2 \cdot l, k)^2 \quad \text{for} \quad l, k < \omega.$$

P. Erdös proved

(3)
$$\omega^{\alpha\cdot 2+1} \rightarrow (\omega^{\alpha+1}, 4)^2 \quad \text{for} \quad \alpha < \omega_1.$$

A. Hajnal proved recently the following theorem. Let $S = \{(n_0, \ldots, n_{k-1}): n_i < \omega \text{ for } i < k\}$ and $[S]^2 = \bigcup_{\nu<l} \mathscr{T}_\nu$ be a 2-partition of type $l < \omega$ of S. Then there exists an infinite set N of integers, for which the partition $(\mathscr{T}_\nu)_{\nu<l}$ is canonical on $S' = \{(n_0, \ldots, n_{k-1}): n_i \in N \text{ for } i < k\}$, i.e. for every pair

$$(n_0, \ldots, n_{2k-1})(m_0, \ldots, m_{2k-1}),\ n_i, m_j \in N$$
$$n_i < n_j \Leftrightarrow m_i < m_j \quad \text{and} \quad n_i = m_j \Leftrightarrow m_i = m_j$$

implies that $\{(n_0, \ldots, n_{k-1})(n_k, \ldots, n_{2k-1})\} \in \mathscr{T}_\nu$ holds iff

$$\{(m_0, \ldots, m_{k-1})(m_k, \ldots, m_{2k-1})\} \in \mathscr{T}_\nu$$

for every $\nu < l$.

ADDED IN PROOF. We learned recently that this result was obtained independently by F. Galvin.

This certainly implies that for every $n < \omega$

(4) $$\omega^n \nrightarrow (\omega^3, f(n))^2 \quad \text{holds for some} \quad f(n) < \omega.$$

If $f(k, n)$ denotes the least integer for which $\omega^n \nrightarrow (\omega^k, f(k, n))^2$ holds for $k \geq 3$, the above mentioned result reduces the determination of $f(k, n)$ to a finite combinatorial problem which is not quite easy to answer. We have computed, e.g. that $f(3, 4) = 5$, but we still do not know whether

Problem 6. $\omega^5 \to (\omega^3, 5)^2$?

However, this is obviously a problem of type (C). The real problem is to determine $f(k, n)$ generally.

None of the results mentioned gives any information about the following problem of type (B):

Problem 7. $\omega^\omega \to (\omega^\omega, 3)^2$?

ADDED IN PROOF (May, 1970). C. C. Chang proved recently $\omega^\omega \to (\omega^\omega, 3)^2$; $\omega^\omega \to (\omega^\omega, 4)^2$ is still open. See C. C. Chang, *A theorem in combinatorial set theory*. Preprint.

3.3. *Symbol*-I *in case of nondenumerable order types and ordinals.* Whenever we have a positive arrow relation $a \to (b_\nu)_c^r$ for cardinals, this obviously implies a corresponding relation for initial ordinals $\alpha \to (\beta_\nu)_c^r$, where α, β_ν are the initial ordinals of a, b_ν, respectively. If $\beta_\nu < \alpha$ for some $\nu < \Omega(c)$ one can ask for what ordinals β'_ν, $|\beta'_\nu| = \beta_\nu$ does the same relation remain true. Usually the method used for the proof of $a \to (b_\nu)_c^r$ yields a slightly stronger result than $\alpha \to (\beta_\nu)_c^r$. There are some results of this type in [**21**], but most of the problems remain unsolved. By Theorem I of [**21**], we have $\aleph_1 \to (\aleph_1, \aleph_0)^2$ and the proof in fact gives $\omega_1 \to (\omega_1, \omega + 1)^2$. A. Hajnal proved, using G.C.H. [**25**] that $\omega_1 \nrightarrow (\omega_1, \omega + 2)^2$ holds.

Problem 8. *Can one prove without using the continuum hypothesis that* $\omega_1 \nrightarrow (\omega_1, \omega + 2)^2$ *holds*?

This might be a problem of type (A).

Using G.C.H. the result of **[25]** gives $\omega_{\alpha+1} \nrightarrow (\omega_{\alpha+1}, \omega_\alpha + 2)^2$ if $\aleph_\alpha$ is regular and, e.g. Theorem I of **[13]** implies $\omega_{\alpha+1} \rightarrow (\omega_{\alpha+1}, \omega_{\mathrm{cf}(\alpha)} + 1)^2$ for every α. These leave the following problem open.

Problem 9. *Assume G.C.H. Does* $\omega_{\omega+1} \rightarrow (\omega_{\omega+1}, \omega + 2)^2$ *hold?*

On the other hand, G.C.H. $\Rightarrow \omega_{\rho+1} \rightarrow (\omega_\rho + 1, \omega_\rho + 1)^2$ for every ρ. There is no counterexample for the following:

Problem 10. *Assume G.C.H. Does then* $\omega_{\rho+1} \rightarrow (\xi, \xi)^2$ *hold for every* $\xi < \omega_{\rho+1}$ *and for every* ρ?

By Theorem 8 of **[25]** we have $\omega_1 \rightarrow (\omega \cdot 2, \omega \cdot k)^2$ for every finite k. Thus the simplest unsolved instances are

Problem 10/A. *Does*

$$\omega_1 \rightarrow (\omega \cdot 2, \omega^2)^2$$

or

$$\omega_1 \rightarrow (\omega \cdot 3, \omega \cdot 3)^2$$

or

$$\omega_1 \rightarrow (\omega + n, \omega + n, \omega + n)^2 \quad \textit{for every } n < \omega$$

hold?

In cases $\rho > 0$ the problem is even more difficult.

ADDED IN PROOF (May, 1970). F. Galvin communicated to us in a letter that he proved $\omega_1 \rightarrow (\omega \cdot 3, \omega \cdot 3)^2$; $\omega_1 \rightarrow (\omega^2, \omega^2)^2$ is still open.

Problem 10/B. *Assume G.C.H. Does* $\omega^2 \rightarrow (\omega_1 + \omega, \omega_1 + \omega)^2$ *hold?*

10/B might be a problem of type (D) but $\omega_2 \rightarrow (\omega_1 \cdot 2, \omega_1 \cdot 2)^2$ (if true) certainly requires new ideas.

It would be easy to formulate Problem 10 without using G.C.H. We formulate one special case.

Assume $a \geq \aleph_0$. Does then $\Omega((2^a)^+) \rightarrow (\xi, \xi)^2$ hold for every $\xi < \Omega(a^{++})$? Let λ denote the type of real numbers. There is no counterexample for

Problem 11. *Does* $\lambda \rightarrow (\alpha)^2_k$ *hold for every* $\alpha < \omega_1$, $k < \omega$?

By Theorem 31 of **[21]** we have $\lambda \rightarrow (\omega + l)^2_3$ for $l < \omega$.

It was proved in **[25]** that $\lambda \rightarrow (\omega \cdot l, \omega \cdot l)^2$ holds for every $l < \omega$ and $\lambda \rightarrow (\eta, \alpha \text{ or } \alpha^*)^2$ holds for every $\alpha < \omega_1$. The simplest unsolved instances are

Problem 11/A.

$$\lambda \rightarrow (\omega^2, \omega^2)^2?$$

$$\lambda \rightarrow (\omega \cdot 2)^2_3?$$

$$\lambda \rightarrow (\omega + l)^2_4 \quad \textit{for} \quad 2 \leq l < \omega?$$

ADDED IN PROOF. F. Galvin proved $\Phi \rightarrow (\alpha)^2_2$ for every real type Φ and $\alpha < \omega_1$. $\lambda \rightarrow (\alpha)^2_3$ is still unsolved and seems to be difficult.

We mention that all the known results remain valid if λ is replaced by a "real type" Φ (Φ is a real type if $\omega_1, \omega_1^* \not\leq \Phi$). This explains our interest in the following problem of type (E).

Problem 12. *Can one prove a relation $\lambda \rightarrow (\Theta_1, \Theta_2)^2$ which does not hold for every real type Φ instead of λ?*

Naturally, one can ask the problems stated in 3.2 for ω_α instead of ω, but there are fewer results.

Specker's result $\omega^3 \nrightarrow (\omega^3, 3)^2$ stated in (1) of 3.2 generalizes easily but the proof of $\omega^2 \rightarrow (\omega^2, k)^2$ uses finitely additive measures and breaks down for ω_1.

In a forthcoming paper **[22]** P. Erdös and R. Rado state

(1) for every ξ and for every finite k, l there exists an $n < \omega$ such that $\omega_\xi \cdot n \rightarrow (\omega_\xi \cdot k, l)^2$,

and A. Hajnal proved

(2) assume G.C.H., then $\omega_{\xi+1} \cdot \rho \nrightarrow (\omega_{\xi+1} \cdot \omega_\xi, 3)^2$ for every $\rho < \omega_{\xi+1}$ and for every ξ.

The following remained unsolved:

Problem 13. *Assume G.C.H.*

$$\omega_1^2 \rightarrow (\omega_1^2, 3)^2?$$

$$\omega_2 \cdot \omega \rightarrow (\omega_2\omega, 3)^2?$$

The first one seems to be of type (B). Note that $\omega_\alpha^2 \nrightarrow (\omega_\alpha + 1, \omega)^2$ is known for every α.

Added in proof (May, 1970). A. Hajnal proved G.C.H. $\Rightarrow \omega_1^2 \nrightarrow (\omega_1^2, 3)^2$. See A. Hajnal, *A negative partition relation*, Proc. Nat. Acad. Sci. U.S.A. (to appear).

4. **Problems for Symbol-II.**

Definition of Symbol-II. Let a, b_ν, r, c have the same meaning as in the definition of Symbol-I.

We write $a \xrightarrow{\text{II}} [b_\nu]_c^r$ if the following statement is true.

Let S be a set if a is a cardinal and S, $\prec$ be a (simply) ordered set if a is an order type such that $|S| = a$ and type $S(\prec) = a$ respectively. Let further $(\mathscr{T}_\nu)_{\nu<\Omega(c)}$ be an r-partition of type c of S. Then there are a subset $S' \subset S$ and an ordinal $\nu_0 < \Omega(c)$ such that $[S']^r \subset \bigcup_{\nu\neq\nu_0<\Omega(c)} \mathscr{T}_\nu$ and $|S'| = b_{\nu_0}$ or typ $S'(\prec) = b_{\nu_0}$ respectively.

We will use the same self-explanatory abbreviations in cases when some of the b_ν's are equal which were introduced for Symbol-I.

Symbol-II was first defined for cardinals in **[13]** and was not yet so thoroughly investigated as Symbol-I. It is obvious that, e.g. $a \nrightarrow [b]_c^r$, $c > 2$ is a much stronger counterexample than $a \nrightarrow (b, b)^r \Leftrightarrow a \nrightarrow [b, b]^r$.

We mention that in **[9]** we have proved that if $a \nrightarrow [a]_a^r$ for some $r < \omega$ then there is a Jónsson algebra of power a.

4.1. *The case of the infinite exponents.* One would expect that as in case of Symbol-I we have a best possible negative result for $r \geq \aleph_0$. There is no counter-example for the following.

Problem 14. *Assume $a, b \geq r \geq \aleph_0$ are cardinals. Then $a \not\rightarrow [b]^r_{b^r}$.*

Note that $b \rightarrow [b]^r_c$ is trival for $c > b^r$. Using an idea of J. Novák described in **[1]** we can prove the following result.

THEOREM (UNPUBLISHED). *Assume $a, r \geq \aleph_0$ are cardinals. Then $a \not\rightarrow [r]^r_{2^r}$.*

The following simple instance of Problem 14 remains unsolved.

Problem 14/A. *Is it true that*

$$a \not\rightarrow [(2^{\aleph_0})^+]^{\aleph_0}_{(2^{\aleph_0})^+}$$

holds for every a? Or assuming G.C.H. is it true that $a \not\rightarrow [\aleph_2]^{\aleph_0}_{\aleph_2}$ holds for every a?

Note that for $a < \aleph_\omega$ this might be a consequence of $\overset{\mathrm{II}}{\not\rightarrow}$ relations with finite exponents. See Problem 17/A, and that assuming G.C.H. we can in fact prove the $\overset{\mathrm{II}}{\not\rightarrow}$ relation for the special case $a = \aleph_2$.

4.2. *Symbol*-II, *in case $r < \omega$.* Theorem 17 of **[13]** states that G.C.H. $\Rightarrow$ $\aleph_{\alpha+1} \not\rightarrow [\aleph_{\alpha+1}]^2_{\aleph_{\alpha+1}}$ for every α and it is well known that $2^{\aleph_\alpha} \not\rightarrow (\aleph_{\alpha+1}, \aleph_{\alpha+1})^2$ i.e. $2^{\aleph_\alpha} \not\rightarrow [\aleph_{\alpha+1}]^2_2$.

Problem 15. *Can one prove without assuming C.H.*

$$2^{\aleph_0} \not\rightarrow [\aleph_1]^2_3 \quad or \quad 2^{\aleph_0} \not\rightarrow [2^{\aleph_0}]^2_3 \quad or \quad \aleph_1 \not\rightarrow [\aleph_1]^2_3?$$

This might be a problem of type (A).

ADDED IN PROOF (May, 1970). We learned from a letter of F. Galvin that he proved $2^{\aleph_0} \rightarrow [2^{\aleph_0}]^2_n$ for $n < \omega$ and $\aleph_1 \not\rightarrow [\aleph_1]^2_4$.

We would like to stress the importance of the following:

Problem 16. *Let a be an inaccessible cardinal for which $a \not\rightarrow (a, a)^2$ holds. Does then $a \not\rightarrow [a]^2_a$ hold?*

We do not know if $a \not\rightarrow [a]^2_a$ holds for the first strongly inaccessible cardinal $a > \aleph_0$.

We think that the problem whether $a \not\rightarrow [a]^2_a$ holds is strongly connected with the following. There exists an a-complete field S of subsets of a set X of power a generated by at most a elements $[X]^{<a} \subseteq S$ in which there is no a-complete proper a-saturated ideal I such that $[X]^{<a} \subset I$.

As to further details of the results concerning $a \rightarrow [b_\nu]^r_c$ a, b_ν cardinals we refer to **[13]**. We formulate only one more problem.

Problem 17 (Erdös, Hajnal, Rado). *Assume $a \geq \aleph_0$, $2 \leq r < \aleph_0$, $b_\nu > r$ are cardinals and $a \not\rightarrow [b_\nu]^r_c$. Does then $2^a \not\rightarrow [b_\nu + 1]^r_c$ hold?*

We cannot give a positive answer even assuming G.C.H.

Problem 17/A. *Does* $2^{2^{\aleph_0}} \nrightarrow [\aleph_1]^3_{\aleph_1}$ *hold? Or does G.C.H.* $\Rightarrow \aleph_2 \nrightarrow [\aleph_1]^3_{\aleph_1}$ *and* $\aleph_{l+1} \nrightarrow [\aleph_1]^{l+2}_{\aleph_1}$ *for* $l < \omega$?

This should be compared with Theorems 17, 25, and Problem 3 of [**13**]. As far as we know no one has investigated Symbol-II for types. There are some very simple problems we cannot answer. Here is one of them.

Problem 18. $\omega^\omega \rightarrow [\omega^\omega]^2_{\aleph_0}$?

Note that $\omega^k \rightarrow [\omega^k]^2_{\aleph_0}$, $k < \omega$ follows from the theorem of A. Hajnal mentioned in subsection 3.2.

5. **Symbol-III and related problems.**

Definition of Symbol-III. Let r, c, d be cardinals, a, b types or cardinals but b should be a cardinal if a is a cardinal.

$a \overset{\text{III}}{\longrightarrow} [b]^r_{c,d}$ denotes that the following statement is true.

Let S be a set if a is a cardinal and let S, $\prec$ be an ordered set if a is a type such that $|S| = a$ or typ $S(\prec) = a$ respectively. Let further $(\mathscr{T}_\nu)_{\nu<\Omega(c)}$ be an r-partition of type c of S. Then there exist an $S' \subseteq S$ and a set N of ordinals less than $\Omega(c)$ such that $|N| \leq d$, $[S']^r \subseteq \bigcup_{\nu\in N} \mathscr{T}_\nu$ and $|S'| = b$ or typ $S'(\prec) = b$ respectively.

Symbol-III is Symbol-V of [**13**] defined in [**13,** 18.3]. We collected a number of results and problems in [**13,** 18] which we do not repeat here; we only point out one of them (Problem 3.1(a) of [**13**]).

Problem 19. *Assume G.C.H. Does then*

$$\aleph_2 \overset{\text{III}}{\nrightarrow} [\aleph_1]^2_{\aleph_1,\aleph_0}$$

hold?

We came to this problem when considering a problem of S. Ulam. Several other people have independently considered this problem and though we were unable to collect references we know by hearsay that both $\aleph_2 \rightarrow [\aleph_1]^2_{\aleph_1,\aleph_0}$ and $\aleph_2 \nrightarrow [\aleph_1]^2_{\aleph_1,\aleph_0}$ are proved to be consistent with the axioms of set theory and G.C.H. E.g. F. Rowbottom proved that the negative relation follows from Gödel's axiom $V = L$.

We formulated this problem because we will formulate a series of other problems related to it and some implications between them.

Let $(A_\nu)_{\nu<\varphi}$ be a sequence of disjoint sets. The set X is said to be a transversal of the sequence $(A_\nu)_{\nu<\varphi}$ if $|A_\nu \cap X| = 1$ for every $\nu < \varphi$.

Problem 19/A. *Assume G.C.H. Let* $(A_\nu)_{\nu<\omega_1}$ *be a sequence of disjoint sets such that* $|A_\nu| = \aleph_0$ *for every* $\nu < \omega_1$.

Does there exist a system $\mathscr{F}$, $|\mathscr{F}| = \aleph_2$ *of almost disjoint transversals (i.e.* $X, \neq Y \in \mathscr{F}$ *implies* $|X \cap Y| \leq \aleph_0$)?

Problem 19/B. *Does there exist under the conditions of Problem* 19/A, *a system* $\mathscr{F}$, $|\mathscr{F}| = \aleph_2$ *of almost disjoint transversals satisfying the following additional* *condition? If $X, Y \in \mathcal{F}$ and $X \cap A_\nu = Y \cap A_\nu$ for some $\nu < \omega_1$ then $X \cap A_\mu = Y \cap A_\mu$ for every $\mu < \nu$.*

It is easy to see that Problem 19/B is equivalent to the well-known Kurepa problem. It is also quite easy to see that Problem 19/B $\Rightarrow$ Problem 19/A $\Rightarrow$ Problem 19. We cannot answer the following problem.

Problem 19/C. *Does either of the implications*

$$\text{Problem 19} \Rightarrow \text{Problem 19/A}, \qquad \text{Problem 19/A} \Rightarrow \text{Problem 19/B}$$

hold?

All these problems are well known.

Let a be a regular cardinal, $a > \aleph_0$. Let X be a set of ordinals less than $\Omega(a)$. A function f defined on X with ordinal values less than $\Omega(a)$ is said to be *regressive* on X if $f(\xi) < \xi$ for every $\xi \in X$. X is said to be stationary (in $\Omega(a)$) if for every regressive function f on X there is a $\rho < \Omega(a)$ such that $(f^{-1}(\rho)) = a$.

It has been recently proved by R. Solovay that every stationary X is the union of a disjoint stationary sets.

Problem 19/D. *Does there exist a system $\mathcal{F}$, $|\mathcal{F}| = \aleph_2$ or $2^{\aleph_1}$ of almost disjoint stationary subsets of ω_1?*

We do not know the answer for any regular a in place of ω_1. It is easy to see that a positive answer to Problem 19/A implies a positive answer to Problem 19/D.

Some problems related to 19/A will be considered in a forthcoming paper by E. C. Milner and the two of us **[17]**.

We mention one more problem of Kurepa type. We do not know if its independence has already been investigated.

Problem 19/E. *Assume G.C.H. Let $|S| = \aleph_\omega$. Does there exist a family $\mathcal{F}$, $|\mathcal{F}| = \aleph_{\omega+1}$, $\mathcal{F} \subseteq [S]^{\aleph_0}$ such that $\mathcal{F} \restriction X = \{F \cap X : F \in \mathcal{F}\}$ has power $\aleph_0$ for every $X \subset S$, $|X| = \aleph_0$?*

We turn to a problem concerning Symbol-III in case of ordinals. We mention that even Symbol-I yields interesting problems for ordinals in case $r = 1$, but these had been solved and completely discussed by E. C. Milner and R. Rado in **[32]**. One of their surprising results states that $\rho \nrightarrow (\omega_\xi^\omega)^1_{\aleph_0}$ holds for every $\rho < \omega_{\xi+1}$ and for every ξ.

This can be formulated in terms of Symbol-III: $\rho \nrightarrow [\omega_\xi^\omega]^1_{\aleph_0, <\aleph_0}$ holds for every $\rho < \omega_{\xi+1}$ and for every ξ.

(Symbol-III was not defined with $<\aleph_0$ in place of d, but has a self-explanatory meaning.)

A straightforward generalization of this would be the following:

Problem 20. *Let $\rho < \omega_{\xi+1}$, $\zeta + 1 < \xi$. Does then*

$$\rho \nrightarrow [\omega_\xi^{\omega_{\zeta+1}}]^1_{\aleph_{\zeta+1}, \aleph_\zeta}$$

hold?

This is certainly a problem of type (A) and we would be interested if a positive answer for it is consistent with the axioms of set theory.

We wish to make some remarks on a special case of it.

Problem. 20/A. *Let $\rho < \omega_3$. Does then $\rho \nrightarrow [\omega_2^{\omega_1}]^1_{\aleph_1,\aleph_0}$ hold for every $\rho < \omega_3$?*

We know the following partial results:

(1) $\rho \nrightarrow [\omega_2^{\omega_1}]^1_{\aleph_1,\aleph_0}$ for $\rho < \omega_2^{\omega_2}$.

(2) If $\omega_2^{\omega_2} \nrightarrow [\omega_2^{\omega_1}]^1_{\aleph_1,\aleph_0}$, then the answer for Problem 19/D is affirmative.

We formulate two more problems of this type.

Problem 21. *Let $\rho < \omega_3$. Does there exist a sequence $(f_\nu^\rho)_{\nu<\rho}$ of type ρ of functions defined on ordinals $< \omega_1$ with values $<\omega_1$ satisfying the following condition? Whenever $\nu_1 < \nu_2 < \rho$ then the set*

(0) $\{\xi < \omega_1 \colon f_{\nu_1}^\rho(\xi) \geq f_{\nu_2}^\rho(\xi)\}$ *is nonstationary in* ω_1.

Problem 21/A. *Does there exist a sequence of functions satisfying the requirements of Problem* 21 *and the stronger conditions?*

If $\nu_1 < \nu_2 < \rho$ *then*

(00) $|\{\xi < \omega_1 \colon f_{\nu_1}^\rho(\xi) \geq f_{\nu_2}^\rho(\xi)\}| < \aleph_1$.

We can prove

(3) A positive answer to Problem 20/A implies that the answer is positive for 21.

(1), (2), (3) will be published in a triple paper with Milner. It is obvious that Problem 21/A $\Rightarrow$ Problem 21.

It is also obvious that a positive answer for Problem 21 and Problem 21/A in the special cases $\rho = \omega_2 + 1$ implies a positive answer to Problem 19/D, Problem 19/A respectively. On the other hand, we do not know if the consistency of Problem 21 or Problem 21/A has already been investigated.

6. Symbol-IV, $a \rightarrow (b)_c^{<\aleph_0}$ and related problems.

Definition of Symbol-IV. Let a, b, c be cardinals. $a \rightarrow (b)_c^{<\aleph_0}$ denotes that the following statement is true. Let S be a set $|S| = a$. Let $(\mathscr{T}_\nu^r)_{\nu<\Omega(c)}$ be an r-partition of type c of S for every $r < \omega$. Then there exist an $r_0 < \omega$, a function $\nu(r) < \Omega(c)$ for $r < \omega$ and a subset $S' \subset S$, $|S'| = b$ such that $[S']^r \subset \mathscr{T}_{\nu(r)}^r$ for every $r_0 \leq r < \omega$.

Symbol-IV is Symbol-II of **[13]**. We have proved in **[1]** that $a \rightarrow (a)_c^{<\aleph_0}$ holds for $c < a$ if a is a measurable cardinal $> \aleph_0$. J. Silver has proved recently **[34]** that $a \nrightarrow (\aleph_0)_2^{<\aleph_0}$ holds for a very large section of cardinals. For other results, history, and references see **[13]** and **[34]**.

We will speak about some strongly related problems.

Definition of Symbol-IV.1. Let a and b be cardinals. $a \Rightarrow (b)^{<\aleph_0}$ denotes that the following statement is true. Let S be a set, $|S| = a$, and let $(\mathscr{T}_0^r, \mathscr{T}_1^r)$ be an r-partition of type 2 of S for every $r < \omega$, such that

(0) $X \subset S$, $|X| = r + 1$ implies $[X]^r \not\subset \mathscr{T}_0$ for every $r < \omega$. Then there exist an $r_0 < \omega$ and $S' \subset S$ such that $|S'| = b$ and $[S']^r \subset \mathscr{T}_1^r$ for every $r_0 \leq r < \omega$.

It is obvious that $a \not\Rightarrow (b)^{<\aleph_0}$ implies $a \nrightarrow (b)_2^{<\aleph_0}$.

We have proved several years ago that all the negative results stated for Symbol-IV (i.e. Symbol-II with $c = 2$ in [**13**]) are valid for Symbol-IV.1 as well.

The following problem (of type (E)) arises.

Problem 22. *Does $a \not\Rightarrow (\aleph_0)^{<\aleph_0}$ hold for the first strongly inaccessible cardinal $a > \aleph_0$ (or for a large section of cardinals)?*

We can prove that if we had defined a symbol $a \Rrightarrow_l (b)^{<\aleph_0}$ by replacing the condition (0) of the definition of Symbol-IV.1 by the stronger condition

(00) $X \subset S$, $|X| = r + 1$ implies $|[X]^r \cap \mathscr{T}_0| < r + 1 - l$ for $r < \omega$ we would have had $2^{\aleph_0} \Rrightarrow_l (\aleph_0)^{<\aleph_0}$ for some $l < \omega$, but we do not know if l can be replaced by 1.

We define a Symbol-IV.2 which is in the same relation to Symbol-IV as Symbol-II is to Symbol-I.

Definition of Symbol-IV.2. Let a, b, c be cardinals.

$a \to [b]_c^{<\aleph_0}$ is said to hold if the following statement is true.

Let S be a set $|S| = a$. Let further $(\mathscr{T}_\nu^r)_{\nu < \Omega(c)}$ be an r-partition of type c of S for every $r < \omega$. Then there exist $S' \subset S$, $r_0 < \omega$ and a function $\nu(r) < \omega$ for $r < \omega$ such that

$$[S_1']^r \subset \bigcup_{\nu \neq \nu(r); \nu < \Omega(c)} \mathscr{T}_\nu^r \quad \text{for every} \quad r < \omega.$$

It is obvious that $a \nrightarrow [b]_c^{<\aleph_0}$ for $c \geq 2$ is a stronger counterexample than $a \nrightarrow (b)_2^{<\aleph_0}$.

We cannot even decide

Problem 23. $\aleph_0 \overset{\text{IV.2}}{\nrightarrow} [\aleph_0]_{\aleph_0}^{<\aleph_0}$?

We always suspected that there is a $\nrightarrow$ relation. We do not know if $a \to [\aleph_0]_{\aleph_0}^{<\aleph_0}$ holds for any "relatively small" cardinal.

This might be a problem of type (D).

Added in proof. Let $a \to (b)^{<\aleph_0}_{c_0,\dots,c_r,\dots,d_0,\dots,d_n\dots}$ denote the following statement: Let S be a set, $|S| = a$. Let further (J_ν^r), $\nu < \Omega(c_r)$ be an r partition of type c_r of S for every $r < \omega$.

Then there are sets B_r of ordinals less than $\Omega(c_r)$ and $S' \subset S$ such that

$$[S]^r \subset \bigcup_{\nu \in B_r} J_\nu^r, \qquad |S'| = b \text{ and } |B_r| \leq d_r \text{ for } r < \omega.$$

J. E. Baumgartner and independently R. Rado and we proved that $\aleph_0 \to [\aleph_0]^{<\aleph_0}_{c_0,\dots,c_r,\dots,d_0,\dots,d_r\dots}$ holds provided $c_r < \omega$ and $d_r \to +\infty$ if $r \to +\infty$. This implies obviously a negative solution of Problem 23.

7. Polarized partition relations and related problems.

Definition of Symbol-V (see [**13**, 3.3]). (1) Let $a, b, c, d_\nu, e_\nu, f_\nu, g_\nu, \nu < \Omega(c)$ be cardinals.

$$\begin{pmatrix} a \\ b \end{pmatrix} \to \begin{pmatrix} d_\nu & & f_\nu \\ & \vee & \\ e_\nu & & g_\nu \end{pmatrix}_c$$

is said to hold if the following statement is true.

Whenever S_0, S_1 are sets such that $|S_0| = a$, $|S_1| = b$ and $(\mathscr{T}_\nu)_{\nu<\Omega(c)}$ is a partition of $S_0 \times S_1$ then there exist $\nu_0 < \Omega(c)$, $S_0' \subset S_0$, $S_1' \subset S_1$ such that $S_0' \times S_1' \subset \mathscr{T}_{\nu_0}$ and either $|S_0'| = d_\nu$ and $|S_1'| = e_\nu$ or $|S_0'| = f_\nu$ and $|S_1'| = g_\nu$.

We write

$$\begin{pmatrix} a \\ b \end{pmatrix} \to \begin{pmatrix} d_\nu \\ e_\nu \end{pmatrix}_c$$

if $d_\nu = f_\nu$, $e_\nu = g_\nu$ for $\nu < \Omega(c)$.

(2) Let $r = r_0 + \cdots + r_{l-1}$, $l < \omega$ for an $r < \omega$.

Let a_i, $b_{i,\nu}$, c be cardinals for $i < l$, $\nu < \Omega(c)$.

$$\begin{pmatrix} a_0 \\ \cdot \\ \cdot \\ \cdot \\ a_{l-1} \end{pmatrix} \to \begin{pmatrix} b_{0,\nu} \\ \cdot \\ \cdot \\ \cdot \\ b_{l-1,\nu} \end{pmatrix}^{r_0,\ldots,r_{l-1}}$$

is said to hold if the following statement is true.

Whenever S_i, $i < l$ are sets such that $|S_i| = a_i$ for $i < l$ and $(\mathscr{T}_\nu)_{\nu<\Omega(c)}$ is a partition of $[S_0]^{r_0} \times \cdots \times [S_{l-1}]^{r_{l-1}}$ then there are subsets $S_i' \subset S_i$ for $i < l$ and $\nu_0 < \Omega(c)$ such that $|S_i'| = b_{i,\nu_0}$ for $i < l$ and

$$[S_0']^{r_0} \times \cdots \times [S_{l-1}']^{r_{l-1}} \subset \mathscr{T}_{\nu_0}.$$

Obviously

$$\begin{pmatrix} a_0 \\ a_1 \end{pmatrix} \to \begin{pmatrix} b_{0\nu} \\ b_{1\nu} \end{pmatrix}_c^{1,1}$$

means the same as

$$\begin{pmatrix} a_0 \\ a_1 \end{pmatrix} \to \begin{pmatrix} b_{0\nu} \\ b_{1\nu} \end{pmatrix}.$$

One could give a definition of Symbol-V for types under obvious restrictions for the entries as in the case of the previous symbols. It is also obvious that as in case of Symbols-I,-II a corresponding "square bracket" symbol can be defined, e.g. in the definition of

$$\begin{pmatrix} a \\ b \end{pmatrix} \to \begin{pmatrix} b_{0\nu} \\ b_{1\nu} \end{pmatrix}_c$$

$S_0' \times S_1' \subset \mathscr{T}_{\nu_0}$ has to be replaced by

$$S_0' \times S_1' \subset \bigcup_{\nu_0=\nu;\nu<\Omega(c)} \mathscr{T}_\nu.$$

In **[13]** we have investigated the symbol as defined in (1) with $c = 2$, and assuming G.C.H. in almost every case.

We mention the following

Problem 24 (Erdös, Hajnal, Rado). *Assume G.C.H. Does then*

$$\begin{pmatrix}\aleph_2\\ \aleph_1\end{pmatrix} \to \begin{pmatrix}\aleph_i & \aleph_j\\ \aleph_1 & , \aleph_1\end{pmatrix}$$

hold for $0 \le i, j \le 1$?

Problem 25 (Erdös, Hajnal, Rado). *Assume G.C.H.*

$$\begin{pmatrix}\aleph_{\omega+1}\\ \aleph_{\omega+1}\end{pmatrix} \to \begin{pmatrix}\aleph_{\omega+1} & \aleph_1\\ \aleph_\omega & \wedge\ \aleph_\omega\end{pmatrix}?$$

See Problems 12 and 14 of **[13]**. For a discussion of the known results see **[13]**.

We think that both problems might be of type (A).

We mention that unlike in the case of Symbol-I most of our results use G.C.H. essentially and even the simplest problems seem to be unsolvable.

We do not know the answer to

Problem 26. *Does*

$$\begin{pmatrix}2^{\aleph_0}\\ \aleph_0\end{pmatrix} \not\to \begin{pmatrix}2^{\aleph_0} & 2^{\aleph_0}\\ \aleph_0 & \wedge\ \aleph_0\end{pmatrix}$$

hold if we do not assume C.H.?

ADDED IN PROOF (May, 1970). We learned from a letter of F. Galvin that Laver proved that

$$\begin{pmatrix}\aleph_1\\ \aleph_0\end{pmatrix} \to \begin{pmatrix}\aleph_1 & \aleph_1\\ \aleph_0 & , \aleph_0\end{pmatrix}$$

is consistent with ZF + AC.

C.H. implies even

$$\begin{pmatrix}\aleph_1\\ \aleph_1\end{pmatrix} \not\to \begin{pmatrix}\aleph_1 & \aleph_0 & \aleph_1 & \aleph_0\\ \aleph_0 & \vee\ \aleph_1 & , \aleph_0 & \vee\ \aleph_1\end{pmatrix}.$$

We did not investigate the case $c \ge \aleph_0$ in detail. We mention that a surprising number of set theoretical problems can be formulated with the help of the polarized partition symbol.

We mention, e.g. that

$$\begin{pmatrix}\aleph_2\\ \aleph_1\end{pmatrix} \not\to \begin{pmatrix}2\\ \aleph_1\end{pmatrix}_{\aleph_0}$$

is equivalent to Problem 19/A.

The following slightly weaker statement seems to be of type (A) as well.

Problem 27. *Assume G.C.H. Does then*

$$\binom{\aleph_2}{\aleph_1} \not\rightarrow \binom{\aleph_0}{\aleph_1}_{\aleph_0}$$

hold?

Note that the methods of **[13]** give the following results: G.C.H. $\Rightarrow$

$$\binom{\aleph_2}{\aleph_1} \not\rightarrow \binom{\aleph_2}{\aleph_1}, \binom{\aleph_0}{\aleph_1}$$

$$\binom{\aleph_2}{\aleph_1} \rightarrow \left(\binom{\aleph_2}{\aleph_1}, \binom{\aleph_2}{\aleph_0}_{\aleph_0}\right)$$

$$\binom{\aleph_2}{\aleph_2} \rightarrow \left(\binom{\aleph_2}{\aleph_1} \vee \binom{\aleph_1}{\aleph_2}, \binom{\aleph_1}{\aleph_1}\right).$$

We never investigated Symbol-V as defined in (2). The methods of **[13]** would give a positive relation if the cardinals $a_0, \ldots, a_{l-1}$ are large and far from each other but the other cases must lead to complicated and involved problems.

We know, e.g. from a result of Sierpiński [35] that

$$\begin{pmatrix}\aleph_1\\ \aleph_1\\ \aleph_1\end{pmatrix} \not\rightarrow \begin{pmatrix}\aleph_0\\ \aleph_0\\ \aleph_0\end{pmatrix}_3^{1,1,1}$$

holds. We state the following problem of type (D).

Problem 28.

$$\begin{pmatrix}\aleph_1\\ \aleph_1\\ \aleph_1\end{pmatrix} \rightarrow \begin{pmatrix}\aleph_0\\ \aleph_0\\ \aleph_0\end{pmatrix}_2^{1,1,1} ?$$

8. **Further problems on partitions.** Let S be a set and let $(S_\xi)_{\xi<\varphi}$ be a disjoint partition of it. Let $\Delta = (\mathscr{T}_\nu)_{\nu<\Omega(c)}$ be an r-partition of S. Δ is said to be canonical with respect to $(S_\xi)_{\xi<\varphi}$, if $X, Y \in [S]^r$, $|X \cap S_\xi| = |Y \cap S_\xi|$ for every $\xi < \varphi$ implies that $X \in \mathscr{T}_\nu$ and $Y \in \mathscr{T}_\nu$ are equivalent for every $\nu < \Omega(c)$.

Problem 29. *Assume G.C.H. Let* $|S| = \aleph_\omega$, $(S_n)_{n<\omega}$ *a disjoint partition of* S, $|S_n| = \aleph_n$. *Let further* $\Delta_r = (\mathscr{T}_0^r, \mathscr{T}_1^r)$ *be an r-partition of type* 2 *of S for every* $r < \omega$.

Does there then always exist an $S' \subset S$, $|S'| = \aleph_\omega$ *such that* Δ_r *is canonical with respect to* $(S'_n)_{n<\omega}$ *for every* $r < \omega$ *(where* $S'_n = S' \cap S_n$*)?*

This should be compared with Lemma 3 of **[13]**. Note that one can ask the problem without assuming G.C.H. for a singular strong limit cardinal instead of

$\aleph_\omega$ and a corresponding version of Lemma 3 [**13**] remains valid for singular strong limit cardinals.

Let

$$a \to \left(\binom{b}{c}, d\right)^2$$

denote that the following statement is true:

Whenever $|S| = a$ and $(\mathscr{T}_0, \mathscr{T}_1)$ is a 2-partition of type 2 of S, then either there exist S', S'', $S' \cap S'' = 0$, $|S'| = b$, $|S''| = c$ such that $x \in S'$, $y \in S''$ implies $\{x, y\} \in \mathscr{T}_0$ or there exists $S' \subset S$, $|S'| = d$ such that $[S']^2 \subset \mathscr{T}_1$. It is obvious that a more general symbol corresponding to Smybol-I can be defined and with the help of the methods of [**13**] we can discuss almost all the problems, e.g.

$$\text{G.C.H.} \Rightarrow \aleph_\omega \to \left(\binom{\aleph_\omega}{\aleph_\omega}, \aleph_\omega\right)^2$$

holds. We mention one which remains unsolved.

Problem 30. *Assume G.C.H. Does*

$$\aleph_{\omega+1} \not\to \left(\binom{\aleph_{\omega+1}}{\aleph_\omega}, \aleph_1\right)^2$$

hold?

Note that as a corollary of Theorems I, 10 of [**13**] we have

$$\text{G.C.H.} \Rightarrow \begin{pmatrix} \aleph_{\omega+1} \to (\aleph_{\omega+1}, \aleph_0)^2 \\ \aleph_{\omega+1} \not\to (\aleph_{\omega+1}, \aleph_1)^2 \end{pmatrix} \quad \text{and} \quad \aleph_{\omega+1} \not\to \left(\binom{\aleph_{\omega+1}}{\aleph_\omega}, \aleph_2\right)^2.$$

Problem 31. *Can one prove without assuming G.C.H. that*

$$2^{\aleph_0} \not\to \left(\binom{\aleph_1}{\aleph_0}, \binom{\aleph_1}{\aleph_0}\right),$$

or at least

$$2^{\aleph_0} \not\to \left(\binom{\aleph_1}{\aleph_1}, \binom{\aleph_1}{\aleph_1}\right)$$

holds?

Problem 32. *Assume G.C.H. Does there exist a graph $\mathscr{G} = \langle g, G\rangle$ with $|g| = \aleph_1$, not containing a subgraph of type $[\![\aleph_0, \aleph_1]\!]$ for which there exist a set $|S| = \aleph_1$ and a 2-partition $(\mathscr{T}_0, \mathscr{T}_1)$ of S such that neither the graph $(S, \mathscr{T}_0)$ nor the graph $(S, \mathscr{T}_1)$ contains a subgraph isomorphic to $\mathscr{G}$*?

This could be expressed by $\aleph_1 \not\to (\mathscr{G}, \mathscr{G})^2$ and would be a further strengthening of the relation

$$\aleph_1 \nrightarrow \left(\binom{\aleph_1}{\aleph_0}, \binom{\aleph_1}{\aleph_0}\right).$$

(For the graph terminology used here see, e.g. [**10**].)

9. **Problems on set mappings.** Let S be a set. A function f with domain $[S]^a$ or $[S]^{<a}$ and with $f(X) \subset S - X$, $|f(X)| < b$ is said to be a set mapping on S of order $\leq b$ and of type a ($<a$) respectively.

$S' \subset S$ is said to be a free subset if $f(X) \cap S' = 0$ for $X \in [S']^a$ (or $X \in [S']^{<a}$), respectively.

DEFINITION OF SYMBOL-VI. $(m, a, b) \xrightarrow{\text{VI}} n$ (or $(m, <a, b) \xrightarrow{\text{VI}} n$) is said to hold if for every S, $|S| = m$ and for every set mapping of type a (of type $<a$) and order $\leq b$ there exists a free subset $S' \subset S$, $|S'| = n$.

We introduced set mappings of type >1 in [**1**]. We will point out some problems stated in [**1**].

In [**1**] Theorem 7 we proved that $(m, <\aleph_0, b) \xrightarrow{\text{VI}} m$ holds for $b < m$ provided m is 0-1 measurable. In view of the recent results one can expect a positive answer to

Problem 33. *Does* $(m, <\aleph_0, 2) \overset{\text{VI}}{\nrightarrow} \aleph_0$ *or at least* $(m, <\aleph_0, \aleph_0) \overset{\text{VI}}{\nrightarrow} \aleph_0$ *hold for those m for which* $m \overset{\text{IV}}{\nrightarrow} (\aleph_0)_2^{<\aleph_0}$ *holds?*

As a matter of fact we could not prove this even for $m = \aleph_\omega$.

See Problem 1 of [**1**].

This problem is also relevant to Jónsson's problem, see [**9**, p. 22].

We know that $m \xrightarrow{\text{I}} (n)_b^k$ implies $(m, k, b^+) \xrightarrow{\text{VI}} n$ but here we do know that the positive results thus obtained are the best possible.

Problem 34.

(A) *Assume that m is regular and* $m \overset{\text{I}}{\nrightarrow} (m, m)^2$. *Does then* $(m, 2, 2) \overset{\text{VI}}{\nrightarrow} m$ *hold?*

(B) *Assume G.C.H. Does then*

$$(\aleph_2, 3, 2) \nrightarrow \aleph_1 \quad or \quad (\aleph_3, 3, \aleph_0) \nrightarrow \aleph_2$$

hold?

(Note that $m \nrightarrow (m, m)^2$ implies $m \nrightarrow (m, 4)^3$ [**27**] and $\aleph_2 \nrightarrow (\aleph_1)_2^3$, $\aleph_3 \nrightarrow (\aleph_2)_{\aleph_1}^3$ holds if G.C.H. is assumed. See [**13**].) If m is a singular strong limit cardinal then $m \nrightarrow (m, m)^2$ but $(m, 2, 2) \rightarrow m$.

Problem 35 (Hajnal). *Assume G.C.H. Let S be a set $|S| = \aleph_{\omega+1}$. Let f be a set mapping on S of type 1 and order $\leq \aleph_{\omega+1}$. Assume further that $|f(x) \cap f(y)| < \aleph_\omega$ for every pair $x \neq y \in S$.*

Does there exist a free subset of power $\aleph_1$?

In [**25**, Theorem 1] assuming G.C.H. this problem is settled in the negative for cardinals $\aleph_{\alpha+1}$ where $\aleph_\alpha$ is regular ($\aleph_1$ stands for $\aleph_{\text{cf}(\omega)+1}$).

Problem 36 (Hajnal). *Let S, $\prec$ be an ordered set of type ω_1. Let f be a set mapping on S of order $\leq \omega_1$ and of type 1. Assume further that $|f(x) \cap f(y)| < \aleph_0$ for every pair $x \neq y \in S$.*

Let $\alpha < \omega_1$. Does there then exist a free subset S', such that typ $S'(\prec) = \alpha$?

Note that the answer is positive if $\alpha < \omega^2 \cdot 2$. (See **[25]**.)

The following problem on almost disjoint sets is strongly connected to the problems mentioned above

Let $|S| = b$; does there exist a system $\mathscr{F} \subset [S]^a$ such that $\mathscr{F}$ is almost disjoint, i.e. $A, B \in \mathscr{F}$, $A \neq B$ implies $|A \cap B| < a$ and such that for $S' \subset S$, $|S'| = a^+$ there is an $A \in \mathscr{F}$, $A \subset S'$? It was proved in **[25]** that the answer is yes if $a \geq \aleph_0$ is regular and $b = a^+$ and G.C.H. holds. The simplest unsolved problems are

Problem 37 (Hajnal). *What is the answer to the above stated problem if G.C.H. holds and*

(A) $a = \aleph_\omega$, $b = \aleph_{\omega+1}$,

(B) $a = \aleph_0$, $b = \aleph_2$?

Many special and difficult problems arise if we consider set mappings of type 1 on the set R of real numbers under different conditions imposed on the sets $f(x)$. Here are some typical unsolved ones.

Problem 38. *Let f be a set mapping of type* 1 *on R.*

(A) *Assume that f is nowhere dense in R. Does there then exist a free subset of power $\aleph_1$?*

(B) *Let f be closed and of measure ≤ 1. Does there then exist a free subset of at least* 3 *points?*

(C) *Let f be bounded and of outer measure ≤ 1. Does there then exist an infinite independent set?*

REMARKS. In case (A) A. Máté **[31]** proved that for every $\alpha < \omega_1$ there exists an independent set of type α. We do not even know the answer in case (A), if $f(x)$ is an ω sequence with limit point x for every x.

In case (B), Gladysz **[24]** proved that there is an independent pair.

In case (C) we proved in **[2]** that for every $k < \omega$ there exists an independent set of k elements, but an independent set of power $\aleph_1$ does not necessarily exist.

10. Problems on families of sets stated in [4].

DEFINITION. A family $\mathscr{F}$ is said to have *property* B if there is a set B such that $A \cap B \neq 0$ and $A \cap -B \neq 0$ for every $A \in \mathscr{F}$. $\mathscr{F}$ is said to *have property* B(s) if there is a set B such that $1 \leq |A \cap B| < s$ for every $A \in \mathscr{F}$.

Problem 39. *Assume G.C.H. Let $|S| = \aleph_{\omega+1}$, $\mathscr{F} \subset [S]^{\aleph_1}$ and assume $|A \cap B| < \aleph_0$ for every pair $A \neq B \in \mathscr{F}$.*

Does then $\mathscr{F}$ possess property B($\aleph_1$) *or at least property* B?

The statement is true if $\aleph_{\omega+1}$ is replaced by a smaller cardinal.

For the background see **[4]**. We think that this is a problem of type (B) if not of type (A).

The following problem seems to be of type (A). There are many possible versions in which to formulate it. One of these is the following.

Problem 40. *Is it true that every $\mathscr{F}$, with $|\mathscr{F}| < 2^{\aleph_0}, \mathscr{F} \subset [S]^{\aleph_0}$ has property* B? (If C.H. holds, the answer is obviously yes.)

In [**4**] we have stated the following problem (7):

Assume $\mathscr{G} = \langle g, G \rangle$ is a graph of $\aleph_2$ vertices. Suppose that every subgraph $\mathscr{G}'$ of it spanned by at most $\aleph_1$ vertices has chromatic number $\leq \aleph_0$. Does then $\mathscr{G}$ have chromatic number $\leq \aleph_0$? Recently in [**12**] we proved that the answer is negative.

We have proved assuming G.C.H. that there exist graphs $\mathscr{G}_{k+1}$ of power $\aleph_{k+1}$ for every $k < \omega$ such that every subgraph of $\mathscr{G}_{k+1}$ spanned by at most $\aleph_k$-vertices has chromatic number $\leq \aleph_0$, but $\mathscr{G}_{k+1}$ has chromatic number greater than $\aleph_0$. (In fact, we prove a more general result which can even be formulated without G.C.H.)

The following problems remain open:

Problem 41. *Assume G.C.H.*

(A) *Does there exist a graph $\mathscr{G}$ of $\aleph_{\omega+1}$ vertices, with chromatic number $> \aleph_0$, such that every subgraph $\mathscr{G}'$ spanned by less than $\aleph_{\omega+1}$ vertices has chromatic number at most $\aleph_0$?*

(B) *Does there exist a graph $\mathscr{G}$ with $\aleph_2$ vertices with chromatic number $\aleph_2$ such that each subgraph spanned by less than $\aleph_2$ vertices has chromatic number $\leq \aleph_0$?*

Note that a corresponding genuine problem can easily be formulated without G.C.H. It is quite possible that the answer to 41/A is yes if $\aleph_{\omega+1}$ is replaced by any regular cardinal a which is not too large ($a \in C_0 \wedge [\aleph_1 a] \subseteq C_1$ of Keisler-Tarski [**30**]).

Our methods of [**12**] break down for some very similar problems stated in [**4**].

Problem 42. *Assume G.C.H.*

(A) *Does there exist a family $\mathscr{F}$, $|\mathscr{F}| = \aleph_2$, $\mathscr{F} \subseteq [S]^{\aleph_0}$ for some S, such that if $\mathscr{F}' \subset \mathscr{F}$, $|\mathscr{F}'| = \aleph_1$ then $\mathscr{F}'$ possesses property* B *and $\mathscr{F}$ does not possess property* B?

(B) *Does there exist a graph $\mathscr{G}$ of power $\aleph_2$ such that every subgraph $\mathscr{G}'$ spanned by less than $\aleph_2$ vertices can be directed so that the number of directed edges emanating from a vertex is finite for every vertex, but this is no longer true for the graph $\mathscr{G}$?*

(C) (W. Gustin) *A family $\mathscr{F}$ is said to have property* G *if there exists a function f with $D(f) = \mathscr{F}$ such that $f(F) \in \mathscr{F}$ and $f(A) \neq f(B)$ for every pair $A \neq B \in \mathscr{F}$.*

Does there exist a family $\mathscr{F}$, $|\mathscr{F}| = \aleph_2$, $\mathscr{F} \subset [S]^{\aleph_0}$ such that every subfamily $\mathscr{F}' \subset \mathscr{F}$, $|\mathscr{F}'| < \aleph_2$ has property G, *but F does not possess property* G?

We mention that a number of related problems are stated in [**4**] which we do not repeat here. Note that in view of the results stated in [**28**] and [**30**] Problem 10 of [**4**] has already been solved. We state one more problem of [**4**] which seems to be of type (A) but of quite different character.

Problem 43. *Let S be the set of ordinals less than ω_1. Does there exist a function f with $D(f) = \omega_1$, $R(f) \subset \omega_1$ such that $f(\xi) < \xi$ for every $\xi < \omega_1$,* *and for every limit number ξ there exists an increasing sequence of ordinals ξ_n, $n < \omega$, $\lim \xi_n = \xi$ such that $\xi_n = f(\xi_{n+1})$ for $n < \omega$?*

ADDED IN PROOF. J. E. Baumgartner proved that the answer to Problem 43 is affirmative, and that it is certainly not of type (A).

11. **Problems on chromatic and colouring numbers of graphs [10], [11].** For the results underlying Problems 44–50 see [**10**] and for the rest see [**11**].

Problem 44. *Assume C.H. Does there exist a graph $\mathcal{G}$ with $\aleph_1$ vertices of chromatic number $\aleph_1$ which does not contain a complete graph of 3 elements and does not contain a complete even graph $[\![\aleph_0, \aleph_0]\!]$?*

ADDED IN PROOF (May, 1970). A positive answer is given in Hajnal's paper mentioned on p. 24.

(A complete even graph $[\![a, b]\!]$ consists of two disjoint sets $|A| = a$, $|B| = b$ and of the edges with one endpoint in A and one endpoint in B.) The answer is affirmative by [**10**, 5.9] if 3 is replaced by $\aleph_0$.

A positive answer would be implied by

Problem 45.

(A) *Assume $\mathcal{G}$ is a graph of chromatic number $a \geq \aleph_0$. Does then $\mathcal{G}$ contain a subgraph $\mathcal{G}'$ of chromatic number a such that $\mathcal{G}'$ does not contain a triangle?*

(This might be a problem of type (D).)

(B) *Does there exist a function $f(k) < \omega$, for $k < \omega$ such that $f(k) \to +\infty$ if $k \to +\infty$ and such that every graph with chromatic number $\geq k$ contains a subgraph of chromatic number $\geq f(k)$ not containing triangles?*

Problem 46. *Let $\mathcal{G}$ be a graph of chromatic number greater than $\aleph_0$. Does then $\mathcal{G}$ contain odd circuits of length $2i + 1$ for $i > j$ for some $j < \omega$?*

The answer is affirmative if $\mathcal{G}$ has chromatic number greater than $\aleph_1$.

Problem 47. *Let $\mathcal{G}$ be a graph of chromatic number $\aleph_0$ and put $N = \{i < \omega:$ there is a circuit of length i contained in $\mathcal{G}\}$. Is it true that*

$$\sum_{i \in N} \frac{1}{i} = +\infty?$$

DEFINITION. The *colouring number of a graph $\mathcal{G}$* is the smallest cardinal b for which the set of vertices has a well-ordering $\prec$ satisfying

$$|\{y \prec x : y \text{ is connected to } x \text{ in } \mathcal{G}\}| < b$$

for every vertex x.

The colouring number of a graph is greater than or equal to its chromatic number.

The problem involved in the Symbol-VII to be defined is due to R. Rado.

DEFINITION OF SYMBOL-VII. $(a, b) \xrightarrow{\text{VII}} (c, d)$ is said to hold if every graph $\mathcal{G}$ with a vertices, all whose subgraphs spanned by a set of power $<b$ have colouring number $\leq c$, has colouring number $\leq d$.

In [**10**] we prove several results concerning this symbol.

Problem 48. *Assume G.C.H.*

(A) $(\aleph_2, \aleph_2) \overset{\text{VII}}{\not\longrightarrow} (\aleph_0, \aleph_0)$?

(B) $(\aleph_{\omega+1}, \aleph_2) \overset{\text{VII}}{\not\longrightarrow} (\aleph_0, \aleph_1)$?

Here $(\aleph_2, \aleph_2) \overset{\text{VII}}{\longrightarrow} (\aleph_0, \aleph_1)$ is true.

This should be compared with Problem 41/A.

We prove in [**10**, Theorems 9.1 and 10.1]

$$(a, \aleph_0) \overset{\text{VII}}{\longrightarrow} (k, 2k-2) \quad \text{for} \quad 2 \le k < \omega, \quad a \text{ arbitrary}$$

but

$$(\aleph_0, \aleph_0) \overset{\text{VII}}{\not\longrightarrow} (k, 2k-3) \quad \text{and} \quad (\aleph_n, \aleph_n) \overset{\text{VII}}{\not\longrightarrow} (k, 2k-3), n < \omega$$

provided G.C.H. holds.

Problem 49. *Assume G.C.H. Is* $(\aleph_{\omega+1}, \aleph_{\omega+1}) \overset{\text{VII}}{\longrightarrow} (k, l)$ *true for some* $l < 2k-2$?

Problem 50. *Assume G.C.H. Is it true that if* $\mathcal{G}$ *has* $\aleph_{\omega+1}$ *vertices and does not contain a complete even graph* $[\![\aleph_0, \aleph_2]\!]$ *then* $\mathcal{G}$ *has colouring number* $\le \aleph_1$?

This is true if $\aleph_{\omega+1}$ is replaced by a smaller cardinal a. This should be compared with Problem 39, since the method for proving the theorems for $a < \aleph_{\omega+1}$ is very similar.

We turn to graph decomposition problems considered in [**11**].

DEFINITION OF SYMBOL-VIII. Let $\mathcal{G} = \langle g, G\rangle$ be a graph with set of vertices g and with set of edges $G \subset [g]^2$.

The sequence $\mathcal{G}_\xi = \langle g_\xi, G_\xi\rangle$, $\xi \prec \varphi$, is said to be an edge decomposition of type $|\varphi|$ of $\mathcal{G}$ if $g_\xi = g$ and $\bigcup_{\xi<\varphi} G_\xi = G$. (An edge decomposition of a complete graph $G = \langle g, [g]^2\rangle$ is a two partition of the set g.) $(a, b) \overset{\text{VIII}}{\longrightarrow} (c, d)$ denotes that every graph $\mathcal{G}$ of a vertices not containing a complete subgraph of power b has an edge decomposition $\mathcal{G}_\xi$, $\xi < \Omega(c)$, of type C where the members $\mathcal{G}_\xi$ do not contain a complete d graph.

Though the problems seem to be quite fundamental our results are very sketchy.

We know, e.g.

$$((2^{(2^{\aleph_0})^+})^+, \aleph_0) \overset{\text{VIII}}{\not\longrightarrow} (\aleph_0, k)$$

holds for every $k < \omega$[i.e. G.C.H. $\Rightarrow (\aleph_4, \aleph_0) \overset{\text{VIII}}{\not\longrightarrow} (\aleph_0, k)$, $k < \omega$], but probably the relation in the following problem is true.

Problem 51. $((2^{\aleph_0})^+, \aleph_0) \overset{\text{VIII}}{\not\longrightarrow} (\aleph_0, k)$ *for* $k < \omega$?

(Note that $(2^{\aleph_0}, (2^{\aleph_0})^+) \overset{\text{VIII}}{\longrightarrow} (\aleph_0, 3)$ is trivial since $2^{\aleph_0} \overset{\text{I}}{\not\longrightarrow} (3)^2_{\aleph_0}$ holds.)

Problem 52. $((2^{\aleph_0})^+, \aleph_1) \overset{\text{VIII}}{\longrightarrow} (\aleph_0, \aleph_0)$?

We do not know if $(m, \aleph_1) \overset{\text{VIII}}{\not\longrightarrow} (\aleph_0, \aleph_0)$ holds for any $m > 2^{\aleph_0}$. So it might be that every graph not containing a complete $\aleph_1$ graph can be decomposed into the union of $\aleph_0$ graphs not containing complete $\aleph_0$ graphs. However, this seems to be very unlikely. We suspect that $((2^{\aleph_0})^+, \aleph_1) \overset{\text{VIII}}{\not\longrightarrow} (\aleph_0, \aleph_0)$ holds or at least assuming G.C.H. one can prove $(\aleph_2, \aleph_1) \overset{\text{VIII}}{\not\longrightarrow} (\aleph_0, \aleph_0)$.

Note that $\aleph_1$ has a special role in Problem 52. We know, e.g. that

$$((2^{\aleph_0})^+, (2^{\aleph_0})^+) \overset{\text{VIII}}{\not\longrightarrow} (\aleph_0, \aleph_1)$$

holds; i.e. assuming G.C.H.

$$(\aleph_2, \aleph_2) \overset{\text{VIII}}{\not\longrightarrow} (\aleph_0, \aleph_1),$$

and trivially

$$(\aleph_2, \aleph_3) \overset{\text{VIII}}{\longrightarrow} (\aleph_1, 3).$$

We do not know the answer for the following.

Problem 53. *Does* $(m, k + 1) \overset{\text{VIII}}{\not\longrightarrow} (\aleph_0, k)$ *hold for any* $m > 2^{\aleph_0}$, $3 \leq k < \omega$? (*Does* $((2^{\aleph_0})^+, 4) \overset{\text{VIII}}{\longrightarrow} (\aleph_0, 3)$ *hold*?)

There is a very interesting finite problem here. It is obvious that $(a, b^+) \overset{\text{VIII}}{\not\longrightarrow} (c, d)$, $a \geq b$ holds if $b \overset{\text{I}}{\longrightarrow} (d)^2_c$ is true. One can ask if this is a best possible condition if a, b, c, d are finite. This is certainly not so, since 6 is the smallest number for which $b \overset{\text{I}}{\longrightarrow} (3)^2_2$ holds but there is an $a(2, 3) = a$ for which $(a, 4) \overset{\text{VIII}}{\not\longrightarrow} (2, 3)$ holds. (This was proved by Volkmann, but the involved proof is still unpublished, a is very large.)

It is reasonable to conjecture

Problem 54. *For every pair of integers c, d there exists an integer $a(c, d)$ such that*

$$(a(c, d), d + 1) \overset{\text{VIII}}{\not\longrightarrow} (c, d)$$

holds.

We cannot make any guess on the order of magnitude of $a(c, d)$.

12. **Problems of [14] and [8].** In [14] we consider several arrow relations of new type. We point out only one problem of the 15 problems stated there.

DEFINITION OF SYMBOL-IX. Let a, b, c, d be types or cardinals, e a cardinal but b, c, d are cardinals if a is a cardinal. Let $S, \prec$ be an ordered set of type a or a set of cardinal a respectively. $a \overset{\text{IX}}{\longrightarrow} [b, c]^d_e$ is said to hold if for every family $\mathscr{F}$, $|\mathscr{F}| = e$ of subsets of S, either there is an S' of typ $S'(\prec) = b$ ($|S'| = b$) such that for every $X \subset S'$, typ $X(\prec) = d$ ($|X| = d$) there is an $A \in \mathscr{F}$, $X \subset A$ or there is an $S'' \subset S$, typ $S''(\prec) = c$ ($|S''| = c$) and an $\mathscr{F}' \subset \mathscr{F}$, $|\mathscr{F}'| = e$ such that $S'' \cap \bigcup \mathscr{F}' = 0$ respectively.

Problem 55. *Assume G.C.H.*

(A) $\aleph_2 \xrightarrow{\text{IX}} [\aleph_2, a]^{\aleph_0}_{\aleph_2}$ $a = \aleph_1 \vee a = \aleph_2$?

(B) $\aleph_{\omega+1} \xrightarrow{\text{IX}} [\aleph_{\omega+1}, \aleph_0]^{\aleph_\omega}_{\aleph_{\omega+1}}$?

(C) $\aleph_{\omega_1} \xrightarrow{\text{IX}} [\aleph_{\omega_1}, \aleph_0]^{\aleph_0}_{\aleph_1}$?

Though these problems seem to be of type (E), because of the involved formulation, they are certainly difficult and, e.g. Problem 55/A might even be of type (A).

As usual the problems for types are more ramified and we do not have the space to discuss them here.

In [**8**] we considered problems of the following type.

Let S be a set $|S| = a$ and let f be a function defined on $[S]^k$, which associates a Lebesgue measurable subset $f(X)$ of $[0, 1]$ of Lebesgue measure $m(X) \geq u$ to every $X \in [S]^k$. For what type of subsets $\mathcal{T}$ of $[S]^k$ does there necessarily exist a $v \in [0, 1]$ such that $v \in \bigcap_{x \in \mathcal{T}} f(X)$?

We define a corresponding *Symbol*-X $(a, u)^k \xrightarrow{\text{X}} \Delta$ where Δ stands for the corresponding class of subsets of $[S]^k$. We have genuine results only in case $k = 2$. We mention two of them:

$$(\aleph_0, u)^2 \xrightarrow{\text{X}} [\![s + 1]\!] \quad \text{iff} \quad u > 1 - \frac{1}{s} \quad \text{for} \quad 2 \leq s < \omega$$

where $[\![s + 1]\!]$ stands for the class of complete subgraphs of $s + 1$ elements.

If $m > \aleph_0$ then $(m, u)^2 \xrightarrow{\text{X}} \aleph_0$ for every positive u.

We can prove $(2^{\aleph_0}, u)^2 \not\xrightarrow{\text{X}} \aleph_1$ for $u \leq \frac{1}{2}$, but our proof for $(2^{\aleph_0}, u)^2 \not\xrightarrow{\text{X}} \aleph_1$ for u arbitrary uses C.H.

Problem 56. *Can one prove* $(2^{\aleph_0}, u)^2 \not\xrightarrow{\text{X}} \aleph_1$ *for some* $u > \frac{1}{2}$ *without using C.H.*?

It is clear from the remarks given in [**8**] that this problem is strongly connected with Problem 15.

Here are two other problems of [**8**] we are interested in.

Problem 57.

(A) $(\aleph_1, u)^3 \xrightarrow{\text{X}} [\![4]\!]$ *for* $u > 0$,

(B) $(\aleph_0, u)^2 \xrightarrow{\text{X}} [\![\aleph_0, \aleph_0]\!]$ *for* $u > \frac{1}{2}$,

where $[\![\aleph_0, \aleph_0]\!]$ *is a complete* $\aleph_0, \aleph_0$ *even graph*?

We know that (B) is false for $u \leq \frac{1}{2}$.

13. **Miscellaneous unpublished problems.**

Problem 58. *Let* S *be a set,* $|S| = a > 2^{\aleph_0}$. *Does there exist a disjoint* $\aleph_0$-*partition* $\bigcup_{\nu < \Omega(2^{\aleph_0})} \mathcal{T}_\nu = [S]^{\aleph_0}$ *of* S *satisfying the following condition*?

Whenever A_n, $n < \omega$ is a sequence of disjoint subsets of S, $|A_n| = 2$ for every $n < \omega$ then for every $\nu < \Omega(2^{\aleph_0})$ there is an $X \in \mathcal{T}_\nu$ such that X is a transversal of the sequence A_n, $n < \omega$.

If the answer is affirmative this is an improvement of the theorem mentioned before Problem 14/A. Note that for $a = 2^{\aleph_0}$ the answer is yes.

For $a > 2^{\aleph_0}$ we do not even know the answer for partitions of type 2.

ADDED IN PROOF. Kunen and F. Galvin proved that the answer to Problem 58 is affirmative for partitions of type 2 and of type $2^{\aleph_0}$ respectively.

DEFINITION OF SYMBOL-XI. Let S, $|S| = a$ be a set and let $\mathscr{A}$, $\mathscr{B}$ be classes of subgraphs of the complete graph with vertices S. $(a, b, \mathscr{A}) \xrightarrow{\text{XI}} (c, \mathscr{B})$ is said to hold if the following statement is true.

Whenever $\mathscr{G}_\xi$, $\xi < \Omega(b)$ is a sequence of graphs $\mathscr{G}_\xi \in \mathscr{A}$ then there is a $\mathscr{G}^* \in \mathscr{B}$ and a set C of ordinals $<\Omega(b)$ $|C| = c$ such that for $\xi \in C$, $\mathscr{G}^*$ and $\mathscr{G}$ have no common edges.

We have several unpublished results on Symbol-XI.

Let $\mathscr{A}(d, a)$ be the set of subgraphs of a complete graph of a vertices not containing complete d-graphs. Let $\mathscr{B}(d, a)$ be the set of complete subgraphs of a complete graph of a vertices spanned by d elements.

We can prove the following results:

(1) $$(\aleph_1, \aleph_0, \mathscr{A}(3, \aleph_1)) \xrightarrow{\text{XI}} (\aleph_0, \mathscr{B}(\aleph_0, \aleph_1))$$

and

(2) $$(\aleph_0, \aleph_1, \mathscr{A}(3, \aleph_1)) \xrightarrow{\text{XI}} (\aleph_0, \mathscr{B}(\aleph_0, \aleph_0))$$

but

(3) $$(\aleph_1, \aleph_0, \mathscr{A}(3, \aleph_1)) \overset{\text{XI}}{\not\longrightarrow} (\aleph_0, \mathscr{B}(\aleph_1, \aleph_1))$$

provided C.H. holds.

The following seems to be an intriguing unsolved case.

Problem 59. *Does* $(\aleph_1, \aleph_0, \mathscr{A}(4, \aleph_1)) \xrightarrow{\text{XI}} (\aleph_0, \mathscr{B}(\aleph_0, \aleph_1))$ *hold or does* $(\aleph_0, \aleph_1, \mathscr{A}(4, \aleph_0)) \xrightarrow{\text{XI}} (\aleph_0, \mathscr{B}(\aleph_0, \aleph_0))$ *hold?*

We also do not know whether the relation in the following problem is true.

Problem 60. *Does* $(\aleph_0, \aleph_1, \mathscr{A}(\aleph_0, \aleph_0)) \overset{\text{XI}}{\not\longrightarrow} (\aleph_0, \mathscr{B}(\aleph_0, \aleph_0))$ *hold?*

ADDED IN PROOF. Using the polarized partition relation V one can express Problems 59 and 60 as follows:

$$\begin{pmatrix}\aleph_0\\ \aleph_1\end{pmatrix} \to \begin{pmatrix}1 & \aleph_0\\ 4 & \aleph_0\end{pmatrix}^{1,2}$$

and

$$\begin{pmatrix}\aleph_1\\ \aleph_0\end{pmatrix} \to \begin{pmatrix}1 & \aleph_0\\ \aleph_0 & \aleph_0\end{pmatrix}^{1,2}$$

respectively.

The answer to both of these problems is affirmative. As to Problem 59 the following is true: If $a \geq \aleph_0$ is 0,1-measurable then

$$\begin{pmatrix}a\\ a^+\end{pmatrix} \to \begin{pmatrix}a & \vee & 1 & a\\ b & & a & a\end{pmatrix}^{1,2}$$

holds for every $b < a$.

The proof of this will be published in a forthcoming paper of A. Hajnal in the Fundamenta Mathematicae.

As to Problem 60 we proved that

$$\binom{a^+}{a} \to \binom{a}{a}_c^{1,2}$$

holds for every 0-1 measurable cardinal $a \geq \aleph_0$, $c < a$.

Then F. Galvin proved that

$$\binom{\aleph_1}{\aleph_0} \to \binom{\aleph_0}{\aleph_0}_c^{1,r}$$

holds for $r, c < \aleph_0$.

He conjectured that

$$\binom{a^+}{a} \to \binom{a}{a}_c^{1,<\aleph_0}$$

will hold for $c < a$ where a is 0-1 measurable cardinal greater than $\aleph_0$.

This was proved by A. Hajnal. For the proof see the above mentioned paper.

We give some more samples of the existing results.

Let $\mathscr{A}([\![a_1, a_2]\!], a)$ be the class of subgraphs of the complete graph $[\![a]\!]$ not containing complete even $[\![a_1, a_2]\!]$ graphs.

We have

$$(4) \qquad (\aleph_0, \aleph_0, \mathscr{A}([\![l, \aleph_0]\!])\aleph_0) \cap \mathscr{A}(k, \aleph_0)) \xrightarrow{\text{XI}} (\aleph_0, \mathscr{B}(\aleph_0, \aleph_0))$$

for every $l, k < \omega$ but a $\overset{\text{XI}}{\not\longrightarrow}$ relation holds with each of the classes standing on the right-hand side for every $l < \omega$ or $k < \omega$ respectively.

Let $\mathscr{B}([\![a_1, a_2]\!], a)$ be the class of complete even $[\![a_1, a_2]\!]$ subgraphs of the complete graph a.

We know

$$(5) \qquad (\aleph_1, \aleph_1, \mathscr{A}(\text{tree}, \aleph_1)) \overset{\text{XI}}{\not\longrightarrow} (\aleph_1, \mathscr{B}([\![\aleph_0, \aleph_0]\!], \aleph_1)).$$

But we do not know

Problem 61. $(\aleph_1, \aleph_0, \mathscr{A}\,(\text{tree}, \aleph_1)) \xrightarrow{\text{XI}} (\aleph_0, \mathscr{B}(\aleph_1, \aleph_1))$

$$\text{or} \xrightarrow{\text{XI}} (\aleph_0, \mathscr{B}([\![\aleph_0, \aleph_1]\!], \aleph_1))?$$

With $\mathscr{B}(\aleph_0, \aleph_1)$ we have an $\xrightarrow{\text{XI}}$ relation because of (1). (We do not even know the answer in case of disjoint trees.)

The results mentioned above will be published in a forthcoming paper by the two of us.

ADDED IN PROOF. Using C.H. the answer to Problem 61 is negative. In fact we have much stronger negative results. See our forthcoming paper mentioned above.

14. **Miscellaneous problems continued.** As a corollary of Lemma 5A of **[13]** we know that

$$2^{2^{\cdot^{\cdot^{2^{\aleph_0}}}}}\ (r-1) = \exp_{r-1}(\aleph_0) \nrightarrow (\aleph_1)^r_2 \quad \text{for} \quad 2 \leq r < \omega.$$

We do not know

Problem 62. *Let* $2 < r < \omega$, $|S| = \exp_{r-1}(\aleph_0)$. *Does there exist an r-partition* $(\mathscr{T}^r_0, \mathscr{T}^r_1)$ *of type* 2 *of S satisfying the following conditions?*

(1) $S' \subset S, |S'| = \aleph_1$ *implies* $[S']^2 \not\subset \mathscr{T}_i$ *for* $i < 2$ *but*

(2) $S' \subset S, |S'| = \aleph_1$ *implies that for every* $n < \omega$ *and for every* $i < 2$ *there is an* $S''_i \subset S', |S''_i| = n$ *such that* $[S''_i]^r \subset \mathscr{T}_i$.

It is possible that (2) can be replaced by the stronger condition

(2′) $S' \subset S$, $|S'| = \aleph_1$ *implies that there are* $S''_i \subset S', |S''_i| = \aleph_0$ *such that* $[S''_i]^2 \subset \mathscr{T}_i$ *for* $i < 2$.

A positive answer to Problem 62 would be an improvement of the theorem of **[13]** already mentioned and it would be useful for the discussion of the following general problem which we formulated from an old result of W. Sierpiński **[36]**.

DEFINITION OF SYMBOL-XII. A family $\mathscr{F}$ of sets is said to have property $B(a, b), b \geq 3$, if $\mathscr{F}' \subset \mathscr{F}$, $|\mathscr{F}'| = a$ implies that for every $b' < b$ there is an $\mathscr{F}'' \subset \mathscr{F}'$, $|\mathscr{F}''| = b'$, $\bigcap \mathscr{F}'' \neq 0$. ($B(a, 3)$ means that $\mathscr{F}$ does not contain a disjointed subfamily of power a.)

$(m, n) \xrightarrow{\text{XII}} (a, b)$ is said to hold if the Cartesian product of two families having property $B(a, b)$ has property $B(m, n)$.

It is easy to see that

$$(1) \qquad (m, b^+) \xrightarrow{\text{XII}} (a, b^+)$$

is equivalent to

$$m \xrightarrow{\text{I}} (a)^b_2 \quad \text{for} \quad b \leq \omega;$$

hence, e.g. $(2^{\aleph_0}, 3) \nrightarrow (\aleph_1, 3)$.

This was proved by Sierpiński and in fact his example gives $(2^{\aleph_0}, 3) \nrightarrow (\aleph_1, \aleph_1)$ but we do not know

Problem 63. *Assume* $3 \leq r < \omega$.
Is it true that

$$(\exp_{r-1}(\aleph_0), r + 1) \nrightarrow (\aleph_1, \aleph_0)$$

or

$$(\exp_{r-1}(\aleph_0), r + 1) \nrightarrow (\aleph_1, \aleph_1)$$

holds?

A positive answer to Problem 62 implies a positive answer to Problem 63.

Problem 64. *Do there exist two families* $\mathscr{F}_1, \mathscr{F}_2$ *both having property* $B(\aleph_1, \aleph_0)$ *and such that* $\mathscr{F} = \mathscr{F}_1 \times \mathscr{F}_2$ *does not possess property* $B(\exp_k(\aleph_0), \aleph_0)$ *for every* $k < \omega$?

We have some more partial results on the Symbol-XII which will be published later.

15. Miscellaneous problems continued.

Problem 65. *Let* $|S| = \aleph_1$ *and let* $(\mathscr{T}_0, \mathscr{T}_1)$ *be a 2 partition of type 2 of S. Let further* S_n, $n < \omega$ *be a disjointed sequence of subsets of S such that* $|S_0| = \aleph_1$, $|S_n| = \aleph_0$ *for* $0 < n < \omega$. *Does there then exist an increasing sequence of integers* $(n_k)_{k<\omega}$, $n_0 = 0$ *and a sequence* B_k *of subsets of S such that* $|B_k| = \aleph_0$, $B_k \subset S_{n_k}$ *and*

$$[B_0, B_k]^{1,1} = \{\{xy\}: x \in B_0 \wedge y \in B_k\} \subset \mathscr{T}_0$$

for some $i < 2$ *and for every* $1 < k < \omega$?

Problem 66 (Erdös, Hajnal, Milner). *Let* $\mathscr{G} = \langle g, G\rangle$ *be a graph and let* $\prec$ *be a well-ordering of the set g such that* $\text{typ}\, g(\prec) = \omega_1^\rho$, $\rho < \omega_2$. *Assume that* $\mathscr{G}$ *does not contain an infinite path. Does there then exist a subset* $g' \subset g$, $\text{typ}\, g'(\prec) = \omega_1^\rho$ *such that* g' *does not contain an edge*?

We know that the answer is affirmative for $\rho \leq \omega + 1$. The first unsolved case is $\rho = \omega + 2$.

In a forthcoming paper with E. Milner we will prove that if the condition that $\mathscr{G}$ does not contain an infinite path is replaced by the condition that $\mathscr{G}$ does not contain a quadrilateral then the answer is affirmative for every ρ even if ω_1 is replaced by ω_α.

The following simple problem seems to be strongly connected with well-known problems concerning denumerable order types.

Problem 67 (Erdös, Milner, Hajnal). *Let* $\mathscr{G} = \langle g, G\rangle$ *be a graph such that* $|g| = \aleph_0$ *and let* $\prec$ *be an arbitrary ordering of g.*

Assume $\mathscr{G}$ *does not contain a quadrilateral. Put* $\Theta = \text{typ}\, g(\prec)$ *and assume that*

(a) $$\text{typ}(g - \{x\})(\prec) \geq \Theta \quad \textit{for every} \quad x \in g.$$

Does then g contain a subset g', $\text{typ}\, g'(\prec) = \Theta$ *such that* g' *contains no edge of* $\mathscr{G}$?

The problem whether for an arbitrary Θ there exist only finitely many vertices x which do not satisfy (a), is equivalent to the well-known problem whether a denumerable order type has only finitely many fixed points.

ADDED IN PROOF (May, 1970). Laver proved that the answer to Problem 67 is affirmative.

Problem 68. *Assume G.C.H. Let S be a set* $|S| = \aleph_1$ *and let* $[S]^2 = \mathscr{T}_0 \cup \mathscr{T}_1 \cup \mathscr{T}_2$ *be a 2-partition of type 3 of S. Assume* $[S'] \subset \bigcup_{i \neq j < 3} \mathscr{T}_j$ *implies* $|S'| \leq \aleph_0$ *for every* $i < 3$. *Does there then exist a subset* $X \subset S$, $|X| = 3$ *such that*

$$[X]^2 \cap \mathscr{T}_i \neq 0 \qquad \textit{for} \quad i < 3?$$

This problem is stated in **[12]** where several similar problems are formulated for the case $|S| = \aleph_1$. On the other hand, using the methods of **[13]** some general results can be proved which we preserve for later publication. We mention only one of them.

If $|S| = \aleph_1$ and $(\mathscr{T}_\nu)_{\nu<\omega_1}$ is a 2-partition of type ω_1 of S such that $[S']^2 \subset \bigcup_{\nu \neq \mu; \mu < \omega_1} \mathscr{T}'_\mu$ implies $|S'| \leq \aleph_0$ for $\nu < \omega_1$ then there is a subset $X \subset S$, $|X| = \aleph_0$ such that every pair of X belongs to different $\mathscr{T}_\nu$'s.

Problem 69. *Let $\mathcal{G} = \langle g, G\rangle$ be a graph $|g| = \aleph_1$. Assume that for every $g' \subset g$ there is a $g'' \subset g'$ such that g'' is finite and each vertex of $g' - g''$ is adjacent to at least one element of g''.*

Does then $\mathcal{G}$ contain a complete $\aleph_1$ graph?

Let $\mathcal{G} = \langle g, G\rangle$ be a graph and let $\mathcal{F}$ be a family of sets. We will briefly say that $\mathcal{F}$ *a-represents* $\mathcal{G}$ if there is a one-to-one mapping φ of g onto $\mathcal{F}$ such that

$$A \neq B \in g \text{ are connected in } \mathcal{G} \text{ iff } |f(A) \cap f(B)| < a.$$

Assuming G.C.H. we can prove that if a is regular then every graph $\mathcal{G}$ of at most a^+ vertices can be a-represented by a family $\mathcal{F}$ of subsets of a set of power a. We cannot answer

Problem 70. *Assume G.C.H. Let $\mathcal{G}$ be a graph of $\aleph_{\omega+1}$ vertices. Can it be $\aleph_\omega$-represented by a family $\mathcal{F}$ of subsets of a set of power $\aleph_\omega$?*

Let A be a set and $\mathcal{F}$ a family of subsets of A. Let $a = \{a_\xi\}_{\xi<\varphi}$ be a sequence of type φ of elements of A. $\mathcal{F}$ *strongly cuts* a if for every $\xi < \varphi$ there exists an $A_\xi \in \mathcal{F}$ such that $A_\xi \cap a = \{a_\zeta\}_{\zeta<\xi}$.

In **[20]** P. Erdös and M. Makkai proved that $|A| \geq \aleph_0$, $|\mathcal{F}| > A$ implies the existence of a sequence of type ω which is either strongly cut by $\mathcal{F}$ or is strongly cut by the family of the complements of $\mathcal{F}$ in A.

The following simple problems remain unsolved.

Problem 71 (Erdös, M. Makkai).

(A) *Assume $|A| = \aleph_1$, $|\mathcal{F}| > \aleph_1$. Does there then exist a sequence of type ω strongly cut by $\mathcal{F}$?*

(B) *Assume $|A| = \aleph_1$, $|\mathcal{F}| > \aleph_1$. Does there exist a sequence of length ξ, $\omega + 2 \leq \xi \leq \omega_1$ which is strongly cut either by $\mathcal{F}$ or by the family of the complements?*

ADDED IN PROOF (May, 1970). Recently S. Shelah obtained a number of results concerning this problem which we do not know yet in detail.

Problem 72. *Assume $\mathcal{G}$ with $|g| = \aleph_1$ does not contain a complete $\aleph_1$ graph. Does then its complement contain a topological complete $\aleph_1$ graph? See* [7].

DEFINITION OF SYMBOL-XIII. $a \xrightarrow{\text{XIII}} (b, c, d)$ is said to hold if the following statement is true. If $|S| = a$ and $\mathcal{F}$ is a family of subsets of S, such that $A \in \mathcal{F}$ implies $|A| < b$ and $A_1 \neq A_2 \in \mathcal{F}$ implies $A_1 \not\subset A_2$ then there are an $S' \subset S$ and $\mathcal{F}' \subset \mathcal{F}$ with $|S'| = c$, $|\mathcal{F}'| = d$ such that $S' \cap (\bigcup \mathcal{F}') = 0$.

Assuming G.C.H. we can give an almost complete discussion of this symbol and many results can be proved without assuming G.C.H. If $a \xrightarrow{\text{I}} (a, a)^2$ holds then $a \xrightarrow{\text{XIII}} (a, a, a)$ holds as well. The only genuine unsolved problem is the following.

Problem 73. *Assume a is strongly inaccessible and* $a \overset{\mathrm{I}}{\not\rightarrow} (a, a)^2$. *Does then* $a \overset{\mathrm{XIII}}{\not\longrightarrow} (a, a, a)$ *hold?*

Problem 74 (Erdös, Rado). *Assume G.C.H. Let* $\mathscr{A}$ *be the class of graphs of at most* $\aleph_\omega$ *vertices such that the valency of every vertex is less than* $\aleph_\omega$. *Does there exist a* $\mathscr{G}_0 = \langle g_0, G_0 \rangle \in \mathscr{A}$ *such that every* $\mathscr{G} \in \mathscr{A}$ *is isomorphic to a subgraph of* $\mathscr{G}_0$ *spanned by some subset of* g_0?

Problem 75 (Erdös, Milner). *Assume G.C.H. Let* $|S| = \aleph_\omega$ *and let* $\mathscr{F}$ *be a family* $\mathscr{F} \subset [S]^{\aleph_0}$, $|\mathscr{F}| = \aleph_{\omega+1}$. *Does there then exist a disjoint partition* $A \cup B \cup C = S$ *of S such that* $|C| \leq \aleph_0$ *and both* $A \cup C$ *and* $B \cup C$ *contain* $\aleph_{\omega+1}$ *elements of* $\mathscr{F}$?

Problem 76 (P. Erdös). *Let* $\mathscr{F}$ *be a family of analytic functions in the unit circle so that for every Z,* $|\{f(Z): f \subset \mathscr{F}\}| \leq a$. *Is it true that* $\mathscr{F}$ *has power* $\leq a$?

This problem was asked for $a = \aleph_0$ by J. Wetzel, and P. Erdös proved that in this case the problem is equivalent to $2^{\aleph_0} > \aleph_1$. If $2^{\aleph_0} > a^+$, then the answer is affirmative in general. The real problem is, e.g. whether $2^{\aleph_0} = \aleph_2$ implies that the answer is negative with $a = \aleph_1$ (see [**19**]).

17. **Some problems in topology; a problem on generalized Ulam matrices.** The second author and I. Juhász considered several problems in general topology where the methods of combinatorial set theory could be applied. We state some of the unsolved problems which seem to be of purely set theoretical character, too.

Problem 77 (J. de Groot, B. A. Efimov, J. Isbell). *Does there exist a Hausdorff space of* $(2^{\aleph_0})^+$ *points not containing a discrete subspace of at least* $\aleph_1$ *points?*

The sharpest result is given in [**29**], a Hausdorff space of $(2^{2^{\aleph_0}})^+$ points contains a discrete subspace of $\aleph_1$ points. For references, see also [**29**].

Problem 77/A (A. Hajnal, I. Juhász). *Assume G.C.H. Let R be an ordered set* $|R| = \aleph_2$ *such that the character of every point of R is* $\aleph_0$. *Does there then exist a disjointed system* $\mathscr{F}$ *of power* $\aleph_2$ *of open intervals of R?*

Note that there is an obvious connection with a special case of the generalized Souslin problem.

Problem 78 (A. Hajnal, I. Juhász). *Does there exist a hereditarily separable Hausdorff space of cardinality greater than that of the continuum?*

Problem 79 (A. Hajnal, I. Juhász). *Assume G.C.H. Does there exist a regular space of power* $\aleph_1$, *such that each subspace of power* $\aleph_1$ *of it has weight* $\aleph_2$?

We proved this for Hausdorff spaces assuming G.C.H. See [**29**].

Problem 80 (A. Hajnal). *Let a be the first weakly inaccessible cardinal* $>\aleph_0$.

Let $|S| = a$. Does there exist a triangular matrix $A_{\xi,\eta}$ of subsets of S for $\xi < \eta < \Omega(a)$ satisfying the following conditions:

(1) *for every $\xi < \Omega(a)$ the family $\{A_{\xi,\eta}\}_{\xi<\eta<\Omega(a)}$ is disjointed?*

(2) *for every $\eta < \Omega(a)$, $|S - \bigcup_{\xi<\eta} A_{\xi,\eta}| < a$?*

This would be a straightforward generalization of Ulam matrices for inaccessible cardinals, and it would give a short direct proof of the fact, that there is an a complete field of sets generated by at most a elements containing $[S]^{<a}$ in which there is no a complete proper a-saturated ideal containing $[S]^{<a}$.

This statement holds for a wide class of weakly and strongly inaccessible cardinals.

In 1950, answering a problem of S. Ulam, L. Alaoglu and P. Erdös proved **[18]** that if $|S|$ is less than the first weakly inaccessible cardinal, then one cannot define $\aleph_0$ σ-additive 0-1 measures on S so that every subset of S is measurable with respect to one of them. It is obvious from their proof that as a corollary of recent results of R. Solovay this would hold if $|S|$ is even larger. Though we did not investigate the problem very closely, it might be worth mentioning that the following simple instance of the problem seems to be still unsolved.

Problem 81 (S. Ulam). *Let $|S| = \aleph_1$. Can one define $\aleph_1$ σ-additive* 0-1 *measures on S so that each subset is measurable with respect to one of them?*

We do not know what happens if 0-1 measure is replaced, e.g., by real valued measure.

Problem 82 (L. Gillman). *Let $|S| = \aleph_1$ and let $\mathcal{T}$ be a nonprincipal prime ideal in the set of subsets of S. Does there exist an $\mathcal{T}' \subset \mathcal{T}$, $|\mathcal{T}'| = \aleph_1$ such that $\bigcup \mathcal{T}'' = S$ for every $\mathcal{T}'' \subset \mathcal{T}'$, $|\mathcal{T}''| \geq \aleph_0$?*

REFERENCES

1. P. Erdös and A. Hajnal, *On the structure of set mappings*, Acta Math. Acad. Sci. Hungar. **9**(1958), 111–131.

2. ———, *Some remarks on set theory*. VIII, Michigan Math. J. **7** (1960), 187–191.

3. ———, *Some remarks on set theory*. VII, Acta Sci. Math. **21** (1960), 154–163.

4. ———, *On a property of families of sets*, Acta. Math. Acad. Sci. Hungar. **12** (1961), 87–123.

5. ———, *Some remarks concerning our paper "On the structure of set mappings"—Nonexistence of a two-valued σ-measure for the first uncountable inaccessible cardinal*, Acta Math. Acad. Sci. Hungar. **13** (1962), 223–226.

6. ———, *On a classification of denumerable order types and an application to the partition calculus*, Fund. Math. **5** (1962), 117–129.

7. ———, *On complete topological subgraphs of certain graphs*, Ann. Univ. Sci. Budapest **7** (1964), 143–149.

8. ———, *Some remarks on set theory*. IX, Michigan Math. J. **11** (1964), 107–127.

9. ———, *On a problem of B. Jónnson*, Bull. Acad. Polon. Sci. Sér. Sci. Math. Astronom. Phys. **14** (1966), 61–99.

10. ———, *On the chromatic number of graphs and set systems*, Acta Math. Acad. Sci. Hungar. **17** (1966), 61–99.

11. ———, *On decomposition of graphs*, Acta. Math. Acad. Sci. Hungar., **18** (1967)

12. ———, *On chromatic numbers of infinite graphs*,, Graph theory symposium held in Tihany, Hungary 1966 83–98.

13. P. Erdös, A. Hajnal, and R. Rado, *Partition relations for cardinals*, Acta Math. Acad. Sci. Hungar. **16** (1965), 193–196.

14. P. Erdös, A. Hajnal, and E. C. Milner, *On the complete subgraph of graphs defined by systems of sets*, Acta Math. Acad. Sci. Hungar., **17** (1966), 159–229.

15. J. Czipszer, P. Erdös, and A. Hajnal, *Some extremal problems on infinite graphs*, Publ. Math. Inst. Hungar. Acad. Sci. Ser. A **7** (1962), 441–457.

16. P. Erdös, G. Födor, and A. Hajnal, *On the structure of inner set mappings*, Acta Sci. Math **20** (1959), 81–90.

17. P. Erdös, A. Hajnal, and E. C. Milner, *On sets of almost disjoint subsets of a set*, Acta Math. Acad. Sci. Hungar, **19** (1968), 209–218.

18. P. Erdös, *Some remarks on set theory*, Proc. Amer. Math. Soc. **1** (1950), 133–137.

19. ———, *An interpolation problem associated with the continuum hypothesis*, Michigan Math. J. **11** (1964), 9–10.

20. P. Erdös and M. Makkai, *Some remarks on set theory.* X, Studia Sci. Math. Hungar. **1** (1966), 157–159.

21. P. Erdös and R. Rado, *A partition calculus in set theory*, Bull. Amer. Math. Soc. **62** (1956), 427–489.

22. ———, Partition relations and transitivity domains of binary relations formed LMS **42** (1967) 624–633.

23. P. Erdös and A. Tarski, *On families of mutually exclusive sets*, Ann. of Math. **44** (1943), 315–429. "On some problems involving inaccessible cardinals" in *Essays on the foundations of mathematics*, Magnes Press, Hebrew Univ., Jerusalem, 1961, pp. 50–82.

24. S. Gladysz, *Bemerkungen über die Unabhängigkeit der Punkte in Bezug auf Mengenwertigen Funktionen*, Acta Math. Acad. Sci. Hungar. **13** (1962), 199–201.

25. A. Hajnal, *Some results and problems on set theory*, Acta Math. Acad. Sci. Hungar. **11** (1960), 277–298.

26. ———, *Proof of a conjecture of S. Ruziewicz*, Fund. Math. **50** (1961), 123–128.

27. ———, *Remarks on a theorem of W. P. Hanf*, Fund. Math. **54** (1964), 109–113.

28. ———, *On the topological product of discrete spaces*, Notices Amer. Math. Soc. **11** (1964).

29. A. Hajnal and I. Juhász, *Discrete subspaces of topological spaces*, Indag. Math. **29** (1967), 343–356; *On hereditary α-Lindelöf and hereditary α-separable spaces*, Acta Math. Acad. Sci. Hungar., to appear.

30. H. J. Keisler and A. Tarski, *From accessible to inaccessible cardinals*, Fund. Math. **53** (1964), 225–308.

31. A. Máté, *On the theory of relations*, Publ. Inst. Hungar. Acad. Sci. **9** (1964), 331–333.

32. F. C. Milner and R. Rado, *The pigeonhole principle for ordinal numbers*, J. London Math. Soc. **15** (1965), 750–768.

33. F. P. Ramsey, *On a problem of formal logic*, Proc. London Math. Soc. (2) **30** (1930), 264–286.

34. J. H. Silver, *Some applications of model theory in set theory*, Ph.D. Thesis, Univ. of California, Berkeley, 1966.

35. W. Sierpiński, *Sur quelques propositions concernant la puissance du continu*, Fund. Math. **36** (1951), 1–13.

36. ———, *Sur un problème de la théorie des relations*, Ann. Scuola Norm. Sup. Pisa **2** (1933), 285–287.

37. E. Specker, *Teilmengen von Mengen mit Relationen*, Comment. Math. Helv. **31** (1957), 302–314.

MATHEMATICAL INSTITUTE OF HUNGARIAN ACADEMY OF SCIENCES

EÖTUÖS LORAND UNIVERSITY CHAIR ANALYSIS I

A MORE EXPLICIT SET THEORY[1]

HARVEY FRIEDMAN

0. **Introduction**. Set theory is usually formulated in terms of closure conditions. A typical example is the power set axiom, which asserts closure under the operation of power set. In Chapter 1, we consider set theory based on closure conditions applied only to definable sets. We formalize this set theory and call it ZF*, and we give a consistency proof of ZF relative to ZF*.

A direct method presents itself for obtaining this relative consistency result; namely, to use a constructible set construction, and prove within ZF* the relativized to the constructible sets of each instance of ZF. This is, of course, in analogy with the method of proof for the consistency of ZF + AxC relative to ZF. However, an examination of the basic principles needed for such a constructible set construction to go through reveals the need for the least counterexample principle for ordinals to be provable in ZF*. By the least counterexample principle for ordinals, we mean the schema $(\exists\alpha)P\alpha \rightarrow (\mu\alpha)P\alpha$, where P is any formula. It does not appear that this schema is derivable in ZF*, even if P is restricted to have only one free variable α. Of course, in ZF, the schema is derivable by means of a closure condition applied to all sets as follows: assume $(\exists\alpha)P\alpha$, and fix such an α. Then form $\{\beta \mid \beta \in \alpha \;\&\; P\beta\}$, and use Foundation to obtain $(\mu\alpha)P\alpha$. This illustrates the basic difference between ZF and ZF*, in that the closure condition, $\{\beta \mid \beta \in \alpha \;\&\; P\beta\}$, is necessarily provable in ZF* only when α is given a definition.

Such a direct attack seems hopeless. An outline of our proof can be found in §4.

Towards the end of this paper, we show how to add elements "on top of" a

---

[1] This paper comprises part of the author's Doctoral Dissertation submitted in partial fulfillment of the requirements for Ph.D. at MIT, August, 1967.

model of ZF to obtain nonstandard models of ZF*. By this means, we obtain results concerning independence from ZF*.

1. **General situation.** Suppose we are given a comprehension axiom $(x_1) \cdots (x_n)(\exists y)Rx_1 \cdots x_ny$. We are interested here in forming the derived schema consisting of the axioms

$$[(\exists !x_1)(\exists !x_2) \cdots (\exists !x_n)(F_1x_1 \,\&\, \cdots \,\&\, F_nx_n)] \rightarrow$$
$$(\exists x_1)(\exists x_2) \cdots (\exists x_n)(F_1x_1 \,\&\, F_2x_2 \,\&\, \cdots \,\&\, F_nx_n \,\&\, (\exists y)Rx_1 \cdots x_ny),$$

where the F_i are formulae with only the free variable x_i.

We are purposely vague about the general situation (what is a comprehension axiom?), since we have only looked at this derived schema when the original axiom is drawn from a natural set of axioms, such as ZF. A report on recent work done on 1st-order arithmetic and 2nd-order arithmetic will occur at the end of this paper.

More specifically, we will look at the schema of schema formed by taking the union of all the schema defined above corresponding to each of the comprehension axioms of ZF. We will not perturb the other (noncomprehension) axioms of ZF, except in minor ways. We call this derived theory ZF*. (We inessentially modify the Replacement schema in ZF for convenience, so that each instance is appropriately placed in the form $(x_1) \cdots (x_n)(\exists y)Rxy$, so that we may pass to the derived schema in the manner above.) The axioms of ZF and ZF* are spelled out in detail, in an elegant form, in the next section.

We are interested in the relation between ZF and ZF* as axiomatic theories. Our main result is that consistency of ZF can be proved in 1st-order arithmetic from the assumption of the consistency of ZF*. It is obvious that ZF* is a subsystem of ZF.

2. **Remarks on terminology and notation.** The only (standard) symbols that can occur in a formula of ZF (or ZF*) are the 2 2-ary relation symbols "$=$" "$\in$"; the 2 quantifiers $(\exists x)$ and (x); the propositional connectives; and variables x_i, y_i, z_i, u_i, v_i, etc. Everything else is nonstandard; when nonstandard symbols occur in a formula, they are meant to be expanded out in such a way that the mere occurrence of a nonstandard symbol implies existence of the corresponding set. For example, $x = \bigcup y$ is an abbreviation for

$$(\exists z)[(w)(w \in z \equiv (\exists u)(w \in u \,\&\, u \in z)) \,\&\, x = z].$$

Also, say, $\varnothing \in x$ would be $(\exists y)[(z)(z \notin y) \,\&\, y \in x]$.

3. **Axioms.** For ZF, we have

0. Axioms for predicate calculus with equality.
1. Extensionality. $(x_0 = x_1) \equiv (x_2)(x_2 \in x_0 \equiv x_2 \in x_1)$.
2. Infinity. $(\exists x_0)(\varnothing \in x_0 \,\&\, (x_1)(x_1 \in x_0 \rightarrow x_1 \cup \{x_1\} \in x_0))$.
3. Power set. $(x_0)(\exists x_1)(x_2)(x_2 \in x_1 \equiv (x_3)(x_3 \in x_2 \rightarrow x_3 \in x_0))$.
4. Sum set. $(x_0)(\exists x_1)(x_2)(x_2 \in x_1 \equiv (\exists x_3)(x_3 \in x_0 \,\&\, x_2 \in x_3))$.

5. Replacement schema. Let Axy be a formula with the free variables x and y and possibly more free variables $x_1, \ldots, x_n$. Then

$$(x_1) \cdots (x_n)(y)[(\exists y_1)(y_2)(y_2 \in y_1 \equiv (\exists y_3)(Ay_3y_2 \mathbin{\&} (y_4)(Ay_3y_4 \rightarrow y_4 = y_2) \mathbin{\&} y_3 \in y))]$$

is an instance. (The domain is y and the axiom asserts the range of the partial function, $A'y_3y_2 = Ay_3y_2 \mathbin{\&} (z)(Ay_3z \rightarrow z = y_2)$, on the domain y, exists.)

6. Foundation. $(x_0)(x_0 \neq \varnothing \rightarrow (\exists x_1)(x_1 \in x_0 \mathbin{\&} (x_2)(x_2 \in x_1 \rightarrow x_2 \notin x_0)))$.

It is clear that by the usual process of making partial functions into total functions axiom schema 5 is the same as the usual formulation in the present context.

Now let Ax be a formula of 1 free variable. Then the formula

$$(\exists y)(x)(x \in y \equiv Ax)$$

is abbreviated as C_A.

For ZF* we have

0. Same as ZF.

1. Same as ZF.

2. Infinity.

$$(\exists x_0)(\varnothing \in x_0 \mathbin{\&} (x_1)(x_1 \in x_0 \rightarrow x_1 \cup \{x_1\} \in x_0) \mathbin{\&} (y)(y \in x_0 \equiv \text{Fin}(y))).$$

Fin(y) will be defined later. Intuitively it means y is a finite ordinal.

3. Power set. The instances are $C_A \rightarrow (\exists x_0)(x_1)(x_1 \in x_0 \equiv (x_2)(x_2 \in x_1 \rightarrow Ax_2))$, A with 1 free variable.

4. Sum set. The instances are $C_A \rightarrow (\exists x_0)(x_1)(x_1 \in x_0 \equiv (\exists x_2)(x_1 \in x_2 \mathbin{\&} Ax_2))$, A with 1 free variable.

5. Replacement schema. The instances are

$$C_A \rightarrow (\exists x_0)(x_1)(x_1 \in x_0 \equiv (\exists x_2)(Ax_2 \mathbin{\&} Bx_2x_1 \mathbin{\&} (x_3)(Bx_2x_3 \rightarrow x_3 = x_1))),$$

for A with 1 free variable, B with exactly 2 free variables.

6. Same as ZF.

REMARKS. Our formulation of Power set in ZF* is seen to be equivalent to the derived schema of Power set in ZF (as given in General Situation) by noticing (1) that if $(\exists! y)Fy$, then C_A, where A is $(\exists y)(Fy \mathbin{\&} x \in y)$ and (2) that if C_A, then $(\exists! x)(y)(y \in x \equiv Ay)$. These latter are obtained by Axiom 1 of ZF*, Extensionality. The same remark applies to Sum set.

Essentially the same idea yields that the union of the derived schema of the instances of Replacement in ZF is equivalent, in the present context, to the schema

$$\begin{aligned}(C_{A_1} \mathbin{\&} \cdots C_{A_n} \mathbin{\&} C_A) \rightarrow (\exists x_0)(x_1 \in x_0 \equiv (\exists x_2)(Ax_2 \mathbin{\&} B(x_2, x_1, \{x \mid A_1x\},\\ \{x \mid A_2x\}, \ldots, \{x \mid A_nx\}) \mathbin{\&} (x_3)(B(x_2, x_3, \{x \mid A_1x\}, \ldots,\\ \{x \mid A_nx\}) \rightarrow x_3 = x_1)))),\end{aligned}$$

where B has $n + 2$ free variables, A and the A_i have 1 free variable. We want to show this schema is contained (in the present context) in Replacement in ZF*. But the above is easily seen to follow from that instance of 5 of ZF*, setting A as

A, Bxy as $(C_{A_1} \& \cdots \& C_{A_n}) \rightarrow B(x_2, x_1, \{x \mid A_1 x\}, \{x \mid A_2 x\}, \ldots, \{x \mid A_n x\})$. (That the above schema contains Replacement in ZF* is obvious.) NOTE: "In the present context" means "using the other axioms of ZF*".

4. **Outline of proof of main theorem.** The main theorem is $\vdash_{\text{ENT}}\text{Con ZF}^* \rightarrow$ Con ZF. The first step in proving this is developing in ZF* an adequate definition of ordinals, which turns out to be a much more delicate matter than for ZF due to the lack of certain key instances of Replacement in ZF*. By an adequate definition of ordinals in ZF*, we mean a definition of ordinals such that, provably in ZF*, members of ordinals are ordinals, and (for a natural definition in ZF* of ω) ω is an ordinal, and ordinals are comparable by ε, and the ε-relation on the ordinals is transitive, and antisymmetric, and antireflexive, and every ordinal has a (natural) successor, except possibly the greatest ordinal. The definition is made and Lemma 1 establishes the above properties for it in the next section; we even obtain more: that, provably in ZF*, the new definition of ordinal coincides with the usual definition given in ZF, on *definable* sets. (This is made precise in Lemma 1(f).)

Next we develop an adequate definition of L within ZF*. Among the properties of the predicate $x \in L$ needed, we must have, provable in ZF*, every member of an $x \in L$ is $\in L$, $\omega \in L$, the new definition of $x \in L$ coincides with the usual definition of constructibility for definable x, and a definable well-ordering of L.

With this machinery, an apparently straightforward "proof" of our main result comes to mind. Namely, just to prove the relativized to L of each instance of ZF in ZF* by taking least counterexamples of various things in ZF*, as Gödel established the relativization of the axioms of ZF to L within ZF. But a moment's reflection will reveal that one can hardly expect that ZF* will prove any general least counterexample principles; i.e., one may well be able to prove in ZF* that $(\exists x)(x \in L \& Px)$, yet not be able to prove $(\exists \mu x)(x \in L \& Px)$ in ZF*, where μ is defined in terms of the definable well-ordering of L. In fact, we do not know how to prove each instance of the relativized of ZF* to L, within ZF*.

In order to get our main result, we form an auxiliary system $\text{ZF}^{*\prime} \subseteq \text{ZF}^*$, whose definition depends on a certain crucial transformation on sentences T. This subsystem has the property that each instance of $\text{ZF}^{*\prime}$ semirelativized (semirelativization is a certain modification of relativization) to L, is provable in ZF*. We form another auxiliary theory ZF′ which is related to $\text{ZF}^{*\prime}$ about as $\text{ZF} + \text{V} = L$ is to ZF*. It turns out that $\text{ZF}' \supseteq \text{ZF}$. It also turns out that in the theory obtained by semirelativizing each comprehension axiom of $\text{ZF}^{*\prime}$ and retaining the other axioms, one can give, for each finite subsystem of ZF, a Skolem hull argument that proves the existence of a (suitably definable) model of this finite subsystem, and hence its consistency.

Putting all this together we get a finitary proof that $(n) \vdash_{\text{ZF}^*}\text{Con}(\text{ZF}'_n)$, where ZF'_n is the first n axioms of ZF′, because the semirelativized of each instance of $\text{ZF}^{*\prime}$ is provable in ZF*. Now since $\text{ZF}' \supseteq \text{ZF}$, we obtain a proof in ENT that Con ZF* $\rightarrow$ Con ZF.

5. **Development of ordinals.** We define $\text{Ord } x = \text{Trans}(x) \& \varepsilon\text{-Conn}(x) \&$

(x is semiclosed under succession) & there are no 3-chains in x, i.e.,

$$\begin{aligned}\mathrm{Ord}(x) = (y)(z)((y \in x \,\&\, z \in y) \rightarrow z \in x) \,\&\\ (y)(z)((y \in x \,\&\, z \in x) \rightarrow y \in z \vee z \in y \vee y = z) \,\&\\ (y)(y \in x \rightarrow (\exists z)[(w)(w \in z \equiv (w \in y \vee w = y)) \,\&\, (z \in x \vee z = x)]) \,\&\\ (y)(z)(w)(\sim[y \in z \,\&\, z \in w \,\&\, (w \in y \vee w = y)]).\end{aligned}$$

We define $\mathrm{Ord}'(x) = \mathrm{Ord}\, x \,\&\, (y)((\mathrm{Ord}\, y \,\&\, y \subseteq x) \rightarrow (y \in x \vee y = x))$. $\mathrm{Ord}''(x) = \mathrm{Ord}'(x) \,\&\, (y)(\mathrm{Ord}'\, y \rightarrow (x \subseteq y \vee y \subseteq x))$. Ord″ will be our notion of ordinal in ZF*. It is obvious that $(x)(y)(\mathrm{Ord}''\, x \,\&\, \mathrm{Ord}''\, y \rightarrow (x \in y \vee y \in x \vee y = x))$. For $\mathrm{Ord}(x)$, we define y *is successor of* x if and only if $y = x \cup \{x\}$.

LEMMA 1. *The following are theorems of* ZF*, *where Ax has* 1 *free variable*:
(a) *If* $\mathrm{Ord}\, x \,\&\, y \in x$, *then* $\mathrm{Ord}\, y$.
(b) *If* $\mathrm{Ord}'\, x$ *and* $y \in x$, *then* $\mathrm{Ord}'\, y$.
(c) *For x with* $\mathrm{Ord}(x)$, *and x not a successor*, $\bigcup x = x$. *For* $\mathrm{Ord}(x)$, *x a successor*, $x = y \cup \{y\}$, *we have* $\bigcup x = y$.
(d) *If* $\mathrm{Ord}''(x)$ *and* $y \in x$, *then* $\mathrm{Ord}''(y)$.
(e) *If* $y = \{x \mid Ax\}$, $\mathrm{Ord}'(y)$, *then* $\mathrm{Ord}''(y)$.
(f) *If* $y = \{x \mid Ax\}$, $\mathrm{Ord}(y)$, *then* $\mathrm{Ord}''(y)$.
(g) $\mathrm{Ord}''(\omega)$. (*Explained below.*)
(h) *If* $y = \{x \mid Ax\}$, *and* $Ax \rightarrow \mathrm{Ord}''(x)$, *then* $\mathrm{Ord}''(\bigcup y)$.

PROOF. (a) Claim Trans(y). Let $z \in y$, $w \in z$. Then by Trans(x), we have $w \in x$ and $y \in x$. By ε-Conn(x), $w \in y \vee y \in w \vee y = w$. We cannot have $y \in w \vee y = w$ because we would have a 3-chain. So $w \in y$. To see ε-Conn(y), let $z \in y$, $w \in y$. Then $z \in x$ and $w \in x$, and so $z \in w \vee w \in z \vee w = z$ by ε-Conn(x). Towards showing y semiclosed, let $z \in y$. Hence $z \in x$, and $z \cup \{z\} \in x$ or $z \cup \{z\} = x$. By ε-Conn(x), we have $z \cup \{z\} \in y$ or $z \cup \{z\} = y$ or $y \in z \cup \{z\}$. But if $y \in z \cup \{z\}$, then $y \in z \vee y = z$. The first yields the 3-chain $z \in y$, $y \in z$, $z = x$; the second yields $y \in y$, which yields the 3-chain $y \in y$, $y \in y$, $y \in y$. So $y \notin z \cup \{z\}$.

Now suppose $a \in b$, $b \in c$, $(c \in a \vee c = a)$, $a, b, c, \in y$. Then $a, b, c \in x$, and we have a 3-chain in x.

(b) By (a), we have Ord(y). Towards showing Ord′(y), let $z \subseteq y$, Ord(z). By Trans(x), $z \subseteq x$. Hence $z \in x \vee z = x$. But $z \neq x$, for if $z = x$, then $x \subseteq y$, and hence $y \in y$. So $z \in x$. By ε-Conn(x), $z \in y \vee z = y \vee y \in z$. But $y \notin z$, since if $y \in z$, then $y \in y$.

(c) Let Ord(x), x with $(y)(y \cup \{y\} \neq x)$. By Trans(x), every member of a member of x is a member of x. So $\bigcup x \subseteq x$, if $\bigcup x$ exists. Now let $y \in x$. By semiclosure of x, $y \cup \{y\} \in x$. But $y \in y \cup \{y\}$, and so $x = \bigcup x$, since also every member of x is a member of a member of x. If $x = z \cup \{z\}$, then again $\bigcup x \subseteq x$, if $\bigcup x$ exists. But $z \notin \bigcup x$, since x has no 3-chains. So $\bigcup x \subseteq z$, if it exists. But every member of z is a member of x, since if $w \in z$, then w is a member of a member of x. So $\bigcup x$ exists and is z.

(d) By (b), Ord$'(y)$. Let Ord$'(z)$. Then $x \subseteq z \vee z \subseteq x$. If $x \subseteq z$, then $y \subseteq z$. If $z \subseteq x$, then $z \in x \vee z = x$. If $z = x$, then $y \subseteq x$. If $z \in x$, then $y \in z \vee z \in y \vee z = y$. Hence $x \subseteq y \vee y \subseteq z$.

(e) Let Ord$'(y)$, $y = \{x \mid Ax\}$. Then either

(1) All elements of y are Ord$''$. Then let z be any Ord$'$. For every $x \in y$ we have $x \subseteq z \vee z \subseteq x$. Either all members of y are $\subseteq z$ or some member of $y \supseteq z$. Suppose the first holds. Then if y is not a successor, by (c) we have $\bigcup y = y$, and so $y \subseteq z$. If $y = u \cup \{u\}$, then $u \subseteq z$. Since Ord$'(z)$, we have $u \in z \vee u = z$. If $u = z$, then $z \subseteq x$. If $u \in z$, then $y \subseteq z$.

Suppose the second holds, i.e., some $x \in y$ contains z. Then clearly $z \subseteq y$. So y is Ord$''$.

(2) Some element of y is not Ord$''$. We have just proved that any Ord$'$ such that every member is Ord$''$ is Ord$''$. The (unique) ε-least element of y which is not Ord$''$ is definable, and has every member an Ord$''$. (There is a least by suitable use of Replacement schema of ZF*, and Foundation.) Just apply (1) to obtain a contradiction.

(f) We merely have to show all definable Ords are Ord$'$. As in (e) we go down to a definable ordinal all of whose members are Ord$''$. Let y be such an ordinal, $y = \{x \mid Bx\}$. Let $z \subseteq y$, Ord z. We wish to show $z \in y \vee z = y$. Suppose $z \notin y \,\&\, y \in z$. Then $\exists w \in y$ with $w \notin z$. But since Ord$''$ w, we have $w \subseteq z \vee z \subseteq w$. If $z \subseteq w$ then $z = w \vee z \in w$ and so $z \in y$. So $w \subseteq z$, $w \notin z$. Now either $w \cup \{w\} \in y \vee w \cup \{w\} = y$. If $w \cup \{w\} = y$ then since $z \subseteq y$, $z = w$, and so $z \in y$. So $w \cup \{w\} \in y$. Now then Ord$''(w \cup \{w\})$. Hence

$$w \cup \{w\} \subseteq z \vee z \subseteq w \cup \{w\}.$$

The first is out, so $z \subseteq w \cup \{w\}$, and since Ord$'(w \cup \{w\})$ we have

$$z = w \cup \{w\} \vee z \in w \cup \{w\},$$

either one implying $z \in y$.

(g) We now explain Axiom 3. Fin(x) = Ord x & $(\exists y)(x = y \cup \{y\})$ *and* $(z)(z \in x \rightarrow (\exists w)(z = w \cup \{w\}))$. *x is a successor* Ord = Ord x & $(\exists y)(y = x \cup \{x\})$. We let ω be the x_0 in Axiom 2 of ZF*.

Now clearly ω is definable, and so we merely have to show Ord(ω).

(1) Trans (ω). Let $n \in \omega$. Let $x \in n$. Then x is a successor Ord or $\varnothing$. Since n is transitive, x is a successor Ord or $\varnothing$ and every member of x is a successor Ord or $\varnothing$, and so x is a finite Ord(i.e. Fin(x)), and so $x \in \omega$.

(2) ε-Conn (ω). Suppose $\exists n \in \omega$ such that for some $m \in \omega$, $n \notin m \,\&\, m \notin n \,\&\, n \neq m$. Take any ε-least such n and call it k. There is an ε-least by Replacement in ZF* and Foundation. Now $k \neq \varnothing$, since $\varnothing \in m \vee \varnothing = m$, for any $m \in \omega$, by Foundation and Trans(m). Hence k is a successor Ord, and $k = l \cup \{l\}$, and by Trans(ω), $l \in \omega$. We must have, for some $m \in \omega$, $k \notin m \,\&\, m \notin k \,\&\, k \neq m \,\&\, (l \in m \vee m \in l \vee m = l)$. But if $m = l$, then $m \in k$. If $m \in l$, then $m \in k$. If $l \in m$, then $l \cup \{l\} = m \vee l \cup \{l\} \in m$, i.e. $k = m \vee k \in m$. Contradiction, by the lack of 3-chains in Ords, and the comparability of l and m.

(3) Semiclosure under succession. This is insured directly from Axiom 2 of ZF*.

(4) No 3-chains. Suppose $n \in m$, $m \in r$, $r \in n \vee r = n$, where n, m, $r \in \omega$. By Trans(r), we have $n \in r$. By Trans(r) again, we have $r \in r$, if $r \in n$. If $r = n$, we also have $r \in r$. These contradict Ord(r).

(h) We have only to show Ord($\bigcup y$).

(1) Trans ($\bigcup y$). If $z \in \bigcup y$, then $z \in w$ for some Ord″(w), $w \in y$; hence any $u \in z$ has $u \in w$, $w \in y$. So $u \in \bigcup y$.

(2) ε-Conn($\bigcup y$). If $z, w \in \bigcup y$, then $z \in z' \in y$, $w \in w' \in y$, z', w' both Ord″. So $z \subseteq w'$ or $w \subseteq z'$. Without loss of generality, assume $z' \subseteq w'$. Then $z \in w'$, $w \in w'$. Hence $z \in w \vee w \in y \vee w = z$.

(3) Semiclosure. Let $z \in \bigcup y$. Then $z \in w \in y$, Ord″(w). So $z \cup \{z\} \in w$ or $z \cup \{z\} \in w$. In the first case, $z \cup \{z\} \in \bigcup y$. In the second suppose:

(A) Some $u \in y$ has $w \subseteq u$, but $w \neq u$. Hence $w \in u$, and so $z \cup \{z\} = w \in \bigcup y$.

(B) Every $u \in y$ has either $w = u$ or $u \subseteq w$. Then clearly $\bigcup y = w = z \cup \{z\}$.

(A) and (B) are exhaustive.

(4) No 3-chains. Suppose $a \in b$, $b \in c$, $(c \in a \vee c = a)$, where $a, b, c \in \bigcup y$. Then a, b, c are Ord″, and $a \in c$. If $c \in a$, then $c \in c$. If $c = a$, $a \in c$, then $c \in c$. Contradicts Ord (c).

6. **Development of** L. We wish to define a class of sets, L, which has a definable well-ordering, and provably so in ZF*. L, of course, will not be an object. We are interested in the predicate $x \in L$.

We let n, m, r, p, q be special variables for elements of ω. We let $\alpha, \beta, \gamma, \ldots$ be special variables for sets x with Ord″ x. We write $\alpha + 1$ for $\alpha \cup \{\alpha\}$. We let λ be a special variable for limit (nonsuccessor and nonnull) Ord″'s.

We say $\underline{x = M(\alpha)} = (\exists f)(\mathrm{Dom} f = \alpha + 1 \;\&\; f(\varnothing) = \varnothing) \;\&\; (\lambda)(\lambda \in \alpha + 1 \to \bigcup_{x \in \lambda} f(x) = f(\lambda)) \;\&\; (\beta)(\beta \in \alpha + 1 \to f(\beta + 1) = \mathrm{Fodo}\,(f(\beta)) \;\&\; x = f(\alpha))$, where $\mathrm{Fodo}(y) = \{x \mid (\exists x_0)(\exists n)(x = \{z \mid z \in y \;\&\; \langle y, \varepsilon_y\rangle \models n(z)[x_0]\} \;\&\; \mathrm{Fin\ Seq}(x_0, y))\}$, where $\langle y, \varepsilon_y\rangle \models n(z)[x_0]$ means the structure $\langle y, \varepsilon_y\rangle$ satisfies the formula with Gödel number n at the sequence of elements of the domain y, (z, x_0), which is the sequence starting with z, followed by the sequence x_0.

REMARK ON FORMALIZATION. We formalize the satisfaction relation in ZF* the same way we do in ZF. Also see §2, Remarks on terminology and notation.

LEMMA 2. *For each Ax with 1 free variable, and for each $Czy_1 \ldots y_n$, and Bzw with only free variables shown, the following is provable in* ZF*: *If $y = \{x \mid Ax\}$ and if Bzw is a well-ordering of $y \cup \{y\}$, and $y_1, \ldots, y_n \in y$, then*

$$\{z \mid z \in y \;\&\; \langle y, \varepsilon_y\rangle \models \bar{n}(z)[y_1, \ldots, y_n]\} = \{z \mid z \in y \;\&\; Czy_1 \cdots y_n\},$$

where n is the Gödel number of C. Thus Fodo(x) *means the set of all sets first-order definable over x, for definable sets x.*

PROOF. Suppose Czy_1 is $(\exists w)(w \in z \;\&\; w \notin y_1)$. We note that both

$$\{z \mid z \in y \;\&\; Czy_1\}$$

and $\{z \mid z \in y \;\&\; \langle y, \varepsilon_y\rangle \models \bar{n}(z)[y_1]\}$ exist by Replacement on the definable y. We want to show $Czy_1 \equiv \langle y, \varepsilon_y \rangle \models \bar{n}[z, y_1]$, for $z \in y$. The proof in the case of ZF is routine. What complicates it in the case of ZF* is that certain sets definable in terms of the members of y may not provably exist in ZF*, and also that theory of Gödel numbering may not be formalizable in ZF*. The latter is not the case, since ω is definable, and hence by Replacement and Foundation, induction on ω provably holds in ZF*, and also we may define $+$ and $\times$, and prove the relevant properties. What comprehension axioms are needed to establish our equivalence? Apparently, what is involved is just that the theory of finitely hereditary sequences of elements of y and natural numbers provably in ZF* have the intended interpretation. For instance, we must verify provability in ZF*, for sentences like "for every $y_1, y_2, y_3 \in y$, $n, m \in \omega$, there exists the sequence $\langle \{y_1, n\}, \{\{n, m\}\}, \{y_2, y_3, \{n\}\} \rangle$". Such sentences can clearly be proved by suitable instances of Replacement in ZF* for definable sets y_1, y_2, y_3, and n, m. One assumes in ZF* that such a sentence is false, and goes to definable counterexamples y_1, y_2, y_3 and n, m via the definable well-ordering of $y \cup \{y\}$, *Bzw*.

LEMMA 3. *Each instance of the following is provable in* ZF*:

(a) $[\alpha = \{x \mid Ax\}$ & $(\beta = \alpha \vee \beta \in \alpha)$ & $(\exists x)(x = M(\beta))] \rightarrow \exists! f$ *satisfying the conditions given in the definition of* $\underline{x = M(\beta)}$.

(b) $\alpha = \{x \mid Ax\} \rightarrow (\exists! x_0)(\underline{x_0 = M(\alpha)})$. *Also*

$$\beta \in \alpha = \{x \mid Ax\} \rightarrow (\exists! x_0)(x_0 = M(\beta)).$$

PROOF. Assume $\alpha = \{x \mid Ax\}$ & $(\exists x)(\underline{x = M(\alpha)})$. Suppose we have 2 functions f, g satisfying conclusion, and $f \neq g$. We take, using Replacement and Foundation in ZF* in the usual way, β to be the ε-least element of $\alpha + 1$ with the property that $\exists f$ and g, $f \neq g$ satisfying definition of $\underline{x = M(\alpha)}$, with $f(\beta) \neq g(\beta)$. Then β is definable. Suppose $\mathrm{Lim}(\beta)$. Then use Replacement on β to get the $\{y \mid (\exists \gamma)(\gamma \in \beta$ & for all f satisfying definition of $\underline{x = M(\alpha)}$, $f(\gamma) = y)\}$. We can apply Sum set in ZF* to get a union U. In any f satisfying the definition of $\underline{x = M(\alpha)}$, clearly $f(\beta) = U$. But this is contrary to hypothesis, $f(\beta) \neq g(\beta)$.

Now suppose $\beta = \delta + 1$. By hypothesis, $f(\delta)$ is fixed when f varies over the functions satisfying the definition of $\underline{x = M(\alpha)}$. Clearly $f(\delta + 1) = \mathrm{Fodo}(f(\delta))$ for any f satisfying the definition of $\underline{x = M(\alpha)}$, and so $f(\delta + 1)$ is also independent of f, again contradicting the definition of β.

Clearly $\beta \neq \varnothing$. So no such β exists.

Now suppose for some $\beta \in \alpha$, part (a) false. Take least such β, and apply above, since least such β is definable. This concludes part (a).

(b) Now suppose $(\exists! x_0)(\underline{x_0 = M(\beta)})$ for all $\beta \in \alpha$, but not for $\beta = \alpha$. We may assume this, without loss of generality, by taking least counterexamples. We conclude the proof of Lemma 3 by obtaining a contradiction. Suppose $\mathrm{Lim}(\alpha)$. Using Replacement in ZF* on α we can get the set of all $M(\beta)$'s, $\beta \in \alpha$. (We write $M(\beta)$ for that x_0 with $\underline{x_0 = M(\beta)}$ if it is unique.) Thus we may take the union by sum set in ZF* and call this U. Now it is not hard to see, under our hypothesis, that using replacement there is an f consisting of only $\langle \beta, M(\beta) \rangle$'s, $\beta \in \alpha$, and $\langle \alpha, U \rangle$, and that this is the required f in the definition of $\underline{U = M(\alpha)}$. So

$$(\exists x_0)(\underline{x_0 = M(\alpha)}).$$

Uniqueness comes from (a). Suppose $\alpha = \gamma + 1$. It is easy to see that γ is definable and by our hypotheses, $f(\gamma)$ is definable. It is obvious that elements of Fodo($f(\gamma)$) are elements of $P(f(\gamma))$ and that $P(f(\gamma))$ exists by power set in ZF*. Furthermore, Fodo ($f(\gamma)$) exists since it can be gotten by replacement on $P(f(\gamma))$. Proceed as above to get an appropriate f to give Fodo($f(\gamma)$) $= M(\alpha)$. Uniqueness follows from (a). The case $\alpha = \varnothing$ is trivial.

We want to insure in ZF* that there is a definable well-ordering of L (among other things). This insurance is easily obtained by a natural definition of L in ZF, but not in ZF*. We have no choice but to complicate the definition of $\in L$ by adding on conditions.

We define $x \in L$, *approximately as* $(\exists \alpha)(\exists y)(\underline{y = M(\alpha)} \;\&\; x \in y)$. But this is not good enough for our purposes. We define 5 extra conditions on this α and y:

(1) $(\beta)(z)([z = M(\beta) \;\&\; x \in z] \rightarrow (\beta \in \alpha \vee \beta = \alpha))$, and

$$(z)(z = M(\alpha) \rightarrow z = y).$$

Whenever x and α have such a y, we say $O(x) = \alpha$.

(2) For all $\beta \in \alpha$, there is a unique corresponding $y = M(\beta)$, and if $\beta, \gamma \leq \alpha$, then $M(\beta) \nsubseteq M(\gamma)$.

(3) The $M(\beta)$'s, $\beta \in \alpha$, and $M(\alpha)$ are transitive sets.

(4) For every $z \in M(\alpha)$, we have $(\exists \beta)(O(z) = \beta)$ and $O(z) \leq \alpha$. Also if $z, w \in M(\alpha)$, then $[z \in M(O(w)) \rightarrow O(z) \leq O(w)]$.

(5) Now, there is a usual definable mapping F in full set theory (identify this with the 2-ary relation $F(x) = y$) mapping the constructible sets 1-1 into ordinals. Of course, ZF* may well not be able to prove $(x)(\exists y)(F(x) = y)$. Condition (5) will be that (the 2-ary relation) F is a 1-1 function when restricted to domain $M(\alpha)$, and $(x)(x \in M(\alpha) \rightarrow \text{Ord}''(F(x)))$. This is the F which, in full set theory, assigns Gödel numbers in the form of ordinals to each constructible set a sequence of ordinals, the first being the rank of the set in the constructible hierarchy $\alpha + 1$; the rest of the sequence codes in, via F on the sets in $M(\alpha)$, how the constructible set in question is first-order defined over $M(\alpha)$.

We define $x \in L = (\exists \alpha)(\exists y)(\underline{y = M(\alpha)} \;\&\; x \in y \;\&\; \alpha$ and y satisfy conditions (1)–(5) above).

We define $x < y \equiv x \in L \;\&\; y \in L \;\&\; F(x) \in F(y)$, for F as in (5).

LEMMA 4. *The following are provable in* ZF*: *if* $x \in L$, *then* $(y)(y \in M(O(x)) \rightarrow y \in L)$. *Also* $(y)(y \in x \rightarrow y \in L)$, *if* $x \in L$.

PROOF. Let $y \in M(O(x))$. By (4) in the definition of $x \in L$, we have y, $M(O(y))$, $O(y)$ satisfy (1), (2), and (3) in the definition of $y \in L$, since $O(y) \leq O(x)$. Towards verifying (4) in the definition of $y \in L$, let $z \in M(O(y))$. Then $z \in M(O(x))$.

Then $(\exists\alpha)(O(z) = \alpha)$. $O(z) \leq O(x)$, for suppose not. Then we get a contradiction via condition (2) in the definition of $x \in L$. The rest of condition (4) follows for y because of condition (4) being satisfied for y, and because $M(O(y)) \subseteq M(O(x))$.

y satisfies (5) (i.e., y, together with $O(y)$, $M(O(y))$ since x does, $M(O(y)) \subseteq M(O(x))$.

To show $(y)(y \in x \rightarrow y \in L)$, notice by Trans$(M(O(x)))$ we have (assuming $y \in x$) that $y \in M(O(x))$, and so by the first part of Lemma 4, $y \in L$.

LEMMA 5. *For each* Ax, *1 free variable*, $\beta = \{x \mid Ax\} \rightarrow (x)(x \in M(\beta) \rightarrow x \in L)$, *is provable in* ZF*. (*Note that* $M(\beta)$ *exists unambiguously by Lemma* 3.) *Also* $\omega \in L$.

PROOF OF LEMMA 5. Form $\{\alpha \mid \alpha \in \beta \,\&\, M(\alpha)$ does not satisfy conclusion$\}$. Take ε-least member, and call it γ.

Case 1. γ is a limit. Now γ is definable in ZF*. Let $x \in M(\gamma)$. Then $x \in M(\alpha)$, for some $\alpha \in y$, and hence by the definition of γ, $x \in L$.

Case 2. $\gamma = \delta + 1$. Let $x \in M(\delta + 1)$. There is a definable well-ordering on $M(\delta)$, $<$, and we may use this to definably well-order, in ZF*, the finite sequences of elements of $M(\delta)$ in the natural way, proving in ZF* that it is a well-ordering. With this well-ordering of $M(\delta + 1)$, we take a least, in $M(\delta + 1)$, x such that there is no α, $M(\alpha)$ satisfying condition 1, assuming there is an x. This least x is definable in ZF*, and so we consequently can form $\{\alpha \mid \alpha \in \delta + 1 \,\&\, x \in M(\alpha)\}$ and take the ε-least member, thereby obtaining a contradiction.

So every $x \in M(\delta + 1)$ possesses a (unique) $O(x)$.

Conditions (2)–(4) are treated similarly, taking definable counterexamples and using definable well-orderings. The proof of (5), after taking least counterexamples, is much like our indication of construction of a definable well-ordering of $M(\delta + 1)$ on the basis of one for $M(\delta)$, above.

To show $\omega \in L$, it suffices to prove $\omega \in M(\omega + 1)$. The proof is like the proof of this fact in ZF.

LEMMA 6. $\vdash_{ZF^*}(x)(x = \omega \equiv (x = \omega)')$, *where* A' *is* A *relativized to the predicate* $\in L$.

PROOF. Left to the reader.

7. **The system** ZF*′. We define a transformation mapping formula in prenex form in the standard notation (described in §2 Remarks on terminology and notation) into formulae which contain the '$<$' symbol. If B is in prenex form define B^- to be the usual prenex form for $\sim B$. Take T to be the identity on formulae with no quantifiers, and take $T((\exists x_i)Bx_i)$ to be $(\exists x_i)(T(Bx_i) \,\&\, (x_j)(x_j < x_i \rightarrow T(B^- x_j)))$. $T((x_i)Bx_i)$ is $(x_i)(T(Bx_i) \vee (\exists x_j)(x_j < x_i \,\&\, T(B^- x_j)))$. It is easily proved by induction that $T(B^-)$ and $\sim T(B)$ are equivalent for any prenex B. Recall that the interpretation of $x_j < x_i$ is $x_j \in L \,\&\, x_i \in L \,\&\, F(x_j) \in F(x_i)$.

We form ZF*′ as follows: Extensionality & Foundation & Infinity & Power set & Sum set & Modified Replacement. The latter is the only difference between

ZF*′ and ZF*. The other axioms are the same. Replacement in ZF*′ is as follows: Any instance of Replacement in ZF* is an instance in ZF*′ provided that the Axy be of the form $T(Bxy)\ \&\ (z)(z < y \rightarrow \sim T(Bxz))$, Bxy having only 2 free variables. It is obvious that ZF*′ $\subseteq$ ZF*.

LEMMA 7. *Extensionality & Foundation & Infinity are theorems of* ZF* *when relativized to L.*

PROOF. For (Infinity)′ take x_0 to be ω. For (Extensionality)′ and (Foundation)′ just note from Lemma 4 that $(x \in L\ \&\ y \in x) \rightarrow y \in L$.

LEMMA 8. *The Power set and Sum set axioms* (*of* ZF*) *are theorems of* ZF* *when relativized to L.*

PROOF. In power set, we have $x = \{y \mid Ay\}$. The relativized to L will be equivalent to $x = \{y \mid y \in L\ \&\ (Ay)'\}\ \&\ x \in L$. Now observe that the relation $(z \subseteq w)'$ is equivalent to $z \subseteq w\ \&\ z \in L\ \&\ w \in L$. So we have to verify that if $x \in L\ \&\ x = \{y \mid y \in L\ \&\ (Ay)'\}$, then there is a set $x_0 \in L$ with x_0 the set of all subsets z of x such that $z \in L$. Now the hypothesis tells us that x is definable, and so x has a definable power set $|P(x)$. We use replacement on $|P(x)$ to get the set of all $O(y)$'s with $y \in L$ and $y \in |P(x)$. This is a definable set of Ord″, and so it has a union U. Now U is definable, and so all $y \subseteq x$ with $y \in L$ have $y \in M(U)$, because if $\beta \in \alpha$, then $M(\beta) \subseteq M(\alpha)$. The required set x_0 of all subsets y of x with $y \in L$ is in $M(U + 1)$, by Lemma 2; hence $x_0 \in L$, by Lemma 5.

The relativized of Sum set is checked similarly.

LEMMA 9. *If* $(\exists y)(y < z\ \&\ Ayx_1 \ldots x_n)$, *and* $O(z)$, $O(x_1), \ldots, O(x_n)$ *all* $\in \alpha = \{x \mid Bx\}$, *then* $\exists\mu O(y)$, *with* $y < z\ \&\ Ayx_1 \cdots x_n$. (*That is, if A any formula with the free variables shown, B any formula with* 1 *free variable, the above is provable in* ZF*.)

PROOF. One just assumes there are counterexamples $z, x_1, \ldots, x_n \in M(\alpha)$ to this lemma, and then goes to definable counterexamples. But we obtain a contradiction, since there is provably a $\mu O(y)$ for these supposed *definable* counterexamples.

LEMMA 10. *The result of relativizing the quantifiers of each instance of Replacement in* ZF*′ *to L* (not *relativizing the* '$<$' *symbols*) *yields theorems of* ZF*. *These resulting sentences are called the semirelativizations.*

PROOF. We take a particular instance of Replacement in ZF*′, e.g., that one whose Axy is $T((\exists z)(w)Czwxy)\ \&\ (u)(u < y \rightarrow \sim T((\exists z)(w)Czwxu))$, where C is quantifier-free. We let D be a definable domain, $D \in L$. We wish to show in ZF* that there is a set $S \in L$ of all $y \in L$ such that for some $x \in L$ with $x \in D$, y is the unique y with $(Axy)'$. This is easily seen to be equivalent to finding a set $S \in L$

of all $y \in L$ such that for some $x \in D$,

$$
(1)\quad \begin{aligned}
&x \in L \;\&\; y \in L \;\&\; (\exists z)_L[(w)_L(Czwxy \vee (\exists w')_L(w' < w \;\&\; {\sim}Czw'xy)) \\
&\qquad \&\; (z')_L(z' < z \rightarrow (\exists w)_L({\sim}Cz'wxy \;\&\; (w')_L(w' < w \rightarrow Cz'w'xy)))] \\
&\&\; (u)_L\{u < y \rightarrow (z)_L[(\exists w)_L({\sim}Czwxu \;\&\; (w')_L(w' < w \rightarrow Czw'xu)) \\
&\qquad \vee (\exists z')_L(z' < z \;\&\; (w)_L(Cz'wxu \vee (\exists w')_L(w' < w \;\&\; {\sim}Cz'w'xu)))]\}.
\end{aligned}
$$

Convenient notation. If X and Y are expressions occurring in (1), then let $[X, Y]$ be the subformula of (1) beginning with X and ending with Y.

Let U = union of the $O(y)$'s such that (1) holds for some $x \in D$. We proceed to place definable bounds on the quantifiers above in such a way that the new formula is equivalent to (1) for $x \in D$, $y \in M(U)$. (Note that for each $x \in D$ there is at most 1 y satisfying (1).)

Let $f_1(\langle x, y\rangle)$ be undefined if 1 is false; be $O(z)$ for $z \in L$ with $[(w)_L, Cz'w'xy]$, otherwise. Define U_1 = union of the range of f_1 on $D \times M(U)$.

Let $f_2(\langle x, y, z\rangle)$ be undefined if $[(w)_L, {\sim}Czw'xy]$; be $O(w)$ for $w \in L$ with ${\sim}[Czwxy, {\sim}Czw'xy]$, otherwise. Define U_2 = union of the range of f_2 on $D \times M(U) \times M(U_1)$.

Let $f_3(\langle x, y, z, w\rangle)$ be undefined if $(w')_L(w' < w \rightarrow Czw'xy)$; otherwise be $\mu O(w')$ for w' with $w' < w \;\&\; {\sim}Czw'xy$ (see Lemma 9). Define U_3 = union of the range of f_3 on $D \times M(U) \times M(U_1) \times M(U_2)$.

Let $f_4(\langle x, y, z\rangle)$ be undefined if $[(z')_L, Cz'w'xy]$; be $\mu O(z')$ with ${\sim}[z' < z, Cz'w'xy]$, otherwise. Define U_4 = union of the range of f_4 on $D \times M(U) \times M(U_1)$.

Let $f_5(\langle x, y, z, z'\rangle)$ be undefined if ${\sim}[(\exists w)_L, Cz'w'xy]$; be $O(w)$ for $w \in L$ with $[{\sim}Cz'wxy, Cz'w'xy]$ otherwise. Define U_5 = union of the range of f_5 on $D \times M(U) \times M(U_1) \times M(U_4)$.

Let $f_6(\langle x, y, z, z', w\rangle)$ be undefined if $[(w')_L, Cz'w'xy]$; otherwise be $\mu O(w')$ with $w' < w \;\&\; {\sim}Cz'w'xy$. Define U_6 = union of the range of f_6 on

$$D \times M(U) \times M(U_1) \times M(U_4) \times M(U_5).$$

Let $f_7(\langle x, y\rangle)$ be undefined if $[(u)_L, \subseteq Cz'w'xu]$; otherwise be $\mu O(u)$ with ${\sim}[u < y, {\sim}Cz'w'xu]$. Define U_7 = union of the range of f_7 on $D \times M(U)$.

Let $f_8(\langle x, y, u\rangle)$ be undefined if $[(z)_L, {\sim}Cz'w'xu]$; otherwise be $O(z)$ for $z \in L$ with ${\sim}[(\exists w)_L, {\sim}Cz'w'xu]$. Define U_8 = union of the range of f_8 on $D \times M(U) \times M(U_7)$.

Let $f_9(\langle x, y, u, z\rangle)$ be undefined if ${\sim}[(\exists w)_L, Czw'xu]$; otherwise be $O(w)$ with $w \in L$ and $[{\sim}Czwxu, Czw'xu]$. Define U_9 = union of the range of f_9 on $D \times M(U) \times M(U_7) \times M(U_8)$.

Let $f_{10}(\langle x, y, u, z, w\rangle)$ be undefined if $[(w')_L, Czw'xu]$; otherwise be $\mu O(w')$ with $w' < w \;\&\; {\sim}Czw'xu$. Define U_{10} = union of the range of f_{10} on

$$D \times M(U) \times M(U_7) \times M(U_8) \times M(U_9).$$

Let $f_{11}(\langle x, y, u, z\rangle)$ be undefined if ${\sim}[(\exists z')_L, {\sim}Cz'w'xu]$; otherwise be $\mu O(z')$ with $[z' < z, {\sim}Cz'w'xu]$. Define U_{11} = union of the range of f_{11} on $D \times M(U) \times M(U_7) \times M(U_8)$.

Let $f_{12}(\langle x, y, u, z, z' \rangle)$ be undefined if $\sim[(w)_L, \sim Cz'w'xu]$; otherwise be $O(w)$ for $w \in L$ with $\sim[Cz'wxu, \sim Cz'w'xu]$. Define U_{12} = union of the range of f_{12} on $D \times M(U) \times M(U_7) \times M(U_8) \times M(U_{11})$.

Let $f_{13}(\langle x, y, u, z, z', w \rangle)$ be undefined if $\sim[(\exists w')_L, \sim Cz'w'xu]$; otherwise be $\mu O(w')$ with $w' < w \,\&\, \sim Cz'w'xu$. Define U_{13} = union of the range of f_{13} on $D \times M(U) \times M(U_7) \times M(U_8) \times M(U_{11}) \times M(U_{12})$.

Note that by suitable instances of Replacement in ZF*, all of the above are provably well defined. Note that each U_i is definable, so that each $M(U_i) \subseteq L$. It is easily seen that for $x \in D$, $y \in M(U)$, it is the case that Axy is equivalent to the predicate Bxy obtained by placing the bounds $M(U_i)$, $1 \leq i \leq 13$, on the appropriate quantifiers in (1).

Now each instance of the following is provable in ZF*: If $\alpha = \{x \mid Ax\}$, and α a limit, then for x and $y \in M(\alpha)$, $x < y$ iff $x < y$ holds when the quantifiers in the definition are relativized to $M(\alpha)$, A of 1 free variable. The proof in ZF* of the schema is like the proof in ZF. Use the definable well-ordering of $M(\alpha)$.

Now let $V = \max(U, U_i, O(D))$. Then relativizing the quantifiers occurring in the expansions of the "$<$"'s that occur in Bxy, to $M(V + \omega)$, we get the same predicate as Bxy, for $x \in D$. Hence we have shown that our S, which we wanted to show originally as $\in L$, is first-order definable over $M(V + \omega)$, and hence

$$\in M(V + \omega + 1), \quad V + \omega + 1 \qquad \text{definable.}$$

8. **The system** ZF′. Making use of the transformation T defined in the previous section, we form ZF′ as follows: First, Extensionality & Foundation & Infinity & Power set and Sum set axioms of ZF. In addition, we have

$$(x)(x \in L) \,\&\, (x)(\text{Ord}\, x \rightarrow \text{Ord}''\,(x)) \,\&\, (x)(x \neq \varnothing \rightarrow (\exists y)(z)(y \in x \,\&\, (z \in x \rightarrow z \not< y))).$$

Replacement in ZF′ will be the following: Let $Bxyy_1 \cdots y_n$ be a formula in prenex form with only the free variables shown. Then

$$(y_1) \cdots (y_n)(x_0)(\exists x_1)(x_2)(x_2 \in x_1 \equiv (\exists x_3)(x_3 \in x_0 \,\&\, T(Bx_3x_2y_1 \cdots y_n) \,\&\, (x_4)(x_4 < x_2 \rightarrow \sim T(Bx_3x_4y_1 \cdots y_n)))),$$

is an instance.

LEMMA 11. $\text{ZF}' \supseteq \text{ZF}$.

PROOF. First, we wish to show in ZF′ each instance of $T(A) \equiv A$. This is trivial for A with no quantifiers.

Suppose $T(A) \equiv A$ is provable in ZF′ for all A in prenex form with n quantifiers. We then wish to show that $T(A) \equiv A$ is provable with A having $n + 1$ quantifiers in prenex form. Then we will have shown $T(A) \equiv A$ provable for any A in prenex form in ZF′.

Let B be in prenex form with $n + 1$ quantifiers. Suppose B is $(\exists x_i)(Ax_i)$. Then $T(B)$ is $(\exists x_i)(T(Ax_i) \,\&\, (x_j)(\sim x_j < x_i \vee T(A^- x)))$. Now $\vdash_{\text{ZF}'} T(Ax_i) \equiv Ax_i$. Since $T(A^- x_j)$ is equivalent to $\sim T(Ax_j)$ we have $\vdash_{\text{ZF}'} T(A^- x_j) \equiv \sim Ax_j$. We have to check that $(\exists x_i)(Ax_i) \equiv (\exists x_i)(Ax_i \,\&\, (x_j)(\sim x_j < x_i \vee \sim Ax_j))$ is provable in ZF′.

Define Cxy to be $y = x \,\&\, T(Ay)$. This is, of course, equivalent with $T(Ay \,\&\, y = x)$, a transformation on a wff of n quantifiers. Now

$(x_0)(\exists x_1)(x_2)(x_2 \in x_1 \equiv (\exists x)(x \in x_0 \,\&\, Cxx_2 \,\&\, (x_4)(x_4 < x_2 \rightarrow \sim T(Ay \,\&\, y = x))))$

is (equivalent to) an axiom of ZF′. But $\vdash_{ZF'} T(Ax_2 \,\&\, x_2 = x_3) \equiv Ax_2 \,\&\, x_2 = x_3$. So $\vdash_{ZF'}(x_0)(\exists x_1)(x_2)(x_2 \in x_1 \equiv Ax_2 \,\&\, x_2 \in x_0)$. Now assume $(\exists x_i)Ax_i$. Choose any such x_i. Take $O(x_i)$, and set $x_0 = M(O(x_i))$, and use the above theorem of ZF′ to get the set of all elements of $M(O(x_i))$ having the property A. ($M(O(x_i))$ is defined and has required properties since $(x)(x \in L)$ is an axiom of ZF′, and $x \in L$ is formalized as in ZF*, previously.) Hence by one of the axioms of ZF′, there is a $<$-least member. Hence we have shown by induction the equivalence between $T(A)$ and A, in ZF′. This has the effect of provably in ZF′ eliminating the T's in the axioms of ZF′, and so ZF′ $\supseteq$ ZF.

9. **The Skolem argument.** We wish to show $(n) \vdash_{ZF^*} \mathrm{Con}(ZF'_n)$, where ZF'_n is the first n axioms of ZF′ in some natural enumeration of them. If we succeed in showing this, then suppose $\sim$Con(ZF). Then $\sim$Con(ZF′). Then

$$(\exists n) \sim \mathrm{Con}(ZF'_n).$$

But then $(\exists n) \vdash_{ENT} \mathrm{Con}(ZF'_n)$. Since ENT is formalizable in ZF*,

$$(\exists n) \vdash_{ZF^*} \mathrm{Con}(ZF'_n).$$

Hence $(\exists n)(\vdash_{ZF^*} \mathrm{Con}\,(ZF'_n) \,\&\, \vdash_{ZF^*} \sim\mathrm{Con}(ZF'_n))$, and so $\sim$Con(ZF*). Hence Con ZF* $\rightarrow$ Con ZF.

We give, without loss of generality, a Skolem closure argument within ZF* to give, provably in ZF*, a set which is a model for (1) Extensionality in ZF, (2) Foundation in ZF, (3) Infinity in ZF, (4) Power set in ZF, (5) Sum set in ZF, (6) $(x)(x \in L)$, (7) $(x)(\mathrm{Ord}\, x \rightarrow \mathrm{Ord}''\, x)$, (8) $(x)(x \neq \varnothing \rightarrow (\exists y)(z)(y \in x \,\&\, z \not< y))$, (9) let $D(xyT)$ be the formula obtained from taking (1) in §7, The system ZF*′ and replacing the 4-place quantifier-free predicate C, with some 5-place quantifier-free predicate $E(zwxyT)$. $(D)(T)(\exists S)(y)(y \in S \equiv (\exists x)(x \in D \,\&\, D(x, y, T)))$.

Thus, (1)–(9) are typical of a finite subsystem of ZF′.

The construction, in ZF*, of the model of these 9 sentences will be much like a Skolem construction in which the initial model is $\varnothing$. At each stage n, we throw in some sets $x \in L$, and we take the union as n ranges over ω.

We simultaneously define α_n and S_n. We are interested in $\bigcup_{n\in\omega} S_n$.

$S_n = M(\alpha_n)$. $S_0 = M(\varnothing) = \varnothing$. $\alpha_0 = \varnothing$.

Consider, for each $x \in S_n$, the $<$-least $y \in x$ with $(z)(z \in y \rightarrow z \notin x)$.

Consider, for each $x \in S_n$, $P_L(x)$ = set of all $y \subseteq x$ with $y \in L$.

Consider, for each $x, y \in S_n$, with $x \nsubseteq y$, the $<$-least element of x not in y.

Consider, for each $x \in S_n$, $\bigcup_L(x)$ = set of all $y \in L$ such that $(\exists z)(z \in x \,\&\, y \in z)$.

Consider, for each $T \in S_n$, the unique $S \in L$ satisfying the semirelativized of (9) to L. (Call this (9)″.)

W continue "considering" through (9)″, closing S_n, in effect, under the "Skolem functions" for (9)″, in such a way that, as in §7, The system ZF*, we have that the Skolem functions produce values definable in terms of the arguments.

We take α_{n+1} = (union of the $O(z)$'s for the z's considered above) + ω. Take $S_{n+1} = M(\alpha_{n+1})$.

We can then use appropriate instances of Replacement in ZF* in combination with the definable well-ordering $<$, to show that if the S_n and α_n are not well defined for each n, then there are definable counterexamples to our construction in the following sense: for some specifically definable sets, the sets corresponding to them that we considered above do not exist. But this is impossible by Lemmas 7, 8, 9, and 10. So our construction is well defined in ZF*.

Now our model $\bigcup_{n\in\omega} S_n$ is an $M(\alpha)$, α definable, α a limit. In particular, it is transitive. It also contains ω. Due to the absoluteness of the definition of L and of the definition of ω in $M(\alpha)$'s, α a limit, it is easily seen in ZF*, putting all this together, that the sentences (1)–(9) are true when the quantifiers range over $M(\alpha)$. Furthermore, since $M(\alpha)$ is definable, the definition of satisfaction and the induction on α are easily developable in ZF*, in order to prove, in ZF*, that Con((1)–(9)).

From the remarks at the beginning of this section, we immediately have

THEOREM 1. $\vdash_{\text{ENT}}$ Con ZF* $\rightarrow$ Con ZF.

10. **Some further results.** We define a sentence of set theory to be *arithmetical* if it is the relativized of some sentence of set theory to ω.

COROLLARY 1. ZF *is a conservative extension of* ZF* *for arithmetical sentences.*

PROOF. Let A be arithmetical, and $\vdash_{\text{ZF}} A$. We can show Con(ZF* + $\sim A$) $\rightarrow$ Con(ZF + $\sim A$) by modifying the proof of Theorem 1 slightly; just redefine the systems ZF*′, ZF′ as ZF*′ + $\sim A$, ZF′ + $\sim A$, respectively. Due to Lemma 6, all of our lemmas carry over. Now since $\sim$Con(ZF + $\sim A$), we have

$$\sim\text{Con}(\text{ZF}^* + \sim A),$$

and so $\vdash_{\text{ZF}^*} A$.

Our next theorem concerns sentences of the form $(x)(\exists!y)Axy$, A arbitrary, with only 2 free variables, that are provable in ZF*. Now in ZF there are many such sentences which define, provably, in ZF, a Skolem function which moves everything and which is 1-1. An example is $(x)(\exists y)(y = |P(x))$. Another is $(x)(\exists y)(y = \{x\})$. Not so in ZF*. Thus,

THEOREM 2. *Let Axy be any formula with only free variables shown, and let* $C = (x)(\exists!y)(Axy \,\&\, y \neq x) \,\&\, (x)(y)(z)((Axy \,\&\, Axz) \rightarrow y = z)$. *Then C is not provable in* ZF*.

PROOF. We let C be a sentence of the above form, and we construct a model for ZF* + $\sim C$, given an arbitrary model for ZF, $\mathscr{A} = \langle X, R\rangle$, where R is a 2-ary relation on X, $X \neq \varnothing$. (All models are assumed to be equality models. Note that ZF* is a first-order theory with equality.)

We define $\mathscr{B}$ as follows: The domain is to be $X \cup Q$, where Q is the rationals. The 2-ary relation, Sxy, is defined as Rxy if $x, y \in X$; $x < y$ if $x, y \in Q$; false, if $x \in Q, y \in X$; true if $x \in X, y \in Q$.

We claim $\mathscr{B}$ satisfies ZF* + $\sim C$.

First, we show that the elements of Q in $\mathscr{B}$ are indistinguishable in the sense that if $Ax_1 \cdots x_n y_1 \cdots y_m$ holds in $\mathscr{B}$ for $x_i \in Q$, $y_j \in X$, then so does $Az_1 \cdots z_n y_1 \cdots y_m$ for $z_i \in Q$ if the two sequences of rationals, x_i, y_i have the same order relations in Q, (i.e., there is a 1-1 order preserving map). To see this, it suffices to show that, given such a pair of similar sequences of rationals, there is an automorphism of $\mathscr{B}$ which keeps the elements of X fixed, and which maps, in an order-preserving way, the sequence x_i onto z_i. And such an automorphism is easily given by any map which fixes the elements of X and maps the rationals 1-1 onto itself, which maps the x_i into the y_j.

Now suppose $\mathscr{B} \models (x)(\exists! y)(Axy)$. Then by indistinguishability, it is clear that for $x \in Q$, we have $\mathscr{B} \models Axy$ for some $y \in X$, for otherwise we would have Axy for $x, y \in Q$, and hence Axz for $z = y + 1$. But now I claim that $\mathscr{B}$ satisfies $A(x + 1, y)$, since $y \in X$, by indistinguishability, assuming $x \in Q$. So $\mathscr{B}$ does not satisfy C.

Clearly $\mathscr{B}$ satisfies Axiom 0 of ZF*.

To verify 1 of ZF*, suppose $\mathscr{B}$ $(x_2)(x_2 \in x_0 \equiv x_2 \in x_1)$, for x_2, x_1 in the domain. Then either x_1 and $x_0 \in X$, or $x_1, x_0 \in Q$. In the first case, we can conclude that $\mathscr{A}$ $(x_2)(x_2 \in x_0 \equiv x_2 \in x_1)$, and so since $\mathscr{A}$ ZF, we have $x_0 = x_1$. In the second case, we have $\frac{1}{2}(x_0 + x_1) \leq x_0$ iff $\frac{1}{2}(x_0 + x_1) \leq x_1$. But then $x_0 = x_1$.

To verify 2 of ZF*, set $x_0 = \omega$ of $\mathscr{A}$. It is easy to see that $\mathscr{B} \models \varnothing \in x_0$, since $\varnothing$ in $\mathscr{B}$ is same as $\varnothing$ in $\mathscr{A}$. Also, $x \cup \{x\}$ remains unchanged for $x \in X$, when we pass from $\mathscr{A}$ to $\mathscr{B}$. Also, the members in $\mathscr{A}$ of x_0 are the same as the members of x_0 in $\mathscr{B}$. Also, the subsets in $\mathscr{A}$ of x_0, or any of its members, are identical with the corresponding elements in $\mathscr{B}$. Putting this together, we see that Axiom 2 of ZF* is satisfied in $\mathscr{A}$ "in the same way" as it is in $\mathscr{B}$.

To see that 3 of ZF* is satisfied by $\mathscr{B}$, suppose $\mathscr{B}$ satisfies C_A. Then the unique element defined in C_A must be $\in X$, by indistinguishability. We let this element be x. What we are looking for is a power set of x in the model $\mathscr{B}$. We claim that $y = P(x)$ in the model $\mathscr{A}$ does the trick. We have to show that $\mathscr{B}$ satisfies $y = P(x)$. But this is obvious, since the only members of y in $\mathscr{B}$ are the members of y in $\mathscr{A}$, and the only subsets of x in $\mathscr{B}$ are the subsets of x in $\mathscr{A}$.

Axiom 4 of ZF* is checked similarly.

To see that Axiom 6 of ZF* is satisfied in $\mathscr{B}$, let $x \in X \cup Q$. If $x \in X$, take x_0 in foundation as an R-least member of x in $\mathscr{A}$. If $x \in Q$, take x_0 to be $\varnothing$ in $\mathscr{A}$ (or $\mathscr{B}$). Obviously $\mathscr{B} \models \varnothing \in x$.

Axiom 5 of ZF* is the most complicated. The set that C_A defines in $\mathscr{B}$ is again $\in X$. Call it D. Then we are interested in the range in $\mathscr{B}$ of the partial function in $\mathscr{B}$, Bx_2x_1 & $(x_3)(Bx_2x_3 \rightarrow x_3 = x_1)$, on the domain D. Now every x_2 with $S(x_2, D)$ has $x_2 \in X$, and so, by indistinguishability, if $S(x_2, D)$, and Bx_2x_1 & $(x_3)(Bx_2x_3 \rightarrow x_3 = x_1)$, then $x_1 \in X$. Now suppose there is a formula Cx_2x_1 such that, for $x_2 \in X$, $x_1 \in X$, Cx_2x_1 holds in $\mathscr{A}$ iff Cx_2x_1 holds in $\mathscr{B}$. Then by Replacement in the model $\mathscr{A}$, we would have (this instance of) Replacement in $\mathscr{B}$, and we would be done. It remains to show that for each formula $Ax_1 \cdots x_n$, with the free variables shown, there is a formula $Bx_1 \cdots x_n$ which holds in $\mathscr{A}$ iff $Ax_1 \cdots x_n$ holds in $\mathscr{B}$, when $x_i \in X$.

It suffices to prove by induction that for any formula $Ax_1 \cdots x_n$, and for any *partial* function f from $\{i \mid 1 \leq i \leq n\}$ into Q, there is a formula $Bx_{i_1} \cdots x_{i_k}$, $\{i_1, \ldots, i_k\} = \{i \mid 1 \leq i \leq n \;\&\; i \notin \mathrm{Dom}(f)\}$, such that for any sequence $x_1 \cdots x_n$ with $x_i \in X$ iff $i \notin \mathrm{Dom}(f)$, $x_i = f(i)$ if $i \in \mathrm{Dom}(f)$, we have $\mathscr{B} \models Ax_1 \cdots x_n$ iff $\mathscr{A} \models Bx_{i_1} \cdots x_{i_k}$.

To see this for $Ax_1 \cdots x_n$ quantifier-free, take B to be the formula obtained from A by (1) replacing all instances of $x_i \in x_j$, $i \notin \mathrm{Dom}(f)$, $j \in \mathrm{Dom}(f)$, by $x_i = x_i$; (2) replacing all instances of $x_i \in x_j$, or $x_i = x_j$, or $x_j = x_i$, $i \in \mathrm{Dom}\,(f)$, $j \notin \mathrm{Dom}(f)$, by $x_j \neq x_j$; (3) replacing all instances of $x_i \in x_j$, or $x_i = x_j$, $i, j \in \mathrm{Dom}(f)$, by $(\exists v)(v = v)$ if $f(i) < f(j)$, $\sim(\exists v)(v = v)$ if not; or $(\exists v)(v = v)$ if $f(i) = f(j)$, $\sim(\exists v)(v = v)$ if not, respectively.

Put $Ax_1 \cdots x_n$ in prenex form, and suppose our claim is true for all formulae with fewer quantifiers.

We may assume that $Ax_1 \cdots x_n$ is $(x_0)Cx_0x_1 \cdots x_n$, since the existential case follows from this case by taking negations. Let f be a partial function from $\{i \mid 1 \leq i \leq n\}$ into Q. Now 2 finite partial functions g, h from $\omega \to Q$ are said to be of the same type if (1) they have the same domain D, (2) $g(x) < g(y)$ iff $h(x) < h(y)$.

Consider the set of partial functions on $\{i \mid 0 \leq i \leq n\}$ which are identical to f on $\{i \mid 1 \leq i \leq n\}$. There are only a finite number of types represented in this set. Pick a representative from each type, and call this set $\{f_1, f_2, \ldots, f_k\}$, $f_1 = f$. Let D_i, $1 \leq i \leq k$, be the formula given by the inductive hypothesis for $Cx_0x_1 \cdots x_n$ for f_i; i.e. each D_i has exactly the free variables x_l for $l \notin \mathrm{Dom}(f_i)$, $0 \leq l \leq n$, and for any $x_0 \cdots x_n$ with $x_l \in X$ for $l \notin \mathrm{Dom}(f_i)$, $x_l = f_i(l)$ for $l \in \mathrm{Dom}(f_i)$, we have $\mathscr{B} \models Cx_0x_1 \cdots x_n$ iff $\mathscr{A} \models D_i x_{p_1} \cdots x_{p_q}$, $\{p_1 \cdots p_q\} = \{r \mid 0 \leq r \leq n \;\&\; r \notin \mathrm{Dom}(f_i)\}$.

Then we take B to be $(x_0)D_1 \;\&\; \bigwedge_{2 \leq i \leq k} D_i$. By indistinguishability, it is easily seen that B and A satisfies the conclusion of our claim for the function f. This concludes the proof of Theorem 2.

Considering for the moment a "more explicit 2nd-order arithmetic," on the basis of the remarks in §1, General situation, we should consider the system Z_2^- (2nd-order arithmetic is Z_2), which is the same as Z_2 except each instance of comprehension is of the form $(\exists x)(n)(n \in x \leftrightarrow An)$, where A has only the free variable shown. Then our Theorem 1 and Corollary 1 hold.

Going to a "more explicit 1st-order arithmetic," we consider ENT* which has the usual recursion equations for $+$ and $\times$, and the usual axioms about 0 and successor. The induction schema is modified. Given two formulae An, Bk, with only one free variable, we take

$$[An \;\&\; (m)(Am \rightarrow m = n) \;\&\; (\exists k)(k \leq n \;\&\; Bk)] \rightarrow (\exists l)(Bl \;\&\; (\forall p)(Bp \rightarrow l \leq p)).$$

Then our Theorem 1 holds. The proofs of these two claims use (and seem to need) the methods of this paper.

STANFORD UNIVERSITY

SETS, SEMISETS, MODELS

PETR HÁJEK

This is an expository article written for two purposes. First, it is to give a survey of works of the members of the Prague seminar on foundations of set theory and a full bibliography of them. Secondly, it is to explicate the matter of interest from the present point of view and, in fact, to summarize the contents of a monograph being written by P. Vopěnka and the author of the present paper. These two purposes are followed simultaneously throughout the paper; the author would be glad if the paper was of some help as a guide for reading papers mentioned in the bibliography.

Speaking on the study of foundations of the set theory in Czechoslovakia we must begin with the name of the late Professor L. Rieger. He was the first Czechoslovak mathematician to work on this field. (See Czech. Math. J. (89) **14** (1964), 629 ff. for a short account of his life and papers.) After his tragic death in 1963, his student and fellow-worker P. Vopěnka founded a seminar and engaged the attention of several young people for the study of foundations. Now, after five years, the seminar consists of the following members: B. Balcar, L. Bukovský, K. Hrbáček, T. Jech, A. Sochor, P. Štěpánek, P. Vopěnka and the author.

Let us begin with a trivial remark. Studying metamathematics, it is not uniquely determined which intuitive concepts are presupposed to be sufficiently known. In the case of the syntax of axiomatic theories, in our case of the set theory (or of theories of sets), the notion of a finite sequence of symbols and that of an effective (decidable) system of these sequences may suffice. There are at least two reasons for such a minimization of means: the metamathematical one, consisting in the fact that the finitary conception of the syntax gives to our metamathematical study more "anthropological" character and enables us to answer adequately the question of what the mathematicians can do (prove, decide) and what they cannot. Secondly, there is a mathematical reason, consisting in the fact that, from the mathematical point of view (i.e. from the point of view of developing a particular axiomatic theory) finitary metamathematical results may be consequently understood as auxiliary principles for obtaining new proofs or notions in the theory we are dealing with. Thus, we keep this finitary intuitive conception, being interested in foundations both as logicians and as mathematicians. E.g. a statement "there is a formula . . ." is demonstrated if and only if an effective method is given for finding such a formula. Speaking on a mapping in the metalanguage we always assume a method to be given which enables to find the image of every element to be mapped.

Even if not formulated explicitly, that which is said above has been our point of view from the beginning. But, because of the simultaneous metamathematico-mathematical interest, parts written in the object language and those written in the metalanguage are sometimes not distinguished clearly in earlier papers from the bibliography. The reader is suggested to read those works on the basis of the present paper.

As we are interested in the study of concrete theories (namely, the Gödel-Bernays set theory and some related theories) we are forced to choose the basic formal system quite rich and thus quite near to actual mathematical techniques. But it need not be explained in details. Imagine simply, we have *variables* and *constants* of various *sorts*, *predicates* of various *ranks*, and *operations* of various *sorts* and *ranks*. One sort is preferred as *universal*. *Terms* and *formulas* are defined in the usual way; any (finite) sequence of formulas may be considered as an *axiomatics*. The *language* of an axiomatics is the (finite) list of all predicates, constants, operations and sorts of variables occurring in the axioms. (A sort may be represented by an arbitrary variable of this sort.) The notion of *logical axioms* (tautologies) and *deductive rules* is defined; thus, we have a formal notion of a *proof* in an axiomatic theory. (*Theory* is given by its axiomatics.) If $\mathcal{T}$ is a theory, then a formula is a *$\mathcal{T}$-formula* iff it is formulated by means of things occurring in the language of $\mathcal{T}$; a $\mathcal{T}$-formula is *$\mathcal{T}$-provable* (denotation: $\mathcal{T} \vdash \varphi$) iff there is a proof of it from the axioms of $\mathcal{T}$. A theory $\mathcal{S}$ is an *extension of $\mathcal{T}$* iff the sequence of axioms of $\mathcal{T}$ is a segment of the sequence of axioms of $\mathcal{S}$. The notion of a *contradictory* and *consistent* theory (introduced in this order) is usual.

A mapping $\mathcal{M}$ of $\mathcal{T}$-formulas into $\mathcal{S}$-formulas is a *syntactic model* of $\mathcal{T}$ in $\mathcal{S}$ iff

(a) $\mathcal{M}$ respects both logical axioms and the axioms of $\mathcal{T}$, i.e. maps these axioms into $\mathcal{S}$-provable formulas,

(b) $\mathcal{M}$ respects deduction rules, i.e. if a $\mathcal{T}$-formula immediately follows from some $\mathcal{T}$-formula(s) (assumption(s)) then the image of the former formula is provable in the extension of $\mathcal{S}$ by the image(s) of the assumption(s);

(c) $\mathcal{M}$ respects the negation, i.e. the negation of the image of a $\mathcal{T}$-formula is provable in the extension of $\mathcal{S}$ by the image of the negation of that formula.

Provability principle. Let $\mathcal{M}$ be a model of $\mathcal{T}$ in $\mathcal{S}$, then the image of every $\mathcal{T}$-provable formula is $\mathcal{S}$-provable.

Consistency principle. If $\mathscr{T}$ has a model in $\mathscr{S}$ and $\mathscr{T}$ is contradictory then $\mathscr{S}$ is too; a fortiori, if $\mathscr{T}$ has a model in $\mathscr{S}$ and $\mathscr{S}$ is consistent then $\mathscr{T}$ is too.

A $\mathscr{T}$-formula φ is said to *hold* in a model $\mathscr{M}$ of $\mathscr{T}$ in $\mathscr{S}$ iff $\mathscr{S} \vdash \varphi^{\mathscr{M}}$ ($\varphi^{\mathscr{M}}$ being the image of φ by $\mathscr{M}$). A model $\mathscr{M}_1$ is *stronger* than $\mathscr{M}_2$ iff every formula holding in $\mathscr{M}_2$ holds in $\mathscr{M}_1$ too. $\mathscr{M}_1$ is *equivalent* to $\mathscr{M}_2$ iff $\mathscr{M}_1$ is stronger than $\mathscr{M}_2$ and $\mathscr{M}_2$ is stronger than $\mathscr{M}_1$. Identical mapping of $\mathscr{T}$-formulas is called the *identical model* of $\mathscr{T}$. If $\mathscr{M}_1$ is a model of $\mathscr{T}_1$ in $\mathscr{T}_2$ and $\mathscr{M}_2$ is a model of $\mathscr{T}_2$ in $\mathscr{T}_3$, then the composed mapping is denoted by $\mathscr{M}_1 * \mathscr{M}_2$ and called $\mathscr{M}_1$ *constructed in* $\mathscr{M}_2$. Theories $\mathscr{T}$ and $\mathscr{S}$ are *equivalent* iff there are models $\mathscr{M}_1$ of $\mathscr{T}$ in $\mathscr{S}$ and $\mathscr{M}_2$ of $\mathscr{S}$ in $\mathscr{T}$ such that $\mathscr{M}_1 * \mathscr{M}_2$ is equivalent to the identical model of $\mathscr{T}$ and $\mathscr{M}_2 * \mathscr{M}_1$ is equivalent to the identical model of $\mathscr{S}$.[1]

EXAMPLES. (1) There may be defined explicitly what a definition of a predicate, constant, operation, sort of variables respectively in a theory is. An extension of a theory by adding such a definition is equivalent to the original theory. (2) More generally for constants: Let $\mathscr{T} \vdash (\exists x_1, \ldots, x_n)\pi(x_1, \ldots, x_n)$. If we add the axiom $\pi(a_1, \ldots, a_n)$ where a's are new constants we obtain an equivalent theory. (In this case we say that we have fixed the parameters with the help of π.)

Now it is possible to formulate the axioms of the fundamental gödelian theory of classes **TC** and describe a very general kind of models of **TC** in itself. We are interested in developing this theory from a unary predicate $\in$ and variables of the universal sort only (the language $(\in, X)$ is called the *fundamental language*); all other notions, including the equality predicate, are defined. (This possibility has been observed and used by several authors.) Instead of doing it explicitly we only give the definitions of the notions we need; the axioms serve only to the fact that the following definitions really are definitions in the sense of the calculus.

$X = Y \equiv (\forall Z)(Z \in X \equiv Z \in Y)$	equality predicate
$(\exists x)(x = X) \equiv (\exists Z)(X \in Z)$	set variables
$z \in \{X, Y\} \equiv .z = X \vee z = Y$	pairing operation
$\langle x, y\rangle = \{\{x\}, \{x, y\}\}$	ordered pair operation
$(\forall x)(x \in V)$	constant for universal class
$x \in \mathfrak{E}(X) \equiv (\exists u, v)(x = \langle u, v\rangle \;\&\; x \in X \;\&\; u \in v)$	$\in$-representation on X (operation)
$x \in X - Y \equiv .x \in X \;\&\; x \notin Y$	difference
$x \in \mathfrak{D}(X) \equiv (\exists y)(\langle y, x\rangle \in X)$	domain
$x \in X \upharpoonright Y \equiv .x \in X \;\&\; (\exists u, v)(x = \langle u, v\rangle \;\&\; v \in Y)$	restriction
$x \in \mathfrak{Cn}(X) \equiv (\exists u, v)(x = \langle u, v\rangle \;\&\; \langle v, u\rangle \in X)$	conversion
$x \in \mathfrak{Cn}_3(X) \equiv (\exists u, v, w)(x = \langle u, v, w\rangle \;\&\; \langle v, w, u\rangle \in X)$	ternary conversion

[1] By the way, if we gave up our finitary point of view, it could be of some interest to deal with the category of theories as objects and (some) syntactic models as morphisms.

The operations $\{\ \}$, $\mathfrak{E}$, $-$, $\mathfrak{D}$, $\restriction$, $\mathfrak{Cn}$, $\mathfrak{Cn}_3$ are called *gödelian operations* and sometimes denoted by $\mathfrak{F}_1(X, Y), \ldots, \mathfrak{F}_7(X, Y)$ (in this order). A term built up from the constant V, universal variables and the operations $\mathfrak{F}_2, \ldots, \mathfrak{F}_7$ is called a *gödelian term.* A formula built up from the predicate $\in$, universal variables (called also class variables), and set variables in which no class variable is bound is called *normal.* A well-known metatheorem on normal formulas may be stated in the following way:

Let $\varphi(x_1, \ldots, x_n, X_1, \ldots, X_m)$ be normal. Then there is a gödelian term $\mathfrak{T}(X_1, \ldots, X_m)$ such that

$$\mathbf{TC} \vdash \langle x_1, \ldots, x_n \rangle \in \mathfrak{T}(X_1, \ldots, X_n) \equiv \varphi(x_1, \ldots, x_n, X_1, \ldots, X_m).$$

Even if we have not written down the axioms of **TC** we shall use some names of them. Besides two auxiliary axioms F1, F2 we have an axiom A1 "justifying the definition of the ordered pair", B1 "justifying the definition of the universal class" and B2 – B7 "justifying the definitions of the operations $\mathfrak{F}_2 - \mathfrak{F}_7$". (E.g. B1 is $(\exists Z)(\forall x)(x \in Z)$ etc.)

A *theory of classes* is any extension of **TC** such that every new axiom either is formulated in the language of the preceding segment (is a proper axiom) or is a definition of a new concept.[2]

Fundamental formulas are formulas of the fundamental language, i.e. those built up from the predicate $\in$ and class variables. Set formulas are formulas built up from $\in$ and set variables.

Fundamentalization principle. Let $\mathscr{T}$ be a theory of classes. Then there is a mapping $\mathscr{F}$ associating with every $\mathscr{T}$-formula φ a fundamental formula $\varphi^{\mathscr{F}}$ deductively equivalent to φ in $\mathscr{T}$; moreover, the sequence of images of proper axioms of $\mathscr{T}$ is a theory $\mathscr{T}^{\mathscr{F}}$ equivalent to $\mathscr{T}$ and $\mathscr{F}$ is a model of $\mathscr{T}$ in $\mathscr{T}^{\mathscr{F}}$. $\varphi^{\mathscr{F}}$ is called the fundamentalization of φ.

Given a theory of classes $\mathscr{T}$ and another theory $\mathscr{S}$, given further a binary predicate $\in^*$ and a sort of variables $X^*, Y^*, \ldots$ in the language of $\mathscr{S}$, we may define a mapping of $\mathscr{T}$-formulas into $\mathscr{S}$-formulas as follows; for every $\mathscr{T}$-formula φ, take its fundamentalization $\varphi^{\mathscr{F}}$ and, in the latter formula, put $\in^*$ instead of $\in$, X^* instead of X, etc. This mapping is denoted by $\mathscr{Im}(\mathscr{T}, \mathscr{L})$, where $\mathscr{L}$ is the language $(\in^*, X^*)$ (read: the *imitation* of the $\mathscr{T}$-formulas given by $\mathscr{L}$); this mapping is a model of $\mathscr{T}$ in $\mathscr{S}$ iff images of proper axioms of $\mathscr{T}$ are provable in $\mathscr{S}$. (This notion is closely related to the Tarski's notion of relative interpretability; cf. also **[18]**, **[42]**.)

Sometimes we do not have such a language $(\in^*, X^*)$ (which may be called an **F**-like language) but we are able to introduce it. In the optimal case, we find two formulas $\chi(X)$ and $\varepsilon(X, Y)$ such that we are able to define X^* as those X that $\chi(X)$ and to define $\in^*$ with help of ε. The couple of formulas χ, ε is called a *nonparametric basis* (of an **F**-like language). A triple $\mathscr{B}$ of formulas $\pi(\boldsymbol{u})$, $\chi(X, \boldsymbol{u})$, $\varepsilon(X, Y, \boldsymbol{u})$

[2] Notions which are used in the usual sense will not be defined here, as the power class, the field of a relation etc.

($\boldsymbol{u}$ being a finite sequence of variables) is called a *parametric basis* iff $\mathscr{S} \vdash (\exists \boldsymbol{u})\pi(\boldsymbol{u})$ and $\mathscr{S} \vdash \pi(\boldsymbol{u}) \rightarrow (\exists X)\chi(X, \boldsymbol{u})$. Having such a basis we may first fix parameters with help of π ("take arbitrary but fixed $\boldsymbol{a}$ such that $\pi(\boldsymbol{a})$") and then define X^* as those X that $\chi(X, \boldsymbol{a})$ and $X^* \in^* Y^*$ as $\varepsilon(X, Y, \boldsymbol{a})$. (We obtain a theory $\mathscr{S}_1$ equivalent to $\mathscr{S}$ in this way.) Let us write $\mathscr{Im}(\mathscr{T}, \mathscr{B})$ instead of $\mathscr{Im}(\mathscr{T}, \mathscr{L})$ where $\mathscr{L}$ is $(\in^*, X^*)$. Or we may associate with every $\mathscr{T}$-formula not containing the variables $\boldsymbol{u}$ (more precisely: with its fundamentalization) a formula $\varphi^{\mathscr{Par}(\mathscr{T},\mathscr{B})}$ by the following induction: $(X \in Y)^{\mathscr{Par}(\mathscr{T},\mathscr{B})}$ is $\varepsilon(X, Y, \boldsymbol{u})$, $(\varphi \ \& \ \psi)^{\mathscr{Par}(\mathscr{T},\mathscr{B})}$ is $\varphi^{\mathscr{Par}(\mathscr{T},\mathscr{B})} \ \& \ \psi^{\mathscr{Par}(\mathscr{T},\mathscr{B})}$ etc., $[(\forall X)\varphi(X)]^{\mathscr{Par}(\mathscr{T},\mathscr{B})}$ is $(\forall X)(\chi(X, \boldsymbol{u}) \rightarrow [\varphi(X)]^{\mathscr{Par}(\mathscr{T},\mathscr{B})})$ etc. (The mapping $\mathscr{Par}$ is called the *parametric imitation* of $\mathscr{T}$-formulas given by $\mathscr{B}$.) If φ is closed then $\pi(\boldsymbol{a}) \rightarrow \varphi^{\mathscr{Par}(\mathscr{T},\mathscr{B})}(\boldsymbol{a})$ is deductively equivalent to $\varphi^{\mathscr{Im}(\mathscr{T},\mathscr{B})}$ in $\mathscr{S}_1$ (one can say even more); but in any case, $\varphi^{\mathscr{Par}(\mathscr{T},\mathscr{B})}$ has at least free variables $\boldsymbol{u}$; to every closed φ a notion concerning $\boldsymbol{u}$ is associated in this way. Given a concrete basis, we write usually φ^* instead of $\varphi^{\mathscr{Im}(\mathscr{T},\mathscr{B})}$ and $\varphi^{\boldsymbol{u}}$ instead of $\varphi^{\mathscr{Par}(\mathscr{T},\mathscr{B})}$. ($\varphi^{\boldsymbol{u}}$ is often read "φ holds in sense of $\boldsymbol{u}$".) We also write $\mathrm{Cls}^{\boldsymbol{u}}(X)$ instead of $\chi(X, \boldsymbol{u})$.

Let $\mathscr{B} = (\pi(\boldsymbol{u}), \chi(X, \boldsymbol{u}), \varepsilon(X, Y, \boldsymbol{u}))$ be a basis in $\mathscr{S}$. A basis $\mathscr{B}' = (\pi')(\boldsymbol{u})$, $\chi(X, \boldsymbol{u}), \varepsilon(X, Y, \boldsymbol{u})$ is called a *specification* of $\mathscr{B}$ (in $\mathscr{S}$) iff $\mathscr{S} \vdash \pi'(\boldsymbol{u}) \rightarrow \pi(\boldsymbol{u})$. If $\mathscr{B}$ determines a model of $\mathscr{T}$ and $\mathscr{B}'$ is a specification of $\mathscr{B}$ then $\mathscr{B}'$ determines a model of $\mathscr{T}$ stronger than the former one.

Very important example for **TC**. Define in **TC**: *Relation* is a class of ordered pairs. *Extension* of x in a relation R (denoted $\mathfrak{Ext}_R(x)$) is the class $\{y; \langle y, x\rangle \in R\}$. R is an *E-like relation* (denotation: $\mathrm{Elk}(R)$) iff R is a nonempty relation, is *internal* (i.e. for any $x, y \in \mathit{field}(R)$, $\mathfrak{Ext}_R(x) = \mathfrak{Ext}_R(y)$ implies $x = y$) and *closed* with respect to the 1st operation (i.e. for every $x, y \in \mathit{field}(R)$, there is a z such that $\mathfrak{Ext}_R(z) = \{x, y\}$).

The triple $\mathscr{N}_{\mathbf{TC}}(R)$ of formulas

$$\mathrm{Elk}(R), \qquad X \subseteq \mathit{field}(R), \qquad (\exists z \in Y)(X = \mathfrak{Ext}_R(z))$$

is a basis which determines a model of **TC** in **TC**. E.g. the relation $E = \mathfrak{E}(V)$ is an E-like relation.

Classes of the model are all subclasses of the field of R; elements of the field of R are codes of sets of the model, the membership is determined by R (X belongs to Y in sense of the model iff the code of X belongs to Y). Now, we are interested in theories stronger than **TC** and their models. Evidently, the basis $\mathscr{N}_{\mathbf{TC}}(R)$ determines a model of **TC** in every theory stronger than **TC**; but if we specify the conditions on R, we obtain more. We are just going to discuss this.

We shall deal with two extensions of **TC**: with the theory of sets **TS** and the theory of semisets **TSS**. The former one is equivalent to Gödel's theory with axiom groups **A**, **B**, **C** and corresponds to **ZF** without the axiom of regularity. The latter one is weaker than **TS** and was formulated by Vopěnka in Summer 1967 in cooperation with the author; this theory is explained here with the permission of P. Vopěnka. *Semisets* are defined in **TC** as subclasses of sets. ($\mathrm{Sm}(X) \equiv (\exists y)(X \subseteq y)$.) This notion is superfluous in **TS**, as all semisets are sets; but it is of a fundamental importance in **TSS**. As we shall see, there are two ways of constructing models of **TS** in **TSS** (with some additional axioms): going below, i.e. omitting some proper classes (nonsets), and going above, i.e. make semisets to sets. It enables us e.g. to demonstrate various forms of the independence of the axiom of constructibility by constructing models of **TSS** in **TS** (which is not difficult) and then working in **TSS**. From this point of view, **TSS** is the key means of the latest Vopěnka's reformulation of Cohen's method; the advantage consists in the fact that main proofs are done not about model-sets etc., say, in the model, but simply in **TSS**. See below for more details.

Relations will play a prominent rôle in the sequel. Let us define some notions about them (in **TC**). R is called nowhere constant iff, for any $x, y \in \mathfrak{D}(R)$, $\mathfrak{Ext}_R(X) = \mathfrak{Ext}_R(Y)$ implies $X = Y$. If R is a relation, then R is called *regular* iff all extensions are semisets, i.e. $(\forall x \in \mathfrak{D}(R))(\exists y)(\mathfrak{Ext}_R(x) \subseteq y)$. Regular nowhere constant relations may be considered as 1-1 associations of semisets to sets; it helps to formulate some axioms. Axioms of **TSS**: those of **TC** plus

(A2)–(A7) $\mathrm{M}(\mathfrak{F}_i(x, y))$ $i = 2, \ldots, 7$ (gödelian operations make sets from sets)

(C1) $(\exists x \neq 0)(\forall y \in x)(\exists z \in x)(y \subset z)$ (infinity)

(C2) R regular nowhere constant $\rightarrow (\mathrm{Sm}(\mathfrak{D}(R)) \equiv \mathrm{Sm}(\mathfrak{W}(R)))$

($\mathfrak{W}(R)$ is the domain of values, i.e. $\{y; (\exists x)(\langle yx \rangle \in R)\}$). Axioms of **TS**: those of **TC** plus (C1), (C2), plus

(C3) every semiset is a set.

The axioms (A2)–(A7) are provable in **TS**, hence **TS** is stronger than **TSS**. Further, the original Gödel's axioms (C2)–(C4) are provable in **TS** and our axioms (C2)–(C3) are provable in Gödel's axiomatics. But we cannot replace (C2) as an axiom of **TSS** by Gödel's (C2), (C3) (power set and sum set axioms). The power class and sum class of a set is proved to be a set in **TSS**. In **TS**, (C2) may be evidently replaced by

(C2′) R regular nowhere constant $\rightarrow (\mathrm{M}(\mathfrak{D}(R)) \equiv \mathrm{M}(\mathfrak{W}(R)))$

(M being the predicate "... is a set").
First axiom of regularity:

(D1) $(\forall X)(\forall x)(\exists y)(\mathfrak{D}(X) \cap x = \mathfrak{D}(X \cap y))$.

Equivalently: For every relation whose domain is a semiset there is a subrelation which is a semiset and has the same domain.

In **TSS** + (D1), the comprehension schema is provable. A class X is said to have *set intersection property* (SIP(X)) iff, for every set x, $X \cap x$ is a set. Define new variables X^* for classes with set intersection property, define $X^* \in^* Y^* \equiv X^* \in Y^*$. In this way, we obtain a model of **TS** + (D1) in **TSS** + (D1) with absolute notion of set. As a consequence, we obtain the

Equiprovability principle. Let φ be a set formula. Then **TSS** + (D1) $\vdash \varphi$ if and only if **TS** + (D1) $\vdash \varphi$.

The model just described is called the natural model and denoted by $\mathcal{N}at$.

Consider the basis $\mathcal{N}_{\mathrm{TC}}(R)$ defined above. We write φ^R instead of $\varphi^{\mathcal{P}ar(\mathbf{TSS},\mathcal{N}_{\mathrm{TC}}(R))}$. Define (in **TSS**): Let R be an E-like relation. (a) R is *almost universal* iff every semiset included in the field of R is included in the extension of a $x \in \mathfrak{D}(R)$; (b) R is *closed* with respect to the ith operation iff $(\mathrm{A}i)^R$ holds ($i = 2, \ldots, 7$; i e. the axiom (Ai) holds in the sense of R); R is *closed* iff it is closed with respect to all 7 operations. (c) R is *relatively infinite* iff $(\mathrm{Cl})^R$ holds. (d) R is a *model relation* (Mrel(R)) iff it is regular, internal, closed, relatively infinite and almost universal.

METATHEOREM. *The basis*

$$\mathrm{Mrel}(R), \qquad X \subseteq \mathit{field}(R), \qquad (\exists z \in Y)(X = \mathfrak{Ext}_R(z))$$

(*it is a specification of* $\mathcal{N}_{\mathrm{TC}}$) *determines a model of* **TSS** *in* **TSS** *and of* **TSS** + (D1) *in* **TSS** + (D1). (*Cf.* [**2**], [**8**]).

In other words, we can prove in **TSS** (+ (D1)) that all axioms of **TSS** (+ (D1)) hold in sense of every model relation. A fortiori, we can prove the same in **TS** (+ (D1)) but we cannot prove (C3)) in sense of every model relation in **TS**. The basis just described is called the *normal basis* for **TSS** and denoted by $\mathcal{N}_{\mathrm{TSS}}(R)$; also the corresponding model is denoted by $\mathcal{N}_{\mathrm{TSS}}(R)$. The composed model $\mathcal{N}at * \mathcal{N}_{\mathrm{TSS}}(R)$ is called the *normal model* of **TS** + (D1) in **TSS** + (D1). The theory **TSS** + (D1) will be denoted by **TSS**$'$.

Model relations of the form $E \cap P$, E being the original membership relation, and P being a transitive class, are of particular interest. We define (in **TSS**): P is a model class (Mcl(P)) iff it is transitive, contains with arbitrary x, y results of all gödelian operations from them and, for every semiset $X \subseteq P$, there is an $x \in P$ such that $X \subseteq x$.

THEOREM (**TSS**). *If P is a model class then $E \cap P$ is a model relation.*[3]

Hence if we want to construct models of **TS** in **TS** + (D1) it suffices to prove the existence of various model relations in the latter theory. Cf. [**34**].

On the other hand, a mathematical theory of model-relations and model-classes may be developed within the theory of (semi)sets; e.g. we may study the structure of model classes as interesting set theoretical objects not speaking about metamathematical problems.

Let us also give a slight generalization of the notion of a model relation. Define (in **TSS**): R is a *weak model relation* (wMrel(R)) iff there is a model relation R_0 and a function F such that F maps the field of R onto the field of R_0 and $(\forall x, y \in \mathit{field}(R))(\langle x, y\rangle \in R \equiv \langle F(x), F(y)\rangle \in R_0)$. An $X \subseteq \mathit{field}(R)$ is called *saturated* iff, for every $x \in X$, $y \in \mathit{field}(R)$, $\mathfrak{Ext}_R(x) = \mathfrak{Ext}_R(y)$ implies $y \in X$.

[3] This theorem may be used also for **ZF** as a metatheorem: let a class P be defined, let **ZF** $\vdash$ (1) $x, y \in P \rightarrow \mathfrak{F}_i(x, y) \in P$, (2) $x \subseteq P \rightarrow (\exists y \in P)(x \subseteq y)$, (3) P is transitive. Then P is a model for **ZF**.

(Denotation: $\mathrm{Sat}_R(X)$.) The basis

$$\mathrm{wMrel}(R), \qquad X \subseteq \mathit{field}(R) \ \&\ \mathrm{Sat}_R(X), \qquad (\exists z \in Y)(X = \mathfrak{Ext}_R(z))$$

determines a model of **TSS** (+ (D1)) in **TSS** (+ (D1)) equivalent to $\mathcal{N}_{\mathbf{TSS}}(R)$.

Ultraproduct-relations (see below) give examples of weak model relations.

Now, let us deal briefly with **TS**. (We shall come back to **TSS** later; the equiprovability principle will supply a lot of theorems of **TSS**.) Ordinal and cardinal numbers are defined in **TS** in the usual way. If the first axiom of choice (E1) is assumed (every set can be well-ordered), cardinal arithmetic may be studied (cf. [**25**]).

Further (stronger) axioms of regularity and choice are formulated as analogous to each other.

(D2) There is a regular relation R such that $\mathfrak{D}(R) = On$ (the class of all ordinal numbers) and $\mathfrak{W}(R) = V$.

(E2) There is a function F such that $\mathfrak{D}(F) = On$ and $\mathfrak{W}(F) = V$.

Evidently, $\mathbf{TS} \vdash \mathrm{E2} \to \mathrm{D2}$, $\mathrm{D2} \to \mathrm{D1}$, $\mathrm{E2} \to \mathrm{E1}$. Both axioms can be strengthened by defining a relation (function) and postulating that it fulfills the conditions of (D2), (E2) respectively.

(D3) is equivalent to the usual regularity axiom; define $p_0 = 0$ (empty set), $p_{\alpha+1} =$ the power set of p_α, $p_\lambda = \bigcup_{\alpha<\lambda} p_\alpha$ for λ limit number, $Ker = \bigcup_{\alpha \in On} p_\alpha$. (D3) is the assumption $Ker = V$. Defining $\langle x\alpha \rangle \in R \equiv x \in p_\alpha$ we obtain a relation about which we prove in **TS** + (D3): R is regular, $\mathfrak{D}(R) = On$, $\mathfrak{W}(R) = V$. Hence $\mathbf{TS} \vdash \mathrm{D3} \to \mathrm{D2}$. It is well known that (D3) is consistent as it is provable in **TS** that Ker is a model class and (D3) holds in the sense of that model class.

In **TS** + (D3), the following theorem concerning model classes is provable [**49**]: Let M, N be model classes, let the axiom (E1) hold in sense of M, let $\mathfrak{P}^M(On) = \mathfrak{P}^N(On)$ (i.e. M and N have the same sets of ordinals). Then $M = N$.

The assumption that (E1) holds in sense of M is essential, see the paper by Jech in this volume.

The consistency of (E2) can be proved by means of the so called effective model. The model class of hereditarily effective sets equals (provably in **TS** + (D3)) to the model class of hereditarily ordinal-definable sets defined by Myhill and Scott; so we do not describe it in this paper (see this volume for the paper of Myhill and Scott). The axiom (E3) is the assumption that the class of all (hereditarily) effective sets equals V. It implies (E2) but need not hold in sense of all hereditarily effective sets. The Gödel axiom of constructibility $V = L$ is stronger than (E3) and holds in sense of the model class of all constructible sets, so all these axioms are consistent (which is well known). More generally, it is provable in **TS** that, for every X, there is the smallest Z such that $\mathrm{Mcl}(Z)$ and $\mathrm{Cls}^Z(X)$; it is denoted by L_X.

On the other hand, the nonprovability of (D3) is an immediate consequence of the following theorem provable in **TS** + (D1): (see [**14**]; the proof can be radically simplified): Let R be a regular internal relation, then there is a model relation S such that $R \subseteq S$, $(\forall x, y)(\langle x, y\rangle \in S \ \& \ y \in \mathit{field}(R) \rightarrow \langle x, y\rangle \in R)$, $u \subseteq \mathit{field}(S) \rightarrow (\exists z \in \mathfrak{D}(S))(u = \mathfrak{Ext}_S(z))$.

E.g. the consistency of the existence of a proper class of urelements (sets such that $x = \{x\}$) follows immediately. Classic Fraenkel-Mostowski permutation model classes may be studied. For some particular results see [**36**], [**40**]. It is also possible to show that not only D1 & ¬E1 is consistent but also ¬D1 & E1.

From now, let **TSS*** be the theory **TSS**′ + (D3) + (E2), **TS*** be **TS** + (D3) + (E2). We deal with **TS***. The *ultraproduct* weak model relation is defined in dependence on a complete boolean algebra and two other parameters. (This generalization of the usual ultraproduct relation is done by Vopěnka.) Let b be a complete boolean algebra; $\mathit{Part}(b)$ is the set of all disjointed partitions of b. (An $x \subseteq b$ is a *disjointed partition* of b iff

$$(1)\ \bigvee x = 1_b \qquad (2)\ (\forall u, v \in x)(u \neq v \rightarrow u \wedge v = 0_b).)$$

$\mathit{Part}(b)$ is a lattice (with respect to the partial ordering by being finer). A set c is a *partitive structure* on b iff it is a filter on $\mathit{Part}(b)$, or c equals to $\mathit{Part}(b)$. Let, moreover, z be an ultrafilter on b; we define $f \in \mathit{Ulc}(b, c)$ (ultraproduct class) iff there is an $x \in c$ such that f is a mapping with the domain x (values arbitrary sets); for $f, g \in \mathit{Ulc}(b, c)$, put $\langle f, g\rangle \in \mathit{Ulr}(b, c, z)$ iff $\bigvee\{u \wedge v; u, v \in b \ \& \ f(u) \in g(v)\} \in z$ (ultraproduct relation).

THEOREM (**TS***). *If b is a complete boolean algebra, z an ultrafilter on it, c a partitive structure on it, then $\mathit{Ulr}(b, c, z)$ is a weak model relation.*

METATHEOREM. *Let $\varphi(x, \ldots, y)$ be a set formula; then the following is provable in* **TS***: *Let b be a complete boolean algebra, z an ultrafilter on it, c a partitive structure on it, let R be $\mathit{Ulr}(b, c, z)$. Then, for every $f, \ldots, g \in \mathit{Ulc}(b, c)$*

$$\varphi^R(\mathfrak{Ext}_R(f), \ldots, \mathfrak{Ext}_R(g)) \equiv V\{u \wedge \cdots \wedge v; u, \ldots, v \in b \ \& \ \varphi(f(u), \ldots, g(v))\} \in z.$$

Considering the power-set of a set x as a complete boolean algebra b with the usual set-theoretical operations, let us write $\mathit{Part}(x)$ instead of $\mathit{Part}(b)$ and speak about partitions of x instead of partitions of b etc. as usual. Functions whose domain is a partition of x can evidently be replaced by functions on x constant on every element of the partition.

It is well known that the ultraproduct model relations may be used for the study of large cardinals (in theories in which we assume such cardinals). We do not mention details here; let us only mention the result of Vopěnka-Hrbáček concerning the fact that the existence of a strongly compact cardinal is inconsistent with the axiom "there is a set a such that $V = L_a$" (see [**41**]). The proof uses two different partitive structures on an appropriate set and corresponding ultraproduct model relations.

We present also two theorems proved by Balcar and Vopěnka (not yet published). Define in **TS***: Let s be an infinite set; an ultrafilter z on s is *uniform* iff every element of z has the cardinality of s. $c_{\aleph_\alpha}$ is the partitive structure on s of all partitions p of s such that the cardinality of p is less than $\aleph_\alpha$. c_z is the partitive structure of all partitions p of s such that $z \cap p \neq 0$. $c_z \dot{\vee} c_{\aleph_\alpha}$ is the partitive structure generated by c_z and $c_{\aleph_\alpha}$. z is *fine* iff $c_z \dot{\vee} c_{\aleph_\alpha}$ is a maximal filter on *Part*(s).

THEOREM. *If* $2^{\aleph_\alpha} = \aleph_{\alpha+1}$ *then, for card*$(s) = \aleph_\alpha$, *there are* $2^{\aleph_{\alpha+1}}$ *fine ultrafilters on* s.

If $f, g \in Ulc(s, c)$ then f is said to be *almost equal* to g iff there is an $x \in z$ such that $f \restriction x = g \restriction x$.

THEOREM. *Let* $\aleph_\alpha$ *be regular*, z *a fine ultrafilter on* $\aleph_\alpha$. *Then every mapping of* $\aleph_\alpha$ *into itself either almost equals to a constant function or there is an* $x \in z$ *such that* $f \restriction x$ *is strictly increasing.*

Let us come back to the theory of semisets. There are two important axioms we shall deal with:

(S1) For every nonempty semiset X, there is a $y \in X$ such that $X \cap y = 0$ (regularity for semisets).

(S2) There is an internal regular relation R such that $\mathfrak{D}(R) = V$ and, for every semiset X, there is a y such that $\mathfrak{Ext}_R(y) = X$ (all semisets can be coded by all sets).

E.g. let M be a constant for a model-class in **TS***, let X^* be a variable for subclasses of M and $\in^*$ the usual membership; then the language $(\in^*, X^*)$ determines a model of **TSS** + (S1) + (S2) in the former theory.

Define in **TSS**: A model relation R is an *extension of the theory* (Eth(R)) iff there is a mapping H of V into the field of R such that

(1) $$\mathrm{Mcl}^R(H''V) \;\&\; (\forall x, y)(x \in y \equiv \langle H(x), H(y)\rangle \in R),$$

(2) for every semiset $X \subseteq$ *field*(R), there is a $y \in$ *field*(R) such that
$$X = \mathfrak{Ext}_R(y).$$

THEOREM (**TSS′** + S1 + S2). *There is an extension* R *of the theory*; *it is uniquely determined in the following sense*: *if* R, S *are extensions of the theory then* Ker^R *is isomorphic to* Ker^S *with respect to* R, S.

Hence, the specification of the normal model of **TS** in **TSS′** + (S1) + (S2) by Eth(R) is a model of **TS** in **TSS′** + (S1) + (S2) such that the identical model of **TSS′** + (S1) + (S2) is equivalent to a (transitive) submodel of the former model. In this way, we obtain a model of **TS** in **TSS′** + (S1) + (S2) "going above".

There is an axiom stronger than (S1) + (S2); it is called the axiom of a support. Define in **TSS**: A semiset X is a *support* iff, for every semiset Y, there is a function f which is a set such that $Y = f^{-1}[X]$.

(Supp) There is a support.

DEFINITION (**TSS**). Let b be a complete boolean algebra (set); a semiset Z is a set-multiplicative ultrafilter on b iff (1) $(\forall x, y \in b)(x \in Z \;\&\; y \geq x \,.\!\rightarrow y \in Z)$, (2) $(\forall x \in b)(x \in Z \vee -x \in Z)$, (3) for every set $a \subseteq Z$, $\bigwedge a \in Z$.

Define: A support is a *boolean support* iff it is a set-multiplicative ultrafilter on a boolean algebra b.

(B Supp) There is a boolean support.

THEOREM (**TSS***). (B Supp) $\equiv$. (Supp) & (S1).

THEOREM (**TSS***). (B Supp) $\rightarrow$. (S1) & (S2).

METATHEOREM. *Let* $\varphi(b)$, ω *be set formulas, let* **TSS***, $\omega \vdash (\forall b)(\varphi(b) \rightarrow b$ *is a complete boolean algebra*) & $(\exists b)\varphi(b)$; *then the formula*

$$(\exists b)(\exists Z)(\varphi(b) \;\&\; Z \textit{ is a set-multiplicative ultrafilter on } b \;\&\; Z \textit{ is a support})$$

is consistent with **TSS*** $+ \, \omega$.

This is proved using ultraproduct model relations.

METATHEOREM. *Let, under the assumptions of the preceding metatheorem,* ψ *be a set formula provable in* **TSS*** $+ \, \omega + (\exists b)(\exists Z)(b$ *is c. b. algebra* & Z *is a set-multiplicative ultrafilter on* b *and is a support*); *then* ψ *is provable in* **TSS*** $+ \, \omega$ (*a fortiori, in* **TS*** $+ \, \omega$).

E.g. a proof of the following theorem may be obtained in this way: (see **[53]**)

THEOREM (**TS***). *Every locally nonseparable metric space is a union of an increasing sequence of* $\aleph_1$ *nowhere dense sets.*

THEOREM (**TSS***). *Let* Z_1, Z_2 *be set-multiplicative ultrafilters on a complete boolean algebra* b, Z_1, Z_2 *supports. Then there is an automorphism* p *of* b *such that* $p''Z_1 = Z_2$.

The theory of supports can be applied very fruitfully to the study of model classes in **TS***, because—as we already have seen—**TSS** with the axioms (S1), (S2) axiomatizes well the power class of an arbitrary model class in sense of which the axiom of choice holds and makes its extension possible. We define in **TS***: Let M be a model class; a set $x \subseteq M$ is a *support over* M iff, for every $y \subseteq M$, there is a function $f \in M$ such that $y = f^{-1}[x]$. The following theorem is obtained immediately:

THEOREM (**TS***). *Let* M *be a model class,* q *a support over* M. *Then there is a* $b \in M$, *which is a complete boolean algebra in sense of* M, *and an ultrafilter* z *on* b *closed under intersections of all systems* $a \subseteq b$, $a \in M$ (*say,* M*-multiplicative*), *which is a support over* M.

Thus, for every definition φ of a complete boolean algebra, we may e.g. suppose consistently in **TS*** (φ a set formula): The algebra defined by φ in sense of the universe of constructible sets has a constructibly multiplicative ultrafilter which is a support over L (and, of course is not constructible). We obtain a lot of consistent axioms in this way; many conditions concerning sets of constructible sets (absoluteness of cardinals, cardinalities of power sets) may be derived from the properties of the algebra b on which we have a support. We may also characterize all model classes with the axiom of choice; they are described by all subalgebras of b. Another result of Vopěnka:

THEOREM (**TS***). *Let $V = L_a$, where $a \subseteq On$, then there is a support over the model class HEf of hereditarily effective sets.*

Now, suppose, we have an arbitrary model class M with the axiom of choice, a $b \in M$ which is a complete boolean algebra in sense of M and an M-multiplicative ultrafilter on b which is a support. The universum $V_M^{(b)}$ of boolean valued functions may be defined in sense of M such that $(\forall f \in M)(f \in V_M^{(b)} \equiv f$ is a mapping of a subset $a \in M$ of $V_M^{(b)}$ into b) (cf. Scott, these Proceedings, part II, and Vopěnka [**50**]). The value $w(f)$ of every $f \in V_M^{(b)}$ in z is defined such that

$$w(f) = \{w(g);\ g \in \mathfrak{D}(f)\ \&\ f(g) \in Z\}.$$

THEOREM (**TS***). *Let z be an M-multiplicative ultrafilter on b (b is a complete boolean algebra in sense of the model class M), let z be a support. Then every set is a value of a boolean valued function from $V_M^{(b)}$. This is to say, boolean valued functions of M code all sets.*

A historical remark should be placed here. After P. J. Cohen had proved the independence of the continuum hypothesis and the axiom of choice in 1963, Vopěnka tried to use his ideas to prove in **TS*** the existence of a model relation in sense of which the continuum hypothesis does not hold; speaking metamathematically, to demonstrate the consistency of the negation of CH with **TS*** in the described straightforward finitary way. It was done in the paper [**12**] published in Russian. Although this paper is closely related to those of Cohen, some new ideas were necessary because—as Sheperdson had proved—it was necessary to construct a non-well-founded relation which is not a set (and is a model relation with the negation of CH) without any assumptions concerning countable models. Then the conception was generalized and simplified in a series of eight papers ([**16**], [**19**], [**20**], [**33**], [**21**], [**31**], [**46**], [**48**]) written partly together with the author of the present paper. It was proved that the number of parameters of the so-called ∇-model relations can be limited to two—a complete boolean algebra and an ultrafilter on it. The whole theory was then presented in the paper [**50**]. The conception of this paper is deeply analogous to that of D. Scott presented at the Los Angeles Summer Institute but, of course, discovered by both authors independently. From the new Vopěnka's point of view, using semisets, boolean valued functions play only auxiliary, even though very important rôle. The advantage consists in the fact that we need only to prove the consistency of the existence of a support on a complete boolean algebra with the theory of semisets (which is done easily using ultraproducts) and then we deal with model classes and supports over them in the set theory, hence in the standard way, having the result about the extension of the theory of semisets. Let us also mention the fact that, in Scott's terminology, the statement "$V^{(B)}$ has a support over $V^{(2)}$" holds in every boolean valued model. (More precisely but less generally: Assume $V = L$. Then, for every complete boolean algebra, the statement "There is a set x of constructible sets such that, for every set y of constructible sets, there is a constructible function f such that y is the counterimage of x by f" has the boolean value 1.)

For consistency proofs of particular statements using the methods just described, see e.g. **[26]**, **[47]**, **[51]**, **[52]**.

Given a normal filter on the group of all automorphisms of b (everything in sense of the model class M) the class of *hereditarily symmetric* boolean valued functions is defined (see **[21]**, **[31]** and Scott, these Proceedings, part II). The values of hereditarily symmetric functions form a model class. In this way, model classes in which the axiom of choice fails can be obtained, and consistency of various statements contradicting axiom of choice may be demonstrated. (See e.g. **[39]**.)

Consistency proofs for some existential formulas using Fraenkel-Mostowski permutation models can be modified using the method just mentioned. A general method comes from Jech and Sochor (see **[37]**, **[38]**). Let us formulate their result in a form modified a little.

Define in **TS**: $p_0(z) = z$, $p_{\alpha+1}(z) =$ the power-set of $p_\alpha(z)$, $p_\lambda(z) = \bigcup_{\alpha<\lambda} p_\alpha(z)$ for λ a limit number, $Ker(z) = \bigcup_{\alpha\in On} p_\alpha(z)$. A definition $\varphi(\alpha)$ of a cardinal number is *good* iff

$$\mathbf{TS} \vdash (\exists!\alpha)(\alpha \text{ cardinal number } \& \ \varphi(\alpha))$$

$$\mathbf{TS},\ Cn = Cn^L \vdash (\forall\alpha)(\forall M)(\mathrm{Mcl}(M) \to \varphi(\alpha) \equiv \varphi^M(\alpha))$$

(Cn being the class of all cardinal numbers, L the model class of all constructible sets). Let α be a constant defined by a good definition of a cardinal number, let $\psi(z)$ be a set formula with only one free variable z. $\psi(z)$ is said to be *α-restricted* iff all quantifiers are restricted onto $p_\alpha(z)$. The formula $(\exists z)\psi(z)$ is said to have a permutation model iff, for a constant λ defined by a good definition of a cardinal number, the following is provable in the theory: $\mathbf{TS} + V = Ker(Ur) + card(Ur) = \lambda$ (Ur is the constant for the class of urelements): "There is a normal filter F on the group of all permutations of urelements such that, if we denote the Fraenkel-Mostowski model class determined by F as M_F, $(\exists z)(z \cap Ur = 0 \ \& \ \psi^{M_F} z)$ holds.

The result of Jech-Sochor: Let α be defined by a good definition, let $\psi(z)$ be an α-restricted set formula, let $(\exists z)\psi(z)$ have a permutation model. Then $(\exists z)\psi(z)$ is consistent with **TS** + (D3).

Bibliography of the Prague Seminar on Foundations of Set Theory

1. P. Vopěnka, *A method of constructing a non-standard model in the Bernays-Gödel axiomatic set theory*, (Russian) Dokl. Akad. Nauk SSSR **143** (1962), 11–12.

2. ———, *Models for set theory*, (Russian, German summary) Z. Math. Logik Grundlagen Math. **8** (1962), 281–292.

3. ———, *Construction of models for set theory by the method of ultra-products*, (Russian, German summary) Z. Math. Logik Grundlagen Math. **8** (1962), 293–304.

4. ———, *Construction of models for set theory by the spectrum method*, (Russian, German summary) Z. Math. Logik Grundlagen Math. **9** (1963), 149–160.
5. ———, *Elementary concepts in set theory*, (Russian, German summary) Z. Math. Logik Grundlagen Math. **9** (1963), 161–167.
6. ———, *Construction of non-standard, non-regular models in set theory*, (Russian, German summary) Z. Math. Logik Grundlagen Math. **9** (1963), 229–233.
7. P. Vopěnka and P. Hájek, *Über die Gültigkeit des Fundierungsaxioms in speziellen Systemen der Mengenlehre*, Z. Math. Logik Grundlagen Math. **9** (1963), 235–241.
8. P. Hájek, *Die durch die schwach inneren Relationen gegebenen Modelle der Mengenlehre*, Z. Math. Logik Grundlagen Math. **10** (1964), 151–157.
9. P. Vopěnka, *Submodels of models of set theory*, (Russian, German summary) Z. Math. Logik Grundlagen Math **10** (1964), 163–172.
10. P. Hájek and A. Sochor, *Ein dem Fundierungsaxiom äquivalentes Axiom*, Z. Math. Logik Grundlagen Math. **10** (1964), 163–172.
11. P. Vopěnka, *Axiome der Theorie endlicher Mengen*, Časopis. Pěst. Mat. **89** (1964), 312–317.
12. ———, *The independence of the continuum hypothesis*, Comment. Math. Univ. Carolinae **5** (1964), Supplement I, 1–48; English transl., Amer. Math. Soc. Transl. (2) **57** (1966), 85–112.
13. P. Vopěnka and L. Bukovský, *The existence of a PCA-set of cardinality* $\aleph_1$, Comment. Math. Univ. Carolinae **5** (1964), 125–128.
14. P. Hájek, *Modelle der Mengenlehre, in denen Mengen gegebener Gestalt existieren*, Z. Math. Logik Grundlagen Math **11** (1965), 103–115.
15. B. Balcar and T. Jech, *Models of the theory of sets generated by a perfect relation*, (Russian, German summary) Časopis. Pěst. Mat. **90** (1965), 413–434.
16. P. Vopěnka, *The limits of sheaves and applications on constructions of models*, Bull. Acad. Polon. Sci. Sér. Sci. Math. Astronom. Phys. **13** (1965), 189–192.
17. K. Přikrý, *The consistency of the continuum hypothesis for the first measurable cardinal*, Bull. Acad. Polon. Sci. Sér. Sci. Math. Astronom. Phys. **13** (1965), 193–197.
18. P. Hájek, *Syntactic models of axiomatic theories*, Bull. Acad. Polon. Sci. Sér. Sci. Math. Astronom. Phys. **13** (1965), 273–278.
19. P. Vopěnka, *On* ∇*-model of set theory*, Bull. Acad. Polon. Sci. Sér. Sci. Math. Astronom. Phys. **13** (1965), 267–272.
20. ———, *Properties of the* ∇*-model*, Bull. Acad. Polon. Sci. Sér. Sci. Math. Astronom. Phys. **13** (1965), 441–444.
21. P. Vopěnka and P. Hájek, *Permutation submodels of the model* ∇, Bull. Acad. Polon. Sci. Sér. Sci. Math. Astronom. Phys. **13** (1965), 611–614.
22. P. Hájek, *Eine Bemerkung über standarde nichtreguläre Modelle der Mengenlehre*, Comment. Math. Univ. Carolinae **6** (1965), 1–6.
23. P. Vopěnka, *Concerning a proof of* $\aleph_1 \leqq 2^{\aleph_0}$ *without the axiom of choice*, Comment. Math. Univ. Carolinae **6** (1965), 111–113.
23a. ———, *Concerning a proof of* $\aleph_1 \leqq 2^{\aleph_0}$ *without the axiom of choice, a correction*, Comment. Math. Univ. Carolinae **6** (1965), 337.
24. P. Vopěnka, A. Pultr, and Z. Hedrlín, *A rigid relation exists on any set*, Comment. Math. Univ. Carolinae, **6** (1965), 149–155.
25. L. Bukovský, *The continuum problem and powers of alephs*, Comment, Math. Univ. Carolinae **6** (1965), 181–197.
26. ———, *The consistency of some theorems concerning Lebesgue measure* (preliminary communication), Comment. Math. Univ. Carolinae **6** (1965), 179–180.
27. P. Vopěnka, *The first measurable cardinal and the generalized continuum hypothesis*, Comment. Math. Univ. Carolinae **6** (1965), 367–370.
28. K. Hrbáček, *Model* $\nabla(\omega_\alpha \to \omega_\beta)$ *in which* β *is a limit number*, Comment. Math. Univ. Carolinae, **6** (1965), 439–442.
29. L. Bukovský, *An elementary proof of normality of the class of accessible cardinals*, Comment. Math. Univ. Carolinae **6** (1965), 409–412.
30. P. Vopěnka, "Construction of a model etc." (summary of [**6**]) in *The theory of models*, North-Holland, Amsterdam, 1965.
31. P. Hájek and P. Vopěnka, *Some permutation submodels of the* ∇*-model*, Bull. Acad. Polon. Sci. Sér. Sci. Math. Astronom. Phys. **14** (1966), 9–14.
32. L. Bukovský and K. Přikrý, *Some metamathematical properties of measurable cardinals*, Bull. Acad. Polon. Sci. Sér. Sci. Math. Astronom. Phys. **14** (1966), 95–99.

33. P. Vopěnka, ∇-*models in which the generalized continuum hypothesis does not hold*, Bull. Acad. Polon. Sci. Sér. Sci Math. Astronom. Phys. **14** (1966), 95–99.

34. L. Bukovský and P. Hájek, *On the standardness and regularity of normal syntactic models of set theory*, Bull. Acad. Polon. Sci. Sér. Sci. Math. Astronom. Phys. **14** (1966), 101–105.

35. P. Vopěnka, *A new proof of Gödel's result on the non-provability of consistency*, Bull. Acad. Polon. Sci. Sér. Sci. Math. Astronom. Phys. **14** (1966), 293–296.

36. T. Jech, *On ordering of cardinalities*, Bull. Acad. Polon. Sci. Sér. Sci. Math. Astronom. Phys. **14** (1966), 293–296.

37. T. Jech and A. Sochor, *On the* ∇-*model of set theory*, Bull. Acad. Polon. Sci. Sér. Sci. Math. Astronom. Phys. **14** (1966), 297–303.

38. ———, *Applications of the* ∇-*model*, Bull. Acad. Polon Sci Sér. Sci. Math. Astronom. Phys., **14** (1966), 351–353.

39. P. Hájek, *The consistency of Church's alternatives*, Bull. Acad. Polon. Sci. Sér. Sci. Math. Astronom. Phys. **14** (1966), 424–430.

40. T. Jech, *On cardinals and their successors*, Bull. Acad. Polon. Sci. Sér. Sci. Math. Astronom. Phys. **14** (1966), 533–537.

41. P. Vopěnka and K. Hrbáček, *On strongly measurable cardinals*, Bull. Acad. Polon. Sci. Sér. Sci. Math. Astronom. Phys. **14** (1966), 587–591.

42. P. Hájek, *Generalized interpretability in terms of models* (note to a paper of R. Montague), Časopis. Pěst. Mat. **91** (1966), 587–591.

43.[4] K. Hrbáček, *Measurable cardinals in some Gödelian set theories*, Comment. Math. Univ. Carolinae **7** (1966), 343–358.

44. T. Jech, *Interdependence of weakened forms of the axiom of choice*, Comment. Math. Univ. Carolinae **7** (1966), 359–371. (Correction appears in Comment. Math. Univ. Carolinae **8** (1967), p. 567.)

45. L. Bukovský, *Consistency theorems connected with some combinatorial problems*, Comment. Math. Univ. Carolinae **7** (1966), 495–499.

46. P. Vopěnka, *The limits of sheaves over extremally disconnected compact Hausdorff spaces*, Bull. Acad. Polon. Sci. Sér. Sci. Math. Astronom. Phys., **15** (1967), 1–6.

47. P. Vopěnka and K. Hrbáček, *The consistency of some theorems on real numbers*, Bull. Acad. Polon. Sci. Sér. Sci. Math. Astronom. Phys. **15** (1967), 107–111.

48. P. Vopěnka and P. Hájek, *Concerning the* ∇-*models of set theory*, Bull. Acad. Polon. Sci. Sér. Sci. Math. Astronom. Phys. **15** (1967), 113–117.

49. P. Vopěnka and B. Balcar, *On complete models of set theory*, Bull. Acad. Polon. Sci. Sér. Sci. Math. Astronom. Phys. **15** (1967), 721–723.

50. P. Vopěnka, *General theory of* ∇-*models*, Comment. Math. Univ. Carolinae **8** (1967), 145–170.

51. T. Jech, *Non-provability of Souslin's hypothesis*, Comment. Math. Univ. Carolinae **8** (1967), 291–305.

52. K. Hrbáček, *A note on the generalized Souslin problem*, Comment Math Univ. Carolinae **8** (1967), 307–310.

53. P. Štěpánek and P. Vopěnka, *Decomposition of metric spaces into nowhere dense sets*, Comment. Math. Univ. Carolinae **8** (1967), 387–403.

54. P. Vopěnka and B. Balcar, *Generator classes in set theory*, Z. Math. Logik Grundlagen Math. **13** (1967), 97–98.

55. L. Bukovský, "Borel subsets of metric separable spaces" in *General topology and its relations to modern analysis and Algebra*, II, Proc. Second Prague Topological Sympos., 1966.

[4] The results of [**43**] were obtained by K. Hrbáček after he had known the paper of Solovay cited there as [**8**]. The priority for the results on "mild extensions" contained in §3 of [**43**] belongs hence fully to Lévy and Solovay. (Note by K. Hrbáček.)

MATHEMATICAL INSTITUTE OF THE
CZECHOSLOVAK ACADEMY OF SCIENCES

THE BOOLEAN PRIME IDEAL THEOREM DOES NOT IMPLY THE AXIOM OF CHOICE

J. D. HALPERN AND A. LÉVY[1]

1. **Introduction, and statement of the results.** One of the most remarkable features of the axiom of choice of *ZF* is that it is equivalent to many different statements taken from various branches of mathematics. A very comprehensive, but by no means exhaustive, list of equivalent forms of the axiom of choice is collected by Rubin and Rubin in [**11**]. This feature of the axiom of choice indicates that it is a very natural axiom of set theory. One of the consequences of the axiom of choice is the Boolean prime ideal theorem, i.e., the statement that every Boolean algebra has a prime ideal. Also the Boolean prime ideal theorem has a considerable number of equivalent forms in several branches of mathematics (see [**4**] for references) although by far not as many as the axiom of choice. Thus also the Boolean prime ideal theorem is indicated to be a natural principle of set theory.

Since the Boolean prime ideal theorem is a consequence of the axiom of choice it is natural to ask whether it is equivalent to that axiom. Mostowski constructed in [**9**] a model of set theory with individuals in which the ordering theorem (i.e., the statement that every set can be ordered) holds, while the axiom of choice fails. Halpern proved in [**4**] that the Boolean prime ideal theorem also holds in this model. By the well-known methods of Mendelson, Shoenfield and Specker this proves also that the Boolean prime ideal theorem is not equivalent to the axiom of choice in the set theory *ZF* without the axiom of foundation. The main result of the present paper is that the Boolean prime ideal theorem is not equivalent to the axiom of choice in full *ZF*.

Our method of proof is as follows. In §3 we shall define a certain set theory *SP* which is an extension of *ZF*. In §4 we shall prove that if *ZF* is consistent so is *SP*.

[1] The research of the first author was supported in parts by a grant from the Institute for Advanced Study and by NSF grant GP-6578. The research of the second author was supported by the U.S. Office of Naval Research, Information Systems Branch, Contract F-61052, 67C0055.

This will be done by Cohen's method of forcing. The proof of the relative consistency of *SP* will be finitary, in the sense that one could prove in primitive recursive arithmetic that if there is a contradiction in *SP* then there is a contradiction in *ZF*. In §3 we shall show that the following statements are theorems of *SP*:

(a) The Boolean prime ideal theorem.

(b) Every set is equinumerous to a subset of the Cartesian product of the set of all real numbers and an ordinal.

(c) There is a Dedekind finite set A of reals which is dense in the real line. Łoś and Ryll-Nardzewski [**8**] prove that (a) implies the ordering theorem; however, (b) is a stronger ordering theorem. In particular (b) implies that every set is equinumerous to a set of sets of ordinals (see Kinna and Wagner [**7**] for a discussion of the latter statement).

(c) implies (in *ZF*) the following three equivalent statements (see Sierpiński [**12**] and Jaegermann [**6**]).

(d) There is a denumerable set of sets of real numbers which has no choice function.

(e) There is a set C of real numbers and a real number $x \in C$ such that x is an accumulation point of C in the topological sense, but is not the limit of a sequence of members of C. (To prove (e) from (c) just choose $C = A$ and $x \notin A$.)

(f) There is a real valued function from the real line which is not continuous at some point x but such that for every sequence $\{x_i\}_{i<\omega}$ converging to x we have $f(x) = \lim_i f(x_i)$. (To prove (f) from (c) take for f the characteristic function of A.)

We know now that if *ZF* is consistent then *SP* is consistent and hence (a)–(c) above are consistent with *ZF*. In particular, we know that in *ZF* even the conjunction of (a) and (b) does not imply that every dense subset of the real line has a denumerable subset (this is the negation of (c)). The latter statement is a weak consequence of the axiom of choice since it follows from the statement that for every denumerable set of sets of reals there is a choice function (the negation of (d)), and it follows, a fortiori, from the existence of a well-ordering of the set of all real numbers.

2. **Relative constructibility.** The most general account of relative constructibility is given by Scarpellini in [**13**]. We shall develop this notion here only in the degree of generality which we need and in a way which suits our particular application of this notion.

All the definitions and theorems of the present section are definitions and theorems of *ZF*. The axiom of choice, which we do not take to be an axiom of *ZF*, is not used here. Even though we use, throughout the paper, only systems of set theory which admit only sets, but no classes, we shall feel free to talk about classes in a way which is well known to be eliminable (see, e.g., Quine [**10**]).

For a function f we shall denote by $\mathfrak{D}(f)$ the *domain* of f (i.e., the set $\{x \mid \exists y(\langle x, y\rangle \in f)\}$) and by $\mathfrak{R}(f)$ the *range* of f (i.e. the set $\{y \mid \exists x(\langle x, y\rangle \in f)\}$). Also, for a set d, $f \mid d$ denotes the function f *restricted to* d, i.e. the function

$\{\langle x, y\rangle \mid \langle x, y\rangle \in f \wedge x \in d\}$. If f and g are functions then $f \circ g$ is the function obtained by their *composition*, i.e., the function $\{\langle x, y\rangle \mid \exists z(\langle x, z\rangle \in g \wedge \langle z, y\rangle \in f)\}$. $f[a]$ denotes the set of all values of f for members of a, i.e. the set $\{f(x) \mid x \in a\} = \mathfrak{R}(f \mid a)$.

A *finite sequence* is a function whose domain is a finite ordinal $m = \{n \mid n < m\}$. m is called the *length* of the sequence. An *m-sequence* is a finite sequence of length m. $\langle t_0, \ldots, t_{m-1}\rangle$ will denote the m-sequence $\{\langle i, t_i\rangle \mid i < m\}$. If f and g are finite sequences with the respective lengths k and l then $f^\frown g$ is the sequence h of length $k + l$ such that $h(i) = f(i)$ for $i < k$ and $h(k + i) = g(i)$ for $i < l$. For a finite set x we shall denote by $|x|$ the cardinality of x. $\sim$ will denote set-theoretical difference, i.e. $a \sim b = \{x \mid x \in a \wedge x \notin b\}$.

$R(\alpha)$ is the function on the ordinals defined by $R(\alpha) = \bigcup_{\beta<\alpha} P(R(\beta))$, where $P(x)$ is the power-set of x. By the axiom of foundation, for every set x there is an ordinal α such that $x \subseteq R(\alpha)$; the least such α is called the *rank* of x, and is denoted by $\rho(x)$.

Throughout the present section b will be a fixed set of subsets of ω and S will be a transitive class (i.e., if $x \in y \in S$ then $x \in S$). We shall refer to the members of S as *standard sets*. We consider now a certain process for creating sets; this process is carried out in steps, indexed by the ordinal numbers. Informally, let us denote by L_α the set of all sets obtained before the αth step. Then at the αth step we create all the sets of the following forms (a)–(d).

(a) $\{x \mid x \in L_\alpha \wedge \phi(x, x_1, \ldots, x_n)\}$, for $x_1, \ldots, x_n \in L_\alpha$, where ϕ is a formula of set theory in which all the quantifiers are restricted to L_α (i.e., all the existential quantifiers are of the form $\exists y(y \in L_\alpha \wedge \cdots$ and all the universal quantifiers are of the form $\forall y(y \in L_\alpha \rightarrow\ \ )$, (b) the standard sets of rank $\leqq \alpha$, (c) the members of b, if $\alpha \geqq \omega$, and (d) b itself if $\alpha > \omega$. To carry out this construction formally it is very convenient to use a certain formal language, which we shall call $\mathscr{L}$. When we want to deal, in algebra, with a finite extension of a field we consider the "language" of the polynomials over that field. (We refer to the polynomials as a "language" because the polynomials are extraneous syntactical objects used to denote the elements of the finite extension.) Our language $\mathscr{L}$ is considerably more complicated than the language of the polynomials, but the relationship between the language and the mathematical structure is basically the same.

DEFINITION 1. (a) The primitive symbols of the language $\mathscr{L}$ are $\neg$ (negation), $\vee$ (disjunction), $\in$ (membership), $=$ (equality), $\exists_\alpha$ (restricted existential quantifier) λ_α (restricted abstraction operator), x_i (set variables), $\mathbf{s}$ (constants for standard sets), a_i (constants for members of b), and b (which is intended to be a name for the set b—here we commit the common transgression of denoting two different objects, the set b and its name, with the same symbol), where α ranges over the ordinals, i over the natural numbers and s over the standard sets. We actually identify these symbols with the sets $\langle 0, 0\rangle$, $\langle 0, 1\rangle$, $\langle 0, 2\rangle$, $\langle 0, 3\rangle$, $\langle 1, \alpha\rangle$, $\langle 2, \alpha\rangle$, $\langle 3, i\rangle$, $\langle 4, s\rangle$, $\langle 5, i\rangle$, $\langle 0, 4\rangle$, respectively (so, e.g., strictly speaking, $\exists_\alpha$ is defined to be $\langle 1, \alpha\rangle$).

(b) Some finite sequences of the primitive symbols of $\mathscr{L}$ will be called *terms*, or *abstraction terms*, or *formulas*, respectively. We define these notions simultaneously and, at the same time, we define a function λ on the terms, the variables x_i and the formulas by induction as follows:

(i) u is a term if and only if u is $\mathbf{s}$ for some standard set s (more precisely, if u is the sequence $\langle \mathbf{s} \rangle$), u is a_i for some $i < \omega$, u is b or u is an abstraction term. (Notice that the variables x_i are not called terms here; what we call terms are usually called constant terms.)

(ii) $\lambda(\mathbf{s}) = \rho(s)$, $\lambda(a_i) = \omega$, $\lambda(b) = \omega + 1$, $\lambda(x_i) = 0$.

(iii) If ϕ, ψ are formulas so are $\neg\phi$, $\phi \vee \psi$ and $\exists_\alpha x_i \phi$. (Formally, $\neg\phi$ is the sequence obtained from ϕ by appending $\neg$ at the beginning, and $\phi \vee \psi$ is the sequence obtained by first putting $\vee$ then appending ϕ and then ψ.) $\lambda(\neg\phi) = \lambda(\phi)$, $\lambda(\phi \vee \psi) = \max(\lambda(\phi), \lambda(\psi))$, $\lambda(\exists_\alpha x_i \phi) = \max(\alpha, \lambda(\phi))$. $\exists_\alpha x_i \phi$ is to be read: There is an x_i in L_α such that ϕ.

(iv) If u, v are terms or variables then $u \in v$ and $u = v$ are formulas and $\lambda(u \in v) = \lambda(u = v) = \max(\lambda(u), \lambda(v)) + 1$.

(v) If ϕ is a formula with the only free variable x_i (where the notion of a free variable is defined as usual) and $\lambda(\phi) \leqq \alpha$ then $⅄_\alpha x_i \phi$ is an abstraction term and $\lambda(⅄_\alpha x_i \phi) = \alpha$. $⅄_\alpha x_i \phi$ is to be read: The set of all x_i in L_α such that ϕ.

The induction is done here as follows. When we define the finite sequence u to be, or not to be, a term or a formula we assume that for every proper subsequence v of u it has already been defined whether v is a term or a formula, and if v is a term or a formula $\lambda(v)$ has been defined too. It is easily seen that for a formula ϕ, $\lambda(\phi)$ is the least ordinal α such that $\alpha \geqq \beta$ for every β such that $\exists_\beta$ occurs in ϕ and $\alpha > \lambda(u)$ for every term or variable u which occurs in ϕ.

(c) Formulas ϕ without free variables will be called *sentences*. The class of all formulas of $\mathscr{L}$ will be denoted by FL, and the class of all terms of $\mathscr{L}$ will be denoted by TL. An *expression* of $\mathscr{L}$ is defined to be a formula or a term or a variable of $\mathscr{L}$. A *closed expression* of $\mathscr{L}$ is defined to be a sentence or a term of $\mathscr{L}$. For an expression ϕ, $\mathrm{occ}(\phi)$ denotes the finite subset of ω which consists of all indices i for which a_i occurs in ϕ. We shall use letters like x, y, z to denote variables x_i, with the understanding that, unless otherwise mentioned, different letters denote different variables.

Informally, for $u \in TL$, $\lambda(u) = \alpha$ means that we can tell, just by looking at the sequence u, that u denotes a member of $L_{\alpha+1}$; and for $\phi \in FL$ $\lambda(\phi) = \alpha$ means that we can tell, just by looking at the sequence ϕ, that ϕ refers only to members of L_α.

It should be noticed that each of the symbols $\neg$, $\vee$, $\in$, $=$, x_i has a double meaning. For example, the symbol $\in$ denotes the membership relation of set theory, but it denotes also the pair $\langle 0, 2 \rangle$. The reader who is disturbed by it can imagine that we use for the "symbols" of $\mathscr{L}$ a different style of fount; it will always be clear from the context which is the intended meaning of each of these symbols which we shall use. Throughout the present paper we shall, unless otherwise mentioned, use the letters ϕ, ψ, χ as variables for formulas in the sense of Definition 1, i.e. as variables for the objects of set theory which belong to the class FL. Formulas in the ordinary sense, that talk about sets but are not necessarily themselves sets, will be denoted by Φ, Ψ, Γ, etc.

At this point we want to correspond to each term u a set which this term denotes. In general, a term contains several of the a_i's, and we cannot say which set the term denotes until we know which members of b the a_i's denote. We shall, therefore, call an *assignment* any function f whose domain is a finite set of natural numbers and whose values are in b. We say that f *assigns* the member z of b to a_i if $f(i) = z$. We say that the assignment f is *sufficient* for the expression u if the domain of f contains all the i's such that a_i occurs in u, i.e. if $\text{occ}(u) \subseteq \mathfrak{D}(f)$. What we want to define now is a function $\text{den}(u, f)$ defined for all terms u and sufficient assignments f. However, we have to define $\text{den}(\phi, f)$ also for sentences ϕ and sufficient assignments f. A sentence, unlike a term, does not denote a set; it is either true or false. Thus $\text{den}(\phi, f)$ should have the values "true" and "false". We choose 1 for "true" and 0 for "false" and define $\text{den}(\phi, f)$ accordingly.

DEFINITION 2. We define $\text{den}(\phi, f)$, where ϕ is a closed expression and f an assignment sufficient for ϕ. We do it by means of a recursive definition the validity of which will be established below.

$$\text{den}(\mathbf{s}, f) = s,$$

$$\text{den}(a_i, f) = f(i),$$

$$\text{den}(b, f) = b \quad \text{(notice that in the left-hand side } b \text{ is } \langle 0, 4\rangle \text{ and in the right-hand side } b \text{ is a set of subsets of } \omega).$$

If u and v are terms then

$$\text{den}(u \in v, f) = 1 \quad \text{if } \text{den}(u, f) \in \text{den}(v, f) \text{ (i.e., if "}u \in v \text{ is true"),}$$
$$= 0 \quad \text{otherwise,}$$

$$\text{den}(u = v, f) = 1 \quad \text{if } \text{den}(u, f) = \text{den}(v, f) \text{ (i.e., if "}u = v \text{ is true"),}$$
$$= 0 \quad \text{otherwise,}$$

$$\text{den}(\neg\phi, f) = 1 - \text{den}(\phi, f),$$

$$\text{den}(\phi \vee \psi, f) = \max(\text{den}(\phi, f), \text{den}(\psi, f)),$$

$$\text{den}(\exists_\alpha x_i \phi(x_i), f) = 1 \quad \text{if for some term } v \text{ with } \lambda(v) < \alpha \text{ and for some assignment } g \text{ sufficient for } \phi(v) \text{ such that } g \mid \text{occ}(\phi(x_i)) = f \mid \text{occ}(\phi(x_i)),\ \text{den}(\phi(v), g) = 1,$$
$$= 0 \quad \text{otherwise,}$$

$\phi(v)$ denotes the formula obtained from $\phi(x_i)$ by substituting v for all free occurrences of x_i.

$$\text{den}(\text{Ӿ}_\alpha x_i \phi(x_i), f) = \{\text{den}(v, g) \mid \lambda(v) < \alpha \wedge g \textit{ is an assignment sufficient for } \phi(v) \wedge g \mid \text{occ}(\phi(x_i)) = f \mid \text{occ}(\phi(x_i)) \wedge \text{den}(\phi(v), g) = 1\}.$$

L_α is defined to be $\{\text{den}(u, f) \mid u \in TL \wedge \lambda(u) < \alpha \wedge f \text{ is an assignment sufficient for } u\}$. It is clear that $\text{den}(\text{Ӿ}_\alpha x_i \phi(x_i), f)$ consists of members x_i of L_α for which $\phi(x_i)$ is true (when the a_i's in it are interpreted according to f). However, it is not clear from the definition that every member $\text{den}(v, g)$ of L_α for which $\phi(v)$ is true is in den($\lambda_\alpha x_i\phi(x_i), f$), since it is possible that $g \mid \text{occ}(\phi(x_i)) \neq f \mid \text{occ}(\phi(x_i))$. But this causes no difficulty, since we can replace the a_i's in v to obtain a new term v' and we can find an assignment g' such that $g' \mid \text{occ}(\phi(x_i)) = f \mid \text{occ}(\phi(x_i))$ and $\text{den}(v', g') = \text{den}(v, g)$. This will follow directly from Corollary 5 below. An analogous remark applies also to $\exists_\alpha x\phi(x_i)$.

This definition defines den(ϕ, f) for all closed expressions ϕ by induction on $\lambda(\phi)$, and for expressions ϕ with the same value of λ by induction on the number of occurrences of the symbols $\neg$, $\wedge$ and $\exists_\alpha$ which are not inside an abstraction term.

We shall use $\phi \wedge \psi$ as an abbreviation of $\neg(\neg\phi \wedge \neg\psi)$, and similarly for the other sentential connectives. $\forall_\alpha x\phi$ is taken to be an abbreviation of $\neg\exists_\alpha x\neg\phi$.

LEMMA 3. *If ϕ is a closed expression and f, g are assignments sufficient for ϕ such that $f \mid \text{occ}(\phi) = g \mid \text{occ}(\phi)$ then* $\text{den}(\phi, f) = \text{den}(\phi, g)$.

PROOF. By induction, as in Definition 2.

The following lemma is an obvious fact, the proof of which is somewhat messy. The reader is advised to skip the proof, at least in the first reading.

LEMMA 4. *Let u be a closed expression and let s be a function on a finite set of natural numbers which includes* occ(u) *into* ω. *(We shall refer to such an s as a substitution function for u.) Let* sub(u, s) *be the expression obtained from u by replacing in it each occurrence of a_i by an occurrence of $a_{s(i)}$, then*

$$\text{den}(u, f \circ s) = \text{den}(\text{sub}(u, s), f)$$

for every assignment f sufficient for sub(u, s), *(and in particular for every assignment f with* $\mathfrak{D}(f) \supseteq \mathfrak{R}(s)$).

PROOF. This is proved by the same induction as in the definition of den. We shall carry it out only for the messiest case, i.e. for $u = \lambda_\alpha x\phi(x)$. Let

$$\begin{aligned} P &= \text{den}(u, f \circ s) \\ &= \{\text{den}(v, h) \mid \lambda(v) < \alpha \wedge \mathfrak{D}(h) \supseteq \text{occ}(\phi(v)) \wedge h \mid \text{occ}(\phi(x)) \\ &\qquad = f \circ s \mid \text{occ}(\phi(x)) \wedge \text{den}(\phi(v), h) = 1\}, \end{aligned}$$

$$\begin{aligned} Q &= \text{den}(\text{sub}(u, s), f) \\ &= \{\text{den}(v', h') \mid \lambda(v') < \alpha \wedge \mathfrak{D}(h') \supseteq \text{occ}(\phi'(v')) \\ &\qquad \wedge h' \mid \text{occ}(\phi'(x)) = f \mid \text{occ}(\phi'(x)) \wedge \text{den}(\phi'(v'), h') = 1\} \end{aligned}$$

where $\phi'(x)$ is sub($\phi(x)$, s).

We have to prove $P = Q$.

(a) $P \subseteq Q$. Let $w \in P$, we shall show that $w \in Q$. Since $w \in P$, $w = \text{den}(v, h)$ where v, h satisfy the requirements given in the definition of P. Let t be a substitution function for $\phi(v)$ such that t coincides with s on occ($\phi(x)$), and t induces a one-one map of $\text{occ}(v) \sim \text{occ}(\phi(x))$ on a subset d of ω disjoint from $s[\text{occ}(\phi(x))]$. Let h' be an assignment which coincides with f on $t[\text{occ}(\phi(x))] = s[\text{occ}(\phi(x))]$ such that

(1) $$h' \circ t \mid (\text{occ}(v) \sim \text{occ}(\phi(x))) = h \mid (\text{occ}(v) \sim \text{occ}(\phi(x))).$$

There is such an h' because t is a one-one map of $\text{occ}(v) \sim \text{occ}(\phi(x))$ on d and $d \cap \text{occ}(\phi(x)) = 0$. Since s and t coincide on $\text{occ}(\phi(x))$, and since h' and f coincide on $t[\text{occ}(\phi(x))]$ we have

(2) $$h' \circ t \mid \text{occ}(\phi(x)) = f \circ s \mid \text{occ}(\phi(x)) = h \mid \text{occ}(\phi(x)).$$

By (1) and (2) $h' \circ t$ and h coincide on all of $\text{occ}(\phi(v))$, and since $\text{den}(\phi(v), h) = 1$ we have, by Lemma 3,

(3) $$\text{den}(\phi(v), h \circ t) = 1.$$

Let us put $v' = \text{sub}(v, t)$. $\text{sub}(\phi(v), t) = \phi'(v')$, since $\text{sub}(\phi(x), t) = \text{sub}(\phi(x), s) = \phi'(x)$ because s and t coincide on $\text{occ}(\phi(x))$. We shall now see that v' and h' satisfy the requirements of Q. Obviously $\lambda(v') = \lambda(v) < \alpha$.

$$\begin{aligned}\mathfrak{D}(h') \supseteq t[\text{occ}(\phi(x))] \cup t[\text{occ}(v) \sim \text{occ}(\phi(x))] &= t[\text{occ}(\phi(x)) \cup \text{occ}(v)] \\ &= t[\text{occ}(\phi(v))] = \text{occ}(\phi'(v')).\end{aligned}$$

$h' \mid \text{occ}(\phi'(x)) = f \mid \text{occ}(\phi'(x))$ since, obviously, $\text{occ}(\phi'(x)) = s[\text{occ}(\phi(x))]$. The last requirement, $\text{den}(\phi'(v'), h') = 1$, follows from (3) by the induction hypothesis. Now, by (1), and (2) and Lemma 3 we have

$$w = \text{den}(v, h) = \text{den}(v, h' \circ t) = \text{den}(\text{sub}(v, t), h') = \text{den}(v', h') \in Q,$$

where we use once again the induction hypothesis.

(b) $Q \subseteq P$. Let $w \in Q$, then $w = \text{den}(v', h')$, where v' and h' satisfy the requirements in the definition of Q. Let t be a substitution function such that t coincides with s on $\text{occ}(\phi(x))$, and for some finite subset d of ω disjoint from $\text{occ}(\phi(x))$, $t \mid d$ is a one-one map of d on $(\text{occ}(v') \sim \text{occ}(\phi'(x')))$, and $\mathfrak{D}(t) = \text{occ}(\phi(x)) \cup d$. Since $\text{occ}(v') \subseteq \mathfrak{R}(t)$ we have $v' = \text{sub}(v, t)$ for some term v such that $\lambda(v) = \lambda(v')$. We set $h = h' \circ t$. We shall now see that v and h satisfy the requirements of P. $\lambda(v) = \lambda(v') < \alpha$. It is easily seen that $\mathfrak{D}(t) = \text{occ}(\phi(x)) \cup \text{occ}(v) = \text{occ}(\phi(v))$, hence $\mathfrak{R}(t) = \text{occ}(\text{sub}(\phi(v), t)) = \text{occ}(\phi'(v'))$. Since $\mathfrak{D}(h') \supseteq \text{occ}(\phi'(v')) = \mathfrak{R}(t)$, by our assumption on h', we have $\mathfrak{D}(h) = \mathfrak{D}(h' \circ t) = \mathfrak{D}(t) = \text{occ}(\phi(v))$. Since t and s coincide on $\text{occ}(\phi(x))$ and h and f coincide on $\text{occ}(\phi'(x)) = s[\text{occ}(\phi(x))]$ we have

$$h \mid \text{occ}(\phi(x)) = h' \circ t \mid \text{occ}(\phi(x)) = f \circ s \mid \text{occ}(\phi(x)).$$

Finally,

$$\text{den}(\phi(v), h) = \text{den}(\phi(v), h' \circ t) = \text{den}(\text{sub}(\phi(v), t), h') = \text{den}(\phi'(v'), h') = 1,$$

by the induction hypothesis. Now we have $w = \text{den}(v', h') = \text{den}(\text{sub}(v, t), h') = \text{den}(v, h' \circ t) = \text{den}(v, h) \in P$, where we use once again the induction hypothesis.

Corollary 5. *If f, g are assignments such that $\mathfrak{R}(f) \subseteq \mathfrak{R}(g)$, then for every term u such that $\text{occ}(u) \subseteq \mathfrak{D}(f)$ there is a term v such that $\text{occ}(v) \subseteq \mathfrak{D}(g)$, $\text{den}(v, g) = \text{den}(u, f)$ and $\lambda(v) = \lambda(u)$.*

PROOF. Let s be a substitution function such that $\mathfrak{D}(s) = \mathfrak{D}(f)$, $\mathfrak{R}(s) \subseteq \mathfrak{D}(g)$ and $g \circ s = f$. Let $v = \text{sub}(u, s)$. Obviously $\text{occ}(v) \subseteq \mathfrak{R}(s) \subseteq \mathfrak{D}(g)$ and $\lambda(v) = \lambda(u)$. By Lemma 4 we have

$$\text{den}(v, g) = \text{den}(\text{sub}(u, s), g) = \text{den}(u, g \circ s) = \text{den}(u, f).$$

COROLLARY 6. *Let $x_1, \ldots, x_n \in L_\delta$. Then there are terms $u_1, \ldots, u_n$, $\lambda(u_i) < \delta$, and an assignment f such that* $\text{occ}(u_i) \subseteq \mathfrak{D}(f)$ *and* $x_i = \text{den}(u_i, f)$ *for* $1 \leqq i \leqq n$.

PROOF. Since $x_1, \ldots, x_n \in L_\delta$ there are terms $v_1, \ldots, v_n$, $\lambda(v_i) < \delta$, and assignments $g_1, \ldots, g_n$ such that $x_i = \text{den}(v_i, g_i)$, $1 \leqq i \leqq n$. Let s_i, $1 \leqq i \leqq n$, be one-one substitution functions for the respective v_i's with pairwise disjoint ranges. Let $u_i = \text{sub}(v_i, s_i)$, let $h = g_i \circ s_i^{-1}$ and let $f = \bigcup_{i=1}^n h_i$. Then, by Lemma 4,

$$\text{den}(u_i, f) = \text{den}(\text{sub}(v_i, s_i), f) = \text{den}(v_i, f \circ s_i) = \text{den}(v_i, g_i) = x_i.$$

LEMMA 7. (a) *If* $\text{occ}(u) \subseteq \mathfrak{D}(f)$ *and* $y \in \text{den}(u, f)$ *then there are v and g such that* $\text{occ}(v) \subseteq \mathfrak{D}(g)$, $\lambda(v) < \lambda(u)$, $g \supseteq f \mid \text{occ}(u)$ *and* $y = \text{den}(v, g)$.

(b) *If* $\text{occ}(u) \subseteq \mathfrak{D}(f)$ *and* $x = \text{den}(u, f)$ *then* $\rho(x) \leqq \lambda(u)$.

PROOF. (a) is an immediate consequence of the definition of den. (b) follows from (a) and the basic properties of the rank function, by transfinite induction on $\lambda(u)$.

We order the subsets of u by first differences, i.e., $x < y$ if the first member of of the symmetric difference $(x \sim y) \cup (y \sim x)$ is in y. Every finite k-sequence q with $\mathfrak{R}(q) \subseteq \{0, 1\}$ determines the subset

$$b_q = \{y \in b \mid (\forall i < k)((q(i) = 1 \to i \in y) \land (q(i) = 0 \to i \notin y))\}$$

of b. We shall refer to these subsets of b as *absolute intervals of b* or just as *absolute intervals*, and to the k-sequences q as *interval designators*. (The use of "interval" is justified by the observation that if $x, y \in b_q$ and $x < z < y$ then also $z \in b_q$.)

3. **The set theory** *SP*. The language of *SP* is the language of *ZF* with the following additional symbols:

S, F—unary predicate symbols; b—an individual constant.

We shall denote by S the class of all sets x such that $S(x)$. The axioms of *SP* are:

(i) All the axioms of *ZF*, where the formulas in the axiom schemas of subsets and replacement may contain also the additional symbols of *SP*.

(ii) b is a "dense" subset of $P(\omega)$, i.e. no absolute interval of b is empty.

(iii) S is a transitive class which contains all ordinals, all finite subsets of itself and all sets $S \cap R(\alpha)$ and $F \cap R(\alpha)$ for every ordinal α.

(iv) F is a class of ordered pairs which is a function from the class *On* of all ordinals onto S.

(v) (The axiom schema of continuity). If $x_1, \ldots, x_n \in S \cup \{b\}$, $m < \omega$ and $y = \langle y_0, \ldots, y_{m-1}\rangle$ is an m-sequence of different members of b and $\Phi(x_1, \ldots, x_n, y)$ holds then there is an m-sequence $\langle c_0, c_1, \ldots, c_{m-1}\rangle$ of pairwise disjoint absolute intervals of b such that $y_i \in c_i$ for $i < m$ and $\Phi(x_1, \ldots, x_n, z)$ holds for every m-sequence $z = \langle z_0, \ldots, z_{m-1} \rangle$ such that $z_i \in b_i$ for $i < m$, where Φ is any formula of SP with no free variables other than $x_1, \ldots, x_n, y$.

(vi) For every x there is a term u and a sufficient assignment f such that $x = \text{den}(u, f)$.

Throughout the rest of the present section the theorems will be proved in the theory SP and therefore they are theorems of SP.

THEOREM 8. *Every absolute interval of b is an infinite set.*

PROOF. By Axiom (ii) no absolute interval of b is empty. Since every absolute interval includes countably many pairwise disjoint absolute intervals the theorem follows.

THEOREM 9. $S \cap b = 0$.

PROOF. Suppose $x \in S \cap b$. Let $y = \langle x \rangle$ and let Φ be the formula $y = \langle x \rangle$. By the axiom schema (v) of continuity and since $x \in S$ there is an absolute interval c_0 such that for every z_0 if $z_0 \in c_0$ then $\langle z_0 \rangle = \langle x \rangle$. Therefore the absolute interval c_0 consists of the only member x, contradicting Theorem 8.

THEOREM 10. *b is Dedekind finite, i.e. b has no denumerable subset.*

PROOF. Suppose that t is an infinite sequence of different members of b. By Axiom (vi) and Corollary 4a, $t = \text{den}(u, f)$ for some term u and some finite sequence f with no reoccurring members. Since $\mathfrak{R}(t)$ is infinite and $\mathfrak{R}(f)$ is finite we have $\mathfrak{R}(t) \sim \mathfrak{R}(f) \neq 0$. Consider the formula "$d$ is the first member of $\mathfrak{R}(\text{den}(u, f)) \sim \mathfrak{R}(f)$, in the sense of $\text{den}(u, f)$". Since $u \in TL$, $u \in S$ by Axiom (iii). Applying Axiom (v) to the sequence $y = f^\frown \langle d \rangle$ we obtain an absolute interval c' of b such that d' is the first member of $\mathfrak{R}(\text{den}(u, f)) \sim \mathfrak{R}(f)$, in the sense of $\text{den}(u, f)$, for every $d' \in c'$. This contradicts Theorem 8.

COROLLARY 11. *There is a Dedekind finite set c dense in the real line.*

PROOF. Since the open real interval (0, 1) is order-isomorphic to the real line, it is enough to obtain a Dedekind finite set c' dense in (0, 1). The function $f: x \to \sum_{n \in x} 2^{-(n+1)}$ for $x \in b$ is a one-one mapping of b into (0, 1]; it is one-one since no $x \in b$ is a finite set because finite subsets of ω are standard by Axiom (iii) whereas no member of b is standard, by Theorem 9. Since the absolute intervals of b are nonempty the image c' of b is dense in [0, 1]. It follows from Theorem 10 that c' is Dedekind finite.

DEFINITION 12. For an assignment f we denote with L_f the class

$$\{\text{den}(u, f) \mid u \in TL \text{ and } \text{occ}(u) \subseteq \mathfrak{D}(f)\}.$$

This should not be confused with the set L_α where α is an ordinal, which is defined in Definition 2. By Lemma 5, L_f depends only on $\mathfrak{R}(f)$.

LEMMA 13. *Let f be an assignment, then there is a relation definable in terms of f which well-orders the class L_f.*

PROOF. Since, by Axiom (iii), $TL \subseteqq S$, F induces, by Axiom (iv), a well-ordering on the class TL of all terms, and this, in turn, induces a well-ordering on L_f.

LEMMA 14. *Let f be an assignment. If $x_1, \ldots, x_n \in L_f$, $m < \omega$, $g' = \langle g'_0, \ldots, g'_{m-1}\rangle$ is an m-sequence of different members of $b \sim \mathfrak{R}(f)$ and $\Phi(x_1, \ldots, x_n, g')$ holds, then there is a sequence $\langle e_0, \ldots, e_{m-1}\rangle$ of absolute intervals of b, pairwise disjoint and disjoint from $\mathfrak{R}(f)$, such that $g'_i \in e_i$ for $i < m$ and $\Phi(x_1, \ldots, x_n, g)$ holds for every m-sequence $g = \langle g_0, \ldots, g_{m-1}\rangle$ such that $g_i \in b_i$ for $i < m$, where Φ is any formula of SP with no free variables other than $x_1, \ldots, x_n, g'$.*

PROOF. By Corollary 5 we can assume, without loss of generality, that f is a finite sequence without reoccurrences. Let u_i be terms such that $\text{occ}(u_i) \subseteqq \mathfrak{D}(f)$ and $x_i = \text{den}(u_i, f)$ for $1 \leqq i \leqq n$, and let $h = f^\frown g'$. If j is the length of f then $f = h \mid j$ and $g' = [h]_j^{j+m}$, where in general for $k \leqq l \leqq$ the length of h $[h]_j^{j+m}$ is the unique finite sequence satisfying $h \mid j^\frown [h]_j^{j+m} = h \mid j + m$. Let $\Psi(h)$ be

$$\Phi(\text{den}(u_1, h \mid j), \ldots, \text{den}(u_n, h \mid j), [h]_j^{j+m}).$$

Since $u_1, \ldots, u_n, j, j + m \in S$ we have, by Axiom (v), a sequence $\langle c_0, \ldots, c_{j+m-1}\rangle$ of pairwise disjoint absolute intervals such that $h_i \in c_i$ for $i < j + m$, and for every $j + m$-sequence t such that $t_i \in c_i$ for $i < j + m$ we have $\Psi(t)$. Let $e_i = c_{j+1}$ for $i < m$, let g be any m-sequence such that $g_i \in e_i$ for $i < m$ and let $t = f^\frown g$. Then $t_i \in c_i$ for $i < j + m$ and hence we have $\Psi(t)$, i.e.

$$\Phi(\text{den}(u_1, t \mid j), \ldots, \text{den}(u_n, t \mid j), [t]_j^{j+m}),$$

which is $\Phi(x_1, \ldots, x_n, g)$. Each of $e_0, \ldots, e_{m-1}$ is disjoint from each of $c_0, \ldots, c_{j-1}$ and hence $e_i \cap \mathfrak{R}(f) = 0$ for $i < m$.

DEFINITION 15. For a subset G of b of $k < \omega$ members we shall say that the sequence $e = \langle e_0, \ldots, e_{k-1}\rangle$ of absolute intervals *distinguishes* G if the intervals e_i, $1 \leqq i \leqq k$, are pairwise disjoint and each e_i contains exactly one member of G.

COROLLARY 16. *Let f be an assignment. If $x_1, \ldots, x_n \in L_f$, $m < \omega$, G' is a subset of $b - \mathfrak{R}(f)$ of cardinality m and $\Phi(x_1, \ldots, x_n, G')$ holds then there is a sequence $\langle e_0, \ldots, e_{m-1}\rangle$ of absolute intervals disjoint from $\mathfrak{R}(f)$ and distinguishing members of G' such that $\Phi(x_1, \ldots, x_n, G)$ holds for every set G of m members which has exactly one member in each set e_i, $i < m$.*

PROOF. Let g' be a finite sequence without reoccurrences such that $\mathfrak{R}(g') = G'$, then we have $\Phi(x_1, \ldots, x_n, \mathfrak{R}(g'))$. Applying Lemma 14 to $\Phi(x_1, \ldots, x_n, \mathfrak{R}(g'))$ we get a sequence $\langle e_0, \ldots, e_{m-1}\rangle$ as above such that $\Phi(x_1, \ldots, x_n, \mathfrak{R}(g))$ holds for every sequence $g = \langle g_0, \ldots, g_{m-1}\rangle$ such that $g_i \in e_i$ for $i < m$. If G is a set with m members and with exactly one member in common with each e_i, $i < m$, then let g be the m-sequence given by $g_i =$ the single member of $G \cap e_i$, then $\mathfrak{R}(g) = G$ and we have $\Phi(x_1, \ldots, x_n, G)$.

DEFINITION 17. If Φ is a formula of the language of SP and δ is a variable over ordinals then $\Phi^{(\delta)}$ denotes the formula obtained from Φ by restricting all the quantifiers in Φ to L_δ, i.e., by replacing each quantifier $\exists x$ by $(\exists x \in L_\delta)$ (universal quantifiers ar; assumed not to be part of the primitive language, but to be given in terms of existential quantifiers). Also, to each formula Φ of SP we correspond a formula $\phi_{\Phi,\delta}$ of $\mathscr{L}$ (which depends on δ) as follows. We replace in Φ each quantifier $\exists x$ by $\exists_\delta x$ and each occurrence of $S(x)$ or $F(x)$ by $x \in \mathbf{s}_\delta$ or $x \in \mathbf{f}_\delta$, respectively where $s_\delta = S \cap R(\delta)$, $f_\delta = F \cap R(\delta)$, $\mathbf{s}_\delta = \langle 4, s_\delta \rangle$, $\mathbf{f}_\delta = \langle 4, f_\delta \rangle$. It should be noticed that $\phi_{\Phi,\delta}$ is obtained from Φ by more than a mere replacement of symbols; e.g., the symbol $\in$, which is apparently not replaced changes its meaning from the membership symbol to $\langle 0, 2 \rangle$. Formally $\phi_{\Phi,\delta}$ is defined by defining, for each formula Φ, a function $H_\Phi(\delta)$ on the ordinals into the class FL of all formulas of $\mathscr{L}$. Notice that we have, for all Φ and δ, $\operatorname{occ}(\phi_{\Phi,\delta}) = 0$.

LEMMA 18. *Let $\Phi(x_1, \ldots, x_n)$ be a formula of SP with no free variables other than $x_1, \ldots, x_n$. If $u_i \in TL$, $\operatorname{occ}(u_i) \subseteq \mathfrak{D}(f)$, $x_i = \operatorname{den}(u_i, f)$ and $\lambda(u_i) < \delta$ for $1 \leqq i \leqq n$, then $\Phi^{(\delta)}(x_1, \ldots, x_n)$ holds if and only if*

$$\operatorname{den}(\phi_{\Phi,\delta}(u_1, \ldots, u_n), f) = 1,$$

where $\phi_{\Phi,\delta}(u_1, \ldots, u_n)$ denotes the sentence of $\mathscr{L}$ obtained from the formula $\phi_{\Phi,\delta}$ of $\mathscr{L}$ by replacing the free occurrences of $x_1, \ldots, x_n$ by $u_1, \ldots, u_n$, respectively.

PROOF. By induction on the length of Φ.

(a) If Φ is $x_i \in x_j$, then $x_i \in x_j$ if and only if $\operatorname{den}(u_i, f) \in \operatorname{den}(u_j, f)$, which holds if and only if $\operatorname{den}(u_i \in u_j, f) = 1$. If Φ is $x_i = x_j$ the proof is similar.

(b) If Φ is $S(x_i)$, then $S(x_i)$ if and only if $S(\operatorname{den}(u_i, f))$ which holds if and only if $\operatorname{den}(u_i, f) \in s_\delta$ (since by Lemma 7(b) $\lambda(u_i) < \delta$ implies $\rho(\operatorname{den}(u_i, f)) < \delta$). The latter holds if and only if $\operatorname{den}(u_i \in \mathbf{s}_\delta, f) = 1$. If Φ is $F(x_i)$ the proof is similar.

(c) If Φ is an atomic formula which contains b then we proceed as in (a) and (b).

(d) If Φ is $\neg\Psi$ or $\Psi \vee \Gamma$, what we need follows directly from the induction hypothesis.

(e) If Φ is $\exists x_0 \Psi(x_0, x_1, \ldots, x_n)$ then we have: $\Phi^{(\delta)}(x_1, \ldots, x_n)$, i.e. $(\exists x_0 \in L_\delta)\Psi^{(\delta)}(x_0, x_1, \ldots, x_n)$, if and only if for some term v with $\lambda(v) < \delta$ and for some assignment g sufficient for v we have $\Psi^{(\delta)}(\operatorname{den}(v, g), \operatorname{den}(u_1, f), \ldots, \operatorname{den}(u_n, f))$. As follows from Corollary 5 and Lemma 3 we can assume that g coincides with f on $\bigcup_{1 \leqq i \leqq n} \operatorname{occ}(u_i)$ (since otherwise we can replace g by an assignment g' which coincides with f on $\bigcup_{1 \leqq i \leqq n} \operatorname{occ}(u_i)$ and such that $\mathfrak{R}(g') \supseteq \mathfrak{R}(f)$) and hence, by Lemma 3, $\Phi^{(\delta)}(x_1, \ldots, x_n)$ if and only if there is an assignment g as above such that $\Psi(\operatorname{den}(v, g), \operatorname{den}(u_1, g), \ldots, \operatorname{den}(u_n, g))$. Since $\operatorname{occ}(\phi_{\Psi,\delta}) = 0$ we have $\operatorname{occ}(\phi_{\Psi,\delta}(x_0, u_1, \ldots, u_n)) = \bigcup_{1 \leqq i \leqq n} \operatorname{occ}(u_i)$. Using now the induction hypothesis for the formula Ψ we get that $\Phi^{(\delta)}(x_1, \ldots, x_n)$ holds if and only if there is a term v and an assignment g sufficient for $\phi_{\Psi,\delta}(v, u_1, \ldots, u_n)$ such that

$$g \mid \operatorname{occ}(\phi_{\Psi,\delta}(x_0, u_1, \ldots, u_n)) = f \mid \operatorname{occ}(\phi_{\Psi,\delta}(x_0, u_1, \ldots, u_n)),$$

$\lambda(v) < \delta$ and $\operatorname{den}(\phi_{\Psi,\delta}(v, u_1, \ldots, u_n))$ and this holds, by the definition of den, just in case that $\operatorname{den}(\phi_{\Phi,\delta}(u_1, \ldots, u_n), f) = 1$.

LEMMA 19. *L_f contains all its finite subsets.*

PROOF. Let $x_1, \dots, x_n \in L_f$. Then for some $u_1, \dots, u_n \in TL$, $\text{occ}(u_1) \subseteqq \mathfrak{D}(f)$ and $\text{den}(u_i, f) = x_i$ for $1 \leqq i \leqq n$. Let $\delta > \lambda(u_1), \dots, \lambda(u_n)$ and let v be $\rightthreetimes_\delta x(x = u_1 \vee \cdots \vee x = u_n)$. v is obviously a term, and one can easily verify that $\text{den}(v, f) = \{x_1, \dots, x_n\}$.

COROLLARY 20. *For every assignment* f, $f \in L_f$.

LEMMA 21. *Let Φ be a formula of SP with no free variables other than $x_1, \dots, x_n, y$. If f is an assignment, $x_1, \dots, x_n \in L_f$ and y is the unique set such that $\Phi(x_1, \dots, x_n, y)$ then $y \in L_f$.*

PROOF. Let $u_i \in TL$ be such that $\text{occ}(u_i) \subseteqq \mathfrak{D}(f)$ and $\text{den}(u_i, f) = x_i$ for $1 \leqq i \leqq n$. We define a sequence α_m of ordinals as follows. α_0 is the least ordinal such that $\alpha_0 > \lambda(u_1), \dots, \lambda(u_n)$ and $y \in L_{\alpha_0}$. The existence of such an α_0 follows from Axiom (vi). Let Δ be the set of all formulas $\Psi(z, z_1, \dots, z_l)$ such that $\exists z \Psi(z, z_1, \dots, z_l)$ is a subformula of Φ (l, of course, depends on Ψ). We define a function * on the ordinals as follows. For $\Psi \in \Delta$ let $H_\Psi(z_1, \dots, z_l)$ be the least ordinal γ such that $(\exists z \in L_\gamma)\Psi(z, z_1, \dots, z_l)$ if $\exists z \Psi(z, z_1, \dots, z_l)$, and $H_\Psi(z_1, \dots, z_l) = 0$ if $\neg \exists z \Psi(z, z_1, \dots, z_l)$. We define

$$\beta^* = \sup\{H_\Psi(z_1, \dots, z_l) \mid \Psi \in \Delta \wedge z_1, \dots, z_l \in L_\beta\}.$$

Thus we have, for every $\Psi \in \Delta$ and every ordinal β

$$z_1, \dots, z_l \in L_\beta \rightarrow [\exists z \Psi(z, z_1, \dots, z_l) \leftrightarrow (\exists z \in L_{\beta^*})\Psi(z, z_1, \dots, z_l)].$$

Let $\alpha_{m+1} = \max(\alpha_m, \alpha_m^*)$ for all $m < \omega$ and let $\delta = \sup\{\alpha_m \mid m < \omega\}$. Then for all $\Psi \in \Delta$

$$z_1, \dots, z_l \in L_\delta \rightarrow [\exists z \Psi(z, z_1, \dots, z_l) \leftrightarrow (\exists z \in L_\delta)\Psi(z, z_1, \dots, z_l)].$$

It is now easily seen, by induction on the length of the subformulas Γ of Φ, that for each such $\Gamma(z_1, \dots, z_l)$ we have

$$z_1, \dots, z_l \in L_\delta \rightarrow [\Gamma(z_1, \dots, z_l) \leftrightarrow \Gamma^{(\delta)}(z_1, \dots, z_l)].$$

In particular, for $\Gamma = \Phi$ we have, since $x_1, \dots, x_n, y \in L_{\alpha_0} \subseteqq L_\delta$, that y is the unique member of L_δ such that $\Phi^{(\delta)}(x_1, \dots, x_n, y)$.

Therefore, for every term w such that $\lambda(w) < \delta$, if g is an assignment sufficient for w such that $g \supseteqq f \mid \bigcup_{1 \leqq i \leqq n} \text{occ}(u_i)$ then we have $\Phi^{(\delta)}(x_1, \dots, x_n, \text{den}(w, g))$ if and only if $\text{den}(w, g) = y$. By Lemma 18 we have, for all such w and g,

$$\text{den}(\phi_{\Phi,\delta}(u_1, \dots, u_n, w), g) = 1$$

if and only if $\text{den}(w, g) = y$. In the sequel let $\chi(z)$ stand for $\phi_{\Phi,\delta}(u_1, \dots, u_n, z)$. We have just shown that

$$\begin{aligned} &\lambda(w) < \delta \wedge \mathfrak{D}(g) \supseteqq \text{occ}(w) \wedge g \supseteqq f \Big| \bigcup_{1 \leqq i \leqq n} \text{occ}(u_i) \\ &\qquad \rightarrow (\text{den}(\chi(w), g) = 1 \leftrightarrow \text{den}(w, g) = y). \end{aligned} \tag{4}$$

Consider the term $v = \gimel_\delta x \exists_\delta y(\chi(y) \wedge x \in y)$. Obviously

$$\operatorname{occ}(v) = \bigcup_{1 \leqq i \leqq n} \operatorname{occ}(u_i) \subseteq \mathfrak{D}(f).$$

We aim to prove $\operatorname{den}(v, f) = y$; once this is proved we have $y \in L_f$, which is what we set out to prove. By definition

$$(5) \qquad \begin{aligned} \operatorname{den}(v, f) = \{\operatorname{den}(u, g) \mid \lambda(u) < \delta \wedge \operatorname{occ}(u) \subseteq \mathfrak{D}(g) \\ \wedge g \supseteq f \mid \operatorname{occ}(v) \wedge \operatorname{den}(\exists_\delta y(\chi(y) \wedge u \in y), g) = 1\}. \end{aligned}$$

(a) $\operatorname{den}(v, f) \subseteq y$. Let $r \in \operatorname{den}(v, f)$, then for some u and g as in (5) $r = \operatorname{den}(u, g)$. In (5) we have for u and g $\operatorname{den}(\exists_\delta y(\chi(y) \wedge u \in y), g) = 1$. Therefore, for some term w such that $\lambda(w) < \delta$ and some assignment g' such that $g' \supseteq g|(\operatorname{occ}(\chi(y)) \cup \operatorname{occ}(u)) \supseteq g|\operatorname{occ}(v) = f|\operatorname{occ}(v)$ we have $\operatorname{den}(\chi(w), g') = 1$ and $\operatorname{den}(u \in w, g') = 1$. $\operatorname{den}(\chi(w), g') = 1$ implies, by (4), $\operatorname{den}(w, g') = y$; $\operatorname{den}(u \in w, g') = 1$ implies, by the definition of den, $r = \operatorname{den}(u, g) = \operatorname{den}(u, g') \in \operatorname{den}(w, g') = y$.

(b) $y \subseteq \operatorname{den}(v, f)$. Since $y \in L_{\alpha_0}$ there is, by Corollary 5, a term w and a sufficient assignment g such that $g \supseteq f \mid \bigcup_{1 \leqq i \leqq n} \operatorname{occ}(u_i)$, $\lambda(w) < \alpha_0$ and $\operatorname{den}(w, g) = y$. By (4) we have $\operatorname{den}(\chi(w), g) = 1$. Let $r \in y$, then there is, by Lemma 7(a) a term u and a sufficient assignment g' such that $\lambda(u) < \lambda(w) < \alpha_0 \leqq \delta$ and $\operatorname{den}(u, g') = r$. By Corollary 5 we can assume that $g' \supseteq g \mid \operatorname{occ}(\chi(w)) \supseteq f \mid \operatorname{occ}(v)$. We have, therefore,

$$\operatorname{den}(\chi(w) \wedge u \in w, g') = \min(\operatorname{den}(\chi(w), g'), \operatorname{den}(u \in w, g')) = 1,$$

and since $\lambda(w) < \delta$ we have, by (5), $r = \operatorname{den}(u, g') \in \operatorname{den}(v, f)$.

COROLLARY 22. *All the absolute intervals are in L_f for every assignment f.*

PROOF. $b \in L_f$ since $b = \operatorname{den}(b, f)$. If q is an interval designator then $q \in S$ by Axiom (iii), b and q determine the corresponding absolute interval b_q uniquely, hence, by Lemma 21, $b_q \in L_f$.

THEOREM 23. *If $\mathfrak{B} \in L_f$ is a Boolean algebra then there is a prime ideal J of $\mathfrak{B}$ such that $J \in L_f$. Therefore, by Axiom* (vi), *the Boolean prime ideal theorem holds.*

PROOF. Let $\mathfrak{B} = \langle B, \cdot, - \rangle$, where $\cdot$ is the product operation and $-$ is the complementation operation. For a nonvoid finite subset a of B we denote by $\prod a$ the product of the members of a. Since $\mathfrak{B} \in L_f$ we have, by Lemma 21, $B, \cdot, - \in L_f$. Our proof will consist of two steps; first we show that among the proper ideals of $\mathfrak{B}$ in L_f there is a maximal one, secondly we show that every such ideal is a prime ideal.

Consider the set of all proper ideals of $\mathfrak{B}$ which are in L_f. Since, by Lemma 13, L_f can be definably well-ordered in terms of f, we can define in terms of f and $\mathfrak{B}$ a function G on some ordinal α such that $G(\gamma)$ is a proper ideal in L_f for $\gamma < \alpha$ and $G(\beta) \subset G(\gamma)$ for $\beta < \gamma < \alpha$, and such that there is no proper ideal in L_f which properly includes $\bigcup_{\beta < \alpha} G(\beta)$. Then $\bigcup_{\beta < \alpha} G(\beta)$ is a proper ideal as desired. It belongs to L_f by Lemma 21, and is maximal in L_f by construction.

As an aid to the reader, we shall use sans serif letters, such as c, e, f, g, h, s for finite sequences. The ith term of the sequence e will be denoted by e_i. For a k-sequence e, $\mathsf{X}\,\mathsf{e}$ is the Cartesian product of e, i.e. $\{\mathsf{c} \mid \mathsf{c}$ *is a* k*-sequence and* $c_i \in e_i$ *for every* $i < k\}$.

From now on we shall assume, merely for the sake of convenience, that the assignment f of the theorem is a finite sequence, i.e. that $\mathfrak{D}(f)$ is a natural number $n = \{m \mid m < n\}$. Anyway, the main point of the theorem holds even if we prove the theorem only for finite sequences f.

Let I be a maximal proper ideal in L_{f}. Suppose I is not prime. Then for some $x \in B$, $x \notin I$ and $-x \notin I$. Let u be a term and h' a k-sequence of different members of $b \sim \mathfrak{R}(\mathsf{f})$ such that $x = \mathrm{den}(u, \mathsf{f}^\frown \mathsf{h}')$. Since f and u are fixed throughout the present proof we shall abbreviate $\mathrm{den}(u, \mathsf{f}^\frown \mathsf{h})$ by $d(\mathsf{h})$ for any k-sequence h. Under the assumption mentioned above we have

$$d(\mathsf{h}') \in B \sim I \quad \text{and} \quad -d(\mathsf{h}') \in B \sim I. \tag{6}$$

By way of an example we shall handle the simple case where $k = 1$ before tackling the general case. Now, $\mathsf{h}' = \langle h_0' \rangle$; we shall write h' for h_0'. It follows from (6) and Lemma 14 that there is an absolute interval c such that

$$h \in c \to d(h) \in B \sim I \wedge -d(h) \in B \sim I. \tag{7}$$

Consider the ideal of $\mathfrak{B}$ generated by $I \cup \{d(h) \mid h \in c\}$. This ideal is in L_f by Lemma 21 and Corollary 22; hence this ideal coincides with B (by the maximality of I). Therefore for some finite set $G_1' \subseteq b$ we have $\prod \{-d(h) \mid h \in G_1'\} \in I$. Similarly, if we consider the ideal generated by $I \cup \{-d(h) \mid h \in c\}$ we obtain a finite set G_2' such that $\prod \{d(h) \mid h \in G_2'\} \in I$. Putting $G' = G_1' \cup G_2'$ we get

$$\prod \{-d(h) \mid h \in G'\} \in I, \qquad \prod \{d(h) \mid h \in G'\} \in I. \tag{8}$$

Let e be a sequence of absolute subintervals of c which distinguishes G' and such that

(9) *If G consists of exactly one member of each e_i, $i < \mathfrak{D}(\mathsf{e})$, then*

$$\prod \{-d(h) \mid h \in G\} \in I, \qquad \prod \{d(h) \mid h \in G\} \in I.$$

The existence of such an e follows from (8) by Corollary 16; $e_i \subseteq c$ since in the consequence of Corollary 16 we can intersect e_i with c, for $i < \mathfrak{D}(\mathsf{e})$.

Let us put $S^1 = \mathfrak{R}(\mathsf{e})$, S^1 is a set of absolute subintervals of $c \subseteq b$. Let $r \in S^1$. Since $r \subseteq c$ (7) holds also with c replaced by r. Replacing c by r in the argument that leads to (9) we obtain here that there is a sequence e^r of pairwise disjoint absolute subintervals of r such that

(10) *If G consists of exactly one member of each $(e^r)_i$, $i < \mathfrak{D}(\mathsf{e}^r)$, then*

$$\prod \{-d(h) \mid h \in G\} \in I, \qquad \prod \{d(h) \mid h \in G\} \in I.$$

Put $S^2 = \bigcup \{\mathfrak{R}(\mathsf{e}^r) \mid r \in S^1\}$. S^2 is, as easily seen, a set of pairwise disjoint absolute subintervals of c. Let y be a set which consists of exactly one member out of each member of S^2. We shall show that for every $z \subseteq y$

$$(11) \qquad \prod \{d(h) \mid h \in z\} \in I \quad or \quad \prod \{-d(h) \mid h \in y \sim z\} \in I.$$

Once (11) is proved we obtain a contradiction as follows. Let 1 be the unit element of $\mathfrak{B}$. Then, since $d(h) + -d(h) = 1$, we have

$$(12) \qquad 1 = \prod_{h \in y} [d(h) + -d(h)] = \sum_{p \in 2^y} \prod_{h \in y} d(h)^{(p(h))} = \sum_{p \in 2^y} a(p)$$

where 2^y is the set of all functions on y into $2(=\{0, 1\})$, $d(h)^{(0)} = d(h)$, $d(h)^{(1)} = -d(h)$ and $a(p) = \prod_{h \in y} d(h)^{(p(h))}$. For a given p put $z = \{h \in y \mid p(h) = 0\}$, then $a(p) = \prod_{h \in z} d(h) \cdot \prod_{h \in y \sim z} -d(h)$. This implies, by (11), that $a(p) \in I$. Since we have $a(p) \in I$ for every $p \in 2^y$ we get, by (12), that also their sum 1 is in I, contradicting our assumption that I is a proper ideal.

We still have to prove (11). If z contains at least one member out of each member e_i of S^1, then, by (9), $\prod \{d(h) \mid h \in z\} \in I$. If, on the other hand, there is a member r of S^1 such that $z \cap r = 0$, then $y \sim z$ contains exactly one member out of each member t of $\mathfrak{R}(\mathrm{e}^r) \subseteq S^2$ since these members are subintervals of c. Hence, by (10),

$$\prod \{-d(h) \mid h \in (y \sim z) \cap \bigcup \mathfrak{R}(\mathrm{e}^r)\} \in I$$

and, a fortiori, $\prod \{-d(h) \mid h \in y \sim z\} \in I$ and (11) is proved.

Having dealt with the case $k = 1$, let us now deal with the general case. For this we need a combinatorial theorem of Halpern and Läuchli [**5**, Theorem 2]. Indeed, it was the present proof that suggested that combinatorial theorem. Before we formulate a particular case of that theorem we have to introduce some notation.

A *tree* $\mathfrak{T} = \langle T, \leqq \rangle$ is a partially ordered set such that for each *node* x (i.e. for each $x \in T$) the set of all predecessors of x, i.e. $\{y \mid y < x\}$, is totally ordered by $\leqq$. The cardinality of this set is called the *order* of x, or the *level* at which x occurs. A *finitistic* tree is a tree with at least one node, all of whose nodes have finite orders and such that each level is a finite set. $T \mid n$ (read: T restricted to n) will denote the set of the nodes of T whose order is $\leqq n$. A set A of nodes *dominates* a set B of nodes if $(\forall x \in B)(\exists y \in A) y \geqq x$; *A supports B* if $(\forall x \in B)(\exists y \in A) y \leqq x$. A set A of nodes is said to be $(m, 1)$-*dense* if there is a node x of order m such that the set of the nodes of order $m + 1$ supported by $\{x\}$ is dominated by A. Given trees $\mathfrak{T}_i = \langle T_i, \leqq_i \rangle$, $i < k$, we shall refer to a Cartesian product $\mathsf{X}_{i<k} A_i$, where A_i is a subset of T_i $(m, 1)$-dense in T_i, as an $(m, 1)$-*matrix*.

THEOREM (HALPERN-LÄUCHLI). *If* $\mathfrak{T}_i = \langle T_i, \leqq_i \rangle$, $i < k$, *are finitistic trees without maximal nodes, then there is a positive integer* n *such that whenever* $\mathsf{X}_{i<k}(T_i \mid n) = Q_1 \cup Q_2$, Q_1 *or* Q_2 *includes an* $(m, 1)$-*matrix for some* $m < n$.

Now we return to our proof for a general k. It follows from (6) and Lemma 14 that there is a k-sequence c of pairwise disjoint absolute intervals such that

$$(13) \qquad \mathsf{h} \in \mathsf{X}\, \mathsf{c} \rightarrow d(\mathsf{h}) \in B \sim I \wedge -d(\mathsf{h}) \in B \sim I.$$

In the case where $k = 1$ it was enough to go through a certain procedure twice to define S^1 and S^2; here we have to define a k-sequence S^n for every $n \geqq 0$. We define S^n by induction on n. First we put $S_i^0 = \{c_i\}$ for $i < k$. While defining the S^n's we shall, simultaneously, prove that they have the following properties (a)–(d).

(a) S_i^m is a finite set of absolute subintervals of the members of S_i^{m-1}, for $m \geqq 1$ and $i < k$. Therefore, by the definition of S^0 and by induction on m, the members of S_i^m are subintervals of c_i, for $i < k$.

(b) Every member of S_i^{m-1} has at least two subintervals in S_i^m, for $m \geqq 1$ and $i < k$.

(c) The members of S_i^m are pairwise disjoint, for $m \geqq 0$, and $i < k$.

(d) If $\mathsf{r} \in \mathsf{X}\, \mathsf{S}^{m-1}$ and G is a finite set which contains exactly one member out of each member of $\bigcup \mathfrak{R}(\mathsf{S}^m)$ then

$$\prod \{d(\mathsf{h}) \mid \mathsf{h} \in \mathsf{X}_{i<k} (r_i \cap G)\} \in I$$

and

$$\prod \{-d(\mathsf{h}) \mid \mathsf{h} \in \mathsf{X}_{i<k} (r_i \cap G)\} \in I.$$

The only one of (a)–(d) which applies in the case $m = 0$ is (c), and it holds since the c_i's are pairwise disjoint. Let us assume now that for $m \leqq n$ S^m is defined and (a)–(d) hold. We shall now define S^{n+1} and prove that (a)–(d) hold for $m = n + 1$. By (a), the members of S_i^n are subintervals of c_i, for $i < k$. Therefore, if $\mathsf{r} \in \mathsf{X}\, \mathsf{S}^n$ then $\mathsf{X}\, \mathsf{r} \subseteq \mathsf{X}\, \mathsf{c}$ and hence, by (13),

$$\text{(14)} \qquad \mathsf{r} \in \mathsf{X}\, \mathsf{S}^n \wedge \mathsf{h} \in \mathsf{X}\, \mathsf{r} \rightarrow d(\mathsf{h}) \in B \sim I \wedge -d(\mathsf{h}) \in B \sim I.$$

Consider the ideal J generated by $I \cup \{d(\mathsf{h}) \mid \mathsf{h} \in \mathsf{X}\, r\}$. Since r is a k-tuple of absolute intervals $\mathsf{r} \in L_f$ by Corollary 22, Axiom (iii) and Lemma 19. Therefore, by Lemma 21, $J \in L_f$. Since I is a maximal proper ideal in L_f, and since by (14) $J \supset I$ we have $1 \in J$. As a consequence there is a finite subset $G_1''(\mathsf{r})$ of $\mathsf{X}\, \mathsf{r}$ such that $\prod \{-d(\mathsf{h}) \mid \mathsf{h} \in G_1''(\mathsf{r})\} \in I$. By (14) $G_1''(\mathsf{r})$ has at least two members. By considering, in the same way, the ideal generated by $I \cup \{-d(\mathsf{h}) \mid \mathsf{h} \in \mathsf{X}\, r\}$ one can show that there is a finite subset $G_2''(r)$ of $\mathsf{X}\, r$ which has at least two members and such that $\prod \{d(\mathsf{h}) \mid \mathsf{h} \in G_2''(\mathsf{r})\} \in I$. Let $G_j'(\mathsf{r}) = \bigcup \{\mathfrak{R}(\mathsf{h}) \mid \mathsf{h} \in G_j''(\mathsf{r})\}$, $j = 1, 2$. Since $G_j''(\mathsf{r})$ is a finite set of k-sequences of members of b, $G_j'(\mathsf{r})$ is a finite subset of $\bigcup \mathfrak{R}(\mathsf{r}) \subseteq b$ and it has at least two members in common with each r_i, $i < k$. If we put

$$G' = \bigcup \{G_1'(\mathsf{r}) \cup G_2'(\mathsf{r}) \mid \mathsf{r} \in \mathsf{X}\, \mathsf{S}^n\}$$

then G' is a finite subset of b which has at least two members in common with each member of $\bigcup (\mathfrak{R}(\mathsf{S}^n))$. Obviously $G_j''(\mathsf{r}) \subseteq \mathsf{X}_{i<k} (r_i \cap G')$ for $j = 1, 2$; hence

$$\text{(15)} \qquad \mathsf{r} \in \mathsf{X}\, \mathsf{S}^n \rightarrow \prod \{-d(\mathsf{h}) \mid \mathsf{h} \in \mathsf{X}_{i<k} (r_i \cap G')\} \in I \wedge \prod \{d(\mathsf{h}) \mid \mathsf{h} \in \mathsf{X}_{i<k} (r_i \cap G')\} \in I.$$

As a consequence of Corollary 22, Axiom (iii) and Lemma 19 $\mathsf{S}^n \in L_f$. Hence, by Corollary 16, there is a finite sequence e which distinguishes G' and such that

$$\text{(16)} \qquad \mathsf{r} \in \mathsf{X}\, \mathsf{S}^n \wedge |G| = \mathfrak{D}(\mathsf{e}) \wedge (\forall j \in \mathfrak{D}(\mathsf{e}))\, |G \cap e_j| = 1 \rightarrow \prod \{-d(\mathsf{h}) \mid \mathsf{h} \in \mathsf{X}_{i<k} (r_i \cap G)\} \in I \wedge \prod \{d(\mathsf{h}) \mid \mathsf{h} \in \mathsf{X}_{i<k} (r_i \cap G)\} \in I.$$

We can assume that each e_j is a subinterval of some member of $\bigcup \mathfrak{R}(\mathsf{S}^n)$ for the following reason. Given e_j, let x be the single member of $G' \cap e_j$. Since $x \in G'$ x belongs to a unique member s_j of $\bigcup \mathfrak{R}(\mathsf{S}^n)$. Thus $x \in s_j \cap e_j$; hence $s_j \cap e_j$ is an absolute interval. The sequence $\langle s_0 \cap e_0, s_1 \cap e_1, \ldots \rangle$ obviously distinguishes G' and can replace e in (16). Moreover, if $s \in \bigcup \mathfrak{R}(\mathsf{S}^n)$ then, as follows easily from our construction of G', $|G' \cap s| \geqq 2$ and therefore s includes at least two subintervals out of the sequence e.

Let us define now the k-sequence S^{n+1} by $S_i^{n+1} = \{e_j \mid j \in \mathfrak{D}(\mathsf{e}) \wedge e_j \textit{ is a subset of a member of } S_i^n\}$, for $i < k$. What we said just now concerning the sequence e shows that S^{n+1} satisfies requirements (a)–(c) above; by (16) S^{n+1} satisfies also requirement (d).

We did not have to worry about using the axiom of choice during our construction of S^{n+1}, since whenever we made arbitrary choices we made only finitely many of them. However, since we define S^n by induction on n, we have to make sure that S^{n+1} itself is not obtained by an arbitrary choice. As a consequence of Corollary 22, Axiom (iii) and Lemma 19 every k-sequence S' which satisfies requirements (a)–(d) on S^{n+1} is in L_f. Our construction of S^{n+1} can be taken to be a proof of the existence of such an S'. Since, by Lemma 13, the class L_f has a well-ordering definable in terms of f, we can now re-define S^{n+1} as the least set S' in that well-ordering which satisfies requirements (a)–(d) for S^{n+1}.

To apply the Halpern-Läuchli theorem we define the following trees. For $i < k$, $T_i = \bigcup_{n<\omega} S_i^n$ and $\leqq_i$ is the converse of the inclusion relation, i.e., $s \leqq_i t \leftrightarrow t \subseteq s$. It follows easily from (a)–(d) above that the nth level of $\mathfrak{T}_i$ is exactly S_i^n. All the requirements of the Halpern-Läuchli theorem hold in the present case. Let n be a positive integer as in the consequence of that theorem.

Let H be a choice function on $W = \bigcup_{i<k, m \leqq n} S_i^m$, i.e. for each $s \in W$, $H(s) \in s$. Let y be the k-sequence given by $y_i = \{H(s) \mid s \in \bigcup_{m \leqq n} S_i^m\}$, for $i < k$. We shall show that for every $z \subseteq \mathsf{X}\,\mathsf{y}$

$$(17) \qquad \prod \{d(\mathsf{h}) \mid \mathsf{h} \in z\} \in I \quad or \quad \prod \{-d(\mathsf{h}) \mid \mathsf{h} \in \mathsf{X}\,\mathsf{y} \sim z\} \in I.$$

Once (17) is proved we obtain a contradiction by exactly the same proof that yields a contradiction from (11) above, replacing there y by $\mathsf{X}\,\mathsf{y}$ and h by h. Let us now prove (17). We partition the set $\mathsf{X}_{i<k}(T_i \mid n)$ as follows.

$$Q_1 = \{\mathsf{g} \in \mathsf{X}_{i<k}(T_i \mid n) \mid \langle H(g_0), \ldots, H(g_{k-1})\rangle \in z\}, \quad Q_2 = \mathsf{X}_{i<k}(T_i \mid n) \sim Q_1.$$

By our choice of n, Q_1 or Q_2 includes an $(m, 1)$-matrix V for some $m < n$. Suppose $V \subseteq Q_1$, $V = \mathsf{X}_{i<k} A_i$ where A_i is $(m, 1)$-dense in T_i. This means that there is a $t_i \in S_i^m$ such that for every $s \in S_i^{m+1}$ if $s \subseteq t_i$ then there is an $r \in A_i$ such that $r \subseteq s$. Let us define a function q on $\bigcup \mathfrak{R}(\mathsf{S}^{m+1})$ by putting, for $s \in S_i^{m+1}$ such that $s \subseteq t_i$, $q(s) = H(r)$, where r is an absolute interval such that $r \subseteq s$ and $r \in A_i$, and by choosing $q(s)$ to be an arbitrary member of s if $s \in S_i^{m+1}$, $s \nsubseteq t_i$, $i < k$. Thus we have for every $s \in \bigcup \mathfrak{R}(\mathsf{S}^{m+1})$ $q(s) \in s$. Let $G = \{q(s) \mid s \in \bigcup \mathfrak{R}(\mathsf{S}^{m+1})\}$. By requirement (d) for $m + 1$ we get

$$(18) \qquad \prod \{d(\mathsf{h}) \mid \mathsf{h} \in \mathsf{X}_{i<k}(t_i \cap G)\} \in I.$$

We shall now prove that $\mathsf{X}_{i<k}(t_i \cap G) \subseteq z$; once this is proved then (18) directly implies $\prod \{d(\mathsf{h}) \mid \mathsf{h} \in z\} \in I$ and (17) holds (under the assumption $V \subseteq Q_1$). To prove $\mathsf{X}_{i<k}(t_i \cap G) \subseteq z$ let $\mathsf{h} \in \mathsf{X}_{i<k}(t_i \cap G)$, then $h_i \in t_i \cap G$ for $i < k$, and hence, by the definition of G, $h_i = q(s_i)$ for some $s_i \in \bigcup \mathfrak{R}(\mathsf{S}^{m+1})$. Since q is a choice function we have $h_i \in s_i$. Since also $h_i \in t_i \in S_i^m$ it follows from (a) and (c) that $s_i \subseteq t_i$. By definition of $q(s)$ for $s \subseteq t_i$ we have $h_i = q(s_i) = H(r_i)$, where $r_i \in A_i$.Therefore $\mathsf{h} = \langle H(r_0), \ldots, H(r_{k-1})\rangle$, where $\mathsf{r} \in \mathsf{X}_{i<k} A_i \subseteq Q_1$. But by definition of Q_1 since $\mathsf{r} \in Q_1$ we have $\langle H(r_0), \ldots, H(r_{k-1})\rangle \in z$, i.e. $\mathsf{h} \in z$, and therefore $\mathsf{X}_{i<k}(t_i \cap G) \subseteq z$.

To deal with the other case, namely where $V \subseteq Q_2$, let us write $z' = \mathsf{X}\, \mathsf{y} \sim z$ and notice that by the definition of y and Q_2

$$Q_2 = \{\mathsf{g} \in \mathsf{X}_{i<k}(T_i \mid n) \mid \langle H(g_0), \ldots, H(g_{k-1})\rangle \in z'\}.$$

We proceed exactly as in the case of $V \subseteq Q_1$, replacing z with z' and Q_1 with Q_2. Where (d) was applied to obtain (18) we use it now to obtain $\prod \{-d(\mathsf{h}) \mid \mathsf{h} \in \mathsf{X}_{i<k}(t_i \cap G)\} \in I$, and finally $\prod \{-d(\mathsf{h}) \mid \mathsf{h} \in z'\} \in I$ so that (17) holds.

LEMMA 24. *Let f, g and h be assignments such that $\mathfrak{R}(h) = \mathfrak{R}(f) \cap \mathfrak{R}(g)$, then $L_h = L_f \cap L_g$.*

PROOF. By Corollary 5, $L_h \subseteq L_f \cap L_g$. Also by Corollary 5, $L_{f''} = L_{g''}$ if $\mathfrak{R}(f'') = \mathfrak{R}(g'')$, for any assignments f'' and g''. Hence we may assume that f, g and h are finite sequences without repetitions and that there are finite sequences f' and g' such that $f = h^\frown f'$, $g = h^\frown g'$. Now suppose that $x \in L_f \cap L_g$. Then $x = \mathrm{den}(u, h^\frown f') = \mathrm{den}(v, h^\frown g')$ for some $u, v \in TL$. Applying Lemma 14 to the formula $\mathrm{den}(u, h^\frown f') = \mathrm{den}(v, h^\frown g')$ (where we notice u, v, h, $f' \in L_f$ by Axiom (iii) and Lemma 19) we obtain a sequence $e = \langle e_0, \ldots, e_{m-1}\rangle$ of absolute intervals such that

$$\mathrm{den}(u, f) = \mathrm{den}(v, h^\frown z) \quad \textit{for every} \quad z \in \mathsf{X}\, e. \tag{19}$$

The right-hand side of (19) determines $x = \mathrm{den}(u, f)$ uniquely in terms of members of L_h (v, h, $e \in L_h$ by Axiom (iii), and Corollaries 20 and 22 and Lemma 19). Hence, by Lemma 21, $x \in L_h$.

COROLLARY 25. *For every set x there is a unique finite subset c of b such that for every assignment g, $x \in L_g$ if and only if $c \subseteq \mathfrak{R}(g)$. We call this c the* support *of x. We shall also refer to the unique f which is the sequence consisting of the members of c, ordered according to first differences, as the support of x.*

THEOREM 26. *There is a one-one map F' of the universe into $On \mathsf{X} R$, where On is the class of all ordinals and R is the set of all real numbers. This establishes statement* (b) *of* §1.

PROOF. There is a one-one map H of the set of all finite sequences of subsets of ω into R, and, by Axioms (iii) and (iv), there is a one-one function G which maps the class TL of all terms into On. For any set x we define $F'(x) = \langle \alpha, H(f)\rangle$, where f is the support of x and α is the least member of $\{G(u) \mid u \in TL \wedge \operatorname{occ}(u) \subseteq \mathfrak{D}(f) \wedge \operatorname{den}(u, f) = x\}$.

4. **The consistency of** ***SP*****.** Our proof of the consistency of SP uses, essentially, Cohen's method of forcing (Cohen [**1**]). We shall use the Boolean-algebraic version of forcing, which is due to Scott and Solovay [**15**]. In some technical points we make use of the presentation of Easton [**2**]. We give Solovay's proof of the Power-set Axiom.

We shall proceed within the set theory ZF strengthened by the axiom of constructibility $V = L$. We now take up again the language $\mathscr{L}$ with some difference in its interpretation. We shall define a function $\|\phi\|$ on the sentences of $\mathscr{L}$ which will be similar to the function $\operatorname{den}(\phi, f)$ above, except that its values will be the elements of a certain given complete Boolean algebra $\mathfrak{B} = \langle B, \cdot, - \rangle$. (A Boolean algebra $\mathfrak{B}$ is said to be *complete* if every set A of its members has a least upper bound $\sum A$ and a greatest lower bound $\prod A$). In addition to the valuation of $\mathscr{L}$ we shall also interpret the language of SP in such a way that the truth values of its sentences will be members of B rather than just "true" and "false". As is usual with Boolean-algebraic interpretations of set theory, $\mathscr{L}$ and the language of SP are understood to talk about a certain universe of imaginary sets which properly includes the universe of all "real" sets. Accordingly, we take now for S, in the definition of $\mathscr{L}$, the class V of all sets, so that every "real" set x will belong to the imaginary universe, being named there by $\mathbf{x}$. For F we take a class which is a mapping of On on V; such an F exists by the axiom of constructibility. In the imaginary universe the constants a_i are taken to have a "fixed" denotation, and b is taken to consist exactly of the sets denoted by the a_i's. We do not need to define $\|u\|$ for the terms u (even though there is a natural definition for it—see Scott and Solovay [**15**]).

DEFINITION 27. Let $\{\mathfrak{a}_{i,n} \mid i, n < \omega\}$ be a fixed double sequence of elements of the complete Boolean algebra $\mathfrak{B}$. We define a function $\|\phi\|$ on the sentences of $\mathscr{L}$ into B by the following inductive definition.

$$\|u \in a_i\| = \sum_{n<\omega} \|u = \mathbf{n}\| \mathfrak{a}_{i,n},$$

$$\|u \in b\| = \sum_{i<\omega} \|u = a_i\|,$$

$$\|u \in \mathbf{s}\| = \sum_{t \in s} \|u = \mathbf{t}\|,$$

$$\|u \in \text{Ӽ}_\alpha x \phi(x)\| = \sum_{\lambda(v)<\alpha} \|u = v\| \, \|\phi(v)\|,$$

$$\|u = v\| = \|\forall_\gamma x (x \in u \leftrightarrow x \in v)\|, \quad \text{where } \gamma = \max(\lambda(u), \lambda(v)),$$

$$\|\neg \phi\| = -\|\phi\|,$$

$$\|\phi \vee \psi\| = \|\phi\| + \|\psi\|,$$

$$\|\exists_\alpha x \phi(x)\| = \sum_{\lambda(v)<\alpha} \|\phi(v)\|.$$

We have now, obviously, $\|\phi \wedge \psi\| = \|\phi\| \cdot \|\psi\|$, $\|\forall_\alpha x\phi(x)\| = \prod_{\lambda(v)<\alpha} \|\phi(v)\|$ and

$$\|u = v\| = \prod_{\lambda(w)<\gamma} \|w \in u\| \cdot \|w \in v\| + (-\|w \in u\|) \cdot (-\|w \in v\|),$$

where $\gamma = \max(\lambda(u), \lambda(v))$.

To justify this definition we define two functions λ_1 and λ_2 on the sentences of $\mathscr{L}$ as follows. If $\lambda(\phi) = \alpha + 1$ and ϕ contains a subformula of the form $u \in v$ where $\lambda(u) = \alpha$ then $\lambda_1(\phi) = 2$; if $\lambda(\phi) = \alpha + 1$, ϕ does not contain a subformula $u \in v$ with $\lambda(u) = \alpha$, but ϕ contains a subformula $u = v$ where $\lambda(u) = \alpha$ or $\lambda(v) = \alpha$ then $\lambda_1(\phi) = 1$; in every other case $\lambda_1(\phi) = 0$. $\lambda_2(\phi)$ is the number of the occurrences of the symbols $\neg$, $\vee$, $\exists_\alpha$ in ϕ which are not inside an abstraction term in ϕ. The definition by induction is justified once the following two facts are noticed. (a) The class of all sentences ϕ of $\mathscr{L}$ with $\lambda(\phi) \leqq \alpha$ is a set. (b) $\|\phi\|$ is defined above in terms of the values $\|\psi\|$, where ψ belongs to a set S_ϕ of sentences such that for every $\psi \in S_\phi$ we have $\lambda(\psi) \leqq \lambda(\phi)$, and if $\lambda(\psi) = \lambda(\phi)$ then either $\lambda_1(\psi) < \lambda_1(\phi)$ or else $\lambda_1(\psi) = \lambda_1(\phi)$ and $\lambda_2(\psi) < \lambda_2(\phi)$. I.e. $\|\phi\|$ is defined by induction on the left lexicographic ordering of the triples $\langle\lambda(\phi), \lambda_1(\phi), \lambda_2(\phi)\rangle$.

DEFINITION 28. With each formula Φ of the language of *SP* (the *SP*-language) with the free variables $x_{i_1}, \ldots, x_{i_n}$ we associate a term τ_Φ of set theory with free variables $u_{i_1}, \ldots, u_{i_n}$ (it has also the free variables $\mathfrak{B}$ and $\mathfrak{a}$, but we shall suppress those) as follows.

$$\tau_{x_i \in x_j}(u_i, u_j) \quad \text{is} \quad \|u_i \in u_j\|,$$

$$\tau_{x_i = x_j}(u_i, u_j) \quad \text{is} \quad \|u_i = u_j\|,$$

$$\tau_{S(x_i)}(u_i) \quad \text{is} \quad \sum_{s \in S} \|u_i = \mathbf{s}\|,$$

$$\tau_{F(x_i)}(u_i) \quad \text{is} \quad \sum_{s \in F} \|u_i = \mathbf{s}\|,$$

τ_Φ, where Φ is an atomic formula containing b is defined in the obvious way, e.g.

$$\tau_{b \in x_i}(u_i) \quad \text{is} \quad \|b \in u_i\|,$$

$$\tau_{\neg\Phi}(u_{i_1}, \ldots, u_{i_n}) \quad \text{is} \quad -\tau_\Phi(u_{i_1}, \ldots, u_{i_n}),$$

$$\tau_{\Phi\vee\Psi}(u_{i_1}, \ldots, u_{i_n}) \quad \text{is} \quad \tau_\Phi(u_{i_1}, \ldots, u_{i_n}) + \tau_\Psi(u_{i_1}, \ldots, u_{i_n}),$$

$$\tau_{\exists x_j \Phi(x_j, x_{i_1}, \ldots, x_{i_n})}(u_{i_1}, \ldots, u_{i_n}) \quad \text{is} \quad \sum_{u_j \in TL} \tau_\Phi(u_j, u_{i_1}, \ldots, u_{i_n}).$$

If Φ is a sentence then τ_Φ is a fixed member of B.

Since the notation $\tau_\Phi(u_{i_1}, \ldots, u_{i_n})$ is cumbersome we shall write, from now on, $\|\Phi(u_{i_1}, \ldots, u_{i_n})\|$ for $\tau_\Phi(u_{i_1}, \ldots, u_{i_n})$. Thus we shall write $\|u \in v \wedge v = u\|$ for $\tau_{x \in y \wedge y = z}(u, v, u)$. Even though this notation introduces some ambiguity, since $\|u \in v \wedge v = u\|$ can also stand for the operation $\|\ \|$ applied to the sentence $u \in v \wedge v = u$ of $\mathscr{L}$, and it can also stand for $\tau_{x \in y \wedge y = x}(u, v)$ this lead to no serious difficulty since, as easily seen, the value of $\|u \in v \wedge v = u\|$ is the same for all of the three interpretations mentioned. We could prove here several lemmas which will guarantee that the value of $\|\Phi(v_1, \ldots, v_n)\|$ is independent of the way we read it, but we prefer not to do it. Therefore the reader is advised to think of one particular interpretation when he encounters an expression of the form $\|\Phi(v_1, \ldots, v_n)\|$, and when such expressions are manipulated he should bear in mind the simple facts about the $\|\Phi\|$'s and the τ_Φ's. If Φ is a formula of *SP* with the free variables $x_{i_1}, \ldots, x_{i_n}$ we shall sometimes write, as an additional abbreviation, $\|\Phi\|$ for $\|\Phi(u_{i_1}, \ldots, u_{i_n})\|$.

LEMMA 29 (SCHEMA). (a) *If Φ is a logically valid formula of the first-order predicate calculus without equality then* $\|\Phi\| = 1$.

(*b*) *If Φ implies Ψ in the first-order predicate calculus without equality then* $\|\Phi\| \leqq \|\Psi\|$.

PROOF. To get (a) take a convenient axiom system for the first-order predicate calculus without equality. Show that the property $\|\Phi\| = 1$ holds for all axioms Φ, and is also preserved under the rules of inference. (b) follows from (a) by means of the deduction theorem.

LEMMA 30. (a) $\|u = u\| = \|\forall x(x = x)\| = 1$. (b) $\|\forall x \forall y(x = y \leftrightarrow y = x)\| = 1$.

PROOF. By direct computation.

LEMMA 31. $t \in s \rightarrow \|\mathbf{t} \in \mathbf{s}\| = 1$.

PROOF. Follows from Lemma 30 (a) by direct computation.

LEMMA 32. (a) $\|\mathbf{t} \in \mathbf{s}\| \neq 0 \rightarrow t \in s$. (b) $\|\mathbf{t} = \mathbf{s}\| \neq 0 \rightarrow t = s$.

PROOF. One proves, by induction on $\rho(s)$

$$
\begin{aligned}
&\forall t[\rho(t) < \rho(s) \wedge \|\mathbf{t} \in \mathbf{s}\| \neq 0 \rightarrow t \in s] \\
(20) \qquad &\wedge \forall t[\rho(t) \leqq \rho(s) \wedge \|\mathbf{t} = \mathbf{s}\| \neq 0 \rightarrow t = s] \\
&\wedge \forall t[\rho(t) \leqq \rho(s) \wedge \|\mathbf{s} \in \mathbf{t}\| \neq 0 \rightarrow s \in t],
\end{aligned}
$$

where, in the induction step, the three parts of (20) are proved in the order in which they are written, using earlier parts in the proof of later ones.

LEMMA 33. (a) $\|w \in u\| \; \|u = v\| \leqq \|w \in u\|$ *if* $\lambda(w) < \max(\lambda(u), \lambda(v))$.

(b) $\|u = v\| \; \|\forall x(x \in v \leftrightarrow x \in w)\| \leqq \|u = w\|$ *if* $\lambda(w) \leqq \max(\lambda(u), \lambda(v))$.

(*The restrictions on $\lambda(w)$ are temporary, they will be removed later on.*)

PROOF. By straightforward computation, using the definitions of $\|u = v\|$ and $\|\forall x(x \in v \leftrightarrow x \in w)\|$.

LEMMA. 34. (a) $\|\forall x \forall y \forall z(x = y \wedge z \in x \rightarrow z \in y)\| = 1$.

(b) $\|\forall x \forall y(x = y \leftrightarrow \forall z(z \in x \leftrightarrow z \in y))\| = 1$.

PROOF. We shall prove

$$(21) \qquad \|u = v\| \; \|w \in u\| \leqq \|w \in v\|$$

and

$$(22) \qquad \|u = v\| \leqq \|\forall z(z \in u \leftrightarrow z \in v)\|$$

jointly by induction on $\gamma = \max(\lambda(u), \lambda(v))$. (21) directly implies (a). To get (b) notice that the inequality $\geqq$ in (22) holds trivially, and hence (22) is indeed an equality. Let us carry out now the induction step in the proof of (21) and (22). If for given u and v and for all w we have (21), and also (21) with u and v interchanged, we get (22) when we prove

$$\|u = v\| \; \|\forall z(z \in u \leftrightarrow z \in v)\| \geqq \|u = v\|$$

by direct computation. Therefore we have only to establish (21). To prove (21), we see that if $\lambda(w) < \max(\lambda(u), \lambda(v))$ this is Lemma 33 (a); hence we can assume $\lambda(w) \geqq \max(\lambda(u), \lambda(v)) = \gamma$.

Case A. u is $\mathbf{s}$.

Case AA. v is $\mathbf{r}$. If $s \neq r$ then by Lemma 32 (b) $\|\mathbf{s} = \mathbf{r}\| = 0$ and (21) holds trivially; if $s = r$ then (21) is outright trivial.

Case AB. v is $\daleth_\alpha x\phi(x)$, then $\alpha \leqq \gamma$. If $t \in s$ then $\|\mathbf{s} = v\| \leqq \|\mathbf{t} \in v\|$ by Lemma 33(a), since $\|\mathbf{t} \in \mathbf{s}\| = 1$ by Lemma 31. Therefore,

$$\begin{aligned}\|\mathbf{s} = v\| \leqq \|\mathbf{t} \in v\| &= \sum_{\lambda(m)<\alpha} \|\mathbf{t} = m\| \; \|\phi(m)\| \\ &\leqq \sum_{\lambda(m)<\alpha} \|\forall x(x \in \mathbf{t} \leftrightarrow x \in m)\| \; \|\phi(m)\| \qquad \text{(by the induction hypothesis).}\end{aligned}$$

Also

$$\begin{aligned}\|\mathbf{s} = v\| \; \|w \in \mathbf{s}\| &= \sum_{t \in s} \|w = \mathbf{t}\| \; \|\mathbf{s} = v\| \\ &\leqq \sum_{t \in s} \sum_{\lambda(m)<\alpha} \|w = \mathbf{t}\| \; \|\forall x(x \in \mathbf{t} \leftrightarrow x \in m)\| \; \|\phi(m)\| \\ &\leqq \sum_{t \in s} \sum_{\lambda(m)<\alpha} \|w = m\| \; \|\phi(m)\|\end{aligned}$$

by Lemma 33(b) since

$$\lambda(m) < \alpha \leqq \gamma \leqq \lambda(w) \leqq \sum_{\lambda(m)<\alpha} \|w = m\| \; \|\phi(m)\| = \|w \in u\|.$$

Case B. u is $\daleth_\beta x\psi(x)$. Then if $\lambda(m) < \beta$ we have

$$\psi(m) \leqq \sum_{\lambda(m)<\beta} \|m = n\| \; \|\psi(n)\| = \|m \in u\|;$$

hence

$$\begin{aligned}\|w \in u\| \; \|u = v\| &= \sum_{\lambda(m)<\beta} \|u = v\| \; \|w = m\| \; \|\psi(m)\| \\ &\leqq \sum_{\lambda(m)<\beta} \|u = v\| \; \|w = m\| \; \|m \in u\|, \quad \text{by the inequality above} \\ &\leqq \sum_{\lambda(m)<\beta} \|m \in v\| \; \|w = m\|, \quad \text{by Lemma 33(a) since } \lambda(m) < \beta \leqq \gamma.\end{aligned}$$

Case BA. v is $\mathbf{r}$. Then

$$\|m \in v\| = \sum_{t \in r} \|m = \mathbf{t}\| = \sum_{t \in r} \|\forall x(x \in m \leftrightarrow x \in \mathbf{t})\|, \quad \text{by the induction hypothesis.}$$

$$\begin{aligned}
\|w \in u\| \; \|u = v\| &\leqq \sum_{\lambda(m) < \beta} \|m \in v\| \; \|w = m\|, \quad \text{shown under “Case B”,} \\
&\leqq \sum_{\lambda(m) < \beta} \sum_{t \in r} \|\forall x(x \in m \leftrightarrow x \in t)\| \; \|w = m\|, \\
&\qquad\qquad \text{by the equality above,} \\
&\leqq \sum_{\lambda(m) < \beta} \sum_{t \in r} \|\mathbf{t} = w\|, \\
&\qquad\qquad \text{by Lemma 33(b) since } \lambda(t) < \lambda(v) \leqq \lambda(w) \\
&\leqq \sum_{t \in r} \|t = w\| = \|w \in \mathbf{r}\|.
\end{aligned}$$

Case BB. v is $\gimel_\alpha x\sigma(x)$. If $\lambda(m) < \beta$ then

$$\|m \in v\| = \sum_{\lambda(n) < \alpha} \|m = n\| \; \|\sigma(n)\| = \sum_{\lambda(n) < \alpha} \|\forall x(x \in m \leftrightarrow x \in n)\| \; \|\sigma(n)\|,$$

by the induction hypothesis.

$$\begin{aligned}
\|w \in u\| \; \|u = v\| &\leqq \sum_{\lambda(m) < \beta} \|m \in v\| \; \|w = m\|, \quad \text{shown under “Case B”,} \\
&\leqq \sum_{\lambda(m) < \beta} \sum_{\lambda(n) < \alpha} \|w = n\| \; \|\forall x(x \in m \leftrightarrow x \in n)\| \; \|\sigma(n)\|, \\
&\qquad\qquad \text{by the equality above} \\
&\leqq \sum_{\lambda(m) < \beta} \sum_{\lambda(n) < \alpha} \|w = n\| \; \|\sigma(n)\|, \\
&\qquad\qquad \text{by Lemma 33(b) since } \lambda(n) < \alpha \leqq \gamma \leqq \lambda(w) \\
&\leqq \sum_{\lambda(n) < \alpha} \|w = n\| \; \|\sigma(n)\| = \|w \in v\|.
\end{aligned}$$

LEMMA 35. (a) $\|\forall x \forall y \forall z(x = y \wedge y = z \rightarrow x = z)\| = 1$.
(b) $\|u = v\| \; \|v = w\| \leqq \|u = w\|$.
(c) $\|\forall x \forall y \forall z(x = y \wedge x \in z \rightarrow y \in z)\| = 1$.
(d) $\|\forall x \forall y(x = y \wedge S(x) \rightarrow S(y))\| = 1$.
(e) $\|\forall x \forall y(x = y \wedge F(x) \rightarrow F(y))\| = 1$.

PROOF. (a) The sentence inside $\| \; \|$ in (a) is a logical consequence of the sentence $\forall x \forall y(x = y \leftrightarrow \forall z(z \in x \leftrightarrow z \in y))$; hence (a) follows from Lemma 34(b) by Lemma 29(b). (b) follows directly from (a). (c)–(e) follow from (b) by the definitions of $\|u \in v\|$, $\|S(u)\|$ and $\|F(u)\|$.

THEOREM 36 (SCHEMA). *If Φ is a logically valid formula of the first-order predicate calculus with equality then $\|\Phi\| = 1$. If Φ logically implies Ψ in the first-order predicate calculus with equality then $\|\Phi\| \leqq \|\Psi\|$.*

PROOF. By Lemmas 29, 34, and 35.

Suppose that for every axiom Φ of *SP* we prove in $ZF + V = L$ that $\|\Phi\| = 1$; then, by Theorem 36, we can prove also in $ZF + V = L$ that $\|\Phi\| = 1$ for every theorem Φ of SP. If SP is inconsistent, then for every sentence Φ, both Φ and $\neg\Phi$ are theorems of SP and hence we can prove in $ZF + V = L$ that $\|\Phi\| = 1$, and also that $\|\neg\Phi\| = 1$, i.e. that also $\|\Phi\| = 0$, which is a contradiction in $ZF + V = L$. By the results of Gödel [3], we can get from the latter contradiction a contradiction in ZF. Therefore if SP is inconsistent so is ZF; i.e. if ZF is consistent so is SP. The rest of this paper consists of showing that for every axiom Φ of SP $\|\Phi\| = 1$.

As we have already mentioned, we can consider our valuation $\|\Phi\|$ as standing for truth in some imaginary universe $\mathcal{N}$. If $\|\Phi\| = 1$ we say that Φ holds, or is true, in $\mathcal{N}$; if $\|\Phi\| = 0$ we say that Φ fails, or is false, in $\mathcal{N}$. It is best to think about $\mathcal{N}$ as an ordinary two-truth-valued universe since, e.g. by Theorem 36 $\Phi \vee \neg\Phi$ is true in $\mathcal{N}$. Our excuse for using Boolean truth values can be that the Boolean truth values just indicate that $\mathcal{N}$ is not uniquely determined but is an arbitrary member of a "class" of possible universes; hence with the information at hand we cannot always say whether a given sentence Φ is true or false in $\mathcal{N}$. The language we use in talking about $\mathcal{N}$ is the language of SP; thus the universe $\mathcal{N}$ has a membership relation $\in$, two unary predicates S and P and a particular member b. We said that we shall consider $\mathcal{N}$ as an ordinary two-truth-valued universe. This is possible because of Theorem 36 which asserts that the first-order predicate calculus with equality "holds" in $\mathcal{N}$. What we have to do now is to prove that all the axioms of SP hold in $\mathcal{N}$.

When we talk about $\mathcal{N}$ we have, actually, at our disposal more than just the language of SP, since we can also regard all the terms of TL as individual constants denoting sets in $\mathcal{N}$. Thus a general statement about $\mathcal{N}$ will be of the form $\Phi(v_1, \ldots, v_n)$, where $v_1, \ldots, v_n \in TL$. We are justified in regarding the terms as individual constants if they behave like such in the semantics of $\mathcal{N}$. To see that this is indeed the case suppose that $\Psi(v_1, \ldots, v_n)$ is a logical consequence of $\Phi(v_1, \ldots, v_n)$ with $v_1, \ldots, v_n$ regarded as constants. Then, by standard arguments of logic, the formula $\Phi(x_1, \ldots, x_n)$ logically implies $\Psi(x_1, \ldots, x_n)$. Hence, by Theorem 36, $\|\Phi(v_1, \ldots, v_n)\| \leqq \|\Psi(v_1, \ldots, v_n)\|$. Therefore if $\Phi(v_1, \ldots, v_n)$ holds in $\mathcal{N}$ then also $\Psi(v_1, \ldots, v_n)$ holds in $\mathcal{N}$.

Our first step will now be to prove that all the axioms of ZF, including the strengthened versions of the axiom schemas of subsets and replacement, hold in $\mathcal{N}$. Once we have done that, if we consider a sentence Ψ of ZF which contains also defined notions of ZF then $\|\Psi\|$ is well defined even if we do not specify a particular procedure for the elimination of the defined notions (e.g. this is the case if Ψ uses the notion of the real numbers even though we have not specified which definition of the real numbers we take up). Suppose we have such a sentence Ψ, and two different procedures for eliminating the defined notions yield the respective sentences Φ_1 and Φ_2 of the primitive language of ZF; then $\Phi_1 \leftrightarrow \Phi_2$ is a theorem of ZF and hence $\|\Phi_1\| = \|\Phi_2\|$. Thus $\|\Psi\|$ can be defined as $\|\Phi\|$, where Φ is any sentence of the primitive language of ZF equivalent to Ψ in ZF. Ψ may, therefore, also contain English expressions.

The *Axiom of Extensionality* holds in $\mathcal{N}$, by Lemma 34(b).

LEMMA 37. $\|u \in v\| = \sum_{\lambda(w) < \lambda(v)} \|u = w\| \; \|w \in v\| \leqq \sum_{\lambda(w) < \lambda(v)} \|u = w\|$.

PROOF. The inequality is obvious. As to the equality, one has to prove both the $\leqq$ and $\geqq$ inequalities. The $\geqq$ inequality follows from $\|u \in v\| \geqq \|u = w\| \, \|w \in v\|$ which is a consequence of Lemma 35(c). The $\leqq$ inequality follows easily from the definition of $\|u \in v\|$.

THEOREM 38. *The Axiom of Union* $\forall x \exists y \forall z \forall t (t \in z \wedge z \in x \rightarrow t \in y)$ *holds in* $\mathcal{N}$.

PROOF. We have to show that for every $u \in TL$, $\|\exists y \forall z \forall t (t \in z \wedge z \in x \rightarrow t \in y)\| = 1$. Let $\alpha \geqq \lambda(u)$ and let $v = \rightthreetimes_\alpha x(x = x)$, then it is enough to show that for all $w, m \in TL$, $\|w \in m\| \, \|m \in u\| \leqq \|w \in v\|$. By Lemma 37

$$\|m \in u\| \leqq \sum_{\lambda(m') < \alpha} \|m = m'\|;$$

hence, by Lemmas 34(a) and 37

$$\|w \in m\| \, \|m \in u\| \leqq \sum_{\lambda(m') < \alpha} \|w \in m\| \, \|m = m'\| \leqq \sum_{\lambda(m') < \alpha} \|w \in m'\|$$

$$\leqq \sum_{\lambda(m') < \alpha} \; \sum_{\lambda(w') < \lambda(m')} \|w = w'\| \leqq \sum_{\lambda(w') < \alpha} \|w = w'\|$$

$$= \sum_{\lambda(w') < \alpha} \|w = w'\| \, \|w' = w'\| = \|w \in v\|.$$

THEOREM 39. *The Axiom of Power-set* $\forall x \exists y \forall z (z \subseteq x \rightarrow z \in y)$, *where* $z \subseteq x$ *is* $\forall t \, (t \in z \rightarrow t \in x)$, *holds in* $\mathcal{N}$.

PROOF. We shall prove that $\|\exists y \forall z (z \subseteq u \rightarrow z \in y)\| = 1$ for every $u \in TL$. Let $\lambda(u) = \alpha$. For any two terms w and w' we define $w \equiv w'$ if for all terms v if $\lambda(v) < \alpha$ then $\|v \in w\| = \|v \in w'\|$ and if also $\|w \subseteq u\| = \|w' \subseteq u\|$. We claim that if $w \equiv w'$ then $\|w \subseteq u\| \, \|w' \subseteq u\| \leqq \|w = w'\|$. This is shown as follows. If $w \equiv w'$ then, as follows from Lemma 37 by direct computation, $\|\forall x \, (x \in u \rightarrow (x \in w \leftrightarrow x \in w'))\| = 1$. Since the formula

$$\forall x \, (x \in u \rightarrow (x \in w \leftrightarrow x \in w')) \wedge w \subseteq u \wedge w' \subseteq u \rightarrow w = w'$$

is a logical consequence of the axiom of extensionality we get, by Theorem 36, that

$$\|w \subseteq u \wedge w' \subseteq u \rightarrow w = w'\| = 1 \text{ and hence } \|w \subseteq u\| \, \|w' \subseteq u\| \leqq \|w = w'\|.$$

Now let $a = \{v \in TL \mid \lambda(v) < \alpha\} \cup \{u\}$ and let c be the set of all functions on a into the Boolean algebra $\mathfrak{B}$. For $f \in c$ let $H(f)$ be the least ordinal β such that there is a $w \in TL$ for which $\lambda(w) = \beta$, $\|w \subseteq u\| = f(u)$ and for all $v \in a \sim \{u\}$, $\|v \in w\| = f(v)$, and let $H(f) = 0$ if there is no such w. Let γ be an ordinal which includes $H[a]$ and let t be $\rightthreetimes_\gamma x \, (x = x)$. Our proof is finished once we show that for every term, w, $\|w \subseteq u\| \leqq \|w \in t\|$. Given $w \in TL$, by definition of γ there is a term w' such that $\lambda(w') < \gamma$, $w' \equiv w$ and $\|w' \subseteq u\| = \|w \subseteq u\|$. We have $\|w \subseteq u\| = \|w \subseteq u\| \, \|w' \subseteq u\|$ since $\|w' \subseteq u\| = \|w \subseteq u\| \leqq \|w = w'\|$, which follows from $w' \equiv w$ by what we have shown before. But since $\lambda(w') < \alpha$ we have $\|w' \in t\| = 1$ by definition of t, hence $\|w \subseteq u\| \leqq \|w = w'\| \, \|w' \in t\| \leqq \|w \in t\|$.

LEMMA 40. (a) *Let $\Gamma(x)$ be a formula of SP, and let T be a subclass of TL and let c be a function on T into B such that for all terms u*

$$\|\Gamma(u)\| = \sum_{v \in T} \|u = v\| \, c(v).$$

Then

$$\|\exists x(\Gamma(x) \wedge \Phi(x))\| = \sum_{v \in T} c(v) \, \|\Phi(v)\|,$$

$$\|\forall x(\Gamma(x) \rightarrow \Phi(x))\| = \prod_{v \in T} (-c(v) + \|\Phi(v)\|).$$

(b) *Let $\Gamma(x)$ be a formula and let T be a subclass of TL such that for all terms u* $\|\Gamma(u)\| = \sum_{v \in T} \|u = v\|$. *Then*

$$\|\exists x(\Gamma(x) \wedge \Phi(x))\| = \sum_{v \in T} \|\Phi(v)\|, \quad \|\forall x(\Gamma(x) \rightarrow \Phi(x))\| = \prod_{v \in T} \|\Phi(v)\|.$$

(c) $\|\exists x(x \in S \wedge \Phi(x))\| = \sum_{s} \|\Phi(\mathbf{s})\|, \; \|\forall x \, (x \in S \rightarrow \Phi(x))\| = \prod_{s} \|\Phi(\mathbf{s})\|.$

(d) $\|\exists x(x \in \mathbf{a} \wedge \Phi(x))\| = \sum_{s \in a} \|\Phi(\mathbf{s})\|, \; \|\forall x \, (x \in \mathbf{a} \rightarrow \Phi(x))\| = \prod_{s \in a} \|\Phi(\mathbf{s})\|.$

PROOF. (a)

$$\|\exists x(\Gamma(x) \wedge \Phi(x))\| = \sum_{u \in TL} \|\Gamma(u)\| \, \|\Phi(u)\| = \sum_{u \in TL} \sum_{v \in T} \|u = v\| \, c(v) \, \|\Phi(u)\|.$$

Since $u = v \wedge \Phi(u)$ is logically equivalent to $u = v \wedge \Phi(v)$ we have $\|u = v\| \, \|\Phi(u)\| = \|u = v\| \, \|\Phi(v)\|$ and therefore

$$\begin{aligned}\|\exists x(\Gamma(x) \wedge \Phi(x))\| &= \sum_{u \in TL} \sum_{v \in T} \|u = v\| \, c(v) \, \|\Phi(v)\| \\ &= \sum_{v \in T} \left(\sum_{u \in TL} \|u = v\| \right) c(v)\Phi(v) \\ &= \sum_{v \in T} c(v)\Phi(v),\end{aligned}$$

since

$$\sum_{u \in TL} \|u = v\| \geqq \|v = v\| = 1.$$

$$\begin{aligned}\|\forall x(\Gamma(x) \rightarrow \Phi(x))\| &= \|\neg \exists x(\Gamma(x) \wedge \neg \Phi(x))\| \\ &= -\sum_{v \in T} c(v)(-\|\Phi(v)\|) \\ &= \prod_{v \in T} (-c(v) + \|\Phi(v)\|).\end{aligned}$$

(b) follows from (a) by taking $c(v) = 1$ for every $v \in T$.

(c) and (d) are immediate consequences of (b).

LEMMA 41. (a) *If* $\|u = \mathbf{s}\| \neq 0$ *then* $\lambda(u) \geqq \rho(s)$,

(b) *If* $\|\mathbf{s} \in u\| \neq 0$ *then* $\lambda(u) > \rho(s)$.

PROOF. By induction on $\lambda(u)$. By Lemma 37,

$$\|\mathbf{s} \in u\| = \sum_{\lambda(v) < \lambda(u)} \|\mathbf{s} = v\| \, \|v \in u\|.$$

Hence, if $\|\mathbf{s} \in u\| \neq 0$ we have $\|\mathbf{s} = v\| \neq 0$ for some v with $\lambda(v) < \lambda(u)$. By the induction hypothesis $\lambda(v) \geqq \rho(s)$ hence $\lambda(u) > \rho(s)$.

If $\|u = \mathbf{s}\| \neq 0$ then, by Theorem 36 and Lemma 34(b), $\|\forall x\,(x \in \mathbf{s} \to x \in u)\| \neq 0$, and hence, by Lemma 40(d), $\prod_{t \in s} \|\mathbf{t} \in u\| \neq 0$. Therefore, by the induction hypothesis, $\lambda(u) > \rho(t)$ for every $t \in s$, hence $\lambda(u) \geqq \rho(s)$.

LEMMA 42. *Let $\Phi(y_1, \ldots, y_n)$ be a formula of SP. For every ordinal α there is an ordinal $\delta \geqq \alpha$ such that for all terms $v_1, \ldots, v_n$ with $\lambda(v_i) < \delta$ for $1 \leqq i \leqq n$ we have*

$$\|\Phi(v_1, \ldots, v_n)\| = \|\phi_{\Phi,\delta}(v_1, \ldots, v_n)\|$$

(*where $\phi_{\Phi,\delta}$ is as in Definition* 17).

PROOF. We define a sequence α_n of ordinals as follows. $\alpha_0 = \alpha$. Let Δ be the set of all formulas $\Psi(z, z_1, \ldots, z_l)$ such that $\exists z \Psi(z, z_1, \ldots, z_l)$ is a subformula of Φ. We define a function $*$ on the ordinals as follows. For $\Psi \in \Delta$ let $F_\Psi(w_1, \ldots, w_l, \mathfrak{a})$, where $w_1, \ldots, w_l \in TL$ and $\mathfrak{a} \in B$, be the least ordinal γ such that there is a term u for which $\lambda(u) = \gamma$ and $\|\Psi(u, w_1, \ldots, w_l)\| = \mathfrak{a}$, and $F_\Psi(w_1, \ldots, w_l, \mathfrak{a}) = 0$ if there is no such u. We define

$$\beta^* = \sup\{F_\Psi(w_1, \ldots, w_l, \mathfrak{a}) + 1 \mid \Psi \in \Delta, w_1, \ldots, w_l \in TL \wedge \lambda(u_1), \ldots, \lambda(u_l) < \beta \wedge \mathfrak{a} \in B\}.$$

Let $\alpha_{n+1} = \max(\alpha, \alpha_n^*)$ for all $n \in \omega$, and $\delta = \sup_{n<\omega} \alpha_n$. We claim that

$$\|\Gamma(w_1, \ldots, w_l)\| = \|\phi_{\Gamma,\delta}(w_1, \ldots, w_l)\| \tag{23}$$

for all $w_1, \ldots, w_l \in TL$, for every subformula Γ of Φ. In particular, this holds when Γ is Φ, which proves the lemma.

(23) is proved by induction on the length of Γ. If Γ is $z \in y$ or $z = y$ then (23) is obvious. If Γ is $S(z)$ then we have to prove $\lambda(v) < \delta \to \|S(v)\| = \|v \in \mathbf{s}_\delta\|$. Using Lemma 41 we get $S(v) = \sum_s \|v = \mathbf{s}\| = \sum_{\rho(s)<\delta} \|v = \mathbf{s}\| = \|v \in \mathbf{s}_\delta\|$. If Γ is $F(z)$ the proof is similar. If Γ is a negation or a disjunction then (23) follows easily from the induction hypothesis. If Γ is $\exists z \Psi(z, z_1, \ldots, z_l)$ then, by the induction hypothesis, we have for all $u, w_1, \ldots, w_l \in TL$ such that $\lambda(u), \lambda(w_1), \ldots, \lambda(w_l) < \delta$,

$$\|\Psi(u, w_1, \ldots, w_l)\| = \|\phi_{\Psi,\delta}(u, w_1, \ldots, w_l)\|.$$

Hence, if $\lambda(w_1), \ldots, \lambda(w_l) < \delta$ then

$$\begin{aligned}\|\exists z \Psi(z, w_1, \ldots, w_l)\| &= \sum_{u \in TL} \|\Psi(u, w_1, \ldots, w_l)\| \\ &\geqq \sum_{\lambda(u)<\delta} \|\Psi(u, w_1, \ldots, w_l)\| \\ &\geqq \sum_{\lambda(u)<\delta} \|\phi_{\Psi,\delta}(u, w_1, \ldots, w_l)\| \\ &\geqq \|\exists_\delta z \phi_{\Psi,\delta}(z, w_1, \ldots, w_l)\|.\end{aligned}$$

This establishes that the left-hand side of (23) is $\geqq$ the right-hand side of (23). To prove the converse inequality, let u be any term and let $\mathfrak{c} = \Psi(u, w_1, \ldots, w_l)$. By definition of α_n and δ we have, for some $n < \omega$, $\lambda(w_1), \ldots, \lambda(w_l) < \alpha_n$. Therefore there is a term u' such that $\lambda(u') < \alpha_n^* \leqq \alpha_{n+1} \leqq \delta$ and $\|\Psi(u', w_1, \ldots, w_l)\| = \mathfrak{c}$. Hence

$$\begin{aligned}\|\Psi(u, w_1, \ldots, w_l)\| = \|\Psi(u', w_1, \ldots, w_l)\| &= \|\phi_{\Psi,\delta}(u', w_1, \ldots, w_l)\| \\ &\leqq \|\exists_\delta z \phi_{\Psi,\delta}(z, w_1, \ldots, w_l)\|.\end{aligned}$$

Therefore

$$\|\exists z \Psi(z, w_1, \ldots, w_l)\| = \sum_{u \in TL} \|\Psi(u, w_1, \ldots, w_l)\| \leqq \|\exists_\delta z \phi_{\Psi,\delta}(z, w_1, \ldots, w_l)\|$$

which means that the left-hand side of (23) is $\leqq$ the right-hand side of (23).

THEOREM 43. *Every instance* $\forall x \exists y \forall z(z \in y \leftrightarrow z \in x \wedge \Psi(z))$ *of the Axiom Schema of Subsets (where* $\Psi(z)$ *is any formula of the language of SP in which* y *is not free) holds in* $\mathcal{N}$.

PROOF. Let $y_1, \ldots, y_n$ be the parameters of $\Psi(z)$. Given $v_1, \ldots, v_n, u \in TL$ let $\alpha > \max(\lambda(v_1), \ldots, \lambda(v_n), \lambda(u))$ then, by Lemma 42, there is an ordinal $\delta \geqq \alpha$ such that for all $w \in TL$ if $\lambda(w) < \delta$ then

$$\|\Psi(w, v_1, \ldots, v_n)\| = \|\phi_{\Psi,\delta}(w, v_1, \ldots, v_n)\|. \tag{24}$$

Let v be the term $\daleth_\delta z\,(z \in u \wedge \phi_{\Psi,\delta}(z, v_1, \ldots, v_n))$. Since $\delta \geqq \alpha > \lambda(u), \lambda(v_1), \ldots, \lambda(v_n)$ and since $\phi_{\Psi,\delta}(z, y_1, \ldots, y_n)$ contains only quantifiers with the subscript δ, v is indeed a term. We verify now

$$\|\forall z(z \in v \leftrightarrow z \in u \wedge \Phi(z, v_1, \ldots, v_n))\| = 1$$

by computation, using (24), Theorem 36 and Lemma 37.

The *Axiom of Infinity* holds in $\mathcal{N}$, as shown in Corollary 54 below.

LEMMA 44.

$$\|\exists x(x \in w \wedge \Phi(x))\| = \sum_{\lambda(v) < \lambda(w)} \|v \in w\|\ \|\Phi(v)\|,$$

$$\|\forall x(x \in w \rightarrow \Phi(x))\| = \prod_{\lambda(v) < \lambda(w)} -\|v \in w\| + \|\Phi(v)\|.$$

PROOF. By Lemma 37, $\|u \in w\| = \sum_{\lambda(v) < \lambda(w)} \|u = v\|\ \|v \in w\|$. Putting $x \in w$ for $\Gamma(x)$, $\{v \mid \lambda(v) < \lambda(w)\}$ for T and $\|v \in w\|$ for $\mathfrak{c}(v)$ in Lemma 40(a) we get the present lemma.

THEOREM 45. *Every instance* $\forall x \exists y \forall z\ [z \in x \rightarrow (\exists t\ \Psi(z, t) \rightarrow \exists t(t \in y \wedge \Psi(z, t)))]$ *of the Axiom Schema of Replacement (where* $\Psi(z, t)$ *is any formula of SP in which* y *is not free) holds in* $\mathcal{N}$. *(Here we use a version of the axiom schema of replacement which, taken together with the axiom schema of subsets, obviously implies the usual version of the former.)*

PROOF. Let $y_1, \ldots, y_n$ be the parameters of $\Psi(z, t)$. Given $v_1, \ldots, v_n, u \in TL$, let $\lambda(u) = \alpha$. For every $v \in TL$ and $\mathfrak{a} \in B$ let $H(v, \mathfrak{a})$ be the least ordinal β such that there is a term w with $\lambda(w) = \beta$ and $\|\Psi(v, w)\| = \mathfrak{a}$, if there is such a w, and $H(v, \mathfrak{a}) = 0$ otherwise. Let γ be an ordinal which includes the set $\{H(v, \mathfrak{a}) \mid \lambda(v) < \alpha, \mathfrak{a} \in B\}$ and let $u' = \text{\reflectbox{$\lambda$}}_\gamma x\,(x = x)$. All we have to do now is to show

$$\|\forall z[z \in u \to (\exists t \Psi(z, t) \to \exists t(t \in u' \wedge \Psi(z, t)))]\| = 1.$$

By Lemma 44 this follows once we prove that for every term v with

$$\lambda(v) < \lambda(u) = \alpha, \qquad \|\exists t \Psi(v, t)\| \leqq \|\exists t(t \in u' \wedge \Psi(v, t))\|,$$

but this follows by easy computation from the definitions of γ and u'.

THEOREM 46. *Every instance* $\exists y \Phi(y) \to \exists y[\Phi(y) \wedge \neg \exists z(\Phi(z) \wedge z \in y)]$ *of the Axiom Schema of Foundation holds in* $\mathcal{N}$.

PROOF. *Let* $x_1, \ldots, x_n$ be the parameters of $\Phi(y)$, and let $u_1, \ldots, u_n \in TL$. We have to show that $\|\exists y \Phi(y)\| \leqq \|\exists y[\Phi(y) \wedge \neg \exists z(\Phi(z) \wedge z \in y)]\|$, where $\Phi(y)$ is short for $\Phi(y, u_1, \ldots, u_n)$. Suppose this fails, then for some $\mathfrak{a} \in B$ we have $0 < \mathfrak{a} \leqq \|\exists y \Phi(y)\|$ but $\mathfrak{a}\|\exists y[(\Phi(y) \wedge \neg \exists z(\Phi(z) \wedge z \in y)]\| = 0$. Since

$$\mathfrak{a} \leqq \|\exists y \Phi(y)\| = \sum_{v \in TL} \|\Phi(v)\|$$

we have

$$0 < \mathfrak{a} = \sum_{v \in TL} \mathfrak{a}\|\Phi(v)\|$$

and hence for some term v we have $\mathfrak{a}\|\Phi(v)\| > 0$. Let w be a term with minimal $\lambda(w)$ such that $\mathfrak{a}\|\Phi(w)\| > 0$. By Lemma 44

$$\|\exists z\,(z \in w \wedge \Phi(z))\| = \sum_{\lambda(w') < \lambda(w)} \|w' \in w \wedge \Phi(w')\|.$$

By the minimality of $\lambda(w)$, if $\lambda(w') < \lambda(w)$ then $\mathfrak{a}\|\Phi(w')\| = 0$; hence

$$\mathfrak{a}\|\exists z(z \in w \wedge \Phi(z))\| = \sum_{\lambda(w') < \lambda(w)} \|w' \in w\|\ \mathfrak{a}\|\Phi(w')\| = 0.$$

Therefore $\mathfrak{a}\|\Phi(w) \wedge \neg \exists z(z \in w \wedge \Phi(z))\| = \mathfrak{a}\ \|\Phi(w)\| > 0$ and a fortiori $\mathfrak{a}\|\exists y[\Phi(y) \wedge \neg \exists z(z \in y \wedge \Phi(z))\| > 0$, contradicting our assumption concerning $\mathfrak{a}$.

Having shown that Axiom (i) of *SP* holds in $\mathcal{N}$ we aim now to show that Axioms (iii) and (iv) hold in $\mathcal{N}$. In order to do this we have to discuss some of the basic facts about absoluteness.

DEFINITION 47. Let M be a unary predicate symbol. We shall also write $x \in M$ for $M(x)$. For a formula Φ of *ZF*, $\Phi^{(M)}$ denotes the formula obtained from Φ by replacing in it every quantifier $\exists x$ by $(\exists x \in M)$ (universal quantifiers are given in terms of existential ones and thus $\forall x$ gets replaced by $(\forall x \in M)$). Let ZF_M be the axiom system consisting of the following axioms

(25) The axioms of ZF, including all the instances of the axiom schemas of subsets and replacement which contain the symbol M.

(26) $$y \in M \wedge x \in y \rightarrow x \in M \text{ (i.e. the class } M \text{ is transitive).}$$

(27) $\Phi^{(M)}$, where Φ ranges over the axioms of ZF. As a consequence of (27), if Φ is a theorem of ZF then $\Phi^{(M)}$ is a theorem of ZF_M.

We shall say that a formula Φ of ZF with the free variables $x_1, \ldots, x_n$ is *absolute* if $ZF_M \vdash x_1, \ldots, x_n \in M \rightarrow (\Phi^{(M)}(x_1, \ldots, x_n) \leftrightarrow \Phi(x_1, \ldots, x_n))$.

DEFINITION 48. The primitive language of ZF has $\in$ as its only extralogical symbol. In practice we use an extended language which is obtained from the primitive language by introduction of new relation- and function-symbols by means of definitions. In the definition of each new symbol we may use symbols already defined. We define, simultaneously, the *relativizations to M*, $R^{(M)}$, $F^{(M)}$, and $\Psi^{(M)}$ of a defined relation symbol R, a defined function-symbol F, and a formula or a term Ψ of the extended language of ZF. (A term of the extended language of ZF should not be confused with a member of TL.) At the same time we also prove that if Γ is the definition of the new symbol then $\Gamma^{(M)}$ is a theorem of ZF_M.

(a) If Ψ is a formula or a term which contains the relation symbols $R_1, \ldots, R_k$ and the function-symbols $F_1, \ldots, F_l$, all of which are already defined, then $\Psi^{(M)}$ is the formula, or term, respectively, obtained from Ψ by the replacements of

$$\exists x \quad \text{by} \quad (\exists x \in M)$$

and

$$R_i, F_i \quad \text{by} \quad R_i^{(M)}, F_i^{(M)}, \quad \text{respectively.}$$

(b) If the relation-symbol R is defined by $\forall x_1 \cdots \forall x_n(R(x_1, \ldots, x_n) \leftrightarrow \Phi(x_1, \ldots, x_n))$, where Φ contains only symbols which have been defined previously, then we define $R^{(M)}$ in ZF_M by

$$\forall x_1 \cdots \forall x_n(R^{(M)}(x_1, \ldots, x_n) \leftrightarrow \Phi^{(M)}(x_1, \ldots, x_n)).$$

If we denote the definition of R given above by Γ, then $\Gamma^{(M)}$, which is

$$(\forall x_1 \in M) \cdots (\forall x_n \in M)(R^{(M)}(x_1, \ldots, x_n) \leftrightarrow \Phi^{(M)}(x_1, \ldots, x_n)),$$

is clearly a theorem of ZF_M.

(c) if F is a function-symbol defined by

$$\forall x_1 \cdots \forall x_n \forall y(F(x_1, \ldots, x_n) = y \leftrightarrow \Phi(x_1, \ldots, x_n, y)),$$

where Φ is a formula which contains only symbols which have been defined previously, and such that

(28) $$ZF \vdash \forall x_1 \cdots \forall x_n \exists y \forall z(\Phi(x_1, \ldots, x_n, z) \leftrightarrow z = y)$$

then we define in ZF_M

$$F^{(M)}(x_1, \ldots, x_n) = y \leftrightarrow x_1, \ldots, x_n, y \in M \wedge \Phi^{(M)}(x_1, \ldots, x_n) \vee \bigvee_{1 \leq i \leq n} x_i \notin M \wedge y = 0.$$

The definition of $F^{(M)}$ is valid since by (28) and Axiom (27) we have

$$ZF_M \vdash \forall x_1 \cdots \forall x_n[x_1, \ldots, x_n \in M \rightarrow (\exists y \in M)(\forall z \in M)(\Phi^{(M)}(x_1, \ldots, x_n, z) \leftrightarrow z = y)].$$

If Γ is the definition of F above then $\Gamma^{(M)}$ is clearly provable in ZF_M.

A formula $\Phi(x_1, \ldots, x_n)$ of the extended language of ZF is said to be absolute if

$$ZF_M \vdash (\forall x_1 \in M) \cdots (\forall x_n \in M)(\Phi^{(M)}(x_1, \ldots, x_n) \leftrightarrow \Phi(x_1, \ldots, x_n)).$$

A term $\tau(x_1, \ldots, x_n)$ of that language is said to be absolute if

$$ZF_M \vdash (\forall x_1 \in M) \cdots (\forall x_n \in M)(\tau^{(M)}(x_1, \ldots, x_n) = \tau(x_1, \ldots, x_n)).$$

Notice that for every function-symbol F we have, in ZF_M, $x_1, \ldots, x_n \in M \rightarrow F^{(M)}(x_1, \ldots, x_n) \in M$. Notice also that, by Axiom (27), two definitions of a relation or function-symbol R which are provably equivalent in ZF lead to two definitions of $R^{(M)}$ which are provably equivalent in ZF_M, and hence we are free, in practice, to use for $R^{(M)}$ any of the definitions of R which suits us.

LEMMA 49. (a) *If the formula Φ is absolute and Ψ is equivalent to Φ in ZF then Ψ is absolute too.*

(b) *If $\tau(x_1, \ldots, x_n)$ is a term then $ZF_M \vdash x_1, \ldots, x_n \in M \rightarrow \tau^{(M)}(x_1, \ldots, x_n) \in M$.*

(c) *$\tau(x_1, \ldots, x_n)$ is an absolute term if and only if $y = \tau(x_1, \ldots, x_n)$ is an absolute formula.*

(d) *If the relation symbol R is defined by $R(x_1, \ldots, x_n) \leftrightarrow \Phi(x_1, \ldots, x_n)$ and Φ is absolute so is $R(x_1, \ldots, x_n)$.*

(e) *If the function symbol F is defined by $F(x_1, \ldots, x_n) = y \leftrightarrow \Phi(x_1, \ldots, x_n, y)$ and Φ is absolute so is $F(x_1, \ldots, x_n)$.*

(f) *If $\tau(x_1, \ldots, x_n)$, $\Phi(x_1, \ldots, x_n)$ and $\tau_i(x_1, \ldots, x_m)$, $1 \leqq i \leqq n$, are absolute so are $\tau(\tau_1(x_1, \ldots, x_m), \ldots, \tau_n(x_1, \ldots, x_m))$ and $\Phi(\tau_1(x_1, \ldots, x_m), \ldots, \tau_n(x_1, \ldots, x_m))$.*

(g) *Every formula obtained from absolute formulas by sentential connection (i.e. by applications of negation and disjunction) and by restricted quantification (i.e. by quantifiers of the type $(\exists x \in y)$ and $(\forall x \in y)$) is absolute.*

(h) *If $\Phi(x)$ and $\Psi(x)$ are absolute formulas (which may have free variables other than x) and $ZF \vdash \forall x \Phi(x) \leftrightarrow \exists x \Psi(x)$ then $\forall x \Phi(x)$ and $\exists x \Psi(x)$ are absolute.*

PROOF. (a) follows directly from Axiom (27).

(b) We have already mentioned that if F is a defined function symbol then $x_1, \ldots, x_n \in M \rightarrow F^{(M)}(x_1, \ldots, x_n) \in M$. Using this, (b) is proved by induction on the length of τ.

(c) is trivial, using (b).

(d) and (e) are immediate consequences of (a) and (c).

(f) follows easily from (b) by comparing the term or formula considered with its relativization to M.

(g) follows easily from its assumptions, using, in the second part, also the transitivity of M (Axiom (26)).

(h) Suppose the free variables of $\Phi(x)$ and $\Psi(x)$, other than x, are $x_1, \ldots, x_n$. One verifies that for $x_1, \ldots, x_n \in M$ we have, by our assumptions and Axiom (27),

$$(\exists x \in M)\Phi^{(M)}(x) \to (\exists x \in M)\Phi(x) \to \exists x\Phi(x) \to \forall x\Psi(x) \to (\forall x \in M)\Psi(x) \\ \to (\forall x \in M)\Psi^{(M)}(x) \to (\exists x \in M)\Phi^{(M)}(x).$$

LEMMA 50. *The following terms and formulas are absolute:* $\{x, y\}$, $\{x\}$, $\langle x, y\rangle$, '*x is an ordered pair*', $p_i(x)$ (= *the ith member of the ordered pair x, if x is an ordered pair, and* 0 *otherwise*) *for* $i = 1, 2$, $x \leqq y$, $x \cup y$, $x \sim y$, '*f is a relation*', '*f is a function*'. '*f is a* 1-1 *function*', $\mathfrak{R}(f)$, $\mathfrak{D}(f)$, $f(x)$ (*as a term with the free variables f and x*), $f \mid x$, 0, 1, 2, 3, 4, 5, '*x is a transitive set*' (*i.e.* $(\forall y \in x)(y \subseteq x)$), '*x is an ordinal*', '*x is a finite ordinal*', $\rho(x)$, '*f is a finite sequence*', '*s is a substitution function*', '*q is an interval designator*', '*f and g are incompatible* (*as functions*)'.

PROOF. Everything is proved by simple-minded applications of the various parts of Lemma 49. We shall only carry out the most difficult case, that of $\rho(x)$. Let $\Phi(f)$ be the formula

$$f \text{ is a function} \wedge \mathfrak{D}(f) \text{ is a transitive set} \wedge \mathfrak{R}(f) \text{ consists of ordinals only} \\ \wedge (\forall x \in \mathfrak{D}(f))[(\forall y \in \mathfrak{D}(f))(y \in x \to f(y) \in f(x)) \\ \wedge \neg(\exists z \in f(x))(\forall y \in \mathfrak{D}(f))(y \in x \to f(y) \in z)].$$

Replace the terms $\mathfrak{D}(f)$, $\mathfrak{R}(f)$, $f(y)$ and $f(x)$ in $\Phi(f)$ by the new variables x', x'', y', y'', respectively. The formula thus obtained is clearly absolute, by the earlier parts of the present lemma and by Lemma 49(g). $\Phi(f)$ is obtained from that formula by resubstituting $\mathfrak{D}(f)$, $\mathfrak{R}(f)$, $f(y)$ and $f(x)$. Therefore $\Phi(f)$ is absolute, by the earlier parts of the present lemma and by Lemma 49(f). As follows from the elementary properties of the function ρ we have

$$y = \rho(x) \leftrightarrow \exists f(\Phi(f) \wedge x \in \mathfrak{D}(f) \wedge f(x) = y) \leftrightarrow \forall f(\Phi(f) \wedge x \in \mathfrak{D}(f) \to f(x) = y).$$

Hence the formula $y = \rho(x)$ is absolute, by earlier parts of the present lemma and by Lemma 49(a) and (b). $\rho(x)$ itself is absolute by Lemma 49(c).

In order to apply absoluteness in our framework we shall show that Axioms (25)–(27) hold in $\mathcal{N}$ with M replaced by S. Axiom (25) becomes now Axiom (i) of SP, which we have already seen to hold in $\mathcal{N}$.

LEMMA 51. $x \in S \wedge y \in x \to y \in S$ (*i.e., S is a transitive class*) *holds in* $\mathcal{N}$.

PROOF. By easy computation.

LEMMA 52.

$$\Phi(x_1, \ldots, x_n) \to \|\Phi^{(S)}(\mathbf{x}_1, \ldots, \mathbf{x}_n)\| = 1,$$
$$\neg\Phi(x_1, \ldots, x_n) \to \|\Phi^{(S)}(\mathbf{x}_1, \ldots, \mathbf{x}_n)\| = 0.$$

PROOF. Jointly, by induction on the length of Φ, using Lemmas 31, 32 and 40 (c).

Lemma 51 asserts that Axiom (26) holds in $\mathcal{N}$ (with S for M). It is an immediate consequence of Lemma 52 that also Axiom (27) holds in $\mathcal{N}$.

LEMMA 53. *If $\Phi(x_1, \ldots, x_n)$ is an absolute formula then*

$$\Phi(x_1, \ldots, x_n) \rightarrow \|\Phi(\mathbf{x}_1, \ldots, \mathbf{x}_n)\| = 1,$$

$$\neg\Phi(x_1, \ldots, x_n) \rightarrow \|\Phi(\mathbf{x}_1, \ldots, \mathbf{x}_n)\| = 0.$$

PROOF. Assume $\Phi(x_1, \ldots, x_n)$ then, by Lemma 52, $\|\Phi^{(S)}(\mathbf{x}_1, \ldots, \mathbf{x}_n)\| = 1$. Since Φ is absolute, and Axioms (25)–(27) hold for S in $\mathcal{N}$ we have

$$\|(\forall x_1 \in S) \cdots (\forall x_n \in S)[\Phi^{(S)}(\mathbf{x}_1, \ldots, \mathbf{x}_n) \leftrightarrow \Phi(\mathbf{x}_1, \ldots, \mathbf{x}_n)]\| = 1.$$

This, together with $\|\mathbf{x}_i \in S\| = 1$ and $\|\Phi^{(S)}(\mathbf{x}_1, \ldots, \mathbf{x}_n)\| = 1$, implies $\|\Phi(\mathbf{x}_1, \ldots, \mathbf{x}_n)\| = 1$. The second part of the lemma is proved similarly.

COROLLARY 54. *The axiom of infinity $\exists x(x$ is an ordinal $\wedge (\exists y \in x) \wedge (\forall z \in x)(\exists t \in x)(z \in t)$ holds in $\mathcal{N}$. Also $\boldsymbol{\omega} = \omega$ holds in $\mathcal{N}$.*

PROOF. Let SF be the set theory ZF without the axiom of infinity. We can replace ZF by SF (and ZF_M by SF_M) in Definitions 47 and 48 to obtain a somewhat stronger notion of absoluteness, which we call SF-absoluteness. Lemmas 49, 50 and 53 hold also for SF-absoluteness, since their proofs never mentioned the axiom of infinity. Let us call the lemmas which we would obtain by this replacement Lemmas 49_{SF}, 50_{SF} and 53_{SF} respectively. Notice that Lemma 53_{SF} does not need the assumption that the axiom of infinity holds in $\mathcal{N}$.

Let $\Phi(x)$ be the formula x *is an ordinal* $\wedge (\exists y \in x) \wedge (\forall z \in x)(\exists t \in x)(z \in t)$. By Lemmas 49_{SF} and 50_{SF} $\Phi(x)$ is SF-absolute. We have, obviously, $\Phi(\omega)$ and hence, by Lemma 53_{SF}, we get $\|\Phi(\boldsymbol{\omega})\| = 1$ and therefore $\|\exists x\Phi(x)\| = 1$.

In ZF we have $\Phi(x) \leftrightarrow x = \omega$; hence this holds also in $\mathcal{N}$. Since $\Phi(\boldsymbol{\omega})$ holds in $\mathcal{N}$ we have also $\boldsymbol{\omega} = \omega$ in $\mathcal{N}$.

THEOREM 55. *"Every ordinal is standard" holds in $\mathcal{N}$.*

PROOF. Let u be any term and let $\alpha > \lambda(u)$. Since every theorem of ZF holds in $\mathcal{N}$ we have

(29) $\quad \|u$ *is an ordinal* $\wedge\ \boldsymbol{\alpha}$ *is an ordinal* $\rightarrow u \in \boldsymbol{\alpha} \vee u = \boldsymbol{\alpha} \vee \boldsymbol{\alpha} \in u\| = 1.$

By Lemmas 50 and 53 $\|\boldsymbol{\alpha}$ *is an ordinal*$\| = 1$, and by Lemma 41 $\|u = \boldsymbol{\alpha}\| = 0$ and $\|\boldsymbol{\alpha} \in u\| = 0$ (since $\rho(\alpha) = \alpha > \lambda(u)$); hence we get from (29) $\|u$ *is an ordinal* $\rightarrow u \in \boldsymbol{\alpha}\| = 1$. From this we get, by $\|\boldsymbol{\alpha} \in S\| = 1$ and by Lemma 51 which asserts that S is transitive in $\mathcal{N}$, $\|u$ *is an ordinal* $\rightarrow u \in S\| = 1$.

THEOREM 56. *"Every finite subset of S is a member of S" holds in $\mathcal{N}$.*

PROOF. To prove the theorem it is enough to show that "$0 \in S$ and for all x, y if $x, y \in S$ also $x \cup \{y\} \in S$" holds in $\mathcal{N}$. $0 \in S$ holds in $\mathcal{N}$ by Lemmas 50 and 49(b). By Lemma 40(a) $\|\forall x \forall y(x, y \in S \rightarrow x \cup \{y\} \in S)\| = 1$ follows if we prove for all s and t $\|\mathbf{s} \cup \{\mathbf{t}\} \in S\| = 1$. Let $r = s \cup \{t\}$. By Lemmas 49(f),

50 and 53 we have $\|\mathbf{r} = \mathbf{s} \cup \{\mathbf{t}\}\| = 1$ and since $\|\mathbf{r} \in S\| = 1$ we get $\|\mathbf{s} \cup \{\mathbf{t}\} \in S\| = 1$.

LEMMA 57. *Let $\Gamma(x)$ be an absolute formula such that $\Gamma(x) \to x \in S$ holds in $\mathcal{N}$.*
(a) *If $u \in TL$ then* $\|\Gamma(u)\| = \sum_{\Gamma(s)} \|u = \mathbf{s}\|$.
(b) $\|\exists x(\Gamma(x) \wedge \Phi(x))\| = \sum_{\Gamma(s)} \|\Phi(\mathbf{s})\|$, $\|\forall x(\Gamma(x) \to \Phi(x))\| = \prod_{\Gamma(s)} \|\Phi(\mathbf{s})\|$.
(c) $\|u$ *is an ordinal*$\| = \sum_{\alpha \in On} \|u = \boldsymbol{\alpha}\|$.
(d) $\|\exists\alpha\Phi(\alpha)\| = \sum_{\alpha \in On} \|\Phi(\boldsymbol{\alpha})\|$, $\|\forall\alpha\Phi(\alpha)\| = \prod_{\alpha \in On} \|\Phi(\boldsymbol{\alpha})\|$.

PROOF. (a) By Lemma 53 $\|\Gamma(\mathbf{s})\| = 1$ if $\Gamma(s)$ and $\|\Gamma(\mathbf{s})\| = 0$ if $\neg\, \Gamma(s)$. Also, since $u = v \wedge \Gamma(u)$ is logically equivalent to $u = v \wedge \Gamma(v)$ we have $\|u = v\|\ \|\Gamma(u)\| = \|u = v\|\ \|\Gamma(v)\|$. Therefore

$$\|\Gamma(u)\| = \|u \in S \wedge \Gamma(u)\| = \sum_{s} \|u = \mathbf{s}\|\ \|\Gamma(u)\| = \sum_{s} \|u = \mathbf{s}\|\ \|\Gamma(\mathbf{s})\| = \sum_{\Gamma(s)} \|u = \mathbf{s}\|.$$

(b) follows from (a) by Lemma 44(b) where we take $T = \{\mathbf{s} \mid \Gamma(s)\}$.

(c) and (d) follow from (a) and (b), respectively, by taking "*x is an ordinal*" for $\Gamma(x)$. The requirements of (a) and (b) are satisfied by Lemma 50 and Theorem 55.

THEOREM 58. (a) *$\forall\alpha(S \cap R(\alpha) \in S)$ holds in $\mathcal{N}$.*
(b) *$\forall\alpha(F \cap R(\alpha) \in S)$ holds in $\mathcal{N}$.*

PROOF. (a) By Lemma 57(d) it is enough to prove that for each α $\|S \cap R(\boldsymbol{\alpha}) \in S\| = 1$. Let $s = R(\alpha)$ then we have $\mathsf{A}x(x \in s \leftrightarrow \rho(x) < \alpha)$ and hence, by Lemma 52, $\|(\forall x \in S)(x \in \mathbf{s} \leftrightarrow \rho^{(S)}(x) < \boldsymbol{\alpha})\| = 1$. By Lemma 50 we have in $\mathcal{N}$ $\forall x(\rho^{(S)}(x) = \rho(x))$, hence $\|(\forall x \in S)(x \in \mathbf{s} \leftrightarrow \rho(x) < \boldsymbol{\alpha})\| = 1$. Since $\|\mathbf{s} \in S\| = 1$ and S is transitive in $\mathcal{N}$ we get $\|\forall x(x \in \mathbf{s} \leftrightarrow x \in S \wedge \rho(x) < \boldsymbol{\alpha})\| = 1$, i.e. $\|\mathbf{s} = S \cap R(\boldsymbol{\alpha})\| = 1$, and hence $\|S \cap R(\boldsymbol{\alpha}) \in S\| = 1$.

(b) is proved like (a), taking $s = F \cap R(\alpha)$ and replacing S in the appropriate places by F.

By Theorems 51, 56, 57, and 58 Axiom (iii) of *SP* holds in $\mathcal{N}$. The next theorem will assert that also Axiom (iv) holds in $\mathcal{N}$.

THEOREM 59. *"F is a function from On onto S" holds in $\mathcal{N}$.*

PROOF. First let us notice that it follows immediately from Lemma 40(b) that

$$\|(\exists x \in F)\Phi(x)\| = \sum_{x \in F} \|\Phi(\mathbf{x})\|, \quad \|(\forall x \in F)\Phi(x)\| = \prod_{x \in F} \|\Phi(\mathbf{x})\|. \tag{30}$$

What we have to prove is

$$\begin{aligned} &\|(\forall x \in F)\ (x \text{ is an ordered pair} \wedge p_1(x) \text{ is an ordinal} \\ &\qquad \wedge p_2(x) \in S \wedge (\forall y \in F)(p_1(y) = p_1(x) \to y = x)) \\ &\qquad \wedge \forall\alpha(\exists r \in F)(p_1(r) = \alpha) \wedge (\forall z \in S)(\exists s \in F)(p_2(s) = z)\| = 1. \end{aligned} \tag{31}$$

By (30) and Lemmas 57(d) and 40(c), (31) follows once we prove that for all

$x, y \in F$, $\alpha \in On$ and $z \in S$ we have

$$(32)\quad \begin{aligned}&\|\mathbf{x} \textit{ is an ordered pair} \wedge p_1(\mathbf{x}) \textit{ is an ordinal} \wedge p_2(\mathbf{x}) \in S\\ &\wedge (p_1(\mathbf{y}) = p_1(\mathbf{x}) \rightarrow \mathbf{y} = \mathbf{x}) \wedge p_1(\mathbf{r}) = \boldsymbol{\alpha} \wedge p_2(\mathbf{s}) = \mathbf{z}\| = 1\end{aligned}$$

where $r = \langle \alpha, F(\alpha)\rangle$ and $s = \langle \beta, z\rangle$ where β is such that $F(\beta) = z$. (32) follows easily from Lemmas 50, 49(c, f, g) and 53.

Our next aim is to show that Axiom (vi) of SP holds in $\mathcal{N}$.

DEFINITION 60. For every finite sequence $\langle u_0, \ldots, u_{n-1}\rangle$ of members of TL with $\max_{i<n} (\lambda(u_i)) + 1 = \alpha$ we define $\{u_0, \ldots, u_n\}^*$ to be

$$\daleth_\alpha x(x = u_0 \vee \cdots \vee x = u_{n-1}),\ \langle u, v\rangle^* = \{\{u\}^*, \{u, v\}^*\}^*.$$

LEMMA 61. (a) $\|v \in \{u_0, \ldots, u_{n-1}\}^*\| = \sum_{i<n} \|v = u_i\|$.

(b) $\|\exists x(x \in \{u_0, \ldots, u_{n-1}\}^* \wedge \Phi(x))\| = \sum_{i<n} \|\Phi(u_i)\|$,

$\|\forall x(x \in \{u_0, \ldots, u_{n-1}\}^* \wedge \Phi(x))\| = \prod_{i<n} \|\Phi(u_i)\|$.

(c) $\|\{u, v\}^* = \{u, v\}\| = 1$.

(d) $\|\langle u, v\rangle^* = \langle u, v\rangle\| = 1$.

PROOF. (a) is proved by direct computation, (b) follows from (a) by Lemma 40(b). (c) $\{u, v\}^* = \{u, v\}$ is equivalent to the statement

$$(\forall x \in \{u, v\}^*)(x = u \vee x = v) \wedge u \in \{u, v\}^* \wedge v \in \{u, v\}^*.$$

The value of the latter statement under $\| \quad \|$ is easily shown to be 1, by direct computation and the use of (b). (d) follows easily from (c).

LEMMA 62. *Let W be the set of all functions h such that $\mathfrak{D}(h)$ is a finite set of finite ordinals and $\mathfrak{R}(h)$ is a set of finite ordinals. For every $h \in W$ let $h^\#$ be the term $\{\langle \mathbf{i}, a_{h(i)}\rangle^* \mid i \in \mathfrak{D}(h)\}^*$, i.e. $h^\#$ is $\{u_0, \ldots, u_{n-1}\}^*$ where the u_j's are the terms $\langle \mathbf{i}, a_{h(i)}\rangle^*$, ordered by the natural order of the i's. We have for all $h, k \in W$*

(a) $\|h^\#$ *is an assignment*$\| = 1$.

(b) *If* $d = \mathfrak{D}(h)$ *then* $\|\mathbf{d} = \mathfrak{D}(h^\#)\| = 1$.

(c) *If* $j \in \mathfrak{D}(h)$ *then* $\|h^\#(\mathbf{j}) = a_{h(j)}\| = 1$.

(d) *If* $i \notin \mathfrak{D}(h)$ *then* $\|\langle \mathbf{i}, a_j\rangle \in h^\#\| = 0$.

(e) *If* $i \in \mathfrak{D}(h)$ *then* $\|\langle \mathbf{i}, a_j\rangle \in h^\#\| = \|a_j = a_{h(i)}\|$.

(f) *If* $h \subseteq k$ *then* $\|h^\# \subseteq k^\#\| = 1$.

(g) $\|(h \circ k)^\# = h^\# \circ \mathbf{k}\| = 1$.

PROOF. (a) Let $\Phi(z, n)$ stand for

$$(\forall x \in z)(\forall y \in z)(x \textit{ is an ordered pair} \wedge p_1(x) \in n \wedge p_2(x) \in b \wedge (p_1(x) = p_1(y) \rightarrow x = y)).$$

"*z is an assignment*" is equivalent in ZF to $(\exists n \in \omega)\Phi(z, n)$. Let n be such that $n \supseteq \mathfrak{D}(h)$, then $\Phi(h^\#, \mathbf{n})$ holds in $\mathcal{N}$, as follows from Lemmas 61(*b*, c), 31 and 32. $\omega = \boldsymbol{\omega}$ holds in $\mathcal{N}$ by Corollary 54. Therefore, since $\Phi(h^\#, \mathbf{n})$ holds in $\mathcal{N}$ also $(\exists n \in \omega)\Phi(h^\#, n)$ holds in $\mathcal{N}$. As a consequence also the equivalent statement "$h^\#$ *is an assignment*" holds in $\mathcal{N}$.

(b) Let $d = \mathfrak{D}(h)$. First let us show that $\mathbf{d} \subseteq \mathfrak{D}(h^\#)$ holds in $\mathcal{N}$.

$$\begin{aligned}\|\mathbf{d} \subseteq \mathfrak{D}(h^\#)\| &= \|(\forall x \in \mathbf{d})(\exists y \in h^\#)(p_1(y) = x)\| \\ &= \prod_{x\in d} \|(\exists y \in h^\#)(p_1(y) = \mathbf{x})\|, \quad \text{by Lemma 40(d)} \\ &= \prod_{x\in d} \sum_{i\in\mathfrak{D}(h)} \|p_1(\langle \mathbf{i}, a_{h(i)}\rangle^*) = \mathbf{x}\|, \quad \text{by Lemma 61(b)} \\ &= \prod_{x\in d} \sum_{i\in\mathfrak{D}(h)} \|\mathbf{i} = \mathbf{x}\|, \quad \text{by Lemma 61(d).}\end{aligned}$$

For $x \in d$ we have $x \in \mathfrak{D}(h)$, and $\|\mathbf{x} = \mathbf{x}\| = 1$, hence we have

$$\|\mathbf{d} \subseteq \mathfrak{D}(h^\#)\| = \prod_{x\in d} \sum_{i\in\mathfrak{D}(h)} \|\mathbf{i} = \mathbf{x}\| = 1.$$

To show that also $\mathfrak{D}(h^\#) \subseteq \mathbf{d}$ holds in $\mathcal{N}$ we write

$$\begin{aligned}\|\mathfrak{D}(h^\#) \subseteq \mathbf{d}\| &= \|(\forall y \in h^\#)(p_1(y) \in \mathbf{d})\| \\ &= \prod_{i\in\mathfrak{D}(h)} \|p_1(\langle \mathbf{i}, a_{h(i)}\rangle^*) \in \mathbf{d}\|, \quad \text{by Lemma 61(b)} \\ &= \prod_{i\in\mathfrak{D}(h)} \|\mathbf{i} \in \mathbf{d}\|, \quad \text{by Lemma 61(d)} \\ &= 1, \quad \text{by Lemma 31.}\end{aligned}$$

Since both $\mathbf{d} \subseteq \mathfrak{D}(h^\#)$ and $\mathfrak{D}(h^\#) \subseteq \mathbf{d}$ hold in $\mathcal{N}$ also $\mathbf{d} = \mathfrak{D}(h^\#)$ holds in $\mathcal{N}$.

(c) If $j \in \mathfrak{D}(h)$ then, by Lemma 61(a), $\|\langle \mathbf{j}, a_{h(j)}\rangle^* \in h^\#\| = 1$. Thus we have in $\mathcal{N}$, $\langle \mathbf{j}, a_{h(j)}\rangle \in h^\#$ (by Lemma 61(d)), and also that $h^\#$ is an assignment and hence a function (by (a)). Therefore we have in $\mathcal{N}$, $h^\#(\mathbf{j}) = a_{h(j)}$.

(d)–(g) are proved by easy computation, using also Lemma 61.

LEMMA 63. (a) $\|u$ *is an assignment*$\| = \sum_{h\in W} \|u = h^\#\|$.
(b) *If* $k \in W$ *then* $\|u$ *is an assignment* $\wedge\ u \supseteq k^\#\| = \sum_{h\in W\wedge h\supseteq k} \|u = h^\#\|$.
(c) $\|\exists x(x$ *is an assignment* $\wedge\ \Phi(x)\| = \sum_{h\in W} \|\Phi(h^\#)\|$,

$$\|\forall x(x \textit{ is an assignment} \to \Phi(x)\| = \textstyle\prod_{h\in W} \|\Phi(h^\#)\|.$$

PROOF. (a) To prove the inequality $\geqq$ it is enough to prove that for every $h \in W$, $u = h^\#$ implies in $\mathcal{N}$ that u is an assignment, but this is a direct consequence of Lemma 62(a).

"*u is an assignment*" is also equivalent in *ZF* to

$$(\exists n \in \omega)[u \textit{ is a function} \wedge \mathfrak{D}(u) \subseteq n \wedge (\forall i \in n)(i \notin \mathfrak{D}(u) \vee i \in \mathfrak{D}(u) \wedge u(i) \in b)].$$

Since $\omega = \boldsymbol{\omega}$ in $\mathcal{N}$ the left-hand side of (a) is equal to

$$(33) \quad \sum_{n<\omega} \|u \textit{ is a function} \wedge \mathfrak{D}(u) \subseteq \mathbf{n} \wedge (\forall i \in \mathbf{n})(i \notin \mathfrak{D}(u) \vee i \in \mathfrak{D}(u) \wedge u(i) \in b)\|.$$

To prove the inequality $\leqq$ we have to show that each of the summands in (33) is

$\leqq \sum_{h \in W} \|u = h^{\#}\|$. Let us denote the nth summand of (33) with $\mathfrak{c}_n$. We have

$$\mathfrak{c}_n = \|u \textit{ is a function}\| \, \|\mathfrak{D}(u) \subseteq \mathbf{n}\| \cdot \prod_{i<n} \|\mathbf{i} \notin \mathfrak{D}(u) \vee (\mathbf{i} \in \mathfrak{D}(u) \wedge u(i) \in b)\|$$

by Lemma 40(d)

$$= \|u \textit{ is a function}\| \, \|\mathfrak{D}(u) \subseteq \mathbf{n}\| \cdot \prod_{i<n} (\|\mathbf{i} \notin \mathfrak{D}(u)\| + \sum_{j<\omega} \|\mathbf{i} \in \mathfrak{D}(u) \wedge u(\mathbf{i}) = a_j\|)$$

since $u(\mathbf{i}) \in b$ is equivalent in ZF to $\exists x(x \in b \wedge u(\mathbf{i}) = x)$

$$= \sum_{d \subseteq n \wedge h \in \omega} \|u \textit{ is a function}\| \, \|\mathfrak{D}(u) \subseteq \mathbf{n}\| \cdot \prod_{i \in n \sim d} \|\mathbf{i} \notin \mathfrak{D}(u)\|$$

$$\cdot \prod_{i \in d} \|\mathbf{i} \in \mathfrak{D}(u) \wedge u(\mathbf{i}) = a_{h(i)}\|$$

by distributing the multiplication over the addition.

To prove $\mathfrak{c}_n \leqq \sum_{h \in W} \|u = h^{\#}\|$ it is now enough to prove that for all $d \subseteq n$ and for every function h on d into ω we have

$$\|u \textit{ is a function}\| \, \|\mathfrak{D}(u) \subseteq \mathbf{n}\| \cdot \prod_{i \in n \sim d} \|\mathbf{i} \notin \mathfrak{D}(u)\|$$

(34)

$$\cdot \prod_{i \in d} \|\mathbf{i} \in \mathfrak{D}(u) \wedge u(\mathbf{i}) = a_{h(i)}\| \leqq \sum_{h \in W} \|h^{\#} = u\|.$$

Let us denote the left-hand side of (34) with $\mathfrak{e}$. We shall prove that indeed $\mathfrak{e} \leqq \|h^{\#} = u\|$ and thereby (34) will be established. By Lemma 62(a) "$h^{\#}$ *is an assignment*" holds in $\mathcal{N}$ and therefore "$h^{\#}$ *is a function*" holds in $\mathcal{N}$, i.e. $\|h^{\#} \textit{ is a function}\| = 1$. By Lemmas 62(b), 49 and 53 $\|\mathfrak{D}(h^{\#}) \subseteq \mathbf{n}\| = 1$. Straightforward computation gives by Lemmas 62(b, c), 31 and 32 $\|\mathbf{i} \notin \mathfrak{D}(h^{\#})\| = 1$ for $i \in n \sim d$, and $\|\mathbf{i} \in \mathfrak{D}(h) \wedge h^{\#}(\mathbf{i}) = a_{h(i)}\| = 1$ for $i \in d$. Therefore we have

$$\mathfrak{c}_n = \|u \textit{ is a function}\| \, \|\mathfrak{D}(u) \subseteq \mathbf{n}\| \cdot \prod_{i \in n \sim d} \|\mathbf{i} \notin \mathfrak{D}(u)\| \, \|\mathbf{i} \notin \mathfrak{D}(h^{\#})\|$$

$$\cdot \prod_{i \in d} \|\mathbf{i} \in \mathfrak{D}(u)\| \, \|\mathbf{i} \in \mathfrak{D}(h^{\#})\| \, \|u(\mathbf{i}) = a_{h(i)}\| \, \|h^{\#}(\mathbf{i}) = a_{h(i)}\|$$

$$\leqq \|u \textit{ is a function}\| \, \|\mathfrak{D}(u) \subseteq \mathbf{n}\|$$

$$\cdot \prod_{i \in n} \|\mathbf{i} \notin \mathfrak{D}(u) \wedge \mathbf{i} \notin \mathfrak{D}(h^{\#}) \vee \mathbf{i} \in \mathfrak{D}(u) \wedge \mathbf{i} \in \mathfrak{D}(h^{\#}) \wedge u(\mathbf{i}) = h^{\#}(\mathbf{i})\|$$

$$\leqq \|u \textit{ is a function} \wedge h^{\#} \text{ is a function} \wedge \mathfrak{D}(u) \subseteq \mathbf{n} \wedge \mathfrak{D}(h^{\#}) \subseteq \mathbf{n}$$

$$\wedge (\forall i \in \mathbf{n})(i \notin \mathfrak{D}(u) \wedge i \notin \mathfrak{D}(h^{\#}) \vee i \in \mathfrak{D}(u) \wedge i \in \mathfrak{D}(h^{\#}) \wedge u(i) = h^{\#}(i))\|$$

$$\leqq \|u = h^{\#}\|,$$

since the statement inside the former $\| \;\; \|$ implies $u = h^{\#}$ in ZF.

(b) By (a),

$$\|u \textit{ is an assignment} \wedge u \supseteq k^{\#}\| = \sum_{h \in W} \|u = h^{\#}\| \, \|u \supseteq k^{\#}\|$$

$$= \sum_{h \in W} \|u = h^{\#}\| \, \|h^{\#} \supseteq k^{\#}\|$$

$$\geqq \sum_{h \in W \wedge h \supseteq k} \|u = h^{\#}\| \, \|h^{\#} \supseteq k^{\#}\| = \sum_{h \in W \wedge h \supseteq k} \|u = h^{\#}\|,$$

by Lemma 62(f). To show that this inequality is indeed an equality it is enough if we show that for every $h \in W$ either $\|h^\# \supseteq k^\#\| = 0$, or else $\|u = h^\#\| \, \|h^\# \supseteq k^\#\| \leqq \|u = l^\#\|$ for some $l \in W$ such that $l \supseteq k$. Let $h \in W$, if $\mathfrak{D}(h) \not\supseteq \mathfrak{D}(k)$ then easy computation shows that $\|h^\# \supseteq k^\#\| = 0$. If, on the other hand $\mathfrak{D}(h) \supseteq \mathfrak{D}(k)$ then easy computation shows that $\|h^\# \supseteq k^\#\| = \prod_{i \in \mathfrak{D}(k)} \|a_{h(i)} = a_{k(i)}\|$. Let $l \in W$ be such that $l(i) = k(i)$ for $i \in \mathfrak{D}(k)$ and $l(i) = h(i)$ for $i \in \mathfrak{D}(h) \sim \mathfrak{D}(k)$. Easy computation shows that $\|h^\# = l^\#\| = \prod_{i \in \mathfrak{D}(k)} \|a_{h(i)} = a_{k(i)}\| = \|h^\# \supseteq k^\#\|$. Therefore we have $\|u = h^\#\| \, \|h^\# \supseteq k^\#\| = \|u = h^\#\| \, \|h^\# = l^\#\| \leqq \|u = l^\#\|$.

(c) follows from (a) by Lemma 40(b).

LEMMA 64. (*Absoluteness of functions defined by induction.*) *Let R be a binary relation such that it can be proved in ZF that R is well founded* (*i.e. every set t has a member x such that $\{y \mid yRx\} \cap t = 0$*) *and that for every x $\{y \mid yRx\}$ is a set and such that*

$$(35) \qquad ZF \vdash (\forall x \in M)\forall y(yRx \to y \in M).$$

If $G(f, x)$ is an absolute function (*i.e. given by an absolute term*) *then the function F defined by $F(x) = G(F \restriction \{y \mid yRx\}, x)$ is also absolute.* (*R and G may have parameters, which become parameters of F.*)

PROOF.

$$\begin{aligned} y = F(x) \leftrightarrow \exists f[x \in \mathfrak{D}(f) \wedge f(x) = y \wedge (\forall t \in \mathfrak{D}(f))\forall z(zRt \to z \in \mathfrak{D}(f)) \wedge \\ (\forall t \in \mathfrak{D}(f))(f(t) = G(f \restriction \{s \mid sRt\}, t))] \\ \leftrightarrow \forall f[x \in \mathfrak{D}(f) \wedge (\forall t \in \mathfrak{D}(f))\forall z(zRt \to z \in \mathfrak{D}(f)) \wedge \\ (\forall t \in \mathfrak{D}(f))(f(t) = G(f \restriction \{s \mid sRt\}, t)) \to y = f(x)] \end{aligned}$$

where $\{s \mid sRt\}$ is absolute because

$$z = \{s \mid sRt\} \leftrightarrow (\forall x \in z)xRt \wedge \forall x(xRt \to x \in y)$$

and by (35) and Lemma 49(c, g), and the absoluteness of the parts inside the square brackets follows from (35) and Lemmas 50 and 49(f, g). The absoluteness of $F(x)$ follows now from Lemma 49(h, c).

LEMMA 65. *The following terms of ZF are absolute.*

(a) $\neg$, $\vee$, $\in$, $=$, x_i, a_i, $\mathbf{s}$, $\exists_\alpha$, λ_α. (*This holds if, say, $\exists_\alpha$ denotes the symbol itself, i.e. $\langle 1, \alpha\rangle$, or the* 1-*termed sequence* $\{\langle 0, \langle 1, \alpha\rangle\rangle\}$ *of this symbol. We shall use* $\neg, \ldots, \lambda_\alpha$ *in the following to denote both things, leaving it to the reader to figure out the right meaning.*)

(b) $s^\frown t$.

(c) ω.

PROOF. (a) follows directly from Lemmas 49 and 50 and Definition 1.

(b) We define $x \dot{-} 1$ by

$$y = x \dot{-} 1 \leftrightarrow x = y \cup \{y\} \vee \neg(\exists z \in x)(x = z \cup \{z\}) \wedge y = x.$$

$x \dot{-} 1$ is absolute by Lemmas 49 and 50. Let $G(f, x)$ be given by

$$z = G(f,x) \leftrightarrow x = 0 \wedge z = s \vee x \neq 0 \wedge z = f(x \dot{-} 1) \cup \{\langle \mathfrak{D}(f(x \dot{-} 1)), t(x \dot{-} 1)\rangle\}.$$

$G(f, x)$ is absolute, by Lemmas 49 and 50. If we define a function $F(x)$ as in Lemma 64, where we take for R the $\in$ well-ordering of the finite ordinals, then F is absolute. As easily seen, if s and t are finite sequences and $n \leqq \mathfrak{D}(t)$ then $F(n) = s^\frown (t \mid n)$; hence $s^\frown t = F(\mathfrak{D}(t))$ and $s^\frown t$ is absolute.

(c) $x = \omega \leftrightarrow x$ *is an ordinal* $\wedge\, x = x \dot{-} 1 \wedge x \neq 0 \wedge (\forall y \in x)(y \dot{-} 1 \in y \vee y = 0)$.

Till now we defined absoluteness for the language of ZF, i.e. for the language whose atomic formulas are of the forms $x \in y$ and $x = y$. Nothing of what we said till now will change if we admit also a third primitive predicate $S(x)$. Relativization to M is defined so as not to affect $S(x)$, i.e. $S(x)^{(M)}$ is just $S(x)$. In addition to (25)–(27) we add to ZF_M the axiom

$$(36) \qquad \forall x(S(x) \rightarrow M(x)).$$

When we define now the language $\mathscr{L}$ we adopt constants $\mathbf{s}$ just for those sets s such that $S(s)$.

LEMMA 66. *The term $\lambda(x)$ is absolute.*

PROOF. Whenever we write in a formula $(\exists x \in \in y)$ it stands for "there is an x which is a member of a member of y", formally $(\exists z \in y)(\exists x \in z)$, where z is a variable which occurs nowhere else in the formula. Similarly $(\exists x \in \in \in y)$ stands for "there is an x which is a member of a member of a member of y". Also $(\forall x \in \in y)$ and $(\forall x \in \in \in y)$ have their obvious meaning.

Let $\Phi(x, y)$ stand for the formula

$$\begin{aligned}
&y \textit{ is an ordinal} \wedge (\forall z \in \mathfrak{D}(x))(\forall s \in \in \in \mathfrak{R}(x)) \\
&[(\langle z, \mathbf{r}\rangle \in x \rightarrow \rho(r) \in y) \wedge (\langle z, x_r\rangle \in x \rightarrow 0 \in y) \\
&\qquad \wedge (\langle z, a_r\rangle \in x \rightarrow (\exists u \in y)\ u \textit{ is not a finite ordinal}) \\
&\qquad \wedge (\langle z, b\rangle \in x \rightarrow (\exists u \in \in y)\ u \textit{ is not a finite ordinal}) \\
&\qquad \wedge (\langle z, \mathsf{Y}_r\rangle \in x \wedge r \textit{ is an ordinal} \rightarrow r \in y) \\
&\qquad \wedge (\langle z, \exists_r\rangle \in x \wedge r \textit{ is an ordinal} \rightarrow r \in y \wedge r = y)].
\end{aligned}$$

$\Phi(x, y)$ is absolute by Lemmas 49, 50, and 64.

$$\begin{aligned}
\lambda(x) = y \leftrightarrow\ &(x \textit{ is not a finite sequence} \vee \mathfrak{D}(x) = 0) \wedge y = 0 \\
&\vee x \textit{ is a finite sequence} \wedge [\mathfrak{D}(x) = \{0\} \\
&\wedge \Phi(x, y + 1) \wedge \neg\, \Phi(x, y) \vee \mathfrak{D}(x) \neq \{0\} \\
&\wedge (\exists r \in \in x(0))(r \textit{ is an ordinal} \wedge x(0) = \mathsf{Y}_r \wedge y = r) \\
&\vee \neg\, (\exists r \in \in x(0))(r \textit{ is an ordinal} \wedge x(0) = \mathsf{Y}_r) \\
&\qquad \wedge \Phi(x, y) \wedge (\forall z \in y) \neg\, \Phi(x, z))].
\end{aligned}$$

$\lambda(x)$ is now absolute by Lemmas 49, 50 and 65(a).

LEMMA 67. (a) *Let T be the function such that for $x \in FL$ $T(x)$ is the set of the subscripts of the free variables of x, for $x \in TL$, $T(x) = \{\{1\}\}$ and for all other sets x $T(x) = \{\{2\}\}$. The term $T(x)$ is absolute.*

(b) *The following formulas are absolute: $x \in TL$, $x \in FL$, x is a sentence of $\mathscr{L}$.*

PROOF. (a) T is defined by induction on the relation R given by

$$\begin{aligned} xRy \leftrightarrow{} & x \text{ and } y \text{ are finite sequences of symbols} \\ & \wedge (\lambda(x) < \lambda(y) \vee \lambda(x) = \lambda(y) \wedge \mathfrak{D}(x) \in \mathfrak{D}(y)). \end{aligned}$$

This relation is obviously absolute. It satisfies the requirement that for every x $\{y \mid yRx\}$ is a set, as we saw in §2, as well as (36). $T(x)$ is defined by means of the function $G(f, x)$, which is given by

$$\begin{aligned} y = G(f, x) \leftrightarrow{} & (\exists i, j \in \omega)[(x = \in^\frown x_i^\frown x_j \vee x = =^\frown x_i^\frown x_j) \wedge y = \{i, j\} \\ & \vee (\exists u, v \in \mathfrak{D}(f))(f(u) = f(v) = \{1\} \wedge (x = \in^\frown x_i^\frown u \\ & \vee x = \in^\frown u^\frown x_i \vee x = =^\frown x_i^\frown u \vee x = =^\frown u^\frown x_i) \wedge y = \{i\} \\ & \vee (x = \in^\frown u^\frown v \vee x = =^\frown u^\frown v) \wedge y = 0)] \\ & \vee (\exists \sigma \in \mathfrak{D}(f))[f(\sigma) \neq \{\{1\}\}, \{\{2\}\} \wedge x = \neg^\frown \sigma \wedge y = f(\sigma)] \\ & \vee (\exists \sigma, \sigma' \in \mathfrak{D}(f))[f(\sigma), f(\sigma') \neq \{\{1\}\}, \{\{2\}\} \wedge x = \vee^\frown \sigma^\frown \sigma' \wedge \\ & \qquad y = f(\sigma) \cup f(\sigma')] \\ & \vee (\exists \sigma \in \mathfrak{D}(f))(\exists i \in \omega)(\exists \alpha \in \in \in \in \in x)[(f(\sigma) \neq \{\{1\}\}, \{\{2\}\} \wedge x = \exists_\alpha{}^\frown x_i^\frown \sigma \\ & \quad \wedge y = f(\sigma) \sim \{i\} \wedge (f(\sigma) = i \wedge \lambda(\sigma) \leqq \alpha \wedge x = \forall_\alpha{}^\frown x_i^\frown \sigma \wedge y = \{\{1\}\})] \\ & \vee [(\exists s \in \in \in \in \in x)(s \in S \wedge x = \mathbf{s}) \vee (\exists i \in \omega)(x = a_i) \vee x = b] \wedge y = \{\{1\}\} \\ \vee{} & [\neg(\exists i, j \in \omega)[x = \in^\frown x_i^\frown x_j \vee x = =^\frown x_i^\frown x_j \vee (\exists u, v \in \mathfrak{D}(f))(f(u) = f(v) = \{\{1\}\} \\ & \wedge (x = \in^\frown x_i^\frown u \vee x = \in^\frown u^\frown x_i \vee x = \in^\frown u^\frown v \vee x = =^\frown x_i^\frown u \\ & \vee x = =^\frown u^\frown x_i \vee x = =^\frown u^\frown v))] \\ & \wedge \neg(\exists \sigma, \sigma' \in \mathfrak{D}(f))[f(\sigma), f(\sigma') \neq \{\{1\}\}, \{\{2\}\} \wedge (x = \neg^\frown \sigma \vee x = \vee^\frown \sigma^\frown \sigma')] \\ & \wedge \neg(\exists \sigma \in \mathfrak{D}(f))(\exists i \in \omega)(\exists \alpha \in \in \in \in \in x)[(f(\sigma) \neq \{\{1\}\}, \{\{2\}\} \\ & \quad \wedge x = \exists_\alpha{}^\frown x_i^\frown \sigma) \vee (f(\sigma) = \{i\} \wedge \lambda(\sigma) \leqq \alpha \wedge x = \forall_\alpha{}^\frown x_i^\frown \sigma)] \\ & \wedge \neg(\exists s \in \in \in \in \in x)(s \in S \wedge x = \mathbf{s}) \wedge \neg(\exists i \in \omega)(x = a_i) \wedge x \neq b] \wedge y = \{\{2\}\}. \end{aligned}$$

$G(f, x)$ is absolute by Lemmas 49, 50, 65 and 66; hence $T(x)$ is absolute, by Lemma 64.

(b) $x \in TL \leftrightarrow T(x) = \{\{1\}\}$,

$x \in FL \leftrightarrow T(x) \neq \{\{1\}\}, \{\{2\}\}$,

x *is a sentence of* $\mathscr{L} \leftrightarrow T(x) = 0$,

and part (b) follows from Lemmas 49 and 50.

LEMMA 68. *Let* $\operatorname{sbt}(\sigma, i, u)$ *be the function such that for $\sigma \in FL$, $i \in \omega$ and $u \in TL$* $\operatorname{sbt}(\sigma, i, u)$ *is the formula obtained from σ by replacing in it all the free* *occurrences of* x_i *by* u. (*We do not care what the value of* $\operatorname{sbt}(\sigma, i, u)$ *is in all other cases.*) $\operatorname{sbt}(\sigma, i, u)$ *is absolute.*

PROOF. We define $\operatorname{sbt}(\sigma, i, u)$ by induction on the same relation R as in the proof of Lemma 67. By what was mentioned there and by Lemma 64, all that has to be done now is to write out the corresponding function $G(f, x)$ for $\operatorname{sbt}(\sigma, i, u)$ and to verify that it is absolute. Looking at the way we write out $G(f, x)$ in the proof of Lemma 67 the reader can now easily write out the function $G(f, x)$ needed here, and deduce its absoluteness from Lemmas 49, 50, 65, 66, and 67.

Before Lemma 65 we introduced the notion of absoluteness also for formulas which contain the unary predicate symbol $S(x)$. We needed this wider notion of absoluteness for Lemma 67, since the formulas $x \in TL$ and $x \in FL$ contain the predicate symbol $S(x)$. In our present application of absoluteness we shall interpret $S(x)$ in set theory (i.e. in the system $ZF + V = L$, which we use) as $x = x$, i.e. $S(x)$ holds for every set; and in $\mathcal{N}$ choose for both $S(x)$ and $M(x)$ the $S(x)$ which we have there. One can verify easily that under these stipulations Lemmas 52 and 53 are also valid for our wider notion of absoluteness.

LEMMA 69. *The term* $\operatorname{occ}(x)$ *is absolute.*

PROOF.

$$
\begin{aligned}
y = \operatorname{occ}(x) \leftrightarrow {} & x \textit{ is not a finite sequence} \wedge y = 0 \\
& \vee\, x \textit{ is a finite sequence} \wedge [(\forall z \in y)(z \in \omega) \\
& \qquad \wedge (\forall i \in \omega)(i \in y \leftrightarrow (\exists t \in \Re(x))(t = a_i)].
\end{aligned}
$$

$\operatorname{occ}(x)$ is absolute by Lemmas 49(c, f, g), 50 and 65(a, c).

LEMMA 70. *The term* $\operatorname{sub}(x, s)$ *is absolute* (*see Lemma* 4).

PROOF.

$$
\begin{aligned}
y = \operatorname{sub}(x, s) \leftrightarrow {} & x \textit{ is not a finite sequence} \wedge y = 0 \\
& \vee\, x \textit{ and } y \textit{ are finite sequences} \wedge \mathfrak{D}(y) = \mathfrak{D}(x) \\
& \wedge (\forall z \in \mathfrak{D}(x))[(\exists i \in \omega)(x(z) = a_i \wedge y(z) = a_{s(i)}) \\
& \qquad \vee \neg(\exists i \in \omega)(x(z) = a_i) \wedge y(z) = x(z)].
\end{aligned}
$$

$\operatorname{sub}(x, z)$ is absolute by Lemmas 49(c, f, g), 50 and 65(a, c).

LEMMA 71. *Let us denote with* t_n *the term* $\{\langle \mathbf{0}, a_0\rangle^*, \langle \mathbf{1}, a_1\rangle^*, \ldots, \langle \mathbf{m}, a_m\rangle^*\}^*$, *where* $m = n - 1$. *For every term* u, *if* $\operatorname{occ}(u) \subseteq n$ *then* $\|\operatorname{den}(\mathbf{u}, t_n) = u\| = 1$, *and for every sentence* ϕ *of* $\mathcal{L}$, *if* $\operatorname{occ}(\phi) \subseteq n$ *then* $\|\operatorname{den}(\boldsymbol{\phi}, t_n) = 1\| = \|\phi\|$.

PROOF. Let $k_n = \{\langle i, i\rangle \mid i < n\}$, then $t_n = k_n^{\#}$. We shall now prove the lemma by induction as in the definition of $\| \ \|$ in Definition 27.

Case 1. $u = \mathbf{s}$ for some $s \in S$. By definition of *den* we have in ZF

(37) $$\forall s \forall f(s \in S \wedge f \textit{ is an assignment} \rightarrow \operatorname{den}(\langle 4, s\rangle, f) = s);$$

hence this holds also in $\mathcal{N}$. Since we have in $\mathcal{N}$ $\mathbf{s} \in S$ and t_n *is an assignment* (by Lemma 62(a)) we have also in $\mathcal{N}$, by (37), $\text{den}(\langle 4, \mathbf{s}\rangle, t_n) = \mathbf{s}$. $u = \mathbf{s} = \langle 4, s\rangle$ implies, by Lemmas 50, 49(c) and 53, that $\mathbf{u} = \langle 4, \mathbf{s}\rangle$ holds in $\mathcal{N}$. We saw that $\text{den}(\langle 4, \mathbf{s}\rangle, t_n) = \mathbf{s}$ holds in $\mathcal{N}$, i.e. $\text{den}(\mathbf{u}, t_n) = u$ holds in $\mathcal{N}$.

Case 2. $u = a_j$ for some $j \in \omega$. By definition of *den* we have in *ZF*

$$(38) \qquad \forall i \forall f(f \textit{ is an assignment} \wedge i \in \mathfrak{D}(f) \rightarrow \text{den}(\langle 5, i\rangle, f) = f(i)),$$

hence this holds also in $\mathcal{N}$. We have in $\mathcal{N}$, t_n *is an assignment* and, for $i < n$, $\mathbf{i} \in \mathfrak{D}(t_n)$ and $t_n(\mathbf{i}) = a_i$, by Lemma 62(a, b, c). Since we assumed $\text{occ}(u) \subsetneqq n$ we have $j < n$ and therefore, by (38) $\text{den}(\langle 5, \mathbf{j}\rangle, t_n) = t_n(\mathbf{j}) = a_j$ in $\mathcal{N}$. Since $u = \langle 5, j\rangle$ we have in $\mathcal{N}$, $\mathbf{u} = \langle 5, \mathbf{j}\rangle$ and thus $\text{den}(\mathbf{u}, t_n) = u$ holds in $\mathcal{N}$.

Case 3. $u = b$. This is handled like Cases 1 and 2.

Case 4. ϕ is $u \in v$. By definition of *den* we have in *ZF*

$$(39) \qquad \begin{aligned} &\forall u \forall v \forall f(u, v \in TL \wedge f \textit{ is an assignment} \wedge \mathfrak{D}(f) \supseteqq \text{occ}(u \in v) \\ &\qquad \rightarrow \text{den}(u \in v, f) = 1 \leftrightarrow \text{den}(u, f) \in \text{den}(v, f), \end{aligned}$$

and hence this holds also in $\mathcal{N}$. Since $u, v \in TL$ and $\phi = u \in v$ we have in $\mathcal{N}$, by Lemmas 67, 65, and 53, $\mathbf{u}, \mathbf{v} \in TL$ and $\boldsymbol{\phi} = \mathbf{u} \in \mathbf{v}$. By Lemma 62 we have in $\mathcal{N}$ t_n *is an assignment* and $\mathfrak{D}(t_n) = \mathbf{n}$. By our assumption, $\text{occ}(u \in v) \subsetneqq n$ and hence we have in $\mathcal{N}$, by Lemmas 69, 65, 50, and 49, $\text{occ}(\mathbf{u} \in \mathbf{v}) \subsetneqq \mathbf{n}$. Substituting $\mathbf{u}$, $\mathbf{v}$ and t_n for u, v and f in (39) we get in $\mathcal{N}$ $\text{den}(\mathbf{u} \in \mathbf{v}, t_n) = 1 \leftrightarrow \text{den}(\mathbf{u}, t_n) \in \text{den}(\mathbf{v}, t_n)$. Since $\text{occ}(u)$, $\text{occ}(v) \subsetneqq \text{occ}(\phi) \subsetneqq n$ we have in $\mathcal{N}$, by the induction hypothesis $\text{den}(\mathbf{u}, t_n) = n$, $\text{den}(\mathbf{v}, t_n) = v$ and thus we have in $\mathcal{N}$ $\text{den}(\mathbf{u} \in \mathbf{v}, t_n) = 1 \leftrightarrow u \in v$. Since $\boldsymbol{\phi} = \mathbf{u} \in \mathbf{v}$ we have in $\mathcal{N}$, $\text{den}(\boldsymbol{\phi}, t_n) = 1 \leftrightarrow u \in v$, i.e.

$$\|\text{den}(\boldsymbol{\phi}, t_n) = 1\| = \|u \in v\| = \|\phi\|.$$

Case 5. ϕ is $u = v$. This is handled like Case 4.

Case 6. $\phi = \psi \vee \chi$. By definition of *den* we have in *ZF*

$$(40) \qquad \begin{aligned} &\forall \psi \forall \chi \forall f(\psi \text{ and } \chi \textit{ are sentences of } L \wedge f \textit{ is an assignment} \\ &\wedge \mathfrak{D}(f) \supseteqq \text{occ}(\psi \vee \chi) \rightarrow \text{den}(\psi \vee \chi, f) = 1 \leftrightarrow \text{den}(\psi, f) = 1 \wedge \text{den}(\chi, f) = 1), \end{aligned}$$

and hence this holds also in $\mathcal{N}$. Since ψ and χ are sentences of L we have in $\mathcal{N}$, by Lemmas 67 and 53, $\boldsymbol{\psi}$ *and* $\boldsymbol{\chi}$ *are sentences of* L. Also, by Lemmas 62, 65, 69, and 53 t_n *is an assignment* and $\mathfrak{D}(t_n) = \mathbf{n} \supseteqq \text{occ}(\boldsymbol{\psi} \vee \boldsymbol{\chi})$ hold in $\mathcal{N}$. Therefore we have in $\mathcal{N}$, by (40), $\text{den}(\boldsymbol{\psi} \vee \boldsymbol{\chi}, t_n) = 1 \leftrightarrow \text{den}(\boldsymbol{\psi}, t_n) = 1 \vee \text{den}(\boldsymbol{\chi}, t_n) = 1$, i.e. we have

$$\|\text{den}(\boldsymbol{\psi} \vee \boldsymbol{\chi}, t_n) = 1\| = \|\text{den}(\boldsymbol{\psi}, t_n) = 1\| + \|\text{den}(\boldsymbol{\chi}, t_n) = 1\|.$$

Since $\text{occ}(\psi)$, $\text{occ}(\chi) \subsetneqq \text{occ}(\psi \vee \chi) \subsetneqq n$ we have by the induction hypothesis, $\|\text{den}(\boldsymbol{\psi}, t_n) = 1\| = \|\psi\|$ and $\|\text{den}(\boldsymbol{\chi}, t_n) = 1\| = \|\boldsymbol{\chi}\|$ and therefore

$$\|\text{den}(\boldsymbol{\psi} \vee \boldsymbol{\chi}, t_n) = 1\| = \|\psi\| + \|\chi\| = \|\psi \vee \chi\|.$$

Since $\phi = \psi \vee \chi$ we have in $\mathscr{N}$, by Lemmas 65, 49, and 53, $\boldsymbol{\phi} = \boldsymbol{\psi} \vee \boldsymbol{\chi}$ and hence $\|\text{den}(\boldsymbol{\phi}, t_n) = 1\| = \|\psi \vee \chi\| = \|\phi\|$.

Case 7. $\phi = \neg\psi$. This is handled like Case 6.

Case 8. $\phi = \exists_\alpha x_i \psi$. By the definition of *den* and Lemma 4 we have in *ZF*

$$\begin{aligned} &\forall\alpha(\forall i \in \omega)\forall\psi\forall f(T(\psi) = \{i\} \wedge f \textit{ is an assignment} \wedge \mathfrak{D}(f) \supseteqq \text{occ}(\psi) \\ (41) \qquad &\to \text{den}(\exists_\alpha x_i \psi, f) = 1 \leftrightarrow (\exists g \supseteqq f)\exists v(g \textit{ is an assignment} \\ &\wedge v \in TL \wedge \lambda(v) < \alpha \wedge \mathfrak{D}(g) \supseteqq \text{occ}(v) \wedge \text{den}(\text{sbt}(\psi, i, v), g) = 1)) \end{aligned}$$

and hence this holds also in $\mathscr{N}$. Since $T(\psi) = \{i\}$ we have in $\mathscr{N}$, by Lemma 67, $T(\boldsymbol{\psi}) = \{\mathbf{i}\}$. We have also in $\mathscr{N}$, as easily seen, $\mathbf{i} \in \omega$, t_n *is an assignment* and $\mathfrak{D}(t_n) = \mathbf{n} \supseteqq \text{occ}(\boldsymbol{\psi})$; therefore, by (41), we have in $\mathscr{N}$

$$\begin{aligned} &\text{den}(\{\langle 0, \langle 1, \boldsymbol{\alpha}\rangle\rangle\}^\frown\{\langle 0, \langle 3, \mathbf{i}\rangle\rangle\}^\frown\boldsymbol{\psi}, t_n) = 1 \\ (42) \qquad &\leftrightarrow (\exists g \supseteqq t_n)\exists v(g \textit{ is an assignment} \wedge v \in TL \wedge \lambda(v) < \boldsymbol{\alpha} \\ &\wedge \mathfrak{D}(g) \supseteqq \text{occ}(v) \wedge \text{den}(\text{sbt}(\boldsymbol{\psi}, \mathbf{i}, v), g) = 1). \end{aligned}$$

The left-hand side of (42) is equivalent in $\mathscr{N}$ to $\text{den}(\boldsymbol{\phi}, t_n) = 1$, since

$$\phi = \exists_\alpha x_i \psi = \{\langle 0, \langle 1, \alpha\rangle\rangle\}^\frown\{\langle 0, \langle 3, i\rangle\rangle\}^\frown\psi$$

implies, by Lemmas 65, 49, and 53, that in $\mathscr{N}$ $\{\langle 0, \langle 1, \boldsymbol{\alpha}\rangle\rangle\}^\frown\{\langle 0, \langle 3, \mathbf{i}\rangle\rangle\}^\frown\boldsymbol{\psi} = \boldsymbol{\phi}$. Since our aim is to prove that $\|\text{den}(\boldsymbol{\phi}, t_n) = 1\| = \|\phi\| = \|\exists_\alpha x_i \psi\|$ all we have to do is to show that $\| \ \|$ applied to the right-hand side of (42) yields $\|\exists_\alpha x_i \psi\|$, i.e.

$$\begin{aligned} (43) \qquad &\|(\exists g \supseteqq t_n)\exists v(g \textit{ is an assignment} \wedge v \in TL \wedge \lambda(v) < \boldsymbol{\alpha} \\ &\wedge \mathfrak{D}(g) \supseteqq \text{occ}(v) \wedge \text{den}(\text{sbt}(\boldsymbol{\psi}, \mathbf{i}, v), g) = 1\| = \|\exists_\alpha x_i \psi\|. \end{aligned}$$

Let us denote the left-hand side of (43) with $\mathfrak{b}$. We have to prove $\mathfrak{b} = \|\exists_\alpha x_i \psi\|$.

First let us prove $\mathfrak{b} \geqq \|\exists_\alpha x_i \psi\|$. $\|\exists_\alpha x_i \psi\| = \sum_{\lambda(v) < \alpha} \|\text{sbt}(\psi, i, v)\|$; therefore it is enough to prove that for each $v \in TL$ with $\lambda(v) < \alpha$, $\mathfrak{b} \geqq \|\text{sub}(\psi, i, v)\|$. Let $\chi = \text{sbt}(\psi, i, v)$ and let m be such that $m \geqq n$ and $\text{occ}(v) \subseteqq m$, then by the induction hypothesis $\|\text{sbt}(\psi, i, v)\| = \|\chi\| = \|\text{den}(\boldsymbol{\chi}, t_m) = 1\|$. Since $\chi = \text{sub}(\psi, i, v)$ we have, by Lemmas 68, 49, and 53 $\boldsymbol{\chi} = \text{sbt}(\boldsymbol{\psi}, \mathbf{i}, \mathbf{v})$ in $\mathscr{N}$; hence

$$\|\text{sbt}(\psi, i, v)\| = \|\text{den}(\text{sbt}(\boldsymbol{\psi}, \mathbf{i}, \mathbf{v}), t_m) = 1\|.$$

Since $m \supseteqq \text{occ}(v)$ we have, by Lemmas 62, 69, 67, 66, and 53 $\|t_m \supseteqq t_n\| = 1$, $\|t_m \textit{ is an assignment}\| = 1$, $\|\mathfrak{D}(t_m) \supseteqq \text{occ}(\mathbf{v})\| = 1$, $\|\mathbf{v} \in TL\| = 1$, $\|\lambda(\mathbf{v}) < \boldsymbol{\alpha}\| = 1$. Therefore we have

$$\begin{aligned} \|\text{sbt}(\psi, i, v)\| &= \|\text{den}(\text{sbt}(\boldsymbol{\psi}, i, v), t_m) = 1\| \\ &= \|t_m \supseteqq t_n \wedge t_m \textit{ is an assignment} \wedge \mathbf{v} \in TL \wedge \lambda(\mathbf{v}) < \boldsymbol{\alpha} \\ &\qquad \wedge \mathfrak{D}(t_m) \supseteqq \text{occ}(\mathbf{v}) \wedge \text{den}(\text{sbt}(\boldsymbol{\psi}, \mathbf{i}, \mathbf{v}), t_m) = 1\| \\ &\leqq \|(\exists g \supseteqq t_n)\exists v(g \textit{ is an assignment} \wedge v \in TL \wedge \lambda(v) < \boldsymbol{\alpha} \\ &\qquad \wedge \mathfrak{D}(g) \supseteqq \text{occ}(v) \wedge \text{den}(\text{sbt}(\boldsymbol{\psi}, \mathbf{i}, v), g) = 1\| = \mathfrak{b}. \end{aligned}$$

Thereby we have finished the proof of $\mathfrak{b} \geqq \|\exists_\alpha x_i \psi\|$.

Now we shall prove $\mathfrak{b} \leqq \|\exists_\alpha x_i \psi\|$. Since Axiom (iii) holds in $\mathscr{N}$ we have in $\mathscr{N}$ $v \in TL \rightarrow v \in S$ and hence

$$\mathfrak{b} = \|\exists g(g \textit{ is an assignment} \wedge g \supseteq t_n \wedge (\exists v \in S)(v \in TL \wedge \lambda(v) < \boldsymbol{\alpha} \wedge \mathfrak{D}(g) \supseteq \operatorname{occ}(v) \wedge \operatorname{den}(\operatorname{sbt}(\boldsymbol{\psi}, \mathbf{i}, v), g))) = 1\|$$

$$= \sum_{h \in W \vee h \supseteq k_n} \sum_{v} \|\mathbf{v} \in TL\| \, \|\lambda(\mathbf{v}) < \boldsymbol{\alpha}\| \cdot \|\mathfrak{D}(h^{\#}) \supseteq \operatorname{occ}(\mathbf{v})\| \, \|\operatorname{den}(\operatorname{sbt}(\boldsymbol{\psi}, \mathbf{i}, \mathbf{v}), h^{\#}) = 1\|,$$

by Lemmas 63(b) and 40(b, c). By Lemmas 67, 66, and 53 $\|v \in TL\| \, \|\lambda(\mathbf{v}) < \boldsymbol{\alpha}\|$ is 1 or 0 according to whether $v \in TL \wedge \lambda(v) < \alpha$ or not. If $\operatorname{occ}(v) = d$ then by Lemmas 69, 49, and 53 we have $\|\operatorname{occ}(\mathbf{v}) = \mathbf{d}\| = 1$ and hence $\|\operatorname{occ}(\mathbf{v}) \subseteq \mathfrak{D}(h^{\#})\| = \|\mathbf{d} \subseteq \mathfrak{D}(h^{\#})\|$ but, by (the proof of) Lemma 62(b), $\|\mathbf{d} \subseteq \mathfrak{D}(h^{\#})\|$ is 1 or 0 according to whether $d \subseteq \mathfrak{D}(h)$ or not. Therefore we have

$$\mathfrak{b} = \sum_{h \in W \wedge h \supseteq k_n \wedge v \in TL \wedge \lambda(v) < \alpha \wedge \mathfrak{D}(h) \supseteq \operatorname{occ}(v)} \|\operatorname{den}(\operatorname{sbt}(\boldsymbol{\psi}, \mathbf{i}, \mathbf{v}), h^{\#}) = 1\|.$$

In order to prove $\mathfrak{b} \leqq \|\exists_\alpha x_i \psi\|$ it is now enough to prove that for each $h \in W$ such that $h \supseteq k_n$ and for each term v with $\lambda(v) < \alpha$ and $\operatorname{occ}(v) \subseteq \mathfrak{D}(h)$ we have $\|\operatorname{den}(\operatorname{sbt}(\boldsymbol{\psi}, \mathbf{i}, \mathbf{v}), h^{\#}) = 1\| \leqq \|\exists_\alpha x_i \psi\|$. Let $m \in \omega$ be such that $m \supseteq \mathfrak{R}(h)$. In ZF we have, by Lemma 4,

$$\begin{aligned} \forall \psi (\forall i \in \omega) \forall v \forall f \forall s [T(\psi) = \{i\} \wedge v \in TL \wedge f \textit{ is an assignment} \\ \wedge\, s \textit{ is a substitution function} \wedge \mathfrak{D}(s) \supseteq \operatorname{occ}(\operatorname{sbt}(\psi, i, v)) \\ \wedge\, \mathfrak{D}(f) \supseteq \mathfrak{R}(s) \wedge (\forall j \in \operatorname{occ}(\psi))(s(j) = j) \\ \rightarrow \operatorname{den}(\operatorname{sbt}(\psi, i, v), f \circ s) = \operatorname{den}(\operatorname{sub}(\operatorname{sbt}(\psi, i, v), s), f) \\ = \operatorname{den}(\operatorname{sbt}(\operatorname{sub}(\psi, s), i, \operatorname{sub}(v, s)), f \circ s) = \operatorname{den}(\operatorname{sbt}(\psi, i, \operatorname{sub}(v, s)), f)], \end{aligned} \tag{44}$$

and therefore this holds also in $\mathscr{N}$. Since we have $T(\psi) = \{i\}$, $v \in TL$, h is a substitution function, $\mathfrak{D}(h) \supseteq \operatorname{occ}(\operatorname{sbt}(\psi, i, v))$, (which follows from our assumption that $\operatorname{occ}(\exists_\alpha x_i \psi) = \operatorname{occ}(\psi) \subseteq n$, $h \supseteq k_n$ and $\operatorname{occ}(v) \subseteq \mathfrak{D}(h)$), $\mathfrak{D}(t_m) = m \supseteq \mathfrak{R}(h)$, $\operatorname{occ}(\psi) \subseteq n$ and $(\forall j \in n)(h(j) = k_n(j) = j)$ we get in $\mathscr{N}$, by Lemmas 53, 54, 67, 50, 62, 69, 68, 49, and 40(d), $\mathbf{i} \in \omega$, $T(\boldsymbol{\psi}) = \{\mathbf{i}\}$, $v \in TL$, t_m *is an assignment*, $\mathbf{h}$ *is a substitution function*, $\mathfrak{D}(\mathbf{h}) \supseteq \operatorname{occ}(\operatorname{sbt}(\boldsymbol{\psi}, \mathbf{i}, \mathbf{v}))$, $\mathfrak{D}(t_m) \supseteq \mathfrak{R}(\mathbf{h})$ and $(\forall j \in \operatorname{occ}(\boldsymbol{\psi}))(\mathbf{h}(j) = j)$. Therefore, substituting $\boldsymbol{\psi}, \mathbf{i}, \mathbf{v}, t_m$ and $\mathbf{h}$ for ψ, i, v, f and s in (44) we get in $\mathscr{N}$

$$\operatorname{den}(\operatorname{sbt}(\boldsymbol{\psi}, \mathbf{i}, \mathbf{v}), t_m \circ \mathbf{h}) = \operatorname{den}(\operatorname{sbt}(\boldsymbol{\psi}, \mathbf{i}, \operatorname{sub}(\mathbf{v}, \mathbf{h})), t_m).$$

By Lemma 62(g) we have $t_m \circ \mathbf{h} = (k_m \circ h)^{\#} = h^{\#}$ in $\mathscr{N}$ and hence we have in $\mathscr{N}$

$$\operatorname{den}(\operatorname{sbt}(\boldsymbol{\psi}, \mathbf{i}, \mathbf{v}), h^{\#}) = \operatorname{den}(\operatorname{sbt}(\boldsymbol{\psi}, \mathbf{i}, \operatorname{sub}(\mathbf{v}, \mathbf{h})), t_m). \tag{45}$$

Since what is left to do is to prove that $\|\operatorname{den}(\operatorname{sbt}(\boldsymbol{\psi}, \mathbf{i}, \mathbf{v}), h^{\#}) = 1\| \leqq \|\exists_\alpha x_i \psi\|$ we have, because of (45), just to show $\|\operatorname{den}(\operatorname{sbt}(\boldsymbol{\psi}, \mathbf{i}, \operatorname{sub}(\mathbf{v}, \mathbf{h})), t_n) = 1\| \leqq \|\exists_\alpha x_i \psi\|$. Let $w = \operatorname{sub}(v, h)$; by Lemmas 70 and 66 we have in $\mathscr{N}$ $\mathbf{w} = \operatorname{sub}(\mathbf{v}, \mathbf{h})$. Thus we have to show $\|\text{den}(\text{sbt}(\boldsymbol{\psi}, \mathbf{i}, \mathbf{w}), t_m) = 1\| \leqq \|\exists_\alpha x_i \psi\|$. Since $\lambda(v) < \alpha$ also $\lambda(w) < \alpha$ and hence, by the induction hypothesis,

$$\|\text{den}(\text{sbt}(\boldsymbol{\psi}, i, w), t_m) = 1\| = \|\text{sbt}(\psi, i, w)\| \leqq \|\exists_\alpha x_i \psi\|.$$

Case 9. $u = \lambda\!\!\!\!\lambda_\alpha x_i \psi$. By definition of *den* and Lemma 4 we have in *ZF*

$$\begin{aligned}\forall \alpha(\forall i \in \omega)\forall \psi \forall f[T(\psi) = \{i\} \wedge f \textit{ is an assignment} \wedge \mathfrak{D}(f) \supseteq \text{occ}(\psi) \\ \rightarrow \forall x(x \in \text{den}(\lambda\!\!\!\!\lambda_\alpha x_i \psi, f) \leftrightarrow \exists v \exists g(v \in TL \wedge \lambda(v) < \alpha \\ \wedge\, g \textit{ is an assignment} \wedge g \supseteq f \wedge \mathfrak{D}(g) \supseteq \text{occ}(v) \\ \wedge\, \text{den}(\text{sbt}(\psi, i, v), g) = 1 \wedge x = \text{den}(v, g))],\end{aligned}$$

and hence, this holds also in $\mathcal{N}$. Substituting $\boldsymbol{\alpha}$, $\mathbf{i}$, $\boldsymbol{\psi}$ and t_n for α, i, ψ and f we get in $\mathcal{N}$, as in Case 8,

$$\begin{aligned}\forall x(x \in \text{den}(\mathbf{u}, t_n) \leftrightarrow \exists v \exists g(v \in TL \wedge \lambda(v) < \boldsymbol{\alpha} \wedge g \textit{ is an assignment} \\ \wedge\, g \supseteq t_n \wedge \mathfrak{D}(g) \supseteq \text{occ}(v) \wedge \text{den}(\text{sbt}(\boldsymbol{\psi}, \mathbf{i}, v), g) = 1 \wedge x = \text{den}(v, g))).\end{aligned}$$

Therefore we have in $\mathcal{N}$, for every $w \in TL$,

$$\begin{aligned}w \in \text{den}(\mathbf{u}, t_n) \leftrightarrow (\exists v \in S)\exists g(g \supseteq t_n \wedge v \in TL \wedge \lambda(v) < \boldsymbol{\alpha} \wedge g \textit{ is an assignment} \\ \wedge\, \mathfrak{D}(g) \supseteq \text{occ}(v) \wedge \text{den}(\text{sbt}(\boldsymbol{\psi}, \mathbf{i}, v), g) = 1 \wedge w = \text{den}(v, g)).\end{aligned}$$

Hence, by Lemmas 63(b) and 40(b, c)

$$\begin{aligned}\|w \in \text{den}(\mathbf{u}, t_n)\| = \sum_v \sum_{h \in W \wedge h \supseteq k_n} \|\mathbf{v} \in TL\| \, \|\lambda(\mathbf{v}) < \boldsymbol{\alpha}\| \, \|\mathfrak{D}(h^\#) \supseteq \text{occ}(\mathbf{v})\| \\ \cdot \|\text{den}(\text{sbt}(\boldsymbol{\psi}, \mathbf{i}, \mathbf{v}), h^\#) = 1\| \, \|w = \text{den}(\mathbf{v}, h^\#)\|\end{aligned}$$

and as in Case 8 this yields

$$\begin{aligned}&\|w \in \text{den}(\mathbf{u}, t_n)\| \\ (46)\quad &= \sum_{h \in W \wedge h \supseteq k_n \wedge v \in TL \wedge \lambda(v) < \alpha \wedge \mathfrak{D}(h) \supseteq \text{occ}(v)} \|\text{den}(\text{sbt}(\boldsymbol{\psi}, \mathbf{i}, \mathbf{v}), h^\#) = 1\| \\ &\qquad \cdot \|w = \text{den}(\mathbf{v}, h^\#)\|.\end{aligned}$$

Let us denote the right-hand side of (46) by $\mathfrak{b}$. We want to show that $u = \text{den}(\mathbf{u}, t_n)$ holds in $\mathcal{N}$. To establish that it is enough to show, since the axiom of extensionality holds in $\mathcal{N}$, that for every term w, $\|w \in u\| = \|w \in \text{den}(\mathbf{u}, t_n)\|$. Therefore, by (46), all we have to show is that for every term w

$$(47) \qquad \|w \in \lambda\!\!\!\!\lambda_\alpha x_i \psi\| = \mathfrak{b}.$$

We shall prove (47) by showing that we have inequalities both ways. First we handle the inequality $\|w \in \lambda\!\!\!\!\lambda_\alpha x_i \psi\| \leqq \mathfrak{b}$.

$$\|w \in \lambda\!\!\!\!\lambda_\alpha x_i \psi\| = \sum_{\lambda(v) < \alpha} \|w = v\| \, \|\text{sbt}(\psi, i, v)\|.$$

Therefore it is enough to prove that if $\lambda(v) < \alpha$ then

(48) $$\|w = v\| \ \|\mathrm{sbt}(\psi, i, v)\| \leqq \mathfrak{b}.$$

By the induction hypothesis we have, for a suitably large $m \geqq n$

$$\|v = \mathrm{den}(\mathbf{v}, t_m)\| = 1$$

and also, for $\chi = \mathrm{sbt}(\psi, i, v)$, $\|\mathrm{den}(\boldsymbol{\chi}, t_m) = 1\| = \|\chi\|$. Because of this, and since $\boldsymbol{\chi} = \mathrm{sbt}(\boldsymbol{\psi}, \mathbf{i}, \mathbf{v})$ in $\mathcal{N}$ (by Lemma 68) (48) becomes

(49) $$\|w = \mathrm{den}(\mathbf{v}, t_m)\| \ \|\mathrm{den}(\mathrm{sbt}(\boldsymbol{\psi}, \mathbf{i}, \mathbf{v}), t_m) = 1\| \leq \mathfrak{b}.$$

But (49) follows directly from the definition of $\mathfrak{b}$, since $t_m = k_m^{\#}$ and $m \geq n$. Now we handle the inequality $\mathfrak{b} \leqq \|w \in \lambda_\alpha x_i \psi\|$. By the definition of $\mathfrak{b}$ all we have to prove is that if $h \in W$, $h \supseteq k_n$, $v \in TL$, $\lambda(v) < \alpha$ and $\mathfrak{D}(h) \supseteq \mathrm{occ}(v)$ then

(50) $$\|\mathrm{den}(\mathrm{sbt}(\boldsymbol{\psi}, \mathbf{i}, \mathbf{v}), h^{\#}) = 1\| \ \|w = \mathrm{den}(\mathbf{v}, h^{\#})\| \leqq \|w \in \lambda_\alpha x_i \psi\|.$$

As in (45) of Case 8 we obtain in $\mathcal{N}$, for a suitably large m

$$\mathrm{den}(\mathrm{sbt}(\boldsymbol{\psi}, \mathbf{i}, \mathbf{v}), h^{\#}) = \mathrm{den}(\mathrm{sbt}(\boldsymbol{\psi}, \mathbf{i}, \mathrm{sub}(\mathbf{v}, \mathbf{h})), t_m)$$

and

$$\mathrm{den}(\mathbf{v}, h^{\#}) = \mathrm{den}(\mathrm{sub}(\mathbf{v}, \mathbf{h}), t_m)$$

and by the induction hypothesis and Lemmas 68 and 70 we have

$$\|\mathrm{den}(\mathrm{sbt}(\boldsymbol{\psi}, \mathbf{i}, \mathrm{sub}(\mathbf{v}, \mathbf{h})), t_m) = 1\| = \|\mathrm{sbt}(\psi, i, \mathrm{sub}(v, h))\|$$

and

$$\|\mathrm{den}(\mathrm{sub}(\mathbf{v}, \mathbf{h}), t_m) = \mathrm{sub}(v, h)\| = 1.$$

Therefore (50) becomes

$$\|\mathrm{sbt}(\psi, i, \mathrm{sub}(v, h))\| \ \|w = \mathrm{sub}(u, h)\| \leqq \|w \in \lambda_\alpha x_i \psi\|,$$

but this does indeed hold by definition of $\|w \in \lambda_\alpha x_i \psi\|$ since $\lambda(\mathrm{sub}(v, h)) < \alpha$.

THEOREM 72. *Axiom* (vi) *of SP holds in* $\mathcal{N}$.

PROOF. Axiom (vi) is

$$\forall x \exists u \exists f (u \in TL \wedge f \textit{ is an assignment} \wedge \mathfrak{D}(f) \supseteq \mathrm{occ}(u) \wedge x = \mathrm{den}(u, f)).$$

To show that it holds in $\mathcal{N}$ it suffices to show that for every term w there is an $m < \omega$ such that

$$\|\mathbf{w} \in TL \wedge t_m \textit{ is an assignment} \wedge \mathfrak{D}(t_m) \supseteq \mathrm{occ}(\mathbf{w}) \wedge w = \mathrm{den}(\mathbf{w}, t_m)\| = 1.$$

We choose m such that $m \supseteq \mathrm{occ}(w)$. $\|\mathbf{w} \in TL\| = 1$ by Lemmas 67 and 53, $\|t_m \textit{ is an assignment}\| = 1$ by Lemma 62, $\|\mathfrak{D}(t_m) \supseteq \mathrm{occ}(\mathbf{w})\| = 1$ by Lemmas 62, 69, 50, and 49, $\|w = \mathrm{den}(\mathbf{w}, t_m)\| = 1$ by Lemma 71.

We have now shown that Axioms (i), (iii), (iv) and (vi) of *SP* hold in $\mathcal{N}$ independently of our choice of the Boolean algebra $\mathfrak{B}$ and the double sequence $\mathfrak{a}$. However, in order to get also Axioms (ii) and (v) of *SP* we shall choose a particular $\mathfrak{B}$ and a particular $\mathfrak{a}$. We consider the set $2^{\omega\times\omega}$ of all functions f on $\omega \times \omega$ into $2 = \{0, 1\}$. We endow this set with the product topology of $\aleph_0$ copies of the two-point discrete space, i.e. the *basic open sets* are of the form $\mathfrak{b}_p = \{f \in 2^{\omega\times\omega} \mid f \supseteq p\}$, where p is a function on a finite subset of $\omega \times \omega$ into 2. We shall refer to such a function p as a *condition.* $\mathfrak{B}$ is taken to be the complete Boolean algebra of all regular open sets of this space, where the inclusion relation $\subseteq$ is taken to be the $\leq$ relation of $\mathfrak{B}$. (See Sikorski **[14]**. An open set is *regular* if it is the interior of its closure.) Every basic open set $\mathfrak{b}_p$ is also closed and is, hence, a regular open set. We take for $\mathfrak{a}_{i,n}$ the set $\{f \in 2^{\omega\times\omega} \mid f(i, n) = 1\}$.

We define now the forcing of an *SP*-formula Φ, with given values $u_{i_1}, \ldots, u_{i_n}$ for the free variables $x_{i_1}, \ldots, x_{i_n}$ of Φ, as follows.

$$p \Vdash \Phi(u_{i_1}, \ldots, u_{i_n}) \quad \text{(read: } p \text{ forces } \Phi(u_{i_1}, \ldots, u_{i_n}))$$
$$\textit{if } \mathfrak{b}_p \leq \|\Phi(u_{i_1}, \ldots, u_{i_n})\|.$$

This notion of forcing is often called weak forcing in the literature.

Let σ be a permutation of ω. We shall now define the action of σ on the members and the subsets of $2^{\omega\times\omega}$ and we shall see that σ induces an automorphism of the Boolean algebra $\mathfrak{B}$. We define $\sigma(f)$, for $f \in 2^{\omega\times\omega}$, by $\sigma(f)(\sigma(i), n) = f(i, n)$, and for a set $U \subseteq 2^{\omega\times\omega}$, $\sigma(U) = \{\sigma(f) \mid f \in U\}$; for a condition p

$$\sigma(p) = \{\langle\langle\sigma(i), n\rangle, l\rangle \mid \langle\langle i, n\rangle, l\rangle \in p\}.$$

It is easily seen that since $\sigma(\mathfrak{b}_p) = \mathfrak{b}_{\sigma(p)}$, σ acts as an homeomorphism of the topological space and hence as an automorphism of the Boolean algebra $\mathfrak{B}$. For the $\mathfrak{a}_{i,n}$'s we have $\sigma(\mathfrak{a}_{i,n}) = \mathfrak{a}_{\sigma(i),n}$. σ can also be defined to act on $TL \cup FL$ by stipulating that for an expression u of $\mathscr{L}$ $\sigma(u)$ is the expression obtained from u by replacing each a_i by $a_{\sigma(i)}$.

LEMMA 73. (a) *For every sentence* ϕ *of* $\mathscr{L}$ $\|\sigma(\phi)\| = \sigma(\|\phi\|)$.
(b) (*schema*). *For every SP-formula* Ψ,

$$\|\Psi(\sigma(v_1), \ldots, \sigma(v_n))\| = \sigma(\|\Psi(v_1, \ldots, v_n)\|).$$

PROOF. (a) is proved by induction as in the definition of $\|\phi\|$.
(b) is proved by induction on the length of Ψ, using also (a).

LEMMA 74. *If* $p \Vdash \Phi(u_1, \ldots, u_n)$ *then also* $p' \Vdash \Phi(u_1, \ldots, u_n)$, *where* p' *is the restriction of* p *to the set of all pairs* $\langle i, n'\rangle$ *such that* a_i *occurs in one of* $u_1, \ldots, u_n$.

PROOF. For a condition q let $\mathfrak{D}^*(q) = \{i \mid \exists n'(\langle i, n'\rangle \in \mathfrak{D}(q))\}$. For p and p' as above, if $\mathfrak{D}^*(p') = \mathfrak{D}^*(p)$ then $p' = p$ and the lemma holds trivially. Otherwise let $\mathfrak{D}^*(p) \sim \mathfrak{D}^*(p') = \{i_1, \ldots, i_l\}$, $l > 0$. Let $m \supseteq \mathfrak{D}^*(p)$. For each integer $k \geq 0$ let σ_k be a permutation of ω which is the identity on $\bigcup_{1\leq j\leq n} \operatorname{occ}(u_j)$ and which maps $i_1, i_2, \ldots, i_l$ on $m + kl + 1, m + kl + 2, \ldots, m + (k + 1)l$, respectively. ($\{i_1, \ldots, i_l\} \cap \bigcup_{1\leq j\leq n} \operatorname{occ}(u_j) = 0$ by definition of p'.) For each $k \geq 0$ and $1 \leq j \leq n$ we have $\sigma_k(u_j) = u_j$, since σ_k is the identity on $\operatorname{occ}(u_j)$. Therefore, by

Lemma 73(b) we have, since $\mathfrak{b}_p \leqq \|\Phi(u_1, \ldots, u_n)\|$, also $\mathfrak{b}_{\sigma_k(p)} \leqq \sigma(\|\Phi(u_1, \ldots, u_n)\|) = \|\Phi(u_1, \ldots, u_n)\|$. We claim that this implies

$$\mathfrak{b}_{p'} \leq \|\Phi(u_1, \ldots, u_n)\| \tag{51}$$

which is what we want to prove. Let us denote $\|\Phi(u_1, \ldots, u_n)\|$ with $\mathfrak{c}$. If the closure of $\mathfrak{c}$ includes $\mathfrak{b}_{p'}$ then (51) follows, since $\mathfrak{c}$, being a regular open set, is the interior of its closure, and also the interior of its closure must include $\mathfrak{b}_{p'}$, which is an open set. If, on the other hand, the closure of $\mathfrak{c}$ does not include $\mathfrak{b}_{p'}$ then there is a $g \in 2^{\omega\times\omega}$ such that $g \in \mathfrak{b}_{p'}$ but g is not in the closure of $\mathfrak{c}$. Then there is a basic open set $\mathfrak{b}_q$ which is a neighborhood of g disjoint from $\mathfrak{c}$. Let k be such that $mk + l + 1 \geqq \mathfrak{D}^*(q)$ then, as easily seen, $q' = \sigma_k(p) \cup q$ is a condition (since $p' \subseteq g, q \subseteq g$ and $(\mathfrak{D}^*(\sigma_k(p)) \sim p') \cap q = 0$) and $\mathfrak{b}_{q'} \subseteq \sigma_k(p) \subseteq \|\Phi(u_1, \ldots, u_n)\| = \mathfrak{c}$, as shown above. On the other hand, since $\mathfrak{b}_{q'} \subseteq \mathfrak{b}_q$, $\mathfrak{b}_{q'}$ is disjoint from $\mathfrak{c}$, which is a contradiction since $\mathfrak{b}_{q'} \neq 0$.

LEMMA 75. (a) *For every* $i \in \omega$, $a_i \subseteq \omega \wedge a_i \in b$ *holds in* $\mathcal{N}$.
(b) $(\forall x \in b)(x \subseteq \omega)$ *holds in* $\mathcal{N}$.

PROOF. (a) and (b) are proved by simple computation, using $\boldsymbol{\omega} = \omega$ in $\mathcal{N}$.

LEMMA 76. *For every interval designator* q

$$\|a_i \in b_{\mathbf{q}}\| = \{f \in 2^{\omega\times\omega} \mid (\forall n \in \mathfrak{D}(q))(f(i, n) = q(n))\}^2$$

(see the end of §2 *for the definition of* b_q*).*

PROOF. Since $a_i \in b$ holds in $\mathcal{N}$ by Lemma 75, $a_i \in b_{\mathbf{q}}$ is equivalent in $\mathcal{N}$, by the definition of b_q, to $(\forall n \in \mathfrak{D}(\mathbf{q}))\,(n \in a_i \leftrightarrow \mathbf{q}(n) = 1)$. Let $\mathfrak{D}(q) = d$, then by Lemmas 50, 49, and 53 we have in $\mathcal{N}$ $\mathfrak{D}(\mathbf{q}) = \mathbf{d}$ and hence

$$\|a_i \in b_{\mathbf{q}}\| = \|(\forall n \in \mathbf{d})(n \in a_i \leftrightarrow \mathbf{q}(n) = 1)\| = \prod_{n \in d} \|\mathbf{n} \in a_i \leftrightarrow \mathbf{q}(\mathbf{n}) = 1\|,$$

by Lemma 40(d). By Lemmas 50, 49, and 53 $\|\mathbf{q}(\mathbf{n}) = 1\|$ is 1 or 0 according to whether $q(n) = 1$ or not. $\|\mathbf{n} \in a_i\| = \mathfrak{a}_{i,n}$ as follows from the definition of $\|\phi\|$. Therefore we have $\|a_i \in b_{\mathbf{q}}\| = \prod_{n\in\mathfrak{D}(q)} \mathfrak{a}_{i,n}^{(q(n))}$, where $\mathfrak{c}^{(0)} = -\mathfrak{c}$, $\mathfrak{c}^{(1)} = \mathfrak{c}$. Since $\mathfrak{a}_{i,n} = \{f \in 2^{\omega\times\omega} \mid f(i, n) = 1\}$ we have $-\mathfrak{a}_{i,n} = \{f \in 2^{\omega\times\omega} \mid f(i, n) = 0\}$ and therefore $\|a_i \in b_{\mathbf{q}}\| = \{f \in 2^{\omega\times\omega} \mid f(i, n) = q(n)\}$.

LEMMA 77. *For every interval designator* q, $b_{\mathbf{q}} \neq 0$ *holds in* $\mathcal{N}$.

PROOF. For $i \in \omega$ put $\mathfrak{c}_i = \|a_i \in b_{\mathbf{q}}\| = \{f \in 2^{\omega\times\omega} \mid (\forall n \in \mathfrak{D}(q))(f(i, n) = q(n))\}$. We shall prove the lemma by showing that $(\exists x \in b)(x \in b_{\mathbf{q}})$ holds in $\mathcal{N}$. We have, by Lemma 40(b), $\|(\exists x \in b)(x \in b_{\mathbf{q}})\| = \sum_{i\in\omega} \|a_i \in b_{\mathbf{q}}\| = \sum_{i\in\omega} \mathfrak{c}_i$. We claim that $\sum_{i\in\omega} \mathfrak{c}_i = 1$. Suppose not, then there is a basic open set $\mathfrak{b}_p$ such that $\mathfrak{b}_p \leqq -\sum_{i\in\omega} \mathfrak{c}_i$. On the other hand, if $i \notin \mathfrak{D}^*(p)$ (where $\mathfrak{D}^*(p) = \{i \mid \exists n(\langle i, n\rangle \in \mathfrak{D}(p))\}$) then, obviously, $\mathfrak{b}_p \cap \mathfrak{c}_i \neq 0$ which contradicts $\mathfrak{b}_p \leqq -\sum_{i\in\omega} \mathfrak{c}_i$.

---

[2] The reader should distinguish between b_q, with the subscript q, and $b_{\mathbf{q}}$, with the boldface subscript which stands for $\mathbf{q}$.

THEOREM 78. *Axiom* (ii) *of SP holds in* $\mathcal{N}$.

PROOF. By Lemma 75 (b) "*b is a set of subsets of* ω" holds in $\mathcal{N}$. In $\mathcal{N}$ we have Axiom (iii) of *SP*; hence every interval designator is in S and therefore

$$\begin{aligned} &\|\forall q(q \text{ is an interval designator} \to b_q \neq 0)\| \\ &= \|(\forall q \in S)(q \text{ is an interval designator} \to b_q \neq 0)\| \\ &= \textstyle\sum_q \|\mathbf{q} \text{ is an interval designator} \to b_{\mathbf{q}} \neq 0\|, \quad \text{by Lemma 40(c).} \end{aligned}$$

By Lemmas 50 and 53 $\|\mathbf{q}$ *is an interval designator*$\|$ is 1 or 0 according to whether q is an interval designator or not; hence

$$\|\forall q(q \text{ is an interval designator} \to b_q \neq 0)\| = \prod_{q \text{ is an interval designator}} \|b_{\mathbf{q}} \neq 0\| = 1,$$

by Lemma 77.

THEOREM 79. *Axiom* (v) *of continuity holds in* $\mathcal{N}$.

PROOF. A general instance of the axiom schema of continuity is obviously equivalent to

$$\begin{aligned} &(\forall x_1, \ldots, x_n \in S \cup \{b\})(\forall m \in \omega)\forall y[y \text{ is an assignment} \\ &\quad \wedge y \text{ is a 1-1 function} \wedge \mathfrak{D}(y) = m \wedge \Phi(x_1, \ldots, x_n, y) \to \exists q(q \text{ is a function} \\ &\quad \wedge \mathfrak{D}(q) = m \wedge (\forall i, j \in m)(q(i) \text{ is an interval designator} \wedge y(i) \in b_{q(i)} \\ &\quad \wedge (i \neq j \to b_{q(i)} \cap b_{q(j)} = 0)) \wedge \forall z(z \text{ is an assignment} \wedge z \text{ is a 1-1 function} \\ &\qquad\qquad \wedge \mathfrak{D}(z) = m \wedge (\forall i \in m)(z(i) \in b_{q(i)}) \to \Phi(x_1, \ldots, x_n, z)))]. \end{aligned}$$

By Lemmas 40(b, d) and 63(c), and since $\|\omega = \boldsymbol{\omega}\| = 1$ and $\|u \in S \cup \{b\}\| = \sum_s \|u = \mathbf{s}\| + \|u = b\|$ the theorem will be proved once we prove that for all $u_1, \ldots, u_n$ such that each u_i is $\mathbf{s}$ for some s or is b, and for all $m \in \omega$ and for all $h \in W$ we have

$$\begin{aligned} &\|h^{\#} \text{ is an assignment} \wedge h^{\#} \text{ is a 1-1 function} \wedge \mathfrak{D}(h^{\#}) = \mathbf{m} \\ &\quad \wedge \Phi(u_1, \ldots, u_n, h)\| \leqq \|\exists q(q \text{ is a function} \wedge \mathfrak{D}(q) = \mathbf{m} \\ &\quad \wedge (\forall i, j \in \mathbf{m})(q(i) \text{ is an interval designator} \wedge h^{\#}(i) \in b_{q(i)} \\ &\quad \wedge (i \neq j \to b_{q(i)} \cap b_{q(j)} = 0)) \wedge \forall z(z \text{ is an assignment} \wedge z \text{ is a 1-1 function} \\ &\quad \wedge \mathfrak{D}(z) = \mathbf{m} \wedge (\forall i \in \mathbf{m})(z(i) \in b_{q(i)}) \to \Phi(u_1, \ldots, u_n, z)))\|. \end{aligned} \tag{52}$$

We denote the right-hand side of (52) with $\mathfrak{e}$. It is an easy consequence of Lemma 62(c) that if h is not a 1-1 function then $\|h^{\#}$ is a 1-1 function$\| = 0$; also by Lemma 62(b) if $\mathfrak{D}(h) \neq m$ then $\|\mathfrak{D}(h^{\#}) = \mathbf{m}\| = 0$. Therefore in order to prove (52) it is enough to show that if h is 1-1 and $\mathfrak{D}(h) = m$ then $\|\Phi(u_1, \ldots, u_n, h^{\#})\| \leqq \mathfrak{e}$. Suppose that $\|\Phi(u_1, \ldots, u_n, h^{\#})\| \nleqq \mathfrak{e}$, then $\|\Phi(u_1, \ldots, u_n, h^{\#})\| - \mathfrak{e} \neq 0$ and there is a basic open set $\mathfrak{b}_{p''}$ such that $\mathfrak{b}_{p''} \leqq \|\Phi(u_1, \ldots, u_n, h^{\#})\| - \mathfrak{e}$. We extend p'' to a condition $p \geqq p''$ such that if $\langle i, j\rangle \in \mathfrak{D}(p)$ and $j' < j$ then $\langle i, j'\rangle \in \mathfrak{D}(p)$ and for all $i, k < m$ with $i \neq k$ there is a j such that $\langle h(i), j\rangle, \langle h(k), j\rangle \in \mathfrak{D}(p)$ and $p(h(i), j) \neq p(h(k), j)$. Obviously $\mathfrak{b}_p \leqq \mathfrak{b}_{p''} \leqq \|\Phi(u_1, \ldots, u_n, h)\| - \mathfrak{e}$. We shall now get a contradiction by proving $\mathfrak{b}_p \leqq \mathfrak{e}$. We define functions $q_0, \ldots, q_{m-1}$ by putting $q_i(j) = p(h(i), j)$ for every pair $\langle h(i), j\rangle \in \mathfrak{D}(p)$. By our choice of $p, q_0, \ldots, q_{m-1}$ are pairwise incompatible interval designators. We put $q = q_0, \ldots, q_{m-1}$. In order to prove $\mathfrak{b}_p \leqq \mathfrak{e}$ it is enough, by the definition of $\mathfrak{e}$, to prove

$$(53)\quad \begin{aligned} \mathfrak{b}_p \leqq \|&\mathbf{q} \textit{ is a function} \wedge \mathfrak{D}(\mathbf{q}) = \mathbf{m} \wedge (\forall i, j \in \mathbf{m})(\mathbf{q}(i) \textit{ is}\\ &\quad \textit{an interval designator} \wedge h^{\#}(i) \in b_{\mathbf{q}(i)} \wedge (i \neq j \rightarrow b_{\mathbf{q}(i)} \cap b_{\mathbf{q}(j)} = 0))\\ &\wedge \forall z(z \textit{ is an assignment} \wedge z \textit{ is a 1-1 function}\\ &\wedge \mathfrak{D}(z) = \mathbf{m} \wedge (\forall i \in \mathbf{m})(z(i) \in b_{\mathbf{q}(i)}) \rightarrow \Phi(u_1, \ldots, u_n, z))\|. \end{aligned}$$

By Lemmas 50, 49 and 53 $\|\mathbf{q}$ *is a function* $\wedge\ \mathfrak{D}(\mathbf{q}) = \mathbf{m}\| = 1$. Let $\mathfrak{d}$ be

$$\|(\forall i, j \in m)(\mathbf{q}(i) \textit{ is an interval designator} \wedge h^{\#}(i) \in b_{\mathbf{q}(i)} \wedge (i \neq j \rightarrow b_{\mathbf{q}(i)} \cap b_{\mathbf{q}(j)} = 0))\|.$$

By Lemma 40(d)

$$(54)\quad \mathfrak{d} = \prod_{i,j\in m} \|\mathbf{q}(\mathbf{i}) \textit{ is an interval designator} \wedge h^{\#}(\mathbf{i}) \in b_{\mathbf{q}(\mathbf{i})} \wedge (\mathbf{i} \neq \mathbf{j} \rightarrow b_{\mathbf{q}(\mathbf{i})} \cap b_{\mathbf{q}(\mathbf{j})} = 0)\|.$$

By Lemmas 50, 49, and 53 we have, for $i < m$, $\|\mathbf{q}(\mathbf{i})$ *is an interval designator*$\| = 1$. By Lemmas 62(c) and 76

$$\|h^{\#}(\mathbf{i}) \in b_{\mathbf{q}(\mathbf{i})}\| = \|a_{h(i)} \in b_{\mathbf{q}(\mathbf{i})}\| = \{f \in 2^{\omega\times\omega} \mid (\forall n \in \mathfrak{D}(q(i)))(f(h(i), n) = q(i)(n))\}.$$

If $i = j$ then obviously $\|\mathbf{i} \neq \mathbf{j} \rightarrow b_{\mathbf{q}(\mathbf{i})} \cap b_{\mathbf{q}(\mathbf{i})} = 0\| = 1$; if $i, j \in m$ and $i \neq j$ then $q(i)$ and $q(j)$ are incompatible and hence, by Lemmas 50, 49, and 53, we have in $\mathcal{N}$ "$\mathbf{q}(\mathbf{i})$ *and* $\mathbf{q}(\mathbf{j})$ *are incompatible*". Since we have also in $\mathcal{N}$, by definition of b_p, "*if* p' *and* p'' *are incompatible interval designators then* $b_{p'} \cap b_{p''} = 0$", we have in $\mathcal{N}$ $b_{\mathbf{q}(\mathbf{i})} \cap b_{\mathbf{q}(\mathbf{j})} = 0$, and therefore $\|\mathbf{i} \neq \mathbf{j} \rightarrow b_{\mathbf{q}(\mathbf{i})} \cap b_{\mathbf{q}(\mathbf{j})} = 0\| = 1$. Thus we get from (54)

$$\begin{aligned} \mathfrak{d} &= \{f \in 2^{\omega\times\omega} \mid (\forall i \in m)(\forall n \in \mathfrak{D}(q(i)))(f(h(i), n) = q(i)(n))\}\\ &= \{f \in 2^{\omega\times\omega} \mid (\forall i \in m)\forall n(\langle h(i), n\rangle \in \mathfrak{D}(p) \rightarrow f(h(i), n) = p(h(i), n))\}\\ &\geqq \{f \in 2^{\omega\times\omega} \mid f \supseteq p\} = \mathfrak{b}_p. \end{aligned}$$

To finish the proof of (53) we still have to prove

$$\mathfrak{b}_p \leqq \|\forall z(z \textit{ is an assignment} \wedge z \textit{ is a 1-1 function} \wedge \mathfrak{D}(z) = \mathbf{m} \wedge (\forall i \in \mathbf{m})(z(i) \in b_{\mathbf{q}(i)}) \rightarrow \Phi(u_1, \ldots, u_n, z))\|.$$

By Lemmas 63(c) and 62(b) this follows once we prove that for every $k \in W$ which is 1-1 with $\mathfrak{D}(k) = m$ we have

$$\mathfrak{b}_p \leq \|(\forall i \in \mathbf{m})(k^{\#}(i) \in b_{\mathbf{q}(i)}) \rightarrow \Phi(u_1, \ldots, u_n, k^{\#})\|$$

or, equivalently

$$(55)\qquad \mathfrak{b}_p \cdot \|(\forall i \in \mathbf{m})(k^{\#}(i) \in b_{\mathbf{q}(i)})\| \leqq \|\Phi(u_1, \ldots, u_n, k^{\#})\|.$$

By Lemmas 40(d), 62(c) we have

$$\begin{aligned} \|(\forall i \in \mathbf{m})(k^{\#}(i) \in b_{\mathbf{q}(\mathbf{i})})\| &= \prod_{i\in m} \|k^{\#}(\mathbf{i}) \in b_{\mathbf{q}(\mathbf{i})}\| = \prod_{i\in m} \|a_{k(\mathbf{i})} \in b_{\mathbf{q}(\mathbf{i})}\| \\ &= \prod_{i\in m} \{f \in 2^{\omega\times\omega} \mid (\forall n \in \mathfrak{D}(q(i)))(f(k(i), n) = q(i)(n))\} \\ &= \bigcap_{i\in m} \{f \in 2^{\omega\times\omega} \mid \forall n(\langle h(i), n\rangle \in \mathfrak{D}(p) \rightarrow f(k(i), n) = p(h(i), n))\} \\ &= \{f \in 2^{\omega\times\omega} \mid f \supseteq r\} = \mathfrak{b}_r, \end{aligned}$$

where r is a condition such that $\mathfrak{D}(r) = \{\langle k(i), n\rangle \mid i < m \wedge \langle h(i), n\rangle \in \mathfrak{D}(p)\}$ and $r(k(i), n) = p(h(i), n)$ for $\langle k(i), n\rangle \in \mathfrak{D}(r)$. Therefore, to prove (55) we have to prove

$$(56)\qquad \mathfrak{b}_p \cdot \mathfrak{b}_r \leq \|\Phi(u_1, \ldots, u_n, k^{\#})\|.$$

Since each u_j, $j < n$, is of the form $\mathbf{s}$ or is b, the only a_i's which occur in $\Phi(u_1, \ldots, u_n, h^{\#})$ are $a_{h(0)}, \ldots, a_{h(m-1)}$. Let p' be the restriction of p to the set of all pairs $\langle h(i), l\rangle$, where $i < m$. Since by our choice of p, $p \Vdash \Phi(u_1, \ldots, u_n, h^{\#})$ we have, by Lemma 74, $p' \Vdash \Phi(u_1, \ldots, u_n, h^{\#})$. Let σ be a permutation of ω such that for each $i < m$ $\sigma(h(i)) = k(i)$, then, by definition of r, $\sigma(p') = r$. Thus we have, by Lemma 73,

$$r = \sigma(p') \Vdash \Phi(\sigma(u_1), \ldots, \sigma(u_n), \sigma(h^{\#})).$$

But since for $1 \leq j \leq n$, u_j is $\mathbf{s}$ or b we have $\sigma(u_j) = u_j$ for $1 \leq j \leq n$, and $\sigma(h^{\#})$ is obviously $k^{\#}$ we get $r \Vdash \Phi(u_1, \ldots, u_n, k^{\#})$, which establishes (56).

Bibliography

1. P. J. Cohen, *Set theory and the continuum hypothesis*, Benjamin, New York, 1966.

2. W. B. Easton, *Powers of regular cardinals*, Ph.D. thesis, Princeton University, 1964.

3. K. Gödel, *The consistency of the continuum hypothesis*, Princeton University Press, Princeton, N.J., 1940.

4. J. D. Halpern, *The independence of the axiom of choice from the Boolean prime ideal theorem*, Fund. Math. **55** (1964), 57–66.

5. J. D. Halpern and H. Läuchli, *A partition theorem*, Trans. Amer. Math. Soc. **124** (1966), 360–367.

6. M. Jaegermann, *The axiom of choice and two definitions of continuity*, Bull. Acad. Polon. Sci. Ser. Sci. Math. Astronom. Phys. **13** (1965), 699–704.

7. W. Kinna and K. Wagner, *Über eine Abschwechung der Auswahlaxioms*, Fund. Math. **42** (1955), 75–82.

8. J. Łoś and C. Ryll-Nardzewski, *Effectiveness of the representation theory for Boolean algebras*, Fund. Math. **41** (1954), 49–56.

9. A. Mostowski, *Über die Unabhängigkeit des Wohlordnungssatz vom Ordnungsprinzip*, Fund. Math. **32** (1939), 201–252.

10. W. V. Quine, *Set theory and its logic*, Harvard University Press, Cambridge, Mass., 1963.

11. H. Rubin and J. E. Rubin, *Equivalents of the axiom of choice*, North-Holland, Amsterdam, 1963.

12. W. Sierpiński, *L'axiome de M. Zermelo et son role dans la théorie des ensembles et l'analyse*, Bull. Acad. Sci. Cracovie, Cl. Sci. Math. e Nat., Série A, 1918, 97–152.

13. B. Scarpellini, *On a family of models of Zermelo-Fraenkel set theory*, Z. Math. Logik Grundlagen Math. **12** (1966), 191–204.

14. R. Sikorski, *Boolean algebras*, Springer-Verlag, Berlin, 1960 and 1964.

15. D. Scott and R. Solovay, *Boolean-valued models for set theory*, these Proceedings, part II.

UNIVERSITY OF MICHIGAN

THE HEBREW UNIVERSITY OF JERUSALEM

ON MODELS FOR SET THEORY WITHOUT AC

TOMÁŠ JECH

1. **Introduction.** The present contribution deals with *symmetric* submodels of ∇-models (Boolean valued models), i.e. with submodels given by a normal filter of groups of automorphisms of the corresponding complete Boolean algebra. We are particularly interested in sets of ordinals and sets of such sets in these models. It can easily be proved (by induction on ranks) that a model for set theory with the axiom of choice is uniquely determined by all its sets of ordinals. We are going to show that this is not the case for models without AC.

By a ∇-model $\nabla(\mathscr{B}, j)$ we understand a model for Gödel-Bernays set theory given by a complete Boolean algebra $\mathscr{B}$ and an ultrafilter j on it (cf. [3]). ∇-models are very closely related to Boolean-valued models. Throughout the paper we use essentially the same notation as in [2] and assume the reader to be familiar with basic concepts of Boolean-valued models.

2. **Symmetric submodels of ∇-models.** A class M (we work within the Gödel-Bernays set theory) is called a *model-class* if it is transitive, almost universal and closed under the fundamental operations, i.e.

(i) $x \in M \rightarrow x \subseteq M$,

(ii) $x \subseteq M \rightarrow (\exists y)(x \subseteq y \;\&\; y \in M)$,

(iii) $x, y \in M \rightarrow F_i(x, y) \in M, \qquad (i = 1, \ldots, 8)$.

Then M with $\in$ restricted (and subclasses properly defined) yields a model for set theory.

In the sequel, let $\mathscr{B}$ be a fixed complete Boolean algebra and j an ultrafilter on it. For each subclass X of the Boolean universe $V^{(\mathscr{B})}$ let us denote

$$\mathrm{Cl}_j(X) = \left\{x : \sum_{y \in X} [\![x = y]\!] \in j\right\}.$$

For each $X \subseteq V^{(\mathscr{B})}$, $\mathrm{Cl}_j(X)$ is a class of the model $\nabla(\beta, j)$.

DEFINITION. Let $\check{V} \subseteq M \subseteq V^{(\mathscr{B})}$ ($\check{V}$ being the image under the natural embedding of the real universe into the Boolean universe). M is called a *$\mathscr{B}$-model-class*, if

(i) $x \in M \to \text{dom}(x) \subseteq M$,

(ii) $x \subseteq M \to (\exists y)(x \subseteq y \,\&\, \check{y} \in M)$,

(iii) $x, y \in M \to F_i^{(\mathscr{B})}(x, y) \restriction M \in M, \qquad (i = 1, \ldots, 8)$,

$F_i^{(\mathscr{B})}$ being the Boolean counterparts of the operations $F_1, \ldots, F_8$.

The following theorem describes the submodels of ∇-models. The proof is a matter of computation.

THEOREM 1. (a) *If $\check{V} \subseteq M \subseteq V^{(\mathscr{B})}$ and if M is a model-class in $\nabla(\mathscr{B}, j)$, then M is a $\mathscr{B}$-model-class.*

(b) *If M is a $\mathscr{B}$-model-class, then* $\text{Cl}_j(M)$ *is a model-class in* $\nabla(\mathscr{B}, j)$, *for every* j.

Under submodels of ∇-models, the symmetric submodels are of particular importance. The basic idea is a modification of the Fraenkel-Mostowski-Specker method and for using it for models with the axiom of regularity we refer to **[1]** and **[4]**.

Each automorphism g of $\mathscr{B}$ can be extended onto the entire Boolean universe $V^{(\mathscr{B})}$:

$$g0 = 0, \quad \text{dom}(gx) = g''\text{dom}(x), \quad (gx)(gy) = g(x(y)).$$

A filter F on the lattice of all groups of automorphisms of $\mathscr{B}$ is *normal*, if

$$G(F) = \{g : (\forall H \in F)(gHg^{-1} \in F)\} \in F.$$

For each $x \in V^{(\mathscr{B})}$, let us denote

$$h(x) = \{g : gx = x\}.$$

If $h(x) \in F$, x is called *F-symmetric*.

The following class (of all *hereditarily F-symmetric* elements of $V^{(\mathscr{B})}$) is a $\mathscr{B}$-model-class:

$$V^{(F)} = \{x \in V^{(\mathscr{B})} : \ \text{dom}(x) \subseteq V^{(F)} \,\&\, x \text{ is } F\text{-symmetric}\}.$$

The proof of $V^{(F)}$ being a $\mathscr{B}$-model-class is easy using the symmetry argument. Denote by $\nabla(\mathscr{B}, F, j)$ the submodel of $\nabla(\mathscr{B}, j)$ given by $V^{(F)}$.

REMARK. If F is principal then AC holds in $\nabla(\mathscr{B}, F, j)$, for every j.

3. **Sets of ordinals in symmetric models.**

DEFINITION 3.1. Let F be normal

$$P_1^F = \{x \in V^{(F)} : \text{dom}(x) \subseteq \check{\text{OR}}\}.$$

Obviously, the class $\text{Cl}_j(P_1^F)$ is in $\nabla(\mathscr{B}, F, j)$ the class of all sets of ordinals. We are going to show that

(i) P_1^F depends only on what closed groups are in F,

(ii) there is a 1-1 correspondence between the classes P_1^F (for various F's) and symmetric ideals on the lattice of all complete Boolean subalgebras of $\mathscr{B}$.

In the sequel, let u, B, g, H denote elements of $\mathscr{B}$, complete subalgebras of $\mathscr{B}$, automorphisms of $\mathscr{B}$ and groups of automorphisms, respectively.

Now, we are going to define two operators h and b. hB is a group of automorphisms for each complete B and bH is a complete subalgebra of $\mathscr{B}$ for each group H. The proofs of the following lemmas are left to the reader.

DEFINITION 3.2.

$$hB = \{g : gu = u \text{ for each } u \in B\},$$

$$bH = \{u : gu = u \text{ for each } g \in H\}.$$

LEMMA 3.3.

(1a) $B_1 \subseteq B_2 \to hB_1 \supseteq hB_2$, (2a) $H_1 \subseteq H_2 \to bH_1 \supseteq bH_2$,

(b) $h(B_1 \vee B_2) = hB_1 \cap hB_2$, (b) $b(H_1 \vee H_2) = bH_1 \cap bH_2$,

$h(B_1 \cap B_2) \supseteq hB_1 \vee hB_2$, $b(H_1 \cap H_2) \supset bH_1 \vee bH_2$,

(c) $h(gB) = g(hB)g^{-1}$, (c) $b(gHg^{-1}) = g(bH)$.

REMARK. gB is the same as $g''B$. Similarly, gI is $\{gB : B \in I\}$, I being an ideal on the lattice $\mathscr{L}_{\mathscr{B}}$ of all complete subalgebras of $\mathscr{B}$.

DEFINITION 3.4.

$$\bar{B} = bhB \qquad (\textit{closure} \text{ of } B),$$

$$\bar{H} = hbH \qquad (\textit{closure} \text{ of } H).$$

LEMMA 3.5.

(a) $hbhB = hB$, $bhbH = bH$,

(b) $B \subseteq \bar{B}$ $H \subseteq \bar{H}$,

(c) $\bar{\bar{B}} = \bar{B}$, $\bar{\bar{H}} = \bar{H}$,

(d) $B_1 \subseteq B_2 \to \bar{B}_1 \subseteq \bar{B}_2$, $H_1 \subseteq H_2 \to \bar{H}_1 \subseteq \bar{H}_2$,

(e) $u \notin B \leftrightarrow (\exists g)(g \in hB \;\&\; gu \neq u)$, $g \notin H \leftrightarrow (\exists u)(u \in bH \;\&\; gu \neq u)$.

Now, we are going to characterize the classes P_1^F by ideals on $\mathscr{L}_{\mathscr{B}}$.

DEFINITION 3.6. $I(F)$ is the ideal on $\mathscr{L}_{\mathscr{B}}$ generated by the collection $\{bH : H \in F\}$.

THEOREM 2. *Let F, F' be two normal filters. Then $P_1^F = P_1^{F'}$ if and only if $I(F) = I(F')$.*

PROOF. With each function from $\check{\mathrm{O}}\mathrm{R}$ into $\mathscr{B}$ we can associate a complete B_x generated by the values of x. On the other hand, for each B there is an x such that $B_x = B$, viz. such a function, the range of which is B. Having a normal filter F, the collection $I = \{B_x : x \in P_1^F\}$ is obviously an ideal on $\mathscr{L}_{\mathscr{B}}$. We shall show that $I = I(F)$. To prove $I(F) \subseteq I$, let $H \in F$. Then there is an x with $B_x = bH$ and we have $h(x) = hB_x = \bar{H}$; hence $x \in P_1^F$. To prove $I \subseteq I(F)$, let $x \in P_1^F$. Then $h(x) = hB_x \in F$; hence $B_x \subseteq \bar{B}_x = bh(x)$.

To make the characterization of sets of ordinals by ideals on $\mathscr{L}_{\mathscr{B}}$ complete, we shall show what ideals correspond to normal F's.

DEFINITION 3.7. Let I be an ideal on $\mathscr{L}_{\mathscr{B}}$.

$$G(I) = \{g : gI = I\}.$$

I is *symmetric*, if

$$B \in I \to b(G(I) \cap hB) \in I.$$

THEOREM 3. (a) *If F is normal then $I(F)$ is symmetric.*
(b) *If I is symmetric then there is a normal F such that $I = I(F)$.*

PROOF. (a) It suffices to prove that $G(I(F)) \supseteq G(F)$. Let $g \in G(F)$, let $B \in I(F)$. We show that $gB \in I(F)$. There is $H \in F$ with $B \subseteq bH$. Hence $gB \subseteq g(bH) = b(gHg^{-1}) \in I(F)$.

(b) Let F be the filter generated by $G(I)$ and $\{hB : B \in I\}$. F is normal, for $G(F) = \{g : (\forall H \in F)(gHg^{-1} \in F)\} \supseteq \{g \in G(I) : (\forall B \in I)(h(gB) \in F)\} = h(I)$. Let us prove $I = I(F)$. If $B \in I$, then $B \subseteq \bar{B} \in I(F)$. If $B \in I(F)$, then $B \subseteq b(G(I) \cap hB_1 \cap \ldots \cap hB_k) = b(G(I) \cap hB')$, where $B_1, \ldots, B_k, B'$ belong to I.

There is another characterization of the classes P_1^F. A group is called *closed* if $\bar{H} = H$.

THEOREM 4. *For F, F' normal, $P_1^F = P_1^{F'}$ holds if and only if F and F' coincide on closed groups.*

The theorem is an immediate consequence of the following lemma:

LEMMA 3.8. *If x is a function from $\check{\mathrm{O}}\mathrm{R}$ into $\mathscr{B}$, then $h(x)$ is closed. For each closed H there is a function x from $\check{\mathrm{O}}\mathrm{R}$ into $\mathscr{B}$ such that $h(x) = H$.*

PROOF. $h(x) = hB_x$, which is closed by Lemma 3.5. If H is closed, then there is x with $B_x = bH$; hence $h(x) = hB_x = \bar{H} = H$.

4. **Two models with the same sets of ordinals.** In this section, two different symmetric models are constructed, which have the same sets of ordinals.

Let $\mathscr{B}$ be the algebra of all regular open sets of the Cantor space $2^{\omega \times \omega}$. We shall often use the fact that $\mathscr{B}$ is generated by the collection

$$A = \{u_s : \mathrm{dom}(s) \subseteq \omega \times \omega \text{ is finite and } \mathrm{rng}(s) \subseteq \{0, 1\}\}$$

where

$$u_s = \{f \in 2^{\omega \times \omega} : f \supseteq s\}.$$

Let G be the group of all permutations of the set $\omega \times \omega$. Every permutation $g \in G$ induces a homeomorphism of $2^{\omega \times \omega}$ (and, a fortiori, an automorphism of $\mathscr{B}$), which is denoted by g as well:

$$(gf)(\langle m, n \rangle) = f(g\langle m, n \rangle).$$

Let H be the group of all permutations of the set ω. Every permutation $g \in H$ induces a permutation of $\omega \times \omega$ (denoted again g):

$$g\langle m, n \rangle = \langle gm, n \rangle.$$

Let $K \subseteq G$ be the group of all the permutations which move with only a finite number of elements. By the *mth column* (of the square $\omega \times \omega$) we understand the set $\{\langle m, n\rangle : n \in \omega\}$. Let, for $m \in \omega$, $G_m \subseteq G$ be the group of all the permutations which keep the first m columns (i.e. the 0th up to $(m-1)$th) fixed.

The groups G, H, K, G_m can be regarded as groups of automorphisms of $\mathscr{B}$. Let us define two filters F, F' as follows:

F is the least filter containing H and all G_m, $m \in \omega$,

F' is the least filter containing K and all G_m, $m \in \omega$.

In the following three lemmas we prove that F and F' are normal and that, for each ultrafilter j on $\mathscr{B}$, the models $\nabla(\mathscr{B}, F, j)$ and $\nabla(\mathscr{B}, F', j)$ are different and have the same sets of ordinals.

LEMMA 4.1. *Both F and F' are normal.*

PROOF. We shall prove that $G(F) \supseteq H$ and $G(F') \supseteq K$. Let us prove, e.g. the second assertion. Let $m \in \omega$, $g \in K$; it suffices to prove that there is $k \in \omega$ such that
$$g G_m g^{-1} \supseteq G_k.$$
Since g moves with only a finite number of elements of $\omega \times \omega$, the number $k \geqslant m$ can be chosen in such a way that g moves only with elements of the first k columns. Given a $g_1 \in G_k$, let us define $g_2 = g^{-1} g_1 g$. It can be easily verified that $g_2 \in G_m$ and that $g g_2 g^{-1} = g_1$.

LEMMA 4.2. $I(F) = I(F')$.

PROOF. It will be proved that $b(G_m \cap H) = b(G_m \cap K) = bG_m$ for each $m \in \omega$ and thus $I(F) = I(F')$ is the ideal generated by $\{bG_m : m \in \omega\}$. Let, for $m \in \omega$, $B(m)$ be the algebra generated by all those $u_s \in A$, where $\mathrm{dom}(s)$ contains elements of the first m columns only. It can be proved that $b(G_m \cap H) = b(G_m \cap K) = bG_m = B(m)$. We prove, e.g. that $b(G_m \cap H) \subseteq B(m)$. Let $u \in b(G_m \cap H)$. There is a collection $\{v_\xi : \xi \in \alpha\} \subseteq A$ of generators such that $u = \sum_{\xi \in \alpha} v_\xi$. For each $\xi \in \alpha$, we shall find $w_\xi = u_s \in A$ such that $v_\xi \leqslant w_\xi \leqslant u$ and that $\mathrm{dom}(s)$ contains elements of the first m columns only. This will prove that $u \in B(m)$. Let $v_\xi = u_{\tilde{s}} \in A$ and let us define w_ξ as $\sum \{g u_{\tilde{s}} : g \in G_m \cap H\}$. It is obvious that $v_\xi \leqslant w_\xi \leqslant u$. We shall prove that $w_\xi = u_s$, s being the restriction of $\tilde{s}$ to the first m columns. Obviously, $u_s \geqslant w_\xi$. To show that $u_s \leqslant w_\xi$, it suffices to prove that for each $u_{\bar{s}} \leqslant u_s$ there is $g \in G_m \cap H$ such that $u_{\bar{s}} \cdot g u_{\tilde{s}} \neq 0$. But $u_{\bar{s}} \leqslant u_s$ is the same as $\bar{s} \supseteq s$ and $u_{\bar{s}} \cdot g u_{\tilde{s}} \neq 0$ holds whenever $g''\mathrm{dom}(\tilde{s}) \cap \mathrm{dom}(\bar{s}) = \mathrm{dom}(s)$. Such $g \in G_m \cap H$ can easily be found.

LEMMA 4.3. *Let j be an ultrafilter on $\mathscr{B}$. There is $a \in V^{(\mathscr{B})}$, which belongs to $\nabla(\mathscr{B}, F, j)$ but not to $\nabla(\mathscr{B}, F', j)$.*

PROOF. Let us define:
$$\mathrm{dom}(y_m) = \check{\omega}, \quad y_m(\check{n}) = \{f \in 2^{\omega \times \omega} : f(m, n) = 1\},$$
$$a = \{y_m : m \in \omega\}.$$

It is obvious that $h(y_m) \supseteq G_m$, $h(\check{a}) \supseteq H$ and hence $y_m \in P_1^F = P_1^{F'}$, $\check{a} \in V^{(F)}$. We shall prove that $\check{a} \notin \mathrm{Cl}_j(V^{(F')})$ for any ultrafilter j. Let, on the contrary, there be $c \in V^{(F')}$ and $u_s \in A$ such that $u_s \leqslant [\![\check{a} = c]\!]$. It means that there is $m \in \omega$ such that $h(c) \supseteq K \cap G_m$ and $\mathrm{dom}(s)$ contains elements of the first m columns only. Let $g \in K \cap G_m$ be the permutation which interchanges $\langle m, 0\rangle$ and $\langle m, 1\rangle$ and identical otherwise. It can be proved that $[\![g\check{a} = \check{a}]\!] = [\![gy_m = y_m]\!] = \{f : f(m, 0) = f(m, 1)\}$. Thus, if choosing $\bar{s} \supseteq s$ such that $\bar{s}(m, 0) \neq \bar{s}(m, 1)$, we obtain $u_{\bar{s}} \leqslant u_s$, $u_{\bar{s}} \leqslant [\![g\check{a} \neq \check{a}]\!]$. This is, along with $gc = c$, the required contradiction.

5. **Families of sets of ordinals.** The result of the present section is Theorem 5, which gives a sufficient condition for two symmetric models to have the same families of sets of ordinals.

DEFINITION 5.1. Let F be normal

$$P_2^F = \{x \in V^{(F)} : \mathrm{dom}(x) \subseteq P_1^F\}.$$

Obviously, $\mathrm{Cl}_j(P)_2^F$ is the class of all families of sets of ordinals in $\nabla(\mathscr{B}, F, j)$.

Let now F_1 and F_2 be two normal filters coinciding on closed groups. Denote by F their restriction to the closed groups.

DEFINITION 5.2.

$$H_1 \leqslant H_2 \leftrightarrow (\forall H \in F)(H_1 \subseteq H \cdot H_2),$$

$$H_1 \simeq H_2 \leftrightarrow H_1 \leqslant H_2 \;\&\; H_2 \leqslant H_1,$$

$$F_1 \simeq H_2 \leftrightarrow (\forall H_1 \in F_1)(\exists H_2 \in F_2)(H_1 \simeq H_2) \;\&\; (\forall H_2 \in F_2)(\exists H_1 \in F_1)(H_1 \simeq H_2).$$

LEMMA 5.3. (a) *The relation $\simeq$ between groups is an equivalence.*

(b) *If $\bar{H}_1, \bar{H}_2 \in F$ and $H_1 \simeq H_2$, then $bH_1 = bH_2$ (and hence $\bar{H}_1 = \bar{H}_2$).*

PROOF. The first part is obvious, let us prove $bH_1 \subseteq bH_2$. Let $u \in bH_1$, $g \in H_2$, let us prove $gu = u$. Since $\bar{H}_1 \in F$, there is $\tilde{g} \in H_1$ with $g\tilde{g} \in \bar{H}_1 = hbH_1$; hence $g\tilde{g}u = u$. Since $\tilde{g}u = u$, we have $gu = u$, Q.E.D.

LEMMA 5.4. *Let x be a function from P_1^F into $\mathscr{B}$ and let $h(x)$ belong either to F_1 or to F_2. Then $h(x)$ is maximal in its equivalence class. In other words, if $H_1 \leqslant h(x)$, then $H_1 \subseteq h(x)$.*

PROOF. Let $H_1 \leqslant h(x)$, let $g \in H_1$. We shall prove that $gx = x$. For $y \in \mathrm{dom}(x)$ it is necessary to prove that $x(gy) = g(x(y))$. Let us denote $u = x(y)$. If $\tilde{g}$ keeps both x and y fixed then $\tilde{g}u = \tilde{g}(x(y)) = (\tilde{g}x)(\tilde{g}y) = x(y) = u$ and thus $h(u) = \{\tilde{g} : \tilde{g}u = u\} \supseteq h(x) \cap h(y)$. Denote

$$H = \overline{h(x) \cap h(y)}.$$

H is closed and is in F_1 or in F_2. Hence $H \in F$. Therefore we can find $\tilde{g} \in h(x)$ such that $g\tilde{g}^{-1} \in H$. Since both $h(u)$ and $h(y)$ are closed, we have $H \subseteq h(u)$, $H \subseteq h(y)$. Hence $gu = \tilde{g}u$ and $gy = \tilde{g}y$. Hence $x(gy) = g(x(y))$, Q.E.D.

THEOREM 5. *Let F_1, F_2 be normal filters which coincide on closed groups. Let $F_1 \simeq F_2$. Then $P_2^{F_1} = P_2^{F_2}$.*

PROOF. Let $h(x) \in F_1$, x being a function from P_1^F into $\mathscr{B}$. There is $H_2 \in F_2$ with $H_2 \leqslant h(x)$ and, by Lemma 5.4., $H_2 \subseteq h(x)$. Hence $h(x) \in F_2$. The same for the other inclusion.

PROBLEM. Are there two different symmetric models $\nabla(\mathscr{B}, F_1, j)$ and $\nabla(\mathscr{B}, F_2, j)$ with the same families of sets of ordinals?

REFERENCES.

1. Paul J. Cohen, *The independence of the axiom of choice*, mimeographed, Stanford, 1963.

2. Dana Scott and Robert Solovay, *Boolean-valued models for set theory.* These Proceedings, part II.

3. Petr Vopěnka, *General theory of ∇-models*, Comment. Math. Univ. Carolinae, **8** (1967), 145–170.

4. Petr Vopěnka and Petr Hájek, *Permutation submodels of the model ∇*, Bull Acad. Polon. Sci **13** (1965), 611–614.

UNIVERSITY OF PRAGUE
PRAGUE, CZECHOSLOVAKIA

PRIMITIVE RECURSIVE SET FUNCTIONS

RONALD B. JENSEN AND CAROL KARP

This paper[1] deals with the primitive recursive operations on the universe of sets. It is well known that the familiar primitive recursive functions of natural numbers can be formulated as functions on HF, the hereditarily finite sets. Such a formulation has technical advantages, for instance, ease of coding.[2] The primitive recursive functions treated here give a perfectly good theory of primitive recursive functions on T when T is HF or any other transitive Prim-closed class. The Prim functions on T are Σ_1^T-definable, their enumerating function is Σ_1 on T and so on. (Theorems 2.10, 2.11.) The advantages of treating primitive recursive functions on sets, rather than just on ordinals, are more than just technical. One really cannot identify the primitive recursive set functions with primitive recursive ordinal functions without making a strong assumption on the enumerability of sets by ordinals, such as an assumption of constructibility.

The present authors arrived at this family of primitive recursive set functions independently and for different reasons. The first author wanted a tool for the study of levels of constructibility, the second, a tool for the classification of systems of infinitary logic. This paper grew out of their meeting during the Logic Colloquium at Leicester in the summer of 1965. The second author had the Stability Theorem of §2 at that time. The growth of a primitive recursive set function in a hierarchy, satisfying the conditions of Theorem 2.9, is bounded and stabilized by a primitive recursive ordinal function. The first author had the equivalence of primitive recursive set functions on sets constructible before α and primitive recursive ordinal functionals on α, giving an early version of the results in §3. He had also established the result in §5 that primitive recursive functions on sets of rank less than α

---

[1] The second author received support from NSF grants GP-5456 and GP-6897 during the preparation of this paper.

[2] See Rödding [14] for details.

are determined by their values on sets constructible below α, provided that α is a limit of admissible ordinals. The lemma used to establish this result is a generalization of Shoenfield's Absoluteness Lemma that is interesting in its own right. The result of §3, giving the equivalence between primitive recursive set functions on sets constructible before α and primitive recursive ordinal functions on α, is a recent result of the first author based on a lemma of his student, Max Schröder. A version of this lemma is given in Appendix II. This is an improvement of the earlier result because the primitive recursive ordinal functions are defined using only ordinal arguments, while the primitive recursive ordinal functionals took both ordinals and sequences of ordinals as arguments. §4 contains an expository treatment of admissible sets and min-recursive functions, material due basically to Kripke and Platek. The equivalence of §3 is extended to min-recursive set functions and recursive ordinal functions. Here again the result seems to be new because earlier versions of this equivalence have dealt with recursive ordinal functionals, functions taking both ordinals and sequences of ordinals as arguments.

The present formulation of primitive recursive set functions is due to Gandy and is simpler than the one the authors had originally. Platek also has an equivalent formulation obtained independently, as part of his study of recursively defined functions in set theory. See **[13]**. There is a formulation of primitive recursive ordinal functions in Takeuti-Kino **[18]**, 1962. Gandy in his review **[2]** points out that their formulation is weaker than the present one.

The work reported here uses the classification of predicates of sets first treated systematically by Lévy in his Memoir **[9]**. By a ZF-formula, we mean a first-order formula in binary predicates $\equiv$, $\in$. Symbols $\wedge$, $\vee$, $\neg$, $\rightarrow$, (w), (Ew) are used for the connectives and quantifiers of the formal language. The corresponding symbols &, $\vee$, $\sim$, $\Rightarrow$, $\forall$, $\exists$ are used informally. Formal variables are v_i, w_i, informal variables are x, y, z, ... A standard Gödel-numbering of the language of set theory is assumed. A Δ_0-formula is one in which every quantifier appears restricted, that is, either in the form $(w)[w \in v \rightarrow \cdots]$ or $(Ew)[w \in v \wedge \cdots]$. A Σ_1-formula is a ZF-formula of the form $(Ew)\mathscr{Q}$ where $\mathscr{Q}$ is Δ_0. A generalized Σ_1-formula is one built up from Δ_0-formulas by connectives $\wedge$, $\vee$, existential and bounded universal quantifiers. Generalized Σ_1-formulas reduce to Σ_1-formulas by equivalences 2.3(3) which hold in many important transitive classes, but not in all of them.

A fixed universe $\langle V, \in\rangle$ of ZF set theory plus the axiom of choice is assumed to underlie the entire discussion. For ZF formulas $\mathscr{A}(v_1, \ldots, v_n)$, $\models \mathscr{A}(x_1, \ldots, x_n)$ means that the sequence $\langle x_1, \ldots, x_n\rangle$ satisfies $\mathscr{A}$ in $\langle V, \in\rangle$. For a class T, $x_1, \ldots, x_n \in T$, $T \models \mathscr{A}(x_1, \ldots, x_n)$ means that $\langle x_1, \ldots, x_n\rangle$ satisfies $\mathscr{A}$ in $\langle T, \in \restriction T\rangle$. A subset $S \subseteq T$ is T-definable just in case there are $x_2, \ldots, x_n \in T$ and a ZF formula $\mathscr{A}(v_1, \ldots, v_n)$ such that $(\forall x \in T)(x \in S \Leftrightarrow T \models \mathscr{A}(x, x_2, \ldots, x_n))$.

Notions of absoluteness, persistence, and stability, are fundamental to the paper. A ZF formula $\mathscr{A}(v_1, \ldots, v_n)$ is T-absolute if

$$(\forall x_1, \ldots, x_n \in T)(\models \mathscr{A}(x_1, \ldots, x_n) \Leftrightarrow T \models \mathscr{A}(x_1, \ldots, x_n)).$$

An important property of Δ_0-formulas is their absoluteness in every transitive class T. A ZF formula $\mathscr{A}(v_1, \ldots, v_n)$ is T-persistent if for any transitive $T' \supseteq T$,

$$(\forall x_1, \ldots, x_n \in T)(T \models \mathscr{A}(x_1, \ldots, x_n) \Rightarrow T' \quad \mathscr{A}(x_1, \ldots, x_n)).$$

An important property of generalized Σ_1-formulas is their T-persistence on every transitive class T. Another property of generalized Σ_1-formulas is their H_κ-absoluteness for regular nondenumerable cardinals κ, where H_κ is the set of all sets hereditarily of power less than κ. This follows from a theorem in Lévy **[9]**, a consequence of the Löwenheim-Skolem Theorem which generalizes Gödel's proof that the generalized continuum hypothesis holds for constructible sets:

$$(\forall x_1, \ldots, x_n)((\exists y) \models \mathscr{Q}(y, x_1, \ldots, x_n)$$
$$\Rightarrow (\exists y)(y \preccurlyeq \omega \cup \mathrm{TC}\{x_1, \ldots, x_n\} \ \& \models \mathscr{Q}(y, x_1, \ldots, x_n)))$$

where $\mathscr{Q}$ is Δ_0. ($\preccurlyeq$ is the cardinal order relation, TC is the transitive closure operation.)

A set function is assumed to operate on the entire universe. The λ-notation is used, $\lambda x_1, \ldots, x_n(\mathscr{T}(x_1, \ldots, x_n))$, $\mathscr{T}$ being a set-theoretical term, for that set function whose value at n-tuple $(x_1, \ldots, x_n)$ is $\mathscr{T}(x_1, \ldots, x_n)$. A set function is Σ_1-definable if its graph is definable on the universe by a Σ_1-formula. The Σ_1-definable set functions make an interesting class which includes the primitive recursive and min-recursive set functions and is closed under Properties 1.3. By the Gödel-Lévy Theorem above, a Σ_1-definable set function cannot raise an infinite cardinality.

1. **Fundamentals.**

1.1 Definition. Let $X_1, \ldots, X_k$ be one-place set functions. Then a set function F is primitive recursive in $X_1, \ldots, X_k$ (written "F is Prim $(X_1, \ldots, X_k)$") just in case it can be obtained from the initial functions by substitution and recursion as follows:

Initial functions:

(1) $F(x) = X_i(x)$, $1 \leq i \leq k$.

(2) $P_{n,i}(\mathbf{x}) = x_i$, $1 \leq n \in \omega$, $\mathbf{x} = (x_1, \ldots, x_n)$, $1 \leq i \leq n$.

(3) $F(x) = 0$.

(4) $F(x, y) = x \cup \{y\}$.

(5) $C(x, y, u, v) = x$ if $u \in v$, y otherwise.

Substitution:

(a) $F(\mathbf{x}, \mathbf{y}) = G(\mathbf{x}, H(\mathbf{x}), \mathbf{y})$, $m, n < \omega$, $\mathbf{x} = (x_1, \ldots, x_n)$, $\mathbf{y} = (y_1, \ldots, y_m)$.

(b) $F(\mathbf{x}, \mathbf{y}) = G(H(\mathbf{x}), \mathbf{y})$.

Recursion:

$F(z, \mathbf{x}) = G(\bigcup \{F(u, \mathbf{x}) \mid u \in z\}, z, \mathbf{x})$, $n < \omega$, $\mathbf{x} = (x_1, \ldots, x_n)$.

A set-theoretical relation is primitive recursive in $X_1, \ldots, X_k$ just in case its characteristic function (with 1 for truth, 0 for falsity) is primitive recursive in $X_1, \ldots, X_k$.

This formulation of primitive recursiveness, due to Gandy, is technically simpler than the earlier ones and has the advantage that it applies equally well to ordinal functions and ordinal functionals. Our original formulation, which is equivalent to the present one, had initial functions (4) and (5) replaced by three: $F(x, y) = \{x, y\}$, $F(x, y) = x - y$, $F(x) = \bigcup x$. The substitution schema was $F(\mathbf{x}) = G(H_1(\mathbf{x}), \ldots, H_p(\mathbf{x}))$ and the recursion schema was $F(z, \mathbf{x}) = G(F \restriction z, z, \mathbf{x})$ where $F \restriction z$ is that function on z whose value at u is $F(u, \mathbf{x})$.

1.2 Definition. Let $X_1, \ldots, X_k$ be one-place functions on ordinals to ordinals. The initial ordinal functions are the restrictions to ordinals of the functions 1.1(1), (2), (3), (5) with (4) replaced by the successor function $F(\alpha) = \alpha + 1 = \alpha \cup \{\alpha\}$. An ordinal function F is ordinally primitive recursive in $X_1, \ldots, X_k$ (written "F is $\mathrm{Prim}_O(X_1, \ldots, X_k)$") just in case it can be obtained from the initial ordinal functions by the substitution and recursion schemata of Definition 1.1. A relation on ordinals is ordinally primitive recursive in $X_1, \ldots, X_k$ just in case its characteristic function is $\mathrm{Prim}_O(X_1, \ldots, X_k)$.

Thus Prim_O functions map ordinals to ordinals and when restricted to natural numbers give the ordinary primitive recursive functions of natural numbers. The Prim_O functions are obviously restrictions to ordinals of Prim functions.

1.3 *Closure properties.* The reader is referred to Appendix I for justification of these fundamental properties of primitive recursive functions.

(1) If $G, H_1, \ldots, H_p$ are $\mathrm{Prim}(X_1, \ldots, X_k)$ ($\mathrm{Prim}_O(X_1, \ldots, X_k)$), then so is $F(\mathbf{x}) = G(H_1(\mathbf{x}), \ldots, H_p(\mathbf{x}))$.

(2) $\mathrm{Prim}(X_1, \ldots, X_k)$ and $\mathrm{Prim}_O(X_1, \ldots, X_k)$ are closed under definition by cases. For example for two cases, if G, H are $\mathrm{Prim}(X_1, \ldots, X_k)$ ($\mathrm{Prim}_O(X_1, \ldots, X_k)$) functions and R is a $\mathrm{Prim}(X_1, \ldots, X_k)$ relation ($\mathrm{Prim}_O(X_1, \ldots, X_k)$-relation), then F is a $\mathrm{Prim}(X_1, \ldots, X_k)$ function ($\mathrm{Prim}_O(X_1, \ldots, X_k)$ function), where $F(\mathbf{x}) = G(\mathbf{x})$ if $R(\mathbf{x})$ and $H(\mathbf{x})$ otherwise.

(3) $\mathrm{Prim}(X_1, \ldots, X_k)$ and $\mathrm{Prim}_O(X_1, \ldots, X_k)$ are closed under Boolean operations on relations.

(4) If G and R are $\mathrm{Prim}(X_1, \ldots, X_k)$ ($\mathrm{Prim}_O(X_1, \ldots, X_k)$) then so is F where $F(z, \mathbf{x}) = \bigcup \{G(u, \mathbf{x}) \mid u \in z \,\&\, R(u, \mathbf{x})\}$.

(5) The $\mathrm{Prim}(X_1, \ldots, X_k)$ and $\mathrm{Prim}_O(X_1, \ldots, X_k)$ relations are closed under bounded quantifiers. Any Δ_0 relation on sets is Prim.

(6) If G and R are $\mathrm{Prim}(X_1, \ldots, X_k)$ then so is F where $F(z, \mathbf{x}) = \{G(u, \mathbf{x}) \mid u \in z \,\&\, R(u, \mathbf{x})\}$.

(7) $\mathrm{Prim}(X_1, \ldots, X_k)$ is closed under recursion with respect to $\in$, i.e., if G is $\mathrm{Prim}(X_1, \ldots, X_k)$ then so is F where $F(z, \mathbf{x}) = G(F \restriction z, z, \mathbf{x})$ and $F \restriction z = (\lambda u F(u, \mathbf{x})) \restriction z$.

(8) $\mathrm{Prim}(X_1, \ldots, X_k)$ and $\mathrm{Prim}_O(X_1, \ldots, X_k)$ are closed under the bounded min rule: $F(z, \mathbf{x}) = (\min \xi)_{\xi \in z} R(\xi, z, \mathbf{x})$ whenever $(\exists \xi \in z) R(\xi, z, \mathbf{x})$.

Note that schemes (6) and (7) do not apply to ordinal functions. Thus the full power of transfinite recursion appears not to be available in Prim_O. It is the main result of §3 that the Prim_O functions really are just the Prim functions that map ordinals to ordinals. The proof involves a coding of constructible sets using only Prim_O functions.

1.4 The following commonly used functions and relations on sets are Prim:

(1) Such relations as $\lambda x, y(x \subseteq y)$, $\lambda x(x$ is transitive), $\lambda x(x$ is an ordinal), $\lambda x(x$ is an ordered pair), $\lambda x(x$ is a relation), $\lambda x(x$ is a function), $\lambda x(x \in \omega)$, $\lambda x(x \subseteq \omega)$, $\lambda x(x$ is finite).

(2) Such functions as $\lambda x(\bigcup x)$, $\lambda x, y(x \cup y)$, $\lambda x, y(x \cap y)$, $\lambda x, y(x - y)$, $\lambda x, y(\{x, y\})$, $\lambda x, y((x, y))$, $\lambda x(\mathrm{Dm}\,(x))$, $\lambda x(\mathrm{Rg}\,(x))$, $\lambda x(x^{-1})$, $\lambda x, y(x \times y)$, $\lambda x, y(x \,\text{‘}\, y)$ where $x \,\text{‘}\, y$ is the value of x at y if x is a function, 0 if not, $\lambda x, y(x \,\text{“}\, y)$, the image of x on y, $\lambda x, y(x \restriction y)$, x restricted to y.

(3) Such other functions as $\lambda x(\mathrm{TC}\,(x))$, the transitive closure of x, $\mathrm{TC}\,(x) = x \cup \bigcup \{\mathrm{TC}\,(y) \mid y \in x\}$, $\lambda x, y(x^y$ if y is finite, 0 if not), and $\lambda x(\mathrm{rank}\,(x))$ where $\mathrm{rank}\,(x) = \bigcup \{\mathrm{rank}\,(y) + 1 \mid y \in x\}$.

(4) The satisfaction relation in a set, $\lambda x, y, z(x$ satisfies y in $\langle z, \in \restriction z\rangle$ and y is Gödel-number of a ZF-formula & x is a mapping of free variables of y to z).

1.5(1) The operations of ordinal arithmetic are Prim_O: $\lambda\alpha\beta(\alpha + \beta)$, $\lambda\alpha\beta(\alpha \cdot \beta)$, $\lambda\alpha\beta(\alpha^\beta)$. Also $\lambda\alpha(\bigcup \alpha)$, $\lambda\alpha\beta(\max\,\{\alpha, \beta\})$, $\lambda\alpha\beta(\alpha \dot{-} \beta)$ where $\alpha \dot{-} \beta$ is $\alpha - \beta$ if $\beta < \alpha$, 0 if not.

(2) If G and R are $\mathrm{Prim}_O\,(X_1, \ldots, X_k)$ then so are F and F', where

$$F(\alpha, \mathbf{x}) = \sum \{G(\xi, \mathbf{x}) \mid \xi < \alpha \;\&\; R(\xi, \mathbf{x})\}$$

and

$$F'(\alpha, \mathbf{x}) = \prod \{G(\xi, \mathbf{x}) \mid \xi < \alpha \;\&\; R(\xi, \mathbf{x})\}.$$

(3) The Gödel pairing function P and its inverses L_1 and L_2 are Prim_O. For $P(\alpha, \beta)$ is the position of (α, β) in the ordering $(\alpha, \beta)R(\gamma, \delta) \Leftrightarrow (\alpha \cup \beta < \gamma \cup \delta)$ or $(\alpha \cup \beta = \gamma \cup \delta \;\&\; \beta < \delta)$ or $(\alpha \cup \beta = \gamma \cup \delta \;\&\; \beta = \delta \;\&\; \alpha < \gamma)$. Then

$$P(\alpha, 0) = \sum_{\gamma < \alpha} (\gamma \cdot 2 + 1), \qquad P(\alpha, \beta) = P(\alpha, 0) + \beta \quad \text{if } \beta < \alpha,$$

$$P(\alpha, \beta) = P(\beta, 0) + \beta + \alpha \quad \text{if } \beta \geq \alpha.$$

Primitive recursive functions are Σ_1-definable and therefore cannot raise cardinality. Examples of Σ_1-definable set functions that are not primitive recursive may be obtained by the usual diagonal methods. The enumerating function for Prim functions is Σ_1-definable but not Prim. In fact it is generalized Σ_1-definable on any Prim-closed transitive T and therefore not Prim on T. See §2.

Closure Properties 1.3 are easy to prove, but 1.6 is more difficult. It is used in §3.

1.6 *Closure properties.* The reader is referred to Appendix I for proofs of the following properties of Prim_O.

(1) If G and H are Prim_O then so is F where $F(0, \mathbf{x}) = G(\mathbf{x})$, $F(n + 1, \mathbf{x}) = H(F(n, \mathbf{x}), n, \mathbf{x})$, $F(\alpha, \mathbf{x}) = 0$ if $\sim(\alpha \in \omega)$.

(2) If G and H are Prim_O then so is F where $F(0) = \gamma$, $F(\alpha) = G(F(H(\alpha), \alpha))$ for $\alpha > 0$, provided that for all $\alpha > 0$, $H(\alpha) < \alpha$.

(3) If R is a Prim_O relation and G is a Prim_O function taking only values 0 and 1, then F is Prim_O where $F(\alpha) = G(\bigcup \{F(\beta) \mid \beta < \alpha \;\&\; R(\beta, \alpha)\}, \alpha)$.

2. **Absoluteness and stability.** Some of the most useful properties of primitive recursive set functions follow from the forms of their defining formulas in set

theory. Let us assume for the moment that the given functions X_i are set-theoretically definable and assign defining formulas $\mathscr{D}_{X_i}(v_0, v_1)$ in ZF to them once and for all. Of course, if the X_i are not ZF-definable we could adjoin one-place function symbols to denote them. We assume that for all sets x, y we have

$$y = X_i(x) \Leftrightarrow \models \mathscr{D}_{X_i}(y, x).$$

Then every Prim $(X_1, \ldots, X_k)$ function F has a defining formula $\mathscr{D}_F$ which is built up from the $\mathscr{D}_{X_i}$ and restricted formulas using only connectives $\wedge$ and $\vee$ with bounded universal and unbounded existential quantifications. They are thus generalized Σ_1 in the $\mathscr{D}_{X_i}$. Moreover the Skolem functions for the unbounded E's in $\mathscr{D}_F$ are themselves Prim $(X_1, \ldots, X_k)$ functions. This is why the Prim $(X_1, \ldots, X_k)$ functions are absolute in transitive Prim $(X_1, \ldots, X_k)$-closed sets.

2.1 Definition. Let T be a transitive class. A set-theoretical formula $\mathscr{A}(v_1, \ldots, v_n)$ is T-absolute just in case for every $\mathbf{x} \in T^n$, $\models \mathscr{A}(\mathbf{x}) \Leftrightarrow T \models \mathscr{A}(\mathbf{x})$. Such a formula is T-persistent just in case for all $\mathbf{x} \in T^n$, all transitive $T' \supseteq T$, $T \models \mathscr{A}(\mathbf{x}) \Rightarrow T' \models \mathscr{A}(\mathbf{x})$. Similarly if $\mathscr{T}$ is a collection of transitive classes then A is $\mathscr{T}$-absolute just in case $\mathscr{A}$ is T-absolute for every T in $\mathscr{T}$, and $\mathscr{T}$-persistent in case $\mathscr{A}$ is T-persistent for every T in $\mathscr{T}$.

A Δ_0-formula is T-absolute for all transitive T, a generalized Σ_1-formula is T-persistent for all transitive T and is $\mathscr{T}$-absolute where $\mathscr{T}$ is the family of sets H_κ, κ a regular nondenumerable cardinal.

2.2 *Defining formulas for the* Prim $(X_1, \ldots, X_k)$ *set functions.* Assume that the given functions $X_1, \ldots, X_k$ have given ZF-defining formulas $\mathscr{D}_{X_i}(v_0, v_1)$ expressing the condition, "$x_0 = X_i(x_1)$". Assume also a standard list of Δ_0-formulas defining the relations $\lambda x(\text{Fcn}(x))$, "x is a function", $\lambda x, y(x \subseteq y)$, $\lambda x(\text{Trans}(x))$, "$x$ is transitive", $\lambda y, x(y = \text{Dm}(x))$, $\lambda z, x, y(z = x \text{`} y)$, $\lambda z, x, y(z = \bigcup x \text{``} y)$. Then to each schema in Definition 1.1 defining a Prim $(X_1, \ldots, X_k)$ set function F is attached a set-theoretical defining formula $\mathscr{D}_F$ for the graph of F as follows:

(1) If F is X_i, then $\mathscr{D}_F(v_0, v_1)$ is $\mathscr{D}_{X_i}(v_0, v_1)$.

(2) If F is $P_{n,i}$, then $\mathscr{D}_F(v_0, \ldots, v_n)$ is $v_0 \equiv v_i$.

(3) If F is the zero function, then $\mathscr{D}_F(v_0, v_1)$ is $(\forall w_0 \in v_0)[\neg w_0 \equiv w_0]$.

(4) If $F(x, y) = x \cup \{y\}$, then $\mathscr{D}_F(v_0, v_1, v_2)$ is

$$(\forall w_0 \in v_0)[w_0 \in v_1 \vee w_0 \equiv v_2] \wedge v_2 \in v_0 \wedge (\forall w_0 \in v_1) w_0 \in v_0.$$

(5) If F is C, then $\mathscr{D}_F(v_0, v_1, v_2, v_3, v_4)$ is $[v_0 \equiv v_1 \wedge v_3 \in v_4] \vee [\neg v_3 \in v_4 \wedge v_0 \equiv v_2]$.

(6) If F arises from G, H by substitution schema (a) then $\mathscr{D}_F(v_0, \ldots, v_{n+m})$ is

$$(Ew_0)[\mathscr{D}_H(w_0, v_1, \ldots, v_n) \wedge \mathscr{D}_G(v_0, v_1, \ldots, v_n, w_0, v_{n+1}, \ldots, v_{n+m})].$$

(b) is similar.

(7) If F arises from G by the recursion schema then $\mathscr{D}_F(v_0, v_1, \ldots, v_{n+1})$ is

$$(Ew_0, w_1)[\text{Fcn}(w_0) \wedge v_1 \subseteq w_1 \wedge \text{Trans}(w_1) \wedge w_1 = \text{Dm}(w_0) \\ \wedge (Ew_2)[w_2 = \bigcup w_0 \text{``} v_1 \wedge \mathscr{D}_G(v_0, w_2, v_1, \ldots, v_{n+1})] \\ \wedge (\forall w_3 \in w_1)(Ew_4, w_5)[w_4 = w_0 \text{`} w_3 \wedge w_5 = \bigcup w_0 \text{``} w_3 \\ \wedge \mathscr{D}_G(w_4, w_5, w_3, v_2, \ldots, v_{n+1})]].$$

(Note: Think of w_1 as TC (v_1) and w_0 as $\lambda u F(u, v_2, \ldots, v_{n+1}) \restriction w_1$.)

2.3 REMARKS. (1) If the $\mathscr{D}_{X_i}$ are generalized Σ_1 then so is $\mathscr{D}_F$ whenever F is Prim $(X_1, \ldots, X_k)$. Thus like all generalized Σ_1-formulas, $\mathscr{D}_F$ is persistent in transitive classes.

(2) If T is transitive and $x, y \in T \,\&\, T \models \mathscr{D}_{X_i}(y, x) \Rightarrow y = X_i(x)$ then if F is Prim $(X_1, \ldots, X_k)$, then $\mathbf{x} \in T^n \,\&\, y \in T \,\&\, T \models \mathscr{D}_F(y, \mathbf{x}) \Rightarrow y = F(\mathbf{x})$. Similarly if the $\mathscr{D}_{X_i}$ are T-persistent, so is $\mathscr{D}_F$.

PROOF. The property $T \models \mathscr{A}(\mathbf{x}) \Rightarrow T' \models \mathscr{A}(\mathbf{x})$ for $\mathbf{x} \in T^n$ and $T \subseteq T'$ is preserved by $\wedge$, $\vee$, E, bounded $\forall$.

(3) A generalized Σ_1-formula reduces to a Σ_1-formula by use of four rules:

$$(Ew)\mathscr{A}(w, \ldots) \wedge (Ew')\mathscr{B}(w', \ldots) \leftrightarrow (Ew, w')[\mathscr{A}(w, \ldots) \wedge \mathscr{B}(w', \ldots)] \text{ if } w \neq w'.$$

$$(Ew)\mathscr{A} \vee (Ew')\mathscr{B} \leftrightarrow (Ew, w')[\mathscr{A} \vee \mathscr{B}].$$

$$(Ew)(Ew')\mathscr{A} \leftrightarrow (Ew'')(Ew \in w'')(Ew' \in w'')\mathscr{A}.$$

$$(\forall w \in v)(Ew')\mathscr{A} \leftrightarrow (Ew'')(\forall w \in v)(Ew' \in w'')\mathscr{A}.$$

These four equivalences are valid in H_κ whenever κ is a regular infinite cardinal. Therefore generalized Σ_1-formulas are H_κ-absolute since Σ_1-formulas are. If the $\mathscr{D}_{X_i}$ are generalized Σ_1 and F is Prim $(X_1, \ldots, X_k)$, then H_κ is closed under F and $\mathscr{D}_F$ is H_κ-absolute.

(4) If T is a transitive class closed under Prim $(X_1, \ldots, X_k)$ functions and if the $\mathscr{D}_{X_i}$ are T-absolute, then $\mathscr{D}_F$ is T-absolute whenever F is Prim $(X_1, \ldots, X_k)$.

PROOF. This will follow from the Stability Theorem, but the reader will find it instructive to prove this first. By (2) it is required only to show $x_1, \ldots, x_n \in T \Rightarrow T \models \mathscr{D}_F(F(\mathbf{x}), \mathbf{x})$ where $\mathbf{x} = (x_1, \ldots, x_k)$. The defining formulas for the initial functions are either Δ_0 or are given T-absolute. Proceeding by induction on the defining schemes, if F arises by substitution, $F(\mathbf{x}, \mathbf{y}) = G(\mathbf{x}, H(\mathbf{x}), \mathbf{y})$, then the outside quantifier on w_0 in $\mathscr{D}_F$ denotes $H(\mathbf{x})$. The induction hypothesis gives $T \models \mathscr{D}_H(H(\mathbf{x}), \mathbf{x})$ and $T \models \mathscr{D}_G(F(\mathbf{x}, \mathbf{y}), \mathbf{x}, H(\mathbf{x}), \mathbf{y})$. Hence $T \models \mathscr{D}_F(F(\mathbf{x}, \mathbf{y}), \mathbf{x}, \mathbf{y})$. If F arises by recursion, $F(x_1, \ldots, x_{n+1}) = G(\bigcup F \text{``} x_1, x_1, \ldots, x_{n+1})$, take the outside existential quantifications on $w_0, w_1, w_2$ in $\mathscr{D}_F$ to be $w_1 = \text{TC}(x_1)$, $w_0 = F \restriction w_1$, $w_2 = \bigcup w_0 \text{``} x_1$. By 1.3 these denote elements of T. The inside parts of $\mathscr{D}_F$ are $\mathscr{D}_G$ and Δ_0-formulas. Thus the T-absoluteness of $\mathscr{D}_F$ follows from the T-absoluteness of $\mathscr{D}_G$.

(5) COROLLARY. *If κ is a regular infinite cardinal, if H_κ is closed under the X_i and each $\mathscr{D}_{X_i}$ is H_κ-absolute, then for any* Prim $(X_1, \ldots, X_k)$ *set function F, H_κ is closed under F and $\mathscr{D}_F$ is H_κ-absolute.*

The notion of stability is similar to absoluteness but it is relative to a hierarchy. The hierarchies we have in mind are the following:

Ramified hierarchy: $\bigcup_\alpha L(\alpha) = L$, the class of constructible sets.	$L(0) = 0$ $L(\alpha + 1) = \mathrm{Def}\,(L(\alpha))$, the $L(\alpha)$-definable subsets of $L(\alpha)$ $L(\lambda) = \bigcup_{\alpha<\lambda} L(\alpha)$, if $\mathrm{Lim}\,(\lambda)$.
Cumulative hierarchy: $\bigcup_\alpha V(\alpha) = V$, the universe of set theory.	$V(0) = 0$ $V(\alpha + 1) = \mathscr{P}(V(\alpha))$ $V(\lambda) = \bigcup_{\alpha<\lambda} V(\alpha)$, if $\mathrm{Lim}\,(\lambda)$.
The hierarchy of sets hereditarily of power $<\kappa$: $\bigcup_\alpha H(\kappa, \alpha) = H_\kappa$.	$H(\kappa, 0) = 0$ $H(\kappa, \alpha + 1) = \mathscr{P}_\kappa(H(\kappa, \alpha))$, the subsets of $H(\kappa, \alpha)$ of power $<\kappa$. $H(\kappa, \lambda) = \bigcup_{\alpha<\lambda} H(\kappa, \alpha)$ if $\mathrm{Lim}\,(\lambda)$.

Note that if κ regular, then $H_\kappa = \{x \mid \mathrm{TC}\,(x) \prec \kappa\}$. If κ is singular, then H_κ is not closed under such Prim functions as $\lambda x(\bigcup x)$ and the stability theorem does not apply. Assume κ regular in this section.

2.4 Definition. Let H be the function defining one of the above hierarchies, let F be an n-place set function with set-theoretical formula $\mathscr{D}_F$ defining its graph. Then we say that the ordinal function b H-stabilizes $\mathscr{D}_F$ just in case for all α and all $\mathbf{x} \in H(\alpha)^n$, the following conditions hold:

(i) $F(\mathbf{x}) \in H(b(\alpha))$.

(ii) $\beta \geq b(\alpha) \;\&\; y \in H(\beta) \Rightarrow (y = F(\mathbf{x}) \Leftrightarrow H(\beta) \models \mathscr{D}_F(y, \mathbf{x}))$.

2.5 Stability Theorem. *Suppose that to each given function X_i is assigned a defining formula $\mathscr{D}_{X_i}$ and an increasing H-stabilizing function b_{X_i}. Then to each function in F* Prim $(X_1, \ldots, X_k)$ *can be assigned an increasing* $\mathrm{Prim}_O\,(b_{X_1}, \ldots, b_{X_k})$-*function which H-stabilizes $\mathscr{D}_F$.*

Before going on with the proof, it might be instructive to point out some examples and simple consequences of the Stability Theorem.

2.6 Corollary. *For ordinals α the following are equivalent:*

(i) α *is* Prim_O-*closed.*

(ii) $L(\alpha)$ *is* Prim-*closed.*

(iii) $V(\alpha)$ *is* Prim-*closed.*

(iv) $V(\alpha)$ *is* Prim $(\mathscr{P})$-*closed.*

Proof. Any one of (ii), (iii), (iv) implies (i) simply because a Prim_O function f is the restriction to ordinals of a Prim function F. Since $\mathrm{Lim}\,(\alpha)$, $\mathbf{x} \in \alpha^n$ implies $\mathbf{x} \in \gamma^n$ for some $\gamma < \alpha$. So $F(\mathbf{x}) \in L(\alpha)$ or $V(\alpha)$ as the case may be. Since $F(\mathbf{x})$ is an ordinal, $F(\mathbf{x}) = f(\mathbf{x}) < \alpha$.

Since $f(\alpha) = \alpha + 1$ V-stabilizes $\mathscr{D}_{\mathscr{P}}$, the defining formula for the graph of the power-set function, the Stability Theorem says that every Prim $(\mathscr{P})$ function F (Prim function F) has a Prim_O V-stabilizing (L-stabilizing) function b_F. Thus if

$\mathbf{x} \in V(\alpha)^n$ ($\mathbf{x} \in L(\alpha)^n$), there is $\gamma < \alpha$ such that $\mathbf{x} \in V(\gamma)^n$ ($\mathbf{x} \in L(\gamma)^n$). Whence $F(\mathbf{x}) \in V(b_F(\gamma)) \subseteq V(\alpha)$ ($F(\mathbf{x}) \in L(b_F(\gamma)) \subseteq L(\alpha)$). Hence (i) implies each of (ii), (iii), (iv).

2.7 COROLLARY. *$F(x) = \bar{\bar{x}}$, the cardinal of x, is not* Prim ($\mathscr{P}$).

PROOF. Let π_1 be the first Prim_O-closed ordinal past ω. Then $V(\pi_1)$ is not closed under F since for example, $\mathscr{P}(V(\omega)) \in V(\pi_1)$ and $F(\mathscr{P}(V(\omega)))$ is $2^{\aleph_0}$ which is not in $V(\pi_1)$. By 2.6, F is not Prim ($\mathscr{P}$).

Note that

$$\begin{aligned}\bar{\bar{x}} &= (\min \xi)(x \approx \xi)\\ &= (\min \xi)(\exists f \in \mathscr{P}(x \times \xi))(\text{Fcn}(f) \,\&\, \text{Dm}(f) = x \,\&\, \text{Rg}(f) = \xi)\end{aligned}$$

would be recursive in $\mathscr{P}$ if the min rule were added.

2.8 COROLLARY. *For regular infinite cardinals κ the following are equivalent for $\alpha \leq \kappa$:*

(i) *α is Prim_O-closed.*

(ii) *$H(\kappa, \alpha)$ is Prim-closed.*

(iii) *$H(\kappa, \alpha)$ is Prim ($\mathscr{P}_{\aleph_1}$)-closed provided that $\sigma < \kappa \Rightarrow \sigma^{\aleph_0} < \kappa$. ($\mathscr{P}_{\aleph_1}(x)$ is the set of at most denumerable subsets of x.)*

PROOF. As $\mathscr{D}_{\mathscr{P}_{\aleph_1}}(v_0, v_1)$ take $(w)[w \in v_0 \leftrightarrow [w \subseteq v_1 \wedge w \preccurlyeq \omega]]$. If $x \in H(\kappa, \alpha)$ then $\mathscr{P}_{\aleph_1}(x) \in H(\kappa, \alpha + 2)$. Moreover if $y \subseteq x$ and $y \preccurlyeq \omega$ by a function f, then $f \in H(\kappa, \alpha + 4)$ for infinite α. It follows that $b(\alpha) = \alpha + 4$ H_κ-stabilizes $\mathscr{P}_{\aleph_1}$. From this point proceed as in the proof of Corollary 2.6.

PROOF OF THE STABILITY THEOREM 2.5. The stabilizing functions b_F are defined as follows:

(1) If F is X_i, then b_F is the given stabilizing function.

(2)–(5) If F is one of the other initial functions let $b_F(\alpha) = \alpha + 1$.

(6) If F arises from G, H by the substitution schema, then $b_F(\alpha) = b_G(b_H(\alpha))$.

(7) If F arises from G by the recursion schema, let p be defined recursively as

$$p(\alpha, 0) = b_G(\alpha), \qquad p(\alpha, \rho) = b_G\Big(\bigcup_{\theta < \rho} p(\alpha, \theta) + 3\Big)$$

for $\rho > 0$. Let

$$b_F(\alpha) = \bigcup_{\rho \leq \alpha} p(\alpha, \rho).$$

An easy induction shows that the b_F are increasing functions and are Prim_O $(b_{X_1}, \ldots, b_{X_n})$. The function H is increasing, every $H(\alpha)$ is transitive, and the values of the initial functions other than the X_i are in $H(\alpha + 1)$ for arguments in $H(\alpha)$. Since their defining formulas are Δ_0, the successor function H-stabilizes the initial functions (2)–(5).

To complete the proof, it suffices to show by induction on the defining schemes for primitive recursive functions that

$$\mathbf{x} \in H(\alpha)^n \Rightarrow F(\mathbf{x}) \in H(b_F(\alpha)) \,\&\, H(b_F(\alpha)) \models \mathscr{D}_F(F(\mathbf{x}), \mathbf{x}).$$

For suppose this has been established and that $\beta \geq b_F(\alpha)$. It is required to show that $\mathbf{x} \in H(\alpha)^n \,\&\, y \in H(\beta) \Rightarrow (y = F(\mathbf{x}) \Leftrightarrow H(\beta) \models \mathscr{D}_F(y, \mathbf{x}))$. If F does not depend on the X_i then $\mathscr{D}_F$ is generalized Σ_1 and the result follows by the persistence of such formulas in transitive classes. If F depends on X_i then since every application of substitution and recursion pushes up b_F, we have $b_F(\alpha) \geq b_{X_i}(\alpha)$. The stability property for the $\mathscr{D}_{X_i}$ give for any γ such that

$$\bigcup_{1 \leq i \leq k} b_{X_i}(\alpha) \leq \gamma \leq \beta,$$

(1) $(x \in H(\alpha) \,\&\, y \in H(\beta) \,\&\, H(\beta) \models \mathscr{D}_{X_i}(y, x)) \Rightarrow y = X_i(x)$.

$(2)_\gamma$ $(x \in H(\alpha) \,\&\, y \in H(\gamma) \,\&\, H(\gamma) \models \mathscr{D}_{X_i}(y, x)) \Rightarrow H(\beta) \models \mathscr{D}_{X_i}(y, x)$.

But properties (1′) and $(2')_\gamma$ of formulas $\mathscr{A}$ are preserved by $\wedge$, $\vee$, E and bounded universal quantification:

(1′) $(\mathbf{x} \in H(\alpha)^n \,\&\, \mathbf{y} \in H(\beta)^m \,\&\, H(\beta) \models \mathscr{A}(\mathbf{y}, \mathbf{x})) \Rightarrow \models \mathscr{A}(\mathbf{y}, \mathbf{x})$.

$(2')_\gamma$ $(\mathbf{x} \in H(\alpha)^n \,\&\, \mathbf{y} \in H(\gamma)^m \,\&\, H(\gamma) \models \mathscr{A}(\mathbf{y}, \mathbf{x})) \Rightarrow H(\beta) \models \mathscr{A}(\mathbf{y}, \mathbf{x})$.

Thus in particular for $\mathscr{D}_F$,

$$\mathbf{x} \in H(\alpha)^n \,\&\, y \in H(\beta) \,\&\, H(\beta) \models \mathscr{D}_F(y, \mathbf{x}) \Rightarrow y = F(\mathbf{x})$$

and

$$\mathbf{x} \in H(\alpha)^n \,\&\, y \in H(b_F(\alpha)) \,\&\, H(b_F(\alpha)) \models \mathscr{D}_F(y, \mathbf{x}) \Rightarrow H(\beta) \models \mathscr{D}_F(y, \mathbf{x}).$$

Thus the result for F follows from $F(\mathbf{x}) \in H(b_F(\alpha)) \,\&\, H(b_F(\alpha)) \models \mathscr{D}_F(F(\mathbf{x}), \mathbf{x})$.

Suppose F arises by substitution from G, G' and assume the induction hypothesis for G, G'. Then for $\mathbf{x}$, $\mathbf{y}$ in $H(\alpha)$, $F(\mathbf{x}, \mathbf{y}) = G(\mathbf{x}, G'(\mathbf{x}), \mathbf{y})$ is in $H(b_G(b_{G'}(\alpha)))$. But $b_F(\alpha) = b_G(b_{G'}(\alpha))$, so $F(\mathbf{x}, \mathbf{y}) \in H(b_F(\alpha))$. Moreover, the outer quantifier on w_0 in $\mathscr{D}_F$ denotes $G'(\mathbf{x})$. Thus $H(b_F(\alpha)) \models \mathscr{D}_F(F(\mathbf{x}, \mathbf{y}), \mathbf{x}, \mathbf{y})$ follows from

$$H(b_{G'}(\alpha)) \models \mathscr{D}_{G'}(G'(\mathbf{x}), \mathbf{x})$$

and $H(b_F(\alpha)) \models \mathscr{D}_G(F(\mathbf{x}, \mathbf{y}), \mathbf{x}, G'(\mathbf{x}), \mathbf{y})$. Substitution (b) is similar.

Suppose finally that F arises by recursion on G and assume the induction hypothesis for G. Then $F(z, \mathbf{x}) = G(\bigcup_{u \in z} F(u, \mathbf{x}), z, \mathbf{x})$. We prove by induction on ρ that z, $\mathbf{x}$ in $H(\alpha)$ and rank $(z) \leq \rho$ implies $F(z, \mathbf{x}) \in H(p(\alpha, \rho))$ and

$$H(p(\alpha, \rho)) \models \mathscr{D}_F(F(z, \mathbf{x}), z, \mathbf{x}).$$

Then the persistence property above and the fact that in all the hierarchies $z \in H(\alpha)$ implies rank $(z) < \alpha$, gives the result $F(z, \mathbf{x}) \in H(b_F(\alpha))$ and

$$H(b_F(\alpha)) \models \mathscr{D}_F(F(z, \mathbf{x}), z, \mathbf{x}).$$

Take the case $\rho = 0$. By the induction hypothesis on G,

$$F(0, \mathbf{x}) = G(0, 0, \mathbf{x}) \in H(b_G(\alpha))$$

and $H(b_G(\alpha)) \models \mathscr{D}_G(F(0, \mathbf{x}), 0, 0, \mathbf{x})$. But then $H(b_F(\alpha)) \models \mathscr{D}_F(F(0, \mathbf{x}), 0, \mathbf{x})$ since the outer quantifiers on w_0, w_1, w_2 in $\mathscr{D}_F$ denote 0, 0, 0, in this case.

In the general case with $z, \mathbf{x} \in H(\alpha)$, rank $(z) \leq \rho \neq 0$, w_0 denotes $F \restriction \mathrm{TC}\,(z)$, w_1 denotes TC (z), w_2 denotes $\bigcup F``\, z$. In all the hierarchies $z \in H(\alpha)$ implies TC $(z) \in H(\alpha + 4)$ and $H(\alpha + 4) \models \mathscr{D}_{\mathrm{TC}}(\mathrm{TC}\,(z), z)$. By induction hypothesis on

ρ, if $\beta = \bigcup_{\theta<\rho} p(\alpha, \theta)$ then $\beta \geq \alpha + 4$ and $u \in \text{TC}(z) \Rightarrow F(u, \mathbf{x}) \in H(\beta)$ and $H(\beta) \models \mathscr{D}_F(F(u, \mathbf{x}), u, \mathbf{x})$. Hence $F \restriction \text{TC}(z) \leq H(\beta + 2)$ and

$$F \restriction \text{TC}(z) = \{v \in H(\beta + 2) \mid (\exists u \in \text{TC}(z)) \\ (\exists v' \in H(\beta))(v = (u, v') \,\&\, H(\beta) \models \mathscr{D}_F(v', u, \mathbf{x}))\} \in H(\beta + 3)$$

because the set is an $H(\beta + 2)$-definable subset of $H(\beta + 2)$. Similarly $F``z \subseteq H(\beta)$ and since all the $H(\beta)$ are transitive, $\bigcup F``z \subseteq H(\beta)$. Then

$$\bigcup F``z \in H(\beta + 1) \subseteq H(\beta + 3).$$

Thus

$$F(z, \mathbf{x}) = G(\bigcup F``z, z, \mathbf{x}) \in H(b_G(\beta + 3)) = H(p(\alpha, \rho)).$$

Also $H(p(\alpha, \rho)) \models \mathscr{D}_G(F(z, \mathbf{x}), \bigcup F``z, z, \mathbf{x})$. The other occurrence of $\mathscr{D}_G$ in $\mathscr{D}_F$ is also satisfied in $H(p(\alpha, \rho))$ by virtue of persistence. Whence

$$H(p(\alpha, \rho)) \models \mathscr{D}_F(F(z, \mathbf{x}), z, \mathbf{x}).$$

A careful study of the proof shows that it can be altered to apply to any function H on ordinals having the properties listed below.

2.9 STABILITY THEOREM. *The Stability Theorem as stated in 2.5 holds for any function H on ordinals having the following properties:*

(1) *H is increasing (i.e., $\alpha < \beta \Rightarrow H(\alpha) \subseteq H(\beta)$) and each $H(\alpha)$ is transitive.*

(2) *For each initial* Prim *set function F there is a* Prim_O *function b such that the values of F are $H(b(\alpha))$ for arguments in $H(\alpha)$.*

(3) $z \in H(\alpha) \Rightarrow \text{rank}(z) < \alpha$.

(4) *There is a* Prim_O *function b such that $z \in H(\alpha) \Rightarrow \text{TC}(z) \in H(b(\alpha))$ and $H(b(\alpha)) \models \mathscr{D}(\text{TC}(z), z)$ where $\mathscr{D}$ is the* ZF*-formula defining the graph of* TC *by expressing the condition, "$y = \text{TC}(z) \Leftrightarrow y$ is the set of all u such that there is a finite $\in$-chain from u to z".*

(5) *For any* ZF*-formula $\mathscr{A}(v_0, v_1, \ldots, v_n)$ there is a* Prim_O *function b such that* $(\forall \mathbf{x} \in H(\alpha)^n)\{y \in H(\alpha) \mid H(\alpha) \models \mathscr{A}(y, \mathbf{x})\} \in H(b(\alpha))$.

2.10 COROLLARY. *If T is transitive and* Prim*-closed and F is an n-place* Prim *set function, then $(\forall \mathbf{x} \in T^n)(\exists t \in T)(t$ is transitive $\&\ t \models \mathscr{D}_F(F(\mathbf{x}), \mathbf{x}))$. In fact, t can be found primitive recursively from $\mathbf{x}$. That is, there is for given F, a* Prim *set function G_F such that for all n-tuples $\mathbf{x}$, $G_F(\mathbf{x})$ is transitive $\&$* $\text{Rg}(\mathbf{x}) \subseteq G_F(\mathbf{x}) \,\&\, F(\mathbf{x}) \in G_F(\mathbf{x}) \,\&\, G_F(\mathbf{x}) \models \mathscr{D}_F(F(\mathbf{x}), \mathbf{x})$. *(Hence $F \restriction T$ is Σ_1^T-definable since the unbounded existential quantifiers in $\mathscr{D}_F$ can be bounded by some $t \in T$.)*

PROOF. The Stability Theorem applies to hierarchies L_t, t transitive, where
$L_t(0) = 0$,
$L_t(\alpha + 1) = \text{Def}(L_t(\alpha)) \cup \{x \in t \cup \{t\} \mid \text{rank}(x) \leq \alpha\}$,
$L_t(\lambda) = \bigcup_{\alpha<\gamma} L_t(\alpha)$.

In fact, the same stabilizing functions as were used in the proof of the Stability Theorem can be used for $H = L_t$. Let $G_F(\mathbf{x}) = L_t(b_F(\alpha + 1))$ where $t = \text{TC}(\text{Rg}(\mathbf{x}))$ and $\alpha = \bigcup \{\text{rank}(y) \mid y \in \text{Rg}(\mathbf{x})\}$. Then each $x_i \in L_t(\alpha + 1)$ and the Stability Theorem gives the result.

2.11 ENUMERATION THEOREM. *The* Prim *set functions have an enumerating function* $E(m, \mathbf{x})$ = *the mth function at* $\mathbf{x}$, *which is generalized* Σ_1-*definable and whose defining formula is T-absolute in every transitive* Prim-*closed class T.*

PROOF. Let us introduce a standard enumeration of Prim set functions: $(2^n \cdot 5^i)^* = P_{n,i}$, $(2 \cdot 7)^*$ is the zero function, $(2^2 \cdot 11)^*$ is $\lambda x, y(x \cup \{y\})$, $(2^4 \cdot 13)^*$ is C. If p^* is G, q^* is H, then $(2^{(p)_0-1} \cdot 17^p \cdot 19^q)^*$ is $\lambda \mathbf{x}, \mathbf{y}(G(\mathbf{x}, H(\mathbf{x}), \mathbf{y}))$ and $(2^{(p)_0+(q)_0-1} \cdot 23^p \cdot 29^q)^*$ is $\lambda \mathbf{x}, \mathbf{y}(G(H(\mathbf{x}), \mathbf{y}))$. If p^* is G then $(2^{(p)_0-1} \cdot 31^p)^*$ is $\lambda z, \mathbf{x}(G(\bigcup_{u \in z} (2^{(p)_0-1} \cdot 31^p)^*(u, \mathbf{x}), z, \mathbf{x}))$. The number-theoretical functions prim (m), "m is the number of a Prim set function", $d(m)$ = the Gödel-number of $\mathcal{D}_{m^*}$, the defining formula for m^*, are primitive recursive functions of natural numbers. Define $E(m, x)$ as $m^*(x)$ if prim (m) & Fcn (x) & Dm $(x) = (m)_0$, 0 if not. Then

$$y = E(m, x) \Leftrightarrow (\sim R(m, x) \,\&\, y = 0) \vee (R(m, x)$$
$$\& (\exists t)(\text{Trans}(t) \,\&\, \text{Rg}(x) \subseteq t \,\&\, y \in t \,\&\, (\langle y \rangle^\frown x \text{ satisfies } d(m) \text{ in } t)))$$

where $R(m, x) \Leftrightarrow$ prim (m) & Fcn (x) & Dm $(x) = (m)_0$. This definition is generalized Σ_1. By Corollary 2.10 the quantifier on t may be restricted to any transitive, Prim-closed T which contains x.

The Stability Theorem can also be made to apply to Gödel's F-function in [3] that enumerates the constructible sets, but in this case the stabilizing functions, though still Prim_O, are not the same as for the other hierarchies.

Gödel's F-function is obtained by iteration and accumulation of his eight fundamental operations:

$$F_1(x, y) = \{x, y\},$$
$$F_2(x, y) = E \cap x = \{(u, v) \mid (u, v) \in x \,\&\, u \in v\},$$
$$F_3(x, y) = x - y,$$
$$F_4(x, y) = x \cap (V \times y),$$
$$F_5(x, y) = x \cap \text{Rg}(y),$$
$$F_6(x, y) = x \cap y^{-1},$$
$$F_7(x, y) = x \cap \text{Cnv}_2(y),$$
$$F_8(x, y) = x \cap \text{Cnv}_3(y),$$

where $\text{Cnv}_2(y) = \{(u, (v, w)) \mid (w, (u, v)) \in y\}$ and

$$\text{Cnv}_3(y) = \{(u, (v, w)) \mid (u, (w, v)) \in y\}.$$

These operations are all Prim. The pairing function P for ordinals and its inverse functions L_1 and L_2 are Prim_O by 1.5 and so is the tripling function $J(i, \alpha, \beta) = 9P(\alpha, \beta) + i$ and its inverse functions K_0, K_1, K_2 which recover i, α, β, respectively. Then the Gödel F-function, called F' here, is defined as

$$F'(\alpha) = F' \text{“} \alpha \quad \text{if} \quad (\exists \gamma \leq \alpha)(\alpha = 9 \cdot \gamma)$$

and

$$F'(\alpha) = F_i(F'K_1(\alpha), F'K_2(\alpha)) \quad \text{if} \quad (\exists \gamma < \alpha)(\alpha = 9 \cdot \gamma + i), \quad i = 1, \ldots, 8.$$

Then F' is clearly a Prim function. The function accumulates when α is divisble by 9. So ac $(\alpha) = (\min \gamma \leq \alpha)(\exists \xi \leq \gamma)(\gamma = 9 \cdot \xi)$ is the largest accumulation point at or below α. The Stability Theorem applies to $H(\alpha) = F'(\text{ac}(\alpha))$.

2.12 THEOREM. *The Stability Theorem as stated in* 2.5 *holds for the function* $H(\alpha) = F'(\text{ac}(\alpha))$.

PROOF. It is required to establish conditions 2.9 (1)–(5). Conditions (1) and (3) are immediate, the others require some calculation. Most of the needed computations can be found in Gödel's monograph. For instance, the proof of the General Existence Theorem shows that for any ZF-formula

$$\mathscr{A}(v_1, \ldots, v_n, v_1', \ldots, v_m')$$

with attached list of variables, there can be attached a natural number j such that the set $\{(x_1, \ldots, x_n) \mid (\bigwedge_{1 \leq i \leq n} x_i \in z) \,\&\, z \models \mathscr{A}(x_1, \ldots, y_1, \ldots)\}$ can be obtained from $z, y_1, \ldots, y_m$ using j iterations of the fundamental operations and Cartesian product operation. This gives (5) and is used for (4) as well. More details may be found in Appendix II.

The result 2.12 is used in §3. Note that it is a corollary that $L(\pi) = F'(\pi)$ when π is Prim_O-closed. One uses the Stability Theorems for L and for F' plus the fact that there is Prim_O G such that for all α, $\alpha \in F'(G(\alpha))$. However, a direct calculation gives a stronger result here. See Linden [**10**].

3. **How primitive recursive ordinal functions and set functions are related.** It is obvious that any Prim_O ordinal function is the restriction to ordinals of a Prim set function. In this section we show the converse. Any Prim set function that maps ordinals to ordinals is Prim_O on the ordinals.

The first step in the proof is to show that the ordinal predicates corresponding to membership and equality via Gödel's F-function, are Prim_O. This gives us a Prim_O model of set theory in the ordinals which allows us to treat Prim set functions on constructible sets as if they were Prim_O ordinal functions. The idea of Lemma 3.1 goes back to Takeuti who showed that these predicates were primitive recursive ordinal functionals. See [**19**]. As for the present version of this lemma, Schröder showed that the predicates can be obtained from Prim_O functions by a use of scheme 1.6 (his formulation differed somewhat because he was not using Gödel's F-function). The first author then showed that Prim_O is itself closed under 1.6.

3.1 LEMMA. $\lambda\alpha\beta(F'(\alpha) = F'(\beta))$ *and* $\lambda\alpha\beta(F'(\alpha) \in F'(\beta))$ *are* Prim_O *predicates.*

The proof is in Appendix II.

3.2 LEMMA. *There is a one-one* Prim *set function N which maps the ordinals onto the constructible sets so that*

(1) $N(\alpha) \subseteq N``\alpha$, $N(0) = 0$, $N(1) = 1$.

(2) *The relation* $\lambda\alpha\beta(N(\alpha) \in N(\beta))$ *is* Prim_O.

(3) *There is a* Prim_O *operation b such that for all* α, $L(\alpha) \subseteq N``b(\alpha)$.

PROOF. N may be obtained by eliminating repetitions in F'. Let

$$\mathrm{Rd}\,(\alpha) = (\min \xi \leq \alpha)(F'(\alpha) = F'(\xi)).$$

Then Rd is Prim_O and $F'(\alpha) = F'(\mathrm{Rd}\,(\alpha))$, $\mathrm{Rd}\,(\alpha) = \mathrm{Rd}\,(\mathrm{Rd}\,(\alpha))$, and $F'(\beta) = F'(\gamma) \Leftrightarrow \mathrm{Rd}\,(\beta) = \mathrm{Rd}\,(\gamma)$. Let $Q(\alpha)$ be the αth fix-point of Rd,

$$Q(\alpha) = (\min \xi)(\xi = \mathrm{Rd}\,(\xi)\ \&\ \xi \geq \bigcup_{\nu<\alpha}(Q(\nu) + 1)).$$

Note that $\mathrm{Rd}\,(9 \cdot \gamma) = 9 \cdot \gamma$ since $9 \cdot \gamma$ is an accumulation point of F', so for every α there is a fix-point of Rd before $\alpha + 9$. Thus the min operator in the definition of Q can be bounded by $(\bigcup_{\nu<\alpha}(Q(\nu) + 1)) + 9$. So Q is Prim_O. The function Q is strictly increasing, $Q(0) = 0, Q(1) = 1$, and for all $\alpha, \alpha \leq Q(\alpha)$ and $Q(\alpha) = \mathrm{Rd}\,(Q(\alpha))$. Let $N(\alpha) = F'(Q(\alpha))$.

Then $N(\alpha) \in N(\beta) \Leftrightarrow F'Q(\alpha) \in F'Q(\alpha)$ is Prim_O. Also $N(\alpha) = N(\beta)$ implies $F'(Q(\alpha)) = F'(Q(\beta))$, so $\mathrm{Rd}\,(Q(\alpha)) = \mathrm{Rd}\,(Q(\beta))$ whence N is one-one. To obtain the bounding function b of (3), first find a Prim_O function G such that $\alpha \in F'(G(\alpha))$ and $G(\alpha)$ is an accumulation point of F'. Then use the Stability Theorem for L in the hierarchy $H(\alpha) = F'(\mathrm{ac}\,(\alpha))$ to find a Prim_O function b' such that

$$\alpha \in H(\beta) \Rightarrow L(\alpha) \in H(b'(\beta)).$$

Let $b(\alpha) = \mathrm{ac}\,(b'(G(\alpha)))$. Then $L(\alpha) \subseteq F'\,\text{``}\,b(\alpha) \subseteq N\,\text{``}\,b(\alpha)$. Hence (3).

For any set function mapping L to L let F^* be the natural transform of F by N, i.e. $F^* = N^{-1} \circ F \circ N$. The conditions $N(0) = 0$ and $N(1) = 1$ guarantee that characteristic functions map into characteristic functions and that the characteristic function of $\lambda\alpha\beta(N(\alpha) \in N(\beta))$ is the *-transform of the characteristic function of $\lambda x, y(x \in y)$.

3.3 THEOREM. *If F is a* Prim *set function then F^* is* Prim_O.

PROOF. Let us do the calculation for 1-place Prim set functions. Let b be the Prim_O function of the preceding lemma and let b_F and b_N be Prim_O stabilizing functions for $\mathscr{D}_F$ and $\mathscr{D}_N$ in the L-hierarchy. Then for any ordinal α, $N(\alpha) \in L(b_N(\alpha + 1))$ by the Stability Theorem. Moreover, if $\gamma = b_F(b_N(\alpha + 1))$ then $F(N(\alpha)) \in L(\gamma)$ and $L(\gamma) \models \mathscr{D}_F(F(N(\alpha)), N(\alpha))$. Then

$$F^*(\alpha) = (\min \beta)(N(\beta) = FN(\alpha)) = (\min \beta < b(\gamma))(\exists \delta < b(\gamma + 1))$$
$$(N(\delta) \models \mathscr{D}_F(N(\beta), N(\alpha))\ \&\ \alpha \cup \beta < \delta\ \&\ \mathrm{Trans}\,(N(\delta))).$$

Thus we have written F^* in the form

$$F^*(\alpha) = (\min \beta < b(\gamma))(\exists \delta < b(\gamma + 1))(\alpha \cup \beta < \delta\ \&\ \models \mathscr{Q}(N(\delta), N(\beta), N(\alpha)))$$

where $\mathscr{Q}$ is restricted. But since $\lambda\alpha\beta(N(\alpha) \in N(\beta))$ is Prim_O, so is

$$\models \mathscr{Q}(N(\delta), N(\beta), N(\alpha))$$

by 1.3(5). Hence F^* is Prim_O.

REMARK. It is also the case that if $X_1, \ldots, X_k$ map constructible sets to constructible sets and if F is Prim $(X_1, \ldots, X_k)$ then F^* is $\mathrm{Prim}_O\,(X^*, \ldots, X^*)$.

3.4 LEMMA. *$N^{-1} \upharpoonright \mathrm{Ord}$ is Prim_O.*

PROOF. Using the recursion schema define

$$F(\alpha) = (\min \beta < b(\alpha + 1))\Big(\mathrm{Ord}\,(N(\beta))\ \&\ \beta \geq \bigcup_{\nu<\alpha} F(\nu)$$

$$\&\ \Big(\sim\mathrm{Lim}\Big(\bigcup_{\nu<\alpha} F(\nu)\Big) \Rightarrow \beta > \bigcup_{\nu<\alpha} F(\nu)\Big)\Big).$$

Since Ord is restricted, 1.3 shows F to be Prim_O. But N^{-1} is increasing on the ordinals, so an easy induction shows that $N(F(\alpha)) = \alpha$ for all α. Hence F is $N^{-1} \upharpoonright \mathrm{Ord}$.

3.5 THEOREM. *If F is* Prim *and maps ordinals to ordinals, then F is* Prim_O *on the ordinals.*

PROOF. $F = N \circ F^* \circ N^{-1} = N^* \circ F^* \circ N^{-1}$ is Prim_O by 3.3 and 3.4.

REMARK. The results of this section also apply to functions on Prim_O-closed ordinals π. If f is $F \upharpoonright L(\pi)$ for some Prim function F, then f^* is $G \upharpoonright \pi$ for some Prim_O function G. If f maps π to π and is $F \upharpoonright \pi$ for some Prim set function F, then f is $G \upharpoonright \pi$ for some Prim_O function G. This applies in particular to $\pi = \omega$ giving the equivalence between Prim set functions on hereditarily finite sets and primitive recursive functions on natural numbers. This case was worked out by Rödding, see **[14]**.

4. **Admissible sets, Σ_1-definable predicates, min-recursive functions.** The early treatments of recursive functions of ordinals dealt only with functions on infinite cardinals. In the work of Takeuti **[17]**, 1960, the κ-recursive functions, κ an infinite cardinal, map κ to κ and are just those functions that can be obtained from functionals using our schemes plus the ordinal min rule bounded by κ. Similarly in the work of Machover **[12]**, 1961, done independently, the κ-recursive functions map κ to κ and are just those whose values are computable in fewer than κ stages by an equation calculus. Again κ is assumed at the outset to be a cardinal.

It has turned out that the assumption that κ is a cardinal was not necessary. What one had to know about κ was that it had closure properties needed to set up the calculus and that whenever $\alpha, \beta < \kappa$ and $\beta = f(\alpha)$ was computable, then $\beta = f(\alpha)$ was computable in fewer than κ stages. This was the approach of Kripke **[8]**, 1964, that led to the concept of admissible ordinals. The τ-recursion theories are perfectly well behaved for any admissible τ and there are many admissible ordinals that are not cardinals. Kripke showed that every infinite cardinal is admissible and that there are κ^+ admissible ordinals of each infinite power κ.

The τ-recursion theories for admissible τ include not only the earlier κ-recursion theories but also the Kreisel-Sacks metarecursion theory, 1963, **[7]**. One takes $\tau = \omega_1^c$, the first nonconstructive ordinal. It should be pointed out, however, that metarecursion theory was obtained not by looking at computations on ω_1^c, but by looking at functions definable in ω-models of an abstract recursion theory.

Platek in [**13**], 1966, independently developed the foundations of recursion theory based on defining schemes for functions not only on ordinals but on sets. He came out with the notion of recursively regular sets, a notion equivalent to admissibility for ordinals.

There has been a parallel development in infinitary logic. In the early treatments of Scott-Tarski [**15**], 1958, and Karp [**4**], 1959, the formulas of the various infinitary languages were restricted by cardinality. The formulas of $\mathfrak{L}_\kappa$ had to be in H_κ, κ an infinite cardinal. Actually there was no good reason for doing things this way as Kreisel rather forcefully pointed out in [**6**], 1963. If, for instance, one is studying proof theory for formulas in a set T, then what one wants to know about T is that it has closure properties necessary for setting up the logic and that when a formula in T is provable, it has a proof in T. Following Kreisel's suggestions, Barwise in [**1**], 1966, developed infinitary logic along these lines and showed that the sets T that are suitable for proving theorems in the familiar deductive systems are precisely the admissible sets and their unions. If a Σ_1^T-set of nonlogical axioms is added, then the sets T suitable for theorem-proving are precisely the admissible T. The family of languages $\mathfrak{L}_T$ with formulas in T, T admissible, comprehend not only the old $\mathfrak{L}_\kappa$ (which are $\mathfrak{L}_{H_\kappa}$ in the present terminology) but the constructive infinitary languages of Takeuti-Kino [**20**], 1963, also Lopez-Escobar [**11**], 1967. Take $T = \mathfrak{L}(\omega_1^c)$. These developments are very important for infinitary languages. The H_κ are admissible but there are many admissible sets besides these, many of the new languages have such interesting properties as compactness and interpolation theorems that the old ones lack.

It may be of interest to the reader to see how the concept of admissibility enters recursion theory when the approach by defining schemes is used. This section is purely expository.

We want to add the ordinal min rule to the defining schemes 1.1:

$$F(\mathbf{x}) = (\min \xi)G(\xi, \mathbf{x}) = 0, \quad \text{provided} \quad (\forall \mathbf{x})(\exists \xi)G(\xi, \mathbf{x}) = 0.$$

The functions obtained by schemes 1.1 plus the ordinal min rule might by called min-recursive in $X_1, \ldots, X_k$. The min-recursive functions are still generalized Σ_1-definable, for the F above is defined by

$$\begin{aligned}\mathscr{D}_F(v_0, v_1, \ldots, v_n) \leftrightarrow \text{Ord}\,(v_0) \wedge (\forall w_0 \in v_0)(Ew_1)[w_1 \neq 0 \\ \wedge \mathscr{D}_G(w_1, w_0, v_1, \ldots, v_n)] \\ \wedge \mathscr{D}_G(0, v_0, v_1, \ldots, v_n).\end{aligned}$$

Remarks 2.3(1), (2), (3), (4), (5) still apply to the min-recursive set functions. The stability lemma fails. The min-recursive set functions grow much faster on the cumulative hierarchy than on the constructible hierarchy.

However, there is something unsatisfactory about these functions. The Prim set functions were defined so that when restricted to $L(\omega)$ they are equivalent to ordinary primitive recursive functions on ω. But the min-recursive set functions on $L(\omega)$ may take values outside of $L(\omega)$. A simple example is

$$F = (\min \xi)(\xi \neq 0 \wedge \xi = \bigcup \xi).$$

Clearly a bounded min-rule ought to be considered. The question arises, what kind of bound? At the very least, the bound α should be chosen so that $L(\alpha)$ is closed under α-min-recursive set functions. Since this implies that α is at least Prim_O-closed, the following theorem tells us exactly what ordinals α should be taken as bounds.

4.1 THEOREM. *Assume* α Prim_O*-closed. Then the following are equivalent:*

(i) $L(\alpha)$ *is closed under the set-functions defined by schemes* 1.1(2), (3), (4), (5), *substitution, recursion, and the* α-min *rule*:

$$F(\mathbf{x}) = (\min \xi)_\alpha G(\xi, \mathbf{x}) = 0, \quad \textit{provided} \quad (\forall \mathbf{x} \in L(\alpha)^n)(\exists \xi < \alpha) G(\xi, \mathbf{x}) = 0.$$

(ii) $L(\alpha)$ *satisfies the* $\Sigma_1^{L(\alpha)}$*-reflection principle: For restricted formulas* $\mathcal{Q}$

$$(\forall x, y_1, \ldots, y_n \in L(\alpha))[(\forall u \in x)(\exists v \in L(\alpha)) \models \mathcal{Q}(u, v, y_1, \ldots, y_n) \Rightarrow$$
$$(\exists w \in L(\alpha))(\forall u \in x)(\exists v \in w) \models \mathcal{Q}(u, v, y_1, \ldots, y_n)].$$

PROOF. Suppose (i), that $\mathcal{Q}$ is restricted, $x, y_1, \ldots, y_n \in L(\alpha)$, and

$$(\forall u \in x)(\exists v \in L(\alpha)) \models \mathcal{Q}(u, v, y_1, \ldots, y_n).$$

Let F be defined on $L(\alpha)$ so that $F(u, \mathbf{y}) = (\min \xi)_\alpha (\exists v \in L(\xi)) \models \mathcal{Q}(u, v, \mathbf{y})$ if $u \in x$, 0 if $u \notin x$. Since L and the satisfaction relation for restricted formulas are both Prim, the relation $\lambda \xi, u, y_1, \ldots, y_n((\exists v \in L(\xi)) \models \mathcal{Q}(u, v, \mathbf{y}))$ is Prim. So F is α-min recursive. Thus $G(x, \mathbf{y}) = \bigcup \{F(u, \mathbf{y}) \mid u \in x\}$ is α-min recursive. By (i), $w = L(G(x, \mathbf{y})) \in L(\alpha)$ and w satisfies the conclusion of the reflection principle. Whence (ii).

Suppose (ii). Since the equivalences 2.3(3) then hold in $L(\alpha)$, generalized Σ_1-formulas reduce to Σ_1-formulas in $L(\alpha)$. Moreover the restricted condition $\mathcal{Q}$ in the Σ_1-reflection principle can be replaced by a $\Sigma_1^{L(\alpha)}$-condition.

To prove that $L(\alpha)$ is closed under α-min-recursive set functions it is necessary to show by induction on the defining schemes that if F is α-min recursive then $L(\alpha)$ is closed under F and its defining formula $\mathcal{D}_F$ is $L(\alpha)$-absolute. The defining formula for F given by the α-min rule is the same as the one for the unbounded ordinal min rule as given above. The closure of $L(\alpha)$ under F and the $L(\alpha)$-absoluteness of $\mathcal{D}_F$ is obvious for the initial functions, and if G and H have these properties then so does F when F arises by substitution from G and H. Suppose G has these properties and $F(x) = (\min \xi)_\alpha G(\xi, \mathbf{x}) = 0$, where

$$(\forall \mathbf{x} \in L(\alpha)^n)(\exists \xi < \alpha) G(\xi, \mathbf{x}) = 0.$$

By definition, $L(\alpha)$ is closed under F. The defining formula for F is

$$\mathcal{D}_F(v_0, v_1, \ldots, v_n) \leftrightarrow \text{Ord}\,(v_0) \wedge \mathcal{D}_G(0, v_0, v_1, \ldots, v_n)$$
$$\wedge\ (\forall w_0 \in v_0)(Ew_1)[w_1 \neq 0 \wedge \mathcal{D}_G(w_1, w_0, v_1, \ldots, v_n)].$$

To show that $\mathcal{D}_F$ is $L(\alpha)$-absolute, we suppose $y, x_1, \ldots, x_n \in L(\alpha)$ and must show $\models \mathcal{D}_F(y, \mathbf{x}) \Leftrightarrow L(\alpha) \models \mathcal{D}_F(y, \mathbf{x})$. But the quantifier on w_1 in $\mathcal{D}_F$ can be restricted to $L(\alpha)$ since $L(\alpha)$ is closed under G, and the $L(\alpha)$-absoluteness of $\mathcal{D}_F$ follows from the $L(\alpha)$-absoluteness of $\mathcal{D}_G$.

Finally suppose F arises by recursion from G,

$$F(z, \mathbf{x}) = G(\bigcup \{F(u, \mathbf{x}) \mid u \in z\}, z, \mathbf{x}),$$

that $L(\alpha)$ is closed under G and that $\mathscr{D}_G$ is $L(\alpha)$-absolute. We show by induction on $z \in L(\alpha)$ that if $\mathbf{x} \in L(\alpha)^n$ then $F(z, \mathbf{x}) \in L(\alpha)$ and $L(\alpha) \models \mathscr{D}_F(F(z, \mathbf{x}), z, \mathbf{x})$. This will suffice since $\mathscr{D}_F$ is generalized Σ_1, hence persistent upward. Suppose $(\forall u \in \mathrm{TC}\,(z))(F(u, \mathbf{x}) \in L(\alpha) \,\&\, L(\alpha) \models \mathscr{D}_F(F(u, \mathbf{x}), u, \mathbf{x}))$. The defining formula $\mathscr{D}_F$ reduces to a Σ_1-formula $(Ew)\mathscr{Q}(w, v_0, \ldots, v_{n+1})$ on $L(\alpha)$. The induction hypothesis shows that $(\forall u \in z)(\exists y \in L(\alpha))L(\alpha) \models (Ew)\mathscr{Q}(w, y, u, \mathbf{x})$. Hence $(\forall u \in z)(\exists v \in L(\alpha))((\exists y \in v)(\exists y' \in v) \models \mathscr{Q}(y', y, u, \mathbf{x}))$. An application of the $\Sigma_1^{L(\alpha)}$-reflection principle gives a set $w \in L(\alpha)$ bounding the v's. Since $L(\alpha)$ is Prim-closed, we may assume w transitive. Thus $(\forall u \in z)(\exists y \in w)L(\alpha) \models \mathscr{D}_F(y, u, \mathbf{x})$. Then $F(u, \mathbf{x}) \in w$ for all $u \in z$. Hence $\bigcup \{F(u, \mathbf{x}) \mid u \in z\} \in L(\alpha)$ and since $L(\alpha)$ is closed under G, $G(\bigcup \{F(u, x) \mid u \in z\}, z, \mathbf{x}) = F(z, \mathbf{x}) \in L(\alpha)$. To see that

$$L(\alpha) \models \mathscr{D}_F(F(z, \mathbf{x}), z, \mathbf{x})$$

where $\mathscr{D}_F$ is as described in 2.2, note that the outer quantifiers $(Ew_0)(Ew_1)$ in $\mathscr{D}_F$ denote $\lambda u F(u, \mathbf{x}) \restriction \mathrm{TC}\,(z)$ and $\mathrm{TC}\,(z)$ respectively. Since $\mathrm{TC}\,(z) \in L(\alpha)$, an application of the $\Sigma_1^{L(\alpha)}$-reflection principle just like the one given above shows that $\lambda u F(u, \mathbf{x}) \restriction \mathrm{TC}\,(z) \in L(\alpha)$. The quantifier on w_2 denotes $\bigcup \{F(u, \mathbf{x}) \mid u \in z\} \in L(\alpha)$ and for any $w_3 \in \mathrm{TC}\,(z)$, w_4, w_5 denote $F(w_3, \mathbf{x})$ and $\bigcup \{F(u, \mathbf{x}) \mid u \in w_3\}$ respectively. All are in $L(\alpha)$. Therefore the $L(\alpha)$-absoluteness of $\mathscr{D}_G$ implies $L(\alpha) \models \mathscr{D}_F(F(z, \mathbf{x}), z, \mathbf{x})$.

The usual definition of admissibility is equivalent to the following:

4.2 Definition. A class T is admissible just in case it is transitive, Prim-closed, and satisfies the Σ_1^T-reflection principle. An ordinal τ is admissible just in case $L(\tau)$ is admissible.

4.3 Examples. (1) (Kripke, Platek). The first admissible ordinal $> \omega$ is ω_1^c, the least nonconstructive ordinal. There are κ^+ admissible ordinals of any infinite power κ.

(2) Whenever the κ_i are regular infinite cardinals, $T = \bigcup_{i \in I} H_{\kappa_i}$ is admissible. For if $\mathscr{Q}$ is Δ_0 and $x, y_1, \ldots, y_n \in T$, then unless $T = H_{\aleph_0}$ there is nondenumerable κ such that $x, y_1, \ldots, y_n \in H_\kappa \subseteq T$. So by the Gödel-Lévy Theorem mentioned in the introduction,

$$(\forall u \in x)(\exists v \in T) \models \mathscr{Q}(u, v, \mathbf{y}) \Rightarrow (\forall u \in x)(\exists v \in H_\kappa) \models \mathscr{Q}(u, v, \mathbf{y}).$$

But any set H_κ is obviously admissible (assuming the axiom of choice). It follows that T is admissible.

(3) (Kripke). Any ordinal α that is an infinite cardinal relative to L is admissible. Outline of proof: Let $C(\alpha) = \{x \mid \mathrm{TC}\,(x) \prec \alpha\}$. Then $C(\kappa) = H_\kappa$ if κ is a regular cardinal, and $C(\kappa) = \bigcup \{H_\gamma \mid \gamma < \kappa \,\&\, \gamma \text{ is a regular cardinal}\}$ if κ is a singular cardinal. In any case, we have proved in ZF + AC the statement

$$(\alpha)[\mathrm{Card}\,(\alpha) \wedge \omega \leq \alpha \rightarrow \mathrm{Adm}\,(C(\alpha))]$$

where $\mathrm{Adm}\,(v_0)$ is the formula expressing the condition "v_0 is admissible" through the coding. Since the satisfaction relation in a set is Prim, $\mathrm{Adm}\,(v_0)$ is L-absolute.

Relativizing the proof to constructible sets, we have $\mathrm{Card}^{(L)}(\alpha)$ and α is infinite implies $C^{(L)}(\alpha)$ is admissible. Using the fact that $L(\xi) \approx \xi$ holds relative to L, it can be seen that $C^{(L)}(\alpha) = L(\alpha)$.

Theorem 4.1 shows that for Prim_O-closed ordinals τ, τ is admissible if and only if the set functions defined by the Prim schemes plus the τ-min rule, map $L(\tau)$ to $L(\tau)$. These functions we call the τ-min recursive set functions. We have seen that these functions restricted to $L(\tau)$ are $\Sigma_1^{L(\tau)}$-definable. The converse also holds and its proof establishes the familiar Normal Form Theorem.

4.4 THEOREM. *Let τ be admissible. Any* τ-min *recursive function restricted to $L(\tau)$ is $\Sigma_1^{L(\tau)}$-definable. Conversely, any $\Sigma_1^{L(\tau)}$-definable function on $L(\tau)$ to $L(\tau)$ is* τ-min *recursive. Moreover, for each n there is a* Prim *relation T_n such that for any n-place* τ-min *recursive set function F there is a restricted formula $\mathscr{Q}$ such that*

$$(\forall \mathbf{x} \in L(\tau)^n)(F(\mathbf{x}) = U(\min \xi)_\tau T_n(\ulcorner \mathscr{Q} \urcorner, \xi, \mathbf{x}))$$

and

$$(\forall \mathbf{x} \in L(\tau)^n)(\exists \xi < \tau) T_n(\ulcorner \mathscr{Q} \urcorner, \xi, \mathbf{x}),$$

where U is the Prim *function $N \circ L_1$. (N is the enumerating function for constructible sets of* §3, *L_1 is the inverse function for the first coordinate of the Gödel pairing function P.)*

PROOF. Suppose there is a restricted formula $\mathscr{Q}$ such that

$$(\forall \mathbf{x} \in L(\tau)^n)(\exists y, z \in L(\tau)) \models \mathscr{Q}(\mathbf{x}, y, z) \,\&$$
$$(\forall y \in L(\tau))(y = F(\mathbf{x}) \Leftrightarrow (\exists z \in L(\tau)) \models \mathscr{Q}(\mathbf{x}, y, z)).$$

Then

$$F(\mathbf{x}) = U(\min \xi)_\tau \models \mathscr{Q}(\mathbf{x}, NL_1(\xi), NL_2(\xi))$$

has the required form.

The Normal Form Theorem can be used to extend to results of §3 on Prim functions to τ-min recursive functions.

4.5 DEFINITION. For admissible τ, the τ-min recursive ordinal functions are those defined by the Prim_O schemes plus the τ-min rule.

Thus a τ-min recursive ordinal function is the restriction to ordinals of a τ-min recursive set function and therefore maps τ to τ.

4.6 THEOREM. *Assume τ admissible. Then*

(1) *If F on $L(\tau)$ is a* τ-min *recursive set function, then $F^* = N^{-1} \circ F \circ N$ is a* τ-min *recursive ordinal function on τ.*

(2) *If F is a* τ-min *recursive set function that maps τ to τ, then F is a* τ-min *recursive ordinal function on τ.*

PROOF. By the Normal Form Theorem, if F is τ-min recursive,

$$F(\mathbf{x}) = NL_1(\min \xi)_\tau \models \mathscr{Q}(\mathbf{x}, NL_1(\xi), NL_2(\xi)).$$

So $NF^*(\alpha_1, \ldots, \alpha_n) = NL_1(\min \xi)_\tau \models \mathscr{Q}(N(\alpha_1), N(\alpha_2), \ldots, NL_1(\xi), NL_2(\xi))$. But L_1 is Prim_O and $\lambda\xi, \ldots \models \mathscr{Q}(N(\xi), \ldots)$ is Prim_O by 1.3(5) and Lemma 3.2. Hence F^* is a τ-min recursive ordinal function on τ.

If moreover F maps τ to τ, then $F = N \circ F^* \circ N^{-1} = N^* \circ F^* \circ N^{-1}$ is a τ-min recursive ordinal function on τ by (1) and Lemma 3.4. Hence (2).

5. **A relativization theorem for ZF-statements.** The main lemma of this section is a sort of "inward" Löwenheim-Skolem Theorem that deals with relativizations from sets to constructible sets.

MAIN LEMMA. *Suppose t is a transitive set closed under the rank function, that a is a constructible set of natural numbers in t, and that τ is the least ordinal not in t. Suppose that $\mathscr{S}$ is a* Prim (a, τ)*-definable set of* ZF*-formulas in one free variable such that a satisfies $\mathscr{S}$ in $\langle t, \in \restriction t\rangle$. Then for admissible α such that $\tau < \alpha$ and $a \in L(\alpha)$ there is a transitive, countable,* Prim (a, τ, α)*-definable set t' such that a satisfies $\mathscr{S}$ in $\langle t', \in \restriction t'\rangle$.*

When we say that a set y is Prim $(X_1, \ldots, X_k)$-definable, the X_i being set functions, we mean that there is a Prim $(X_1, \ldots, X_k)$ function F such that $y = F(0)$. A set-function is Prim $(x_1, \ldots, x_k)$, the x_i being sets, just in case it is Prim $((\lambda x)x_1, \ldots, (\lambda x)x_k)$. This is the same as saying that the x_i may be taken as parameters. Thus to say that a set is Prim $(x_1, \ldots, x_k)$-definable is to say that there is a Prim function F such that $y = F(0, x_1, \ldots, x_k)$. Clearly then a Prim $(x_1, \ldots, x_k)$ definable set belongs to any Prim-closed set containing the x_i. Thus the set t' above is constructible and in fact belongs to $L(\pi)$ where π is the first Prim-closed ordinal above α.

To say that a set of ZF formulas is arithmetic comes to the same thing as saying that it is definable relative to $L(\omega)$ by a ZF-formula. Thus an arithmetic set is Prim (ω)-definable. Moreover ω is Prim (τ) definable if τ is infinite. Thus we have as a corollary:

THEOREM. *Suppose that t is a transitive set closed under the rank function and that τ is the least ordinal not in t. Suppose that $\mathscr{S}$ is an arithmetic set of* ZF *statements true in $\langle t, \in \restriction t\rangle$. Then for admissible α such that $\tau < \alpha$, there is a* Prim (τ, α)*-definable set t' such that $\mathscr{S}$ is true in $\langle t', \in \restriction t'\rangle$.*

Thus, for example, any such set $\mathscr{S}$ that has a model, has a constructible model.

If $\mathscr{S}$ consists of Σ_1-formulas then the persistence of such formulas upward can be used to replace t' by an $L(\alpha_0)$, thus obtaining the theorem stated below. It sharpens Lévy's Theorem 43 in **[9]** which states that if $\mathscr{A}$ is a Σ_1-statement and $\models \mathscr{A}$, then $L(\omega_1^{(L)}) \models \mathscr{A}$. It also is a generalization of Shoenfield's Absoluteness Lemma in **[16]**. This lemma states that Σ_2^1-properties of constructible functions on ω to ω are L-absolute. But Σ_2^1-properties in the analytic hierarchy reduce to Σ_1-predicates in the set-theoretical hierarchy by a well-known reduction of Kleene **[5]**.

THEOREM. *Suppose that α_0 is a limit of admissible ordinals and that a is a constructible set of natural numbers in $L(\alpha_0)$. Then*

(1) *If $\mathscr{A}$ is a Σ_1-formula in one free variable such that $V(\alpha_0) \models \mathscr{A}[a]$, then $L(\alpha_0) \models \mathscr{A}[a]$.*

(2) *If $\mathscr{A}$ is a generalized Σ_1-formula in one free variable such that $V(\tau) \models \mathscr{A}[a]$ for some $\tau < \alpha_0$, then $L(\alpha_0) \models \mathscr{A}[a]$.*

The first statement follows from the lemma because if $V(\alpha_0) \models \mathscr{A}[a]$ and $\mathscr{A}$ is Σ_1 then there is $\tau < \alpha_0$ such that $V(\tau) \models \mathscr{A}[a]$. If $\mathscr{A}$ is generalized Σ_1 and $V(\alpha_0)$ is not admissible then it may not be possible to reduce $\mathscr{A}$ to a Σ_1-formula in $V(\alpha_0)$. So the hypothesis must be strengthened to give the τ.

COROLLARY. *If α_0 is a limit of admissible ordinals, and if F and G are* Prim *set functions that agree on $L(\alpha_0)$ then they agree on $V(\alpha_0)$.*

PROOF. Suppose that F and G are Prim functions of one variable which do not agree on $V(\alpha_0)$. Then there is $x \in V(\alpha_0)$ such that $F(x) \neq G(x)$. Now $V(\alpha_0)$ need not be admissible nor even a limit of admissibles, but it surely is Prim-closed (by Corollary 2.6). So $V(\alpha_0) \models \mathscr{A}$ where $\mathscr{A}$ is

$$(Ew_0)(Ew_1)(Ew_2)[\mathscr{D}_F(w_0, w_2) \wedge \mathscr{D}_G(w_1, w_2) \wedge w_1 \neq w_0].$$

By Lemma 2.12 there is transitive $t \in V(\alpha_0)$ such that x, $F(x)$, $G(x) \in t$ and $t \models \mathscr{A}$. If then τ is the least ordinal not in t, $V(\tau) \models \mathscr{A}$. By theorem, $L(\alpha_0) \models \mathscr{A}$. But then F and G do not agree on $L(\alpha_0)$.

REMARKS. (1) The theorem implies that Σ_1-properties of constructible sets of natural numbers relativize from V to L. But Σ_1-properties of other constructible sets need not relativize. For example, "x is a countable ordinal" is a Σ_1-property of x but an ordinal may be countable without being countable relative to L.

(2) The theorem above need not apply to admissible α_0. Similarly in the Main Lemma, even if τ is admissible, α might have to be taken above τ. For example, consider the relation $\leq_O$ on natural numbers used by Kleene to encode the constructive ordinals. Though not Prim (ω)-definable, this relation has an implicit Prim (ω) definition, meaning that there is a Prim function F such that

$$(\forall x)(x = \leq_O \Leftrightarrow F(x, \omega) = 0).$$

Then ω_1^c, the first nonconstructive ordinal, is admissible (Kripke-Platek) and

$$V(\omega_1^c) \models (Ew_0)\mathscr{D}_F(0, w_0, \omega)$$

since $\leq_O \in V(\omega_1^c)$ and $V(\omega_1^c)$ is Prim-closed. By Lemma 2.12 again, there is $\tau < \omega_1^c$ such that $V(\tau) \models \mathscr{A}$ where $\mathscr{A}$ is the generalized Σ_1-statement, $(Ew_0)\mathscr{D}_F(0, w_0, \omega)$. But not $L(\omega_1^c) \models \mathscr{A}$ since if $\mathscr{A}$ were true in $L(\omega_1^c)$ that would mean $\leq_O \in L(\omega_1^c)$. The valuation function for the codings is $\Sigma_1^{L(\omega_1{}^c)}$-definable so this would put $\omega_1^c \in L(\omega_1^c)$.

We now proceed to the proof of the Main Lemma. Suppose that $\langle t, a, \tau, \mathscr{S}\rangle$ are given in accordance with the hypothesis and that α is admissible, $\tau < \alpha$, and $a \in L(\alpha)$. Given that a satisfies $\mathscr{S}$ in $\langle t, \in \restriction t\rangle$, we are to find a Prim (a, τ, α)-definable, countable transitive set t' such that a satisfies $\mathscr{S}$ in t'. Several first-order languages extending the language of ZF set theory are needed. First there is $\mathfrak{L}(\Theta)$ for $\Theta \subseteq \tau$ which has in addition to the predicate symbols $\equiv$, ϵ, a one-place function symbol $\mathbf{r}$ and individual constants c and c_ν for $\nu \in \Theta$. The language

$\mathfrak{L}'(\Theta)$ is like $\mathfrak{L}(\Theta)$ except that individual constants d_n, $n \in \omega$, are added. The language $\mathfrak{L}''$ is like $\mathfrak{L}'(0)$ except that individual constants e_n, $n \in \omega$, are added. It is assumed that the languages are coded so that Fmla (Θ), the set of (Gödel-sets of) formulas of $\mathfrak{L}(\Theta)$, and Fmla$'$ (Θ), the set of (Gödel-sets of) formulas of $\mathfrak{L}'(\Theta)$, are Prim (ω, Θ)-definable. Fmla$''$ is Prim (ω)-definable. By 1.3, 1.4, the relation $\lambda x, y(x \vdash y)$ is Prim (ω) where $\vdash$ denotes provability in first-order predicate logic with identity. Similarly, the function $(\lambda x, y)$ Con (x, y), the set of consequences of x in y, is Prim (ω).

PROOF OF THE MAIN LEMMA. Let $\mathscr{C} \subseteq$ Fmla (τ) be the set

$$\begin{aligned}\{\ulcorner\mathscr{A}(c)\urcorner\ \ \ulcorner\mathscr{A}\urcorner \in \mathscr{S}\} &\cup \{\text{Axiom of Extensionality}, \ulcorner(w_0)[w_0 \mathrel{E} c \rightarrow \mathrm{Ord}\,(w_0)]\urcorner,\\ &\qquad \ulcorner(w_0)(w_1)[w_0 \mathrel{E} w_1 \rightarrow \mathbf{r}(w_0) \mathrel{E} \mathbf{r}(w_1) \wedge \mathrm{Ord}\,\mathbf{r}(w_0)]\urcorner\}\\ &\cup \{\ulcorner c_\mu \mathrel{E} c_\nu\urcorner \mid \mu < \nu < \tau\} \cup \{\ulcorner\neg c_\mu \mathrel{E} c_\nu\urcorner \mid \mu, \nu \in \tau, \mu \not< \nu\}\\ &\cup \{\ulcorner(w_0)[w_0 \mathrel{E} c_\nu \rightarrow \mathrm{Ord}\,(w_0)]\urcorner \mid \nu \in \tau\} \cup \{\ulcorner c_\mu \mathrel{E} c\urcorner \mid \mu \in a\}\\ &\cup \{\ulcorner\neg c_\mu \mathrel{E} c\urcorner \mid \mu \in \tau, \mu \notin a\}.\end{aligned}$$

The ordinal τ may be assumed infinite since the lemma is trivial if τ finite, so ω is Prim (τ) definable and $\mathscr{C}$ is Prim (a, τ)-definable. Moreover, $\mathscr{C}$ is consistent since $\langle t, \in \restriction t, \mathrm{rank} \restriction t, K\rangle \models \mathscr{C}$ where $K(\ulcorner c_\nu\urcorner) = \nu$, $K(\ulcorner c\urcorner) = a$.

DEFINITION. For $\Theta \subseteq \tau$ and $\mathscr{X}, \mathscr{Y} \subseteq$ Fmla (τ), we say that $\mathscr{X}$ satisfies the Θ-rule relative to $\mathscr{Y}$ if for all statements of the form $\ulcorner(w_0)[\mathrm{Ord}\,(w_0) \rightarrow \mathscr{B}]\urcorner \in \mathscr{Y}$,

$$(\forall \nu \in \Theta)(\mathscr{X} \vdash \mathscr{B}(w_0/c_\gamma)) \Rightarrow \mathscr{X} \vdash (w_0)[\mathrm{Ord}\,(w_0) \rightarrow \mathscr{B}].$$

We make two extensions of $\mathscr{C}$, the first being to a Prim (a, τ, α)-definable set $\mathscr{C}'$ which is consistent and satisfies the τ-rule relative to Fmla (τ). To define $\mathscr{C}'$ begin with the Prim (τ)-function G such that $G(x) = \{y \in \mathrm{Fmla}\,(\tau) \mid y$ is a statement of form $\ulcorner(w_0)[\mathrm{Ord}\,(w_0) \rightarrow \mathscr{B}]\urcorner$ and $(\forall \nu \in \tau)\ulcorner\mathscr{B}(w_0/c_\nu)\urcorner \in x\}$. Let $\mathscr{C}' = \bigcup_{\beta<\alpha} F(\beta)$ where $F(0) = \mathscr{C}$ and for $\beta > 0$,

$$F(\beta) = \mathrm{Con}\left(\bigcup_{\xi<\beta} F(\xi), \mathrm{Fmla}\,(\tau)\right) \cup G\left(\bigcup_{\xi<\beta} F(\xi)\right).$$

Then $\mathscr{C}'$ is Prim (a, τ, α)-definable. The admissibility of α is used to show that $\mathscr{C}'$ satisfies the τ-rule. Suppose $(w_0)[\mathrm{Ord}\,(w_0) \rightarrow \mathscr{B}]$ is a statement of Fmla (τ) and $(\forall \nu \in \tau)\mathscr{C}' \vdash \mathscr{B}(w_0/c_\nu)$. Then $(\forall \nu \in \tau)(\exists \xi \in \alpha)F(\xi) \vdash \mathscr{B}(w_0/c_\nu)$. But F is a Prim (a, τ) set function, so the condition $(\lambda \nu, \xi)(F(\xi) \vdash \mathscr{B}(w_0/c_\gamma))$ is $\Sigma_1^{L(\alpha)}$ with parameters a, $\ulcorner\mathscr{B}\urcorner \in L(\alpha)$. By the $\Sigma_1^{L(\alpha)}$-reflection principle there is $\beta < \alpha$ such that $(\forall \nu \in \tau)(\exists \xi \in \beta)F(\xi) \vdash \mathscr{B}(w_0/c_\nu)$. So $\ulcorner(w_0)[\mathrm{Ord}\,(w_0) \rightarrow \mathscr{B}]\urcorner$ is in $F(\beta + 2)$. Hence $\mathscr{C}'$ satisfies the τ-rule relative to Fmla (τ). An induction on β shows that the same model that we had for $\mathscr{C}$ is a model of $\mathscr{C}'$. Hence $\mathscr{C}'$ is consistent.

Now pass to the language $\mathfrak{L}'(\tau)$ with additional constants d_n, $n < \omega$, to serve as witnesses for the quantifications. Since the d_n do not appear in $\mathscr{C}'$, $\mathscr{C}'$ still satisfies the τ-rule relative to Fmla$'$ (τ). Moreover, $(*)_{\mathscr{C}',\tau}$ where

$(*)_{\mathscr{X},\Theta}$: For all statements $(w_0)\mathscr{B} \in$ Fmla$'$ (Θ), $(\forall n \in \omega)\mathscr{X} \vdash \mathscr{B}(w_0/d_n)$ implies $\mathscr{X} \vdash (w_0)\mathscr{B}$.

The second extension of $\mathscr{C}$ is to a Prim (a, α, τ)-definable set $\mathscr{C}'' \subseteq$ Fmla$'$ (τ) such that for some countable, Prim (a, τ, α)-definable set Θ, $\omega \subseteq \Theta \subseteq \tau$, $\mathscr{C}''$ is maximal consistent relative to Fmla$'$ (Θ), and $\mathscr{C}''$ satisfies the Θ-rule relative to Fmla$'$ (Θ) and $(*)_{\mathscr{C}'',\Theta}$. The usual canonical model made from maximal consistent sets of formulas with witnessing constants for the quantifiers, will reduce to the set t' that we seek.

To define $\mathscr{C}''$, Θ, we use the language $\mathfrak{L}''$ with none of the c_ν's but with new constants e_n, $n < \omega$, that will serve to stand for the chosen c_ν's. Let E be a Prim function enumerating the statements of Fmla$''$ of form $\ulcorner(w_0)\mathscr{B}\urcorner$ on ω in such a way that the e_i appearing in $E(n)$ have $i < n$. Define by simultaneous recursion on n sets $H(n) \subseteq$ Fmla$'$ (τ) and ordinals $J(n) \in \tau$ such that $H(n)$ satisfies the τ-rule relative to Fmla$'$ (τ) and $(*)_{H(n),\tau}$, and $H(n)$ is consistent and contains $\mathscr{C}'$, as follows:

(1) $H(0) = \mathscr{C}'$, $J(0) = 0$.

Suppose $H(n)$ satisfies the τ-rule relative to Fmla$'$ (τ) and $(*)_{H(n),\tau}$. If $m \leq n$ and $E(m) = \ulcorner(w_0)\mathscr{B}\urcorner$, let Sub $E(m)$ be the result of replacing the e_i in $\mathscr{B}$ by $c_{J(i)}$. Define $H(n+1)$ as follows:

(2) If $n + 1 = 3m$, let $H(n+1) = H(n)$, $J(n+1) = m$.

(3a) If $n + 1 = 3m + 1$ and $H(n) \vdash$ Sub $E(m)$, let $H(n+1) = H(n)$, $J(n+1) = 0$.

(3b) If not $H(n) \vdash$ Sub $E(m)$, let k be the least natural number such that not $H(n) \vdash$ Sub $\ulcorner\mathscr{B}(w_0/d_k)\urcorner$, and let

$$H(n+1) = H(n) \cup \{\neg\text{Sub } \ulcorner\mathscr{B}(w_0/d_k)\urcorner\},\ J(n+1) = 0.$$

(4a) If $n + 1 = 3m + 2$ and $H(n) \vdash$ Sub $E(m)$ or if $E(m)$ does not have form $\ulcorner(w_0)[\text{Ord}\,(w_0) \to \mathscr{B}']\urcorner$, let $H(n+1) = H(n)$, $J(n+1) = 0$.

(4b) If not $H(n) \vdash$ Sub $E(m)$ and $E(m) = \ulcorner(w_0)[\text{Ord}\,(w_0) \to \mathscr{B}']\urcorner$, let $J(n+1)$ be the least $\nu < \tau$ such that not $H(n) \vdash$ Sub $\ulcorner\mathscr{B}'(w_0/c_\nu)\urcorner$ and let

$$H(n+1) = H(n) \cup \{\neg\text{Sub } \ulcorner\mathscr{B}'(w_0/c_{J(n+1)})\urcorner\}.$$

Note that $H(n+1)$ still satisfies the τ-rule relative to Fmla$'$ (τ) and $(*)_{H(n+1),\tau}$. For example in case 3b, suppose $(\forall p \in \omega)H(n+1) \vdash \ulcorner\mathscr{C}(w_0/d_p)\urcorner$ with $\ulcorner(w_0)\mathscr{C}\urcorner \in$ Fmla$'$ (τ). Then $(\forall p \in \omega)H(n) \vdash \ulcorner\neg\text{Sub } \mathscr{B}(w_0/d_k) \to \mathscr{C}(w_0/d_p)\urcorner$ with k as in 3b. Then $(*)_{H(n),\tau}$ implies $H(n) \vdash \ulcorner(w_0)[\neg\text{Sub } \mathscr{B}(w_0/d_k) \to \mathscr{C}(w_0)]\urcorner$. So $H(n+1) \vdash \ulcorner(w_0)\mathscr{C}\urcorner$. The other cases are similar. By induction on n, each $H(n)$ is consistent. Thus so is $H(\omega) = \bigcup_{n\in\omega} H(n)$. Let $\Theta = \{J(n) \mid n \in \omega\}$. Then $H(\omega)$ and Θ are Prim (a, τ, α)-definable sets. Let $\mathscr{C}''$ be the extension of $H(\omega)$ to a maximal consistent set relative to Fmla$'$ (Θ). The usual construction shows that $\mathscr{C}''$ is Prim (a, τ, α)-definable because the Prim (a, τ, α) enumeration J of Θ can be used to obtain a Prim (a, τ, α) function F that enumerates statements of Fmla$'$ (Θ) on ω. Thus the function that begins with $H(\omega)$, and at stage m adds $F(m)$ if the result is consistent, is Prim (a, τ, α). Thus $\mathscr{C}''$ is Prim (a, τ, α)-definable.

We may not have $(*)_{\mathscr{C}'',\tau}$, but because witnesses to quantifiers were put into $\mathscr{C}''$ at step 3b of the definition of H, we will have $(*)_{\mathscr{C}'',\Theta}$. This follows because every statement $\ulcorner(w_0)\mathscr{B}\urcorner \in$ Fmla$'$ (Θ) is Sub $E(m)$ for some $m \in \omega$. So either $\mathscr{C}'' \vdash \ulcorner(w_0)\mathscr{B}\urcorner$

or there is k such that $\mathscr{C}'' \vdash \ulcorner \neg \mathscr{B}(w_0/d_k) \urcorner$. For the same reason, C'' satisfies the Θ-rule relative to Fmla$'$ (Θ).

Now form the canonical model on constants d_n, $n \in \omega$, modulo provability from $\mathscr{C}''$. Thus $d_m \approx d_n \Leftrightarrow \mathscr{C}'' \vdash d_m \equiv d_n$ and the model is $\mathfrak{A} = \langle A, E, r, K \rangle$ where

$A = \{d_n/\approx \mid n \in \omega\}$ where $d_n/\approx \; = \{d_m \mid m \in \omega \,\&\, d_m \approx d_n\}$.

E is the relation $(\lambda x, y)((\exists m \in \omega)(\exists n \in \omega)(d_m \in x \,\&\, d_n \in y \,\&\, \mathscr{C}'' \vdash d_m \mathrel{\epsilon} d_n))$.

r is the operation (λx)(the equivalence class $d_n/\approx$ such that

$$(\exists m \in \omega)(d_m \in x \,\&\, \mathscr{C}'' \vdash \mathbf{r}(d_m) \equiv d_n)).$$

For $n \in \omega$, $K(\ulcorner d_n \urcorner) = \ulcorner d_n \urcorner/\approx$.

For $\nu \in \Theta$, $K(\ulcorner c_\nu \urcorner)$ the equivalence class $\ulcorner d_n \urcorner/\approx$ such that $\mathscr{C}'' \vdash c_\nu \equiv d_n$. Similarly for c.

Condition $(*)_{\mathscr{C}'',\Theta}$ tells us that $r(x)$ and $K(\ulcorner c_\nu \urcorner)$ are always defined. The sets A, E, r, K are all Prim (a, τ, α)-definable. The usual induction on statements $\mathscr{B}$ of Fmla$'$ (Θ) shows that $\mathfrak{A} \models \mathscr{B} \Leftrightarrow \mathscr{C}'' \vdash \mathscr{B}$. Therefore all the statements of $\mathscr{C} \cap$ Fmla (Θ) are true in $\mathfrak{A}$ and $\mathfrak{A} \models \ulcorner (w_0)\text{Ord}\,(w_0) \rightarrow \mathscr{B} \urcorner \Leftrightarrow (\forall \nu \in \Theta)$ $\mathfrak{A} \models \ulcorner \mathscr{B}(w_0/c_\nu) \urcorner$.

The set t' that we seek is the result of collapsing $\langle A, E \rangle$ onto a transitive set. It is well known that the function $\varphi(x) = \{\varphi(y) \mid y \in A \,\&\, yEx\}$ is an isomorphism of $\langle A, E \rangle$ to a transitive set $\langle t', \in \restriction t' \rangle$ provided that $\langle A, E \rangle$ is a well-founded model of the axiom of extensionality. This is where the fact that $\mathfrak{A}$ satisfies the Θ-rule is used. Since $\mathfrak{A}$ satisfies (w_0) Ord $(\mathbf{r}(w_0))$, for any $x \in A$ there is $\nu \in \Theta$ such that $r(x) = K(\ulcorner c_\nu \urcorner)$. Since $\mathfrak{A}$ satisfies $\neg c_\mu \equiv c_\nu$ whenever $\mu \neq \nu \in \Theta$, this ν is unique. Call it $\rho(x)$. Since $\mathfrak{A}$ satisfies $(w_0)(w_1)[w_0 \mathrel{\epsilon} w_1 \rightarrow \mathbf{r}(w_0) \mathrel{\epsilon} \mathbf{r}(w_1)]$ and also satisfies the diagram of Θ, $x\,E\,y \Rightarrow \rho(x) \in \rho(y)$. Hence E is well founded. Moreover, ρ is a Prim (a, τ, α)-definable set because

$$\rho = \{(x, \nu) \mid x \in A \,\&\, \nu \in \Theta \,\&\, r\,\text{‘}\,x = K\,\text{‘}\,\ulcorner c_\nu \urcorner\}$$

and A, Θ, r, K are all Prim (a, τ, α)-definable. So the function φ above defined by recursion on E can be defined by ordinal recursion. In fact,

$$\varphi = \bigcup_{\theta \in \tau} G(\theta)$$

where

$$G(\theta) = \bigcup_{\xi \in \theta} G(\xi) \cup \left\{ \left(x, \left\{ \bigcup_{\xi \in \theta} G(\xi)\,\text{‘}\,y \;\middle|\; yEx \;\&\; y \in A \right\} \right) \;\middle|\; x \in A \;\&\; \rho(x) \leq \theta \right\}.$$

Thus φ is itself a Prim (a, τ, α)-definable set and thus so is $t' = \text{Rg}\,(\varphi)$. Then $\langle t', \in \restriction t', \varphi \circ r, \varphi \circ K \rangle$ is a model of $\mathscr{C} \cap$ Fmla (Θ) and the Θ-rule still holds. To show that a satisfies $\mathscr{S}$ in t' it suffices to show that $\varphi K(\ulcorner c \urcorner) = a$. But $\omega \subseteq \Theta$ and for $n \in \omega$, $\ulcorner (w_0) w_0 \mathrel{\epsilon} c \rightarrow \text{Ord}\,(w_0) \urcorner$ is true in $\langle t', \in \restriction t', K \rangle$. Thus for $u \in t'$, $u \in \varphi K(\ulcorner c \urcorner)$ implies $u = \varphi K(\ulcorner c_\nu \urcorner)$ for some $\nu \in \Theta$. Since the diagram of Θ is true in the model, this implies $u = \varphi K(\ulcorner c_i \urcorner)$ for some $i < n$. Hence an induction on n shows that $\varphi K(\ulcorner c_n \urcorner) = n$ for all $n \in \omega$. Since the diagram of a and the statement $(w_0)[w_0 \mathrel{\epsilon} c \rightarrow \text{Ord}\,(w_0)]$ also holds in the model, $\varphi K(\ulcorner c \urcorner) = a$. Since statements

$\ulcorner\mathscr{A}(c)\urcorner$ hold in the model for $\ulcorner\mathscr{A}\urcorner \in \mathscr{S}$, a satisfies $\mathscr{S}$ in t'. Thus the proof is complete.

APPENDIX I

CLOSURE PROPERTIES OF Prim AND Prim_O

PROOF OF 1.3. (1) Proof is by induction on p. For $p = 1$, $F(x) = G(H(x))$ is Prim $(X_1, \ldots, X_k)$ by substitution scheme (b). For $p \geq 2$, $F_1(\mathbf{x}, y_2, \ldots, y_p) = G(H_1(\mathbf{x}), y_2, \ldots, y_p)$ is Prim $(X_1, \ldots, X_k)$ by substitution (b). Then

$$F_2(\mathbf{x}, y_3, \ldots, y_p) = F_1(\mathbf{x}, H_2(\mathbf{x}), y_3, \ldots, y_p) = G(H_1(\mathbf{x}), H_2(\mathbf{x}), y_3, \ldots, y_p)$$

is Prim $(X_1, \ldots, X_k)$ by substitution (a). Continue till the y's vanish.

1.3. (2) $F(\mathbf{x}) = C(G(\mathbf{x}), H(\mathbf{x}), 0, cR(\mathbf{x}))$ where cR is the characteristic function of R (1 for truth, 0 for falsity).

1.3. (3) The characteristic function of $\sim R$ is N applied to cR where $N(x) = C(0, 1, 0, x)$. The characteristic function of the disjunction $R \vee S$ is the function $\max(x, y) = C(y, x, x, y)$ applied to cR and cS.

1.3. (4) First consider the function $F(z, \mathbf{x}) = \bigcup \{G(u, \mathbf{x}) \mid u \in z\}$. Let

$$J(y, u, z, \mathbf{x}) = C(G(u, \mathbf{x}), y, u, z).$$

Then $J(y, u, z, \mathbf{x}) = G(u, \mathbf{x})$ if $u \in z$, y if not. Let

$$H(v, z, \mathbf{x}) = J(\bigcup \{H(u, z, \mathbf{x}) \mid u \in v\}, v, z, \mathbf{x})$$

as defined by the recursion scheme. Then

$$H(z, z, \mathbf{x}) = \bigcup \{H(u, z, \mathbf{x}) \mid u \in z\} = \bigcup \{G(u, \mathbf{x}) \mid u \in z\}.$$

Then 1.3(4) follows because if $F(z, \mathbf{x}) = \bigcup \{G(u, \mathbf{x}) \mid u \in z \,\&\, R(u, \mathbf{x})\}$, then $F(z, \mathbf{x}) = \bigcup \{C(G(u, \mathbf{x}), 0, 0, cR(u, \mathbf{x})) \mid u \in z\}$.

1.3. (5) We show the closure of Prim $(X_1, \ldots, X_k)$ and Prim_O $(X_1, \ldots, X_k)$ relations under bounded quantifiers. The characteristic function of $(\exists u \in z)R(u, \mathbf{x})$ is $\bigcup \{cR(u, \mathbf{x}) \mid u \in z\}$.

1.3. (6) As before it suffices to consider $F(z, \mathbf{x}) = \{G(u, \mathbf{x}) \mid u \in z\}$. $F(z, \mathbf{x}) = \bigcup \{\{G(u, \mathbf{x})\} \mid u \in z\}$ where $\lambda x\{x\}$ is Prim since $\{x\} = 0 \cup \{x\}$.

1.3. (7) Consider $F(z, \mathbf{x}) = G(F \upharpoonright z, z, \mathbf{x})$. Note that $\lambda x, y(x \upharpoonright y)$ is Prim by 1.3(6) since $x \upharpoonright y = \{u \mid u \in x \,\&\, (\exists v, w \in \cup\cup u)(u = (v, w) \,\&\, v \in y)\}$ and the condition on u, y is Δ_0. Thus it suffices to show that

$$H(z, \mathbf{x}) = \{(v, F(v, \mathbf{x})) \mid v \in \text{TC}(z)\} = F \upharpoonright \text{TC}(z)$$

is Prim, for $F(z, \mathbf{x}) = G(H(z, \mathbf{x}) \upharpoonright z, z, \mathbf{x})$. $H(z, \mathbf{x})$ can be written in the form of the recursion scheme as $H(z, \mathbf{x}) = y \cup \{(u, G(y \upharpoonright u, u, \mathbf{x})) \mid u \in z\}$ where

$$y = \bigcup \{H(u, \mathbf{x}) \mid u \in z\}.$$

1.3. (8) Consider $F(z, \mathbf{x}) = (\min \xi)_{\xi \in z} R(\xi, z, \mathbf{x})$ assuming $(\exists \xi \in z)R(\xi, z, \mathbf{x})$. Prim and Prim_O must be treated separately.

The ordinal sum and product operations are Prim by 1.3(7). Using 1.3(7) it is easy to see that Prim $(X_1, \ldots, X_k)$ is closed under schemes $F(\alpha, \mathbf{x}) = \sum \{G(\xi, x) \mid \xi < \alpha\}$, $F(z, \mathbf{x}) = 0$ if $\sim$Ord (z), and $F(\alpha, \mathbf{x}) = \prod \{G(\xi, x) \mid \xi < \alpha\}$, $F(z, \mathbf{x}) = 0$ if $\sim$Ord (z) provided that G is ordinal valued.

$$\mathrm{TC}\,(x) = x \cup \bigcup \{\mathrm{TC}\,(u) \mid u \in z\}$$

is Prim and so is $O(x) = \{u \in \mathrm{TC}\,(x) \mid \mathrm{Ord}\,(u)\}$, the least ordinal greater than all the ordinals in x. So it suffices to consider $G(\beta, \mathbf{x}) = (\min \xi)_{\xi<\beta} S(\xi, \mathbf{x})$ where S is Prim $(X_1, \ldots, X_k)$ since

$$F(z, \mathbf{x}) = (\min \xi)_{\xi \in z} R(\xi, z, \mathbf{x}) = (\min \xi)_{\xi < O(z)} (R(\xi, z, \mathbf{x}) \,\&\, \xi \in z)$$

has the form of G. But $G(\beta, \mathbf{x}) = \sum_{\xi<\beta} (N(\xi, x) \cdot \prod_{\mu<\xi} N(\mu, x))$ where N is the characteristic function of $\sim S$.

Consider Prim_O. The ordinal sum and product operations are usually written by a recursion of the form

$F(0, \mathbf{x}) = G(\mathbf{x})$,
$F(\alpha + 1, \mathbf{x}) = H(F(\alpha, \mathbf{x}), \alpha, \mathbf{x})$,
$F(\lambda, x) = \bigcup_{\alpha<\lambda} F(\alpha, \mathbf{x})$ if $\lambda = \bigcup \lambda > 0$.

But $\mathrm{Prim}_O\,(X_1, \ldots, X_k)$ is closed under this scheme if F is increasing since $F(\alpha, \mathbf{x}) = G(\mathbf{x})$ if $\alpha = 0$, $H(\gamma, \alpha, x)$ if $\bigcup \alpha < \alpha, \gamma$ if $\bigcup \alpha = \alpha > 0$ where $\gamma = \bigcup \{F(\xi, x) \mid \xi < \alpha\}$. Actually $\mathrm{Prim}_O\,(X_1, \ldots, X_k)$ is always closed under this scheme as we see in 1.6, but this is harder to prove if F is not increasing.

Since these are increasing functions with recursive definitions of this form, the ordinal sum and product operations are Prim_O and $\mathrm{Prim}_O\,(X_1, \ldots, X_k)$ is closed under schemes $F(\alpha, \mathbf{x}) = \sum \{G(\xi, \mathbf{x}) \mid \xi < \alpha\}$ and $F(\alpha, \mathbf{x}) = \prod \{G(\xi, \mathbf{x}) \mid \xi < \alpha\}$. Hence $\mathrm{Prim}_O\,(X_1, \ldots, X_k)$ is closed under the bounded min scheme.

Proof of 1.4. The relations of 1.4(1) are all Δ_0 except $\lambda x(x$ is finite). The rank function set for sets is Prim since $\mathrm{rk}\,(x) = \bigcup \{\mathrm{rk}\,(u) + 1 \mid u \in x\}$. Let $\Omega(x) = (\min \xi)_{\xi<\mathrm{rk}(x)} (\xi > 0 \,\&\, \xi = \bigcup \xi)$. Then if $\mathrm{rk}\,(x)$ is infinite, $\Omega(x) = \omega$. Define $F(u, x) =$ the set of u-element subsets of x if $u \in \omega$, by recursion as follows:

$F(0, x) = \{0\}$,
$F(\alpha + 1, x) = \{y \cup \{w\} \mid y \in F(\alpha, x) \,\&\, w \in x - y\}$,
$F(\lambda, x) = 0$ if $\lambda = \bigcup \lambda$,
$F(z, x) = 0$ if $\sim$ Ord (z).

Then x is finite $\Leftrightarrow \mathrm{rk}\,(x) \in \omega$ or $(\exists u \in \Omega(x))(x \in F(u, x))$.

The proofs of 1.4(2) and 1.4(3) are similar to the proofs of 1.3 and 1.4(1).

Proof of 1.4(4) We will only outline a proof that the satisfaction relation for ZF formulas in a set is Prim because the calculation is so similar to the one in Lévy [**9**]. One begins with a suitable Gödel-numbering of ZF formulas so that if $\mathscr{A}_n$ is the formula with Gödel-number n, the predicates $\lambda n(\mathrm{Fmla}\,(n))$, "$n$ is the Gödel-number of a formula", $\lambda m, n(v_m$ is free in formula $\mathscr{A}_n)$, $\lambda n(\text{rank of } \mathscr{A}_n)$, are primitive recursive in the sense of number theory. Then they are restrictions to ω of Prim set functions. Also $\mathrm{FV}\,(n) = \{m \mid v_m \text{ is free in } \mathscr{A}_n\}$ is a Prim function. Let $G(m) = \{n \mid n < m \,\&\, \mathrm{Fmla}\,(n) \,\&\, \mathrm{rank}\,(\mathscr{A}_n) < m\}$. Then G is Prim and is closed under subformulas. Then also F is Prim where F is defined by recursion as

$$F(m, z) = \{((n, f), 1) \mid n \in G(m) \,\&\, f \in z^{\mathrm{FV}(n)} \,\&\, f \text{ satisfies } \mathscr{A}_n \text{ in } \langle z, \in \restriction z\rangle\}$$

$$\cup \{((n, f), 0) \mid n \in G(m) \,\&\, f \in z^{\mathrm{FV}(n)} \,\&\, f \text{ does not satisfy } \mathscr{A}_n \text{ in } \langle z, \in \restriction z\rangle\}.$$

Two cases would be for example $\mathscr{A}_n = [v_i \mathrel{\mathsf{E}} v_j]$, where f satisfies $\mathscr{A}_n$ in $\langle z, \in \restriction z\rangle \Leftrightarrow f\text{ ` } i \in f \text{ ` } j$, or $\mathscr{A}_n = (Ev_i)\mathscr{A}_k$ where f satisfies $\mathscr{A}_n$ in $\langle z, \in \restriction z\rangle \Leftrightarrow (\exists u \in z)(((k, f \cup \{(i, u)\}), 1) \in \bigcup_{j\in m} F(j, z))$. Finally, x satisfies y in $\langle z, \in \restriction z\rangle \Leftrightarrow$ Fmla (y) & $x \in z^{\mathrm{FV}(y)}$ & $((y, x), 1) \in F((y \cup \operatorname{rank}(y)) + 1, z)$.

PROOF of 1.5. Similar to proof of 1.3.

PROOF OF 1.6. Suppose F is defined by recursion as $F(0, \mathbf{x}) = G(\mathbf{x})$, $F(n + 1, \mathbf{x}) = H(F(n, \mathbf{x}), n, \mathbf{x})$, $F(\alpha, \mathbf{x}) = 0$ if $\sim(\alpha \in \omega)$. The relation $\lambda\alpha(\alpha \in \omega)$ is Δ_0, hence Prim_O. Using the pairing function P and its inverses L_1, L_2 define K by recursion as

$K(0, \mathbf{x}) = G(\mathbf{x})$, $K(1, \mathbf{x}) = P(G(\mathbf{x}), H(G(\mathbf{x}), 0, \mathbf{x}))$,
$K(n + 1, \mathbf{x}) = P(K(n, \mathbf{x}), H(L_2K(n, \mathbf{x}), n, \mathbf{x}))$ if $n > 0$,
$K(\alpha, x) = 0$ if $\sim(\alpha \in \omega)$.

Then K is increasing and Prim_O by proof of 1.3(8). Then F is Prim_O since $F(0, \mathbf{x}) = G(\mathbf{x})$, $F(n, \mathbf{x}) = L_2K(n, \mathbf{x})$ if $n \in \omega \,\&\, n \neq 0$.

Suppose F has the form of 1.6(2), $F(0) = \gamma$, $F(\alpha) = G(F(H(\alpha)), \alpha)$ for $\alpha > 0$ where $H(\alpha) < \alpha$ for all $\alpha > 0$. Use 1.6(1) to show that N is Prim_O where $N(0) = 0$, $N(\alpha) = (\min k < \omega \cap \alpha + 1)(H^k(\alpha) = 0)$ for $\alpha > 0$. Then $F(\alpha) = K(N(\alpha), \alpha)$ where $K(0, \alpha) = \gamma$, $K(k + 1, \alpha) = G(K(k, \alpha), H^{N(\alpha)-k}(\alpha))$. Using 1.6(1) again, F is Prim_O. (What happens is that we write

$$\begin{aligned} F(\alpha) = G(FH(\alpha), \alpha) &= G(G(FH^2(\alpha), H(\alpha)), \alpha) \\ &= G(G(G(FH^3(\alpha), H^2(\alpha)), H(\alpha)), \alpha) \end{aligned}$$

and so on until $H^n(\alpha) = 0$.)

Now suppose that G and R are Prim_O and

$$F(\alpha) = G(\bigcup \{F(\beta) \mid \beta < \alpha \,\&\, R(\beta, \alpha)\}, \alpha).$$

Since G is assumed to take only values 0 and 1, F does also, and we may think of F as being defined recursively by cases as

$F(\alpha) = G(0, \alpha)$ if $\sim(\exists\beta < \alpha)(R(\beta, \alpha) \,\&\, F(\beta) = 1)$,
$F(\alpha) = G(1, \alpha)$ if $(\exists\beta < \alpha)(R(\beta, \alpha) \,\&\, F(\beta) = 1)$.

The first step is to spread out the values of F to leave spaces for coding. A suitable spacing function is the increasing function D defined by recursion as

$$D(\alpha) = \bigcup_{\beta<\alpha} D(\beta) \quad \text{if} \quad \alpha = \bigcup \alpha, \qquad D(\alpha + 1) = D(\alpha) \cdot 2 + 1.$$

Then D is Prim_O by the proof of 1.3(8). Using the bounded min operator we see that E_1 and E_2 are also Prim_O where $E_1(\gamma)$ is the unique α, $E_2(\gamma)$ the unique

$\nu \leq D(\alpha)$ such that $\gamma = D(\alpha) + \nu$. Let $\bar{G}(\eta, \alpha) = G(\eta \dot{-} D(\alpha) \cdot 2, \alpha)$. Define $\bar{F}$ by cases as follows for $\gamma = D(\alpha) + \nu$, $\nu \leq D(\alpha)$:

(i) $\bar{F}(\gamma) = \bar{G}(\bigcup_{\xi<\gamma} \bar{F}(\xi), \alpha)$ if $\nu = D(\alpha)$.

If $\nu < D(\alpha)$, let $\nu = D(\beta) + \mu$, $\mu \leq D(\beta)$.

(ii) $\bar{F}(\gamma) = D(\alpha) + \bar{F}(\nu)$ if $\bigcup_{\xi<\gamma} \bar{F}(\xi) < D(\alpha + 1)$ & $\sim(\nu = D(\beta) \cdot 2$ & $R(\beta, \alpha)$ & $\bar{F}(\nu) = 1)$.

(iii) $\bar{F}(\gamma) = D(\alpha + 1)$ if (ii) fails.

What $\bar{F}$ does in the interval from $D(\alpha)$ to $D(\alpha) \cdot 2$ is to copy $\bar{F}(\nu)$ translated upward to $D(\alpha) + \bar{F}(\nu)$ at each point $D(\alpha) + \nu$, until it comes to the first point (if any) where $\nu = D(\beta) \cdot 2$ & $R(\beta, \alpha)$ & $\bar{F}(\nu) = 1$. From then on $F(D(\alpha) + \nu) = D(\alpha + 1)$. An easy induction shows

(1) $\gamma < D(\delta) \Rightarrow \bar{F}(\gamma) \leq D(\delta)$,

(2) $\bar{F}(D(\alpha) \cdot 2) = F(\alpha)$,

(3) for $\nu < D(\alpha)$, $\bar{F}(D(\alpha) + \nu) = D(\alpha) + \bar{F}(\nu) \leq D(\alpha) \cdot 2$ unless

$$(\exists \beta < \alpha)(D(\beta) \cdot 2 \leq \gamma \;\&\; R(\beta, \alpha) \;\&\; F(\beta) = 1).$$

Note that $\bar{F}(D(\alpha) \cdot 2) = \bar{G}(\bigcup_{\nu<D(\alpha)} \bar{F}(D(\alpha) + \nu), \alpha)$ and the sup is at most $D(\alpha) \cdot 2$ unless $(\exists \beta < \alpha)(R(\beta, \alpha)$ & $\bar{F}(D(\beta) \cdot 2) = 1)$, otherwise the sup is $D(\alpha + 1)$. So $\bar{F}(D(\alpha) \cdot 2) = F(\alpha)$ by induction on α.

$\bar{F}(\gamma)$ can be defined in terms of $\eta = \bigcup_{\xi<\gamma} \bar{F}(\xi)$ as $\bar{F}(\gamma) = H(\eta, \gamma)$ where H is defined by cases as follows for $\gamma = D(\alpha) + \nu$, $\nu \leq D(\alpha)$:

(i)′ $H(\eta, \gamma) = \bar{G}(\eta, \alpha)$ if $\nu = D(\alpha)$.

If $\nu < D(\alpha)$, let $\nu = D(\beta) + \mu$, $\mu \leq D(\beta)$.

(ii)′ $H(\eta, \gamma) = D(\alpha) + H(\eta \dot{-} D(\alpha), \nu)$ if $\eta < D(\alpha + 1)$ & $\sim(\nu = D(\beta) \cdot 2$ & $R(\beta, \alpha)$ & $\bar{G}(\eta \dot{-} D(\alpha), \beta) = 1)$.

(iii)′ $H(\eta, \gamma) = D(\alpha + 1)$ if (ii)′ fails.

An induction on γ shows $\bar{F}(\gamma) = H(\bigcup_{\xi<\gamma} \bar{F}(\xi), \gamma)$ using the fact that

$$\bigcup_{\xi<D(\alpha)+\nu} \bar{F}(\xi) = \bigcup_{\xi<\nu} \bar{F}(D(\alpha) + \xi) = D(\alpha) + \bigcup_{\xi<\nu} \bar{F}(\xi)$$

if

$$\sim(\exists \beta < \alpha)(D(\beta) \cdot 2 \leq \nu \;\&\; R(\beta, \alpha) \;\&\; F(\beta) = 1).$$

So $\eta \dot{-} D(\alpha) = \bigcup_{\xi<\nu} \bar{F}(\xi)$ in this case.

It remains only to show that H is Prim_O. But this follows from 1.6(2). The form of the recursive definition of $H(\eta, \gamma)$ is

$$H(\eta, \gamma) = G_1(\eta, \gamma) \quad \text{if} \quad R_1(\eta, \gamma),$$

$$H(\eta, \gamma) = G_2(H(F_1(\eta, \gamma), F_2(\eta, \gamma)), \eta, \gamma) \quad \text{otherwise},$$

where G_1, G_2, F_1, F_2, R_1 are all Prim_O. (Recall that α, ν, β, μ are all Prim_O functions of γ.) Moreover $F_1(\eta, \gamma) < \eta$, $F_2(\eta, \gamma) < \gamma$ whenever $\eta, \gamma > 0$. So by use of the pairing function P, $H(\eta, \gamma)$ can be reduced to a function $\bar{H}$, where $\bar{H}(P(\eta, \gamma)) = H(\eta, \gamma)$ and $\bar{H}$ has a definition of form 1.6(2). Therefore $\bar{H}$ is Prim_O, whence $\bar{F}$ is Prim_O, and finally F is Prim_O.

APPENDIX II

CALCULATIONS WITH GÖDEL'S F-FUNCTION

PROOF OF THEOREM 2.12. Computations in Gödel [3] have to be only slightly modified. He defines the ordered n-tuple recursively as $(x_1) = x_1$, $(x_1, \ldots, x_n) = (x_1, (x_2, \ldots, x_n))$ for $n > 1$. Let Exp (x, n) be the set of ordered n-tuples with coordinates in x. So Exp $(x, 1) = x$, Exp $(x, n+1) = x \times$ Exp (x, n). The fundamental operations were chosen so that they would generate the definable subsets of a set.

THEOREM. *For any* ZF *formula* $\mathscr{A}(v_1, \ldots, v_n, v_1', \ldots, v_m')$ *with attached list of variables containing all those free in* $\mathscr{A}$, *a set of the form*

$$\{(x_1, \ldots, x_n) \in \operatorname{Exp}(z, n) \mid z \models \mathscr{A}(x_1, \ldots, x_n, y_1, \ldots, y_m)\}$$

is obtainable from $z, y_1, \ldots, y_m$ *by finitely many iterations of the fundamental operations and the Cartesian product operation.*

PROOF. Gödel [3, pp. 8–11].

LEMMA. *There is an increasing* Prim_O *operation* $\times_F$ *such that for all* α, β, $F'(\alpha \times_F \beta) = F'(\alpha) \times F'(\beta)$.

PROOF. The least accumulation point of F' above α is ac $(\alpha) + 9$. So $(F'(\alpha) \times F'(\beta)) \cup (F'(\beta) \times F'(\alpha)) \subseteq F'(\nu)$ where

$$\nu = \mathrm{ac}\left(\bigcup_{\gamma,\delta<\alpha\cup\beta} J(1, J(1, \gamma, \delta), J(1, \gamma, \gamma))\right) + 9.$$

Then ν is an increasing Prim_O function of α, β and

$$F'(\alpha) \times F'(\beta) = (F'(\nu) \cap (V \times F'(\beta))) \cap (F'(\nu) \cap (V \times F'(\alpha)))^{-1}.$$

This is obtained by an iteration of the fundamental operations on ν, α, β, so $\alpha \times_F \beta = J(4, \nu, \beta) \cap_F J(6, \nu, J(4, \nu, \alpha))$ where $\gamma \cap_F \delta = J(3, \gamma, J(3, \gamma, \delta))$.

To complete the proof of Theorem 2.12 it is required to check conditions 2.9(1)-(5) for the function $H(\alpha) = F'(\mathrm{ac}(\alpha))$, where ac (α) is the largest accumulation point of F' less than or equal to α. Conditions (1) and (3) are immediate. To check (2) it suffices to produce Prim_O functions b for each of the initial Prim set functions so that function values are in $H(b(\alpha))$ for arguments in $H(\alpha)$. For the projections and for the function C take $b(\alpha) = \alpha + 8$. For the zero take $b(\alpha) = 9$. For the function $F(x, y) = x \cup \{y\}$ take

$$b(\alpha) = 9 \cdot \bigcup_{\delta,\epsilon<\nu} J(3, \nu, J(3, \nu, \delta) \cap_F J(3, \nu, \epsilon))$$

where $\nu = 9 \cdot J(1, \alpha, \alpha)$. Hence (2).

For (5) it is required to show that any ZF formula $\mathscr{A}(v_0, v_1, \ldots, v_n)$ there is an increasing Prim_O function b such that

$$(\forall \mathbf{x} \in H(\alpha)^n)\{y \in H(\alpha) \mid H(\alpha) \models \mathscr{A}(y, \mathbf{x})\} \in H(b(\alpha)).$$

Choose j so that the required subset of $H(\alpha)$ is obtainable from $H(\alpha), x_1, \ldots, x_n$ by j iterations of the Cartesian product and fundamental operations. Then take $b(\alpha) = g(j, \alpha)$ where $g(0, \alpha) = \mathrm{ac}(\alpha) + 9$, the least accumulation point of F' above α, and $g(i + 1, \alpha) = \mathrm{ac}(\nu) + 9$ where

$$\nu = \bigcup \{\gamma \times_F \delta \mid \gamma, \delta < g(i, \alpha)\} \cup \bigcup \{J(i, \gamma, \delta) \mid \gamma, \delta < g(i, \alpha), i < 9\}.$$

Then the subset of $H(\alpha)$ is in $H(b(\alpha))$ because all of $x_1, \ldots, x_n$ and $H(\alpha)$ are in $H(g(0, \alpha))$, any set arising by one use of the operations on $x_1, \ldots, x_n, H(\alpha)$ is in $H(g(1, \alpha))$ and so on.

For (4) we need a Prim_O function G such that for all α, $\alpha \in F'(G(\alpha))$. For G we can just iterate the function b above for the formula Ord (v_0), "v_0 is an ordinal". The formula is Δ_0, so $\mathrm{Ord} \cap H(\alpha) \in H(b(\alpha))$. Thus we may take $G(0) = 9$, $G(\alpha) = b(\bigcup_{\beta<\alpha} G(\beta))$ for $\alpha > 0$. Note that each $G(\alpha)$ is an accumulation point of F'.

For (4) we must show that there is a Prim_O function b such that

$$(\forall x \in H(\alpha))(\mathrm{TC}(x) \in H(b(\alpha))\ \&\ H(b(\alpha)) \models \mathscr{D}(\mathrm{TC}(x), x))$$

where $\mathscr{D}(v_0, v_1)$ defines the condition $v_0 = \mathrm{TC}(v_1)$ in the form "v_0 is the set of all u such that there is a $\in$-chain from u to v_1". Let $\mathscr{A}(v_1, v_2)$ be the generalized Σ_1-formula expressing the condition "there is a $\in$-chain from v_1 to v_2". Then $\mathscr{D}(v_0, v_1)$ is $(w_0)[w_0 \mathrel{E} v_0 \leftrightarrow \mathscr{A}(w_0, v_1)]$. We claim that it suffices to show that there is a Prim_O function c such that $\alpha < c(\alpha)$ and for all $x \in H(\alpha)$ and $u \in \mathrm{TC}(x)$, $H(c(\alpha))$ contains a $\in$-chain from u to x together with its length. For suppose we have such a function c. Then $H(c(\alpha)) \models \mathscr{A}(u, x)$ because the unbounded E-quantifiers in $\mathscr{A}$ denote u, x, and $\in$-chains from u to x and their lengths. Applying (5) to $c(\alpha)$ and the formula $\mathscr{A}$, $\mathrm{TC}(x) = \{u \in H(c(\alpha)) \mid H(c(\alpha)) \models \mathscr{A}(u, x)\} \in H(b'(c(\alpha)))$ where b' is Prim_O. Whence taking $b(\alpha) = b'(c(\alpha))$ we have

$$H(b(\alpha)) \models \mathscr{D}(\mathrm{TC}(x), x).$$

To obtain c, note that for $x \in H(\alpha)$, $u \in \mathrm{TC}(x)$, the length of a $\in$-chain from u to x is finite and at most rank $(x) + 1 \leq \alpha$. So with G as above, the length of a $\in$-chain is in $H(G(\alpha)) = F'(G(\alpha))$. Then a $\in$-chain, being of form $\{(0, u), \ldots, (n, x)\}$, is a finite subset of $F'(G(\alpha) \times_F G(\alpha))$ having at most α elements. Pairs of elements of $\in$-chains are in $F'(c'(\alpha))$ where

$$c'(\alpha) = \bigcup \{J(1, \gamma, \delta) \mid \gamma, \delta < G(\alpha) \times_F G(\alpha)\}.$$

So Prim_O function c can be found so that each $c(\alpha)$ is an accumulation point of F' (so that $H(c(\alpha)) = F'(c(\alpha))$) and the unions of finitely many, but at most α, elements of $F'(c'(\alpha))$ are in $F'(c(\alpha))$.

Proof of Lemma 3.1. It is required to show that the predicates

$$\lambda\alpha\beta(F'(\alpha) = F'(\beta)) \quad \text{and} \quad \lambda\alpha\beta(F'(\alpha) \in F'(\beta))$$

are Prim_O. But in order to prove this we have to prove something stronger. If symbols α^* are interpreted as $F'(\alpha)$, we must show that not only do atomic statements $\alpha^* = \beta^*$ and $\alpha^* \in \beta^*$ define Prim_O predicates of ordinals, but so do statements of a simple quantifier-free language based on individual constants α^*. The statements have to have a special Gödel-numbering in order for the recursion to go through.

Let $h(\alpha) = 7 \cdot P(\alpha, \alpha) + 1$ where P is Gödel's pairing function enumerating pairs by the maximum coordinates, then antilexicographically (1.5(3)). Let $r(\alpha) = \bigcup_{\beta<\alpha} r(\beta)$ if $\alpha = \bigcup \alpha$, $r(\alpha + 1) = h^{100}r(\alpha)$. Then h and r are increasing and Prim_O. Also r^{-1} is Prim_O since $r^{-1}(\alpha) = (\min \beta \leq \alpha)(\alpha = r(\beta))$. The terms and statements of the language will be identified with their Gödel-numbers, so $\alpha^* = 6 \cdot r(\alpha)$ and so on.

Terms t	Interpretations $I(t)$
$\alpha^* = 6 \cdot r(\alpha)$	$F'(\alpha)$
$[t, t'] = 6P(t, t') + 1$	$\{I(t), I(t')\}$
Defined terms	
$p(t, t') = [[t, t], [t, t']]$	$(I(t), I(t'))$
$p(t, t', t'') = p(t, p(t', t''))$	$(I(t), I(t'), I(t''))$

Atomic terms have rank 0, the rank of $[t, t']$ is $(\mathrm{rk}\,(t) \cup \mathrm{rk}\,(t')) + 1$.

LEMMA 1. $\lambda\alpha$ Term (α) *is* Prim_O.

Term $(\alpha) \Leftrightarrow (\exists\beta < \alpha)(\alpha = 6 \cdot r(\beta))$ or $(\exists\beta, \gamma < \alpha)$(Term (β) & Term (γ) & $\alpha = 6 \cdot P(\beta, \gamma) + 1)$. Since P has Prim_O inverse functions, L_1, L_2, the characteristic function of the term predicate has a recursive definition of form 1.6(3).

LEMMA 2. *There are* Prim_O *functions* F_1, F_2, F_3 *such that* $F_1(\alpha^*) = \alpha$, $F_2([t, t']) = t$, $F_3([t, t']) = t'$.

PROOF. $F_1(\beta) = r^{-1}(\beta/6)$, $F_2(\beta) = L_1((\beta - 1)/6)$, $F_3(\beta) = L_2((\beta - 1)/6)$.

LEMMA 3. *If* $\mu < \beta$ *for individuals* μ^* *of term* t *and if* t *has rank less than* 100, *then* $t < \beta^*$.

PROOF. We show by induction that if the rank of t is at most n, then $h^{100-(n+1)}(t) < \beta^*$. If t has rank 0, $t = \mu^*$ for some $\mu < \beta$. So $\mu + 1 \leq \beta$ and $r(\mu + 1) = h^{100}(r(\mu)) \leq r(\beta)$. So $h^{99}(\mu^*) < h^{100}(r(\mu)) \leq r(\beta) < \beta^*$. If $t = [t', t'']$, suppose $t' \leq t''$ and that the larger rank is n. Then

$$h^{100-(n+2)}([t', t'']) \leq h^{100-(n+2)}(7 \cdot P(t', t'')) < h^{100-(n+1)}(t'') < \beta^*.$$

Statements A	Interpretation
$t \neq t' = 6 \cdot r(P(t, t') \cup P(t', t)) + 2$	$I(t) \neq I(t')$
$t \in t' = 6 \cdot r(P(t', t)) + 3$	$I(t) \in I(t')$
$\neg A = 6 \cdot A + 4$	negation of A
$A \vee B = 6P(A, B) + 5$	disjunction of A, B

The other Boolean connectives are introduced by their customary definitions. Note that our statements are not statements in quite their usual sense because we have found it convenient to make $t \mathrel{\ddagger} t' = t' \mathrel{\ddagger} t$. We write $\models \alpha$ for the predicate "α is a statement that is true for the interpretation". Statements $t \mathrel{\ddagger} t'$, $t \in t'$ have rank 0. The rank of $\neg A$ is rk $(A) + 1$, the rank of $A \vee B$ is (rk $(A) \cup$ rk $(B)) + 1$. The proof of Lemma 4 is like that of Lemma 1.

LEMMA 4. *$\lambda\alpha$ Statement (α) is* Prim_O.

LEMMA 5. *There are* Prim_O *functions* $F_1, \ldots, F_6$ *such that* $F_1(\neg A) = A$, $F_2(A \vee B) = A$, $F_3(A \vee B) = B$, $F_4(t \mathrel{\ddagger} t') = t \cup t'$, $F_5(t \in t') = t$, $F_6(t \in t') = t'$.

PROOF. Similar to Lemma 2.

LEMMA 6. (1) $\beta < \gamma \Rightarrow P(\alpha, \beta) < P(\alpha, \gamma)$ *and* $P(\alpha, \beta) < P(\gamma, \alpha)$.

(2) *If A is a statement of rank* < 100 *which is a Boolean combination of atomic statements of the form* $t \in t'$, $t' \in t$, $t \mathrel{\ddagger} t'$, $t' \in t''$, $t' \mathrel{\ddagger} t''$ *where* $t', t'' < s$, *then* $A < t \in s$.

(3) *If A is a statement of rank* < 100 *which is a Boolean combination of atomic statements of the form* $t' \in t_2$, $t'' \in t_1$, $t' \in t''$, $t' \mathrel{\ddagger} t''$, *where t, t' are terms such that* $t' < t_1$ *and* $t'' < t_2$, *then* $A < t_1 \mathrel{\ddagger} t_2$, $A < t_1 \in t_2$, *and* $A < t_2 \in t_1$.

PROOF. (1) is a property of the ordering of pairs enumerated by P. The proofs of (2) and (3) are similar. For (2) first show that if A atomic then $h^{99}(A) < t \in s$, for (3) show that $h^{99}(A) < t_1 \mathrel{\ddagger} t_2 \cap t_1 \in t_2 \cap t_2 \in t_1$. To see this in the case of (2), use (1) to see that $P(t', t'') \cup P(t, t') \cup P(t', t) < P(s, t)$. So atomic A has form $6 \cdot r(\gamma) + 2$ or $6 \cdot r(\gamma) + 3$ with $\gamma + 1 \leq P(s, t)$. So

$$r(\gamma + 1) = h^{100}r(\gamma) \leq r(P(s, t)).$$

Then $h^{99}(A) < h^{100}(r(\gamma)) \leq r(P(s, t)) < t \in s$. For (3) the argument is similar. Then continue as in the proof of Lemma 3 to complete the proof.

THEOREM. *The predicate $\lambda\alpha(\models \alpha)$ is* Prim_O. *Therefore so are predicates* $\lambda\alpha\beta(F'(\alpha) \in F'(\beta))$ *and* $\lambda\alpha\beta(F'(\alpha) = F'(\beta))$.

PROOF. Consider the characteristic function of the predicate $\lambda\alpha(\ \alpha)$. We will write it in the form $S(\alpha) = G(\bigcup \{S(\beta) \mid \beta < \alpha \,\&\, R(\beta, \alpha)\}, \alpha)$ with Prim_O G and R, so 1.6 will show that it is Prim_O. Define $G(i, \alpha) = 0$, if $\sim$ Statement (α), $G(0, \alpha) = 1$ if α has form $\neg A$, 0 if not, and $G(1, \alpha) = 0$ if α has form $\neg A$, 1 if statement α is not a negation. Then G is Prim_O and it remains to define a Prim_O relation $R(\beta, \alpha)$ such that

(i) $R(\beta, \alpha) \Rightarrow \beta < \alpha$.

(ii) If $\alpha = \neg A$, then $\models \alpha \Leftrightarrow (\forall\beta)(R(\beta, \alpha) \Rightarrow \sim \models \beta)$.

(iii) If α is a statement not of form $\neg A$, then $\models \alpha \Leftrightarrow (\exists\beta)(R(\beta, \alpha) \,\&\, \models \beta)$.

Then an easy induction shows that $S(\alpha) = 1 \Leftrightarrow \models \alpha$. Assuming α is a statement, the relation $R(\beta, \alpha)$ is defined by cases as follows:

α	$R(\beta, \alpha)$
(1) $\neg A$	$\beta = A$
(2) $A \vee B$	$\beta = A$ or $\beta = B$
(3) $\gamma^* \neq \delta^*$	$(\exists \nu < \gamma)(\beta = (\nu^* \in \gamma^* \wedge \neg\, \nu^* \in \delta^*))$ or $(\exists \nu < \delta)(\beta = (\nu^* \in \delta^* \wedge \neg\, \nu^* \in \gamma^*))$
(4) $\gamma^* \neq [t, t']$	$\beta = \neg(t \in \gamma^*)$ or $\beta = \neg(t' \in \gamma^*)$ or $(\exists \nu < \gamma)(\beta = (\nu^* \in \gamma^* \wedge \nu^* \neq t \wedge \nu^* \neq t'))$
(5) $[s, s'] \neq [t, t']$	$\beta = (\gamma \neq \delta \wedge \gamma' \neq \delta')$ for some $\gamma, \gamma' \in \{s, s'\}$, $\delta, \delta' \in \{t, t'\}$
(6) $t \in [s, s']$	$\beta = \neg(t \neq s)$ or $\beta = \neg(t \neq s')$
(7) $t \in (9 \cdot \gamma)^*$	$(\exists \nu < 9 \cdot \gamma)(\beta = \neg(t \neq \nu^*))$
(8) $t \in (9 \cdot \gamma + 1)^*$	$\beta = (t \in [L_1(\gamma)^*, L_2(\gamma)^*])$
(9) $t \in (9 \cdot \gamma + 2)^*$	$(\exists \mu, \nu < \gamma)\beta = (t \in L_1(\gamma)^* \wedge \neg t \neq p(\mu^*, \nu^*) \wedge \mu^* \in \nu^*)$
(10) $t \in (9 \cdot \gamma + 3)^*$	$\beta = (t \in L_1(\gamma)^* \wedge \neg t \in L_2(\gamma)^*)$
(11) $t \in (9 \cdot \gamma + 4)^*$	$(\exists \mu, \nu < \gamma)(\beta = (\neg t \neq p(\mu^*, \nu^*) \wedge \nu^* \in L_2(\gamma)^* \wedge t \in L_1(\gamma)^*))$
(12) $t \in (9 \cdot \gamma + 5)^*$	$(\exists \mu < \gamma)(\exists \nu < L_1(\gamma))(\beta = (t \in L_1(\gamma)^* \wedge \neg t \neq \nu^* \wedge p(\mu^*, \nu^*) \in L_2(\gamma)^*))$
(13) $t \in (9 \cdot \gamma + 6)^*$	$(\exists \mu, \nu < \gamma)(\beta = (t \in L_1(\gamma)^* \wedge \neg t \neq p(\mu^*, \nu^*) \wedge p(\nu^*, \mu^*) \in L_2(\gamma)^*))$
(14) $t \in (9 \cdot \gamma + 7)^*$	$(\exists \lambda, \mu, \nu < \gamma)(\beta = (t \in L_1(\gamma)^* \wedge \neg t \neq p(\lambda^*, \mu^*, \nu^*) \wedge p(\nu^*, \lambda^*, \mu^*) \in L_2(\gamma)^*))$
(15) $t \in (9 \cdot \gamma + 8)^*$	$(\exists \lambda, \mu, \nu < \gamma)(\beta = (t \in L_1(\gamma)^* \wedge \neg t \neq p(\lambda^*, \mu^*, \nu^*) \wedge p(\lambda^*, \nu^*, \mu^*) \in L_2(\gamma)^*))$

Lemmas 2 and 5 are used to show that R is Prim_O. Lemmas 3 and 6 show $R(\beta, \alpha) \to \beta < \alpha$. Then properties (ii) and (iii) are proved by induction on α. This completes the proof.

NOTE ADDED IN PROOF. The editor has reminded us that the initial function $F(x) = 0$ is not needed in the formulation of primitive recursive set functions. It can be defined by the recursion schema as $F(x) = \bigcup_{u \in x} F(u)$. We had noted this in some correspondence with Gandy in 1968 but had forgotten about it.

REFERENCES

1. J. Barwise, *Infinitary logic and admissible-sets*, Ph.D. Dissertation, Stanford University, Stanford, Calif., 1966.

2. R. Gandy, Review #4677, Math. Reviews **29** (1965), 882.

3. K. Gödel, *The consistency of the continuum hypothesis*, Princeton Univ. Press, Princeton, N.J., 1940.

4. C. Karp, *Languages with expressions of infinite length*, Ph.D. Dissertation, University of Southern California, Los Angeles, Calif., 1959.

5. S. C. Kleene, *On the forms of predicates in the theory of constructive ordinals*, Amer. J. Math. **77** (1955), 405–428.

6. G. Kreisel, "Model-theoretic invariants: Applications to recursion theory and hyperarithmetic operations" in *The theory of models*, Proc. Sympos. Berkeley, 1963, North-Holland, Amsterdam, 1965, pp. 190–206.

7. G. Kreisel and G. Sacks, *Metarecursive sets* (abstract), J. Symbolic Logic **28** (1963), 304–305.

8. S. Kripke, *Transfinite recursions on admissible ordinals* (abstract), J. Symbolic Logic **29** (1964), 161.

9. A. Lévy, *A hierarchy of formulas in set theory*, Mem. Amer. Math. Soc., No. 57 (1965), 76 pp.

10. T. Linden, "Equivalences between Gödel's definitions of constructibility" in *Sets, models and recursion theory*, Proc. Tenth Logic Colloquium, Leicester, 1965, North-Holland, Amsterdam, 1967, pp. 33–43.

11. E. G. K. Lopez-Escobar, *Remarks on an infinitary language with constructive formulas*, J. Symbolic Logic **32** (1967) 305–318.

12. M. Machover, *The theory of transfinite recursion*, Bull. Amer. Math. Soc. **67** (1961), 575–578.

13. R. Platek, *Foundations of recursion theory*, Ph.D. Dissertation, Stanford University, Stanford, Calif., 1966.

14. D. Rödding, *Primitiv-rekursive Funktionen über einem Bereich endlicher Mengen*, Arch. Math. Logik. Grundlagenforsch. **10** (1967), 13–29.

15. D. Scott and A. Tarski, *The sentential calculus with infinitely long expressions*, Colloq. Math. **6** (1958), 165–170.

16. J. Shoenfield, "The problem of predicativity" in *Essays on the foundations of mathematics*, Magnes Press, The Hebrew University, Jerusalem, 1961, pp. 132–139.

17. G. Takeuti. *On the recursive functions of ordinal numbers*, J. Math. Soc. Japan **12** (1960), 119–128.

18. G. Takeuti and A. Kino, *On hierarchies of predicates of ordinal numbers*, J. Math. Soc. Japan **14** (1962), 199–232.

19. G. Takeuti, *A formalization of the theory of ordinal numbers*, J. Symbolic Logic **30** (1965), 295–317.

20. G. Takeuti and A. Kino, *On predicates with infinitely long expressions*, J. Math. Soc. Japan **15** (1963), 176–190.

Seminar für Logik und Grundlagenforschung der Universität Bonn

University of Maryland

END EXTENSIONS OF MODELS OF SET THEORY

H. JEROME KEISLER AND JACK H. SILVER

1. **Introduction.** Mac Dowell and Specker [**8**] proved the following theorem for models of Peano arithmetic:

Every model $\mathfrak{A}$ *of Peano arithmetic has a proper elementary extension* $\mathfrak{B}$ *such that* $\mathfrak{A}$ *is an initial segment of* $\mathfrak{B}$. (In the terminology of Gaifman [**1**], $\mathfrak{B}$ is an *end extension of* $\mathfrak{A}$.)

In this paper we shall study the corresponding question for models of set theory: *Which models of Zermelo-Fraenkel set theory have end elementary extensions*?

In order to state the above question more precisely, we need some notation. ZF denotes Zermelo-Fraenkel set theory (including the axiom of regularity), and ZFC denotes ZF plus the axiom of choice. We shall let $\mathfrak{A} = \langle A, E\rangle$ and $\mathfrak{B} = \langle B, F\rangle$ be models, where A, B are nonempty sets and E, F are binary relations. The symbol $\mathfrak{A} \prec \mathfrak{B}$ means that $\mathfrak{B}$ is a *proper* elementary extension of $\mathfrak{A}$. (See Tarski and Vaught [**12**].) If $\mathfrak{A}$ is a model of ZF, we denote by Ord($\mathfrak{A}$) the linearly ordered structure $\langle$Ordinals of $\mathfrak{A}$, $E\rangle$. The *cofinality* of Ord($\mathfrak{A}$) is the least cardinal κ such that Ord($\mathfrak{A}$) has an unbounded subset of power κ. We define an end extension of $\mathfrak{A}$ in the following way. $\mathfrak{B}$ is an *end extension* of $\mathfrak{A}$, if and only if

(i) $\mathfrak{B}$ is a proper extension of $\mathfrak{A}$, and

(ii) $b \in B$, $a \in A$, $b\,F\,a$ implies $b \in A$.

If $\mathfrak{B}$ is an end extension of $\mathfrak{A}$ and $\mathfrak{A} \prec \mathfrak{B}$, we call $\mathfrak{B}$ an *end elementary extension* of $\mathfrak{A}$.

We remark that if $\mathfrak{B}$ is an end extension of $\mathfrak{A}$ and if $\mathfrak{A}, \mathfrak{B}$ are models of ZF, then Ord($\mathfrak{A}$) is an initial segment of Ord($\mathfrak{B}$). Conversely, if $\mathfrak{A}$ is a model of ZFC, $\mathfrak{A} \prec \mathfrak{B}$, and Ord($\mathfrak{A}$) is an initial segment of Ord($\mathfrak{B}$), then $\mathfrak{B}$ is an end extension of $\mathfrak{A}$.

Two positive results about end extensions are known from the literature. We state them here without proof.

THEOREM 1.1 (KEISLER AND MORLEY [**6**]). *If* $\mathfrak{A}$ *is a model of* ZF *and* Ord($\mathfrak{A}$) *has cofinality* ω, *then* $\mathfrak{A}$ *has an end elementary extension. In particular, every countable model of* ZF *has an end elementary extension.*

THEOREM 1.2. *If* κ *is a weakly compact inaccessible cardinal then the natural model* $\langle R(\kappa), \in\rangle$ *has an end elementary extension.*

Weakly compact cardinals are defined in Tarski [**11**]. The above result is an almost trivial consequence of the definition of a weakly compact cardinal (see [**10**, pp. 24–25]). The above two theorems also hold for the theory ZF with a finite or countable number of extra predicates. By ZF with the extra predicates $P_0, P_1, P_2, \ldots$ we mean the theory, in a language L with the predicate symbols $\in, P_0, P_1, P_2, \ldots$, whose axioms are the usual axioms of ZF but with the replacement scheme for *all* formulas of the language L. When extra predicates are allowed, Theorem 1.2 above has the following converse.

THEOREM 1.3 (KEISLER [**4**]). *An inaccessible cardinal* κ *is weakly compact if and only if for every set* $S \subset R(\kappa)$, *the model* $\langle R(\kappa), \in, S\rangle$ *has an end elementary extension.*

The proof of the above theorem was never published, although the proof of an analogous theorem for measurable cardinals was given in [**5**]. In this paper we shall state and prove a stronger result, Theorem 2.2, from which Theorem 1.3 above will follow as a corollary.

In §2 of this paper we shall prove some negative results about end elementary extensions. In particular, we show that the model $\langle R(\theta_1), \in\rangle$, where θ_1 is the first (uncountable) inaccessible cardinal, has no end elementary extensions. Most of the results of §§2 and 3 were announced in the authors' joint abstract [7]. In §3 we refine the results of §2 to give information about well-founded end extensions. §4 contains some positive results about end extensions, which are partial converses to the results of §2. The results of §4 were obtained by the first author since the publication of the abstract [7].

2. **Necessary conditions for a model to have end extensions.** To state our first lemma we need a notion which is weaker than the elementary extension. Let $p < \omega$. By a p-formula we mean a (first order) formula φ which has at most p blocks of universal and existential quantifiers; that is

$$\varphi \in \Sigma_p^0 \cup \Pi_p^0.$$

We shall say that $\mathfrak{B}$ is a *p-extension* of $\mathfrak{A}$, in symbols $\mathfrak{A} \prec_p \mathfrak{B}$, if

(i) $\mathfrak{B}$ is a proper extension of $\mathfrak{A}$, and

(ii) for every p-formula $\varphi(v_0, \ldots, v_n)$ and every $a_0, \ldots, a_n \in A$, $a_0, \ldots, a_n$ satisfies φ in $\mathfrak{A}$ if and only if it satisfies φ in $\mathfrak{B}$.

Thus we have $\mathfrak{A} \prec \mathfrak{B}$ if and only if for all $p < \omega$, $\mathfrak{A} \prec_p \mathfrak{B}$.

For any structure $\mathfrak{B}$ and any ordinal α of $\mathfrak{B}$, let $\mathfrak{B}_\alpha = \langle B_\alpha, F\rangle$ be the substructure of $\mathfrak{B}$ such that

$$B_\alpha = \{b \in B : \mathfrak{B} \models \text{“}b \in R(\alpha)\text{”}\}.$$

We shall use the notation $\mathfrak{B}_\alpha$ in the proof below and again later on.

THEOREM 2.1. *Suppose $\mathfrak{A}$ is a well-founded model of* ZFC, Ord($\mathfrak{A}$) *has cofinality $> \omega$, and for all $p < \omega$, $\mathfrak{A}$ has an end p-extension (not necessarily a model of* ZFC). *Then for all $p < \omega$, $\mathfrak{A}$ has a well-founded end p-extension.*

PROOF. We shall prove the lemma first under the assumption that $\mathfrak{A}$ has an end elementary extension $\mathfrak{B}$. $\mathfrak{B}$ is a model of ZFC. Let $p < \omega$. By the reflection principle, there exists an ordinal α of $\mathfrak{B}$ such that $\mathfrak{B}_\alpha \prec_p \mathfrak{B}$, and $\mathfrak{B}_\alpha$ is an end extension of $\mathfrak{A}$. Let $b \in B_\alpha - A$. Using the axiom of choice within the model $\mathfrak{B}$, we may well-order the set B_α and choose Skolem functions for the model $\mathfrak{B}_\alpha$. Let $\mathfrak{C} = \langle C, F\rangle$ be the Skolem hull of the set $A \cup \{b\}$ with respect to these Skolem functions. ($\mathfrak{C}$ can only be defined outside $\mathfrak{B}$, not within $\mathfrak{B}$.) Then $\mathfrak{C} \prec \mathfrak{B}_\alpha$. From

$$\mathfrak{A} \prec \mathfrak{B}, \quad \mathfrak{C} \prec \mathfrak{B}_\alpha, \quad \mathfrak{B}_\alpha \prec_p \mathfrak{B}, \quad A \subset C,$$

we conclude that $\mathfrak{A} \prec_p \mathfrak{C}$ and $\mathfrak{C}$ is an end extension of $\mathfrak{A}$.

Suppose $\mathfrak{C}$ is not well founded. Then there is a countable decreasing F-chain

$$\cdots c_2 \, F \, c_1 \, F \, c_0$$

of elements of $\mathfrak{C}$. Since Ord($\mathfrak{A}$) has cofinality $>\omega$, there is an ordinal γ of $\mathfrak{A}$ such that $c_0, c_1, c_2, \ldots$ all belong to the Skolem hull H of $A_\gamma \cup \{b\}$. The set H “belongs” to $\mathfrak{B}$, and in the sense of $\mathfrak{B}$, $\langle H, F\rangle$ is an $\in$-structure which satisfies the axiom of extensionality. Therefore, in $\mathfrak{B}$, $\langle H, F\rangle$ is isomorphic to a transitive $\in$-structure $\langle G, F\rangle$. In $\mathfrak{B}$, the sets H and G have the same power as the set $R(\gamma) \cup \omega$, and this cardinality belongs to $\mathfrak{A}$. It follows that G belongs to $\mathfrak{A}$, and therefore the structure $\langle G, F\rangle$ is well founded. But $\langle H, F\rangle$ is not well founded. This contradiction shows that $\mathfrak{C}$ is well founded.

To complete the proof we need only point out that there exists a $q < \omega$ (which depends on p) such that the above argument still goes through if the assumption $\mathfrak{A} \prec \mathfrak{B}$ is weakened to $\mathfrak{A} \prec_q \mathfrak{B}$. (Here $\mathfrak{B}$ need not be a model of all of ZFC.)

We now obtain a necessary condition for a model of ZFC to have an end elementary extension. We shall say that a model $\mathfrak{A}$ of ZF has the Π^1_1 *reflection property* if for all $a \in A$ and every Π^1_1 (second order universal) formula $\varphi(v_0)$, if $\varphi(a)$ holds in $\mathfrak{A}$ then the formula

$$\text{“}(\exists\alpha)(a \in R(\alpha) \quad \text{and} \quad \varphi(a) \quad \text{holds in} \quad \langle R(\alpha), \in\rangle)\text{”}$$

holds in $\mathfrak{A}$. This definition may also be applied to models of ZF with extra predicates $P_0, P_1, \ldots$, when we replace $\langle R(\alpha), \in\rangle$ by $\langle R(\alpha), \in, R_0, R_1, \ldots\rangle$.

Notice that if $\mathfrak{A}$ is a natural model, say $\mathfrak{A} = \langle R(\beta), \in\rangle$, then $\mathfrak{A}$ has the Π^1_1 reflection property if and only if for all $a \in A$ and all Π^1_1 formulas $\varphi(v_0)$, if $\varphi(a)$ holds in $\mathfrak{A}$, then $\varphi(a)$ holds in $\mathfrak{A}_\alpha$ for some $\alpha < \beta$.

For example, if θ_1 is the first inaccessible cardinal, then the model $\langle R(\theta_1), \in\rangle$ clearly does *not* have the Π_1^1 reflection property. If μ_1 is the first Mahlo number, then $\langle R(\mu_1), \in\rangle$ does not have the Π_1^1 reflection property.

We remark that a cardinal κ is Π_1^1 indescribable in the sense of Hanf and Scott [3] if and only if $\langle R(\kappa), \in, S\rangle$ has the Π_1^1 reflection property for all $S \subset R(\kappa)$. It is known from [3] that κ is weakly compact if and only if κ is Π_1^1 indescribable (a proof is given in Silver [10]).

THEOREM 2.2. *Suppose* $\mathfrak{A}$ *is a well-founded model of* ZFC, Ord($\mathfrak{A}$) *has cofinality* $>\omega$, *and* $\mathfrak{A}$ *has an end elementary extension. Then* $\mathfrak{A}$ *has the* Π_1^1 *reflection property.*

PROOF. Suppose $\mathfrak{A}$ does not have the Π_1^1 reflection property. Then there is an $a \in A$ and a Π_1^1 formula $\varphi(v_0)$ such that $\varphi(a)$ holds in $\mathfrak{A}$, but the formula $\psi(v_0)$,

$$(\forall\alpha)(v_0 \in R(\alpha) \rightarrow \varphi(v_0) \quad \text{fails in} \quad \langle R(\alpha), \in\rangle),$$

is satisfied by a in $\mathfrak{A}$. For some $p < \omega$, ψ is a p-formula. By Theorem 2.1, $\mathfrak{A}$ has a well-founded end p-extension $\mathfrak{B}$. Let β be the least ordinal of $\mathfrak{B}$ which does not belong to $\mathfrak{A}$. Then $\mathfrak{A} = \mathfrak{B}_\beta$. Since $\psi(a)$ holds in $\mathfrak{A}$, $\psi(a)$ holds in $\mathfrak{B}$. Therefore in $\mathfrak{B}$, $\varphi(a)$ does not hold in $\langle R(\beta), \in\rangle$. But φ is a Π_1^1 formula, so $\varphi(a)$ really does not hold in $\mathfrak{B}_\beta = \mathfrak{A}$. This contradiction completes our proof.

The following result, which has both a weaker hypothesis and a stronger conclusion than Theorem 2.2, can be proved by a straightforward modification of the above proof.

THEOREM 2.3. *Suppose* $\mathfrak{A}$ *is a well-founded model of* ZFC, Ord($\mathfrak{A}$) *has cofinality* $>\omega$, *and for all* $p < \omega$, $\mathfrak{A}$ *has an end p-extension. Then for every set* $S \subset A$ *which is definable in* $\mathfrak{A}$ *the model* $\langle A, E, S\rangle$ *has the* Π_1^1 *reflection property.* (S is definable in $\mathfrak{A}$ if there is a first order formula $\theta(v_0)$ such that $S = \{a_0 \in A : \mathfrak{A} \models \theta(a_0)\}$.)

Since $\langle R(\theta_1), \in\rangle$ does not have the Π_1^1 reflection property, Theorem 2.2 shows that $\langle R(\theta_1), \in\rangle$ has no end elementary extension. Similarly, $\langle R(\mu_1), \in\rangle$ has no end elementary extension.

By Theorem 1.1, every model of ZF of power ω has an end elementary extension. The next example shows that ω is the only reasonable cardinal with this property.

COROLLARY 2.4. *Let* κ *be a cardinal such that* $\omega < \kappa$ *and there exists a weakly inaccesible cardinal* $\lambda > \kappa$. *Then there is a well-founded model* $\mathfrak{A}$ *of* ZF *plus the axiom of constructibility* (*briefly*, ZF + V = L), *such that A has power* κ *and* $\mathfrak{A}$ *has no end elementary extension.*

PROOF. The model $\langle L_\lambda, \in\rangle$ is a model of ZF + V = L in which the following sentence θ holds:

"There exists a least ordinal α such that $\kappa < \alpha$, cf(α) $\geqq \Omega$, and $\langle L_\alpha, \in\rangle$ is a model of ZF + V = L."

In the above formula Ω is a constant denoting the first uncountable ordinal in the absolute sense (not in the sense of the constructible universe).

The ordinal α is easily seen to have cardinality κ, in view of the results of

Gödel [2]. $\langle L_\alpha, \in\rangle$ is a well-founded model of ZF + V = L. In $\langle L_\alpha, \in\rangle$ it can be shown that $cf(\alpha) = \Omega$, so $cf(\alpha) = \Omega$ in the absolute sense. The Π_1^1 sentence

$$\text{"cf(OR)} \geqq \Omega \quad \text{and} \quad \text{ZF + V = L"}$$

holds in $\langle L_\alpha, \in\rangle$. However, θ does not hold in $\langle L_\alpha, \in\rangle$. Therefore $\langle L_\alpha, \in\rangle$ does not have the Π_1^1 reflection property, and has no end elementary extension.

The above corollary may be improved by controlling the cofinality of $\text{Ord}(\mathfrak{A})$ as well as the power of $\mathfrak{A}$. For instance, if μ is any regular uncountable cardinal $\leqq\kappa$, the model $\mathfrak{A}$ may be chosen so that $\text{Ord}(\mathfrak{A})$ has cofinality μ. Thus the natural generalization of Theorem 1.1 to larger cofinalities fails.

Theorems 2.1 and 2.2 (and 2.3) also hold for models of ZFC with finitely many predicates. Indeed, the same proofs go through. Theorem 1.3 follows at once from Theorem 2.2 with extra predicates. One direction is already given by Theorem 1.2. For the other direction, if κ is not weakly compact, then κ is Π_1^1 describable. Hence for some $S \subset R(\kappa)$, the model $\langle R(\kappa), \in, S\rangle$ does not have the Π_1^1 reflection property, and therefore has no end elementary extension.

3. **Well-founded end extensions.** We shall now state a number of other results which, like Theorem 2.1, say that if $\mathfrak{A}$ has an end elementary extension then $\mathfrak{A}$ has an end extension of a certain kind.

A model $\mathfrak{A}$ is said to be *κ-well founded* if there is no decreasing E-chain of length κ, that is, there is no sequence a_α, $\alpha < \kappa$, of elements of $\mathfrak{A}$ such that

$$\beta < \alpha \quad \text{implies} \quad a_\alpha \, E \, a_\beta.$$

It is obvious that well founded is equivalent to ω-well founded, and the notion of κ-well founded becomes weaker as κ increases.

THEOREM 3.1. *Suppose κ is an infinite cardinal, $\mathfrak{A}$ is a κ-well founded model of* ZFC, *and the cofinality of* $\text{Ord}(\mathfrak{A})$ *is different from the cofinality of κ. Then*

(i) *If for all $p < \omega$, $\mathfrak{A}$ has an end p-extension, then for all $p < \omega$, $\mathfrak{A}$ has a κ-well founded end p-extension.*

(ii) *If κ has cofinality $>\omega$ and $\mathfrak{A}$ has an end elementary extension, then $\mathfrak{A}$ has κ-well founded end elementary extension.*

The proof of part (i) is exactly like the proof of Theorem 2.1. To get (ii), we use the proof of (i) to obtain a chain

$$\mathfrak{A} \prec_0 \mathfrak{B}_1 \prec_1 \mathfrak{B}_2 \prec_2 \mathfrak{B}_3 \cdots,$$

such that for each $p < \omega$, $\mathfrak{B}_p$ is an end p-extension of $\mathfrak{A}$ and is κ-well founded. It follows that $\bigcup_{p<\omega} \mathfrak{B}_p$ is a κ-well founded end elementary extension of $\mathfrak{A}$.

We now turn to a condition which is stronger than well foundedness. A model $\mathfrak{A}$ is said to be *κ-closed* if for all cardinals $\lambda < \kappa$, $\mathfrak{A}$ satisfies the infinitely long sentence

$$(\forall x_\alpha : \alpha < \lambda)(\exists y)(\forall z)\Big(z \in y \leftrightarrow \bigvee_{\alpha<\lambda} z = x_\alpha\Big).$$

That is, every subset of A of power λ "belongs" to the model $\mathfrak{A}$. For $\lambda = 2$, this is the axiom of pairs. Every model of ZF is ω-closed. Every ω_1-closed model is well founded, and κ-closed becomes stronger as κ increases. If κ is a singular cardinal then a model $\mathfrak{A}$ of ZF is κ-closed if and only if it is κ^+-closed. So the notion of κ-closed is interesting only when κ is regular.

Consider a model $\mathfrak{B}$ and a set $X \subset B$. We shall say that $\mathfrak{B}$ *contains* X if there exists $b \in B$ such that

$$X = \{a \in B : a\, F\, b\}.$$

Thus $\mathfrak{A}$ is κ-closed if and only if $\mathfrak{A}$ contains every subset of A of power less than κ. We say that $S(\lambda)$ *is standard* in $\mathfrak{B}$ if there is a λth ordinal l in $\mathfrak{B}$ and $\mathfrak{B}$ contains every set of its ordinals less than l.

LEMMA 3.2. *Let $p < \omega$ and let κ be a regular cardinal. Suppose $\mathfrak{B}$ is a model of* ZFC, *$S(\lambda)$ is standard in $\mathfrak{B}$ for every cardinal $\lambda < \kappa$, $X \subset B_\alpha$ for some ordinal α of $\mathfrak{B}$ and $\mathfrak{B}$ contains every subset of X of power less than κ. Then there is a κ-closed structure $\mathfrak{A} \prec_p \mathfrak{B}$ such that $X \subset A$.*

PROOF. We may assume that $\kappa > \omega$, and also that α contains an initial segment of order type κ. There is an ordinal β of $\mathfrak{B}$ of cofinality $>\alpha$ such that $\mathfrak{B}_\beta \prec_p \mathfrak{B}$. In $\mathfrak{B}$ we may formalize the first order logic with the predicates $=$, $\in$, and with Skolem functions. Using the axiom of choice in $\mathfrak{B}$, the model $\mathfrak{B}_\beta$ may be expanded to a model $\mathfrak{B}_\beta^*$ with Skolem functions. Then in $\mathfrak{B}$ there is a function $\mathrm{val}(n, x)$ such that for each natural number n and each sequence x in B_β, $\mathrm{val}(n, x) = t_n(x)$ where t_n is the nth Skolem term.

Let λ be a cardinal, $\lambda < \kappa$. Then $\mathfrak{B}$ has a λth ordinal l. In $\mathfrak{B}$, let H be the set of all finite sequences of elements of l. Since $S(\lambda)$ is standard in $\mathfrak{B}$, $\mathfrak{B}$ contains every subset of H.

Working outside of $\mathfrak{B}$, consider the least set T_λ of formal expressions such that:

For each $\gamma < \lambda$, the constant $c_\gamma \in T_\lambda$;

If $\tau_n(v_0, \ldots, v_p)$ is a Skolem term and

$$t_0, \ldots, t_p \in T_\lambda,$$

then

$$\tau_n(t_0, \ldots, t_p) \in T_\lambda.$$

If $\gamma \leq \lambda$ and $t_\delta \in T_\lambda$ for all $\delta < \gamma$, then

$$\langle t_\delta : \delta < \lambda \rangle \in T_\lambda.$$

We may use subsets of H to code up the set T_λ of expressions with the following "coding function" g.

$$g(c_\gamma) = \{\langle \gamma, 0 \rangle\};$$

$$g(\tau_n(t_0, \ldots, t_p)) = \{\langle n, 1 \rangle\} \cup \bigcup_{i=0}^{p} \{i^\frown s : s \in g(t_i)\};$$

$$g(\langle t_\delta : \delta < \lambda \rangle) = \bigcup_{\delta < \lambda} \{\delta^\frown s : s \in g(t_\delta)\}.$$

The range G_λ of the function g is definable in $\mathfrak{B}$, hence is contained in $\mathfrak{B}$. Since $S(\lambda)$ is standard in $\mathfrak{B}$ and $\mathfrak{B}$ contains every subset of X of power $\leq\lambda$, it follows that $\mathfrak{B}$ contains every subset of $\lambda \times X$ of power $\leq\lambda$. Therefore $\mathfrak{B}$ contains every function x on λ into X.

For each such function x, define $\text{val}_\lambda(u, x)$, $u \in G_\lambda$, by

$$\text{val}_\lambda(g(c_\gamma), x) = x_\gamma;$$
$$\text{val}_\lambda(g(\tau_n(t_0, \ldots, t_p), x)) = \text{val}(n, \langle \text{val}_\lambda(g(t_0), x), \ldots, \text{val}_\lambda(g(t_p), x)\rangle);$$
$$\text{val}_\lambda(g(\langle t_\delta : \delta < \gamma\rangle), x) = \langle \text{val}_\lambda(t_\delta, x) : \delta \leq \gamma\rangle.$$

It is clear that if $\lambda < \lambda' < \kappa$, $u \in G_\lambda$, and x is a function on λ' into X, then

$$\text{val}_\lambda(u, x \restriction \lambda) = \text{val}_{\lambda'}(u, x).$$

For each λ and x, the function $\text{val}_\lambda(u, x)$ is a function in $\mathfrak{B}$ which maps G_λ into B_β.

Let

$$C = \{\text{val}_\lambda(u, x) : \lambda < \kappa, u \in G_\lambda, x \in X^\lambda\}.$$

Then $\mathfrak{C} = \langle C, F\rangle \prec \mathfrak{B}_\beta$, hence $\mathfrak{A} \prec_p \mathfrak{C} \prec_p \mathfrak{B}$. Since κ is regular, we see that $\mathfrak{C}$ is κ-closed.

It follows from the above lemma that $\mathfrak{B}$ is the union of a directed family of κ-closed models $\mathfrak{C} \prec_p \mathfrak{B}$ such that $X \subset C$. For if C_1, C_2 are κ-closed and $b \in B$, then $\mathfrak{B}$ contains every subset of $C_1 \cup C_2 \cup \{b\}$ of power $<\kappa$.

The assumption $X \subset B_\alpha$ may be weakened to $X \subset B$ if the model $\mathfrak{B}$ has a definable well-ordering of the universe, e.g., if $\mathfrak{B}$ is a model of the axiom of constructibility. For then we can choose in $\mathfrak{B}$ Skolem functions for all p-formulas.

In Lemma 3.2, the hypothesis that $\mathfrak{B}$ is a model of ZFC can be weakened somewhat. In fact, there is a $q < \omega$ (depending on p) such that "$\mathfrak{B}$ is a model of ZFC" can be weakened to "every q-sentence provable in ZFC holds in $\mathfrak{B}$." Thus we have the following theorem.

THEOREM 3.3. *Suppose $\mathfrak{A}$ is a κ-closed model of* ZFC, *and for all $p < \omega$, $\mathfrak{A}$ has an end p-extension. Then for all $p < \omega$, $\mathfrak{A}$ has a κ-closed end p-extension.*

PROOF. If $\mathfrak{B}$ is an end q-extension of $\mathfrak{A}$, then $S(\lambda)$ is standard in $\mathfrak{B}$ for all $\lambda < \kappa$. By 3.2 there is a κ-closed $\mathfrak{C}$ such that $A \subset C$ and $\mathfrak{C} \prec_p \mathfrak{B}$. Then $\mathfrak{C}$ is an end p-extension of $\mathfrak{A}$.

Of course, Theorem 3.3 also holds for ZFC with finitely many extra predicates.

We conclude this section with a brief discussion of another version of Theorems 2.1 and 3.3. By ZFC + Sat we mean the theory ZFC with an extra binary predicate symbol Sat, and extra axioms which make $\text{Sat}(n, x)$ hold if and only if the nth formula (involving only $=$ and $\in$) is satisfied by x. It is shown by Montague and Vaught [9] that if $\langle B, F, S\rangle$ is a model of ZFC + Sat, then there is a closed unbounded class of ordinals α of $\mathfrak{B} = \langle B, F\rangle$ such that $\mathfrak{B}_\alpha \prec \mathfrak{B}$. Using this strong reflection principle, we can prove

THEOREM 2.1a. *Suppose $\mathfrak{A}$ is a well-founded model of* ZFC, Ord($\mathfrak{A}$) *has cofinality* $>\omega$, *and $\mathfrak{A}$ has an end elementary extension $\mathfrak{B}$ which can be made into a model* $\langle B, F, S\rangle$ *of* ZFC + Sat. *Then $\mathfrak{A}$ has a well-founded end elementary extension.*

THEOREM 3.3a. *Suppose $\mathfrak{A}$ is a κ-closed model of* ZFC *and $\mathfrak{A}$ has an end elementary extension $\mathfrak{B}$ which can be made into a model* $\langle B, F, S\rangle$ *of* ZFC + Sat. *Then $\mathfrak{A}$ has a κ-closed end elementary extension.*

4. **Sufficient conditions for a model to have end extensions.** Theorems 1.1 and 1.2 give sufficient conditions for a model to have end extensions. In this section we shall obtain a more general sufficient condition. Our result will give an improvement of Theorem 1.2, and also a converse of Theorem 2.3.

THEOREM 4.1. *Suppose $\mathfrak{A}$ is a model of* ZFC *and for every set $S \subset A$ which is definable in $\mathfrak{A}$, the model $\langle A, E, S\rangle$ has the Π^1_1 reflection property. Then for all $p < \omega$, $\mathfrak{A}$ has an end p-extension.*

PROOF. In the model $\mathfrak{A}$, formalize the first order logic for a language with the identity and $\in$ symbols and with a Skolem function $f_{\varphi,n}$ for each formula φ and variable v_n.

Our proof will go through whether or not ω is standard in the model $\mathfrak{A}$. When ω is not standard, however, our notation is made more difficult because there will be nonstandard formulas in the logic formalized within $\mathfrak{A}$. So we shall assume temporarily that ω is standard in $\mathfrak{A}$ and present our proof under this assumption.

Let $p < \omega$, and let F be the set of all p-formulas in which no Skolem functions occur. The set

$$S = \{\langle\varphi, x\rangle : \varphi \in F \text{ and } x \text{ satisfies } \varphi \text{ in } \mathfrak{A}\}$$

is definable in $\mathfrak{A}$. By an *interpretation* we shall mean a function $I(\varphi, n, x)$ defined for all $\varphi \in F$, $n < \omega$, and $x \in A$, such that

$$\exists v_n \varphi(x, v_n) \rightarrow \varphi(x, I(\varphi, n, x))$$

always holds in $\mathfrak{A}$. By a *term*, $t(v_0)$, we shall mean a term built up from Skolem functions $f_{\varphi,n}$ with $\varphi \in F$, from constants for the elements of the model $\mathfrak{A}$, and from the variable v_0. By an *expression* we shall mean an open formula built up from terms of the above kind and the predicates $=, \in$.

An interpretation I gives a value for each Skolem function, namely

$$f_{\varphi,n}(x) = I(\varphi, n, x).$$

Consequently, I also gives us a value for each term $t(v_0)$ and a truth value for each expression $\varphi(v_0)$.

The main step of our proof is to show

(*) there exists an interpretation I and a set D of expressions such that:

(1) D is closed under finite conjunction;

(2) For each expression φ, either $\varphi \in D$ or $(\neg\varphi) \in D$;

(3) For each $\varphi(v_0) \in D$ there exists $y \in A$ such that $\varphi(y)$ is true under the interpretation I;

(4) For each $x \in A$, $(\neg v_0 = x) \in D$;

(5) For each term $t(v_0)$ and each $x \in A$, if $(t(v_0) \in x) \in D$, then there exists $y\, E\, x$ such that $(t(v_0) = y) \in D$.

Once (*) is established, we may construct the end extension B in the following way. First we choose an ultrafilter D^* over the set A such that for each expression $\varphi(v_0) \in D$, the set

$$\varphi^* = \{y \in A : \varphi(y) \quad \text{is true under the interpretation } I\}$$

belongs to D^*. The conditions (1)–(3) guarantee the existence of such a D^*. Now form the ultrapower

$$D^*\text{-prod } \mathfrak{A}.$$

Each term $t(v_0)$ gives rise to an element

$$t^* \in D^*\text{-prod } A,$$

namely, t^* is the equivalent class modulo D^* of the function T where $T(y)$ is the value of $t(y)$ under the interpretation I. Let

$$B = \{t^* : t(v_0) \text{ is a term}\},$$

and let $\mathfrak{B}$ be the submodel of D^*-prod $\mathfrak{A}$ with universe B. Since $\mathfrak{B}$ is closed under the "ultrapowers" of the Skolem functions $f_{\varphi,n}$, D^*-prod $\mathfrak{A}$ is a p-extension of $\mathfrak{B}$. It follows that the natural embedding of $\mathfrak{A}$ into D^*-prod $\mathfrak{A}$ maps $\mathfrak{A}$ isomorphically onto a p-submodel $\mathfrak{A}^*$ of $\mathfrak{B}$. $\mathfrak{B}$ is a proper p-extension of $\mathfrak{A}^*$ because of (4); indeed, the element $(v_0)^*$, coming from the term v_0, belongs to $\mathfrak{B}$ but not to $\mathfrak{A}^*$. Finally, the condition (5) on D insures that $\mathfrak{B}$ is an end extension of $\mathfrak{A}^*$.

We now verify (*). We first observe that (*) may be expressed as a Σ_1^1 sentence σ in the model $\mathfrak{A}$. Moreover, there is a first-order sentence θ involving the predicates $=, \in, S$ such that:

(i) θ holds in $\langle A, E, S\rangle$; and

(ii) for each ordinal α of $\mathfrak{A}$, θ holds in $\langle A_\alpha, E, S \cap A_\alpha\rangle$ if and only if A is a p-extension of $\mathfrak{A}_\alpha$.

Notice that, although S is definable in $\mathfrak{A}$, $S \cap A_\alpha$ is not necessarily definable in $\mathfrak{A}_\alpha$.

We now consider the Σ_1^1 sentence $\theta \rightarrow \sigma$ involving $=, \in, S$. We claim that the following sentence holds in $\langle A, E, S\rangle$:

"For every ordinal α, $\langle R(\alpha), \in, S \cap R(\alpha)\rangle$ satisfies $\theta \rightarrow \sigma$".

To establish our claim, let α be an ordinal of $\mathfrak{A}$ such that in $\mathfrak{A}$, $\langle R(\alpha), \in, S \cap R(\alpha)\rangle$ satisfies θ. Let z be an arbitrary element of $A - A_\alpha$. Working in the model $\langle A, E, S\rangle$, there is an ordinal $\beta > \alpha$ such that $z \in R(\beta)$ and $\langle R(\beta), \in, S \cap R(\beta)\rangle$ is a model of θ. Then $\langle R(\beta), \in, S \cap R(\beta)\rangle$ is a p-extension of $\langle R(\alpha), \in, S \cap R(\alpha)\rangle$. By the axiom of choice there is a function I on $(R(\beta) \cap F) \times \omega \times R(\beta)$ into $R(\beta)$ which is an interpretation in the sense of $\langle R(\beta), \in\rangle$. Moreover, the restriction of I to

$R(\alpha)$ will be an interpretation in the sense of $\langle R(\alpha), \in\rangle$. Let D be the set of all expressions in $R(\alpha)$ which are satisfied by z in $\langle R(\beta), \in\rangle$ under the interpretation I. Then the sentence σ is satisfied in $\langle R(\alpha), \in\rangle$ when we take this choice of I and D. This verifies our claim.

Since $\langle A, E, S\rangle$ has the Π_1^1 reflection property, the Σ_1^1 sentence $\theta \rightarrow \sigma$ must also hold in $\langle A, E, S\rangle$. Since θ holds in $\langle A, E, S\rangle$, it follows that σ holds in $\mathfrak{A}$. This verifies (*) and completes our proof (at least for the case that ω is standard in $\mathfrak{A}$).

In the case ω is not standard in $\mathfrak{A}$, we need only carry out the above proof with the provision that satisfaction of p-formulas is always understood in the sense of the model $\mathfrak{A}$. The model $\mathfrak{B}$ will then be a p-extension of $\mathfrak{A}$ with respect to nonstandard p-formulas as well as standard p-formulas.

Combining Theorems 4.1 and 2.3, we obtain

COROLLARY 4.2. *If $\mathfrak{A}$ is a well-founded model of* ZFC *and* Ord($\mathfrak{A}$) *has cofinality $>\omega$, then the following are equivalent:*

(i) *For all $p < \omega$, $\mathfrak{A}$ has an end p-extension;*

(ii) *For every set $S \subset A$ definable in $\mathfrak{A}$, the model $\langle A, E, S\rangle$ has the Π_1^1 reflection property.*

Our next result is a sufficient condition for $\mathfrak{A}$ to have an end elementary extension.

THEOREM 4.3. *Suppose $\langle A, E, S\rangle$ is a model of* ZFC + Sat *which has the Π_1^1 reflection property. Then $\mathfrak{A} = \langle A, E\rangle$ has an end elementary extension.*

PROOF. Similar to the proof of Theorem 4.1. The only difference is that instead of p-formulas we consider arbitrary formulas (involving $\in$ and $=$), and instead of p-extensions we always have elementary extensions.

The results of this section also hold for models of ZFC with extra predicates. We may thus combine Theorem 4.4 above with Theorems 2.1a and 3.3a. Let ZFC + Sat + Sat′ be the theory ZFC + Sat plus a second extra predicate symbol Sat′ expressing the satisfaction of formulas of ZFC + Sat.

COROLLARY 4.4. *Let $\langle A, E, S, S'\rangle$ be a model of* ZFC + Sat + Sat′ *which has the Π_1^1 reflection property. Then*

(i) *If $\mathfrak{A} = \langle A, E\rangle$ is well founded and* Ord($\mathfrak{A}$) *has cofinality $>\omega$, then $\mathfrak{A}$ has a well-founded end elementary extension.*

(ii) *If $\mathfrak{A}$ is a κ-closed model, then $\mathfrak{A}$ has a κ-closed end elementary extension.*

PROOF. By Theorem 4.3 with the extra predicate Sat, the model $\langle A, E, S\rangle$ has an end elementary extension $\langle B, F, T\rangle$, which is a model of ZFC + Sat. Theorems 2.1a and 3.3a now apply for (i), (ii) respectively.

Notice that (ii) generalizes the fact that if κ is weakly compact then $\langle R(\kappa), \in\rangle$ has a κ-closed end elementary extension.

References

1. H. Gaifman, *Uniform extension operators for models and their applications*, Tech. Report No. 21, U.S. Office of Naval Research, 1965.

2. K. Gödel, *The consistency of the continuum hypothesis*, Ann. of Math. Studies, No. 3, Princeton Univ. Press, Princeton, N.J., 1940.

3. W. Hanf and D. Scott, *Classifying inaccessible cardinals* (abstract), Amer. Math. Soc. Notices, **8** (1961), 445.

4. H. J. Keisler, *The equivalence of certain problems in set theory with problems in the theory of models* (abstract), Amer. Math. Soc. Notices, **9** (1962), 339.

5. ——— *Limit ultraproducts*, J. Symbolic Logic **30** (1965), 212–233.

6. H. J. Keisler and M. Morley, *Elementary extensions of models of set theory*, Israel J. Math **6** (1968), 49–65.

7. H. J. Keisler and J. H. Silver, *Well-founded extensions of models of set theory*, Notices Amer. Math. Soc. **14** (1967), 256.

8. R. Mac Dowell and E. Specker, *Modelle der Arithmetik. Infinitistic methods*, Proc. Sympos. on the Foundations of Math. at Warsaw, 1959. Oxford, 1961, pp. 257–263.

9. R. Montague and R. L. Vaught, *Natural models of set theory*, Fund. Math. **47** (1959), 219–242.

10. J. H. Silver, *Some applications of model theory in set theory*, Doctoral thesis, University of California, Berkeley, 1967.

11. A. Tarski, *Some problems and results relevant to the foundations of set theory*. Logic, Methodology, and Philosophy of Science. (Proc. 1960 Internat. Congr.) Stanford Univ. Press, Stanford, Calif., 1962, pp. 125–136.

12. A. Tarski and R. L. Vaught, *Arithmetical extensions of relational systems*, Compositio Math. **13** (1957), 81–102.

UNIVERSITY OF WISCONSIN

UNIVERSITY OF CALIFORNIA, LOS ANGELES AND BERKELEY

OBSERVATIONS ON POPULAR DISCUSSIONS OF FOUNDATIONS

G. KREISEL

1. **Introduction.** Mathematical logicians usually say they prefer to prove 'hard' theorems instead of formulating carefully philosophical distinctions and views. Naturally, one can't expect too much when they do publish on philosophical, foundational matters. Equally naturally, since they are not accustomed to the discipline of foundational analysis, when they see a careful foundational discussion such as **[G1]** or **[G2]**, they read it less closely than mathematical arguments (corresponding, *mutatis mutandis*, to old fashioned philosophers, who are convinced they understand the results of mathematical logic without close attention to formulations or proofs). The aim of this note is to select a few simple foundational observations and distinctions which provide some orientation on *progress in foundations* without, I hope, overtaxing the reader's patience.

Above all, there is the distinction between logical and mathematical foundations or, better, between the *foundation* and *organization* of mathematics. The distinction is useful for analyzing the nature of the problems presented by (existing) category theory, and, more generally, for analyzing the role of foundations for working mathematicians.

A central component of current popular thinking is the *formalist conception* of mathematics in one of several forms. In its most extreme form it rejects all foundational problems (and hence of course the distinction between foundation and organization). Another, perhaps more common, version does not reject these problems, but simply assumes that a foundation must consist in a formalist reduction; thus failure of *such* a reduction is interpreted as constituting a difficulty for foundations *tout court*.

At the end of this note I go over some of the more extravagant 'arguments' which have been spawned by recent formal independence results in set theory.

2. **Foundational progress.** Naturally, as in any subject, the open problems of foundations get the expert's main attention. But one should not forget the remarkable progress compared to the situation 50 years ago, or, equivalently, compared to the state of *common* knowledge. Not only progress in the sense of internal growth, of theories of one's own theories, but old notions have been analyzed (formulated) and old problems answered. For example, by way of analysis of concepts (in words!) we have the description of the cumulative type structure and the realization (cf. e.g. [**Z**, p. 47] or [**G1**, pp. 262–263]) that, for *this* structure, the contradictory form of the comprehension axiom is really due to a specified error. As an example of a precise formulation, in terms of adequacy conditions,[1] of a foundational conception, we have Hilbert's program. Or as an example of a technical result we have Gödel's incompleteness theorem which establishes the limitations of a conception (in fact, of Hilbert's). At the same time it reestablishes the interest of a rival view which seemed unattractive, because *unnecessarily* complicated, as long as the simpler conception appeared tenable; in this way intuitionistic mathematics, which goes beyond Hilbert's finitist-formalist conception, has become a natural object of attention.

Insisting on the objective fact of progress in foundations is not just a matter of tribal honour! On the contrary, foundational discussions which neglect this progress are worthless; objectively when they rehash dead issues[2] (like new proposals for squaring the circle) and subjectively for one of the following two common reactions. One is to be heavyhearted because one feels one has nothing to build on; the other is to become light headed, because one feels that 'anything goes' since, whatever one proposes, it can't be worse than nothing at all. In fact, the impression that there has been no progress may lead one to invent such negative doctrines as positivism to explain the impression (cf. [**K3**, pp. 140–142]).[3]

---

[1] For a pedantic exposition of the all-important adequacy conditions in the case of set theoretic and formalist foundations (Hilbert's program), see, e.g. [**KK**, pp. 166–168, respectively 206–212].

[2] For instance, A. Robinson's 'Formalism-64' [**Ro1**] or Cohen's paper (in this volume) which, as Cohen points out, was influenced by Robinson's. The number '64' is supposed to refer to 1964, but the paper might just as well have been written in 1864 since it presents merely a crude version of formalism. It neither takes into account Frege's classical critique of that kind of formalism, nor, worse still, the massive studies of the one genuine attempt to pursue formalism, namely Hilbert's program, nor even Hilbert's analysis of adequacy conditions. So, not only does it not add to our knowledge, but it ignores existing results. Presumably, both Robinson's and Cohen's papers *reflect* pretty well what a number of people think. (As Cohen points out the views of these papers are not consistent in *detail*.)

[3] There are, of course, some subsidiary reasons for the impression. One is certainly the frivolous tradition of quoting snippets of Greek philosophy, often even textually doubtful, as if nothing had happened since then! It is right not to pay too much attention to contemporary fashions, but quite wrong not to use the *penetrating* and *detailed* analysis by the pioneers of modern logic (such as Frege, cf. Footnote 2). Another is that there are some who merely want a reason for not bothering about foundations. Why bother to find a reason? Is it to convince oneself (or others) that one is not 'losing' *anything* by concentrating on one's own work? This seems too much to hope for. Is it to follow, unconsciously of course, Oxford's Dr. Jowett? (It isn't knowledge if I don't know it.)

3. **Formalism: an obstacle to recognizing progress when one sees it.** First, let us not forget the positive aspect of formalist foundations, namely the remarkable discovery that a great deal of mathematics which does not present itself as 'a formal game with symbols' nevertheless permits an adequate formalist foundation in the sense of Hilbert's program. But let us note at the same time that this fact is remarkable just because it is not plausible!

The negative side of formalism, already mentioned in the introduction, is the requirement that a foundation, to deserve this name, must be a formalist reduction. If this requirement is accepted, the popular impression that there has been little foundational progress is surely right since all the evidence goes to show that no *formalist* foundation is possible! But a much more far reaching effect of accepting the requirement is this: it becomes *genuinely* difficult to appreciate work in foundations since the formalist restriction affects the idea of *precision* itself; a notion will be rejected as 'imprecise' not only when it is really vague (and hence should be rejected or at least analyzed) but simply because it does not fit into the formalist scheme. But inasmuch as this scheme has failed, *further progress will use notions and methods outside the scheme;* in fact, the principal fruitful questions need not even be formulable in terms accepted in this scheme. So, having accepted the formalist scheme, one is not merely *unwilling* to appreciate (or study!) further work in foundations, but simply *unable* to do so without reexamining one's ideas.

The attraction of formalism, particularly for mathematicians, is surely no accident. In a sense it aims to provide a *theoretical* reason for not bothering about foundations (cf. Footnote 3). For, as Hilbert stressed, formalism was to provide a *final solution* of all foundational problems by showing that mathematical practice loses nothing (as far as elementary conclusions are concerned) by *getting rid* of 'abstract nonsense'. Naturally, final solutions by segregation have appeal: they just aren't always right. In particular, while Hilbert thought he would *eliminate* the notions and methods that strike one as problematic, research has shown the need for these notions. One has to *analyze* and study them. Note that, theoretically, this switch constitutes very real progress: one may have suspected the conclusion, but, before Gödel's incompleteness theorem, there was no precise analysis of what they were needed for. See Footnote 12 for further elaboration.

4. **Organization and foundations: a distinction.** Obviously, the most striking difference is this. Foundations are concerned with the *validity*, constructive or nonconstructive, of mathematical principles; in practice one hardly looks at principles unless one is convinced of their validity, and one's principal interest is to make them *efficient*. Put differently, foundations provide reasons *for* axioms, practice is concerned with deductions *from* axioms, and the organization of practice wants to make these deductions as intelligible as possible. Some practical consequences stare one in the face.

For foundations it is important to know what we are talking about; we make the subject as *specific* as possible. In this way we have a chance of making *strong* assertions. For practice, to make a proof intelligible, we want to eliminate all properties which are not relevant to the result proved, in other words, we make the subject matter less specific.

Foundations provide an *analysis* of practice. To deserve this name, foundations must be expected to introduce notions which do *not* occur in practice. Thus in foundations of set theory, *types* of sets are treated explicitly while in practice they are generally absent; and in foundations of constructive mathematics, the analysis of the logical operations involves (intuitive) *proofs* while in practice there is no explicit mention of the latter.

Foundations and organization are similar in that both provide some sort of more systematic exposition. But a step in this direction may be crucial for organization, yet foundationally trivial, for instance a new *choice of language* when (i) old theorems are simpler to state but (ii) the primitive notions of the new language are defined in terms of the old, that is if they are logically dependent on the latter. Quite often, (i) will be achieved by using new notions with more 'structure', that is less analyzed notions, which is a step in the *opposite* direction to a foundational analysis. In short, foundational and organizational aims are liable to be actually contradictory.

Here it may be remarked that existing practice of category theory definitely raises organizational problems, above all a good choice of language for the formulation of category theory. It is then a separate matter to what extent a proposed theory of categories, formulated in this hypothetical language, raises foundational problems for an adequate reduction to set theory (either in the strict sense of [**KK**, pp. 166–168], or the generalized sense, ibid., p. 169). Existing formulations, e.g. by Lawvere in this volume, patently do not raise such a problem. Also, the mere use of the phrase 'category of all categories' does not imply a set theoretic foundation in which $x \in x$ for some x! for, if the hypothetical language of category theory is independent of set theory there is no compelling reason why the *of* in 'category of categories' should be interpreted as $\in$.

Before going further into the relation between mathematical practice and foundations, it is worth noting the obvious distinction between (i) foundational analysis (which is specifically concerned with validity) and (ii) general conceptual analysis (which, in the traditional sense of the word, is certainly a philosophical activity). As mentioned above, the working mathematician is rarely concerned with (i), but he does engage in (ii), for instance when establishing *definitions* of such concepts as length or area or, for that matter, natural transformation. For this activity to be called an *analysis* the principal issue must be whether the definitions are *correct*, not merely, for instance, whether they are useful technically for deriving results not involving the concepts (when their correctness is irrelevant). In short, it's not (only) what you do it's the way that you do it.

5. **Theory and the practical man: an analogy.** Once the distinction between foundations and organization is made, it is natural to ask to what extent foundations affect mathematical practice. Clearly, certain[4] foundational

[4] Not necessarily the most striking ones; for example, Gödel's incompleteness theorem has, so far, hardly affected mathematical practice at all (and, certainly not as much as lesser results in logic).

discoveries (logical languages, the axiomatic method) have altered practice out of all recognition. But, I think, *day-to-day foundational research does not affect day-to-day practice*. This conviction may well be behind the mathematician's reluctance to get involved in foundations although the reasons usually given are different and often, weak (cf. Footnote 3).

To see the relation between foundational and 'technical' work in more detail, and with much needed detachment, let us look at a superficially quite different matter, namely the way in which the 'practical' man uses and views theoretical work on his own subject.

(i) The practical man will generally dismiss the theoretical questions concerning matters that he handles well; rightly so because, on *familiar* ground, his understanding is not only more efficient, but simply less dubious than any theoretical analysis.

(ii) *Fundamental* differences in rival theories do not 'disturb' practical experience. (Not surprisingly, since a theoretical view would not have survived if it were in conflict with *familiar* experience.) As is well known, one often has to think long and hard, and, possibly, has to *extend* experience by novel kinds of experiments, to find an *experimentum crucis*, before one can decide between two theories. Consequently, to the practical man fundamental theoretical problems may often appear simply ludicrous.

(iii) The *sensible* practical man realizes that the questions which he dismisses may be the key to a theory. Further, since he doesn't have a good theoretical analysis of familiar matters, sometimes not even the concepts needed to frame one, he will not be surprised if a novel situation turns out to be genuinely problematic. He will use working hypotheses and *ad hoc* experience, or simply wait till theoretical analysis is simple enough for his purposes.

(iv) We have the following very *general pattern* of events. First a theory is built up on some startling discovery, that is something quite contrary to intuitive impressions. When such a theory has lasted some time and then fails, the popular reaction is typically journalistic; being preoccupied with current events, it forgets that the original discovery was itself contrary to naive convictions.

Now let us compare the working mathematician's view of foundations with the practical man's view of theory.[5] As to (i) we need only recall the introduction. As to (ii) consider the role of two such fundamentally different *notions* as the cumulative hierarchy generated by the power set operation and the ramified (constructible) hierarchy; they both satisfy the formal laws actually used in current practice. So mathematicians ask, 'What is wrong with simply postulating $V = L$?' More generally, in connection with foundational conceptions, amateurish formulations usually produce simply ludicrous associations; this is familiar from the so-called 'realist' or 'platonist' conception which is presented as if it involved huge (material)

[5] Amusingly, even 'defects' have their parallel, such as the dangerous effect of a little learning; compare, for instance, the sound discussion of the continuum hypothesis by Bing **[Bi]** who takes a robust, genuinely practical view, to the 'arguments' in the next section.

substances.[6] As to (iii) I think mathematicians take this line with measurable cardinals.[7] As to (iv), recall such startling discoveries as the formalization (of the bulk) of mathematics or the explicit definition of (the bulk of) mathematical notions in the language of set theory. And, in connection with failures, the reaction to the failure of Hilbert's program fits very well into that vulgar journalist pattern; but probably an extreme example is the reaction to formal independence proofs, considered in the next section.

6. **Formal independence results.** To the best of my knowledge a number of people are troubled by the 'arguments' quoted below on the significance of independence proofs, for instance of the continuum hypothesis CH. (This malaise is not surprising since preposterous suggestions often leave one speechless.)

Argument from impotence. The independence results support the formalist position against the realist position. (The fact that a position is *unable* to decide something makes it better.)

Argument from omnipotence. CH is such a 'simple' and 'fundamental' matter; there is something radically wrong if we cannot settle it. (There are a lot of subsets of ω; there are a lot of 1-1 mappings from ω to initial segments of the ordinals; we certainly do not *know*, even in the wide sense of having a definition in the usual language of set theory, a listing of the subsets of ω by means of ordinals. So why should we expect to *know* the answer to this particular 'simple' matter?)

Argument from the exception. For Gödel's original formally undecided arithmetic formula we know axioms to decide it; not so for CH; so numbers are OK and sets not. (Gödel's formula is $\prod_1^0$, or $\sum_1^0$ depending on how you look at it.

[6] In the first place, these formulations are slipped in by critics of a position, but then they stick. (Ultimately, the critic is ridiculing *himself* for not finding a sensible formulation.) A good formulation of the realist position will, presumably, be related to some popular current formulations as a good statement of atomic theory is related to Eddington's (who compared matter to a swarm of flies—without even postulating that the flies are held together by, say, bonds of love).

[7] Cohen [*] expresses some reservations. I find them *prima facie* very understandable! Thus, to my amazement, I learned the other day that, in contrast to [**K2**, p. 113, 1.721], people working on measurable cardinals generally *define* κ to be measurable only for *uncountable* κ. So, on that definition, one doesn't know *any* measurable κ! No wonder that nonspecialists find the subject suspicious. On the most natural definition, $\kappa = 2$ and $\kappa = \omega$ are *provably* measurable; and just as the first κ which is additive for unions of 2 sets, namely ω, is also κ-additive, *i.e.*, finitely additive; so, if there is an uncountable κ which has a countably additive measure, then the least such κ is κ-additive. Put differently, the first uncountable measurable κ would be a solution to the equation:

$$2 \text{ is to } \omega \text{ (w.r.t. measurability) as } \omega \text{ is to } x.$$

From this point of view, nonexistence of an uncountable measurable cardinal would support the (somewhat) vague idea that *nothing* (uncountable) *is quite as different from* ω *as* ω *is from finite.* But Cohen's specific charge of 'opportunism' hardly applies universally. Of course, only people who have thought about the subject of large cardinals are competent to judge axioms about them; such people *may* have a vested interest. But they may also have a genuine scientific interest in finding out about the subject. In any case, isn't the situation in connection with independence results similar? Certainly Gödel, who proved the first basic theorems in this area, has always stressed limitations of formal independence results. However, it must be admitted, there have been opportunistic aberrations in connection with *recent* independence results, which seem to need a good dose of logical hygiene; see Footnotes 9, 11 below.

For formulae of *this* syntactic form a formal independence proof always provides a decision and this is where it stops. It is no longer true for Δ^0_2.)[8]

Argument for public consumption. The business of the parallel axiom.[9] Of course, there are similarities with CH. By the time the independence *proofs* were discovered, people were convinced of the independence anyway [**C**, p. 39] (except that, in the case of the parallel axiom, the mathematical formulation of the idea of independence solved a problem of the time). In a more technical direction, the easiest way of defining a model for the unfamiliar case, that is non-Euclidean space, $\neg$CH respectively, is to go 'up': to a higher dimensional Euclidean space, to a higher type in the type structure.[10] Also the models of Klein and Poincaré, used in the independence proofs of the parallel axiom, are of great interest in various parts of mathematics (for instance complex function theory); so are Gödel's L (used in the independence proof of $\neg$CH) and several Cohen-type models. More important, all the models mentioned of Klein and Poincaré, and of Cohen are patently *different* from the intended notions: the latter are quite clear enough to decide that much![11] One *difference* is this. CH is (provably) *not* independent of the full (second order) version of Zermelo's axioms (we know this much without knowing which way it is decided; the example is not empty since, for instance, the replacement axiom *is* second-order independent). In contrast, as for most independence results of 'ordinary' mathematics, the parallel axiom is independent of

---

[8] Specifically, more than 25 years ago I found a perfectly natural formula of this form (and published it later in [**Kl**]) which we know to be formally undecidable in set theory. But nobody has any idea whether it is true or false.

[9] I believe this comparison first appeared in print in one of the San Francisco dailies after a news conference in 1963, soon after Cohen's discovery. (Actually, I believe, the public would have been impressed enough by the (in my opinion) less dubious comparison with the geometric construction problems of doubling the cube or squaring the circle.) Almost five years later, A. Robinson [**Ro2**] returns to this comparison, incidentally once again (cf. Footnote 2) without a second thought for its weaknesses (such as those mentioned below, though several of them had been pointed out in the meantime). The main claim (p. 192) is not well considered: The fundamental importance of the advent of non-Euclidean geometry is that by contradicting the axiom of parallels it denied the uniqueness of geometrical concepts and, hence, their reality; cf. the text below and Footnote 11. *Digression.* For those of us who love the sinner even when we abhor the sin, it is interesting to ask what is behind Robinson's tactics. Could it be that some universal principle of indefiniteness of basic, standard concepts strikes him as a plug for the *philosophical importance* of nonstandard models in place of standard concepts (blithely disregarding the fact that his nonstandard models are *defined* by use of the set of statements which are true in the *standard* model)?

[10] When one uses the ramified set theoretic hierarchy (Gödel's contructible sets L) as in Cohen's original proof, the natural thing to do is to start with a model L_α of set theory and define the generic sets required in some L_β where β is of *higher* level than α. If one uses boolean-valued models, the realizations of 'real numbers' are *functions* of real numbers, which are one type *higher* than real numbers.

[11] This could be used (if it were necessary) to dispose of such claims as in Footnote 9, that these independence proofs show the intuitive notions considered to be indefinite. But a much more illuminating step at this point is to remember *examples which really do show that one has overlooked something*, that one's intended 'notion' was mistaken. (It is possible to be honest about such things!) For instance, one may have spoken of *the* ordering of some given field, compatible with the field operations, thinking that there was only one, and an example turns up which shows that there are many.

the remaining geometric axioms even if the continuity principle is treated as a second order axiom.[12] But the real *deception* produced by the comparison between the two independence proofs comes from this: it wasn't the *mathematical independence proofs of the parallel axiom which led to the physically important and for this reason well-known non-Euclidean geometries, but quite different considerations*, for instance Riemann's on the nature of geometry, which preceded the formal independence proofs of the parallel axiom. In fact, it is the *decision* between the parallel axiom and its negation (for the physical interpretation of geometry) which, probably, is the outstanding contribution in this area. Thus the comparison is trying to cash in on this famous geometric work, and, at the same time, it denies that, in the case of set theory, there is anything objective to consider!

To end on a more positive note we can summarize the situation as follows. CH *is* decided by the full (second order) axioms of Zermelo; by the above this is already something though we don't know which way (as, until recently [**S**], we only knew there were 9 or 10 complex quadratic fields of class number one but didn't know if there are 10). Our *present* analysis of Zermelo's axioms, that is the first order schemata in the usual language of set theory, is not sufficient to decide CH. Put succinctly: not the notion of set, but our analysis (present knowledge) of this notion is at fault.

To decide CH the obvious questions to ask are:

(i) Can we formulate the principles involved in the discovery of the existing axioms of set theory? (Axioms which, after all, also had to be found.) Do these principles, when formulated as generally as we can, decide CH?

(ii) Can we extend the language of set theory to compensate for the fact that

---

[12] Zermelo's axioms or their extension by Fränkel present themselves in the first place as *second* order axioms (just as Peano's axioms for arithmetic); for instance the comprehension axiom or the axiom of regularity; see [**KK**, p. 175]. The corresponding *first order schemata* are then obtained by replacing the second order variables by a list of explicitly defined predicates, and, possibly, adding some first order formulae which are (second order) consequences of the *full*, i.e., second order axioms (for a systematic analysis see, e.g., [**K4**, §4]). Though the following remarks on *higher order notions* are classical, it is necessary to repeat them for the sake of a reader who is influenced, consciously or unconsciously, by formalist doctrines (otherwise he will be ill at ease if he is thoughtful, or commit the absurdity of simply rejecting such notions, as does the extreme formalist in Section 1, if he is thoughtless). Higher order notions are of course, *not* part of formalist foundations; but, pursued seriously as in Hilbert's program, the latter aim to *show* that (in appropriate contexts) higher order notions can be replaced; for instance *accepting* the semantic notion of logical validity one proves the completeness theorem for certain classes of formulae (which wouldn't even make sense if the notions involved in validity were not accepted). Serious formalist foundations do not try to follow Procrustes in simply rejecting any notion outside its framework. Granted this we still have the following *simple-minded puzzle:* is it not circular to use second order notions which involve the concept of set (or subset) in axiomatizations of this concept? Sure, it would be circular if one were looking for a *reduction*, a definition of this concept in, say, more elementary terms; but reduction is certainly not the only way to study a notion: one can reflect on the notion itself! And it is by no means circular to use a concept in order to *state the facts* about it; as one uses the concept *finite* to state properties of such notions as 'hereditarily finite set', or the satisfaction relation which, of course, is defined by use of the logical operations, in order to state properties of these operations. To suspect a *vicious* circle we should have to have independent reasons, for instance actual ambiguities.

any first order schema in a fixed language, such as the comprehension schema in the usual language of set theory, is only an approximation to the second order axiom from which the schema is derived? And decide CH by means of axioms (which are evident for the intended interpretation) of the extended language?

Most people seem to be pessimistic about (i) and (ii); at least they do not pursue these questions. There has been a great deal of work on a much more specific suggestion **[G2]** to try and use axioms of infinity for deciding CH; but so far all concrete proposals have turned out to be, demonstrably, insufficient. Personally, I don't think that this kind of 'statistical' negative evidence has much weight: if one overlooks an important principle in one context, say in connection with CH, there is at least a chance one will overlook it in another context (axioms of infinity) too, if these contexts are related at all.

For elaborations of points in this note see, e.g., **[K5]**.[13]

Bibliography

Bi. R. H. Bing, *Challenging conjectures*, Amer. Math. Monthly **74** (1967), 56–64.

C. P. J. Cohen, "Independence results in set theory" in *The theory of models*, North-Holland, Amsterdam, 1965.

G1. K. Gödel, "What is Cantor's continuum problem" in *Philosophy of mathematics*, Prentice-Hall, Englewood Cliffs, N.J., 1964.

G2. ———, "Remarks on problems in mathematics" in *The undecidable*, Raven Press, New York, 1965.

K1. G. Kreisel, *Note on arithmetic models for consistent formulae of the predicate calculus*, Fund. Math. **37** (1950), 265–285.

K2. ———, "Mathematical logic" in *Lectures on modern mathematics*, vol. 3, Wiley, New York, 1965.

K3. ———, "Informal rigour and completeness proofs" in *Problems in the philosophy of mathematics*, North-Holland, Amsterdam, 1967.

K4. ———, *A survey of proof theory*, J. Symbolic Logic **33** (1968), 321–388.

K5. ———, *Two notes on the foundations of Set-Theory*, Dialectica **23** (1969), 93–114.

KK. G. Kreisel and J. L. Krivine, *Elements of mathematical logic. Model theory*, North-Holland, Amsterdam, 1967.

---

[13] Also I go into the most obvious question raised by the present point of view: Granted that formal independence of an assertion, in general, tells us nothing about its truth or even its definiteness, have formal independence results no positive use? Of course they do, and the use is not dissimilar to that of various consistency proofs for assuming $\sqrt{-1}$ relative to the theory of real numbers. Such proofs allow us to use complex numbers, in addition to *certain* properties ('axioms') of the field of real numbers, for deriving valid theorems of *suitable syntactic form* about reals (e.g. universal statements); depending on the sophistication of the consistency proofs we may extend the class of theorems or the properties of real numbers to which such *conservative extension results* apply (concerning the addition of CH or $\neg$CH see, e.g., Footnotes 35 and 22 on pp. 376 and 364 resp. of **[K4]**). Evidently it is not the mere existence of these conservative extension results that is useful; but the empirical fact that often it is easier to *find* the desired proof by a detour *via* the unwanted (and unnecessary) hypothesis. Note also that, in terms of conservative extension results, we can give a good answer to the question: *What is there to choose at the present stage of knowledge between A and $\neg A$ when* (i) *we do not know whether A is true but* (ii) *we do know that A is formally independent of currently recognized principles of set theory*? The answer is: *Extract conservative extension results from your formal independence proof and look at the syntactic form of the conclusions that you have drawn from A and $\neg A$ respectively.*

Ro1. A. Robinson, "Formalism 64" in *Logic, Methodology and Philosophy of Science, Dialectica*, North-Holland, Amsterdam, 1965, pp. 228–246.

Ro2. ———, *Some thoughts on the history of mathematics*, Compositio Math. **20** (1968), 188–193.

S. H. M. Stark, *A complete determination of the complex quadratic fields of class number one*, Michigan Math. J. **14** (1967), 1–27.

Z. E. Zermelo, *Über Grenzzahlen und Mengenbereiche*, Fund. Math. **16** (1930), 29–47.

STANFORD UNIVERSITY

INDESCRIBABILITY AND THE CONTINUUM[1]

KENNETH KUNEN

0. **Introduction.** Garland [**1**], whose notation we follow here, points out that, by a theorem of Zykov, the cardinal $2^{\aleph_0}$ is $\bigvee_2^1$ characterizable. This means that there is a $\bigvee_2^1$ sentence φ of second order logic which is true in a model $\langle M\rangle$ iff $\bar{\bar{M}} = 2^{\aleph_0}$ (for example, let φ say that there exist relations on M which make M into a complete ordered field). Also, Garland proves that $\aleph_\alpha$ for α small (for example, a Δ_2^1 ordinal) is both $\bigvee_2^1$ and $\bigwedge_2^1$ characterizable, as are the first weak inaccessible, the first fixed point of weak inaccessibles, etc. Thus, if $2^{\aleph_0}$ happens to be one of these cardinals, it will happen to be $\bigwedge_2^1$ characterizable. The main purpose of this paper is to show (Theorem 1) that there is no $\bigwedge_2^1$ sentence which "naturally" characterizes $2^{\aleph_0}$; i.e. that it is consistent with ZF to assume that $2^{\aleph_0}$ is not $\bigwedge_2^1$ characterizable. We shall also briefly discuss describability properties of cardinals below the continuum.

1. A reflection property

DEFINITION. *A cardinal κ has the* $\bigwedge_n^1(\bigvee_n^1)$ reflection property *iff for all* $\bigwedge_n^1$ $(\bigvee_n^1)$ *sentences φ in the binary relation symbol $<$,*

$$\langle\kappa; <\rangle \models \varphi \rightarrow \exists\lambda < \kappa[\langle\lambda; <\rangle \models \varphi].$$

Of course, such a κ is not $\bigwedge_n^1$ $(\bigvee_n^1)$ characterizable. Note also that since λ being a cardinal is a $\bigwedge_1^1$ property of $\langle\lambda; <\rangle$, λ can be restricted to being a cardinal in the definition (except for $\bigvee_1^1$, which we shall not consider here).

THEOREM 1. *If* ZF *is consistent, so is* ZF + AC + $2^{\aleph_0} < \aleph_{\omega_1}$ + $2^{\aleph_0}$ *has the* $\bigwedge_2^1$ *reflection property.*

[1] Work supported by a NSF Graduate Fellowship, and by NSF Grant GP-8569. Most of the results here appeared in the author's doctoral dissertation [**2**, § 16].

REMARKS. Since $2^{\aleph_0}$ is $\bigvee_2^1$ characterizable, it is the largest cardinal in a $\bigwedge_2^1$ spectrum; thus, its having the $\bigwedge_2^1$ reflection property is in fact equivalent to its not being $\bigwedge_2^1$ characterizable.

In [2, § 17], we show, as a curiosity, that the relational system $\langle \boldsymbol{R}; <\rangle$, where $<$ is the usual ordering of the real numbers $\boldsymbol{R}$, is always $\bigwedge_1^1$ characterizable.

The theorem will be proved by starting with ZF + AC + GCH, finding a cardinal κ which has the $\bigwedge_2^1$ reflection property, and then using a Boolean extension (see Scott-Solovay [3]) to blow up $2^{\aleph_0}$ to equal κ. The main problem will be to check that κ retains its $\bigwedge_2^1$ reflection property in the extension. Note that since $2^{\aleph_0}$ is $\bigvee_2^1$ characterizable this argument could not go through with the $\bigvee_2^1$ reflection property.

LEMMA 1. *There is a regular cardinal $\kappa < \aleph_{\omega_1}$ which has the $\bigwedge_n^1$ reflection property for all n.*

PROOF. If not, then for each $\xi < \omega_1$, there is second order sentence φ_ξ such that $\langle \aleph_{\xi+1}; <\rangle \models \varphi_\xi$ but for all $\alpha < \aleph_{\xi+1}$, $\langle \alpha; <\rangle \models \neg\varphi_\xi$. Then the φ_ξ are all distinct, which is impossible, since there are only $\aleph_0$ sentences.

Now for each ordinal α, let $\mathscr{B}_\alpha$ be the Boolean algebra of regular open sets of the topological space ${}^\alpha 2$ (where 2 has the discrete topology and ${}^\alpha 2$ has the usual product topology). If $\alpha < \beta$, identify $\mathscr{B}_\alpha$ with a subalgebra of $\mathscr{B}_\beta$ by identifying the space ${}^\beta 2$ with ${}^\alpha 2 \times {}^{\beta-\alpha}2$. $V^{\mathscr{B}_\alpha}$ is the Boolean-valued universe formed with $\mathscr{B}_\alpha$, and $[\![\Phi]\!]^{\mathscr{B}_\alpha}$ is the Boolean value of the formula Φ in $V^{\mathscr{B}_\alpha}$.

If s is a finite function from a subset of α into 2, let N_s be the clopen set determined by s:

$$N_s = \{F \in {}^\alpha 2 : F \restriction \mathrm{Dom}\,(s) = s\}.$$

Let $\langle s_\xi : \xi < \alpha\rangle$ be a primitive recursive coding of all such s; assume $s_0 = 0$ (the empty function). If R is a $\mathscr{B}_\alpha$-valued m-place relation on α, define $\tilde{R}$ on α by

$$\tilde{R}(\xi, \eta_1, \ldots, \eta_m) \leftrightarrow N_{s_\xi} \subseteq [\![R(\eta_1, \ldots, \eta_m)]\!]^{\mathscr{B}_\alpha}.$$

LEMMA 2. *For every $\bigwedge_n^1(\bigvee_n^1)$ formula $\varphi(X_1, \ldots, X_p)$, there is another $\bigwedge_n^1(\bigvee_n^1)$ formula $\varphi^*(X_1, \ldots, X_p, x)$ such that for any cardinal, α, any sequence $R_1, \ldots, R_p$ of $\mathscr{B}_\alpha$-valued relations on α, and any $\xi < \alpha$,*

$$\langle \alpha; <\rangle \models \varphi^*(\tilde{R}_1, \ldots, \tilde{R}_p, \xi) \leftrightarrow N_{s_\xi} \subseteq [\![\langle \alpha; <\rangle \models \varphi(R_1, \ldots, R_p)]\!]^{\mathscr{B}_\alpha}.$$

PROOF. By induction on φ. Thus, for example, suppose $\varphi(Y)$ is $\exists X\psi(X, Y)$ and we have that

$$\langle \alpha; <\rangle \models \psi^*(\tilde{R}, \tilde{S}, \xi) \leftrightarrow N_{s_\xi} \subseteq [\![\langle \alpha; <\rangle \models \psi(R, S)]\!]^{\mathscr{B}_\alpha}.$$

Then $N_{s_\eta} \subseteq [\![\langle \alpha; <\rangle \models \varphi(S)]\!]^{\mathscr{B}_\alpha}$ iff $\forall\xi\exists\zeta\exists R[s_\xi \supseteq s_\eta \rightarrow s_\zeta \supseteq s_\xi \wedge \psi^*(\tilde{R}, \tilde{S}, \zeta)]$, so take this last formula for $\varphi^*(\tilde{S}, \eta)$. Note φ^* has just one more existential second order quantifier than ψ^*. Similarly for universal second order quantification and for first order quantification.

Now to prove Theorem 1, assume ZF + AC + GCH, and let $\kappa < \aleph_{\omega_1}$ be regular and have the $\bigwedge_2^1$ reflection property. Then in $V^{\mathscr{B}_\kappa}$, we have ZF + AC + $2^{\aleph_0} = \kappa < \aleph_{\omega_1}$, so it remains only to show that in $V^{\mathscr{B}_\kappa}$, κ has the $\bigwedge_2^1$ reflection property.

Let φ be $\bigwedge_2^1$. By symmetry of $\mathscr{B}_\kappa$, $[\![\langle\kappa; <\rangle \models \varphi]\!]^{\mathscr{B}_\kappa}$ is either 1 or 0. Suppose it is 1. Then $\langle\kappa; <\rangle \models \varphi^*$ (0). Thus, there is a cardinal $\lambda < \kappa$ such that $\langle\lambda; <\rangle \models \varphi^*$ (0), so $[\![\langle\lambda; <\rangle \models \varphi]\!]^{\mathscr{B}_\lambda} = 1$. Now we want to get $[\![\langle\lambda; <\rangle = \varphi]\!]^{\mathscr{B}_\kappa} \models 1$. Let φ be, for example, $\forall X \exists Y \psi(X, Y)$. Take any $\mathscr{B}_\kappa$-valued subset A of λ. A depends only on the values $[\![\xi \in A]\!]$ for $\xi < \lambda$, and, by Bockstein's theorem, each $[\![\xi \in A]\!]$ depends only on a countable number of coordinates. Thus, all the values $[\![\xi \in A]\!]$ for $\xi < \lambda$ depend only on λ coordinates; without loss of generality we may assume these are the first λ. Then, with the identification $\mathscr{B}_\lambda \subset \mathscr{B}_\kappa$, A is actually a $\mathscr{B}_\lambda$-valued subset of λ, so, since $[\![\langle\lambda; <\rangle \models \varphi]\!]^{\mathscr{B}_\lambda} = 1$, there is a $\mathscr{B}_\lambda$-valued subset, B, of λ such that $[\![\langle\lambda; <\rangle \models \psi(A, B)]\!]^{\mathscr{B}_\lambda} = 1$. Since ψ involves only quantification over ordinals $[\![\langle\lambda; <\rangle \models \psi(A, B)]\!]^{\mathscr{B}_\kappa} = 1$. A was arbitrary, so $[\![\langle\lambda; <\rangle \models \varphi]\!]^{\mathscr{B}_\kappa} = 1$, proving Theorem 1.

2. **Stronger reflection properties.** Since the reflection property considered above does not even imply that $2^{\aleph_0}$ is weakly inaccessible, it is reasonable to look for something stronger.

DEFINITION. *Let κ be a cardinal, R a relation on κ. κ has the $\bigwedge_n^1(\bigvee_n^1)$ R-*reflection property *iff for all $\bigwedge_n^1$ ($\bigvee_n^1$) sentences, φ, in R and $<$,*

$$\langle\kappa; <, R\rangle \models \varphi \rightarrow \exists\lambda < \kappa[\langle\lambda; <, R \restriction \lambda\rangle \models \varphi].$$

*κ is $\bigwedge_n^1$ ($\bigvee_n^1$)-*indescribable *iff it has the $\bigwedge_n^1$ ($\bigvee_n^1$) R-reflection property for all R.*

Note that the definition used here of $\bigwedge_n^1$ ($\bigvee_n^1$) indescribability does not include being strongly inaccessible.

If κ is $2^{\aleph_0}$, κ is not even $\bigwedge_1^1$ indescribable, since if $\langle a_\xi : \xi < \kappa\rangle$ enumerates $\mathscr{P}(\omega)$, and $R = \{\omega \cdot \xi + n : n \in a_\xi\}$, the sentence $\forall X \exists \xi \forall n_{<\omega}[X(n) \leftrightarrow R(\omega \cdot \xi + n)]$ cannot reflect to a smaller ordinal.

There is a reflection property related to $\bigwedge_2^1$ indescribability which $2^{\aleph_0}$ can satisfy. This is formulated in the following theorem.

THEOREM 2. *Suppose* ZF + AC + $\exists\kappa$ [*κ is $\bigwedge_2^1$ indescribable*] *is consistent. Then so is* ZF + AC + $2^{\aleph_0}$ *has the $\bigwedge_2^1$ R-reflection property for all R which are ordinal definable from a set of ordinals of cardinality* $<2^{\aleph_0}$.

Note that if κ just has the $\bigwedge_1^1$ reflection property for such R, it is weakly inaccessible. For if $\kappa = \lambda^+$, reflection fails for $\langle\kappa; <, \{\lambda\}\rangle$, while if $\operatorname{cof}(\kappa) = \lambda < \kappa$, reflection fails for $\langle\kappa; <, F, \{\lambda\}\rangle$, where F maps λ cofinally into κ. Similarly, κ is the κth weak inaccessible, etc.

Theorem 2 is proved similarly to Theorem 1. Assume ZF + AC + κ is $\bigwedge_2^1$ indescribable. Since $\bigwedge_2^1$ indescribability relativizes to L, assume also GCH. In $V^{\mathscr{B}_\kappa}$, we have ZF + AC + $2^{\aleph_0} = \kappa$, so it again remains only to check the reflection property.

By the usual symmetry argument, it is sufficient to check that if φ is Λ^1_2, R a $\mathscr{B}_\kappa$-valued subset of κ such that $[\![R$ is ordinal-definable from a set of ordinals of cardinality σ and $\langle\kappa; <, R\rangle \models \varphi]\!]^{\mathscr{B}_\kappa} = 1$, where $\sigma < \kappa$, then for some $\lambda < \kappa$, $[\![\langle\lambda; <, R \upharpoonright \lambda\rangle]\!]^{\mathscr{B}_\kappa} = 1$. R can depend on only σ coordinates of $\mathscr{B}_\kappa$, so, without loss of generality, we can consider it to be a $\mathscr{B}_\sigma$-valued subset of κ. By Lemma 2, $\langle\kappa; <\rangle \models \varphi^*(0, \tilde{R})$. Let λ be such that $\sigma < \lambda < \kappa$ and $\langle\lambda; <\rangle \models \varphi^*(0, \tilde{R} \upharpoonright \lambda)$. Then $[\![\langle\lambda; <, R \upharpoonright \lambda\rangle \models \varphi]\!]^{\mathscr{B}_\lambda} = 1$ (since $\tilde{R} \upharpoonright \lambda = (R \upharpoonright \lambda)^{\sim}$), and by the same argument as in the proof of Theorem 1, $[\![\langle\lambda; <, R \upharpoonright \lambda\rangle \models \varphi]\!]^{\mathscr{B}_\kappa} = 1$, proving Theorem 2.

3. **Indescribability below the continuum.** As we have seen, there are limits on the indescribability of $2^{\aleph_0}$. There does not seem to be any such limits on the indescribability of smaller cardinals. Thus, for example,

THEOREM 3. *Suppose* ZF + AC + $\exists\kappa$ [κ *is* Λ^1_n *indescribable for all* n] *is consistent. Then so is* ZF + AC + $\exists\kappa < 2^{\aleph_0}$ [κ *is* Λ^1_n *indescribable for all* n].

The proof uses the fact that once we have decided to blow up the continuum larger than κ, it does not matter how much larger, as far as second order properties of κ are concerned. Thus, similarly to Lemma 2, one can prove

LEMMA 3. *For every* Λ^1_n (V^1_n) *formula* $\varphi(X_1, \ldots, X_p)$, *there is another* Λ^1_n (V^1_n) *formula* $\varphi^\#(X_1, \ldots, X_p, x)$ *such that for any cardinals* $\alpha < \beta$, *any sequence* $R_1, \ldots, R_p$ *of* $\mathscr{B}_\alpha$*-valued relations on* α, *and any* $\xi < \alpha$,

$$\langle\alpha; <\rangle \models \varphi^\#(\tilde{R}_1, \ldots, \tilde{R}_p, \xi) \leftrightarrow N_{s_\xi} \subseteq [\![\langle\alpha; <\rangle \models \varphi(R_1, \ldots, R_p)]\!]^{\mathscr{B}_\beta}.$$

Note that we are using the identification $\mathscr{B}_\alpha \subset \mathscr{B}_\beta$, so the R_i are also $\mathscr{B}_\beta$-valued relations.

Theorem 3 now follows by starting with a cardinal, κ, which is Λ^1_n indescribable for all n, and passing to the Boolean extension $V^{\mathscr{B}_\beta}$ for any cardinal $\beta > \kappa$.

Stronger results are also possible. For example, Solovay has pointed out that if κ is measurable and u a normal ultrafilter on κ, then

$$\{\alpha < \kappa: [\![\forall m, n\ [\alpha \text{ is } \Lambda^m_n \text{ indescribable}]\!]^{\mathscr{B}_\kappa} = 1\} \in \mathscr{U}.$$

4. **Concluding remarks.** The consistency of certain reflection properties of the continuum having been established, it is natural to ask whether these properties follow from any other assumption about the continuum being large.

One such assumption is that $2^{\aleph_0} = \kappa$, where κ carries an $\aleph_1$-saturated non-trivial κ-complete ideal. It might be expected, in analogy with Scott's theorem that measurable cardinals are Λ^2_1 indescribable, that κ must have some reflection properties. Unfortunately, very little can be said about κ. Methods due to Rowbottom do show that any V^1_1 property of $\langle\kappa; <\rangle$ is shared by all $\langle\lambda; <\rangle$ such that λ is a cardinal and $\omega < \lambda < \kappa$. However, in [2] we show that it is consistent that $\langle\kappa; <\rangle$ be Λ^1_1 characterizable.

It is not known whether this last consistency result holds with κ real-valued measurable.

BIBLIOGRAPHY

1. S. Garland, *Second-order cardinal characterizability*, these Proceedings, part II.
2. K. Kunen, *Inaccessibility properties of cardinals*, Doctoral dissertation, Stanford, 1968.
3. D. Scott and R. Solovay, *Boolean-valued models of set theory*, these Proceedings, part II.

UNIVERSITY OF WISCONSIN

THE SIZES OF THE INDESCRIBABLE CARDINALS

AZRIEL LÉVY

The aim of the present paper is to provide some additional information on the size of the indescribable cardinals of Hanf and Scott [**1**]. The major tool which we shall use will be the notion of an enforceable class which is closely related to the notions of a normal class and a strongly normal class of Keisler and Tarski [**2**]. Unlike [**2**], the present paper will use mostly metamathematical ideas; therefore we shall be able to use simpler notions and to obtain also simpler proofs of the results in [**2**] concerning the sizes of the weakly compact cardinals and the measurable cardinals.

Let $\langle A, U_1, \ldots, U_k\rangle$ be a structure, i.e. A is a set and $U_1, \ldots, U_k$ are relations of any finite number of places. Contrary to the usual terminology we do not require that if U_i is an n-ary relation then $U_i \subseteq A^n$; the semantics of the appropriate language will, of course, be indifferent to the replacement of U_i by $U_i \cap A^n$. For the structure $\langle A, U_1, \ldots, U_k\rangle$ we consider the language which has constants $U_1, \ldots, U_k$ for the relations $U_1, \ldots, U_k$ (we use the same symbol for the object and for its name) and variables of every finite type. The variables of type 1 range over the objects of type 1 over A, i.e. over the members of A; the variables of type 2 range over the objects of type 2 over A, i.e. over the subsets of A; the variables of type 3 range over the objects of type 3 over A, i.e. over all sets of subsets of A, and so on. Π^n_m (Σ^n_m) will denote the set of all formulas which are logically equivalent to formulas of the form $\forall X_1 \exists X_2 \forall X_3 \cdots X_m \varphi$ (resp. $\exists X_1 \forall X_2 \exists X_3 \cdots X_m \varphi$) where $X_1, \ldots, X_m$ are variables of type $n + 1$, and φ contains only variables of type $\leq n + 1$ and quantifiers on variables of type $\leq n$. Since we do not distinguish between logically equivalent formulas we have $\Sigma^n_m, \Pi^n_m \subseteq \Sigma^{n'}_{m'}, \Pi^{n'}_{m'}$ if $n < n'$ or if $n = n'$ and $m < m'$. Ω will denote in the following any set of formulas of this language. Capital letters will denote variables of type 2, unless otherwise mentioned. If R is a unary relation, i.e. a set, we shall write the statement that x is in R as Rx or as $x \in R$; if R is a binary relation we shall write the statement that $\langle x, y\rangle$ is in R as Rxy or as xRy. We identify each ordinal number with the set of all smaller ordinals. The cardinals are identified with the initial ordinals. $|x|$ is the cardinal of the set x.

DEFINITION 1. (a) Let θ be an ordinal. A class X of ordinals is said to be *Ω-enforceable at θ* if there are relations $U_1, \dots, U_k$ and a sentence $\varphi \in \Omega$ such that φ is true in $\langle\theta, U_1, \dots, U_k\rangle$ but is false in $\langle\alpha, U_1, \dots, U_k\rangle$ for every $\alpha \in \theta - X$, $\alpha > 0$.

(b) An ordinal θ is said to be *Ω-describable* if $\theta = 0$ or if the null-class 0 is enforceable at θ. θ is said to be *Ω-indescribable* if it is not Ω-describable, i.e. if $\theta > 0$ and for every structure $\langle\theta, U_1, \dots, U_k\rangle$ and every $\varphi \in \Omega$ if φ is true in $\langle\theta, U_1, \dots, U_k\rangle$ then there is an ordinal $0 < \alpha < \theta$ such that φ is true also in $\langle\alpha, U_1, \dots, U_k\rangle$.

(c) A class X is said to be *Ω-enforceable* if it contains all the Ω-indescribable ordinals and is Ω-enforceable at every Ω-indescribable ordinal.

Our indescribable ordinals differ from those of **[1]** only in that they are not required to be inaccessible. (It will, however, be shown, that the Π^1_1-indescribable ordinals are weakly inaccessible, and therefore, if we assume the generalized continuum hypothesis, they are inaccessible.) Hanf and Scott state in [**1** (vi)] that the inaccessible Π^1_1-indescribable ordinals are the weakly compact cardinals. Our Ω-enforceable classes are, in a sense, the duals of the (strongly) normal classes of [**2**]. Our Theorems 2, 9, 10(b), 11, 12, 13 and 15 correspond to Theorems 1.11, 3.24, 1.19, 3.27, 1.21, 3.29, 1.33, 3.35, 1.27, 3.32, 1.31, and 3.33 of [**2**]. The exact relationship between the Ω-enforceable classes and the strongly normal and the normal classes of [**2**] will be discussed towards the end of the paper. At the end of the paper we shall also discuss the relationship between the notions of enforceability and describability as given here, and between the respective notions obtained by replacing θ and α in Definition 1 by $R(\theta)$ and $R(\alpha)$ respectively, where $R(\alpha)$ is defined by $R(\alpha) = \bigcup_{\beta<\alpha} P(R(\beta))$, where $P(x)$ is the power-set of x.

We shall prove that even the Π^1_1-indescribable ordinals are very big by showing that certain classes which contain only very big ordinals are Π^1_1-enforceable. E.g., we shall show that the class of all weakly inaccessible cardinals is Π^1_1-enforceable, and from this we can, of course, conclude that all the Π^1_1-indescribable ordinals are weakly inaccessible cardinals.

LEMMA 2. (a) *If X is Ω-enforceable (at θ) and $Y \supseteq X$ then Y is Ω-enforceable (at θ).*

(b) *Suppose $\Omega \subseteq \Omega'$; if X is Ω-enforceable (at θ) then X is also Ω'-enforceable (at θ), and if θ is Ω'-indescribable then θ is also Ω-indescribable.*

By Lemma 2(a) if θ is Ω-describable then *every* class X is Ω-enforceable at θ. Therefore an Ω-enforceable class is Ω-enforceable at every ordinal θ.

LEMMA 3. *If $\beta < \theta$ then the interval (β, θ) $(= \{\gamma \mid \beta < \gamma < \theta\})$ is Σ^0_1-enforceable at θ.*

PROOF. Take $U = \{\beta\}$ and $\varphi = \exists x U x$.

COROLLARY 4. *Every successor ordinal is* Σ^0_1*-describable.*

LEMMA 5. ω *is* Π^0_2*-describable.*

PROOF. Take $U = <$ and $\varphi = \forall y \exists z (y < z)$.

THEOREM 6. θ *is* Π^1_0*-indescribable if and only if* θ *is regular and* $>\omega$. (*This is, essentially* [**1**, (iv)].)

PROOF. If θ is 0 or ω or a successor ordinal then θ is Π^1_0-describable as we saw above. If θ is a singular limit number $>\omega$ let f be a function on some $\gamma < \theta$ onto a confinal subset of θ. Let $U_1 = <$, $U_2 = \{\langle \delta, f(\delta)\rangle \mid \delta < \gamma\}$, $U_3 = \{\gamma\}$ and $\varphi = \exists x[U_3 x \wedge \forall y(y < x \rightarrow \exists z U_2 yz)]$. U_1, U_2, U_3 and φ obviously enforce 0 at θ, hence θ is Π^1_0-describable. Thus we have shown that if θ is not regular or if $\theta = \omega$ then θ is Π^1_0-describable. We shall now show that if θ is regular and $>\omega$ then θ is Π^1_0-indescribable. Suppose that $\varphi \in \Pi^1_0$ and φ is true in $\langle \theta, U_1, \ldots, U_k\rangle$. Let $f_0, f_1, \ldots, f_i, \ldots, i < \omega$, be all the (first-order) Skolem functions of the structure $\langle \theta, U_1, \ldots, U_k\rangle$, and let this sequence be such that each Skolem function occurs in it infinitely many times. We define a sequence α_n of ordinals as follows. $\alpha_0 = 0$; let a_n be the set of all the values of f_n for arguments $<\alpha_n$; then we set $\alpha_{n+1} = \max(\alpha_n, \bigcup a_n)$. We assume, as an induction hypothesis, that $\alpha_n < \theta$, hence the cardinality $|a_n|$ of a_n is $\max(\aleph_0, |\alpha_n|) < \theta$. Since θ is regular also $\bigcup a_n < \theta$ and thus $\alpha_{n+1} < \theta$. Put $\alpha = \bigcup_{n<\omega} \alpha_n$. Since θ is regular and $>\omega$ we have $\alpha < \theta$. α is obviously closed under all Skolem functions, hence $\langle \alpha, U_1, \ldots, U_k\rangle$ is an elementary substructure of $\langle \theta, U_1, \ldots, U_k\rangle$. Since φ is true in the latter structure it is also true in the former.

Since every regular ordinal is a cardinal we can refer to the Π^n_m-indescribable ordinals, for $n > 0$, as the Π^n_m*-indescribable cardinals.*

Let $p(x, y)$ be the pairing function of the ordinals given by

$$\forall\alpha\exists\beta\exists\gamma(p(\beta, \gamma) = \alpha) \wedge \forall\alpha\forall\beta\forall\gamma\forall\delta(p(\alpha, \beta) < p(\gamma, \delta) \leftrightarrow [\max(\alpha, \beta) < \max(\gamma, \delta) \vee (\max(\alpha, \beta) = \max(\gamma, \delta) \wedge (\alpha < \gamma \vee (\alpha = \gamma \wedge \beta < \delta)))]).$$

It is easily seen that for every regular ordinal θ if $\alpha, \beta < \theta$ then $p(\alpha, \beta) < \theta$. Let P be the ternary relation given by $P\alpha\beta\gamma \leftrightarrow p(\alpha, \beta) = \gamma$.

LEMMA 7. *If* $n > 0$, θ *is closed under the function* p, *and* X *is* Π^n_m*-enforceable* (Σ^n_m*-enforceable*) *at* θ, *then* X *can be enforced at* θ *by a sentence* $\varphi \in \Pi^n_m$ ($\varphi \in \Sigma^n_m$) *together with the relations* $<$, P *and a single unary relation* U.

PROOF. Suppose X is enforced at θ by a sentence $\varphi \in \Pi^n_m$ and relations $U_1, \ldots, U_k, <, P$. Suppose that, for $1 \le i \le k$, U_i is an l_i-ary relation. We shall show how to reduce the number $\sum_{i=1}^k l_i + 2k$ by at least one if it is >3, and thus after a finite number of steps one is left with $<$, P and a single unary relation U.

If for some i, $l_i > 1$ then we replace U_i by

$$U_i' = \{\langle \alpha_1, \ldots, \alpha_{l_i-2}, p(\alpha_{l_i-1}, \alpha_{l_i})\rangle \mid \langle \alpha_1, \ldots, \alpha_{l_i}\rangle \in U_i\}$$

which is an $(l_i - 1)$-ary relation and we obtain from φ a sentence φ' by replacing each occurrence of $U_i x_1 \cdots x_{l_i}$ in φ by $\exists z(Px_{l_i-1}x_{l_i}z \wedge U_i'x_1 \cdots x_{l_i-2}z)$. Since $n > 0$, $\varphi' \wedge \forall x \forall y \exists z Pxyz \in \Pi_m^n$. It is easily seen that the latter sentence together with the relations $U_1, \ldots, U_i', \ldots, U_k, <, P$ enforces X at θ; thus we reduce $\sum_{i=1}^k l_i + 2k$ by 1. If $k > 1$ and for every $1 \leq i \leq k$, $l_i = 1$ then we replace the unary relations $U_1, \ldots, U_k$ by a single binary relation $U = \bigcup_{i=1}^k \{i\} \times U_i$. Let $N_i(x)$ be the formula

$$\exists x_0 \cdots \exists x_{i-1} \left(\bigvee_{j=0}^{i-2} x_j < x_{j+1} \wedge x_{i-1} < x \wedge \forall \left(z(z < x \rightarrow \bigvee_{j=0}^{i-1} z = x_j \right) \right)$$

which asserts that $x = i$. We replace in φ every occurrence of $U_i x$, $1 \leq i \leq k$, by $\exists y(N_i(y) \wedge Uyx)$ and we obtain a sentence φ'. Since $n > 0$, $\varphi' \wedge \exists y N_k(y) \in \Pi_m^n$. It is easily seen that the latter sentence, together with the relations U, $<$ and P enforces X at θ; thus we have reduced $\sum_{i=1}^k l_i + 2k$ by at least one. The same proof works for Σ_m^n.

From now on we shall denote with $*\Pi_m^n$ $(*\Sigma_m^n)$ the set of all formulas in Π_m^n (Σ_m^n) which contain no constants other than $<$, P, and the unary relation symbol U.

THEOREM 8. *For $n, m > 0$ there is a formula $\Phi(x) \in \Pi_m^n$ (Σ_m^n), with the only free variable x, which is a universal formula for all sentences $\varphi \in *\Pi_m^n$ (resp. $\varphi \in *\Sigma_m^n$) with respect to the structures $\mathfrak{A} = \langle \alpha, P, <, S, T, K, U\rangle$ where $P, <, S, T, K$ are fixed classes and α is an ordinal $\geq \omega$ closed under p; i.e. for every sentence $\varphi \in *\Pi_m^n$ $(\varphi \in *\Sigma_m^n)$ there is a finite ordinal r such that for every structure $\mathfrak{A}$ as above φ is true in $\mathfrak{A}$ if and only if r satisfies $\Phi(x)$ in $\mathfrak{A}$.*

PROOF. Let

$$S = \{\langle \beta, \gamma, \delta\rangle \mid \beta, \gamma < \omega \wedge \beta + \gamma = \delta \vee (\beta \geq \omega \vee \gamma \geq \omega) \wedge \delta = 0\},$$
$$T = \{\langle \beta, \gamma, \delta\rangle \mid \beta < \omega \wedge \beta \cdot \gamma = \delta \vee \beta \geq \omega \wedge \delta = 0\}.$$

For $\beta = p(\gamma, \delta)$ let $K_1(\beta) = \gamma$, $K_2(\beta) = \delta$. Each ordinal β is taken to represent the sequence $K_2(\beta), K_2(K_1(\beta)), K_2(K_1(K_1(\beta))), \ldots$. Since for every $\gamma > 0$, $K_1(\gamma) < \gamma$, the terms of this sequence, except finitely many, are zeros. Conversely, any sequence of ordinals all of whose terms, except finitely many, are zeros is represented in this way by some ordinal β. Let $K(l, \beta)$ denote the lth term, $l \geq 0$, of the sequence represented by β, and let $K = \{\langle l, \beta, \gamma\rangle \mid K(l, \beta) = \gamma\}$. If $\alpha \geq \omega$ and α is closed under $p(\beta, \gamma)$ then α is a limit number and is closed also under $\beta + \gamma$ and $\delta \cdot \lambda$ for finite β, γ, δ. α is also closed under $K(l, \beta)$ since for all l and β, $K(l, \beta) \leq \beta$.

First let us consider the following lemma.

(*) For $n > 0$, $m \geq 0$ there is a formula $\Psi(x, X_1, \ldots, X_m) \in \Pi_1^n \cap \Sigma_1^n$ with free variables only as indicated, where $X_1, \ldots, X_m$ are variables of type $n + 1$, such that for every formula $\psi(X_1, \ldots, X_m) \in {}^*\Pi^n_0$ with free variables only as indicated there is a finite ordinal r such that for all structures $\mathfrak{A} = \langle\alpha, <, P, S, T, K, U\rangle$, where $\mathfrak{A}\alpha$ is $\geq\omega$ and closed under p, and for all objects $X_1, \ldots, X_m$ of type $n + 1$ over α, ψ is satisfied in $\mathfrak{A}$ by $X_1, \ldots, X_m$ if and only if Ψ is satisfied in $\mathfrak{A}$ by $r, X_1, \ldots, X_m$.

The general ideas as to how to obtain the formula Ψ of (*), which is a universal formula for ${}^*\Pi^n_0$, are well known. Therefore we shall do it, at the end of the proof of the theorem, just for $n = 1$. In the meanwhile, we assume that (*) is proved and proceed to prove the theorem.

Let $\varphi \in {}^*\Pi^n_m$ be $\forall X_1 \exists X_2 \cdots X_m \psi(X_1, \ldots, X_m)$, let Ψ and r be as in (*) and let Φ be $\forall X_1 \exists X_2 \cdots X_n \Psi(x, X_1, \ldots, X_m)$. It is an immediate consequence of (*) that if $\alpha \geq \omega$ and α is closed under p then φ is true in $\mathfrak{A}$ if and only if r satisfies $\Phi(x)$ in $\mathfrak{A}$. Therefore the theorem will be proved once we show that $\Phi \in \Pi^n_m$.

We have to deal separately with the cases where m is even and where m is odd. Suppose m is even then, since $\Psi \in \Pi^n_1 \cap \Sigma^n_1$ we can take $\Psi = \exists Y\chi$, where $\chi \in \Pi^n_0$ and Y is a variable of type $n + 1$, and thus $\Phi = \forall X_1 \exists X_2 \cdots \forall X_{m-1} \exists X_m \exists Y \chi$. In order to prove $\Phi \in \Pi^n_m$ we have only to show that the existential quantifiers $\exists X_m$ and $\exists Y$ can be collapsed to a single existential quantifier without introducing quantifiers of type $n + 1$ into χ. Let us define the operations odd_n and $even_n$ on all objects of type n as follows. $\mathrm{even}_1(\beta) = 2 \cdot \beta$, $\mathrm{odd}_1(\beta) = 2 \cdot \beta + 1$, and if X is an object of type $n > 1$ then $\mathrm{even}_n(X) = \{\mathrm{even}_{n-1}(Y) \mid Y \in X\}$, $\mathrm{odd}_n(X) = \{\mathrm{odd}_{n-1}(Y) \mid Y \in X\}$. It is easily seen that if α is a limit number (and in particular, if α is closed under p) then if X is an object of type n over α then also $\mathrm{even}_n(X)$ and $\mathrm{odd}_n(X)$ are objects of type n over α, that $even_n$ and odd_n are one-one functions with disjoint ranges, and that these functions can be defined by formulas which involve only variables of type $\leq n$. Since $\chi \in \Pi^n_0$ the variables X of type $n + 1$ occur in χ only in subformulas of the form $W \in X$. Therefore the formula $\exists X_m \exists Y \chi(x, X_1, \ldots, X_m, Y)$ is equivalent to the formula $\exists Z \chi'(x, X_1, \ldots, X_{m-1}, Z)$, where χ' is obtained from χ by first replacing each subformula of the form $W \in X_m$ or $W \in Y$ by $\mathrm{even}_n(W) \in Z$ or $\mathrm{odd}_n(W) \in Z$, respectively, and by subsequent elimination of the terms $\mathrm{even}_n(W)$ and $\mathrm{odd}_n(W)$ thus introduced. As we have mentioned above, this elimination introduces no variables of type $>n$. The cases where m is odd and where we deal with Σ^n_m instead of Π^n_m are similar.

Now we are left with the task of proving (*) for $n = 1$.

When we write formulas of the language we shall use freely the terms $0, 1, 2, \ldots, p(x, y), u + v, u \cdot x, K(u, x)$. We know that the instances of their uses can be eliminated in the usual way by means of the relations $<, P, S, T$ and K for structures $\mathfrak{A} = \langle\alpha, <, P, S, T, K, \ldots\rangle$ where α is closed under p. (We do not care if these terms obtain "wrong" values for infinite u or v.) One should notice that the elimination of those terms is uniform for all structures $\mathfrak{A}$ as above and that in the elimination process the only variables added are variables of type 1. We shall also use $x < \omega$ as an abbreviation for $x = 0 \vee \exists y(x = y + 1) \wedge$

$\forall z(z < x \to z = 0 \vee \exists y(z = y + 1))$. Now that we have addition and multiplication, as well as the predicate $x < \omega$, we have at our disposal all the elementary arithmetical formulas, as formulas which use only variables of type 1.

Let $^*\Pi^1_{0,m}$ denote the set of all formulas in $^*\Pi^1_0$ with no variables of type 2 other than $X_1, \ldots, X_m$. We shall also write X_0 for U. The variables of type 1 are $v_0, v_1, \ldots$. We attach Gödel numbers to the formulas of $^*\Pi^1_{0,m}$. The Gödel number of φ is:

$3^i \cdot 5^j$	if φ is $v_i = v_j$;
$2 \cdot 3^i \cdot 5^j$	if φ is $v_i < v_j$;
$2^2 \cdot 3^i \cdot 5^j \cdot 7^k$	if φ is $Pv_iv_jv_k$;
$2^3 \cdot 3^i \cdot 5^j$	if φ is $v_i \in X_j$;
$2^4 \cdot 3^i$	if φ is $\neg\psi$ and the Gödel number of ψ is i;
$2^5 \cdot 3^i \cdot 5^j$	if φ is $\psi \wedge \chi$ and the Gödel numbers of ψ and χ are i and j respectively;
$2^6 \cdot 3^i \cdot 5^j$	if φ is $\exists v_i\psi$ and the Gödel number of ψ is j.

(We assume that the sentential connectives other than $\neg$ and $\wedge$ are obtained from $\neg$ and $\wedge$, and that $\forall v_i$ is short for $\neg\exists v_i\neg$.)

$\sigma(R)$ will be a formula which asserts that R is a satisfaction class for the set $^*\Pi^1_{0,m}$, i.e. $p(p(x, y), z) \in R$ if and only if x is the Gödel number of a formula φ of $^*\Pi^1_{0,m}$ and $z = 0$ or $z = 1$ according to whether the sequence $K(0, y), K(1, y), K(2, y), \ldots$ satisfies φ or not, where the free variable v_i assumes the value $K(i, y)$.

$$\begin{aligned}
\sigma(R) = \forall x\forall y\forall z[&(p(p(x, y), z) \in R \leftrightarrow (z = 0 \vee z = 1) \wedge \exists i\exists j\exists k(i, j, k < \omega \\
&\wedge [(x = 3^i \cdot 5^j \wedge (z = 1 \leftrightarrow K(i, y) = K(j, y))) \\
&\vee (x = 2 \cdot 3^i \cdot 5^j \wedge (z = 1 \leftrightarrow K(i, y) < K(j, y))) \\
&\vee (x = 2^2 \cdot 3^i \cdot 5^j \cdot 7^k \wedge (z = 1 \leftrightarrow p(K(i, y), K(j, y)) = K(k, y))) \\
&\vee \bigvee_{i=0}^{m} (x = 2^2 \cdot 3^i \cdot 5^j \wedge (z = 1 \leftrightarrow K(i, y) \in X_i)) \\
&\vee (x = 2^4 \cdot 3^i \wedge \exists t(p(p(i, y), t) \in R \wedge (z = 1 \leftrightarrow t = 0))) \\
&\vee (x = 2^5 \cdot 3^i \cdot 5^j \wedge \exists s\exists t(p(p(i, y), s) \in R \wedge p(p(j, y), t) \in R \wedge z = s \cdot t)) \\
&\vee (x = 2^6 \cdot 3^i \cdot 5^j \wedge \exists s(p(p(j, y), s) \in R \\
&\wedge (z = 1 \leftrightarrow \exists u(\forall l(l < \omega \wedge l \neq i \to K(l, u) = K(l, y)) \wedge p(p(j, u), 1) \in R))))])].
\end{aligned}$$

Since for a given structure $\mathfrak{A} = \langle \alpha, <, P, S, T, K, U\rangle$ where $\alpha > \omega$ and α is closed under p, and for given subsets $X_1, \ldots, X_m$ of α there is exactly one subset R of α such that $\sigma(R)$ the formulas $\forall R(\sigma(R) \to p(p(x, 0), 1) \in R)$ and $\exists R(\sigma(R) \wedge p(p(x, 0), 1) \in R)$ are equivalent. The former is obviously in Π^1_1 and the latter is in Σ^1_1. We denote either of these formulas with Ψ. To see that Ψ is as required in (*), let $\psi \in {}^*\Pi^1_{0,n}$, then if we take for r the Gödel number of ψ then $r, X_1, \ldots, X_n$ satisfy Ψ in $\mathfrak{A}$ if and only if $X_1, \ldots, X_n$ satisfy ψ in $\mathfrak{A}$.

THEOREM 9. *If $m, n > 0$ and if for each α ($\alpha < \theta$) the class X_α is Π^n_m-enforceable (at θ) then the diagonal intersection $X^D = \{\alpha \mid \alpha \in \bigcap_{\beta<\alpha} X_\beta\}$ of the classes X_α is Π^n_m-enforceable (at θ). The same holds also for Σ^n_m-enforceability.*

PROOF. It is enough to prove the local version (at θ) and to prove it only for a Π^n_m-indescribable θ. By Lemma 7, X_α is enforced at θ by $<$, P, a unary relation U_α and a formula $\varphi_\alpha \in \Pi^n_m$. Let r_α be a finite ordinal corresponding to φ_α as in Theorem 8 (i.e. a Gödel number of φ_α). Let $U = \bigcup_{\alpha<\theta} \{\alpha\} \times U_\alpha$, $V = \{\langle \alpha, r_\alpha \rangle \mid \alpha < \theta\}$. Let $\Phi(x)$ be the formula of Theorem 8 and let us write it as $\Phi(x, U)$. We denote with $\Phi(x, U_x)$ the formula obtained from it by replacing each subformula of the form $z \in U$ by Uxz. Let τ be the sentence $\forall y \exists z (y < z) \wedge \forall x \forall y \exists z Pxyz$. It is now easily seen, as a consequence of Theorem 6, Lemma 2 and Theorem 8, that X^D is enforced at θ by the relations $<$, P, S, T, K, U, V and the sentence $\tau \wedge \forall x \forall y (Vxy \rightarrow \Phi(y, U_x))$. This sentence is seen to be in Π^n_m by the standard arguments of higher-order logic. The proof for the case of Σ^n_m-enforceability is obtained from the proof we gave by replacing everywhere Π^n_m by Σ^n_m.

COROLLARY 10. (a) *If $n, m > 0$, $\rho < \theta$ and for each $\alpha < \rho$ X_α is Π^n_m-enforceable (Σ^n_m-enforceable) at θ then also $\bigcap_{\alpha<\rho} X_\alpha$ is Π^n_m-enforceable (Σ^n_m-enforceable) at θ.*

(b) *If, $n, m > 0$, ρ is less than the first Π^n_m-indescribable (Σ^n_m-indescribable) cardinal and for each $\alpha < \rho$, X_α is Π^n_m-enforceable (Σ^n_m-enforceable) then also $\bigcap_{\alpha<\rho} X_\alpha$ is Π^n_m-enforceable (Σ^n_m-enforceable).*

PROOF. (a) For $\rho \le \alpha < \theta$ put $X_\alpha = \mathrm{On}$, where On is the class of all ordinals, then $X^D = \{\alpha \mid \alpha \in \bigcap_{\beta<\alpha} X_\beta\} = \{\alpha \mid \alpha < \rho \wedge \alpha \in \bigcap_{\beta<\alpha} X_\beta\} \cup \bigcap_{\alpha<\rho} X_\alpha$. By Theorem 9 X^D is Π^n_m-enforceable (Σ^n_m-enforceable) at θ. Suppose that X^D is enforced at θ by the sentence $\varphi \in \Pi^n_m$ ($\varphi \in \Sigma^n_m$) and the relations $U_1, \ldots, U_k$. Let $V = \{\rho\}$, then $X^D \cap \{\alpha \mid \rho < \alpha\} \subseteq \bigcap_{\alpha<\rho} X_\alpha$ is enforced at θ by the sentence $\varphi \wedge \exists x (x \in V)$ and the relations $U_1, \ldots, U_k, V$. The sentence $\varphi \wedge \exists x (x \in V)$ is easily seen to be in Π^n_m (Σ^n_m). (b) follows immediately from (a).

THEOREM 11. *The class* Rg *of all regular ordinals is Π^1_1-enforceable.*

PROOF. By Theorem 6 and Lemma 2 it is enough to prove that Rg is Π^1_1-enforceable at every regular ordinal θ. Rg is enforced at θ by the relations $<$ and P and the sentence $\forall F[\forall x \forall y \exists z Pxyz \wedge (\forall x \forall y \forall z (p(x, y) \in F \wedge p(x, z) \in F \rightarrow y = z) \rightarrow \forall x \exists y \forall z (z < x \rightarrow \forall u (p(z, u) \rightarrow u < y)))] \in \Pi^1_1$.

THEOREM 12. *The class* IN *of all inaccessible cardinals is Π^1_1-enforceable at every inaccessible cardinal.*

PROOF. Let θ be inaccessible and Π^1_1-indescribable. The class In is enforced at θ by the relations P and $<$ and the sentence

$$\exists y[\exists u(u < y) \wedge \forall z(z < y \rightarrow \exists t(z < t < y))] \wedge \forall R \forall x \exists y \forall z \exists u(u < y \wedge \forall t(t < x \rightarrow (p(z, t) \in R \leftrightarrow p(u, t) \in R))) \in \Pi^1_1.$$

This is easily seen once we regard R, for a given x, as a mapping of the ordinals z on the subsets $\{t \mid t < x \wedge p(z, t) \in R\}$ of x. The second part of our sentence says that for every ordinal x and every mapping R as above there is an ordinal y such that no new subsets of x are obtained from R for values of the argument z which are $\geq y$.

THEOREM 13. *Let $n, m > 0$ and let F be an operation on classes which preserves Π^n_m- (Σ^n_m-) enforceability (at θ). Let $F^{(0)}(X) = X$ and, for $\alpha > 0$, $F^{(\alpha)}(X) = X \cap \bigcap_{\beta<\alpha} F(F^{(\beta)}(X))$, and let $F^{\Delta}(x) = \{\alpha \mid \alpha \in \bigcap_{\beta<\alpha} F^{(\beta)}(X)\}$, then also F^{Δ} preserves Π^n_m- (Σ^n_m-) enforceability (at θ).*

PROOF. First we prove, by induction on $\alpha < \theta$, that if X is Π^n_m- (Σ^n_m-) enforceable at θ then so is $F^{(\alpha)}(X)$. This follows immediately from Corollary 10(a). By Theorem 9, also $F^{\Delta}(X)$ is Π^n_m- (Σ^n_m-) enforceable at θ.

LEMMA 14. *If $X \subseteq \theta$, $\bigcup X = \theta$ and $X \cup \{\theta\}$ is a closed set then X is Π^0_2-enforceable at θ.*

PROOF. X is enforced at θ by the classes $<$ and X and the sentence

$$\forall x \exists y (y > x \wedge y \in X).$$

Let H be the operation given by $H(X) = \{\alpha \mid \alpha \in X \wedge \forall y (y \subseteq \alpha \wedge \bigcup y = \alpha \wedge y \cup \{\alpha\} \textit{ is closed} \rightarrow y \cap X \neq 0\}$. This can be called Mahlo's operation and *is* the dual of the operation M of [**2**, Definition 1.29].

THEOREM 15. *If $n > 1$ or $n = 1$ and $m > 0$ and X is Π^n_m-enforceable (at θ) so is $H(X)$. If $n > 1$ or $n = 1$ and $m > 1$ and X is Σ^n_m-enforceable (at θ) so is $H(X)$.*

PROOF. Let X be Π^n_m-enforceable at a Π^n_m-indescribable cardinal θ. We shall prove first

(*) If $y \subseteq \theta$, $\bigcup y = \theta$ and $y \cup \{\theta\}$ is a closed set then $y \cap X \neq 0$.

If y is as in (*) then, by Lemmas 14 and 2(b) y is Π^n_m-enforceable at θ and hence, by Corollary 10(a), $y \cap X \neq 0$. To prove the local version of the theorem we assume that X is enforced at θ by the sentence $\varphi \in \Pi^n_m$ and the relations $U_1, \ldots, U_k$. $H(X)$ is now enforced at θ by the relations $<, U_1, \ldots, U_k, X$ and the sentence

$$\varphi \wedge \forall Y[\forall x \exists y (y > x \wedge y \in Y) \wedge \forall z (\forall x (x < z \rightarrow \exists y (x < y < z \wedge y \in Y)) \rightarrow z \in Y) \rightarrow \exists z (z \in Y \wedge z \in X)],$$

which is seen to be in Π^n_m by the standard arguments. To prove the global version of the theorem we must still show that if X is Π^n_m-enforceable then $H(X)$ contains all the Π^n_m-indescribable cardinals. If θ is Π^n_m-indescribable then, since X is Π^n_m-enforceable, $\theta \in X$ and hence, by (*), $\theta \in H(X)$. The proof of the second half of the theorem is similar.

$H(\mathrm{Rg})$, where Rg is the class of all regular ordinals, is the class of all ρ_0-numbers of Mahlo [**3**], $H^{(2)}(\mathrm{Rg})$ is the class of all ρ_1-numbers of [**3**], and so on. Therefore, Theorems 11 and 15 entail that every Π^1_1-indescribable cardinal is a ρ_0-number (and is, a fortiori, weakly inaccessible) and is even a ρ_1-number and so on. Therefore the first Π^1_1-indescribable cardinal is greater than the first ρ_0-number (which is not a ρ_1-number), and is also greater than the first ρ_1-number (which is not a ρ_2-number), and so on. If we also apply Theorem 13 we get that all the Π^1_1-indescribable cardinals are, e.g., in the class $(H^\Delta)^\Delta(\mathrm{Rg})$, which tells us even more about how big the Π^1_1-indescribable cardinals are.

THEOREM 16. (a) *The class X is Σ^1_{m+1}-enforceable (at θ) if and only if it is Π^1_m-enforceable (at θ). θ is Σ^1_{m+1}-indescribable if and only if it is Π^1_m-indescribable.*

(b) *The class of all Π^1_m-indescribable cardinals is Π^1_{m+1}-enforceable.*

(c) *For $n > 1$ both respective classes of all the Π^n_m-indescribable ordinals and all the Σ^n_m-indescribable ordinals are Π^n_{m+1}-enforceable and Σ^n_{m+1}-enforceable.*

(d) *For $n > 1$ the class of all the Π^n_m-indescribable (Σ^n_m-indescribable) cardinals is Σ^n_m-enforceable (resp., Π^n_m-enforceable) at every inaccessible cardinal θ which is both Σ^n_m-indescribable and Π^n_m-indescribable.*

PROOF. (a) Since $\Pi^1_m \subseteq \Sigma^1_{m+1}$ we have, by Lemma 2, to prove only that if X is Σ^1_{m+1}-enforceable at θ then X is also Π^1_m-enforceable at θ. Suppose X is enforced at θ by the sentence $\exists Y\varphi(Y)$, $\varphi(Y) \in \Pi^1_m$, and the relations $U_1, \ldots, U_k$, then X is obviously enforced at θ by the sentence $\varphi(U_0) \in \Pi^1_m$ and the relations $U_0, U_1, \ldots, U_k$, where U_0 is any subset of θ which satisfies $\varphi(Y)$ in $\langle\theta, U_1, \ldots, U_k\rangle$.

(b) Let θ be Π^1_{m+1}-indescribable, let $*\Pi^1_m$ be the set of all sentences in Π^1_m which contain only the constants $<$, P and U, and let $\varphi \in *\Pi^1_m$. An ordinal $\alpha > 0$ is $\{\varphi\}$-indescribable (where $<$ and P assume the meaning agreed upon) if and only if the sentence $\forall U[\varphi \to \exists x(x > 0 \wedge \varphi^{(x)})]$, which we denote with ψ, is true in $\langle\alpha, <, P\rangle$, where $\varphi^{(x)}$ is the sentence obtained from φ by replacing the quantifiers $\exists y$, $\forall y$, $\exists Y$, $\forall Y$ by $\exists y(y < x \wedge$, $\forall y(y < x \to$, $\exists Y(Y \subseteq x \wedge$ and $\forall Y(Y \subseteq x \to$, respectively, where $Y \subseteq x$ stands for $\forall z(z \in Y \to z < x)$. By the standard arguments concerning the possibility of moving quantifiers of lower type across quantifiers of higher type, etc., we see that $\varphi^{(x)} \in \Pi^1_m$ hence $\psi \in \Pi^1_{m+1}$. Since θ itself, being Π^1_{m+1}-indescribable, is $\{\varphi\}$-indescribable the class X_φ of all $\{\varphi\}$-indescribable cardinals is enforced at θ by the sentence $\psi \in \Pi^1_{m+1}$. By Lemma 7, the class X of all Π^1_m-indescribable cardinals is just $\bigcap_{\varphi \in *\Pi^1_m} X_\varphi$. Since $*\Pi^1_m$ is denumerable we get, by Corollary 10(a), that X is Π^1_{m+1}-enforceable at θ.

(c) Here we go through arguments corresponding to those in (b) with the following differences. When we get to $\forall U[\varphi \to \exists x(x > 0 \wedge \varphi^{(x)})]$ we notice first that $\varphi^{(x)}$ is now the formula obtained from φ by replacing quantifiers on variables of types 1 and 2 as mentioned in (b); quantifiers $\exists Z$ and $\forall Z$ on a variable Z of type 3 are replaced by $\exists Z[\forall Y \forall y(Y \in Z \wedge y \in Y \to y < x) \wedge$ and $\forall Z[\forall Y \forall y(Y \in Z \wedge y \in Y \to y < x) \to$, respectively, and so on. Since $\varphi \in \Pi^n_m$ ($\varphi \in \Sigma^n_m$) the standard arguments of higher-order logic show that $\varphi^{(x)} \in \Pi^n_m$ (resp. $\varphi^{(x)} \in \Sigma^n_m$) and since U is only a variable of type 2 and $n > 1$ we get $\forall U[\varphi \to \exists x(x > 0 \wedge \varphi^{(x)})] \in \Pi^n_{m+1} \cap \Sigma^n_{m+1}$.

(d) For an inaccessible θ the set $\{b \mid \exists \gamma(\gamma < \theta \wedge b \subseteq \gamma)\}$ is of cardinality θ. Let F be a one-one function mapping θ on this set and let $R = \{\langle\alpha\beta\rangle \mid \beta \in F(\alpha)\}$. The proof of (d) proceeds now like that of (c) except that in $\varphi^{(x)}$ instead of referring to objects of types 2, 3, ... over x we refer to the ordinals which represent these objects by means of R, i.e. $\varphi^{(x)}$ is obtained from φ by replacing the quantifiers $\exists y, \forall y, \exists Y, \forall Y, \exists Z, \forall Z, \ldots$, where Z is a variable of type 3, by

$$\exists y(y < x \wedge, \forall y(y < x \rightarrow, \exists y(\forall u(Ryu \rightarrow u < x) \wedge, \forall y(\forall u(Ry \rightarrow u < x) \rightarrow,$$
$$\exists z(\forall u \forall v(Rzu \wedge Ruv \rightarrow v < x) \wedge, \forall z(\forall u \forall v(Rzu \wedge Ruv \rightarrow v < x) \rightarrow, \ldots,$$

and by replacing the atomic subformulas $u \in Y$, $Y \in Z, \ldots$ of φ by Ryu, $Rzy, \ldots$. Now $\varphi^{(x)}$ contains only variables of type 1, and therefore $\forall U(\varphi \rightarrow \exists x(x > 0 \wedge \varphi^{(x)}))$ is in Σ^n_m or Π^n_m according to whether $\varphi \in \Pi^n_m$ or $\varphi \in \Sigma^n_m$. The same applies also to the sentence

$$\forall x \exists y \forall A(\forall z(z \in A \rightarrow z < x) \rightarrow \exists u(u < y \wedge \forall z(z \in A \leftrightarrow Ruz)))$$
$$\wedge \forall U(\varphi \rightarrow \exists x(x > 0 \wedge \varphi^{(x)}))$$

which enforces the class X_φ of all $\{\varphi\}$-indescribable ordinals at θ.

Let us denote with π^n_m (σ^n_m) the least Π^n_m-indescribable (resp., Σ^n_m-indescribable) cardinal. Let $\bar{\pi}^n_m$ ($\bar{\sigma}^n_m$) be the least inaccessible Π^n_m-indescribable (Σ^n_m-indescribable) cardinal, and let $\bar{\tau}^n_m$ be the least inaccessible cardinal which is both Π^n_m-indescribable and Σ^n_m-indescribable. The existence of these cardinals cannot be proved in set theory, since the classes of which they are supposed to be the least members may be void. Therefore in the following Corollary 17 wherever it is asserted that $\beta < \gamma$ this means that if γ exists β exists and $\beta < \gamma$.

COROLLARY 17. (*This is, with minor additions*, [**1**, (v)].)
(a) $\sigma^1_{m+1} = \pi^1_m$, $\bar{\sigma}^1_{m+1} = \bar{\pi}^1_m$
(b) $\pi^1_m < \pi^1_{m+1} < \pi^2_0$, $\bar{\pi}^1_m < \bar{\pi}^1_{m+1} < \bar{\pi}^2_0$.
(c) *For* $n > 1$ $\pi^n_m, \sigma^n_m < \pi^n_{m+1}, \sigma^n_{m+1} < \pi^{n+1}_0$.
(d) *For* $n > 1$ $\bar{\pi}^n_m, \bar{\sigma}^n_m < \bar{\tau}^n_m < \bar{\pi}^n_{m+1}, \bar{\sigma}^n_{m+1} < \bar{\pi}^{n+1}_0$.

The inequalities in Corollary 17 are very strong inequalities, in the sense that wherever it is asserted that $\beta < \gamma$ it is indeed true that γ is much bigger than β. For example, in order to compare π^1_2 and π^1_1 notice that the class X of all Π^1_1-indescribable cardinals is Π^1_2-enforceable at π^1_2 and therefore we have, e.g. $\pi^1_2 \in (H^\Delta)^\Delta(X)$ and that $(H^\Delta)^\Delta(X)$ has even π^1_2 members below π^1_2.

To obtain in our framework the results of Keisler and Tarski [**2**] concerning the strongly normal and the normal classes we have to generalize Definition 1(c) as follows.

DEFINITION 18. Let Ω be a set of sentences and let A be a class of ordinals. A class X of ordinals is said to be Ω-enforceable in A if X contains all the Ω-indescribable ordinals which are in A and X is Ω-enforceable at every Ω-indescribable ordinal θ which is in A.

The following Theorem 19 generalizes some previous theorems concerning classes which are Ω-enforceable to obtain statements concerning classes which are Ω-enforceable in A. The proofs of the generalizations require only trivial changes in the proofs of the original theorems. Next to each part of Theorem 19 we shall mention which earlier theorem it generalizes.

THEOREM 19. (a) *If X is Ω-enforceable in A and $Y \supseteq X$ then Y also is Ω-enforceable in A (Lemma* 2(a)).

(b) *If $\Omega \subseteq \Omega'$ and X is Ω-enforceable in A then X is also Ω'-enforceable in A (Lemma* 2(b)).

(c) *If $n, m > 0$ and for each α the class X_α is Π^n_m-enforceable (Σ^n_m-enforceable) in A then the diagonal intersection $X^D = \{\alpha \mid \alpha \in \bigcap_{\beta<\alpha} X_\beta\}$ of the classes X_α is Π^n_m-enforceable (Σ^n_m-enforceable) in A (Theorem* 9).

(d) *If $n, m > 0$, ρ is less than the first Π^n_m-indescribable (Σ^n_m-indescribable) cardinal in A and for each $\alpha < \rho$, X_α is Π^n_m-enforceable (Σ^n_m-enforceable) in A then also $\bigcap_{\alpha<\rho} X_\alpha$ is Π^n_m-enforceable (Σ^n_m-enforceable) in A (Corollary* 10(b)).

(e) *Let $n, m > 0$ and let F be an operation on classes which preserves Π^n_m-enforceability (Σ^n_m-enforceability) in A, then also F^Δ preserves Π^n_m-enforceability (Σ^n_m-enforceability) in A (Theorem* 13).

(f) *If $n > 1$ or $n = 1$ and $m > 0$ and X is Π^n_m-enforceable in A so is $H(X)$. If $n > 1$ or $n = 1$ and $m > 1$ and X is Σ^n_m-enforceable so is $H(X)$ (Theorem* 15).

(g) *A class X is Σ^1_{m+1}-enforceable in A if and only if it is Π^1_m-enforceable in A (Theorem* 16(a)).

(h) *If A is Π^1_{m+1}-enforceable in A (i.e. if A is enforceable at every Π^1_{m+1}-indescribable cardinal which is in A) then the class of all Π^1_m-indescribable ordinals of A is Π^1_{m+1}-enforceable in A (Theorem* 16(b)).

(i) *For $n > 1$ both respective classes of all the Π^n_m-indescribable cardinals of A and of all the Σ^n_m-indescribable cardinals of A are Π^n_{m+1}-enforceable in A, if A is Π^n_{m+1}-enforceable in A, and are Σ^n_{m+1}-enforceable in A, if A is Σ^n_{m+1}-enforceable in A (Theorem* 16(c)).

It seems that the closest notion in our framework to the notion of a strongly normal class of [**2**] is the notion of a complement of a class Π^1_1-enforceable in the class IN of all inaccessible cardinals. One can prove that the complement of every class X which is Π^1_1-enforceable in IN (and, a fortiori, the complement of every Π^1_1-enforceable class X) is strongly normal. The author does not know whether the complement of every strongly normal class is Π^1_1-enforceable in IN. Theorem 12 asserts that IN is Π^1_1-enforceable in IN. When we put IN for A and Π^1_1 for Ω in Theorem 19 we get Theorems 3.24, 3.27, 3.29, 3.35, 3.32, 3.33 and 3.36 of [**2**], with "the complement of a class which is Π^1_1-enforceable in IN" replacing "strongly normal". Keisler has given a characterization of the strongly normal classes similar in its idea to that of a class Π^1_1-enforceable in IN (this has not been published yet).

A cardinal θ is said to be *measurable* if $\theta > \omega$ and there is a nonprincipal θ-complete (i.e. closed under union of $<\theta$ members) prime ideal on the power-set of θ. Every measurable cardinal is inaccessible and Π^2_1-indescribable ([**1**, (vii)]—but not every inaccessible Π^2_1-indescribable cardinal is measurable—see Vaught [**4**]). It seems that the closest notion in our framework to that of a normal class of [**2**] is the notion of the complement of a class Π^2_1-enforceable in the class M of all measurable cardinals. It can be shown that the complement of every class which is

Π_1^2-enforceable in M is normal; the author does not know whether one can prove also the converse. When we put M for A and Π_1^2 for Ω in Theorem 19 we get Theorems 1.11, 1.19, 1.21, 1.27 and 1.31 of **[2]** with "the complement of a class which is Π_1^2-enforceable in M" replacing "normal". If a class is Π_2^1-enforceable in IN then it is, a fortiori, Π_1^2-enforceable in M, which is the analogue of the fact that every strongly normal class is normal. The analogue of Theorem 3.4 of **[2]** is the fact that by Theorems 19(h) and 12 the class of all weakly compact cardinals is Π_2^1-enforceable in IN and therefore, a fortiori, Π_1^2-enforceable in M. Vaught announced in **[4]** that the least measurable cardinal is greater than all the cardinals $\bar{\pi}_m^n$ and $\bar{\sigma}_m^n$. Therefore Theorem 18 gives us some lower estimates on the size of the least measurable cardinal.

Other, closely related, notions of enforceable classes and indescribable cardinals are as follows. Let R be the function on the ordinals defined by

$$R(\alpha) = \bigcup_{\beta<\alpha} P(R(\beta)),$$

where $P(x)$ is the power set of x. Definition 20 is now a repetition of Definition 1 with $R(\theta)$ and $R(\alpha)$ replacing θ and α, respectively.

DEFINITION 20. (a) Let θ be an ordinal. A class X of ordinals is said to be *weakly-Ω-enforceable at θ* if there are relations $U_1, \ldots, U_k$ and a sentence $\varphi \in \Omega$ such that φ is true in $\langle R(\theta), U_1, \ldots, U_k\rangle$ but false in $\langle R(\alpha), U_1, \ldots, U_k\rangle$ for every $\alpha \in \theta - X$, $\alpha > 0$.

(b) An ordinal θ is said to be *weakly-Ω-describable* if $\theta = 0$ or if the null-class 0 is weakly enforceable at θ.

(c) A class X is said to be *weakly-Ω-enforceable* if it contains all the ordinals which are not weakly-Ω-describable and is weakly-Ω-enforceable at every (not weakly-Ω-describable) ordinal.

THEOREM 21. (a) *For $n \geq 1$, if X is Π_m^n-enforceable (Σ_m^n-enforceable) at θ then X is also weakly-Π_m^n-enforceable (Σ_m^n-enforceable) at θ.*

(b) *If θ is not inaccessible then θ is weakly-Π_0^1-describable.*

(c) *If Ω is Π_m^n with $n = 1$ and $m > 0$ or with $n > 1$, or if Ω is Σ_m^n with $n = 1$ and $m > 1$ or with $n > 1$, then for every inaccessible θ, X is weakly-Ω-enforceable at θ if and only if X is Ω-enforceable at θ.*

(d) *If Ω is as in* (c) *then a class X is weakly-Ω-enforceable if and only if X is Ω-enforceable in* IN.

PROOF. (a) Let Ord (x) be the formula

$$\forall y \forall z((y \in z \in x \rightarrow y \in x) \wedge (y, z \in x \rightarrow y = z \vee y \in z \vee z \in y)).$$

As is well known we have

(i) For every structure $\mathfrak{A} = \langle R(\alpha), \in, \ldots\rangle$, a member x of $R(\alpha)$ satisfies Ord (x) in $\mathfrak{A}$ if and only if x is an ordinal less than α.

For every sentence φ let $\varphi^{(\mathrm{Ord})}$ be the relativization of φ to Ord(x), i.e. the sentence obtained from φ by replacing the quantifiers $\exists y, \forall y, \exists Y, \forall Y, \exists Z, \forall Z, \ldots$

where Z is a variable of type 3, by

$$\exists y(\mathrm{Ord}(y)\wedge, \forall y(\mathrm{Ord}(y)\rightarrow, \exists Y(\forall u(u \in Y \rightarrow \mathrm{Ord}(u))\wedge,$$
$$\forall Y(\forall u(u \in Y \rightarrow \mathrm{Ord}(u))\rightarrow, \exists Z(\forall Y \forall u(u \in Y \in Z \rightarrow \mathrm{Ord}(u))\wedge,$$
$$\forall Z(\forall Y \forall u(u \in Y \in Z \rightarrow \mathrm{Ord}(u))\rightarrow, \ldots.$$

(Notice that in $\varphi^{(\mathrm{Ord})}$ the symbol $\in$ stands for two different things. In $u \in Y$, $\in$ stands for the membership relation of type theory, and in such a case we can write Yu for $u \in Y$; the $\in$ symbol inside the formula $\mathrm{Ord}(u)$ is the name of the binary relation $\in$ of the structure $\langle R(\alpha), \in, \ldots\rangle$. There is no need to use two different symbols for $\in$, since we can distinguish between the two senses of $\in$ by seeing whether it is flanked by two variables of type 1 or not.) As is easily seen we have

(ii) For every ordinal α and sentence φ, φ is true in $\langle \alpha, U_1, \ldots, U_k\rangle$ if and only if $\varphi^{(\mathrm{Ord})}$ is true in $\langle R(\alpha), \in, U_1, \ldots, U_k\rangle$. If X is Π^n_m-enforceable at θ then there is a sentence $\varphi \in \Pi^n_m$ and classes $U_1, \ldots, U_k$ such that φ is true in $\langle \theta, U_1, \ldots, U_k\rangle$ but false in $\langle \alpha, U_1, \ldots, U_k\rangle$ for every $\alpha \in \theta - X$, $\alpha > 0$. By (ii) $\varphi^{(\mathrm{Ord})}$ is true in $\langle R(\theta), \in, U_1, \ldots, U_k\rangle$ but false in $\langle R(\alpha), \in, U_1, \ldots, U_k\rangle$ for every $\alpha \in \theta - X$, $\alpha > 0$. Since, as easily seen by standard arguments of type theory, also $\varphi^{(\mathrm{Ord})} \in \Pi^n_m$, we get that X is weakly-Π^n_m-enforceable at θ. The proof for Σ^n_m is similar.

(b) By Definition 1, Corollary 4 and Lemma 5, if $\theta \leq \omega$ or θ is not regular then θ is Π^1_0-describable and hence, by (a), θ is also weakly-Π^1_0-describable. We have still to prove that if θ is such that for some cardinal $\beta < \theta$, $2^\beta \geq \theta$, where 2^β denotes cardinal exponentiation, then θ is weakly Π^1_0-describable. Since $|R(\beta)| \geq \beta$ and $R(\beta + 1) = P(R(\beta))$ we have $|R(\beta + 1)| \geq 2^\beta \geq \theta$. Let S be a one-one function on a subset u of $R(\beta + 1)$ onto θ, and let $V = \{u\}$. The void class 0 is enforced at θ by the relations $\in$, S, V and the sentence

$$\exists u[u \in V \wedge \forall x(x \in u \rightarrow \exists y Sxy)].$$

(c) If X is Π^n_m-enforceable at θ then X is weakly-Π^n_m-enforceable at θ, by (a). To prove the other direction we proceed as follows: if α is inaccessible then, as is well known, $|R(\alpha)| = \alpha$. Let F be a one-one function mapping $R(\theta)$ on θ such that for every inaccessible $\alpha < \theta$ F maps the members of $R(\alpha)$ onto the set of all ordinals $< \alpha$. Such a function F is easily constructed by induction. Suppose that X is weakly enforced at θ by $\varphi \in \Pi^n_m$ and the relations $U_1, \ldots, U_k$. For $1 \leq i \leq k$ let U'_i denote the relation which U_i induces on θ by means of F, i.e. if U_i is an l-ary relation then for all $\alpha_1, \ldots, \alpha_l$, $\langle \alpha_1, \ldots, \alpha_l\rangle \in U'_i \leftrightarrow \alpha_1, \ldots, \alpha_l < \theta \wedge \langle F(\alpha_1), \ldots, F(\alpha_l)\rangle \in U_i$. If $\alpha \leq \theta$ is inaccessible then, since F maps $R(\alpha)$ on α, the structures $\alpha, U'_i, \ldots, U'_k$ and $\langle R(\alpha), U_1, \ldots, U_k\rangle$ are isomorphic, and therefore φ is true in one of them if and only if it is true in the other. Thus φ and the relations $U'_1, \ldots, U'_k$ enforce at θ the class $X \cup (\theta - \mathrm{IN})$. By Theorem 12 the class IN is Π^1_1-enforceable at θ, hence by Corollary 10(a) also the classes

$$(X \cup (\theta - \mathrm{IN})) \cap \mathrm{IN} = X \cap \mathrm{IN} \subseteq X$$

are Π^n_m-enforceable at θ. For Σ^n_m the proof is similar.

(d) This follows directly from (b) and (c).

BIBLIOGRAPHY

1. W. P. Hanf and D. Scott, *Classifying inaccessible cardinals*, Notices Amer. Math. Soc. **8** (1961), p. 445.

2. H. J. Keisler and A. Tarski, *From accessible to inaccessible cardinals*, Fund. Math. **53** (1964), 225–308. Corrections, ibid. **57** (1965), p. 119.

3. P. Mahlo, *Zur Theorie und Anwendung der ρ_0-Zahlen*. Berichte über die Verhandlungen der Königlich Sächsischen Gesellschaft der Wissenschaften zu Leipzig, Mathematisch-Physische Klasse, **64** (1912), 108–112, and **65** (1913), 268–282.

4. R. L. Vaught, *Indescribable cardinals*, Notices Amer. Math. Soc. **10** (1963), p. 126.

THE HEBREW UNIVERSITY OF JERUSALEM

ON THE LOGICAL COMPLEXITY OF SEVERAL AXIOMS OF SET THEORY

AZRIEL LÉVY[1]

1. **Introduction and statement of the results.** We shall discuss here the question of the logical complexity of the axiom of choice, the simple and the generalized continuum hypotheses, the axiom of constructibility and some related statements of set theory. To determine the logical complexity of a statement of set theory means to find out whether the statement is "purely universal", "purely existential" and so on. The question of the logical complexity of the axiom of choice was on the mind of many mathematicians interested in the metamathematics of set theory; however, it was not much discussed, partly because there was not even a good formulation of this question, but mainly because no good tools for solving this question were available. As we shall see, our formulation of the problems mentioned above will be very simple. However, in solving these problems, we shall often make use of Cohen's method of forcing (Cohen [**1**]).

Let ZF be the set theory whose only extralogical primitive symbol is the membership symbol $\in$ and which consists of the axioms of extensionality, union, power-set, infinity, replacement and foundation [**4**, Axioms I, III, IV, VII, VIII and IX]. Let AC denote the local axiom of choice [**4**, Axiom VI]. The way to find out the logical complexity of AC is to look for the "least complicated" sentence equivalent to AC in ZF. Which measure of complexity shall we use? We could use the quantifier hierarchy of the first-order predicate calculus, but this hierarchy does not seem to be the right one for the present problem. AC can be formulated as

(1) $\forall x \exists f(f \textit{ is a function} \wedge \textit{domain}\ (f) = x \wedge (\forall y \in x)(y \neq 0 \rightarrow f(y) \in y))$.

[1] Research sponsored by the Information Systems Branch, Office of Naval Research, Washington, D.C., under Contract F-61052-67-C-0055.

Once we classify the formula inside the outer parentheses of (1) as being of the lowest degree of complexity, then we get that the degree of complexity of AC is at most $\forall\exists$, i.e. AC is at most a universal-existential statement. We shall now see that there are indeed good reasons for considering the formula inside the outer parentheses of (1) as being of the lowest degree of complexity. In the ordinary quantifier hierarchy what causes a formula to be complicated are references to objects other than the "given" objects, i.e. other than the objects denoted by the free variables, or parameters, of the formula. The formula "*f is a function*" of (1) is equivalent to the formula

$$(2)\quad (\forall y \in f)[(\exists u, v)(y = \langle u, v\rangle) \wedge \forall u \forall v \forall w(\langle u, v\rangle \in f \wedge \langle u, w\rangle \in f \rightarrow v = w)].$$

In (2) we refer to objects other than f, namely members y of f and members u, v, w of members of members of f. However, in set theory when f is "given", one considers also as given the members of f and hence also the members of the members of f and so on. Thus the question whether f is a function or not is determined only by "looking at given objects", hence the formula "*f is a function*" is of the least possible complexity. The same argument holds also for the other parts of the formula inside the outer parentheses of (1).

Using this basic idea to define a hierarchy of formulas in set theory, we proceed as follows.

DEFINITION 1. (a) Any quantifier of the type $\forall x(x \in y \rightarrow$ or $\exists x(x \in y \wedge$ is called a *restricted quantifier*. A quantifier which is not restricted is said to be *unrestricted*.

(b) A formula of the primitive language of ZF which contains no unrestricted quantifiers is said to be a *restricted formula*.

(c) Σ_n is defined to be the set of all formulas equivalent in ZF to formulas of the form $\exists x_1 \forall x_2 \exists x_3 \cdots x_n \psi$, where ψ is a restricted formula. Π_n is defined to be the set of all formulas equivalent in ZF to $\forall x_1 \exists x_2 \forall x_3 \cdots x_n \psi$, where ψ is a restricted formula.

(d) Λ_n is defined as follows. Let R be the function on the ordinals defined by $R(\alpha) = \bigcup_{\beta<\alpha} P(R(\beta))$, where $P(x)$ is the power-set of x. $R(\omega)$ is thus the set of all hereditarily finite sets. Λ_0 is taken to be the set of all formulas equivalent in ZF to $\psi(R(\omega), x_1, \ldots, x_n)$ where ψ is a restricted formula. Λ_{n+1} is the least set Γ of formulas such that

(i) If $\phi \in \Lambda_n$, then ϕ, $\exists x \phi$, $\forall x \phi \in \Gamma$.

(ii) If $\phi, \psi \in \Gamma$, then $\neg \phi$, $\phi \rightarrow \psi \in \Gamma$.

(iii) If $\phi \in \Gamma$, then $(\exists x \in y)\phi$, $(\exists x \in R(\omega))\phi$, $(\forall x \in y)\phi$, $(\forall x \in R(\omega))\phi \in \Gamma$.

(iv) If $\phi \in \Gamma$ and ψ is equivalent to ϕ in ZF, then also $\psi \in \Gamma$.

Our discussion above justifies our regarding the members of Σ_0 $(= \Pi_0)$ as the simplest formulas of set theory. The hierarchy defined here is studied in detail in [**9**]. (There Σ_n, Π_n and Λ_n are denoted with Σ_n^{ZF}, Π_n^{ZF} and $\Lambda_{R(\omega),n}^{\mathrm{ZF}}$, respectively.) It is easily seen that we have $\Sigma_0 = \Pi_0$ and $\Lambda_n \subseteq \Sigma_{n+1} \cap \Pi_{n+1} \subseteq \Sigma_{n+1}, \Pi_{n+1} \subseteq \Lambda_{n+1}$ for all $n \geq 0$ [**9**, Lemmas 1, 3, 4]. While Σ_n and Π_n are very natural, Λ_n is somewhat artificial.

Let AC_0, CH, GCH, $P(\omega) \subseteq L$ and $V = L$ denote the statements "*$P(\omega)$ can be well ordered*", the simple continuum hypothesis, the generalized continuum hypothesis, "*all subsets of ω are constructible*" and the axiom of constructibility, respectively. It can be seen, by easy checking [**9**, Theorem 34], that AC, GCH, $V = L \in \Pi_2$, AC_0, CH, $P(\omega) \subseteq L \in \Sigma_2 \cap \Pi_2$. One asks, naturally, whether these are the strongest results, e.g. whether one has also AC, GCH, $V = L \in \Sigma_2$ or not. A more far-reaching question is, for example, the following. Which are the simplest sentences ϕ which nontrivially imply AC or nontrivially follow from AC in ZF? These questions are answered by Theorems 1 and 2 below.

Let WAC denote the countable axiom of choice for pairs, i.e. the statement that for every countable set A of unordered pairs there exists a function f on A such that $f(x) \in x$ for every $x \in A$. Let DC denote the axiom of (countably many) dependent choices.

THEOREM 1. *Let ϕ be any sentence in Σ_2.*
(a) *If* $ZF \vdash \phi \rightarrow WAC$, *then* $ZF + AC \vdash \neg\phi$; *in particular, if* $ZF \vdash \phi \rightarrow AC$, *then* $ZF \vdash \neg\phi$.
(b) *If* $ZF + AC \vdash \phi \rightarrow GCH$, *then* $ZF + AC \vdash \neg\phi$.
(c) *If* $ZF + GCH \vdash \phi \rightarrow V = L$, *then* $ZF + GCH \vdash \neg\phi$.

Theorem 1 asserts that in ZF no sentence $\phi \in \Sigma_2$ implies AC, GCH or $V = L$ unless it does so for the trivial reason that ϕ is refutable. Assuming the consistency of ZF, which by Gödel [**5**] implies the consistency of $ZF + V = L$, it follows immediately from Theorem 1 that

(a) WAC, $AC \notin \Sigma_2$.
(b) GCH is not equivalent in ZF + AC to any sentence in Σ_2 (and a fortiori $GCH \notin \Sigma_2$).
(c) $V = L$ is not equivalent in ZF + GCH to any sentence in Σ_2 (and a fortiori $V = L \notin \Sigma_2$).

THEOREM 2. *Let ϕ be any sentence in Λ_1.*
(a) *If* $ZF \vdash \phi \rightarrow AC_0$, *then* $ZF + DC \vdash \neg\phi$.
(b) *If* $ZF + AC \vdash \phi \rightarrow CH$, *then* $ZF + AC \vdash \neg\phi$.
(c) *If* $ZF + GCH \vdash \phi \rightarrow P(\omega) \subseteq L$, *then* $ZF + GCH \vdash \neg\phi$.

Part (a) of Theorem 2 would be stronger if the conclusion were $ZF \vdash \neg\phi$ rather than $ZF + DC \vdash \neg\phi$, but the author does not know whether this stronger statement is true. Theorem 2 establishes that if ZF is consistent, then

(a) $AC_0 \notin \Lambda_1$.
(b) CH is not equivalent in ZF + AC to any sentence in Λ_1.
(c) $P(\omega) \subseteq L$ is not equivalent in ZF + GCH to any sentence in Λ_1.

When we ask now whether the various statements considered above *imply* statements simpler than themselves, we see first that each of $V = L$, GCH and AC implies the simpler statement AC_0, which is not a theorem of ZF (by Cohen [**1**]); each of $V = L$ and GCH implies the simpler statement CH, which is not even a theorem of ZF + AC; and $V = L$ implies the simpler statement $P(\omega) \subseteq L$ which is not even a theorem of ZF + GCH. What we can still get is the following theorem.

THEOREM 3 [**9**, THEOREM 44]. *Let ϕ be any sentence in Λ_1. If* ZF $\vdash V = L \rightarrow \phi$, *then* ZF + DC $\vdash \phi$.

As a consequence of this theorem, if $\phi \in \Lambda_1$ and ZF $\vdash A \rightarrow \phi$, where A is any one of AC, GCH, $V = L$, AC_0, CH, $P(\omega) \subseteq L$, then ZF + DC $\vdash \phi$. We know how to strengthen the conclusion to ZF $\vdash \phi$ only for $\phi \in \Sigma_1$ [**9**, Theorem **44**].

Results which are closely related to Theorem 3, and which improve the results of this theorem in several directions, are those of Platek [**12**]. Platek classifies ϕ according to its position in the quantifier hierarchy of type theory. In comparing Platek's results to Theorem 3, the reader should notice that using methods like those of [**9**, Theorems 36 and 43] one can easily show that every sentence in Λ_0 is equivalent in ZF + DC to a sentence in the Boolean closure of Σ_2^1 of type theory.

The hierarchy we considered is based on the idea that once a set is "given" so are all its members. But this is by no means the only acceptable point of view. It is also common to regard the axiom of constructibility as being "universal", while according to the point of view discussed above it is "universal-existential". The idea behind regarding the axiom of constructibility as "universal" is that in addition to regarding all the members of a "given" set as "given", one regards also all the ordinals as "given". To deal with such questions we use model-theoretic ideas. We get in §4 that from the point of view which we have just described $V = L$ is "universal" but none of GCH and AC is even "existential-universal".

2. **Proof of Theorem 1.** We shall follow, in a great part of the proof, [**10**, §2]. As was done there, we shall first consider a set theory which we shall call ZFM. The present ZFM is analogous to, but not identical with, the system ZFM of [**10**]. The language of ZFM is the language of ZF with the additional individual constants M, X_0, θ and f_0. For a formula ϕ of ZF, we denote with $\phi^{(z)}$ the relativization of ϕ to z, i.e. the formula obtained from ϕ by replacing all the quantifiers $\forall x$ and $\exists x$ in it by $\forall x(x \in z \rightarrow$ and $\exists x(x \in z \wedge$, respectively. The axioms of ZFM are the axioms of ZF + AC together with (3)–(7) below.

(3) $\phi \leftrightarrow \phi^{(M)}$, for every sentence ϕ of ZF.

(4) *M is transitive* $(x \in y \in M \rightarrow x \in M)$.

(5) *θ is a regular cardinal* $\wedge\ \theta \in M \wedge |M| = \theta$ (where $|x|$ denotes the cardinality of x).

(6) $f_0 \in M \wedge f_0$ *is a one-one function which maps θ onto the set* $\{b \mid (\exists \gamma < \theta)(b \subseteq \gamma)\}$.

(7) $x_0 \in M \wedge \chi(x_0)$, where $\chi(x)$ is a fixed formula of ZF with no free variables other than x.

LEMMA 4. *If ψ is a sentence of* ZF *such that* ZFM $\vdash \psi$, *then* ZF + AC $\vdash$ $\exists \alpha(\alpha$ *is a regular cardinal* $\wedge (\forall \beta < \alpha)(2^\beta \leq \alpha) \wedge \exists x(|Tc(x)| \leq \alpha \wedge \chi(x))) \rightarrow \psi$, *where* 2^β *denotes cardinal exponentiation (i.e.* 2_β *is the cardinality of the set of all functions on* β *into* 2), *and* $Tc(x)$ *denotes the transitive closure of* x, *i.e. the set whose members are* x, *the members of* x, *the members of the members of* x *and so on.*

PROOF. Assume ZFM $\vdash \psi$. Then ψ follows in ZF + AC from finitely many of the additional axioms of ZFM, therefore for some sentences $\phi_1, \ldots, \phi_k$ of ZF we have ZF + AC $\vdash [\bigwedge_{i=1}^k(\phi_i \leftrightarrow \phi_i^{(M)}) \wedge M \textit{ is transitive} \wedge \theta \textit{ is a regular cardinal} \wedge \theta \in M \wedge |M| = \theta \wedge f_0 \in M \wedge f_0 \textit{ is a one-one function which maps } \theta \textit{ on } \{b \mid (\exists\gamma < \theta)(b \subseteq \gamma)\} \wedge x_0 \in M \wedge \chi(x_0)] \rightarrow \psi$.

Since M, x_0, θ and f_0 do not occur in ψ we have, by the rules of logic, ZF + AC $\vdash$ $\exists z\exists x\exists\alpha\exists f\,[\bigwedge_{i=1}^k(\phi_i \leftrightarrow \phi_i^{(z)}) \wedge z \textit{ is transitive} \wedge \alpha \textit{ is a regular cardinal} \wedge \alpha \in z \wedge |z| = \alpha \wedge f \in z \wedge f \textit{ is a one-one function which maps } \alpha \textit{ on } \{b \mid (\exists\gamma < \alpha)(b \subseteq \gamma)\} \wedge x \in z \wedge \chi(x)] \rightarrow \psi$. The conclusion of the lemma follows now readily once we prove

$$\text{(8)} \qquad \begin{aligned} &\text{ZF + AC} \vdash \forall\alpha\forall x[\alpha \textit{ is a regular cardinal} \\ &\wedge (\forall\beta < \alpha)(2^\beta \leq \alpha) \wedge |Tc(x)| \leq \alpha] \rightarrow \exists z\exists f \\ &\left[\bigwedge_{i=1}^k (\phi_i \leftrightarrow \phi_i^{(z)}) \wedge z \textit{ is transitive} \wedge \alpha \in z\right. \\ &\wedge |z| = \alpha \wedge f \in z \wedge f \textit{ is a one-one function} \\ &\left.\textit{which maps } \alpha \textit{ on } \{b \mid (\exists\gamma < \alpha)(b \subseteq \gamma)\}\right]. \end{aligned}$$

It follows easily from Montague [**11**, Theorem 1] (see the proof of [**8**, Theorem 2]), which is a version of the Skolem-Löwenheim theorem, that

$$\text{(9)} \qquad \text{ZF} \vdash \forall a\exists u\left(a \subseteq u \wedge \bigwedge_{i=1}^k (\phi_i \leftrightarrow \phi_i^{(u)})\right).$$

Let α and x be as in (8); then there is a one-one function f which maps α on $\{b \mid (\exists\gamma < \alpha)(b \subseteq \gamma)\}$. Let us choose for $\underline{a}$ in (9) the set $\alpha \cup \{\alpha\} \cup Tc(x) \cup Tc(f)$; by what we know about α, x and f the cardinality of this set is α. By the extended Skolem-Löwenheim theorem the structure $\langle u, \epsilon\rangle$ has an elementary substructure $\langle v, \epsilon\rangle$ such that $a \subseteq v$ and $|v| = \alpha$. Since $a \subseteq v$ we have $\alpha, f, x \in v$ and since $\langle v, \epsilon\rangle$ in an elementary substructure of $\langle u, \epsilon\rangle$ we have, by (9), $\bigwedge_{i=1}^k(\phi_i \leftrightarrow \phi_i^{(v)})$. We define now a function g on v by $g(t) = \{g(y) \mid y \in t \wedge y \in v\}$; by the axiom of foundation there is a unique such g. We can assume, without loss of generality, that ϕ_1 is the axiom of extensionality, and therefore, since $\phi_1^{(v)}$ holds, g is one-one. We put $z = \{g(t) \mid t \in v\}$. It is easily seen that since $\underline{a}$ is transitive, $g(t) = t$ for each $t \in a$, hence $\alpha \cup \{\alpha\} \cup Tc(x) \cup Tc(f) \subseteq z$. Since $|v| = \alpha$ also $|z| = \alpha$. z is obviously transitive. For $y_1, y_2 \in v$, $y_1 \in y_2 \leftrightarrow g(y_1) \in g(y_2)$, therefore g is an isomorphism of $\langle v, \epsilon\rangle$ on $\langle z, \epsilon\rangle$ and thus we have for every sentence ϕ of ZF $\phi^{(v)} \leftrightarrow \phi^{(z)}$. We now have established the conclusion of (8), which is what was left to be proved.

We shall first prove part (c) of Theorem 1. Let ϕ be a sentence in Σ_2, i.e. ϕ is equivalent to $\exists x\,\forall y\phi_0(x, y)$, where $\phi_0(x, y)$ is a restricted formula. We assume that ZF + GCH $\vdash \phi \to V = L$. Let us denote $\forall y \phi_0(x, y)$ with $\chi(x)$. We shall prove that for this $\chi(x)$ there is a contradiction in ZFM + GCH. Once we have done that we get, by Lemma 4,

$$\text{ZF} + \text{AC} \vdash \exists\alpha(\alpha \textit{ is a regular cardinal} \wedge (\forall\beta < \alpha)(2^\beta \leq \alpha) \\ \wedge \exists x(|Tc(x)| \leq \alpha \wedge \chi(x))) \to (\text{GCH} \to \exists x(x \neq x)).$$

Since

$$\text{ZF} + \text{GCH} \vdash \forall x \, \exists\alpha(\alpha \textit{ is a regular cardinal} \wedge (\forall\beta < \alpha)(2^\beta \leq \alpha) \wedge |Tc(x)| \leq \alpha)$$

and since GCH implies AC we get ZF + GCH $\vdash \exists x\chi(x) \to \exists x(x \neq x)$, i.e. ZF + GCH $\vdash \neg\exists x\chi(x)$, ZF + GCH $\vdash \neg\phi$.

Now we have to prove a contradiction in ZFM + GCH. Let us outline the idea of the proof. We start with the "model" M of ZF in ZFM and extend it, by Cohen's method, to a "model" N of ZF + GCH in which $V = L$ fails.[2] $x_0 \in M \subseteq N$ and since we have $\forall y \phi_0(x_0, y)$ we have, a fortiori, $(\forall y \in N)\phi_0(x_0, y)$. $\phi_0(x, y)$, being a restricted formula, in absolute with respect to the transitive set N (see, e.g. [**9**, Lemma 34]), i.e. we have $\phi_0(x, y) \leftrightarrow \phi_0^{(N)}(x, y)$, and thus we have $(\exists x \forall y \phi_0(x, y))^{(N)}$. Since N is a model of ZF + GCH $(\exists x \forall y \phi_0(x, y))^{(N)}$ implies $(V = L)^{(N)}$, which contradicts our earlier statement that $V = L$ fails in N.

In constructing N and in establishing its properties we apply Cohen's method as in [**10**, §2]. There, equality was taken to be a defined notion, but if the reader wishes he can consider equality as primitive and introduce slight changes in the treatment, as is apparent from the treatment of [**6**, §4]. In [**10**, §2] new subsets of ω were introduced, but here, if $\theta < \omega$, M contains already all the subsets of ω as well as all the subsets of all ordinals $\gamma < \theta$, and therefore we shall introduce here a new subset of θ. Since we introduce now only a single subset of θ we need only a *unary* predicate A (which means that we choose $\tau = 1$ in [**10**]). We now take up Definitions 15–18 of [**10**]. A *condition* is now a function on a subset of θ of cardinality $<\theta$ into $\{0, 1\}$. If $p \subseteq p'$ we say that p' is an *extension* of p. The notion of forcing is defined as in [**10**, Definition 20–22] with θ replacing ω in the appropriate places in [**10**, Definition 21 (f)]. The treatment proceeds now as in [**10**]. When we come to prove the analogue of [**10**, Lemma 44] that every condition p can be extended to a generic function P, we make use of the facts that since $|M| = \theta$ there are only θ formulas in the M-language, and also that the union of an ascending sequence of $<\theta$ conditions is again a condition, by the regularity of θ. We finally get a transitive set $N \supseteq M$ which is a model of ZF.

Let us now show that GCH holds in N. Let α be an infinite cardinal of N. Since the notion of an infinite ordinal is absolute with respect to transitive models, α is an infinite ordinal (of set theory) and α is also an infinite ordinal of M (by the analogue of [**10**, Lemma 51] which says that M and N have the same ordinals).

---

[2] The idea of using forcing to extend general transitive models M, rather than only denumerable models M as done by Cohen [**1**] was known to several people and is not due to the author. The author originally obtained the results of this paper by a somewhat different method, without using nondenumerable models M.

Since M contains fewer mappings than N α is also a *cardinal* of M. Let β be the cardinal of M which immediately follows α. In proving that in N the power set of α is of the cardinality which immediately follows α we shall deal separately with the cases $\alpha < \theta$ and $\alpha \geq \theta$. In the first case, where $\sigma < \theta$, we have, by (4) and (6), $P(\alpha) \subseteq M$; and since GCH holds in M, by GCH (of ZFM + GCH) and (3), M contains a function g which is a one-one mapping of β on $P(\alpha)$. Since $M \subseteq N$ we also have $g \in N$, i.e. $P(\alpha)$ is equinumerous in N to β which is the cardinal following α in M. Since in N the cardinality of $P(\alpha)$ is at least the next cardinal in N, which is $\geq \beta$, we have that β is the next cardinal in N and that $2^\alpha = \beta$ in N. Now let us consider the case where $\alpha \geq \theta$. The power-set of α in N is $P(\alpha) \cap N$. Since the valuation function $\text{val}_P(u)$ ([**10**, Definition 45]) is absolute with respect to N we can construct in N the set t such that for each member x of $P(\alpha) \cap N$ it contains all the terms u of least rank such that $\text{val}_P(u) = x$. Since all the terms are in M, $t \subseteq M \subseteq N$ and $t \subset N$. The ranks of the members of t are bounded in N by an ordinal γ of N; and by the absoluteness of the rank function, if R_γ is the set of M which contains all members of M whose rank is $<\gamma$, then $t \subseteq R_\gamma$. Since AC is a theorem of ZFM + GCH, R_γ can be well ordered in M and, a fortiori, in N. This gives, of course, a well ordering of t and of $P(\alpha) \cap N$ in N. Let us define now, in M, for every term w

$$[w] = \{\langle p, \lambda\rangle \mid \lambda < \alpha \wedge p \Vdash \boldsymbol{\lambda} \in w\}.$$

If $\text{val}_P(w_1) \cap \alpha \neq \text{val}_P(w_2) \cap \alpha$, then $[w_1] \neq [w_2]$ (since for some $\lambda < \alpha$ and some $p \subseteq P$ we have $p \Vdash \boldsymbol{\lambda} \in w_1$ but not $p \Vdash \boldsymbol{\lambda} \in w_2$, or $p \Vdash \boldsymbol{\lambda} \in w_2$ but not $p \Vdash \boldsymbol{\lambda} \in w_1$). Therefore the function h of N on the set $W = \{[w] \mid w \textit{ is a term}\}$ of M given by $h([w]) = \text{val}_P(w) \cap \alpha$ is a well-defined function of N. Let us estimate the cardinality of W in M. $W \subseteq P(C \times \alpha) \cap M$, where C is the set of all conditions. In M we have, by (6), $|C| = \theta$; and since $\alpha \geq \theta$ we get $|C \times \alpha| = \alpha$ in M. We have the GCH in M, therefore $|P(C \times \alpha) \cap M| = \beta$ in M, where β is the cardinal following α in M, hence $|W| \leq \beta$ in M. Using now the function h above we get in N a mapping of β on the set $P(\alpha) \cap N$, hence $|P(\alpha) \cap N| \leq \beta$ in N, which establishes that β is the cardinal successor of α also in N and that $2^\alpha = \beta$ also in N.

Finally, one proves as in [**10**, (56) and Lemma 55] that in N θ has a subset which is not constructible, namely the set $\text{val}_P(\lambda_\theta x A(x))$. Therefore $V = L$ does not hold in N and this yields a contradiction in ZFM + GCH, as mentioned above.

Now we shall prove part (b) of Theorem 1. Let ϕ be $\exists x \forall y \phi_0(x, y)$, where ϕ_0 is a restricted formula. We assume that ZF + AC $\vdash \phi \rightarrow$ GCH. We denote, again, $\forall y \phi_0(x, y)$ with $\chi(x)$, and we shall prove, for this $\chi(x)$, a contradiction in ZFM + GCH. We shall construct a set $N \supseteq M$ similar to the set N which we constructed above. We shall prove that N is a model of ZF + AC. We shall also prove that GCH does not hold in N, by showing that 2^θ in N is not the cardinal successor of θ in N. On the other hand, we have $\exists x \forall y \phi_0(x, y)$ in N, as above, which is a contradiction since ZF + AC $\vdash \phi \rightarrow$ GCH. The construction of N

proceeds as follows. Let γ be a cardinal of M which is greater than the cardinal successor of θ in M. The notion of forcing will be as in [**10**] except that we replace ω by θ and choose β for τ (in [**10**, Definition 19]). The proof that N is a model of ZF is as usual. The proof that AC holds in N which we gave above for part (c) of Theorem 1 holds also in our present case. As in Cohen [**1**, Chapter IV, §10] one can prove that all the cardinals of M which are $\geq\theta$ (and, in fact, *all* the cardinals of M) are also cardinals of N. By the very construction of N it is obvious that $2^\theta = \beta$ in N. By what we have said just now about the cardinals of N, β is not the cardinal successor of θ in N.

To prove part (a) of Theorem 1 we start as in parts (b) and (c) and proceed to show, for the same choice of $\chi(x)$, that ZFM is inconsistent. This is done by constructing a set N, similar to those which we constructed above, and by showing that N is a model of ZF and yet WAC does not hold in N. We shall indicate briefly how we define the ramified language and the forcing relation in order to obtain N. The ramified language contains constants $a^i_{n,\mu}$, b^i_n, c_n and d, where $i < 2$, $n < \omega$ and $\mu < \theta$. The $a^i_{n,\mu}$ stand for generic subsets of θ, b^i_n stands for $\{a^i_{n,\mu} \mid \mu < \theta\}$, c_n for $\{b^0_n, b^1_n\}$ and d for $\{\langle n, c_n\rangle \mid n < \theta\}$. The forcing conditions are the function on subsets of $\omega \times \theta \times 2$ of cardinality $<\theta$ into $\{0, 1\}$. The forcing relation is defined in the appropriate way. We prove that in N the sequence denoted by d is a sequence of length ω of pairs for which there is no choice function, by the methods of Cohen [**1**] and Feferman [**3**]. Thus WAC fails in N.

3. **Proof of Theorem 2.** We proved the three parts of Theorem 1 by constructing in ZFM (or in ZFM + GCH) a model N in which ϕ held and yet the assumed consequence of ϕ failed. Since the model $N \supseteq M$ had to contain x_0 we could not control the behavior of N at small cardinals, and we could make $V = L$, or GCH, or AC, fail only at θ, which was enough for our purpose. When we come to prove Theorem 2 we have to make those axioms fail already at ω, and therefore we have to deal here with denumerable models M. We required that $x_0 \in M$ so that ϕ will hold in N; here we shall make sure that ϕ holds in N by similar means.

Now ZFM will denote a theory whose language is the language of ZF with the additional individual constant M, and whose axioms are the axioms of ZF together with the axiom DC of dependent choices as well as (10)–(12) below.

(10) $\phi \leftrightarrow \phi^{(M)}$, for every sentence ϕ of ZF.

(11) M *is transitive* $\wedge\ |M| = \aleph_0$.

(12)
$$\forall x_1 \forall x_2 \cdots \forall x_{k_i}(x_1, \ldots, x_{k_i} \in R(\omega) \wedge \exists x \psi_i(x, x_1, \ldots, x_{k_i}) \rightarrow \exists x(x \in M \wedge \psi_i(x, x_1, \ldots, x_{k_i}))), \quad i = 1, \ldots, l,$$
where $\psi_1, \ldots, \psi_l$ are l fixed restricted formulas with no free variables other than indicated.

LEMMA 5. *If ψ is a sentence of* ZF *such that* ZFM $\vdash \psi$, *then* ZF + DC $\vdash \psi$.

PROOF. Since ZFM $\vdash \psi$ we have, for some sentences $\phi_1, \ldots, \phi_m$ of ZF,

$$\mathrm{ZF} + \mathrm{DC} \vdash \bigwedge_{i=1}^{m} (\phi_i \leftrightarrow \phi_i^{(M)}) \wedge M \textit{ is transitive} \wedge |M| = \aleph_0 \wedge$$

$$\bigwedge_{i=1}^{l} x_1 \cdots x_{k_i}(x_1, \ldots, x_{k_i} \in R(\omega) \wedge \exists x \psi_i(x, x_1, \ldots, x_{k_i})$$

$$\rightarrow \exists x(x \in M \wedge \psi_i(x, x_1, \ldots, x_{k_i}))) \rightarrow \psi$$

and therefore

$$\mathrm{ZF} + \mathrm{DC} \vdash \exists z\Big[\bigwedge_{i=1}^{m} (\phi_i \leftrightarrow \phi_i^{(z)}) \wedge z \textit{ is transitive} \wedge |z| = \aleph_0 \wedge$$

$$(13) \qquad \bigwedge_{i=1}^{l} \forall x_1 \cdots \forall x_{k_i}(x_1, \ldots, x_{k_i} \in R(\omega) \wedge \exists x \psi_i(x, x_1, \ldots, x_{k_i})$$

$$\rightarrow \exists x(x \in z \wedge \psi_i(x, x_1, \ldots, x_{k_i})))\Big] \rightarrow \psi.$$

Thus the lemma will be proved once we prove in ZF + DC the existence of a z as in (13). By [**9**, Theorem 43] we have $x_1, \ldots, x_{k_i} \in R(\omega) \wedge \exists x \psi_i(x, x_1, \ldots, x_k) \rightarrow \exists x(x \in L \wedge \mathrm{Od}(x) < \omega_{1,L} \wedge \psi_i(x, x_1, \ldots, x_{k_i}))$ where $\mathrm{Od}(x)$ is the constructible order of x (Gödel [**5**, 9.42]) and $\omega_{1,L}$ is the ordinal which is ω_1 in the model L. Since $x_1, \ldots, x_{k_i} \in R(\omega) \subseteq L$ and since the $\psi_i(x, x_1, \ldots, x_{k_i})$ are absolute, the function $h_i(x_1, \ldots, x_{k_i})$ defined by

$$h_i(x_1, \ldots, x_{k_i}) = \textit{the least } \lambda \textit{ such that } \psi_i(F(\lambda), x_1, \ldots, x_{k_i}) \textit{ if there is such a } \lambda, \textit{ and } 0 \textit{ otherwise,}$$

where F is the function of Gödel [**5**, 9.3], is a constructible function. Since AC holds in the model L the range of h_i has an upper bound $\gamma_i < \omega_{1,L}$. Let $\gamma = \max(\gamma_1, \ldots, \gamma_l)$. Let $F''\gamma = \{F(\delta) \mid \delta < \gamma\}$; by [**5**, 9.5] $F''\gamma$ is a countable transitive set and we have

$$(14) \qquad \forall x_1 \cdots \forall x_{k_i}(x_1, \ldots, x_{k_i} \in R(\omega) \wedge \exists x \psi_i(x, x_1, \ldots, x_{k_i}) \rightarrow \exists x(x \in F''\gamma \wedge \psi_i(x, x_1, \ldots, x_{k_i}))) \quad i = 1, \ldots, l.$$

By the construction of Montague [**11**] mentioned in §2, and by the Skolem-Löwenheim theorem [**9**, Theorem 59] we obtain a set u which includes $F''\gamma$ and which satisfies all the requirements for z in (13) other than the transitivity of z. We can assume, without loss of generality, that ϕ_1 is the axiom of extensionality, and therefore there is a unique ϵ-isomorphism of u on a transitive set z. This isomorphism is the identity on the transitive set $F''\gamma$. The set z thus obtained satisfies the requirements of (13) by what we have just said, and by (14).

We shall now prove part (c) of Theorem 2. We assume that $\phi \in \Lambda_1$ and ZF + GCH $\vdash \phi \rightarrow P(\omega) \subseteq L$. We assume also that all the universal quantifiers $\forall x$ in ϕ which are not of the form $\forall x(x \in y \rightarrow$ or $\forall x(x \in R(\omega) \rightarrow$ were replaced by $\neg\exists x\neg$. Every subformula of ϕ of the form $\exists x\chi$, where χ does not start with $x \in y \wedge$ or with $x \in R(\omega) \wedge$ is such that $\chi \in \Lambda_0$. As mentioned in [**9**, bottom of p.

59] $\exists x\chi$ is equivalent in ZF to a formula $\exists x\chi'$, where χ' is a restricted formula, therefore we can assume that all these $\exists x\chi$ have been replaced by the appropriate formulas $\exists x\chi'$. We choose now for the formulas $\exists x\psi_i$, $i = 1, \ldots, l$, in (12) all the subformulas of ϕ of the form $\exists x\chi'$ as mentioned above. We extend M to a model of N by adding a generic subset of ω; this subset of ω is nonconstructible (see Cohen [**1**, IV, §6] or the proof of Theorem 1(c) here). One shows that all the axioms of ZF, as well as GCH, hold in N. Since $P(\omega) \subseteq L$ fails in N and since we assume that ZF + GCH $\vdash \phi \rightarrow P(\omega) \subseteq L$, we have ZFM + GCH $\vdash \neg\phi^{(N)}$. It follows easily from (12) that ZFM $\vdash \phi \leftrightarrow \phi^{(N)}$, hence ZFM + GCH $\vdash \neg\phi$. Thus we have, by Lemma 5, ZF + GCH $\vdash \neg\phi$.

To prove part (b) of Theorem 2 we make the same choice of ψ_i's as above, and then we construct within ZFM + AC a model N in which $2^{\aleph_0} > \aleph_1$, as done by Cohen in [**1**, IV, §8] or in the proof of Theorem 1(b). We now get as above, using our assumptions on ϕ, ZFM + AC $\vdash \neg\phi$. By Lemma 5 this implies ZF + AC $\vdash \neg\phi$.

To prove part (a) of Theorem 2 we again make the same choice of the ϕ_i's. We get N from M by adding to M denumerably many generic subsets of ω as in Feferman [**3**, §4]. We obtain now as above, using our present assumptions on ϕ, ZFM $\vdash \neg\phi$ and hence, by Lemma 5, ZF + DC $\vdash \neg\phi$.

4. **Considering all ordinals as given.** In this section we shall refer to the formulas which we consider to be analogues of formulas equivalent to universal, existential, and existential-universal formulas as $\forall$-formulas, $\exists$-formulas and $\exists\forall$-formulas, respectively.

The characteristic property of statements ϕ equivalent to universal sentences in a first-order theory T is that whenever ϕ holds in a structure $\mathfrak{A}$ which is a model of T it holds also in every substructure $\mathfrak{B}$ of $\mathfrak{A}$ which is a model of T (A. Robinson [**13**, 3.3.2]). Our present approach is that whenever a set is given also all its members are given, and also that all the ordinals are given; therefore we have to consider here only transitive substructures $\mathfrak{B}$ of $\mathfrak{A}$ which contain all ordinals of $\mathfrak{A}$ and which are models of ZF; we shall refer to these substructures as *admissible substructures*. (A substructure $\mathfrak{B} = \langle B, S\rangle$ of $\mathfrak{A} = \langle A, R\rangle$ is said to be *transitive* if whenever $x \in B$ and $y\, R\, x$, then also $y \in B$.) Thus the $\forall$-sentences are the sentences ϕ such that whenever ϕ holds in a model $\mathfrak{A}$ of ZF, ϕ holds also in every admissible substructure of $\mathfrak{A}$.

We shall now see that $V = L$ is indeed an $\forall$-statement. If $\mathfrak{B} = \langle B, S\rangle$ is an admissible substructure of $\mathfrak{A} = \langle A, R\rangle$, then the class L of $\mathfrak{A}$ is included in B (by Shepherdson [**15**, 3.3.2]). If $V = L$ holds in $\mathfrak{A}$ then $A = L$ and hence also $B = L$; i.e. $\mathfrak{B}$ coincides with $\mathfrak{A}$ and $V = L$ holds in $\mathfrak{B}$ too. Thus $V = L$ is an $\forall$-statement for the trivial reason that if $V = L$ holds in $\mathfrak{A}$, $\mathfrak{A}$ has no admissible proper substructures.

The characteristic property of statements ϕ equivalent in a first-order theory T to existential-universal sentences of T is the following [**13**, 3.4.6 and 3.4.8]. For every sequence $\mathfrak{A}_n$ of models of T such that $\mathfrak{A}_n$ is a substructure of $\mathfrak{A}_{n+1}$ and such that the union $\mathfrak{A}$ of the $\mathfrak{A}_n$'s is also a model of T, if ϕ holds in $\mathfrak{A}$ then for some $k < \omega$ ϕ holds in all the $\mathfrak{A}_n$'s with $n \geq k$. We shall see that the corresponding property of being an $\exists\forall$-sentence, where $\mathfrak{A}_n$ is required to be an admissible substructure of $\mathfrak{A}_{n+1}$, fails for GCH and AC (if ZF is consistent). GCH and AC are of course $\forall\exists$-statements since they are in Π_2.

Let us see that GCH is not an $\exists\forall$-statement, not even in ZF + AC. If ZF is consistent so is ZF + $V = L$, and ZF + $V = L$ has a denumerable model. Easton [**2**] showed that a denumerable model of ZF + $V = L$ can be extended to a denumerable model $\mathfrak{B}$ of ZF + AC + $\forall\alpha(\aleph_\alpha$ is regular $\rightarrow 2^{\aleph_\alpha} = \aleph_{\alpha+2})$. By the construction of Jensen [**7**] one can extend $\mathfrak{B}$ to a model $\mathfrak{A}$ of ZF + GCH by adding to $\mathfrak{B}$ generic maps which collapse some of the cardinals in $\mathfrak{B}$. Since $\mathfrak{B}$ is denumerable there is an ascending sequence α_n of ordinals of $\mathfrak{B}$ such that for every ordinal β of $\mathfrak{B}$ there is an $n < \omega$ such that $\alpha_n > \beta$ in $\mathfrak{B}$. Let $\mathfrak{A}_n$ be the substructure of $\mathfrak{A}$ which consists of all the members of A which are constructible in $\mathfrak{A}$ from the members of B and from the collapsing maps for ordinals $<\aleph_{\alpha_n}$ (see Scarpellini [**14**] for the notion of constructibility relative to given sets). Each $\mathfrak{A}_n$ is a model of ZF + AC. It is easily seen that $\mathfrak{A}$ is the union of the $\mathfrak{A}_n$'s, and that for every ordinal β of $\mathfrak{B}$ such that $\beta > \alpha_n$ in $\mathfrak{B}$ $2^{\aleph_\beta} = \aleph_{\beta+2}$ holds in $\mathfrak{A}_n$. Thus GCH fails in each of the $\mathfrak{A}_n$'s but holds in their union $\mathfrak{A}$.

We shall see now that AC is not an $\exists\forall$-statement. Let $\mathfrak{B}$ be a denumerable model of ZF + $V = L$. Let $\mathfrak{A}$ be the model obtained from $\mathfrak{B}$ by the construction of Easton [**2**], with the difference that instead of adding many generic subsets to a regular $\aleph_\alpha$ we add only $\aleph_0$ such subsets. Then $\mathfrak{A}$ is a model of ZF + GCH. Let α_n be an ascending sequence as above with the additional requirement that each α_n is regular. Let $\mathfrak{A}_n$ be the substructure of $\mathfrak{A}$ which consists of all members of A which are constructible in the function $F(\gamma, m)$ defined on $\alpha_n \times \omega$ such that for a regular $\gamma < \alpha_n$ $F(\gamma, m)$ is the mth new generic subset of $\aleph_\gamma$, and also in the generic subsets $a_m^{\alpha_n}$, $m < \omega$, of $\aleph_{\alpha_n}$. It can be easily shown, by the methods of Feferman [**3**], that AC does not hold in $\mathfrak{A}_n$ since the power-set of $\aleph_{\alpha_n}$ in $\mathfrak{A}_n$ cannot be well ordered in $\mathfrak{A}_n$. It is also clear that the union of the $\mathfrak{A}_n$'s is $\mathfrak{A}$.

Finally let us remark that an $\exists$-statement is now, obviously, a statement ϕ such that whenever $\mathfrak{A}$ is a model of ZF and $\mathfrak{B}$ is an admissible substructure of $\mathfrak{A}$ if ϕ is true in $\mathfrak{B}$ then it is also true in $\mathfrak{A}$. If an $\exists$-statement ϕ follows from $V = L$ in ZF then ϕ is already provable in ZF, as seen from Gödel's theorem in [**5**] that $V = L$ holds in the "structure" $\langle L, \in\rangle$. If an $\exists$-statement ϕ implies AC_0 in ZF it is refutable in ZF + AC, as follows from the construction of Cohen [**1**] of a model in which AC_0 fails.

BIBLIOGRAPHY

1. P. J. Cohen, *Set theory and the continuum hypothesis*, Benjamin, 1966.

2. W. B. Easton, *Powers of regular cardinals*, Ph.D. Thesis, Princeton University, 1964.

3. S. Feferman, *Some applications of the notions of forcing and generic sets*, Fund. Math. **56** (1965), 325–345.

4. A. A. Fraenkel and Y. Bar-Hillel, *Foundations of set theory*, North-Holland, Amsterdam, 1958.

5. K. Gödel, *The consistency of the continuum hypothesis*, Princeton University Press, Princeton, N.J., 1940.

6. J. D. Halpern and A. Lévy, *The Boolean prime ideal theorem does not imply the axiom of choice*, these Proceedings.

7. R. B. Jensen, *Ramsey cardinals and the general continuum hypothesis* (abstract), Notices Amer. Math. Soc. **14** (1967), p. 253.

8. A. Lévy, *Axiom schemata of strong infinity in axiomatic set theory*, Pacific J. Math. **10** (1960), 223–238.

9. ———, *A hierarchy of formulas in set theory*, Mem. Amer. Math. Soc. No. 57 (1965).

10. ———, *Definability in axiomatic set theory*. I, Logic, Methodology and Philos. Sci. (Proc. 1964 Internat. Congr.), pp. 127–151, North-Holland, Amsterdam, 1965.

11. R. Montague, "Fraenkel's addition to the axioms of Zermelo" in *Essays on the foundations of mathematics*, Magnes Press, The Hebrew Univ., Jerusalem, 1961, 91–114.

12. R. A. Platek, *Eliminating the continuum hypothesis* (abstract), Notices Amer. Math. Soc. **14** (1967), p. 366.

13. A. Robinson, *Introduction to model theory and to the metamathematics of algebra*, North-Holland, Amsterdam, 1963.

14. B. Scarpellini, *On a family of models of Zermelo-Fraenkel set theory*. Z. Math. Logik Grundlagen Math. **12** (1966), 191–204.

15. J. C. Shepherdson, *Inner models for set theory—Part* I, J. Symbolic Logic **16** (1951), 161–190.

THE HEBREW UNIVERSITY OF JERUSALEM

CATEGORICAL ALGEBRA AND SET-THEORETIC FOUNDATIONS

SAUNDERS MAC LANE[1]

1. Introduction. This note will speculate on some of the ways in which the current practice of category theory may suggest or require revisions in the use of axiomatic set theory as a foundation for mathematical practice.

Categorical algebra has developed in recent years as an effective method of organizing parts of mathematics. Typically, this sort of organization uses notions such as that of the category **G** of all groups. This category consists of two collections: The collection of all groups G and the collection of all homomorphisms $t: G \to H$ of one group G into another one; the basic operation in this category is the composition of two such homomorphisms. To realize the intent of this construction it is vital that this collection **G** contain *all* groups; however, if "collection" is to mean "set" in any one of the usual axiomatizations of set-theory, this intent cannot be directly realized. This raises the problem of finding some axiomatization of set theory—or of some foundational discipline like set theory—which will be adequate and appropriate to realizing this intent. This problem may turn out to have revolutionary implications vis-a-vis the accepted views of the role of set theory.

2. Categories via axioms. The definition and elementary properties of a category involve no real foundational problems. This may be indicated by reformulating in axiomatic terms, free of set theory, one of the standard descriptions of a category. (An introductory exposition of categories is given in an algebra text by Mac Lane-Birkhoff [**15**]; the more extensive standard expositions are

[1] The studies discussed here were supported in part under a grant from the Office of Naval Research.

those by Freyd [**2**], Mac Lane [**14**], and Mitchell [**16**].) A *category* **C** is a two-sorted system, the sorts being called *objects* A of **C** and *morphisms* f of **C**. The undefined terms, in context, are "A is the *domain* (or, the *codomain*) of f", "k is the *composite* of g with f", and "t is the *identity* morphism of A". The first axiom asserts that every morphism f has exactly one object A as domain and one object B as codomain; when this is the case, we write $f: A \to B$. Next, for all morphisms g and f, there exists a composite k of g with f if and only if domain $(g) =$ codomain (f); when this is the case, the composite k is unique, and is written as $k = gf$. Another axiom requires that this composition be associative $[h(gf) = (hg)f]$ whenever possible. Every object B has one and only one identity morphism, with domain and codomain B, which is written as $1_B: B \to B$. Finally, $1_B f = f$ for every $f: A \to B$ and $g 1_B = g$ for every $g: B \to C$.

A prime example is "the" category **S** of sets (more exactly, the category defined from any (model of *a*) suitable set theory). In this category the objects are the sets S and the morphisms are the functions $f: S \to T$, with the usual domain, identity functions and composition. Note explicitly that the notion of function is *not* that customary in axiomatic set theory; indeed, for categorical purposes a morphism (function) has *both* a given domain *and* a given codomain. This means that a function must be described as a suitable set of ordered pairs *plus* a suitable set as codomain. For example, let S be a proper subset of T and $D \subset S \times S$ the set of all ordered pairs (s, s) for $s \in S$. Then (S, D, S) is (the graph of) the usual identity function $1_S: S \to S$, while (S, D, T) is (the graph of) the injection ("insertion") $i_{S,T}: S \to T$ (that function which maps S as a subset into T). For categorical purposes 1_S and $i_{S,T}$ must be treated as *different* functions, essentially because they behave differently under functors (for a detailed explanation, see Mac Lane-Birkhoff [**15**, p. 325]).

A morphism of categories is called a *functor*. In detail, a functor $F: \mathbf{C} \to \mathbf{D}$ assigns to each object A of **C** an object $D = F(A)$ of **D** and to each morphism $: A \to B$ of **C** a morphism $d = F(f): F(A) \to F(B)$ of **D**. Here "assigns", viewed axiomatically, is a locution for the following undefined terms, "D is the image of A under F" and "d is the image of f under F". The axioms for a functor F require that $F(1_B) = 1_{FB}$ for every B and that $F(gf) = F(g)F(f)$ whenever the composite gf is defined.

A morphism of functors is called a *natural transformation*. In detail, if **C** and **D** are categories and $F: \mathbf{C} \to \mathbf{D}$ and $F': \mathbf{C} \to \mathbf{D}$ are functors, then a natural transformation $\theta: F \to F'$ "assigns" to each object A of **C** a morphism $\theta_A: F(A) \to F'(A)$ in **D** in such fashion that the equality $\theta_B F(f) = F'(f)\theta_A$ (of composites in **D**) holds for every morphism $f: A \to B$ of **C**. This equality is usually pictured by the statement that the following square diagram commutes:

$$\begin{array}{ccc} F(A) & \xrightarrow{F(f)} & F(B) \\ {\scriptstyle \theta_A}\downarrow & & \downarrow{\scriptstyle \theta_B} \\ F'(A) & \xrightarrow[F'(f)]{} & F'(B) \end{array}$$

This commutative diagram expresses the "naturality" of θ.

A final basic notion is that of adjoint functor, due to Kan [**8**]. It may be described in terms of the closely related notion of a "universal construction", due to Samuel [**18**] and Bourbaki. Given a function $U: \mathbf{C} \to \mathbf{X}$, and an object X of $\mathbf{X}$, a *universal morphism* from X under U is a morphism $m: X \to UR$, with R an object of $\mathbf{C}$, such that every morphism $f: X \to UA$ in $\mathbf{X}$ can be written as a composite $f = U(f')m$ for exactly one morphism $f': R \to A$ of $\mathbf{C}$. The corresponding commutative diagram is

$$\begin{array}{ccc} X & \xrightarrow{\ m\ } & UR \\ & \searrow^{f} & \downarrow{\scriptstyle U(f')} \\ & & UA \end{array}$$

A left adjoint of the given functor $U: \mathbf{C} \to \mathbf{X}$ is then described as a functor $F: \mathbf{X} \to \mathbf{C}$ (in the opposite direction) for which there is a natural transformation $\eta: I \to UF$, $I: \mathbf{X} \to \mathbf{X}$ the identity functor, and such that each $\eta_X: X \to UFX$ is universal from X under U. The multiplicity of working examples of adjoint functors is matched by the protean forms of their definitions, as given in the standard references cited above.

3. Categories via sets and classes. In their original introduction of the notion of a category [**1**], Eilenberg-Mac Lane noted that the category of all sets could not be described legitimately as a set, and so proposed to describe it as a class within a Gödel-Bernays set theory (in which *both* "set" and "class" are primitive notions, as in Rubin [**17**]). Call such a category "large". In detail, then, a *large* category is a class of objects and a class of morphisms, with domain, composition, etc. all such as to satisfy the axioms given above for a category. Take the set theory in the version where sets are regarded as special classes, and call a category $\mathbf{C}$ *small* when the class of all its morphisms is a set; since every object has an identity morphism, this also implies that the class of all objects of $\mathbf{C}$ is a set. More generally, for any pair of objects A and B in a large category the comprehension axiom for classes constructs the class

$$\hom_{\mathbf{C}}(A, B) = \{f \mid f \text{ a morphism of } \mathbf{C} \text{ and } f: A \to B\}.$$

A (large) category is said to be locally small (Mitchell's different and aberrant use of this term is to be rejected) when each class $\hom_{\mathbf{C}}(A, B)$ is a set. Usually, the definition of a category is given in terms of such "hom-sets", and so defines only these locally small categories. In particular, the category $\mathbf{S}$ of all sets (of our underlying Gödel-Bernays set theory) is a large category which is locally small (given sets S and T, $\hom(S, T) = T^S$ is a set).

The same small-large distinction may be drawn for other, more familiar, algebraic objects. Thus a "large group" is a *class* G equipped with a function $G \times G \to G$ which satisfies the usual axioms for multiplication in a group, while a "small group" is one for which this class G is actually a set. Now the comprehension axiom of Gödel-Bernays set theory allows one to form the large category of all small groups, in which the class of objects is the class of all (small) groups.

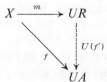

Again, this is a locally small category. There are many other useful examples of locally small categories with objects all (small) Mathematical systems of a fixed sort and morphisms all homomorphisms of these systems. A detailed indication of the construction of these categories by Gödel-Bernays set theory is given in Mac Lane-Birkhoff [**15**, Chapter XV].

Two such constructions of large categories are especially useful. One is **Cat:** The category of all small categories; with objects all small categories and morphisms all functors. For many purposes, as indicated below, one would like to have a bigger **cat**, say the category of all large categories or of all categories *überhaupt*. Another one is the *functor category* $\mathbf{A}^{\mathbf{B}}$ determined by a large category **A** and a small category **B**. Its objects are all functors $F: \mathbf{B} \to \mathbf{A}$ and its morphisms $F \to F'$ are all natural transformations $\theta: F \to F'$. Since the domain category **B** is small, each functor $F: \mathbf{B} \to \mathbf{A}$ can be represented by a set, say the usual graph of F; hence one may form the class of all such F, making $\mathbf{A}^{\mathbf{B}}$ a locally small category. There are, however, many properties of large categories **A** and **B** which can be effectively visualized in the (superlarge (?)) category $\mathbf{A}^{\mathbf{B}}$.

4. Categories in universes. Grothendieck and his followers have made extensive use of categories in algebraic geometry and in this connection have proposed an alternative way of treating large categories within Zermelo-Fraenkel set theory. Define a *universe* to be a set U whose elements $x \in U$ themselves form a model of ZF (under the given membership relation). Add the axiom that every set is a member of some universe (since the cardinal of a universe is inaccessible, this amounts to assuming the existence of many inaccessible cardinals). A category is now always described by sets; that is, it is a set of objects together with a set of morphisms with the added structures which we have already described axiomatically. If U is a universe then a category *within* U is a category whose set of objects is a member of the universe U. Thus a category within U is much like a small category. Moreover, the usual comprehension axiom of ZF allows us to form $\mathbf{Cat}_U$, the category of all categories within U. This will be a category within some larger universe.

There are a number of variants of this idea. In Gabriel [**3**] and Sonner [**19**] a universe is described not as a model of ZF, but as a set closed under union, power set, and several more elementary operations. The essential point is that one may always form the functor category $\mathbf{A}^{\mathbf{B}}$ of two given categories—in general, by going to a larger universe U'. All these large and superlarge categories are sets, subject to the familiar manipulations of set theory. Given any universe U', one can always form the category of all categories within U'. This is still not that will-of-the-wisp, the category of all categories *überhaupt*. There are more subtle questions of the following sort: If R is a ring within the universe U and U' is some larger universe, how is the category of all left R-modules which are elements of U related to the category of all left R-modules which are elements of U'? One would like the relation to be close, but little seems to be known on this point.

Instead of using universes, set-theorists may prefer to work with the familiar levels V_α (all sets of rank less than α) of the cumulative hierarchy of sets (see e.g. Kreisel-Krivine [**8**]). Both the approach by small and large categories and the approach via universes can evidently be described within these levels. The **cats** multiply; there appears to be a category $\mathbf{Cat}_\alpha$ whose objects are all categories **C** in which the set of morphisms is an element of V_α.

5. Basic questions. Our fundamental observation is just this: There is an appreciable body of results about categories (a few indications are given below, in §7) but the received methods of defining categories in terms of sets do not provide any *one* single context (i.e. any one model of a standard set theory) within which one can comprehensively state these results. Using universes, all the functor categories are there, but there is no category of all groups. Using Gödel-Bernays, one has the category of all (small) groups, but only a few functor categories. This situation raises a technical question and a philosophical question. The technical question is that of rearranging categories or sets (or both) to get this body of theorems whole. This question was explicitly formulated in Mac Lane [**15**]. Then Lawvere in [**10**] and [**11**] raised the basic possibility that the attempted explication of categories (and of other Mathematical objects) by sets might be replaced by an explication in terms of some other "fundamental" concepts. In detail, he proposed first an axiomatization of sets not in terms of the usual membership relation but in terms of the category of sets, and he provided an elementary such axiomatization. Going further, he proposed a list of axioms on **Cat** (the category of all categories) as a potential foundation of Mathematics. This involves both technical difficulties (Isbell [7]) and some exciting possibilities.

This also suggests a general philosophical question: Why should the "official" foundations of Mathematics necessarily be couched in terms of set-theory axiomatized by a membership relation? This perhaps inconvenient question, once clearly faced, raises the inescapable possibility of many alternative views of foundations. For instance, "function" rather than "set" might be taken as the fundamental notion. Both combinatory logic and von Neumann's original version of Gödel-Bernays may be regarded as axiomatizations of "functions"; there is good reason to suppose that other and more efficient such axiomatizations exist. Foundations in terms of sets make the most "primitive" Mathematical notion the starting point; there is considerable reason to suppose that the foundation would fit the facts better if it started with some more highly structured notions (set and function, category, or Mathematical structure). The set-theoretic approach is often described in terms of the intuitively constructed hierarchy V_α of cumulative types; one can easily provide similar hierarchies built on different structural concepts. In these ways, many alternative directions for investigation of the foundations are opened.

The prior situation in the foundations of Mathematics had in one respect a very simple structure. One could produce one formal system, say Zermelo-Fraenkel set theory, with the property that all the ordinary operations of practising Mathematicians could be carried out within this one system and on objects of this system. Explicitly, this meant that "every" Mathematical object was or could be defined to be a set, and that all of the arguments about these objects could be reduced to the axioms of ZF set theory. In practise, this did not mean that the arguments were actually so reduced; the working Mathematicians usually thought in terms of a naive set theory (probably one more or less equivalent to ZF) and was trained in describing everything as a set of this naive sort. This one-formal-system "monolithic" approach has also been convenient for specialists on foundations ever since Frege and Whitehead-Russell. On the one hand, all the classical nineteenth century problems of foundations (the construction of integers, real numbers, analysis; the properties of ordinal numbers; the axiom of choice and the continuum hypotheses) could be stated in this one system. On the other hand, alternative formal system could then be tested by comparison (as to strength or relative consistency) with this one system.

This happy situation no longer applies to the practice of category theory. Here the Mathematician is working with a variety of objects (categories of all groups, functor categories) which cannot all be described simultaneously as objects of any one foundational system. What follows? One might hope for some one new foundational system (the category of categories?) within which all the desired objects live; a practical requirement could be that this system could be used "naively" by Mathematicians not sophisticated in foundational research. The alternative would be the simultaneous use of several formal systems. This alternative is only foreshadowed by the Grothendieck use of universes. If the alternative is taken seriously, it has drastic consequences; the next section will give only a very small sample.

6. Multiple systems for foundations. Consider a formal foundation of Mathematics which is *multiple* in the sense that there are a number of different basic axiom systems T, T' which are collectively such that the objects of concern to the working Mathematician can be interpreted in one or the other of these systems. Recall also that the working mathematician usually regards the objects of his study as if they were really *there* (in some realistic or platonistic sense). "Really there" could be read as objects within a fixed interpretation M of one of the basic axiom systems. Here "interpretation" has the intuitive force of "model" (but do not take "model" in the usual technical sense of a model within some universe of sets). For example, this suggests that one would speak of a set-system M, meaning *an* interpretation (model) of ZF in exactly the same way that one speaks of an "abstract group" as a model of the axioms of group theory. Inevitably, there will be discussion of pseudo-set-systems and quasi-set-systems satisfying a variety of weaker axioms or different axioms (some not stated in terms of $\in$). The dogma that there is just *one* set theory disappears.

To handle categories, one must speak of "all so and so's" in some M. The language necessary to do this might be a very simple two sorted language with sorts *items* x, y, z (for things in M) and *classes* A, B, C (for collections of items). Primitive notions might be ordered pairs $\langle x, y\rangle$ of items and membership $x \in A$

(only for an item x in a class A). Axioms should include the expected axioms on the equality of ordered pairs, extensionality for classes, the existence of the null class, the singleton $\{x\}$, and the cartesian product $A \times B$ of classes, as well as a comprehension axiom for classes. These axioms can be regarded as simple axioms for a rudimentary fragment of class (or set) theory. In this language a category (such as a category of all so-and-so's in M) may be described as a class of objects and a class of morphisms with the usual structural properties. For such categories, we may then readily carry out the operations usual in elementary category theory; for example, one may construct the opposite of a category $\mathbf{C}^{\mathrm{op}}$ (reverse the direction of all morphisms) and the cartesian product $\mathbf{C} \times \mathbf{D}$ of two categories. More advanced constructions, such as that of functor categories, would require more powerful axioms (say, for each class B, the existence of a right adjoint to $A \mapsto A \times B$). The essential point is that the original system T, with model M, has been expanded by adjoining classes, in a way analogous to the passage from Zermelo-Fraenkel to Gödel-Bernays set theory. This passage illustrates what we mean by multiple systems of foundations, with varied notions of category (small, large, or otherwise) within these systems.

7. The uses of large categories. We turn from speculation to a few indications of the utility of extra large categories.

The Yoneda Lemma (Freyd [**2**, p. 112], Mac Lane [**14**, p. 54]) has become a fundamental tool of categorists. Let the large category $\mathbf{C}$ be locally small, so that each hom (A, B) is a set. Hold the object A fixed; then the function sending each object to the set hom (A, B) can be made into a functor from $\mathbf{C}$ to the (large) category $\mathbf{S}$ of (small) sets. This functor is usually written

$$h_A = \mathrm{hom}\,(A, -) \colon \mathbf{C} \to \mathbf{S}$$

and is called the covariant hom functor. Now let $F \colon \mathbf{C} \to \mathbf{S}$ be any other functor, and consider the class Nat (h_A, F) of all natural transformations $\theta \colon h_A \to F$. By definition, each such transformation assigns to every object B of $\mathbf{C}$ a morphism $\theta_B \colon \mathrm{hom}\,(A, B) \to F(B)$ in the category $\mathbf{S}$. In particular, the function $\theta_A \colon \mathrm{hom}\,(A, A) \to F(A)$ sends the identity morphism 1_A onto an element $\theta_A(1_A) \in F(A)$. This correspondence $\theta \mapsto \theta_A(1_A)$ is itself a function

$$\psi \colon \mathrm{Nat}\,(h_A, F) \to F(A). \tag{1}$$

An easy argument shows that this function ψ is a bijection (is one-one onto). In fact the Yoneda lemma states simply that ψ is a bijection and a natural transformation.

The statement that ψ is natural means intuitively that the definition we have given for ψ depends on no artificial choices. It should also be interpreted as a natural transformation between functors. Indeed, if we regard the ordered pair (A, F) of the object A and the functor F as an object of the product category $\mathbf{C} \times \mathbf{S}^{\mathbf{C}}$, then both sides $F(A)$ and Nat (h_A, F) of (1) can be regarded as the values on (A, F) of suitable functors $\mathbf{C} \times \mathbf{S}^{\mathbf{C}} \to \mathbf{S}$; it is then routine to check that ψ as we have defined it is a natural transformation between these functors. The only trouble is that the functor category $\mathbf{S}^{\mathbf{C}}$ used here is not legitimate (for $\mathbf{C}$ large). If we employ universes instead of classes, the category $\mathbf{S}$ does not stay put.

Completeness is a basic property of a category, and is defined by closure under products and equalizers. First we define these terms. If f, $g: A \to B$ are two morphisms with the same domain A and the same codomain B, an *equalizer* of f, g is a morphism $e: R \to A$ such that (i) $fe = ge$; (ii) whenever $h: C \to A$ has $fh = gh$, there is a unique $h': C \to R$ such that $h = eh'$. This description clearly makes e (the opposite of) a universal. Let I be a set and A_i for $i \in I$ an I-indexed family of objects of $\mathbf{C}$ (in other words, A is a function on I to the objects of $\mathbf{C}$). A *product* of the family $\{A_i \mid i \in I\}$ consists of an object P of $\mathbf{C}$ and an I-indexed family $p_i: P \to A_i$, $i \in I$, of morphisms of $\mathbf{C}$ with the following (opposite of a) universal property: Whenever $f_i: C' \to A_i$ for $i \in I$, there is a unique $f: C \to P$ with $p_i f = f_i$ for all $i \in I$. For example, in the category of sets, the usual cartesian product ΠA_i of the sets A_i, equipped with its projections $p_i: \Pi A_i \to A_i$ to each of the factors, is readily seen to be a product in the categorical sense we have described. The corresponding cartesian products are also such products in the category of all small groups or of all small topological spaces.

A large category $\mathbf{C}$ is said to be *complete* whenever (a) it contains an equalizer for any two of its morphisms $f, g: A \to B$; (b) it contains a product for any set-indexed family of its objects. The importance of these two conditions is that together they imply that $\mathbf{C}$ contains (inverse) limits for all set-indexed systems (directed or not); technically, if $\mathbf{J}$ is any small category, every functor $F: \mathbf{J} \to \mathbf{C}$ has a limit (see Freyd [**2**], where "limit" is called "left root"). The discussion also indicates that the categories of sets, of groups, and of topological spaces (like many similar categories) are all complete. Observe that the set-class distinction enters vitally here. Completeness is defined using sets (of indices) by way of products or limits over sets; the familiar examples of complete categories are *large* categories hence are classes (see Freyd [**2**, p. 78]).

Completeness is a vital assumption in the fundamental existence theorem for adjoint functors due to Freyd. Let $\mathbf{C}$ be a complete locally small category and $U: \mathbf{C} \to \mathbf{X}$ a functor which carries products and equalizers (in $\mathbf{C}$) to products and equalizers, respectively, in $\mathbf{X}$. Then U has a left-adjoint F *provided* U satisfies the following "solution-set condition": For each object X of $\mathbf{X}$ there is a set J, a J-indexed family of objects R_j of $\mathbf{C}$ and a J-indexed family of morphisms $m_j: X \to UR_j$ of $\mathbf{X}$, such that to each $f: X \to UA$ there is a $j \in J$ and an $f': R_j \to A$ such that $f = U(f')m_j: X \to UA$. (Note that this solution-set condition is like the definition of an adjoint by a universal morphism, except that it specifies a *set* m_j of such morphisms which are "collectively" weakly universal.) For the proof of this adjoint functor theorem (with some needless added hypotheses) see Freyd [**2**, p. 84].

This and related theorems raise the question: If a given category $\mathbf{C}$ is not complete, can it be embedded in a complete category? Now $\mathbf{C}$ has a natural (Yoneda) embedding

$$Y: \mathbf{C} \to \mathbf{S}^{\mathbf{C}^{\mathrm{op}}}, \qquad Y(A) = h^A$$

into a functor category, where $\mathbf{C}^{\mathrm{op}}$ is the opposite of $\mathbf{C}$ and h^A is the contravariant hom-functor, with $h^A(B) = \hom(B, A)$. The completeness of $\mathbf{S}$ readily implies that $\mathbf{S}^{\mathbf{C}^{\mathrm{op}}}$ is complete, at least for $\mathbf{C}$ small. Hence a natural way to complete $\mathbf{C}$ is to embed $\mathbf{C}$ via Y in the closure of $Y(\mathbf{C})$ under products and equalizers. This procedure, however, is not immediately applicable if $\mathbf{C}$ is large. The best current theorems in this direction (Isbell [7]) make sophisticated use of set-theoretical considerations.

To illustrate the use of **Cat** we discuss fibered categories. Consider for instance the category of all (small) modules, where by a module we mean a pair (R, A) with R a ring, and A a left R-module. A morphism $(R, A) \to (R', A')$ of modules is a pair (s, f) consisting of a ring homomorphism $s: R \to R'$ and a homomorphism $f: A \to A'$ of (additive) abelian groups such that $f(ra) = (sr)(fa)$ for all $r \in R$, $a \in A$. Clearly this gives a category, and the assignments $(R, A) \mapsto R$, $(s, f) \mapsto s$ give a functor P on this category to the category of rings. We call P a fibered category, and observe that for any one ring R the fiber $P^{-1}(R)$ over R is itself a category.

In general a fibered category is a functor $P: \mathbf{A} \to \mathbf{C}$ satisfying certain lifting conditions about "cleavages", too complex to state here (see Gray, [4]). Since P is a functor, we have for each object C of $\mathbf{C}$ a category, the *fiber over* C, consisting of all objects A with $P(A) = C$ and all the morphisms f with $P(f) = 1_C$. Call this category $F(C)$. The cleavage conditions suffice to insure that each morphism $f: C \to C'$ on the "base" C yields a functor $F(f): F(C') \to F(C)$ backwards between the corresponding fibers; moreover, F is a (contravariant) functor. Indeed, F is a functor $\mathbf{C}^{\mathrm{op}} \to \mathbf{Cat}$. Effective treatment of the theory of fibered categories requires a systematic use of these functors to **Cat**, where **Cat** ought to be at least the category of all large categories. For this—and for many similar cases such as Benabou's "profunctors"—we need a working foundation which will handle the category of large categories.

Bibliography

1. Samuel Eilenberg and Saunders Mac Lane, *General theory of natural equivalences*, Trans. Amer. Math. Soc. **58** (1945), 231–294.

2. Peter Freyd, *Abelian categories. An introduction to the theory of functors*, Harper's Series in Modern Math., Harper and Row, New York, 1964.

3. Pierre Gabriel, *Des catégories abéliennes*, Bull. Soc. Math. France **90** (1962), 323–448.

4. John W. Gray, *Fibred and cofibred categories*, Proc. Conference Categorical Algebra, (La Jolla, 1965), Springer-Verlag, New York, 1966, pp. 21–83.

5. Alexander Grothendieck, *Catégories fibrées et descente*, Seminaire de Géometrie Algébrique de l'Institut des Hautes Etudes Scientifiques, Paris, 1961.

6. John R. Isbell, *Small subcategories and completeness*, Math. Systems Theory **2** (1968), 27–50.

7. ———, Review of [**11**] in Math. Reviews **34** #7332.

8. D. M. Kan, *Adjoint functors*, Trans. Amer. Math. Soc. **87** (1958), 294–329.

9. George Kreisel and J. L. Krivine, *Elément de logique mathématique; Théorie des modèles*, Dunod, Paris, 1967.

10. F. W. Lawvere, *An elementary theory of the category of sets*, Proc. Nat. Acad. Sci. USA **52** (1964), 1506–1511.

11. F. W. Lawvere, *The category of categories as a foundation of mathematics*, Proc. Conference Categorical Algebra (La Jolla 1965), Springer-Verlag, New York, 1966, pp. 1–20.

12. Saunders Mac Lane, *Duality for groups*, Bull. Amer. Math. Soc. **56** (1950), 485–516.

13. ———, *Locally small categories and the foundations of set theory*, Proc. Sympos. Foundations Math., Infinitistic Methods, Warsaw, 1959, pp. 25-43.

14. ———, *Categorical algebra*, Bull. Amer. Math. Soc. **71** (1965), 40–106.

15. Saunders Mac Lane and Garrett Birkhoff, *Algebra*, Macmillan, New York, 1967.

16. Barry Mitchell, *Theory of categories*, Academic Press, New York and London, 1965.

17. Jean E. Rubin, *Set theory for the mathematician*, Holden-Day, San Francisco, 1967.

18. Pierre Samuel, *On universal mappings and free topological groups*, Bull. Amer. Math. Soc. **54** (1948), 591–598.

19. Johann Sonner, *On the formal definition of categories*, Math. Z. **80** (1962), 163–176.

THE UNIVERSITY OF CHICAGO

THE SOLUTION OF ONE OF ULAM'S PROBLEMS CONCERNING ANALYTIC RECTANGLES

R. MANSFIELD[1]

Ulam has asked [**9**, p. 9] whether every analytic subset of the plane belongs to the σ-field generated by the analytic rectangles (i.e. the cartesian products of two linear analytic sets). Recently, Solovay has shown [**8**] that the assumption of the existence of large cardinal numbers (e.g., measurable cardinals) can be used to settle several long outstanding problems of descriptive set theory: for example, he proves that every complement of an analytic set (CA-set) is either countable or contains a perfect subset. Methods suggested by this result of Solovay shall be used to prove from the axioms of set theory that the standard universal analytic subset of $N \times N$ (N is the Baire space N^N, where N is the set of integers) does *not* lie in the σ-field generated by the analytic rectangles.

L is, of course, Gödels universe of constructible sets and, for any α in N^N, $L(\alpha)$ is the class of sets hereditarily constructible from α. For any α, $L(\alpha)$ is a transitive model for the axioms of set theory together with the axiom of choice and the generalized continuum hypothesis [**2**]. A subset P of N^N is in $\Sigma_n^1(\alpha)$ if the proposition $\beta \in P$ is expressible by a Σ_n^1 formula whose parameters are all in $L(\alpha)$. It is well known [**3**, p. 454] that the analytic sets are just those sets that are in $\Sigma_1^1(\alpha)$ for some α in N^N. Rowbottom has shown [**4**] that if there is a measurable cardinal, then for any α in N^N, Ω_1, the first uncountable ordinal, is inaccessible in $L(\alpha)$. Solovay's theorem and thus the result of this paper follows from just this technical property of Ω. Indeed Solovay's result can be proven by using the assumption that $\Omega_1^{(L(\alpha))}$ is countable to show that if P is in $\Pi_1^1(\alpha)$ and contains an element not in $L(\alpha)$, then it contains a perfect set. But now, $\Omega_1^{(L(\alpha))}$ (that ordinal which in

[1] The author wishes to thank Professor Dana Scott for much valuable assistance and advice.

the set theory $L(\alpha)$ plays the role of the first uncountable ordinal) is less than the first ordinal inaccessible in $L(\alpha)$ and is therefore countable; thus any subset of $L(\alpha) \cap N^N$ is countable. We can therefore make the desired conclusion about CA-sets ($\Pi_1^1(\alpha)$-sets) which does not mention the constructible universe.

Any finite sequence of integers can be uniquely coded by a single integer through the device that

$$\#(\langle s_1, s_2, \ldots, s_n \rangle) = 2^{s_1+1} \cdot 3^{s_1+1} \cdot \ldots \cdot p_n^{s_n+1}$$

where p_n is the nth odd prime. The empty sequence is given the number 1. The relation $n \subset m$ holds if and only if n and m are both codes for finite sequences and the sequence coded by n is a subsequence of the sequence coded by m. $n \subset m$ is clearly a primitive recursive relation. If $\beta \in N^N$, $\bar{\beta}(n)$ is to be the number for the sequence $\langle \beta(0), \beta(1), \ldots, \beta(n) \rangle$.

A *tree* is a set of sequence numbers that is closed under the subsequence relation. Given F an arbitrary subset of N^N there is a tree naturally associated with it.

$$\mathrm{Tr}(F) = \{s : \exists \alpha, n[\alpha \in F \wedge \bar{\alpha}(n) = s]\}.$$

And with any tree T there is the associated set,

$$[T] = \{\beta : \forall n[\bar{\beta}(n) \in T\}.$$

The closure of F in the product topology (N itself is given the discrete topology) is just $[\mathrm{Tr}(F)]$. Thus the function Tr gives a one-to-one correspondence between closed subsets of N^N and trees. A closed set is perfect if and only if every sequence in its tree has at least two distinct extensions in the set. The tree of a perfect set will also be called perfect. A closed set is *α-constructibly coded* if its tree is in $L(\alpha)$.

The concept of an elementary subtree is also needed. Given any sequence number s, $\{t : t \subseteq s \vee s \subseteq t\}$ is a perfect tree, call it E_s. Then S is an elementary subtree of T if there is a sequence number s such that $[S] = [T] \cap [E_s]$. If T is perfect and $s \in T$, then $[T] \cap [E_s]$ is also perfect. It is equally trivial to see that for any perfect tree T and any integer n, there are sequence numbers s_0, s_1 satisfying the following three conditions:

1. $s_0, s_1 \in T$,
2. length $(s_i) \geq n, \quad i = 0, 1$,
3. $[E_{s_0}] \cap [E_{s_1}] = 0$.

Solovay has pointed out that his result can be slightly extended. If P is in $\Pi_1^1(\alpha)$, then the statement that it contains a perfect set can be expressed as,

$$\exists T[T \text{ is a perfect tree} \wedge T \subseteq P].$$

This formula is $\Sigma_2^1(\alpha)$ and thus, by Shoenfield's absoluteness lemma **[6]**, if it is true in the universe, it is true in $L(\alpha)$, i.e., if P is uncountable, it contains an α-constructibly coded perfect set. It has been known for a long time that any uncountable $\Sigma_1^1(\alpha)$ set contains an α-constructibly coded perfect subset **[3,** p. 461**]**.

Let us define for $\alpha \in N^N$

$$D(\alpha) = \{\beta \in N^N : \forall P[P \in \Sigma_1^1(\alpha) \to \exists T[T \in L(\alpha) \wedge T \text{ is a perfect tree} \wedge \beta \in [T] \wedge [[T] \subseteq P \vee [T] \subseteq N^N - P]]]\}.$$

Intuitively D is the set of β such that for any Σ_1^1 set P, the proposition $\beta \in P$ is decided by a constructively coded perfect set. This concept was first used by Gandy in connection with hyperarithmetically coded perfect sets [**1**].

LEMMA. *For every α in N^N, if $\Omega_1^{(L(\alpha))}$ is countable, then $D(\alpha)$ contains a perfect set; thus the Borel hierarchy induced on $D(\alpha)$ as a subtopological space of N^N does not stop at any countable ordinal.*

PROOF. Let S_n be a predicate universal for the countable collection of predicates $\Sigma_1^1(\alpha)$. Then there is a function F whose domain is the set of finite sequences of 0's and 1's and whose range is a subset of the α-constructible perfect trees and which satisfies the following recursion relations:

(1) $[F(\langle\cdot\rangle)] \subseteq S_0$ or $[F(\langle\cdot\rangle)] \subseteq N^N - S_0$.

(2) If s is a sequence number of length n, let s_0, s_1 be chosen so that $\min\{\text{length}(s_0), \text{length}(s_1)\} \geq n$ and $s_0, s_1 \in F(s)$ and $E_{s_0} \cap E_{s_1} = 0$. Then, for $i < 2$,

$$F(s * \langle i\rangle) \subseteq F(s) \cap E_{s_i} \cap S_n$$

or

$$F(s * \langle i\rangle) \subseteq F(s) \cap E_{s_i} \cap (N^N - S_n)$$

since either $F(s) \cap E_{s_i} \cap S_n$ or $F(s) \cap E_{s_i} \cap (N^N - S_n)$ is uncountable and since both are in $\Sigma_1^1(\alpha)$ or $\Pi_1^1(\alpha)$ there must be a value for $[F(s * \langle i\rangle)]$ which satisfies this condition.

Let γ be an arbitrary element of 2^N. Then, by definition, for every n,

$$[F(\bar{\gamma}(n+1))] \subseteq [F((\bar{\gamma}n))] \cap [E_{\bar{\gamma}(n)_{\gamma(n+1)}}].$$

The increasing sequence

$$\langle\cdot\rangle_{\gamma(0)} \subseteq \bar{\gamma}(0)_{\gamma(1)} \subseteq \cdots \subseteq \bar{\gamma}(n)_{\gamma(n+1)} \subseteq \cdots$$

defines a unique function, namely that function β such that

$$\forall k \exists n[\bar{\beta}(n) = \bar{\gamma}(k)_{\gamma(k+1)}].$$

Furthermore, since for every n and m, $\bar{\gamma}(n)_{\gamma(n+1)} \in [F(\bar{\gamma}(m))]$, it is seen that β is the only element of $\bigcap_n F(\bar{\gamma}(n))$. Let

$$G = \bigcup_{\gamma \in 2^N} \bigcap_n [F(\bar{\gamma}(n))], \qquad G_n = \bigcup \{F(s) : \text{length}(s) = n\}.$$

G_n is a finite union of closed sets and is therefore closed. The above discussion is easily seen to yield that $G = \bigcup_n G_n$; thus G is closed. The fact that G is perfect follows immediately from the condition that for every s, E_{s_0} and E_{s_1} were chosen to be disjoint. This completes the proof of the lemma.

Let $\Sigma(\alpha, \beta)$ be the standard Σ_1^1 predicate such that for any analytic set S there is an α_0 with $S = \Sigma(\alpha_0, \cdot)$. Suppose, by way of contradiction, that Σ is a Borel combination of analytic rectangles. That is to say suppose there are sequences of analytic sets $\langle A_n \rangle_{n<\omega}$, $\langle B_n \rangle_{n<\omega}$ and a Borel operator $\boldsymbol{B}$ such that $\Sigma = \boldsymbol{B}(\langle A_n \times B_n \rangle_{n<\omega})$. This means that for any α_0, $\Sigma(\alpha_0, \cdot) = \boldsymbol{B}(\langle B_n^* \rangle)$ where

$$\begin{aligned} B_n^* &= B_n \quad \text{if } \alpha_0 \in A_n, \\ &= 0 \quad \text{if } \alpha_0 \notin A_n. \end{aligned}$$

Any function α can be made to represent an infinite sequence of functions by the device that $\alpha_n(m) = \alpha(2^n \cdot 3^m)$. So let α be chosen so that

$$A_n = \Sigma(\alpha_{2n}, \cdot), \qquad B_n = \Sigma(\alpha_{2n+1}, \cdot).$$

Then, since $\boldsymbol{\Delta}_2^1$ is closed under Borel operations, there is a function γ such that $\alpha \in L_\gamma$ and the set $\boldsymbol{B}(\langle A_n \times B_n \rangle)$ is in $\boldsymbol{\Delta}_2^1(\gamma)$. Then, of course, each B_n is in $\Sigma_1^1(\gamma)$ and thus, relative to $D(\gamma)$, is a union of γ-constructibly coded perfect sets. At this point we would like to be able to call in the lemma, but unless $\Omega^{(L(\alpha))}$ is countable, it does not apply. So let $\mathscr{B}$ be the Boolean algebra which collapses Ω_1 **[5]**. Then within the $\mathscr{B}$-valued set theory $V^{(\mathscr{B})}$, $L(\gamma) \leq V$, thus the sentence ⌜$\Omega_1{}^{(L(\gamma))}$ is countable⌝ is $\mathscr{B}$-valid. And so the sentence ⌜$\forall_n[B_n$ is an F_γ set⌝ is also $\mathscr{B}$-valid, since there are now only countably many γ-constructibly coded perfect sets. Let τ be the level of the Borel operator $\boldsymbol{B}$. Then within $V^{(\mathscr{B})}$ τ is still the level of $\boldsymbol{B}$. So that it is $\mathscr{B}$-valid that, relative to $D(\gamma)$, *every* analytic set and *a fortiori* every Borel set is a Borel set of level $\leq \tau + 1$, contradicting the lemma. Thus it is $\mathscr{B}$-valid that

$$\Sigma \neq B(\langle A_n \times B_n \rangle).$$

But this last statement is in $\Sigma_2^1(\gamma)$ and furthermore the extension $V \subseteq V^{(\mathscr{B})}$ satisfies all the conditions of Shoenfield's absoluteness lemma; it is therefore also true in V, contradicting the assumption that Σ was in the γ-field generated by the analytic rectangles.

It should be pointed out, for those readers unfamiliar with the Boolean valued set theories, that the above argument makes perfect sense if it is assumed from the beginning that for every $\gamma \in N^N$, $\Omega^{(L(\gamma))}$ is countable, and all reference to $V^{(\mathscr{B})}$ is eliminated by replacing all statements of the form ⌜Φ⌝ is $\mathscr{B}$-valid by a simple affirmation of Φ. The assumption can be justified either on the grounds that it follows from the existence of a measurable cardinal, or that it can be proven consistent with the aid of an inaccessible cardinal, or that the above method eliminates it altogether from the argument.

A slight modification of the above proof can be used to show that Σ is not in the σ-field generated by the PCA rectangles. This is to be done by modifying the definition of $D(\alpha)$ to say that for any P in $\Sigma_2^1(\alpha)$ the proposition $\beta \in P$ is decided on $V - L$ by an α-constructibly coded perfect set, and making corresponding modifications throughout the course of the proof. The author suspects that it is consistent to assume that Σ is not in the σ-field generated by arbitrary rectangles.

BIBLIOGRAPHY

1. R. O. Gandy, *Hyperdegrees in* Σ_1^1 *sets*, mimeographed notes, Stanford University, 1967.

2. K. Gödel, *The consistency of the continuum hypothesis*, Princeton University Press, Princeton, N.J., 1940.

3. K. Kuratowski, *Topology*, vol. 1, Academic Press, New York and London, 1966.

4. F. Rowbottom, *Large cardinals and small constructible sets*, Doctoral dissertation, University of Wisconsin, 1964.

5. D. S. Scott and R. Solovay, *Boolean valued models for set theory*, these Proceedings, part II.

6. J. R. Shoenfield, *The problem of predicativity*, Essays on the foundations of Mathematics, pp. 132–139, Magnes Press, Hebrew Univ., Jerusalem, 1961.

7. J. Silver, Doctoral dissertation, University of California, Berkeley, 1966.

8. R. Solovay, *The cardinality of* Σ_2^1 *sets*, Foundations of Mathematics, Springer-Verlag, New York, 1968.

9. S. M. Ulam, *Problems in modern mathematics*, Science Editions, Wiley, New York, 1964.

UNIVERSITY OF MANCHESTER

PREDICATIVE CLASSES

YIANNIS N. MOSCHOVAKIS[1]

When we formulate precisely Zermelo's axiom of subsets or Fraenkel's axiom of replacement, we must make explicit which *classes* (collections of sets that need not be sets) we accept as legitimate. In his original (informal) axiomatization of set theory, Zermelo called these objects *definite conditions* and left this term essentially undefined. It was apparently Skolem who first suggested that the definite conditions should be precisely the *first-order-definable* (relations or) collections of sets. (See the introductions to the relevant papers of Zermelo, Fraenkel and Skolem in **[8]** for a more careful history of this development.) In a language with set (lower case) and class (upper case) variables, this amounts to postulating the schema

$$(\Delta^0_\infty\text{-Comp}) \qquad \exists S \forall x[x \in S \leftrightarrow \alpha(x)],$$

where $\alpha(x)$ has no bound class variables but may have free variables of either sort, excluding S.

Schema Δ^0_∞-Comp introduces an element of impredicativity in set theory, since in order to determine whether $\alpha(x)$ is true or false we must understand quantification over the universe of sets. This much impredicativity has long been recognized as essential to the development of set theory and is probably implicit in our intuitive conception of arbitrary (infinite) sets as completed totalities.

Let $\alpha_1, \alpha_2, \ldots$ be an effective enumeration of the sentences which have no class variables. Any person who accepts quantification over the universe of sets as meaningful must admit the class

$$\{n : \alpha_n \text{ is true}\}$$

[1] The author is a Guggenheim fellow. The preparation of this paper was sponsored in part by an NSF Grant.

as well defined. This class, however, is not first-order definable and cannot be proved to exist in Gödel-Bernays set theory—apparently because of our artificial restriction to finite first-order formulas in Δ^0_∞-Comp.

We study here set theories whose classes are just those that are *predicatively definable* in terms of sets and set quantification. In our explication of predicative definability these are precisely the *hyperprojective* (in $\in$) subclasses of the universe of sets as defined and studied in AFOC[2]. According to the main result of ACID, we assure the existence of these classes (in a model-theoretic sense) if we add to the first-order axioms of whichever set theory we are studying the schema of Δ^1_1-Comprehension,

$$(\Delta^1_1\text{-Comp})\quad \forall x[\exists T\alpha(x, T) \leftrightarrow \forall T\beta(x, T)] \rightarrow \exists S\forall x[x \in S \leftrightarrow \exists T\alpha(x, T)],$$

where $\alpha(x, T)$, $\beta(x, T)$ have no bound class variables but may have free variables of either sort, excluding S.

The adjunction of schema Δ^1_1-Comp (together with some weak induction axioms, natural for a two-sorted theory) to the axioms of Gödel-Bernays results in a significant strengthening of these axioms—e.g. we can prove that there are natural and countable elementary submodels of the universe. Thus this theory can be considered a predicative (relative to set quantification) justification of the natural, informal constructions in Professor Shoenfield's lectures at this Institute.

Since we are concerned with classes rather than sets, our results are to some extent independent of the set-existence axioms that we assume. We shall formulate them in the context of a weak, predicative theory of sets and classes, whose set-existence axioms are those of PZF, S. Kripke's theory in [3]. Most of our results hold for arbitrary extensions of this theory by set (first-order) axioms.[3]

The key results of this paper are Theorems 3(c) and 4 and they are proof-theoretic in nature. In a future paper we hope to study the standard models of predicative theories with classes and their relevance for the theory of admissible sets. That interconnections exist is obvious from our results here and the main theorem of the joint paper [1]: *Let A be a transitive set closed under pairing, let A^+ be the intersection of all admissible sets which contain A as an element. Then the subsets of A that are elements of A^+ are precisely the hyperprojective* (*in* $\in$) *sets.*

In §1 we give the definitions and state the results and in the remainder of the paper we sketch the proofs. These are based mostly on a formalization of the theory developed in AFOC and ACID and so we omit all details and computations; we hope that the outline given is sufficiently clear to convince a person familiar with this theory that he could supply all details in the unlikely case that he should wish to.

[2] We refer to [5] and [6] collectively as AFOC and to [7] as ACID.

[3] I am indebted to Jon Barwise and Robin Gandy for many discussions during 1967–1968 which made me appreciate set theories without the power set axiom in general and admissible sets in particular. The original draft of this paper written in the summer of 1967 concerned extensions of Gödel-Bernays set theory and had a very different flavor.

1. **Definitions and results.** We use a two-sorted language with set variables $x, y, z, \ldots$, class variables $S, T, U, \ldots, \in, =$ and the usual logical symbols. It is convenient to include a constant 0 for the empty set and a function symbol $\{x, y\}$ for the unordered pair; the ordered pair

$$(x, y) = \{\{x, y\}, \{x\}\}$$

is defined in the usual way.

Greek letters α, $\beta(x)$, $\alpha(x, S)$ etc. will denote *first-order formulas*, i.e. formulas with no class quantifiers. A formula Φ is $\Sigma^1_n(\Pi^1_n)$ if it is of the form $\exists S_1 \forall S_2 \exists S_3 \cdots S_n \alpha (\forall S_1 \exists S_2 \forall S_3 \cdots S_n \alpha)$.

Relations will be represented in the usual way by sets (or classes) of pairs. Put

$$x \in \mathit{Field}(S) \leftrightarrow \exists y[(x, y) \in S \vee (y, x) \in S],$$

(1.1) $$\mathrm{WF}(S) \leftrightarrow \forall x[x \neq 0 \rightarrow (\exists y \in x)(\forall z \in x) \neg (z, y) \in S];$$

if WF(S), then every nonempty set has a member which is minimal for S, i.e. the class of pairs in S is a well-founded relation.

If $\Phi(x)$ is any formula, then

(Φ-Ind) $$\{\mathrm{WF}(S) \mathbin{\&} \forall x\{\forall y[(y, x) \in S \rightarrow \Phi(y)] \rightarrow \Phi(x)\}\} \rightarrow (\forall x \in \mathit{Field}(S))\Phi(x)$$

asserts that we can prove $\Phi(x)$ for x's in the field of a well-founded relation by transfinite induction. For any class Γ of formulas, Γ-Ind is the schema obtained from Φ-Ind by letting Φ vary over all the formulas in Γ—we are particularly interested in Σ^1_1-Ind and Π^1_1-Ind.

The basic theory we study is PZF_1, the natural extension to a theory with class variables of Kripke's PZF in [3]. The axioms of PZF_1 are the defining axioms for 0 and $\{x, y\}$, extensionality for both sets and classes, $\forall x \exists S[x = S]$, union, replacement

(Repl) $$\forall u \exists !\, v(u, v) \in S \rightarrow \exists z \forall v[v \in z \leftrightarrow (\exists u \in x)(u, v) \in S],$$

regularity

(Reg) $$\forall u \forall v[(u, v) \in S \leftrightarrow u \in v] \rightarrow \mathrm{WF}(S)$$

and the axiom schemata of Δ^1_1-Comp, Σ^1_1-Ind and Π^1_1-Ind.

Another axiom schema that will come into play is Σ^1_1-Collection,

(Σ^1_1-Coll) $$\forall x \exists S \alpha(x, s) \rightarrow \exists S \forall x \exists y \alpha(x, S^y),$$

where

(1.2) $$S^y = \{u : (y, u) \in S\};$$

we let PZF^+_1 be the theory $\mathrm{PZF}_1 + \Sigma^1_1$-Coll.

By taking $\Phi(x) \leftrightarrow x \notin T$ in Σ^1_1-Ind, it is easy to obtain

(Reg′) $$\mathrm{WF}(S) \leftrightarrow \forall T[T \neq 0 \rightarrow (\exists y \in T)(\forall z \in T) \neg (z, y) \in S];$$

thus a relation is well founded with respect to sets if and only if it is well founded with respect to classes. Using this it is easy to prove the *axiom of infinity* in PZF_1,

(Inf) $$\exists x[0 \in x \;\&\; \forall y[y \in x \rightarrow y \cup \{y\} \in x]].$$

Extensions of the Gödel-Bernays theory are obtained by adding the axioms of power set and choice to PZF_1 or PZF_1^+. If by "choice" we mean the local axiom (*every set can be well ordered*), then these theories are extensions of PZF_1 or PZF_1^+ by first-order axioms.

The axiom schema Δ_1^1-Comp allows us to define in PZF_1 *satisfaction* and *truth* for structures whose domains are proper classes and whose relations are represented by proper classes of tuples.

THEOREM 1. (a) *In* PZF_1 *we can prove: if* $\langle U, R_1, \ldots, R_n\rangle$ *is a structure,* T *a binary relation that well orders* U *and* $x \subseteq U$, *then there exists a set* $u \supseteq x$ *and relations* $r_1, \ldots, r_n$ *such that* $\langle u, r_1, \ldots, r_n\rangle$ *is an elementary substructure of* $\langle U, R_1, \ldots, R_n\rangle$.

(b) *In* PZF_1^+ *axiom of power set we can prove: there is a natural model* V_ξ (= *the set of sets of rank* $<\xi$) *so that the structure* $\langle V_\xi, \in\rangle$ *is an elementary substructure of the universe* $\langle V, \in\rangle$.

In §3 we shall argue that the theory of computability on a structure $\langle U, R_1, \ldots, R_n\rangle$ of AFOC can be formalized in PZF_1. An easy consequence will be

THEOREM 2. *Both* PZF_1 *and* PZF_1^+ *are finitely axiomatizable.*

Let U be a class defined by a formula $x \in U$, $\mathscr{D}$ a collection of classes defined by a formula $S \in \mathscr{D}$ (formulas $x \in U$, $S \in \mathscr{D}$ may contain set or class parameters, in particular U may be a variable). The relativization

$$\langle U, \mathscr{D}\rangle \models \Phi$$

of a formula Φ to the structure $\langle U, \mathscr{D}\rangle$ is obtained by replacing in Φ all set quantifiers $\forall x$, $\exists x$ by $(\forall x \in U)$, $(\exists x \in U)$ and all class quantifiers $\forall S$, $\exists S$ by $(\forall S \in \mathscr{D})$, $(\exists S \in \mathscr{D})$. We say that $\langle U, \mathscr{D}\rangle$ is a (transitive $\in$-) *model* of a theory $\mathscr{T}$ if U is a transitive class and all relativizations $\langle U, \mathscr{D}\rangle \models \Phi$ of axioms Φ of $\mathscr{T}$ hold. For finitely axiomatizable $\mathscr{T}$ (like PZF_1, PZF_1^+) we can express this by a single formula of PZF_1.

A formula Φ is *absolute for* $\langle U, \mathscr{D}\rangle$ if we can prove in PZF_1 that for values of the free variables of Φ in $\langle U, \mathscr{D}\rangle$, Φ is equivalent to its relativization.

Let $S \in \mathbf{HP}(U)$ be the formal version of the relation "*S is a subclass of U, hyperprojective in* $\in$" as defined in AFOC—this will be a Σ_1^1 formula—let V be the universe of sets. Theorems 3-5 will follow immediately from the formalization of the theory in AFOC, ACID.

THEOREM 3. (a) *The formula* $S \in \mathbf{HP}(V)$ *is absolute for* $\langle V, \mathbf{HP}(V)\rangle$.

(b) *In* PZF_1 *we can prove*: $\langle V, \mathbf{HP}(V)\rangle$ *is a model of* PZF_1^+.

(c) *If* $\mathscr{T}$ *is any extension of* PZF_1 *by first-order axioms,* Φ *a* Σ_1^1 *formula and* $\mathscr{T} + \Sigma_1^1\text{-Coll} \vdash \Phi$, *then* $\mathscr{T} \vdash \Phi$.

THEOREM 4. *Let $\mathcal{T}$ be an extension of* PZF_1 *by first-order axioms, let* $\alpha(T)$, $\beta(T)$ *be first-order formulas, assume that*

$$\mathcal{T} + \Sigma_1^1\text{-Coll} \vdash \forall T\alpha(T) \to \exists T\beta(T).$$

There exist first-order formulas $\alpha_1(T)$, $\beta_1(T)$ *such that*

$$\mathcal{T} \vdash \forall T\alpha(T) \to \forall T\alpha_1(T),$$
$$\mathcal{T} \vdash \forall T\alpha_1(T) \leftrightarrow \exists T\beta_1(T),$$
$$\mathcal{T} \vdash \exists T\beta_1(T) \to \exists T\beta(T).$$

(Δ_1^1-*Interpolation.*)

THEOREM 5. *Assume that* $S \in \mathcal{D}$ *is a* Σ_n^1 *formula. Then in* $\mathrm{PZF}_1 + \Sigma_n^1$-Ind *we can prove: if* $\langle U, \mathcal{D}\rangle$ *is a model of* PZF_1, *then* $\langle U, \mathbf{HP}(U)\rangle$ *is a model of* PZF_1^+ *and* $\mathbf{HP}(U) \subseteq \mathcal{D}$.

Some of the work in §§2,3 could be simplified if we worked in PZF_1^+ instead of PZF_1, but we consider part (c) of Theorem 3 sufficiently interesting to be worth the small additional work—of course (c) follows trivially from (b) of Theorem 3. What makes proofs in PZF_1^+ a little easier than proofs in PZF_1 is the possibility of advancing existential class quantifiers past set quantifiers to the front of a formula with first-order scope. Thus in PZF_1^+ every formula in which only existential class quantifiers occur and these only positively (these have been called *extended* Σ_1^1 formulas) is equivalent to a Σ_1^1 formula.

2. **Induction in** PZF_1. We show here how to formalize in PZF_1 inductive definitions with respect to first-order formulas. This is the key to the formalization of computability theory that we shall outline in §3.

Let

$$\Phi \textit{ is } \Sigma_1^1$$

abbreviate the universal closure of

$$\Phi \leftrightarrow \Phi_1,$$

where Φ_1 is any Σ_1^1 formula and similarly for Φ *is* Π_1^1; then

$$\Phi \textit{ is } \Delta_1^1 \leftrightarrow \Phi \textit{ is } \Sigma_1^1 \;\&\; \Phi \textit{ is } \Pi_1^1.$$

Repeated use of the transformation

$$\exists S\exists T\Phi(S, T) \leftrightarrow \exists U\Phi_1(U), \tag{2.1}$$

where $\Phi_1(U)$ comes from $\Phi(S, T)$ by replacing each $t \in S$ by $(t, 0) \in U$ and each $t \in T$ by $(t, 1) \in U$ allows us to prove in PZF_1 Φ *is* Σ_1^1 for every Φ of the form $\exists S_1 \cdots \exists S_n\alpha$, and similarly for a multiple universal prefix.

Let us temporarily use German letters $\mathfrak{a}$, $\mathfrak{b}$, $\mathfrak{c}$ for arbitrary lists of variables of both sorts. If for a formula $f(\mathfrak{a}, \mathfrak{b})$ we know that $\forall\mathfrak{a}\exists!\,\mathfrak{b}f(\mathfrak{a}, \mathfrak{b})$, then we let $f(\mathfrak{a})$ denote the function that $f(\mathfrak{a}, \mathfrak{b})$ defines and as usual we abbreviate

$$\Phi(f(\mathfrak{a})) \leftrightarrow \exists\mathfrak{b}[f(\mathfrak{a}, \mathfrak{b}) \;\&\; \Phi(\mathfrak{b})]. \tag{2.2}$$

It is easy to verify in PZF_1 that

(I) $$\Phi(\mathfrak{b}) \text{ is } \Delta_1^1 \;\&\; f(\mathfrak{a}, \mathfrak{b}) \text{ is } \Sigma_1^1 \;\&\; \forall \mathfrak{a} \exists!\, \mathfrak{b} f(\mathfrak{a}, \mathfrak{b}) \to \Phi(f(\mathfrak{a})) \text{ is } \Delta_1^1.$$

One common way of obtaining such Σ_1^1 $f(\mathfrak{a}, S)$ is by definitions of the form

$$f(\mathfrak{a}, S) \leftrightarrow \forall x[x \in S \leftrightarrow \alpha(\mathfrak{a}, x)]$$

which we abbreviate

$$f(\mathfrak{a}) = \{x : \alpha(\mathfrak{a}, x)\};$$

e.g. the definition of S^y in (1.2) is of this type.

(II) *In* PZF_1: $\Phi(x, S)$ *is* Σ_1^1 & $(\forall x \in y)\exists!\, S\Phi(x, S) \to \exists T(\forall x \in y)\Phi(x, T^x)$.

PROOF. Put

$$T = \{(x, u) : \exists S[\Phi(x, S) \;\&\; u \in S]\}$$
$$= \{(x, u) : \forall S[\Phi(x, S) \to u \in S]\}$$

using Δ_1^1-Comp. (More precisely this definition should be

$$T = \{y : \exists x \exists u[y = (x, u) \;\&\; \exists S[\Phi(x, S) \;\&\; u \in S]]\}$$

and a similar one for the Π_1^1 version.)

(III) *In* PZF_1: *if* $f(S, T)$ *is* Σ_1^1 *and* $\forall S \exists!\, Tf(S, T)$, *then for each* S_0 *there is a* U *such that* $U = \bigcup_{n \in \omega} f^n(S_0)$, *where* $f^0(S) = S$, $f^{n+1}(S) = f(f^n(S))$.

PROOF. First we define

$$T = f^n(S) \leftrightarrow n \in \omega \;\&\; \exists U[(\forall i \in n) U^{i+1} = f(U^i) \;\&\; U^0 = S \;\&\; T = U^n]$$

which is easily proved from the hypothesis to be Σ_1^1 and such that $(\forall n, S)\exists!\, T[T = f^n(S)]$. Applying this to $S = S_0$ and taking by (II) some T such that

$$(\forall n \in \omega)[T^n = f^n(S_0)],$$

it is enough to put $U = \bigcup_n T^n$.

As usual

$$\mathit{Function}(S) \leftrightarrow \forall x \exists!\, y(x, y) \in S.$$

(IV) *In* PZF_1: *if for every* $n \in \omega$, *Function*(S^n) *and if* x *is a set, then there exists a set* y *such that* $x \subseteq y$ *and for every* $n \in \omega$, $S^n[y] \subseteq y$ (*i.e.* $t \in y \;\&\; (t, s) \in S^n \to s \in y$).

PROOF. Put by induction $x_0 = x$,

$$x_{n+1} = x_n \cup \bigcup_{k=0}^{\infty} S^k[x_n]$$

and take $y = \bigcup_n x_n$; y is a set by replacement, since ω is a set.

PROOF OF THEOREM 1(a). To define truth for a given structure $\langle U, R_1, \ldots, R_n \rangle$, we first settle on some canonical coding of all sentences in the language appropriate for the structure (with constants from U) and then define an operator $f(S)$

such that if S is the class of codes of true sentences of length n, then $f(S)$ is the class of codes of true sentences of length $n + 1$; then the class of true sentences will be $\bigcup_{n\in\omega} f^n(0)$ by (III). Next we assign to each sentence in prenex form (with no constants) Skolem functors, using the given well-ordering T and using (II) we code all these Skolem functors into one class T_0. Now (IV) gives us a closure of any given $x \subseteq U$ by these Skolem functors which then defines the elementary substructure that we seek in the usual way.

Theorem 1(b) can be proved in the same fashion by one of the known classical methods, e.g. that of Montague-Vaught **[4]**.

A formula $\alpha(S)$ is *positive in S* if prime subformulas of the form $t \in S$ occur only positively. For such formulas we can easily show

(V) $$in\ \mathrm{PZF}_1: \alpha(S)\ \&\ S \subseteq T \to \alpha(T).$$

A formula $\alpha(S)$ is *positive existential in S* if it is positive in S and no subformula of the form $t \in S$ is within the scope of a universal quantifier. For such formulas we can easily show

(VI) $$in\ \mathrm{PZF}_1: \alpha(S) \to \exists u[u \subseteq S\ \&\ u\ is\ finite\ \&\ \alpha(u)].$$

Let $\alpha(x, S)$ be positive in S, first-order, perhaps with free variables other than x, S. We say that a formula $\alpha^\infty(x)$ *expresses the property defined inductively by* $\alpha(x, S)$ if

(VII) $$in\ \mathrm{PZF}_1: \{\alpha(x, S)\ \&\ \forall y[y \in S \to \alpha^\infty(y)]\} \to \alpha^\infty(x)$$

and

(VIII) $$in\ \mathrm{PZF}_1: \forall x\{[\alpha(x, S)\ \&\ \forall y[y \in S \to \Phi(y)]] \to \Phi(x)\} \to \forall x[\alpha^\infty(x) \to \Phi(x)]$$

for each Σ_1^1 or Π_1^1 formula $\Phi(x)$. Intuitively, $\{x : \alpha^\infty(x)\}$ is then the least (under $\subseteq$) stationary class of the monotone operator

$$\Gamma(S) = \{x : \alpha(x, S)\}$$

defined by $\alpha(x, S)$. We cannot in general prove in PZF_1 that this least stationary class exists, since we cannot in general find a Δ_1^1 formula $\alpha^\infty(x)$ that satisfies (VII) and (VIII). We show in the remainder of this section that if $\alpha(x, S)$ is first-order positive in S, we can choose $\alpha^\infty(x)$ to be Σ_1^1 and if $\alpha(x, S)$ is in addition existential in S, we can choose $\alpha^\infty(x)$ to be first-order.

Suppose first that $\alpha(x, S)$ is positive, existential in S, put

(2.3) $$\alpha^\infty(x) \leftrightarrow \exists f[f\ proves\ \alpha^\infty(x)]$$

where

$$\begin{aligned} f\ proves\ \alpha^\infty(x) \leftrightarrow\ & Function(f)\ \&\ domain(f) = n + 1 \in \omega \\ & \&\ (\forall i \leq n)(\forall y \in f(i))\alpha(y, \textstyle\bigcup_{j<i} f(j)) \\ & \&\ x \in f(n). \end{aligned}$$

To prove (VII) for this choice of $\alpha^\infty(x)$, assume $\alpha(x, S)$ and choose a finite $u \subseteq S$

such that $\alpha(x, u)$ by (VI). Since $(\forall y \in u)\alpha^\infty(y)$, let us assign to each $y \in u$ some f_y such that f_y *proves* $\alpha^\infty(y)$. Let $k = \mathit{supremum}\{\mathit{domain}(f_y) : y \in u\}$, define f on $k + 1$ by

$$f(i) = \bigcup\{f_y(j) : j \leq i, y \in u\} \quad \text{if} \quad i < k,$$
$$f(k) = x.$$

Now the fact that $u \subseteq \bigcup\{f(i) : i < k\}$ and (V) immediately imply f *proves* $\alpha^\infty(x)$. To prove (VIII) assume $\alpha^\infty(x)$, choose f such that f *proves* $\alpha^\infty(x)$; now an easy finite induction on $i \in \mathit{domain}(f)$ shows that $(\forall y \in f(i))\Phi(y)$, so that we have $\Phi(x)$.

Sometimes it is easier to do induction on ω rather than induction on the definition of $\alpha^\infty(x)$ as formalized by (VII) and (VIII). If $\alpha^\infty(x)$, defined by (2.3) holds, put

$$|x|_\alpha = \mathit{infimum}\{n : \exists f[f \text{ proves } \alpha^\infty(x) \;\&\; \mathit{domain}(f) = n + 1\},$$

(for definiteness we set $|x|_\alpha = \omega$ if $\neg\alpha^\infty(x)$). *Proof by induction on* $|x|_\alpha$ is formalized by

(IX) *in* $\mathrm{PZF}_1 : \forall x\{[\alpha^\infty(x) \;\&\; \forall y[\,|y|_\alpha < |x|_\alpha \rightarrow \Phi(y)]] \rightarrow \Phi(x)\}$
$$\rightarrow \forall x[\alpha^\infty(x) \rightarrow \Phi(x)],$$

where again $\Phi(x)$ is an arbitrary Σ_1^1 or Π_1^1 formula. This is easily verified.

In the case when $\alpha(x, S)$ is first-order positive but not necessarily existential, we wish to follow a similar method and also to define $|x|_\alpha$ so that (IX) will hold. A difficulty arises in that the ordinals required to "close" the induction defined by $\alpha(x, S)$ not only go beyond ω but may go beyond the ordinals of the system; we need to allow for *pre-well-ordered classes* to represent ordinals larger than any well-ordering of a set. Put

$$\begin{aligned}(2.4)\qquad \mathrm{PWO}(W) \leftrightarrow\ & (\forall x, y, z)[(x, y) \in W \;\&\; (y, z) \in W \rightarrow (x, z) \in W] \\ & \;\&\; \mathrm{WF}(W) \\ & \;\&\; (\forall x, y \in \mathit{Field}(W))\{[(x, y) \notin W \;\&\; (y, x) \notin W] \\ & \quad \rightarrow \forall u[[(u, x) \in W \leftrightarrow (u, y) \in W] \\ & \qquad \;\&\; [(x, u) \in W \leftrightarrow (y, u) \in W]].\end{aligned}$$

It is easy to check that if $\mathrm{PWO}(W)$, then the relation

$$(2.5)\quad \{(x, y) : (x, y) \in W\} \cup \{(x, y) : x, y \in \mathit{Field}(W) \;\&\; (x, y) \notin W \;\&\; (y, x) \notin W\}$$

is reflexive, transitive, connected and well founded—from being a well-ordering it only lacks antisymmetry. It is convenient for us to deal with the "strict pre-well-orderings" as defined by (2.4) rather than with the more familiar associated relations of (2.5). It is essential that we go beyond well-orderings to represent ordinals, since we have no global axiom of choice to well order V.

Pre-well-orderings are compared for length via *similarity relations*. Put

$$\begin{aligned} S: W_1 \cong W_2 \leftrightarrow (\forall x, y)\{(x, y) \in S \leftrightarrow [&x \in Field(W_1) \,\&\, y \in Field(W_2) \\ &\& \, (\forall u: (u, x) \in W_1)(\exists v: (v, y) \in W_2)(u, v) \in S \\ &\& \, (\forall v: (v, y) \in W_2)(\exists u: (u, x) \in W_1)(u, v) \in S]\} \\ &\& \, (\forall x \in Field(W_1))(\exists y \in Field(W_2))(x, y) \in S \\ &\& \, (\forall y \in Field(W_2))(\exists x \in Field(W_1))(x, y) \in S. \end{aligned}$$

A simple transfinite induction shows in PZF_1 that similarity relations are unique, when they exist:

$$PWO(W_1) \,\&\, S: W_1 \cong W_2 \,\&\, T: W_1 \cong W_2 \to S = T.$$

The initial segments of pre-well-orderings are given by

$$W^{[u]} = \{(x, y) \in W: (y, u) \in W\}.$$

If we abbreviate

$$\begin{aligned} |W_1| = |W_2| &\leftrightarrow (\exists S)[S: W_1 \cong W_2], \\ |W_1| < |W_2| &\leftrightarrow (\exists u)(\exists S)[S: W_1 \cong W_2^{[u]}], \\ |W_1| \leq |W_2| &\leftrightarrow [|W_1| = |W_2| \vee |W_1| < |W_2|], \end{aligned}$$

then a simple Σ_1^1 induction shows in PZF_1 that

$$PWO(W_1) \,\&\, PWO(W_2) \to |W_1| < |W_2| \vee |W_1| = |W_2| \vee |W_2| < |W_1|,$$

so that to a large extent pre-well-orderings behave like well-orderings. We intuitively refer to the similarity classes under $\cong$ as *long ordinals*.

Let $\alpha(x, S)$ be first-order positive in S, put

$$\begin{aligned} R^\alpha(W, T) \leftrightarrow PWO(W) \,\&\, Relation(T) \,\&\, (\forall u \notin Field(W))T^u = 0 \\ \& \, (\forall u \in Field(W))T^u = \{x: \alpha(x, \bigcup\{T^v: (v, u) \in W\})\}. \end{aligned}$$

A trivial induction on W shows that

(X) *in* $PZF_1: PWO(W_1) \,\&\, S: W_1 \cong W_2 \,\&\, R^\alpha(W_1, T_1) \,\&\, R^\alpha(W_2, T_2) \,\&\, (u, v) \in S \to T_1^u = T_2^v$;

in particular, when $W_1 = W_2 = W$ and $S: W \cong W$, this gives

(XI) *in* $PZF_1: PWO(W) \,\&\, R^\alpha(W, T_1) \,\&\, R^\alpha(W, T_2) \to T_1 = T_2$.

Now a Σ_1^1 induction on W using (II) and (XI) shows

$$PWO(W) \to \forall u \exists T R^\alpha(W^{[u]}, T)$$

which with (XI) gives

(XII) *in* $PZF_1: PWO(W) \to \exists!\, T R^\alpha(W, T)$.

Hence, if $PWO(W)$, then

$$\exists T[R^\alpha(W, T) \,\&\, \alpha(x, \bigcup_u T^u)] \leftrightarrow \forall T[R^\alpha(W, T) \to \alpha(x, \bigcup_u T^u)],$$

so that by Δ_1^1-Comp we can put

$$F^\alpha(W) = \{x : \exists T[R^\alpha(W, T) \mathbin{\&} \alpha(x, \bigcup_u T^u)]\} \quad \text{if} \quad \mathrm{PWO}(W),$$
$$= 0 \quad \text{if} \quad \neg\, \mathrm{PWO}(W).$$

Now (XII) again implies that $F^\alpha(W)$ depends only on the long ordinal of W, i.e.

(XIII) *in* $\mathrm{PZF}_1 : \mathrm{PWO}(W_1) \mathbin{\&} |W_1| = |W_2| \to F^\alpha(W_1) = F^\alpha(W_2)$.

Intuitively, $F^\alpha(W)$ consists just of those x such that $\alpha^\infty(x)$ is realized in $|W|$ or fewer steps of the inductive process determined by $\alpha(x, S)$. Thus we can easily show in PZF_1,

$$\mathrm{PWO}(W) \mathbin{\&} R^\alpha(W, T) \to (\forall u \in \mathit{Field}(W))[T^u = F^\alpha(W^{[u]})].$$

We can now set

$$\alpha^\infty(x) \leftrightarrow \exists W[\mathrm{PWO}(W) \mathbin{\&} x \in F^\alpha(W)].$$

In order to prove (IX) in addition to (VII) and (VIII) for this definition of $\alpha^\infty(x)$, we must assign to each x a class $|x|_\alpha$ so that

(XIV) *in* $\mathrm{PZF}_1 : \alpha^\infty(x) \to [\mathrm{PWO}(|x|_\alpha) \mathbin{\&} x \in F^\alpha(|x|_\alpha)]$,

(XV) *in* $\mathrm{PZF}_1 : \mathrm{PWO}(W) \mathbin{\&} x \in F^\alpha(W) \to |x|_\alpha \leq |W|$.

Put

$$s <^\alpha t \leftrightarrow \exists W[\mathrm{PWO}(W) \mathbin{\&} s \in F^\alpha(W) \mathbin{\&} t \notin F^\alpha(W)];$$

if $\alpha^\infty(t)$, it follows immediately that

$$s <^\alpha t \leftrightarrow \forall W[\mathrm{PWO}(W) \mathbin{\&} t \in F^\alpha(W) \to (\exists u) s \in F(W^{[u]})],$$

so that when $\alpha^\infty(t)$, the relation $s <^\alpha t$ is Δ_1^1. By Δ_1^1-Comp now, put

$$|x|_\alpha = \{(s, t) : t <^\alpha x \mathbin{\&} s <^\alpha t\} \quad \text{if} \quad \alpha^\infty(x),$$
$$= V \quad \text{if} \quad \neg\alpha^\infty(x).$$

Intuitively, the field of $|x|_\alpha$ consists of all t such that $\alpha^\infty(t)$ is recognized before $\alpha^\infty(x)$ and $|x|_\alpha$ preorders such t's precisely by the order in which $\alpha^\infty(t)$ is recognized. Formally,

(XVI) *in* $\mathrm{PZF}_1 : \mathrm{PWO}(W) \mathbin{\&} R^\alpha(W, T) \mathbin{\&} x \in T^u \mathbin{\&} (\forall v : (v, u) \in W) x \notin T^v$
$\to (\exists S) S : W^{[u]} \cong |x|_\alpha$

which is proved by taking

$$S = \{(v, t) : (v, u) \in W \mathbin{\&} t \in T^v \mathbin{\&} (\forall z : (z, v) \in W) t \notin T^z\}$$

and showing $S : W^{[u]} \cong |x|_\alpha$ by transfinite induction. From (XVI) now, (XIV) and (XV) follow easily. Proofs of (VII), (VIII) and (IX) with these definitions are easy and we only outline that of (IX). Assume $\alpha^\infty(x)$ and show by Φ-Ind on $|x|_\alpha$, using the hypothesis of (IX) that $(\forall u \in \mathit{Field}(|x|_\alpha))\Phi(u)$. Since $\mathit{Field}(|x|_\alpha) = \{y : |y|_\alpha < |x|_\alpha\}$, the hypothesis of (IX) gives then immediately $\Phi(x)$.

3. **Formalization of abstract computability theory in** PZF_1. Let B^* be a fixed class containing 0 and closed under the formation of ordered pairs, put

$$B = \{x \in B^* : x \neq 0 \ \& \ (\forall u, v) x \neq (u, v)\}.$$

The definitions of §1 of AFOC apply and in particular yield n-tuple operations $\langle x_1, \ldots, x_n \rangle$ on B^* to B^* which are represented by terms of PZF_1. We represent n-ary partial multiple valued (p.m.v.) functions on B^* to B^* by classes of $n + 1$-tuples: to $f(x_1, \ldots, x_n)$ we assign

$$\{\langle x_1, \ldots, n_n, z \rangle \in B^* : f(x_1, \ldots, x_n) \to z\}.$$

In order to formalize the theory of primitive computability we introduce the class of *primitive computable formulas*. We think of $B^*, \boldsymbol{\varphi} = \varphi_1, \ldots, \varphi_l$ as fixed class terms (variables or defined classes) that represent a domain and a sequence of p.m.v. functions of $n_1, \ldots, n_l$ arguments respectively and we give a syntactical definition of a class of formulas that represent the functions defined by C1–C7 of AFOC. More precisely, we define by induction the predicate *the formula* $\alpha(\mathbf{u}, z) = \alpha(u_1, \ldots, u_k, z)$ *is primitive computable in* $B^*, \boldsymbol{\varphi}$ *with variables* $\mathbf{u}$ *and value* z—when this is the case, the formula $\alpha(\mathbf{u}, z)$ will have no class variables other than those in $B^*, \boldsymbol{\varphi}$ but may have set variables other than $\mathbf{u}$, z. Some of the cases in the definition are:

Case C0. *For each* $1 \leq i \leq l$, *the formula* $\langle t_1, \ldots, t_{n_i}, x_1, \ldots, x_n, z \rangle \in \phi_i$ *is primitive computable in* $B^*, \boldsymbol{\varphi}$ *with variables* $t_1, \ldots, t_{n_i}, x_1, \ldots, x_n$ *and value* z.

Case C1. *The formula* $z = y$ *is primitive computable in* $B^*, \boldsymbol{\varphi}$ *with variables* $x_1, \ldots, x_n$ *and value* z.

Case C5. *If* $\alpha(\mathbf{x}, t) = \alpha(x_1, \ldots, x_n, t)$ *and* $\beta(t, \mathbf{x}, z)$ *are primitive computable in* $B^*, \boldsymbol{\varphi}$ *with the indicated variables and values, then*

$$\exists t[\alpha(\mathbf{x}, t) \ \& \ \beta(t, \mathbf{x}, z)]$$

is primitive computable in $B^*, \boldsymbol{\varphi}$ *with variables* $\mathbf{x}$ *and value* z.

The remaining cases C2–C4, C6, C7 can be constructed similarly by formalizing the corresponding clauses in the definitions of primitive computability—only *Case* C6 (induction) is not completely trivial and requires a simple analysis of induction in PZF_1.

To each formula primitive computable in $B^*, \boldsymbol{\varphi}$ we assign a term f, its *primitive computable index* by induction on the definition of primitive computable formulas, following the indexing in §2 of AFOC. For example, a formula primitive computable by C0 above has index $\langle 0, d_i + n, i \rangle$ (recall that as in AFOC, $0 = 0, n + 1 = (n, 0)$), a formula primitive computable by C1 above has index $\langle 1, n, y \rangle$ (this term has y as a free variable) etc. To agree with the notation in AFOC, we abbreviate the primitive computable formula with index f by

$$\{f\}_{\mathrm{pr}}(B^*, \boldsymbol{\varphi}, \mathbf{u}) \to z$$

or even

$$\{f\}_{\mathrm{pr}}(\mathbf{u}) \to z$$

when $B^*, \boldsymbol{\varphi}$ are clear in context.

Of particular interest will be those formulas which are primitive computable in B^* alone (no $\boldsymbol{\varphi}$) and whose index is a closed term, built up from 0 alone. These define the functions that we called combinatorial in AFOC and we call them *combinatorial formulas.*

It is easy to prove in PZF_1 for each primitive computable formula

$$\{f\}_{\mathrm{pr}}(B^*, \boldsymbol{\varphi}, \mathbf{u}) \to z$$

that if each φ_i is single-valued and totally defined, then

$$(\forall \mathbf{u} \in B^*)\exists!\, z\{f\}_{\mathrm{pr}}(B^*, \boldsymbol{\varphi}, \mathbf{u}) \to z;$$

this is just the formal version of Lemma 1 of AFOC and can be verified by induction on the definition of primitive computable formula. In particular this holds for combinatorial formulas, and if we agree by convention that

$$\neg [u_1, \ldots, u_k \in B^*] \to [\{f\}_{\mathrm{pr}}(B^*, \mathbf{u}) \to z \leftrightarrow z = 0],$$

then for each combinatorial $\{f\}_{\mathrm{pr}}(B^*, \mathbf{u}) \to z$, the formula $\forall \mathbf{u} \exists!\, z\{f\}_{\mathrm{pr}}(B^*, \mathbf{u}) \to z$ is a theorem. This allows us to "substitute" combinatorial "terms" $\{f\}_{\mathrm{pr}}(B^*, \mathbf{u})$ in formulas by the convention established by (2.2).

Formalization of §3-§4 of AFOC along these lines is completely trivial. As an example, here is the formal version of Lemma 14 of AFOC, the very useful recursion theorem for primitive computable functions: *for each list* $B^*, \varphi_1, \ldots, \varphi_l$ *and each* k *there is a fixed combinatorial function* $rc_{\mathrm{pr}}(f)$ *such that for each primitive computable formula* $\{f\}_{\mathrm{pr}}(B^*, \boldsymbol{\varphi}, u_1, \ldots, u_k) \to z$ *we can prove in* PZF_1:

$$(\forall u_1, \ldots, u_k, z \in B^*)[\{f\}_{\mathrm{pr}}(B^*, \boldsymbol{\varphi}, rc_{\mathrm{pr}}(f), \mathbf{u}) \to z \leftrightarrow \{rc_{\mathrm{pr}}(f)\}_{\mathrm{pr}}(B^*, \boldsymbol{\varphi}, \mathbf{u}) \to z].$$

Classes of computable functions which are closed under the enumeration schema C8 of §5 of AFOC (first introduced by Kleene in [**2**]) cannot be defined independently of their indexing. Instead, there is a basic relation in the theory, $\{f\}_i(B^*, \boldsymbol{\varphi}, \mathbf{u}) \to z$ (where $i = p$ for *prime computability*, $i = v$ for *search computability* and $i = h$ for *hyperprojectivity*) which is defined inductively and which enumerates the relevant class as f ranges over B^*. Put

$$R_i(B^*, \boldsymbol{\varphi}, x) \leftrightarrow \mathrm{Seq}(x) \,\&\, lh(x) \geq 2$$
$$\& \{(x)_1\}_i(B^*, \boldsymbol{\varphi}, (x)_2, \ldots, (x)_{lh(x)-1}) \to (x)_{lh(x)};$$

now for each fixed $B^*, \boldsymbol{\varphi}$, $R_i(B^*, \boldsymbol{\varphi}, x)$ is a relation defined inductively by those clauses among C0′–C9′ which are specified for $i = p, v$ or h. If we follow the method of formalizing induction outlined in §2, we can set

$$R_i(B^*, \boldsymbol{\varphi}, x) \leftrightarrow \alpha_i^{\infty}(x),$$

where $\alpha_i(x, S)$ is positive in S and can be constructed easily from the informal inductive definitions given in AFOC. When $i = p, v$, then $\alpha_i(x, S)$ is actually positive existential in S, so that $R_i(B^*, \boldsymbol{\varphi}, x)$ is first-order; when $i = h$, then $\alpha(x, S)$ is first-order but not existential in S, so that $R_h(B^*, \boldsymbol{\varphi}, x)$ is Σ_1^1. We abbreviate formally

$$\{f\}_i(B^*, \boldsymbol{\varphi}, u_1, \ldots, u_k) \to z \leftrightarrow R_i(B^*, \boldsymbol{\varphi}, \langle f, u_1, \ldots, u_k, z\rangle)$$

for each k, so that these formulas again are first-order when $i = p, \nu$ and Σ_1^1 when $i = h$.

The basic theory of the relations $\{f\}_i(B^*, \boldsymbol{\varphi}, \mathbf{u}) \to z$ is developed in Lemmas 15–24 (for $i = p$), 25–32 (for $i = \nu$) and 37–43 (for $i = h$) of AFOC. Proofs of these lemmas can be formalized immediately in PZF_1 using (VII), (VIII), and (IX), since they are basically inductive. As an example consider Lemma 43, the Transitivity Lemma for hyperprojective relations. Its formal version is: *there is a combinatorial function* $\mathrm{tr}_h(f, c)$ *such that in* PZF_1 *we can prove*

$$(\forall \mathbf{u}', z \in B^*)[(\mathbf{u}', z) \in \chi \leftrightarrow \{c\}_h(B^*, \boldsymbol{\varphi}, \mathbf{u}') \to z]$$
$$\to (\forall \mathbf{u}, z \in B^*)[\{f\}_h(B^*, \chi, \boldsymbol{\varphi}, \mathbf{u}) \to z \leftrightarrow \{\mathrm{tr}_h(f, c)\}_h(B^*, \boldsymbol{\varphi}, \mathbf{u}) \to z].$$

Here there is a separate consequent for each k corresponding to the number of variables in the list $u_1, \ldots, u_k$; the single consequent which we prove and from which all the above follow by substitution is

$$(3.1)\quad (\forall f, u, z \in B^*)[R_h(B^*, \chi, \boldsymbol{\varphi}, \langle f, u, z\rangle) \leftrightarrow R_h(B^*, \boldsymbol{\varphi}, \langle \mathrm{tr}_h(f, c), u, z\rangle)].$$

We first define $\mathrm{tr}_h(f, c)$ using the instructions given in the proof of Lemma 43. Formally this gives us a fixed combinatorial formula, so that (3.1) at least makes sense. The proof of

$$(\forall f, u, z \in B^*)[R_h(B^*, \chi, \boldsymbol{\varphi}, \langle f, u, z\rangle) \to R_h(B^*, \boldsymbol{\varphi}, \langle \mathrm{tr}_h(f, c), u, z\rangle)]$$

is by induction on the definition of $R_h(B^*, \chi, \boldsymbol{\varphi}, x)$, i.e. (VIII) with $\alpha^\infty(x) \leftrightarrow R_h(B^*, \chi, \boldsymbol{\varphi}, x)$. In establishing the hypothesis of (VIII) we make several applications of (VII), now taking $\alpha^\infty(x) \leftrightarrow R_h(B^*, \boldsymbol{\varphi}, x)$. Proof of the converse implication is by ordinal induction on the definition of $R_h(B^*, \boldsymbol{\varphi}, x)$, i.e. (IX). Both inductions are allowed since the consequences are Σ_1^1 in both cases.

PROOF OF THEOREM 2. Let $\mathscr{T}$ be the theory obtained from PZF_1 by omitting the schemata Δ_1^1-Comp, Σ_1^1-Ind and Π_1^1-Ind and adding the Gödel-Bernays comprehension schema Δ_∞^0-Comp. It is well known that $\mathscr{T}$ is finitely axiomatizable. Our analysis of inductive definitions relative to positive existential formulas in §2 makes clear that the theories of prime and search computability can be formalized in $\mathscr{T}$. Let

$$(3.2)\qquad \exists S\alpha(S, \mathbf{T}, \mathbf{u}, x) \leftrightarrow \exists S\alpha(S, T_1, \ldots, T_n, u_1, \ldots, u_k, x)$$

be a fixed Σ_1^1 formula, where we have exhibited all the free variables. Repeated application of the "multiple-valued Skolem function" operation

$$\forall x \exists y \beta(x, y) \leftrightarrow \exists S[\forall x \exists y(x, y) \in S \;\&\; \forall x \forall y[(x, y) \in S \to \beta(x, y)]]$$

and the contraction of class variables given in (2.1) reduces (3.2) in PZF_1 to an equivalent formula of the form

$$(3.3)\qquad \exists S \forall \mathbf{s} \exists \mathbf{t} \gamma(S, \mathbf{T}, \mathbf{s}, \mathbf{t}, \mathbf{u}, x),$$

where $\mathbf{s}$, $\mathbf{t}$ are lists of set variables and $\gamma(S, \mathbf{T}, \mathbf{s}, \mathbf{t}, \mathbf{u}, x)$ is quantifier free. Take $B^* = V$ and $\boldsymbol{\varphi} = \in', S', T_1', \ldots, T_n'$, where

$$\in = \{(x, y) : x \in y\}$$

and for any class U,

$$U' = \{(x, 0) : x \in U\} \cup \{(x, 1) : x \notin U\}. \tag{3.4}$$

It is trivial to verify that for each quantifier free $\gamma(S, \mathbf{T}, \mathbf{s}, \mathbf{t}, \mathbf{u}, x)$ there is a closed term f such that in PZF_1,

$$\begin{aligned}\gamma(S, \mathbf{T}, \mathbf{s}, \mathbf{t}, \mathbf{u}, x) &\leftrightarrow \{f\}_{\mathrm{pr}}(V, \in', S', \mathbf{T}', \mathbf{s}, \mathbf{t}, \mathbf{u}, x) \to 0\\ &\leftrightarrow \{f\}_p(V, \in', S', \mathbf{T}', \mathbf{s}, \mathbf{t}, \mathbf{u}, x) \to 0,\end{aligned}$$

where we have used Lemma 16 of AFOC. An application of the Transitivity Lemma for prime computable functions will now yield a closed term f' so that in PZF_1 (3.3) becomes equivalent to

$$\exists S \forall s \exists t \{f'\}_p(V, \in', S', \langle T_1, \ldots, T_n\rangle', s, t, \langle u_1, \ldots, u_k\rangle, k) \to 0, \tag{3.5}$$

where

$$\langle T_1, \ldots, T_n\rangle = \{(1, y) : y \in T_1\} \cup \cdots \cup \{(n, y) : y \in T_n\}.$$

This easily implies that the single instance of Σ_1^1-Coll

$$\begin{aligned}&\forall x \exists S \forall s \exists t[\{f\}_p(V, \in', S', T', s, t, u, x) \to 0]\\ &\qquad\to \exists S \forall x \exists y \forall s \exists t[\{f\}_p(V, \in', S^{y\prime}, T', s, t, u, x) \to 0]\end{aligned}$$

implies every instance of Σ_1^1-Coll in $\mathscr{T}$. Similar applications of the reduction of (3.2) to (3.5) allows us to substitute single axioms for Δ_1^1-Comp, Σ_1^1-Ind, and Π_1^1-Ind, so that both PZF_1 and $\mathrm{PZF}_1^{\dagger}$ are finitely axiomatizable.

We proceed to the theory of hyperprojective relation which is our main interest here. The only case that concerns us is when

$$\boldsymbol{\varphi} = \in' \cap B^*, U_1' \cap B^*, \ldots, U_l' \cap B^*,$$

where U' is defined by (3.4), so for this $\boldsymbol{\varphi}$ let us abbreviate

$$\{f\}_i(\mathbf{U}, \mathbf{u}) \to z \leftrightarrow \{f\}_i(B^*, \boldsymbol{\varphi}, \mathbf{u}) \to z.$$

Since the identity relation is clearly hyperprojective in this $\boldsymbol{\varphi}$, we do not need to formalize those parts of the theory dealing with the problems arising when the relation $\sim$ of (10.8) of AFOC differs from $=$.

The relation

$$x \in H(\mathbf{U}) \leftrightarrow x \in H(B^*, \in' \cap B^*, U_1' \cap B^*, \ldots, U_l' \cap B^*)$$

is defined inductively by (11.7) of AFOC, so we can express it by a Σ_1^1 formula in PZF_1 using the technique of §2. Lemma 47 and Theorem 6 of AFOC are easy to prove formally in PZF_1, by induction as above. The ordinals $|s| = |s|\,(\mathbf{U})$ that are attached to each $x \in H(\mathbf{U})$ by (12.1) of AFOC are precisely the ordinals that we attached to elements satisfying an inductive definition in §2—formally these are pre-well-orderings. To prove formally the key Theorem 7 of AFOC, we first compute an index p of the required hyperprojective function $p(x, y)$ and then show by formal inductions on $|x|(\mathbf{U})$, i.e. (IX) that

$$\begin{aligned} &x \in H(\mathbf{U}) \mathbin{\&} \neg\, (|y|(\mathbf{U}) < |x|(\mathbf{U})) \to \{p\}_h(\mathbf{U}, x, y) \to 0,\\ &y \in H(\mathbf{U}) \mathbin{\&} \neg\, (|x|(\mathbf{U}) \leq |y|(\mathbf{U})) \to \{p\}_h(\mathbf{U}, x, y) \to \mathbf{1},\\ &x \in H(\mathbf{U}) \mathbin{\&} \neg\, (|y|(\mathbf{U}) < |x|(\mathbf{U})) \to (\forall z \neq 0) \neg\, \{p\}_h(\mathbf{U}, x, y) \to z,\\ &y \in H(\mathbf{U}) \mathbin{\&} \neg\, (|x|(\mathbf{U}) \leq |y|(\mathbf{U})) \to (\forall z \neq 1) \neg\, \{p\}_h(\mathbf{U}, x, y) \to z. \end{aligned}$$

These inductions are legitimate since the consequents of the first two are Σ^1_1 and the consequents of the last two are Π^1_1.

For a given formula $\Phi(\mathbf{u}, \mathbf{T}) \leftrightarrow \Phi(u_1, \ldots, u_k, T_1, \ldots, T_n)$ (which may have free variables other than $\mathbf{u}, \mathbf{T}$), let us abbreviate

$$\Phi(\mathbf{u}, \mathbf{T}) \textit{ is } s\mathbf{HP}(\mathbf{U})$$

any equivalence of the form

$$(\forall \mathbf{u} \in B^*)(\forall \mathbf{T} \subseteq B^*)[\Phi(\mathbf{u}, \mathbf{T}) \leftrightarrow \{f\}_h(\mathbf{T}, \mathbf{U}, \mathbf{u}) \to 0],$$

where f is a term in which none of the variables in the list $\mathbf{u}$ occur. It is trivial to verify in PZF_1

$$\begin{aligned} \Phi(-) \textit{ is } \mathbf{sHP}(\mathbf{U}) \mathbin{\&} \psi(-) \textit{ is } \mathbf{sHP}(\mathbf{U}) &\to \Phi(-) \mathbin{\&} \psi(-) \textit{ is } \mathbf{sHP}(\mathbf{U}),\\ \Phi(x, -) \textit{ is } \mathbf{sHP}(\mathbf{U}) &\to (\exists x \in B^*)\Phi(x, -) \textit{ is } \mathbf{sHP}(\mathbf{U}) \end{aligned}$$

etc. for all the first-order operations on B^* and for substitution of hyperprojective functions in a formula—this is essentially the formal version of Lemma 48 of AFOC.

For a given formula $\Phi(\mathbf{u}, \mathbf{T})$ as above, put

$$\begin{aligned} \Phi(\mathbf{u}, \mathbf{T}) \textit{ is } \mathbf{HP}(\mathbf{U}) \leftrightarrow (\exists f)(\forall \mathbf{u} \in B^*)(\forall \mathbf{T} \subseteq B^*)&\\ \{[\Phi(\mathbf{u}, \mathbf{T}) \leftrightarrow \{f\}_h(\mathbf{T}, \mathbf{U}, \mathbf{u}) \to 0]&\\ \mathbin{\&} [\neg\, \Phi(\mathbf{u}, \mathbf{T}) \leftrightarrow \{f\}_h(\mathbf{T}, \mathbf{U}, \mathbf{u}) \to 1]\}.& \end{aligned} \tag{3.6}$$

Finally put

$$\begin{aligned} S \in \mathbf{HP}(\mathbf{U}) &\leftrightarrow x \in S \textit{ is } \mathbf{HP}(\mathbf{U})\\ &\leftrightarrow (\exists f)(\forall x \in B^*)\{[x \in S \leftrightarrow \{f\}_h(\mathbf{U}, x) \to 0]\\ &\qquad \mathbin{\&} [x \notin S \leftrightarrow \{f\}_h(\mathbf{U}, x) \to 1]\}. \end{aligned}$$

Now (3.6) with $\Delta\text{-}^1_1\mathrm{Comp}$ imply in PZF_1

$$\Phi(x) \textit{ is } \mathbf{HP}(\mathbf{U}) \to (\exists S \in \mathbf{HP}(\mathbf{U}))(\forall x \in B^*)[\Phi(x) \leftrightarrow x \in S].$$

The formal version of Lemma 49 of AFOC that we need is

$$\neg\, [x \in H(\mathbf{U}) \textit{ is } \mathbf{HP}(\mathbf{U})]$$

and is very simple to show in PZF_1. From it follows immediately the formal version of the key Theorem 8.

(a) *There is a fixed combinatorial function* $eh(m, \mathbf{u})$ *such that in* PZF_1:

$$\Phi(\mathbf{u}) \textit{ is } \mathbf{sHP}(\mathbf{U}) \rightarrow (\exists m)(\forall \mathbf{u} \in B^*)[\Phi(\mathbf{u}) \leftrightarrow eh(m, \mathbf{u}) \in H(\mathbf{U})].$$

(b) *We can prove in* PZF_1:

$$(\forall \mathbf{u} \in B^*)(\exists ! z)\{f\}_h(\mathbf{U}, \mathbf{u}) \rightarrow z \ \& \ (\forall \mathbf{u} \in B^*)[\Phi(\mathbf{u}) \leftrightarrow \{f\}_h(\mathbf{U}, \mathbf{u}) \in H(\mathbf{U})]$$
$$\rightarrow \{\Phi(\mathbf{u}) \textit{ is } \mathbf{HP}(\mathbf{U}) \leftrightarrow (\exists x \in H(\mathbf{U}))(\forall \mathbf{u} \in B^*)$$
$$[\Phi(\mathbf{u}) \rightarrow |\{f\}_h(\mathbf{U}, \mathbf{u})| \ (\mathbf{U}) \leq |x| \ (\mathbf{U})]\}.$$

After this point, formalization of the theory in AFOC is direct and there is no need to reproduce any details here. The key results are Lemma 50 ($x \in H(\mathbf{U})$ *is* $\mathbf{sHP}(\mathbf{U})$), Lemma 51 (reduction for semihyperprojective formulas), Lemma 53 (normal branch lemma), Lemma 55 (boundedness for p.m.v. hyperprojective functions into $H(\mathbf{U})$) and Lemma 65, whose formal version asserts that we can prove in PZF_1

$$S \in \mathbf{HP}(\mathbf{U}) \textit{ is } \mathbf{sHP}(\mathbf{U})$$

from which follows that $S \in \mathbf{HP}(\mathbf{U})$ is equivalent to a Σ_1^1 formula, since the formula $\{f\}_h(S, \mathbf{U}) \rightarrow 0$ is Σ_1^1.

PROOF OF THEOREM 3. Let B^* be a fixed class closed under pairing, consider the structure $\langle B^*, \mathbf{HP}(B^*)\rangle = \langle B^*, \mathbf{HP}(B^*, \in')\rangle$. In §1 we defined the relativization $\langle B^*, \mathbf{HP}(B^*)\rangle \models \Phi$ of any formula Φ to $\langle B^*, \mathbf{HP}(B^*)\rangle$.

Assume that the formula $WF(S)$ of (1.1) is absolute for $\langle B^*, \mathbf{HP}(B^*)\rangle$. This is immediate for $\langle V, \mathbf{HP}(V)\rangle$ since $WF(S)$ is first-order. We show here that for such B^*, the structure $\langle B^*, \mathbf{HP}(B^*)\rangle$ satisfies Σ_1^1-Ind, Π_1^1-Ind, Δ_1^1-Comp and Σ_1^1-Coll, and that $S \in \mathbf{HP}(B^*)$ is absolute, from which (a) and (b) of Theorem 3 will follow immediately.

That $x \in H(B^*)$ is absolute for such $\langle B^*, \mathbf{HP}(B^*)\rangle$ follows from the formal version of Theorem 14 of AFOC and this implies immediately that $S \in \mathbf{HP}(B^*)$ is absolute since it is first-order definable on B^* in terms of $x \in H(B^*)$.

Since $S \in \mathbf{HP}(B^*)$ is Σ_1^1, the relativization of every Σ_1^1 formula to $\langle B^*, \mathbf{HP}(B^*)\rangle$ is also Σ_1^1 and similarly for Π_1^1 formulas. This, together with the assumed absoluteness of $WF(S)$ easily prove the relativizations of Σ_1^1-Ind and Π_1^1-Ind using Σ_1^1-Ind and Π_1^1-Ind in the universe.

To prove the relativization of Δ_1^1-Comp let $\exists T\alpha(T, \mathbf{U}, x)$, $\forall T\beta(T, \mathbf{U}, x)$ be the given Σ_1^1 and Π_1^1 formulas where we have exhibited all the free class variables. The hypothesis is $U_1, \ldots, U_l \in \mathbf{HP}(B^*)$ and

$$\forall x\{\exists T \in \mathbf{HP}(B^*))\alpha(T, \mathbf{U}, x) \leftrightarrow (\forall T \in \mathbf{HP}(B^*))\beta(T, \mathbf{U}, x)\}.$$

Now the formal version of Lemma 63 implies that both formulas

$$(\exists T \in \mathbf{HP}(B^*))\alpha(T, \mathbf{U}, x), \qquad (\exists T \in \mathbf{HP}(B^*)) \neg \beta(T, \mathbf{U}, x)$$

are $\mathbf{sHP}(B^*, \mathbf{U})$, hence $\{x \in B^* : (\exists T \in \mathbf{HP}(B^*))\alpha(T, \mathbf{U}, x)\}$ is in $\mathbf{HP}(B^*, \mathbf{U})$, since these relations are complementary on B^*. The formal version of the Transitivity Lemma for hyperprojective functions implies now that this class actually is in $\mathbf{HP}(B^*)$.

To show the relativization of Σ_1^1-Coll to $\langle B^*, \mathbf{HP}(B^*)\rangle$ we use in a similar manner the formal version of Lemma 61, (17.6) of AFOC.

Finally, to show part (c) of Theorem 3, assume that $\mathscr{T} + \Sigma_1^1\text{-Coll} \vdash \exists S\alpha(S)$. From part (b) then, $\mathscr{T} \vdash (\exists S \in \mathbf{HP}(V))\alpha(S)$, since first-order formulas are surely absolute for $\langle V, \mathbf{HP}(V)\rangle$, hence $\mathscr{T} \vdash \exists S\alpha(S)$.

Proof of Theorem 4. The proof of Theorem 3(b) that we outlined actually shows a relativized theorem, i.e. we can prove in PZF_1 that for any given classes $U_1, \ldots, U_l$ $\langle V, \mathbf{HP}(V, \mathbf{U})\rangle$ is a model of PZF_1^+. Suppose then that $U_1, \ldots, U_l, T$ are the only class variables in $\alpha(T), \beta(T)$; the hypothesis of Theorem 4 implies then that

$$\mathscr{T} \vdash (\exists T \in \mathbf{HP}(V, \mathbf{U})) \neg\, \alpha(T) \vee (\exists T \in \mathbf{HP}(V, \mathbf{U}))\beta(T).$$

It is easier to visualize the proof if we think of $\alpha(T), \beta(T)$ as having a free variable x, so that the disjuncts above define relations. These relations are $\mathbf{sHP}(V, \mathbf{U})$, so by the formal version of the Reduction Lemma 51 of AFOC we can find $\mathbf{sHP}(V, \mathbf{U})$ formulas Φ, Ψ such that

$$\mathscr{T} \vdash \Phi \rightarrow (\exists T \in \mathbf{HP}(V, \mathbf{U})) \neg\, \alpha(T),$$
$$\mathscr{T} \vdash \Psi \rightarrow (\exists T \in \mathbf{HP}(V, \mathbf{U}))\beta(T),$$
$$\mathscr{T} \vdash \neg\, (\Phi \,\&\, \Psi),$$
$$\mathscr{T} \vdash \Phi \vee \Psi.$$

Now the formal version of Theorem 14 of AFOC implies that for suitable first-order $\alpha_1(T), \beta_1(T)$,

$$\Phi \leftrightarrow (\exists T \in \mathbf{HP}(V, \mathbf{U})) \neg\, \alpha_1(T),$$
$$\Psi \leftrightarrow (\exists T \in \mathbf{HP}(V, \mathbf{U}))\beta_1(T)$$

from which the conclusion of Theorem 4 follows trivially.

Proof of Theorem 5. For this we must formalize the theory given in §5 of ACID. Since this is routine if one uses the preliminary work of this section, we give here only the formal statement of Theorem 4 of ACID, the key result that we need.

For given B^*, $\mathbf{U}$ and $x \in H(\mathbf{U})$, the class $H_x(\mathbf{U})$ is defined by (15.1) of AFOC—formally the relation $t \in H_x(\mathbf{U})$ is represented by a fixed Σ_1^1 formula which is equivalent to a fixed Π_1^1 formula when $x \in H(\mathbf{U})$. Put

$$S \in \mathbf{HP}_x(\mathbf{U}) \leftrightarrow \exists y[|y|\ (\mathbf{U}) < |x|\ (\mathbf{U}) \;\&\; S \textit{ is search computable in } H_y(\mathbf{U})];$$

again the formula $S \in \mathbf{HP}_x(\mathbf{U})$ is easily seen to be Σ_1^1 and equivalent to a fixed Π_1^1 formula when $x \in H(\mathbf{U})$.

(a) *$t \in H_1(\mathbf{U})$ is first-order on $B^*(|1|\,(\mathbf{U}) = 0)$.*

(b) *There is a fixed first-order on B^* formula $\alpha(x, y, S, t)$ such that in* PZF_1: *if $|x|\,(\mathbf{U}) = |y|\,(\mathbf{U}) + 1$, then*

$$t \in H_x(\mathbf{U}) \leftrightarrow \alpha(x, y, H_y(\mathbf{U}), t).$$

(c) *There are fixed first-order on B^* formulas $\beta_1(x, T, s)$, $\beta_2(x, T, s)$, $\gamma(x, S, t)$ such that in* PZF_1: *if $|x|\,(\mathbf{U})$ is a limit ordinal, then*

$$\begin{aligned}(\forall s \in B^*)(\exists T \beta_1(x, T, s) &\leftrightarrow ([\exists T \in \mathbf{HP}_x(\mathbf{U}))\beta_1(x, T, s)\\ &\leftrightarrow (\forall T\ \beta_2(x, T, s)\\ &\leftrightarrow (\forall T \in \mathbf{HP}_x(\mathbf{U}))\beta_2(x, T, s)]\end{aligned}$$

and if

$$S_0 = \{s \in B^* : \exists T \beta_1(x, T, s)\},$$

then

$$t \in H_x(\mathbf{U}) \leftrightarrow \gamma(x, S_0, t).$$

Now to prove Theorem 5. The proof that we outlined for (IX) in §2 shows that in $\mathrm{PZF}_1 + \Sigma^1_n$-Ind we can prove (IX) for each Σ^1_n formula $\Phi(x)$. Now using the hypothesis of Theorem 5, this fact and the formal version of Theorem 4 of ACID, it is trivial to show that

$$(\forall x \in H(U))H_x \in \mathscr{D}$$

from which $\mathbf{HP}(U) \subseteq \mathscr{D}$ follows. If

$$\mathscr{E} = \{S \subseteq U : \langle U, \mathscr{D}\rangle \models S \in \mathbf{HP}(V)\},$$

then the relativization of Theorem 3 to $\mathscr{D}$ shows that $\langle U, \mathscr{E}\rangle$ is a model of PZF_1^+. Applying what we just proved to this model, we obtain $\mathbf{HP}(U) \subseteq \mathscr{E}$. Since $\mathscr{E} \subseteq \mathbf{HP}(U)$ is immediate (the formula $S \in \mathbf{HP}(U)$ is Σ^1_1, hence preserved under extensions), we have $\mathscr{E} = \mathbf{HP}(U)$ which shows that $\langle U, \mathbf{HP}(U)\rangle$ is a model of PZF_1^+.

References

1. J. Barwise, R. O. Gandy and Y. N. Moschovakis, *The next admissible set*, J. Symbolic Logic (to appear).

2. S. C. Kleene, *Recursive functionals and quantifiers of finite types*. I, Trans. Amer. Math. Soc. **91** (1959), 1–52.

3. S. Kripke, *Transfinite recursion, constructible sets and analogues of cardinals*, Lecture Notes, Axiomatic Set Theory (University of California, Los Angeles, 1967) pp. IV-T-1—IV-T-7.

4. R. Montague and R. L. Vaught, *Natural models of set theories*, Fund. Math. **47**, (1959), 219–242.

5. Yiannis N. Moschovakis, *Abstract first order computability*. I, Trans. Amer. Math. Soc. **138** (1969), 427–464.

6. ———, *Abstract first order computability*. II, Trans. Amer. Math. Soc. **138** (1969), 465–504.

7. ———, *Abstract computability and invariant definability*, J. Symbolic Logic **34** (1969), 605–633.

8. Jean von Heijenoort, *From Frege to Gödel*, Harvard Univ. Press, Cambridge, Mass., 1967.

University of California, Los Angeles

ON SOME CONSEQUENCES OF THE AXIOM OF DETERMINATENESS

JAN MYCIELSKI

R. Solovay has shown that the axiom of determinateness implies the proposition

$\mathscr{Q}$. *Every well ordered family of countable sets of reals has a countable union.*

We want to improve this result here by showing that $\mathscr{Q}$ follows already from the following two propositions which are known to follow from the axiom of determinateness (see **[2]**).

$\mathscr{C}$. *Every countable family of sets of reals has a selector.*

$\mathscr{P}$. *Every uncountable set of reals has a perfect subset.*

In fact, we will not even use $\mathscr{C}$, but one of its consequences due to A. Church **[1]** (see also **[2]**):

$\mathscr{R}$. ω_1 *is a regular ordinal.*

Notice that $\mathscr{C}$ (and even the principle of dependent choices) and $\mathscr{P}$ hold in a model of Zermelo-Fraenkel set theory without the axiom of choice produced by Solovay [3] and hence $\mathscr{Q}$ holds also in this model. This model satisfies also the proposition that all sets of reals have the property of Baire and are Lebesgue measurable.

Let us put $\mathfrak{k} = |\mathfrak{P}(\omega)/\equiv|$, where $\mathfrak{P}$ is the power set operation and $A \equiv B \leftrightarrow |A \cup B \sim A \cap B| < \aleph_0$. Solovay has remarked that $\mathscr{Q}$ implies the following facts, excluding the existence of several one-to-one maps,

$$\aleph_1 \text{ in } 2^{\aleph_0}, \qquad \aleph_1 \text{ in } \mathfrak{k},$$

where $\mathfrak{m}$ in $\mathfrak{n} \leftrightarrow$ non $(\mathfrak{m} \leq \mathfrak{n} \vee \mathfrak{n} \leq \mathfrak{m})$. This completes a set of similar facts about cardinal numbers deduced in [**2**] from $\mathscr{C}$, $\mathscr{P}$ and the property of Baire (or Lebesgue measurability) of all sets of reals.

THEOREM. *$\mathscr{P}$ and $\mathscr{R}$ implies $\mathscr{Q}$.*

LEMMA. *$\mathscr{P}$ and $\mathscr{R}$ implies that a countable union of countable sets of reals is countable.*

PROOF. Consider the partition of Lebesgue of the real line into $\aleph_1$ nonempty sets S_ξ $(\xi < \omega_1)$ (see e.g. [**2**]). Suppose that T_n $(n < \omega)$ are countable sets of reals but $\bigcup_{n<\omega} T_n$ is uncountable. Then, by $\mathscr{P}$, $\bigcup_{n<\omega} T_n$ contains a perfect set. Since every perfect set of reals is of power $2^{\aleph_0}$ it follows that there exists a partition of the real line into countably many countable sets T_n^* $(n < \omega)$. Now each set T_n^* intersects only countably many sets S_ξ and ω_1 would have to be a countable union of countable sets which contradicts $\mathscr{R}$. Q.E.D.

PROOF OF THEOREM. Suppose that $\mathscr{Q}$ fails. Then there exists a sequence X_ξ $(\xi < \omega_1)$ of disjoint countable sets of reals. By $\mathscr{P}$ the set $\bigcup_{\xi<\omega_1} X_\xi$ contains a perfect set. Since every perfect set of reals is of power $2^{\aleph_0}$ it follows that there exists a partition of the real line into $\aleph_1$ disjoint countable sets A_ξ $(\xi < \omega_1)$ and a similar partition into disjoint countable sets $\mathbf{P}_\xi$ $(\xi < \omega_1)$ of the set $\mathbf{P}$ of all perfect sets of reals (because $|\mathbf{P}| = 2^{\aleph_0}$ also).

Now we will define a partition of ω_1 into two disjoint sets M, N such that both sets $\bigcup_{\xi\in M} A_\xi$ and $\bigcup_{\xi\in N} A_\xi$ intersect every set $P \in \mathbf{P}$. This of course contradicts $\mathscr{P}$ and proves our theorem.

M and N will be unions of two sequences M_ξ, N_ξ $(\xi < \omega_1)$ of segments of ω_1 which we define by induction as follows. Suppose that all M_ξ and N_ξ for all $\xi < \tau$ are already defined, where $\tau < \omega_1$. Let α be the first ordinal larger than all those of $\bigcup_{\xi<\tau}(M_\xi \cup N_\xi)$ and β the first ordinal such that $\bigcup_{\alpha\leq\xi<\beta} A_\xi$ intersects every set $P \in \mathbf{P}_\tau$. It follows by our Lemma that $\beta < \omega_1$. We put $M_\tau = \beta \sim \alpha$.[1] Let γ be the first ordinal such that $\bigcup_{\beta\leq\xi<\gamma} A_\xi$ intersects every $P \in \mathbf{P}_\tau$. Again by $\mathscr{R}$ and our Lemma, γ exists and $\gamma < \omega_1$. We put $N_\tau = \gamma \sim \beta$. It is obvious that both sets $M = \bigcup_{\xi<\omega_1} M_\xi$ and $N = \bigcup_{\xi<\omega_1} N_\xi$ have the required property. Q.E.D.

REFERENCES

1. A. Church, *Alternatives to Zermelo's assumption*, Trans. Amer. Math. Soc. **29** (1927), 178–208.

2. J. Mycielski, *On the axiom of determinateness*, Fund. Math. **53** (1964), 205–224; part II, ibidem **59** (1966), 203–212.

3. R. Solovay, *The measure problem*, Notices Amer. Math. Soc. **12** (1965), 217.

INSTITUTE OF MATHEMATICS
POLISH ACADEMY OF SCIENCES AND
UNIVERSITY OF COLORADO
BOULDER, COLORADO

[1] $\sim$ denotes the difference of sets; every ordinal α is the set of previous ordinals.

EMBEDDING CLASSICAL TYPE THEORY IN 'INTUITIONISTIC' TYPE THEORY

JOHN MYHILL

A long time ago Gödel proved that classical first-order arithmetic could in an appropriate sense be embedded in intuitionistic first-order arithmetic. The sense is this: let $\boldsymbol{Z}_{\mathrm{C}}$ be classical first-order arithmetic and let $\boldsymbol{Z}_{\mathrm{I}}$ be first-order arithmetic with Heyting logic. Then if a formula containing only $\forall$, $\neg$, $\wedge$ amongst its connectives is provable in $\boldsymbol{Z}_{\mathrm{C}}$, it is also provable in $\boldsymbol{Z}_{\mathrm{I}}$. This makes one believe that intuitionistic arithmetic is only apparently weaker than classical; in fact it is a richer system in that it makes more distinctions than $\boldsymbol{Z}_{\mathrm{C}}$. For example it distinguishes between constructive disjunction $A \vee B$ and nonconstructive disjunction $\neg(\neg A \wedge \neg B)$ where classical arithmetic considers only the latter: likewise with constructive existential quantification $(\exists x)$ and nonconstructive $\neg(\forall x)\neg$. There has been some discussion at this conference of more or less liberal intuitionisms, based on Zermelo-Fraenkel set theory or some fragment thereof plus intuitionistic logic. It seems a natural question to see how far Gödel's result can be extended: this paper is a small contribution to that problem.

The formalism of the classical theory of types can be described as follows. *Terms* are numerical variables $x_0, y_0, z_0; \ldots; 0; f(\mathfrak{s}_1, \ldots, \mathfrak{s}_n)$ where $\mathfrak{s}_1, \ldots, \mathfrak{s}_n$ are numerical terms and f is drawn from a prescribed stock of recursive functions (including successor of course); and set-variables $x_i, y_i, z_i, \ldots$ of type $i > 0$. Atomic *formulas* are $\mathfrak{s}_1 = \mathfrak{s}_2$ where $\mathfrak{s}_1$ and $\mathfrak{s}_2$ are terms of the same type ≥ 0, and $\mathfrak{s}_1 \in \mathfrak{s}_2$ where $\mathfrak{s}_1$ and $\mathfrak{s}_2$ are of consecutive ascending types. The other formulas are formed from these by means of $\forall x_i$, $\wedge$, $\neg$ in the usual way. The logical axioms are the usual ones of a many-sorted classical predicate calculus.

The identity axioms are

C0. $x_\iota = x_i$,

C1. $x_0 = y_0 \rightarrow_C \mathfrak{s}(x_0) = \mathfrak{s}(y_0)$ ($\mathfrak{s}(x_0)$ of type 0),

C2. $x_i = y_i \rightarrow_C [x_i \in z_{i+1} \leftrightarrow_C y_i \in z_{i+1}]$ $(i \geq 0)$,

C3. $(x_{i+1} = y_{i+1}) \rightarrow_C [z_i \in x_{i+1} \leftrightarrow_C z_i \in x_{i+1}]$ $(i \geqq 0)$,

where $A \rightarrow_C B$ means $\neg(A \wedge \neg B)$ and $A \leftrightarrow_C B$ means $[A \rightarrow_C B] \wedge [B \rightarrow_C A]$.

For sets we have set-existence

C4. $(\exists_C x_{i+1})(\forall y_i)(y \in x \leftrightarrow_C A(y))$

where $(\exists_C x)$ is $\neg(\forall x)\neg$, and extensionality

C5. $(\forall x_i)(x_i \in y_{i+1} \leftrightarrow_C x_i \in z_{i+1}) \leftrightarrow_C y = z$.

Finally we have the usual Peano axioms and recursive definitions.

The intuitionistic system has the same notation together with $\rightarrow \vee \exists$ (which we shall write sometimes $\rightarrow_I \vee_I \exists_I$ for clarity). The axioms are the Heyting axioms for logic, the Peano axioms and axioms C1–C5 obtained from the above by changing the subscript 'C' to 'I' throughout. $A \leftrightarrow_I B$ is here short for $[A \rightarrow_I B] \wedge [B \rightarrow_I A]$.

Let us write $\vdash_C$ and $\vdash_I$ for provability in the classical and intuitionistic systems respectively, and $\vdash_{C_0}$ and $\vdash_{I_0}$ for provability in the arithmetical fragments. Gödel's proof that if $\vdash_{C_0} A$ then $\vdash_{I_0} A$ proceeds roughly as follows. Call a formula A *stable*, if

$$\vdash_{I_0} \neg\neg A \rightarrow A;$$

then (1) by induction on the length of the proof, if $\vdash_{C_0} A$, then $\vdash_{I_0} \neg\neg A$; (2) by induction on the complexity of the formula, if A contains only $\forall \wedge \neg$, then A is stable. We need to modify this proof for our own purposes.

The only special feature of $\mathbf{I}_0$ and $\mathbf{C}_0$ needed for (2) is that for each atomic formula $\mathfrak{s}_1 = \mathfrak{s}_2$ of $\mathbf{I}_0$, $\neg\neg[\mathfrak{s}_1 = \mathfrak{s}_2] \rightarrow [\mathfrak{s}_1 = \mathfrak{s}_2]$ is provable. We would have to have

(I) $$\vdash_I \neg\neg [x = y] \rightarrow [x = y],$$

and

(II) $$\vdash_I \neg\neg [x \in y] \rightarrow [x \in y],$$

to carry out a similar proof, and these we do not have. Taking our cue from (II), we *restrict our attention to sets which satisfy double negation*. Specifically, we define *stable sets* inductively by

$$\mathrm{St}(x_1) \leftrightarrow_I (\forall x_0)[\neg\neg x_0 \in x_1 \rightarrow_I x_0 \in x_1]$$

$$\mathrm{St}(x_{i+1}) \leftrightarrow_I (\forall x_i)[\neg\neg x_i \in x_{i+1} \rightarrow_I x_i \in x_{i+1} \wedge x_i \in x_{i+1} \rightarrow_I \mathrm{St}(x_i)].$$

I.e. a stable set is a set which satisfies double negation and so do all its members, the members of its members, etc.

Now we define a formula A with free set variables $\mathfrak{y}^{(1)}, \ldots, \mathfrak{y}^{(n)}$ to be stable if

$$\vdash_{\mathrm{I}} \mathrm{St}(\mathfrak{y}^{(1)}) \wedge \cdots \wedge \mathrm{St}(\mathfrak{y}^{(n)}) \wedge \neg\neg A \to_{\mathrm{I}} A,$$

i.e. if it obeys double negation for all stable values of its variables. It is trivial that atomic formulas $\mathfrak{x} \in \mathfrak{y}$ are stable, and it is an easy exercise to prove the same for formulas $\mathfrak{x} = \mathfrak{y}$. Here by exactly Gödel's proof we have

LEMMA I. *If A contains only $\forall \wedge \neg$ as its connectives, and if A_S is obtained by restricting all (universal) quantifiers of A to stable sets, then A_S is stable.*

For our next lemma, observe that $A \to_{\mathrm{C}} B$ and $A \to_{\mathrm{I}} B$ are equivalent (for stable values of the variables) for stable formulas A and B. Hence if A is one of the formulas C0–C3, C5, $\vdash_{\mathrm{I}} A_S$. Likewise for C4, because A contains only $\forall \wedge \neg$ and so is stable by Lemma I. Consequently $\vdash_{\mathrm{I}} B_S$ and so $\vdash_{\mathrm{I}} \neg\neg B_S$. By induction on the length of the proof we get

LEMMA II. *For each theorem T of* **C**, $\vdash_{\mathrm{I}} \neg\neg T_S$.

Combining the two lemmas we get the

THEOREM. *If $\vdash_{\mathrm{C}} T$, then $\vdash_{\mathrm{I}} T_S$.*

Thus *the intuitionistic theory of types contains the classical*, which can be regarded as more restricted than it in two respects—it does not make the distinction of two kinds of existence or disjunction mentioned above, and it deals only with stable sets.

What is the bearing of all this on set theory? There have been a couple of attempts during this conference to propose a 'liberal' intuitionism as a foundation for mathematics. If someone were to propose Zermelo plus Heyting logic as such a foundation, how could our result affect our judgement of this proposal? It is clear that if a system S_1 is satisfactory for an intuitionistic standpoint, and if S_2 can be embedded in S_1, then S_2 is likewise satisfactory (modulo a reinterpretation). Thus because Heyting arithmetic is intuitionistically acceptable, Gödel's result shows that classical arithmetic is too, if we read the connectives and quantifiers aright. On the other hand we can reason in the opposite direction. If we think that classical type theory is intuitionistically unacceptable, we shall interpret the above result to show that intuitionistic type theory and a fortiori intuitionistic Zermelo is intuitionistically unacceptable. Since the logic is acceptable the difficulty must lie in the set theory itself, and specifically in the use of impredicative definitions. By any odds the avoidance of such definitions has turned out to be a much more significant aspect of the intuitionistic restriction, than the change in the underlying logic.

However one might still want to keep intuitionistic Zermelo or intuitionistic theory of types as a refinement and extension of classical mathematics—namely, one which treats sets that are not necessarily stable and which distinguishes two kinds of existence and two kinds of disjunction, etc. The classical theory of types would be a special case of the intuitionistic. One would like from this point of view to extend our earlier theorem say, to embed classical **ZF** in intuitionistic **ZF**. One would then have an intuitionistic **ZF** system that is at once powerful and refined. But there seem to be insuperable technical difficulties in the latter embedding, at least if we construe 'set' in the classical system as meaning 'hereditarily stable set' in the intuitionistic. It is however quite conceivable that another interpretation may work—for example taking 'set' as 'hereditarily stable constructible set' or 'hereditarily stable ordinal definable set'.

STATE UNIVERSITY OF NEW YORK, BUFFALO

ORDINAL DEFINABILITY

JOHN MYHILL AND DANA SCOTT

Consider Zermelo-Fraenkel set theory (which we call ZF) in one of its usual formalizations. We will allow *terms* as well as *formulas*, where these terms can be formed with the aid of, say, a description operator which by convention denotes the empty set in the improper cases. In such a theory we have a perfectly clear conception of what it means to be a *definable set*: these are the sets denoted by terms without free variables. Of course this is not a definition within the theory; in fact, a formal definition of definable set is impossible. Suppose we had such a definition. Certainly we believe that there are only countably many definable sets and uncountably many ordinal numbers. Thus if the definition of definability were adequate, then we could prove the existence of an undefinable ordinal. But the *least* undefinable ordinal is definable, and a contradiction is reached. Instead of a contradiction (which is only informal) we can rephrase this argument as the following more positive and precise statement:

METATHEOREM. *Every definable class with a definable well-ordering is included in every definable class which contains all definable sets.*

Here by a definable class we mean the collection represented by a formula of set theory with one free variable; while the definable well-ordering is represented by a formula with two free variables in such a way that the property of well-ordering of the class is provable as a schema and it is provable that every initial segment is a *set*. We are going to show that there is a definable class which has *both* of the properties mentioned in the metatheorem: it has a definable well-ordering and it contains all definable sets. Therefore it is *uniquely determined*; only it is not clear that it *exists*. The class will be identified in §1 where it will be shown to have the desired properties. In §2 the class will be used to give a simple relative consistency proof for the axiom of choice. In §3 the notion will be related to Gödel's constructible sets. And in §4 brief historical remarks will be given.

1. Ordinal-definable sets. From the remarks in the introduction it is clear that the class we are looking for must contain all the ordinal numbers. Further, if $\tau(\alpha_0, \ldots, \alpha_{n-1})$ is any term with ordinal parameters and no other free variables, then for any particular ordinal values of the parameters the term $\tau(\alpha_0, \ldots, \alpha_{n-1})$ must denote a set in the class. The reason is that the set of all values of the term form a definable class with a definable well-ordering (from a proper kind of lexicographical ordering of the parameters). Note that every element of a definable class with a definable well-ordering can be denoted by a term $\tau(\alpha)$ (the αth element of the class). Hence the class we want must be exactly the class of all sets of the form $\tau(\alpha_0, \ldots, \alpha_{n-1})$ for all ordinals $\alpha_0, \ldots, \alpha_{n-1}$ and *all* terms τ. Thus the class (which is obviously called the class of all *ordinal-definable sets*) is identified, but it has not yet been given a formal definition *within* ZF. It is not even at once obvious that a formal definition is possible—in view of the impossibility of defining the class of definable sets (those definable outright without parameters).

The reason that the class of ordinal definable sets is definable rests with the so-called *Reflection Principle*. This principle shows us that if $\tau(\alpha_0, \ldots, \alpha_{n-1})$ is any term, then the following statement is provable in ZF:

$$\forall \alpha_0, \ldots, \alpha_{n-1} \exists \beta [\alpha_0, \ldots, \alpha_{n-1} < \beta \wedge \forall \xi_0, \ldots, \xi_{n-1} < \beta [\tau(\xi_0, \ldots, \xi_{n-1}) = \tau^{(V_\beta)}(\xi_0, \ldots, \xi_{n-1})]],$$

where $V_\beta = \bigcup_{\gamma<\beta} \{x : x \subseteq V_\gamma\}$ is the set of all sets of rank less than β. Thus for given ordinals $\alpha_0, \ldots, \alpha_{n-1}$ we can find an ordinal β such that

$$\tau(\alpha_0, \ldots, \alpha_{n-1}) = \tau^{(V_\beta)}(\alpha_0, \ldots, \alpha_{n-1}),$$

where the inscription on the right indicates that in the given term all the quantifiers—and the description operators—have their bound variables restricted to V_β. In other words, the set in question is definable—first-order definable—in the relational system $\langle V_\beta, \varepsilon_{V_\beta}, \alpha_0, \ldots, \alpha_{n-1}\rangle$, where ε_{V_β} denotes the restriction of the membership relation to V_β. Now the class of all such relational systems is clearly definable in ZF. Every set definable in such a relational system is ordinal definable. The notion of being definable in a relational system (whose domain is a *set*) is definable. Hence, the class of ordinal-definable sets is in fact definable.

We shall improve this result by showing that the additional parameters can be eliminated by choosing the β so that all the $\alpha_0, \ldots, \alpha_{n-1}$ are *definable* in $\langle V_\beta, \varepsilon_{V_\beta}\rangle$. Some notation is useful. If A is any set, we let $\mathrm{Df}(A)$ stand for the set of all first-order definable elements of the relational system $\langle A, \varepsilon_A\rangle$. We assume that the reader can supply the formal definition of this notion as well as prove in ZF the transitivity of definability which we can formulate as:

LEMMA. (i) $0, a_0, \ldots, a_{n-1}, A \in \mathrm{Df}(B) \rightarrow \tau^{(A)}(a_0, \ldots, a_{n-1}) \in \mathrm{Df}(B)$.
(ii) $0, A \in \mathrm{Df}(B) \wedge A \subseteq B \rightarrow \mathrm{Df}(A) \subseteq \mathrm{Df}(B)$.

We shall also need the definability of the sets V_β:

LEMMA. $\alpha \in \mathrm{Df}(V_\beta) \rightarrow V_\alpha \in \mathrm{Df}(V_\beta)$.

The proof of this last statement is easily obtained from the formal definition of the subclass $\{V_\xi : \xi < \beta\} \subseteq V_\beta$ which is mentioned in [7]. We next establish:

THE EXTENDED REFLECTION PRINCIPLE.

$$\forall \alpha_0, \ldots, \alpha_{n-1} \exists \beta [\alpha_0, \ldots, \alpha_{n-1} \in \mathrm{Df}(V_\beta) \\ \wedge \forall \xi_0, \ldots, \xi_{n-1} < \beta[\tau(\xi_0, \ldots, \xi_{n-1}) = \tau^{(V_\beta)}(\xi_0, \ldots, \xi_{n-1})]]$$

PROOF. Notice first that the ordinary reflection principle as stated above at once generalizes to several terms (hint: use one of the variable parameters to index the terms). Using this remark we establish our extension for the case $n = 2$ which illustrates the method. Proceeding by contradiction, assume the negation of the formula and consider the terms

$$\sigma_0 = \bigcap_{\alpha_0} [\neg \forall \alpha_1 \Phi(\alpha_0, \alpha_1)], \qquad \sigma_1 = \bigcap_{\alpha_1} [\neg \Phi(\sigma_0, \alpha_1)],$$

where $\bigcap$ is intersection (i.e., the least-number operator for ordinals) and

$$\Phi(\alpha_0, \alpha_1) \Leftrightarrow \exists \beta[\alpha_0, \alpha_1 \in \mathrm{Df}(V_\beta) \wedge \forall \xi_0, \xi_1 < \beta[\tau(\xi_0, \xi_1) = \tau^{(V_\beta)}(\xi_0, \xi_1)]].$$

By assumption $\neg \Phi(\sigma_0, \sigma_1)$ holds. But by the reflection principle there is a β such that

$$[\sigma_0, \sigma_1 < \beta \wedge \sigma_0 = \sigma_0^{(V_\beta)} \wedge \sigma_1 = \sigma_1^{(V_\beta)} \wedge \forall \xi_0 \xi_1 < \beta[\tau(\xi_0, \xi_1) = \tau^{(V_\beta)}(\xi_0, \xi_1)]].$$

We thus see that $\sigma_0, \sigma_1 \in \mathrm{Df}(V_\beta)$, so that $\Phi(\sigma_0, \sigma_1)$ holds, and a contradiction has been reached. (Clearly we have just used here once more the idea of the least undefinable ordinal.)

By the way, there is really no need to restrict parameters $\xi_0, \ldots, \xi_{n-1}$ to *ordinals* $< \beta$ in the extended reflection principle; we can just as well use arbitrary *sets* $x_0, \ldots, x_{n-1} \in V_\beta$ by the same proof. Likewise, in place of n we can have some other integer m.

With these formal matters out of the way, we can now define and establish the adequacy of the following notion of ordinal definable sets:

DEFINITION. $\mathrm{OD} = \{x : \exists \alpha[x \in \mathrm{Df}(V_\alpha)]\}$. We shall now freely use class abstraction to define (virtual) classes that are not sets, even though our quantifiers are always restricted to *sets*. In particular, V is the class of all sets and L is the class of constructible sets. Any formulas involving abstracts can be rewritten equivalently without them; and likewise for any expression denoting a *set*. The reader can consult Quine [6] for details. Here we shall continue to use the word *term* for those expressions that always denote sets.

THEOREM. $a_0, \ldots, a_{n-1} \in \mathrm{OD} \rightarrow \tau(a_0, \ldots, a_{n-1}) \in \mathrm{OD}$.

PROOF. In other words we must show that the ordinal definable sets are closed under definability. Now by using the lemmas and the extended reflection principle we can find β with $\alpha, \alpha_0, \ldots, \alpha_{n-1} \in \mathrm{Df}(V_\beta)$ and $a_i \in \mathrm{Df}(V_{\alpha_i})$ for $i < n$ and

$$\tau(\alpha_0, \ldots, \alpha_{n-1}) = \tau^{(V_\alpha)}(a_0, \ldots, a_{n-1}).$$

But then we see that $\tau(a_0, \ldots, a_{n-1}) \in \mathrm{Df}(V_\beta)$, again by use of the lemmas.

Note that $\alpha \in \mathrm{Df}(V_{\alpha+1})$ so that the class of all ordinals $\mathrm{OR} \subseteq \mathrm{OD}$; thus OD does in fact contain all the ordinal-definable sets.

To be able to define a well-ordering of OD we must imagine the formal definition of satisfaction in relational systems together with a Gödel numbering of expressions. Then we can see how to define the term $\mathrm{df}(A, t)$ which denotes the element of A defined by the term with Gödal number t in the relational system $\langle A, \varepsilon_A \rangle$. We can therefore define:

Definition. (i) $\mu(x) = \bigcup_\alpha [x \in \mathrm{Df}(V_\alpha)]$,

(ii) $\nu(x) = \bigcup_{t<\omega} [\mathrm{df}(V_{\mu(x)}, t) = x]$,

(iii) $x \prec y \Leftrightarrow x, y \in \mathrm{OD} \wedge [\mu(x) < \mu(y) \vee [\mu(x) = u(y) \wedge \nu(x) < \nu(y)]]$.

The idea is that $x \prec y$ iff x can be defined in an earlier V_α than y, or else in the same V_α by a definition with a smaller Gödel number than any defining y. We can leave to the reader the verification of the fact that the definable relation $\prec$ really well-orders OD; thus making OD the largest definable class with a definable well ordering.

Since it is obvious that $\mathrm{L} \subseteq \mathrm{OD}$, we see that if ZF is consistent, then ZF + [V = OD] is consistent, because

$$\mathrm{V} = \mathrm{L} \rightarrow \mathrm{V} = \mathrm{OD}.$$

The axiom V = OD may very well be a more reasonable axiom than V = L, but be that as it may, this extension of ZF has an interesting property. Note that $\mathrm{V} = \mathrm{L} \leftrightarrow \mathrm{L} = \mathrm{OD}$ is easily provable in ZF.

Metatheorem. ZF + [V = OD] *is the weakest extension of* ZF *to a theory with the* selection property: *every definable nonempty class contains a definable element.*

Proof. In view of the definable well-ordering of OD, it is clear that ZF + [V = OD] has the selection property. Suppose now that T extends ZF and for every formula $\Phi(x)$ with one free variable, there is a term φ (with no free variables) such that

$$\vdash_T \exists x\, \Phi(x) \rightarrow \Phi(\varphi).$$

Suppose we apply this condition to the formula $\Phi(x) \Leftrightarrow [x \in \mathrm{V} \wedge x \notin \mathrm{OD}]$. Then we obtain a term φ such that

$$\vdash_T \mathrm{V} \neq \mathrm{OD} \rightarrow [\varphi \in \mathrm{V} \wedge \varphi \notin \mathrm{OD}].$$

But we already know that

$$\vdash_{\mathrm{ZF}} \varphi \in \mathrm{V} \rightarrow \varphi \in \mathrm{OD}.$$

Hence, V = OD must be provable in T, which is thus an extension of ZF + [V = OD].

Note that this theorem shows that in any extension of ZF with the selection property, a *uniform selector* is definable by means of the well-ordering relation $\prec$. This easy result would not be so obvious without the use of the class OD.

Another application of the class OD concerns the definable sets. Let us describe the situation in terms of models. Let $\mathcal{M}$ be any model for ZF. Let $\mathcal{OD}^{(\mathcal{M})}$ be the submodel of elements satisfying the definable predicate of being *ordinal definable*. Let $\mathcal{D}^{(\mathcal{M})}$ be the submodel (not generally first-order definable) of elements *definable* in $\mathcal{M}$. From what we have noted about selectors we see:

METATHEOREM. *For any model $\mathcal{M}$ of* ZF, *the submodel $\mathcal{D}^{(\mathcal{M})}$ is an elementary subsystem of $\mathcal{OD}^{(\mathcal{M})}$.*

Hence, for example, the theory of $\mathcal{D}^{(\mathcal{M})}$ is recursively reducible to that of $\mathcal{M}$ because of the definability of OD. It is not quite clear what this really means, but it is curious that we can discover through the formal theory all the formal properties of the definable sets *without* being able to define them directly in the theory.

As a final remark, let us consider the ordinal-definable portion of the continuum. If we identify the continuum with $P\omega = \{x : x \subseteq \omega\}$, then this portion is $P\omega \cap \mathrm{OD}$. Without some assumption (even the axiom of choice is of no help) we cannot prove that $P\omega \cap \mathrm{OD}$ is *uncountable*. All we know is that $P\omega \cap \mathrm{OD}$ is the largest portion of the continuum having a *definable* well-ordering (largest in the sense that it includes all other such definable subsets). Since the question of which reals are or are not definable is difficult, we can make the cardinality of this set slightly more definite. For any set A, let $\mathrm{Th}(A)$, the *theory* of A, be the set of all Gödel numbers of sentences true in $\langle A, \varepsilon_A \rangle$. Then we can show:

THEOREM. *$P\omega \cap \mathrm{OD}$ has the same cardinality as $\{\mathrm{Th}(V_\alpha) : \alpha \in \mathrm{OR}\}$.*

PROOF. This sharpens a result of Possel and Fraïssé [5]. The proof is fairly straightforward. Note first that

$$\{\mathrm{Th}(V_\alpha) : \alpha \in \mathrm{OR}\} \subseteq P\omega \cap \mathrm{OD}.$$

Next, for each $n < \omega$ let

$$I_n = \{\mu(x) : x \in P\omega \cap \mathrm{OD} \wedge \nu(x) = n\}.$$

Clearly $P\omega \cap \mathrm{OD}$ has the same cardinality as

$$\bigcup_{n<\omega} \{n\} \times I_n.$$

Also for $\alpha, \beta \in I_n$, $\alpha \neq \beta$, it is obvious that

$$\mathrm{Th}(V_\alpha) \neq \mathrm{Th}(V_\beta)$$

because the same term (namely, the one with Gödel number n) defines different sets in the two systems. Hence each I_n has cardinality at most that of $\{\mathrm{Th}(V_\alpha) : \alpha \in \mathrm{OD}\}$ and the result follows.

2. The relative consistency of the axiom of choice. Without the aid of the axiom of choice we showed in §1 that there is a definable well-ordering of OD and that the class has interesting closure properties. This certainly indicates that it is likely that we can define a submodel which satisfies all the axioms of ZF including choice. The desired submodel is *not* OD itself because a nonempty definable set need not have any ordinal definable elements (e.g. the set $P\omega \sim \mathrm{OD}$, which may very well be nonempty.) The solution is to take just those sets that are *hereditarily ordinal-definable* in the sense that they and all their ancestors in the membership relation are ordinal definable. Formally we have:

DEFINITION.
$$\mathrm{HOD} = \{x : \exists y \subseteq \mathrm{OD}[x \in y \subseteq Py]\}.$$
If we let
$$C(x) = \bigcap_y [x \subseteq y \subseteq Py]$$
be the *hereditary* (often: *transitive*) *closure* of x, then the definition could also be written:
$$\mathrm{HOD} = \{x \in \mathrm{OD} : C(x) \subseteq \mathrm{OD}\}.$$
We can easily prove by induction on the rank of sets the useful:

LEMMA. $\mathrm{HOD} = \{x \in \mathrm{OD} : x \subseteq \mathrm{HOD}\}$.

COROLLARY. $\mathrm{OR} \subseteq \mathrm{HOD}$

The main result about HOD is the

METATHEOREM. *All the axioms of* ZF *including choice are provable when relativized to* HOD. *Hence if* ZF *is consistent, then so is* ZF + AC.

PROOF. (I) Extensionality. This is obvious because HOD is a hereditary class. (II) Comprehension. Consider a formula $\Phi(x, w_0, \ldots, w_{n-1})$. If we assume $u, w_0, \ldots, w_{n-1} \in \mathrm{HOD}$, then
$$v = \{x \in u : \Phi^{(\mathrm{HOD})}(x, w_0, \ldots, w_{n-1})\}$$
exists, and $v \subseteq \mathrm{HOD}$. But since all parameters are in OD, we have $v \in \mathrm{OD}$ too; therefore $v \in \mathrm{HOD}$. Thus the relativized instance of the comprehension axiom corresponding to Φ is in fact provable. (III) Replacement. The argument is similar. (IV) Power set. If $u \in \mathrm{HOD}$, then $\{x \in \mathrm{HOD} : x \subseteq u\} \in \mathrm{HOD}$. (V) Union. If $u \in \mathrm{HOD}$, then $\bigcup u \in \mathrm{HOD}$. (VI) Infinity. $\omega \in \mathrm{HOD}$. (VII) Foundation. Any restriction of a well-founded relation, like ε, is obviously also well founded: (VIII) Choice. If $u \in \mathrm{HOD}$, then
$$\{\langle x, y\rangle \in u \times u : x \prec y\} \in \mathrm{HOD},$$
and this relation well-orders u.

Note that the equivalences $\mathrm{HOD} = \mathrm{OD} \Leftrightarrow \mathrm{HOD} = \mathrm{V} \Leftrightarrow \mathrm{OD} = \mathrm{V}$ are easily provable in ZF. Hence among these interesting classes we have only these problematical equalities: $\mathrm{L} = \mathrm{HOD}$ and $\mathrm{HOD} = \mathrm{V}$. The known consistency results are as follows, relative to ZF,

ZF + [L = HOD] + [HOD = V],	Gödel **[1]**
ZF + [L = HOD] + [HOD ≠ V],	Lévy [3]
ZF + [L ≠ HOD] + [HOD = V],	McAloon **[4]**
ZF + [L ≠ HOD] + [HOD ≠ V],	McAloon **[4]**

Also due to McAloon is the consistency result that $\text{OD}^{(\text{HOD})} \neq \text{HOD}$ is consistent with ZF. This would seem to clear up most of the basic questions concerning the relationship between these notions. Also in the above ZF can be replaced by ZF + AC.

3. Second-order constructibility. Let $\text{DF}^0(A)$ be the set of all *subsets* of A definable by *first-order* formulas in $\langle A, \varepsilon_A\rangle$ using *parameters* from A; that is, all sets of the form

$$\{x \in A : \Phi^{(A)}(x, a_0, \ldots, a_{n-1})\}$$

where $a_0, \ldots, a_{n-1} \in A$. As is well known, the constructible sets of Gödel can be defined by recursion as follows

$$\text{L}^0_\alpha = \bigcup_{\beta<\alpha} \text{DF}^0(\text{L}^0_\beta),$$

and

$$\text{L}^0 = \bigcup_{\alpha\in\text{OR}} \text{L}^0_\alpha,$$

where the superscript 0 is usually dropped. (As in hierarchy theory, we use the superscript 0 to indicate that the quantifiers are first-order.) By the way, from this definition it is clear why $\text{L}^0 \subseteq \text{HOD}$.

An extension of the notion would be to allow *second-order* formulas instead of just first-order definitions. So let $\text{DF}^{-1}(A)$ be the set of definable subsets where the *quantifiers* in the formulas are restricted *either* to A *or* to PA (or even $P(A \times A)$, etc.) and where the *parameters* are still from A. We define

$$\text{L}^1_\alpha = \bigcup_{\beta<\alpha} \text{DF}^1(\text{L}^1_\beta),$$

$$\text{L}^1 = \bigcup_{\alpha\in\text{OR}} \text{L}^1_\alpha.$$

Again $\text{L}^1 \subseteq \text{HOD}$ is obvious by induction. Clearly we could go on to define $\text{L}^2, \text{L}^3, \ldots$, all of them included in HOD. It is not interesting to do so, however, if we are willing to assume the axiom of choice.

THEOREM. $\text{AC} \to \text{L}^1 = \text{HOD}$.

PROOF. It is enough to show

$$a \subseteq \text{L}^1 \wedge a \in \text{HOD} \to a \in \text{L}^1.$$

Thus let $a \in \text{Df}(V_\alpha)$. We choose $\beta \in \text{OR}$ so large that not only is $a \subseteq \text{L}^1_\beta$, but the cardinality of L^1_β is greater than that of V_α. This is possible by the axiom of choice because the cardinalities of the L^1_β are unlimited. Now we transcribe into a second-order formula the condition:

There is an isomorphism f that carries $\langle C(\{x\}), \varepsilon_{(C\{x\})}\rangle$ over to an hereditary subsystem of a copy of $\langle V_\alpha, \varepsilon_{V_\alpha}\rangle$ on a subset of L^1_β in such a way that $f(x)$ satisfies in the copy the formula defining a as a subset of V_α.

The only parameter we need here is α, and the only problem is seeing that a relation's being a copy of $\langle V_\alpha, \varepsilon_{V_\alpha}\rangle$ can be expressed in a second-order condition. But that is really clear from the recursive definition of V_α. (We leave the details to the reader.) In any event, $a \in L^1_{\beta+1}$.

4. Historical remarks. The notion of ordinal definability was due to Gödel [**2**] who observed that it enabled one to give an easier proof of the consistency of the axiom of choice than that of [**1**], but presumably not of the consistency of the continuum hypothesis. The idea was then rediscovered independently by one of the present authors, G. Takeuti [**8**], and Post [**9**]. Martin Davis lent us Post's 1952 notebook, wherein several of the results of the present paper are contained. In addition, Hájek informed us that in joint work with Vopěnka and Balcar the notion was rediscovered recently by closing the class of sets $\{V_\alpha : \alpha \in \mathrm{OR}\}$ under Gödel's eight operations to obtain the class OD, and they also noted the proof of consistency of the axiom of choice.

BIBLIOGRAPHY

1. K. Gödel, *The consistency of the continuum hypothesis*, Ann. of Math. Studies no. 3, Princeton Univ. Press, Princeton, N. J., 1940.

2. ———, "Remarks before the Princeton Bicentennial Conference" in *The undecidable*, edited by M. Davis, Raven Press, 1964, pp. 84–88.

3. A. Lévy, *Definability in axiomatic set theory*. I, Proc. 1964 Internat. Congr. for Logic, Philosophy, and Methodology of Science, Jerusalem, North-Holland, Amsterdam, 1965.

4. K. McAloon, Thesis, Berkeley, 1966.

5. R. de Possel and R. Fraïssé, *Hypothèse de la théorie des relations qui permittent d'associer, à un bon ordre d'un ensemble, un bon ordre, défini sans ambiguité, de l'ensemble de ses parties. Le raisonnement en mathématiques et en sciences expérimentales*, Colloq. Internat. du Centre Nat. de la Recherche Sci., 70. Editions du Centre National de la Recherche Scientifique, Paris, 1958, pp. 51–55.

6. W. V. Quine, *Set theory and its logic*, Harvard University Press, Cambridge, Mass., 1965.

7. D. Scott, *Axiomatizing set theory*, these Proceedings, part II.

8. G. Takeuti, *Remarks on Cantor's absolute*, J. Math. Soc. Japan. **13** (1961), 197–206.

9. E. L. Post, *A necessary condition for definability for transfinite von Neumann-Gödel set theory sets, with an application to the problem of the existence of a definable well-ordering of the continuum*, Preliminary Report, Bull. Amer. Math. Soc. **59** (1953), p. 246.

10. ———, *Solvability, definability, provability; History of an error*, Bull. Amer. Math. Soc., **59** (1953), p. 245.

STATE UNIVERSITY OF NEW YORK, BUFFALO

STANFORD UNIVERSITY

AN AXIOM OF STRONG INFINITY AND ANALYTIC HIERARCHY OF ORDINAL NUMBERS

KANJI NAMBA

Introduction. As many axioms of strong infinity in set theory are proposed, it remains unknown whether many of these axioms are consistent with the axioms of set theory. Therefore, many problems and conjectures concerning the axiom of strong infinity have been proposed by many mathematicians, among them are the existence of strongly inaccessible cardinals, of hyper-inaccessible cardinals of arbitrary type, arithmetical and both 1-transcendency of cardinals, existence of Ramsey cardinals, existence of measurable cardinals, Π^1_1-transcendency of measurable cardinals, etc.

To consider such kinds of axioms of strong infinity, it seems to be adequate to consider the analytic hierarchy of ordinal numbers. It is very interesting that the axioms of strong infinity which seem to be somewhat constructible, such as the existence of strongly inaccessible cardinals, the existence of hyper-inaccessible cardinals of arbitrary type, etc., are expressed by formulas of the form $x\exists P(x)$, where $P(a)$ is a Π^1_1-formula; and the axioms of strong infinity which seem to be somewhat nonconstructible, such as the existence of Ramsey cardinals, the existence of measurable cardinals, etc., are expressed by formulas of the form $\exists x P(x)$, where $P(a)$ is a Δ^1_2-formula. And there are also axioms of strong infinity of the form $\exists x P(x)$, where $P(a)$ is a Σ^1_1-formula. Existence of a model for an axiom system, existence of standard models on some ordinal numbers, etc., have this form. There is a very interesting contrast between Σ^1_1-infinity and Π^1_1-infinity and this will be discussed later.

Now we shall explain about the concept of transcendency, which is proposed by

G. Takeuti in his paper **[15]**. Arithmetical (Δ_1^1-) transcendency of cardinals, which is abbreviated as ATC (Δ_1^1-TC), is stated as follows:

ATC (Δ_1^1-TC). For every arithmetical (Δ_1^1-) function $f(a_1, \ldots, a_n)$, we have

$$\overline{\overline{f(a_1, \ldots, a_n)}} \leq \max(\aleph_0, a_1, \ldots, a_n).$$

This is slightly generalized as follows:

f-ATC (f-Δ_1^1-TC). Let f_i $(i = 1, \ldots, n)$ be functions such that

(1) $f_i(a_1, \ldots, a_n) \leq c$ for all $a_1, \ldots, a_n$,

(2) $f_i(a_1, \ldots, a_n) = 0$ if $\max(a_1, \ldots, a_n) \geq c$.

And let $g(f_i, \ldots, f_m, a_1, \ldots, a_s)$ be a function arithmetical (Δ_1^1-) in $f_1, \ldots, f_m$. Then we have

$$\overline{\overline{g(f_1, \ldots, f_m, a_1, \ldots, a_n)}} \leq \max(\aleph_0, c, a_1, \ldots, a_n).$$

Let $\aleph_\tau$ be a cardinal number. If there is a nontrivial 2-valued measure on $\aleph_\tau$, or equivalenty, if there is a nonprincipal $\aleph_0$-complete ultrafilter over $\aleph_\tau$, then $\aleph_\tau$ is called a 2-valued measurable cardinal or $\aleph_0$-complete cardinal. It is well known that if there exists an $\aleph_0$-complete cardinal, then there is a cardinal $\aleph_\tau$ with the property that there is a nonprincipal ultrafilter $F_{\aleph_\tau}$ with the character $\aleph_\tau > \aleph_0$. The smallest $\aleph_0$-complete cardinal numbers have this property, and in such a case $\aleph_\tau$ is called a normally measurable cardinal number.

Let us assume that there is a normally measurable cardinal. And let $\aleph_\tau$ be any normally measurable cardinal. Then Δ_1^1-TC and f-Δ_1^1-TC hold if we consider $\aleph_\tau$ as universe. From this we obtain that the predicates 'a is a cardinal', 'a is a regular cardinal', 'a is a strongly inaccessible cardinal', etc., are Π_1^1 but not Σ_1^1. The following result means that the defining formula of Δ_1^1-function depends essentially on its form of defining formula and not essentially on its universe.

Let $f'(a_1, \ldots, a_n)$ and $f(a_1, \ldots, a_n)$ be two Δ_1^1-functions defined by the formula of the same form considering $\aleph_\theta$ $(<\aleph_\tau)$ and $\aleph_\tau$ as universe, respectively. Then we have the following

$$\forall x_1 \cdots \forall x_n(x_1, \ldots, x_n < \aleph_\theta \rightarrow f'(x_1, \ldots, x_n) = f(x_1, \ldots, x_n)).$$

Now we consider the cardinal number $\aleph_2$ and the analytic hierarchy considering $\aleph_2$ as universe. Then the predicate $a > \omega \wedge \mathrm{Card}(a)$, which means that a is a non-countable cardinal less than $\aleph_2$, is expressible by a Π_1^1-formula, and it has exactly one solution, $\aleph_1$. Therefore the function f defined by

$$f(a) = b \equiv b > \omega \wedge \mathrm{Card}(b)$$

is a Δ_1^1-function. So Δ_1^1-TC does not hold for this function. But it should be noted that such a formula does not define a Δ_1^1-function if we consider a suitable cardinal number $\aleph_\tau$ as universe.

Let $A(a, a_1, \ldots, a_n)$ be a Δ_1^1-formula considering $\aleph_\tau$ as universe. And let $\aleph_0 < \aleph_{c_1} < \cdots < \aleph_{c_n} < \aleph_\tau$ and $\aleph_0 < \aleph_{d_1} < \cdots < \aleph_{d_n} < \aleph_\tau$ be two sequences of cardinals. Then we have the following

$$\forall x(x < \aleph_{c_1}, \aleph_{d_1} \rightarrow (A(x, \aleph_{c_1}, \ldots, \aleph_{c_n}) \leftrightarrow A(x, \aleph_{d_1}, \ldots, \aleph_{d_n}))).$$

The same kind of extension as in the case of ATC to f-ATC is also true in this case. This theorem shows that, in the case of a Δ_1^1-formula, its truth depends on order of cardinality in the formula and not on its magnitude.

Now we consider some properties of Π_1^1- and Σ_1^1-formulas. Let $F_{\aleph_\tau}$ be a nonprincipal $\aleph_0$-complete ultrafilter over $\aleph_\tau$ with the character $\aleph_\tau$. By $\mathrm{On}^{\aleph_\tau}/F_{\aleph_\tau}$ we denote the ultraproduct of On (the class of all ordinal numbers) determined by $F_{\aleph_\tau}$. And by a^* we denote the representative element of ath equivalence class of $\mathrm{On}^{\aleph_\tau}/F_{\aleph_\tau}$. Let a^* be $(a_1, \ldots, a_\nu, \ldots)$ $(\nu < \aleph_\tau)$. By $(a)_\nu$ we denote the νth component a_ν of a^*. According to this notation, for every Π_1^1-formula $P(a)$, we have $P(a) \rightarrow \{\nu : P((a)_\nu)\} \in F_{\aleph_\tau}$ for all a. So if we substitute $\aleph_\tau$ for a in the formula, we have $P(\aleph_\tau) \rightarrow \{\nu : P((\aleph_\tau)_\nu)\} \in F_{\aleph_\tau}$. And of course we have $\{\nu : (\aleph_\tau)_\nu < \aleph_\tau\} \in F_{\aleph_\tau}$. Therefore, we obtain that the first element of Π_1^1-class of ordinals cannot be a normally measurable cardinal number.

Let $P_1(a), P_2(a), \ldots, P_n(a), \ldots$ be an enumeration of all Π_1^1-formulas such that $P_n(\aleph_\tau)$. Then clearly we have $\{\nu : P_n((\aleph_\tau)_\nu)\} \in F_{\aleph_\tau}$ for all n, and

$$\{\nu : (\aleph_\tau)_\nu < \aleph_\tau\} \in F_{\aleph_\tau}.$$

Since $F_{\aleph_\tau}$ is $\aleph_0$-complete, we obtain that

$$\{\nu : P_n((\aleph_\tau)_\nu) \text{ for all } n, (\aleph_\tau)_\nu < \aleph_\tau\} \in F_{\aleph_\tau}.$$

By this, we have a ν_0 such that $P_n((\aleph_\tau)_{\nu_0})$ for all n, and $(\aleph_\tau)_{\nu_0} < \aleph_\tau$. By putting $a = (\aleph_\tau)_{\nu_0}$ we obtain that there is a cardinal number $a < \aleph_\tau$ such that a has all Π_1^1-properties of $\aleph_\tau$.

Let $\aleph_\sigma$ and $\aleph_\tau$ $(<\aleph_\sigma)$ be two normally measurable cardinals. Let $P(a)$ be a Σ_1^1-formula considering $\aleph_\sigma$ as universe. Then $P(\aleph_\tau)$ implies $P(\aleph_\alpha)$ for all $\aleph_\alpha$ such that $\aleph_0 < \aleph_\alpha \leq \aleph_\tau$; of course $\aleph_\alpha$ may be a singular cardinal number. Especially, putting $\aleph_\alpha$ as $\aleph_1$, we have $P(\aleph_\tau) \rightarrow P(\aleph_1)$. If we consider a Π_1^1-formula $P(a)$, then we have $P(\aleph_\alpha) \rightarrow P(\aleph_\tau)$ for all $\aleph_0 < \aleph_\alpha \leq \aleph_\tau$. Namely, we have that if there is a cardinal number $\aleph_\alpha$ $(\aleph_0 < \aleph_\alpha \leq \aleph_\tau)$ such that $P(\aleph_\alpha)$, then the normally measurable cardinal number $\aleph_\tau$ satisfies $P(\aleph_\tau)$.

Now we consider the Σ_1^1-formula $P(a)$ considering $\aleph_{\tau+1}$ as universe defined by $P(a) \equiv \forall x(a < x \rightarrow \neg \mathrm{Card}\,(x))$ where Card (a) is a Π_1^1-formula which means that a is a cardinal number. Then we have

$$P(\aleph_\tau) \quad \text{but} \quad \neg P(\aleph_\alpha) \quad \text{for all} \quad \aleph_\alpha < \aleph_\tau.$$

This means that the above theorem does not hold for $\aleph_{\tau+1}$. Namely, some conditions on $\aleph_\sigma$ is necessary to prove the theorem.

The above theorem means that the normally measurable cardinal number has all Π_1^1-properties of smaller uncountable cardinals. Combining this theorem with the theorem stated before, we have a cardinal $\aleph_\alpha$ ($<\aleph_\tau$) which has exactly those Π_1^1-properties that are properties of $\aleph_\tau$. Namely, there is neither Σ_1^1- nor Π_1^1-formula which distinguishes these two different cardinals $\aleph_\alpha$ and $\aleph_\tau$, one of which is a normally measurable cardinal and the other is not. And, moreover, for any Σ_1^1- and Π_1^1-formula $P(a)$, we have

$$P(\aleph_\tau) \leftrightarrow \{\nu : P((\aleph_\tau)_\nu)\} \in F_{\aleph_\tau}.$$

Let $F_{\aleph_\tau}$ be a nonprincipal $\aleph_0$-complete ultrafilter. And let $\{\nu : (\aleph_\tau)_\nu = \nu\} \in F_{\aleph_\tau}$, then $F_{\aleph_\tau}$ is called normal. Let $F_{\aleph_\tau}$ and $G_{\aleph_\tau}$ be two normal ultrafilters. Then we have $\{\nu : P(\nu)\} \in F_{\aleph_\tau} \leftrightarrow \{\nu : P(\nu)\} \in G_{\aleph_\tau}$ for every Σ_1^1- and Π_1^1-formula. Assume that $\{\nu : \text{Meas}\ (\nu)\} \in F_{\aleph_\tau}$ for some normal ultrafilter where Meas (a) is a Δ_2^1-formula which means that a is a normally measurable cardinal. Then there is a normal ultrafilter $G_{\aleph_\tau}$ such that $\{\nu : \text{Meas}\ (\nu)\} \notin G_{\aleph_\tau}$. Therefore this property does not hold for such large cardinal numbers and the Δ_2^1-formula.

Let $P(a, a_1, \ldots, a_n)$ be a Σ_1^1-formula considering $\aleph_\tau$ as universe. Then we have for any uncountable cardinal $\aleph_c < \aleph_\tau$

$$\forall x_1 \cdots \forall x_n(x_1, \ldots, x_n < \aleph_c \wedge \exists x P(x, x_1, \ldots, x_n) \\ \rightarrow \exists x(x < \aleph_c \wedge P(x, x_1, \ldots, x_n))).$$

This theorem is called Σ_1^1-transcendency of cardinal numbers, which is of course a generalization of Δ_1^1-transcendency of cardinals. And it is considered as a generalized form of Löwenheim-Skolem theorem.

A similar theorem is also true for Π_1^1-formulas. But this time the concept of uncountable cardinals is replaced by 2-valued measurable cardinals.

Let $P(a, a_1, \ldots, a_n)$ be a Π_1^1-formula considering $\aleph_\sigma$ as universe. Then we have for any 2-valued measurable cardinal $\aleph_\tau < \aleph_\sigma$

$$\forall x_1 \cdots \forall x_n(x_1, \ldots, x_n < \aleph_\tau \wedge \exists x P(x, x_1, \ldots, x_n) \\ \rightarrow \exists x(x < \aleph_\tau \wedge P(x, x_1, \ldots, x_n))).$$

This theorem is called Π_1^1-transcendency of normally measurable cardinals.

Let $P(a)$ be a Π_1^1-formula such that $\exists x(x > \omega \wedge \text{Card}\ (x) \wedge P(x))$. Then by the above theorem for some uncountable cardinal $\aleph_\alpha$ ($<\aleph_\tau$) we have $P(\aleph_\alpha)$. Therefore we get $P(\aleph_\tau)$. Namely

$$\exists x(x > \omega \wedge \text{Card}\ (x) \wedge P(x)) \leftrightarrow P(\aleph_\tau).$$

This is not true for Σ_1^1-formulas because if we take $P(a)$ as a Σ_1^1-formula, which means that a is not a regular ordinal, then we have

$$\exists x(x > \omega \wedge \text{Card}\ (x) \wedge P(x)) \quad \text{but } \neg P(\aleph_\tau).$$

Let $A(a, a_1, \ldots, a_n)$ be a Π_1^1-formula considering $\aleph_\sigma$ as universe. And let $\aleph_{\tau_1} < \cdots < \aleph_{\tau_n} < \aleph_\sigma$ and $\aleph_{\sigma_1} < \cdots < \aleph_{\sigma_n} < \aleph_\sigma$ be two sequences of normally measurable cardinals. Then we have

$$\forall x(x < \aleph_{\tau_1}, \aleph_{\sigma_1} \rightarrow (A(x, \aleph_{\tau_1}, \ldots, \aleph_{\tau_n}) \leftrightarrow A(x, \aleph_{\sigma_1}, \ldots, \aleph_{\sigma_n}))).$$

This theorem shows that in the case of Σ_1^1- and Π_1^1-formula, its truth depends on order of 2-valued measurable cardinals in the formula and not on its magnitude.

The concept of transcendency is now stated as follows: Let $Q(a)$ be a formula such that $\forall x \exists y(x \leq y \wedge Q(y))$ and

$$\forall x(Q(x) \to \forall x_1 \cdots \forall x_n(x_1, \ldots, x_n < x \wedge \exists y P(y, x_1, \ldots, x_n) \\ \to \exists y(y < x \wedge P(y, x_1, \ldots, x_n)))).$$

Then $Q(a)$ is said to be $P(a, a_1, \ldots, a_n)$ transcendental. By this terminology we have that there is a Π_1^1-formula, $a > \omega \wedge \text{Card}\,(a)$, which is transcendental for every Σ_1^1-formulas.

Let A and B be two classes of formulas. And let a formula $Q(a)$ in B satisfy the condition that for any formula $P(a, a_1, \ldots, a_n)$ in A, $Q(a)$ is $P(a, a_1, \ldots, a_n)$ transcendental. Then we say $A < B$. By assuming a suitable axiom of strong infinity, we have $A < \Delta_1^1 \leq \Sigma_1^1 < \Pi_1^1 < \Delta_2^1$, where A is the class of all arithmetical formulas. The following problems are very important: Is it true that $\Delta_1^1 < \Sigma_1^1$? Which of $\Sigma_2^1 < \Pi_2^1$, $\Pi_2^1 < \Delta_2^1$ is true, or are they incomparable? Anyhow, there are many problems in degrees of transcendency.

1. Analytic hierarchy of ordinal numbers. In this paper, we shall use the following notations:

Number variables:	$a, b, c, \ldots, a_1, a_2, a_3, \ldots$ (range over all ordinal numbers).
Number constants:	$0, 1, 2, \ldots, \omega, \omega + 1, \omega + 2, \ldots$.
Function variables:	$\alpha, \beta, \gamma, \ldots, \alpha_1, \alpha_2, \alpha_3, \ldots$ (range over all functions of ordinal numbers).
Function constants:	$'$, max, Iq, $g^1, g^2, j, \ldots$.
Predicates constants:	$=, <, \in$.
Logical symbols:	$\neg$ (not), $\vee$ (or), $\exists$ (there exists) (the symbols $\wedge$ (and), $\to$ (implies), $\leftrightarrow$ (equivalent to), $\forall$ (for all) are considered as a combination of $\neg, \vee, \exists$).

The function constants $'$, max, Iq, g^1, g^2, and j are defined by the following: a' is the successor of a; max (a, b) is the maximum of a and b;

$$\text{Iq}\,(a, b) = \begin{cases} 0 & \text{if } a < b, \\ 1 & \text{otherwise;} \end{cases}$$

$$j(g^1(a), g^2(a)) = a, \qquad g^1(j(a, b)) = a, \qquad g^2(j(a, b)) = b,$$

$$j(a, b) < j(c, d) \leftrightarrow \max\,(a, b) < \max\,(c, d) \\ \vee (\max\,(a, b) = \max\,(c, d) \wedge (b < d \vee (b = d \wedge a < c))).$$

Let $\alpha(a_1, \ldots, a_n, a)$ be a function. Then the operations $\mu x \alpha(a_1, \ldots, a_n, x)$

and $\alpha^a(a_1, \ldots, a_n, b)$ are defined by the following:

$$\mu x\alpha(a_1, \ldots, a_n, x) = \begin{cases} \text{the least } x \text{ such that } \alpha(a_1, \ldots, a_n, x) = 0 \text{ if} \\ \exists x(\alpha(a_1, \ldots, a_n, x) = 0), \\ 0 \quad \text{otherwise.} \end{cases}$$

$$\alpha^a(a_1, \ldots, a_n, b) = \begin{cases} \alpha(a_1, \ldots, a_n, b) & \text{if } b < a, \\ 0 & \text{otherwise.} \end{cases}$$

A function $f(\alpha_1, \ldots, \alpha_m, a_1, \ldots, a_n)$ is called an arithmetical function in $\alpha_1, \ldots, \alpha_m$ if it is defined by a series of applications of the following schemata, provided that $g, g_1, \ldots, g_e$ are given functions arithmetical in the functions indicated in the functions:

(I) $f(\alpha_1, \ldots, \alpha_m, a_1, \ldots, a_n) = \mathrm{Iq}\,(a_i, a_j) \quad i, j = 1, \ldots, n;$

(II) $f(\alpha_1, \ldots, \alpha_m, a_1, \ldots, a_n) = j(a_i, a_j) \quad i, j = 1, \ldots, n;$

(III) $f(\alpha_1, \ldots, \alpha_m, a_1, \ldots, a_n) = g^1(a_i),$
$\qquad = g^2(a_i) \quad i = 1, \ldots, n;$

(IV) $f(\alpha_1, \ldots, \alpha_m, a_1, \ldots, a_n) = \alpha_i(a_1, \ldots, a_j) \quad i = 1, \ldots, m,$
$\qquad j = 1, \ldots, n;$

(V) $f(\alpha_1, \ldots, \alpha_m, a_1, \ldots, a_n)$
$\qquad = g(\alpha_1, \ldots, \alpha_m, g_1(\alpha_1, \ldots, \alpha_m, a_1, \ldots, a_n), \ldots,$
$\qquad g_e(\alpha_1, \ldots, \alpha_m, a_1, \ldots, a_n));$

(VI) $f(\alpha_1, \ldots, \alpha_m, a_1, \ldots, a_n) = \mu x g(\alpha_1, \ldots, \alpha_m, a_1, \ldots, a_n, x);$

(VII) $f(\alpha_1, \ldots, \alpha_m, a_1, \ldots, a_n, a) = g(\alpha_1, \ldots, \alpha_m, f^a, a_1, \ldots, a_n, a)$

where $g(\alpha_1, \ldots, \alpha_m, \alpha, a_1, \ldots, a_n, a)$ is an arithmetical function in $\alpha_1, \ldots, \alpha_m, \alpha$ and f^a is an abbreviation for the function $\lambda x f^a(\alpha_1, \ldots, \alpha_m, a_1, \ldots, a_n, x)$.

The functions $a_1', a_i, 0, \omega, \max\,(a_i, a_j)$ are defined as follows:

$$a_1' = \mu x\,\mathrm{Iq}\,(a_i, x),$$
$$a_i = \mu x\,\mathrm{Iq}\,(a_i, x'),$$
$$0 = \mu x\,\mathrm{Iq}\,(x, x'),$$
$$\omega = \mu x j(\mathrm{Iq}\,(j(0, x), x'), \mathrm{Iq}\,(0, x)),$$
$$\max\,(a_i, a_j) = \mu x j(\mathrm{Iq}\,(a_i, x'), \mathrm{Iq}\,(a_j, x')).$$

And the function $\lambda x f^a(\alpha_1, \ldots, \alpha_m, a_1, \ldots, a_n, x)$ defined by the schema (VII) has the following property:

$$f^a(\alpha_1, \ldots, \alpha_m, a_1, \ldots, a_n, b) = \begin{cases} f^{b+1}(\alpha_1, \ldots, \alpha_m, a_1, \ldots, a_n, b) & \text{if } b + 1 < a, \\ g(\alpha_1, \ldots, \alpha_m, f^b, a_1, \ldots, a_n, b) & \text{if } b + 1 = a, \\ 0 \quad \text{if } b \geq a. \end{cases}$$

A function $f(\alpha_1, \ldots, \alpha_m, a_1, \ldots, a_n)$ which is arithmetical in $\alpha_1, \ldots, \alpha_m$ is called recursive in $\alpha_1, \ldots, \alpha_m$ if the schema (VI) is applied only to the case where the following condition is satisfied:

$$\forall x_1 \cdots \forall x_n \exists x(g(\alpha_1, \ldots, \alpha_m, x_1, \ldots, x_n, x) = 0).$$

A formula $A(\alpha_1, \ldots, \alpha_m, a_1, \ldots, a_n)$ is called arithmetical in $\alpha_1, \ldots, \alpha_m$ if there is an arithmetical function in $\alpha_1, \ldots, \alpha_m$ such that

$$\forall x_1 \cdots \forall x_n(A(\alpha_1, \ldots, \alpha_m, x_{10} \ldots, x_n) \leftrightarrow f(\alpha_1, \ldots, \alpha_m, x_1, \ldots, x_n) = 0).$$

A formula $A(\alpha_1, \ldots, \alpha_m, a_1, \ldots, a_n)$ is called analytic in $\alpha_1, \ldots, \alpha_m$ if it is equivalent to a formula obtained from an arithmetical formula by prefixing a sequence of number-quantifiers and function-quantifiers and has no free function variables except $\alpha_1, \ldots, \alpha_m$. A formula $A(\alpha_1, \ldots, \alpha_m, a_1, \ldots, a_n)$ is called a Π^1_1-formula in $\alpha_1, \ldots, \alpha_m$ if it is equivalent to a formula of the form

$$\forall \alpha B(\alpha_1, \ldots, \alpha_m, \alpha, a_1, \ldots, a_n)$$

where $B(\alpha_1, \ldots, \alpha_m, \alpha, a_1, \ldots, a_n)$ is arithmetical in $\alpha_1, \ldots, \alpha_m, \alpha$. A formula is called a Σ^1_1-formula in $\alpha_1, \ldots, \alpha_m$ if its negation is a Π^1_1-formula in $\alpha_1, \ldots, \alpha_m$. And a formula is called a Δ^1_1-formula in $\alpha_1, \ldots, \alpha_m$ if it is both Π^1_1- and Σ^1_1-formula in $\alpha_1, \ldots, \alpha_m$. In the above definition if $m = 0$, then the formula $A(a_1, \ldots, a_n)$ is called simply arithmetical, recursive, Π^1_1-, etc. The concept of Π^1_2, Σ^1_2, $\Delta^1_2, \ldots$ are defined similarly.

A function $f(\alpha_1, \ldots, \alpha_m, a_1, \ldots, a_n)$ is called Π^1_1-, Σ^1_1-, Δ^1_1-, Π^1_2-, ... in $\alpha_1, \ldots, \alpha_m$ if there is a formula $A(\alpha_1, \ldots, \alpha_m, a_1, \ldots, a_n, a)$ which is Π^1_1-, Σ^1_1-, Δ^1_1-, Π^1_2-, $\cdots$ in $\alpha_1, \ldots, \alpha_m$ such that

$$\forall x_1 \cdots \forall x_n \forall x(f(\alpha_1, \ldots, \alpha_m, x_1, \ldots, x_n) = x \leftrightarrow A(\alpha_1, \ldots, \alpha_m, x_1, \ldots, x_n, x)).$$

Let a_0 be an ordinal number. We can define analytic hierarchy of ordinal numbers by considering a_0 as universe. For the convenience of description we use $A^{(a_0)}$ to stand for the relativization of notion A to a_0. And $A^{(a_0)}$ is sometimes referred to as A in (a_0).

2. Measurable cardinals, ultraproducts. A subset F of the powerset $\mathfrak{P}(A)$ of a set A is called a filter over the set if the following conditions are satisfied:

(1) $B \in F$ and $A \supset C \supset B$ imply $C \in F$.

(2) $B, C \in F$ implies $B \cap C \in F$.

(3) $\varnothing \notin F$.

A filter F over a set A is called an ultrafilter if for any subset B of A we have $B \in F$ or $A - B \in F$. An ultrafilter F is called principal if there is an element p of A such that $B \in F$ if and only if $p \in B$.

A filter F is said to be $\aleph_a$-complete if $I \leq \aleph_a$ and $A_\nu \in F$ for all $\nu \in I$ imply $\bigcap_{\nu \in I} A_\nu \in F$. A cardinal number $\aleph_b$ is said to be $\aleph_a$-complete if there is a non-principal $\aleph_a$-complete ultrafilter over $\aleph_b$. The character of a filter F is defined to be the smallest cardinal number $\aleph_a$ such that F is not $\aleph_a$-complete if there exists such a cardinal number. The character of F is denoted by ch (F). A cardinal number $\aleph_\tau$ is called normally measurable if there is a nonprincipal ultrafilter $F_{\aleph_\tau}$ over $\aleph_\tau$ with ch $(F_{\aleph_\tau}) = \aleph_\tau > \aleph_0$. It is well known that the smallest $\aleph_0$-complete (2-valued measurable) cardinal number is a normally measurable cardinal. Therefore, if there is an $\aleph_0$-complete cardinal number, then there is a normally measurable cardinal number.

Let $\aleph_\tau$ be a normally measurable cardinal number. Then there is an ultrafilter $F_{\aleph_\tau}$ over $\aleph_\tau$ such that ch $(F_{\aleph_\tau}) = \aleph_\tau$. It is well known that $\aleph_\tau$ is regular, strongly inaccessible, and so on.

Let On be the class of all ordinal numbers and we consider the ultraproduct of On determined by $F_{\aleph_\tau}$ which is denoted by $\mathrm{On}^{\aleph_\tau}/F_{\aleph_\tau}$. The equivalence-relation $=^*$ and the order-relation $<^*$ between two elements u and v of $\mathrm{On}^{\aleph_\tau}$ are defined by

$$u =^* v \equiv \{\nu : u(\nu) = v(\nu)\} \in F_{\aleph_\tau},$$
$$u <^* v \equiv \{\nu : u(\nu) < v(\nu)\} \in F_{\aleph_\tau}.$$

Since $=^*$ is an equivalence-relation, we can classify $\mathrm{On}^{\aleph_\tau}$ into equivalence classes. Let $u \in \mathrm{On}^{\aleph_\tau}$. Then by $\bar{u}$ we denote the equivalence class which includes the function u. Now we choose a representative element from each equivalence class $\bar{u}$ in such a way that if $(a, a, \ldots, a, \ldots) \in \bar{u}$ then the representative element of $\bar{u}$ is $(a, a, \ldots, a, \ldots)$, otherwise the representative element of $\bar{u}$ is an arbitrary but fixed element of $\bar{u}$. By T we denote the class of all representative elements defined above. Since $F_{\aleph_\tau}$ is $\aleph_0$-complete, we have that T is well ordered by the relation $<^*$. Let the ath element of T by the well-ordering $<^*$ be a^*. Then we have $T = \{a^* : a \in \mathrm{On}\}$. And moreover if $a < \aleph_\tau$ then $a^* = (a, a, \ldots, a, \ldots)$.

Let a be an ordinal number and let $a^* = (a_1, a_2, \ldots, a_\nu, \ldots)$. Then by $(a^*)_\nu$ we denote the νth component a_ν of a^*. And $(a^*)_\nu$ is sometimes written as $(a)_\nu$.

An ultrafilter $F_{\aleph_\tau}$ is said to be a normal ultrafilter if $\{\nu : (\aleph_\tau)_\nu = \nu\} \in F_{\aleph_\tau}$. Let $F_{\aleph_\tau}$ be an $\aleph_0$-complete ultrafilter. Then the ultrafilter $G_{\aleph_\tau}$ defined by

$$B \in G_{\aleph_\tau} \equiv \{\nu : (\aleph_\tau)_\nu \in B\} \in F_{\aleph_\tau}$$

is a normal ultrafilter.

An ordinal number a is said to be the ultraproduct of a sequence of ordinal numbers $a_1, a_2, \ldots, a_\nu, \ldots$ if $\{\nu : (a)_\nu = a_\nu\} \in F_{\aleph_\tau}$. A function $f(a_1, \ldots, a_n)$ is said to be the ultraproduct of a sequence of functions $f_1(a_1, \ldots, a_n)$, $f_2(a_1, \ldots, a_n), \ldots, f_\nu(a_1, \ldots, a_n), \ldots$ if for every $a_1, \ldots, a_n$

$$\{\nu : (f(a_1, \ldots, a_n))_\nu = f_\nu((a_1)_\nu, \ldots, (a_n)_\nu)\} \in F_{\aleph_\tau}.$$

An ordinal number a is said to be U-invariant if $\{\nu : (a)_\nu = a\} \in F_{\aleph_\tau}$. A function $f(a_1, \ldots, a_n)$ is said to be U-invariant if for every $a_1, \ldots, a_n$

$$\{\nu : (f(a_1, \ldots, a_n))_\nu = f((a_1)_\nu, \ldots, (a_n)_\nu)\} \in F_{\aleph_\tau}.$$

Let $\Phi(\alpha_1, \ldots, \alpha_m, a_1, \ldots, a_n)$ be an analytic formula. Then by

$$\Phi^*(\alpha_1, \ldots, \alpha_m, a_1, \ldots, a_n)$$

we denote the formula obtained from $\Phi(\alpha_1, \ldots, \alpha_m, a_1, \ldots, a_n)$ restricting all function quantifiers to the function which is an ultraproduct of a sequence of functions.

LEMMA 1. *Let $f(\alpha_1, \ldots, \alpha_m, a_1, \ldots, a_n)$ be an arithmetical function in $\alpha_1, \ldots, \alpha_m$. And let $\{\nu:(h(a_1, \ldots, a_n))_\nu = h_{i\nu}((a_1)_\nu, \ldots, (a_n)_\nu)\} \in F_{\aleph_\tau}$ for all $i = 1, \ldots, m$. Then we have*

$$\{\nu:(f(h_1, \ldots, h_m, a_1, \ldots, a_n))_\nu = f(h_{1\nu}, \ldots, h_{m\nu}, (a_1)_\nu, \ldots, (a_n)_\nu)\} \in F_{\aleph_\tau}.$$

PROOF. In the case where the defining schema is one of (I) to (III), we have

$$\{\nu:(\mathrm{1q}\,(a, b))_\nu = \mathrm{1q}\,((a)_\nu, (b)_\nu)\} \in F_{\aleph_\tau}.$$
$$\{\nu:(j(a, b))_\nu = j((a)_\nu, (b)_\nu)\} \in F_{\aleph_\tau}.$$
$$\{\nu:(g^1(a))_\nu = g^1((a)_\nu)\} \in F_{\aleph_\tau}.$$
$$\{\nu:(g^2(a))_\nu = g^2((a)_\nu)\} \in F_{\aleph_\tau}.$$

In the case where the last schema is (IV), the function $f(\alpha_1, \ldots, \alpha_m, a_1, \ldots, a_n)$ is defined by

$$f(\alpha_1, \ldots, \alpha_m, a_1, \ldots, a_n) = \alpha_i(a_1, \ldots, a_n).$$

By the assumption of the lemma, we have

$$\{\nu:(h_i(a_1, \ldots, a_n))_\nu = h_{i\nu}((a_1)_\nu, \ldots, (a_n)_\nu)\} \in F_{\aleph_\tau}.$$

Namely, we have

$$\{\nu:(f(h_i, \ldots, h_m, a_1, \ldots, a_n))_\nu = f(h_{i\nu}, \ldots, h_{m\nu}, (a_1)_\nu, \ldots, (a_n)_\nu)\} \in F_{\aleph_\tau}.$$

The case (V) is treated as usual.

In the case where the last schema is (VI), the function is defined by

$$f(\alpha_1, \ldots, \alpha_m, a_1, \ldots, a_n) = \mu x g(\alpha_1, \ldots, \alpha_m, a_1, \ldots, a_n, x).$$

By the assumption of the induction, we have

$$\{\nu:(g(h_1, \ldots, h_m, a_1, \ldots, a_n, a))_\nu = g(h_{1\nu}, \ldots, h_{m\nu}, (a_1)_\nu, \ldots, (a_n)_\nu, (a)_\nu)\} \in F_{\aleph_\tau}.$$

And it is easily proved that $\exists x(g(h_1, \ldots, h_m, a_1, \ldots, a_n, x) = 0)$ if and only if

$$\{\nu:\exists x(g(h_{1\nu}, \ldots, h_{m\nu}, (a_1)_\nu, \ldots, (a_n)_\nu, x) = 0)\} \in F_{\aleph_\tau}.$$

In the case where $\neg\exists x(g(h_1, \ldots, h_m, a_1, \ldots, a_n, x) = 0)$, we have

$$\{\nu:(f(h_1, \ldots, h_m, a_1, \ldots, a_n))_\nu = 0\} \in F_{\aleph_\tau},$$
$$\{\nu:f(h_{1\nu}, \ldots, h_{m\nu}, (a_1)_\nu, \ldots, (a_n)_\nu) = 0\} \in F_{\aleph_\tau}.$$

In the case where $\exists x(g(h_1, \ldots, h_m, a_1, \ldots, a_n, x) = 0)$, we consider an ordinal

$$d = \mu x g(h_1, \ldots, h_m, a_1, \ldots, a_n, x).$$

From this we obtain

$$\{\nu : g(h_{1\nu}, \ldots, h_{m\nu}, (a_1)_\nu, \ldots, (a_n)_\nu, (d)_\nu) = 0\} \in F_{\aleph_\tau}.$$

Therefore we have

$$\{\nu : f(h_{1\nu}, \ldots, h_{m\nu}, (a_1)_\nu, \ldots, (a_n)_\nu) \geq (d)_\nu\} \in F_{\aleph_\tau}.$$

Assume that $\{\nu : f(h_{1\nu}, \ldots, h_{m\nu}, (a_1)_\nu, \ldots, (a_n)_\nu) < (d)_\nu\} \in F_{\aleph_\tau}$, then we have $f(h_1, \ldots, h_m, a_1, \ldots, a_n) < d$. This contradicts the definition of d. Namely, we have

$$\{\nu : (f(h_1, \ldots, h_m, a_1, \ldots, a_n))_\nu = f(h_{1\nu}, \ldots, h_{m\nu}, (a_1)_\nu, \ldots, (a_n)_\nu)\} \in F_{\aleph_\tau}.$$

In the case where the last schema is (VII), the function $f(\alpha_1, \ldots, \alpha_m, a_1, \ldots, a_n, a)$ is defined by

$$f(\alpha_1, \ldots, \alpha_m, a_1, \ldots, a_n, a) = g(\alpha_1, \ldots, \alpha_m, f^a, a_1, \ldots, a_n, a).$$

Now assume that there is an ordinal number a such that

$$\{\nu : (f(h_1, \ldots, h_m, a_1, \ldots, a_n, a))_\nu \neq f(h_{1\nu}, \ldots, h_{m\nu}, (a_1)_\nu, \ldots, (a_n)_\nu, (a)_\nu)\} \in F_{\aleph_\tau}.$$

Let c be the smallest such ordinal number. Then by the definition of c, we have

$$\{\nu : (f^c(h_1, \ldots, h_m, a_1, \ldots, a_n, a))_\nu = f^{(c)_\nu}(h_{1\nu}, \ldots, h_m, (a_1)_\nu, \ldots, (a_n)_\nu, (a)_\nu)\} \in F_{\aleph_\tau}.$$

Hence by the assumption of the induction, we obtain

$$\{\nu : (g(h_1, \ldots, h_m, f^c, a_1, \ldots, a_n, c))_\nu = g(h_{1\nu}, \ldots, h_{m\nu}, f^{(c)_\nu}, (a_1)_\nu, \ldots, (a_n)_\nu, (c)_\nu)\} \in F_{\aleph_\tau}.$$

This means that

$$\{\nu : (f(h_1, \ldots, h_m, a_1, \ldots, a_n, c)) = f(h_{1\nu}, \ldots, h_{m\nu}, (a_1)_\nu, \ldots, (a_n)_\nu, (c)_\nu)\} \in F_{\aleph_\tau}.$$

This contradicts the assumption on c. Namely we have

$$\{\nu : (f(h_1, \ldots, h_m, a_1, \ldots, a_n, a))_\nu = f(h_{1\nu}, \ldots, h_{m\nu}, (a_1)_\nu, \ldots, (a_n)_\nu, (a)_\nu)\} \in F_{\aleph_\tau}.$$

Thus we complete the proof of the lemma.

THEOREM A. *Let $\Phi(\alpha_1, \ldots, \alpha_m, a_1, \ldots, a_n)$ be an anayltic formula in $\alpha_1, \ldots, \alpha_m$. And let $\{\nu : (h_i(a_1, \ldots, a_n))_\nu = h_{i\nu}((a_1)_\nu, \ldots, (a_n)_\nu)\} \in F_{\aleph_\tau}$ for all $i = 1, \ldots, m$. Then we have*

$$\Phi^*(h_1, \ldots, h_m, a_1, \ldots, a_n) \leftrightarrow \{\nu : \Phi(h_{1\nu}, \ldots, h_{m\nu}, (a_1)_\nu, \ldots, (a_n)_\nu)\} \in F_{\aleph_\tau}.$$

PROOF. The case where $\Phi(\alpha_1, \ldots, \alpha_m, a_1, \ldots, a_n)$ is of the form

$$f(\alpha_1, \ldots, \alpha_m, a_1, \ldots, a_n) = g(\alpha_1, \ldots, \alpha_m, a_1, \ldots, a_n).$$

By Lemma 1, we have the following equivalence:

$$f(h_1, \ldots, h_m, a_1, \ldots, a_n) = g(h_1, \ldots, h_m, a_1, \ldots, a_n)$$
$$\leftrightarrow \{\nu: (f(h_1, \ldots, h_m, a_1, \ldots, a_n))_\nu = (g(h_1, \ldots, h_m, \ldots, a_1, a_n))_\nu\} \in F_{\aleph_\tau}$$
$$\leftrightarrow \{\nu: f(h_{1\nu}, \ldots, h_{m\nu}, (a_1)_\nu, \ldots, (a_n)_\nu) = g(h_{1\nu}, \ldots, h_{m\nu}, (a_1)_\nu, \ldots, (a_n)_\nu)\} \in F_{\aleph_\tau}.$$

Namely, we obtain that

$$\Phi^*(h_1, \ldots, h_m, a_1, \ldots, a_n) \leftrightarrow \{\nu: \Phi(h_{1\nu}, \ldots, h_{m\nu}, (a_1)_\nu, \ldots, (a_n)_\nu)\} \in F_{\aleph_\tau}.$$

The case where the outermost symbol is one of $<$, $\neg$, $\vee$ is treated as usual.

The case where $\Phi(\alpha_1, \ldots, \alpha_m, a_1, \ldots, a_n)$ is of the form

$$\exists x \Psi(\alpha_1, \ldots, \alpha_m, a_1, \ldots, a_n, x).$$

Assume that $\exists x \Psi^*(h_1, \ldots, h_m, a_1, \ldots, a_n, x)$. Then there is an ordinal number a such that $\Psi^*(h_1, \ldots, h_m, a_1, \ldots, a_n, a)$. By the assumption of the induction, we have

$$\{\nu: \Psi(h_{1\nu}, \ldots, h_{m\nu}, (a_1)_\nu, \ldots, (a_n)_\nu, (a)_\nu)\} \in F_{\aleph_\tau}.$$

Hence we have

$$\{\nu: \exists x \Psi(h_{1\nu}, \ldots, h_{m\nu}, (a_1)_\nu, \ldots, (a_n)_\nu, x)\} \in F_{\aleph_\tau}.$$

Conversely, we assume the above formula. Then there is an ordinal number b such that $\{\nu: \Psi(h_{1\nu}, \ldots, h_{m\nu}, (a_1)_\nu, \ldots, (a_n)_\nu, (b)_\nu)\} \in F_{\aleph_\tau}$. By the assumption of the induction, we have $\Psi^*(h_1, \ldots, h_m, a_1, \ldots, a_n, b)$. Hence we have

$$\exists x \Psi^*(h_1 \ldots, h_m, a_1, \ldots, a_n, b).$$

The case where $\Phi(\alpha_1, \ldots, \alpha_m, a_1, \ldots, a_n)$ is of the form

$$\exists \alpha \Psi(\alpha_1, \ldots, \alpha_m, \alpha, a_1, \ldots, a_n).$$

Now we assume that $\exists \alpha^* \Psi^*(h_1, \ldots, h_m, \alpha^*, a_1, \ldots, a_n)$ where the range of α^* is restricted to the function which is an ultraproduct of a sequence of functions. Then there is a function $h(a_1, \ldots, a_s)$ and a sequence of functions such that

$$\Psi^*(h_1, \ldots, h_m, h, a_1, \ldots, a_n),$$
$$\{\nu: (h(a_1, \ldots, a_s))_\nu = h_\nu((a_1)_\nu, \ldots, (a_s)_\nu)\} \in F_{\aleph_\tau}.$$

By the assumption of the induction, we have

$$\{\nu: \Psi(h_{1\nu}, \ldots, h_{m\nu}, h_\nu, (a_1)_\nu, \ldots, (a_n)_\nu)\} \in F_{\aleph_\tau}.$$

Hence we obtain that

$$\{\nu: \exists \alpha \Psi(h_{1\nu}, \ldots, h_{m\nu}, \alpha, (a_1)_\nu, \ldots, (a_n)_\nu)\} \in F_{\aleph_\tau}.$$

Conversely, we assume the above formula. Then there is a sequence of functions $h_1(a_1, \ldots, a_s), h_2(a_1, \ldots, a_s), \ldots, h_\nu(a_1, \ldots, a_s), \ldots$ and a function $h(a_1, \ldots, a_s)$ such that

$$\{\nu : (h(a_1, \ldots, a_n))_\nu = h_\nu((a_1)_\nu, \ldots, (a_n)_\nu)\} \in F_{\aleph_\tau},$$

$$\{\nu : \Psi(h_{1\nu}, \ldots, h_{m\nu}, h_\nu, (a_1)_\nu, \ldots, (a_n)_\nu)\} \in F_{\aleph_\tau}.$$

By the assumption of the induction, we have $\Psi^*(h_1, \ldots, h_m, h, a_1, \ldots, a_n)$. Therefore we have

$$\exists \alpha^* \Psi^*(h_1, \ldots, h_m, \alpha^*, a_1, \ldots, a_n).$$

Thus we complete the proof of the theorem.

As simple applications of this theorem we have that if $\Phi(h_1, \ldots, h_m, a_1, \ldots, a_n)$ is a Π^1_1-formula in $h_1, \ldots, h_m$, where $h_1, \ldots, h_m$ are U-invariant functions, then

$$\Phi(h_1, \ldots, h_m, a_1, \ldots, a_n) \to \{\nu : \Phi(h_1, \ldots, h_m, (a_1)_\nu, \ldots, (a_n)_\nu)\} \in F_{\aleph_\tau}.$$

Therefore if $\Phi(h_1, \ldots, h_m, a_1, \ldots, a_n)$ is a Σ^1_1-formula in U-invariant functions, then

$$\{\nu : \Phi(h_1, \ldots, h_m, (a_1)_\nu, \ldots, (a_n)_\nu)\} \in F_{\aleph_\tau} \to \Phi(h_1, \ldots, h_m, a_1, \ldots, a_n).$$

If it is a Δ^1_1-formula in U-invariant functions, then

$$\Phi(h_1, \ldots, h_m, a_1, \ldots, a_n) \leftrightarrow \{\nu : \Phi(h_1, \ldots, h_m, (a_1)_\nu, \ldots, (a_n)_\nu)\} \in F_{\aleph_\tau}.$$

Let $f(h_1, \ldots, h_m, a_1, \ldots, a_n)$ be a Δ^1_1-function in U-invariant functions, then

$$\{\nu : (f(h_1, \ldots, h_m, a_1, \ldots, a_n))_\nu = f(h_1, \ldots, h_m, (a_1)_\nu, \ldots, (a_n)_\nu)\} \in F_{\aleph_\tau}.$$

From this we obtain that every function which is Δ^1_1- in U-invariant functions is U-invariant.

Now we shall give some examples of Π^1_1-formulas, and it should be noted that some well-known axioms of strong infinity have the form $\exists x P(x)$ where $P(a)$ is a Π^1_1-formula.

$$\bar{\bar{a}} < \bar{\bar{b}} \equiv \forall \alpha \exists c(c < b \wedge \forall d(d < a \to \alpha(d) = 0)).$$

$$a < \mathrm{cf}(b) \equiv \forall \alpha \exists c(c < b \wedge \forall d(d < a \wedge \alpha(d) < b \to \alpha(d) < c)).$$

$$2^{\bar{\bar{a}}} < \bar{\bar{b}} \equiv \forall \alpha \forall c(\forall d(\alpha(d) \leq 1) \wedge \forall d \forall e(d < c \wedge e < c \wedge d \neq e \to \exists f(f < a \wedge \alpha(j(d,f)) \neq \alpha(j(e,f)))) \to c < b).$$

$$\mathrm{Card}\ (a) \equiv \forall b(b < a \to \bar{\bar{b}} < \bar{\bar{a}}).$$

$$\mathrm{Reg}\ (a) \equiv \forall b(b < a \to b < \mathrm{cf}(a)).$$

$$\mathrm{We}\ (a) \equiv \mathrm{Reg}\ (a) \wedge \forall b(b < a \to \exists c(\bar{\bar{b}} < \bar{\bar{c}} \wedge \bar{\bar{c}} < \bar{\bar{a}})).$$

$$\mathrm{St}\ (a) \equiv \mathrm{Reg}\ (a) \wedge \forall b(b < a \to 2^{\bar{\bar{b}}} < \bar{\bar{a}}).$$

$$\mathrm{Norm}\ (\alpha) \equiv \forall a \forall b(a < b \to \alpha(a) < \alpha(b)) \wedge \forall a(\forall c(a \neq c + 1) \to \forall b(b < \alpha(a) \to \exists c(c < a \wedge b < \alpha(c)))).$$

Norm (α) means that α is a normal function.

$$\begin{aligned}&\text{In }(a, R)\\ &\equiv \forall\alpha(\text{Norm }(\alpha) \wedge \forall b(b < a \rightarrow \alpha(b) < a) \rightarrow \exists c(c < a \wedge \alpha(c) = c \wedge R(c))).\end{aligned}$$

In (a, R) means that a is R-inaccessible.

$$\text{Hyp}_1(a) = \text{St}(a)\,\text{In}(a, \text{St}), \qquad \text{Hyp}_2(a) = \text{Hyp}_1(a)\,\text{In}(a, \text{Hyp}_1), \ldots .$$

$\text{Hyp}_1(a)$ means that a is hyper-inaccessible of type 1.

$$\begin{aligned}\text{Hyp}(a, b, T) \equiv \forall\beta[\{&\forall c(T(c) \rightarrow \beta(0, c) = 0)\\ &\wedge \forall c\forall d(\text{In}(c, \lambda x\beta(d, x) = 0) \wedge \beta(d, c) = 0 \rightarrow \beta(d+1, c) = 0)\\ &\wedge \forall c\forall d(\forall e(d \neq e+1) \wedge \forall e(e < d \rightarrow \beta(e, c) = 0) \rightarrow \beta(d, c) = 0)\}\\ &\rightarrow \beta(a, b) = 0].\end{aligned}$$

Hyp (a, b, T) means that b is hyper-inaccessible of type a with respect to T. Hyp (a, b, St) means that b is hyper-inaccessible of type a, and Hyp (a, a, St) means that a is hyper-inaccessible of its own type.

We can also define Π_1^1-formulas such as Hyp $(a, a, \lambda x$ Hyp $(x, x, \text{St}))$ or even Hyp $(a, a, \lambda x$ Hyp $(x, x, \lambda y$ Hyp $(y, y, \text{St})))$, . . . , the first cardinal satisfying such formulas is tremendously large.

Let $P(a)$ be a Π_1^1-formula. Then we obtain that

$$P(\aleph_\tau) \rightarrow \{\nu : P((\aleph_\tau)_\nu)\} \in F_{\aleph_\tau}.$$

And by $\{\nu : (\aleph_\tau)_\nu < \aleph_\tau\} \in F_{\aleph_\tau}$, we have

$$P(\aleph_\tau) \rightarrow \{\nu : P((\aleph_\tau)_\nu) \wedge (\aleph_\tau)_\nu < \aleph_\tau\} \in F_{\aleph_\tau}.$$

This means that $\aleph_\tau$ cannot be the least element of Π_1^1-class of ordinals.

Let $f(a_1, \ldots, a_n)$ be a function such that $f(a_1, \ldots, a_n) = 0$ if $\max(a_1, \ldots, a_n) \geq \aleph_\tau$. Now we define the functions $f_\nu(a_1, \ldots, a_n)$ $(\nu < \aleph_\tau)$ by the following:

$$\begin{aligned}f_\nu(a_1, \ldots, a_n) &= (f(a_1, \ldots, a_n))_\nu \quad \text{if } (\max(a_1, \ldots, a_n))_\nu < (\aleph_\tau)_\nu,\\ &= 0 \quad \text{otherwise}.\end{aligned}$$

In the case where $\max(a_1, \ldots, a_n) < \aleph$, we have

$$\{\nu : (f(a_1, \ldots, a_n))_\nu = f_\nu(a_1, \ldots, a_n)\} \in F_{\aleph_\tau},$$

$$\{\nu : (a_1)_\nu = a_1, \ldots, (a_n)_\nu = a_n\} \in F_{\aleph_\tau}.$$

Therefore we obtain that

$$\{\nu : (f(a_1, \ldots, a_n))_\nu = f_\nu((a_1)_\nu, \ldots, (a_n)_\nu)\} \in F_{\aleph_\tau}.$$

In the case where $\max(a_1, \ldots, a_n) \geq \aleph$, we have $\{\nu : \max((a_1)_\nu, \ldots, (a_n)_\nu) \geq (\aleph_\tau)_\nu\} \in F_{\aleph_\tau}$. Therefore we have $\{\nu : f_{\nu i}((a_1)_\nu, \ldots, (a_n)_\nu) = 0\} \in F_{\aleph_\tau}$. By

$$f(a_1, \ldots, a_n) = 0,$$

we have

$$\{\nu : (f(a_1, \ldots, a_n))_\nu = f_\nu((a_1)_\nu, \ldots, (a_n)_\nu)\} \in F_{\aleph_\tau}.$$

Namely, $f(a_1, \ldots, a_n)$ is an ultraproduct of a sequence of functions $f_1(a_1, \ldots, a_n)$, $f_2(a_1, \ldots, a_n), \ldots, f_\nu(a_1, \ldots, a_n), \ldots$.

Let $\mathrm{Bd}(\alpha, a)$ be the formula defined by

$$\mathrm{Bd}\,(\alpha, a) \equiv \forall x_1 \cdots \forall x_n(\max\,(x_1, \ldots, x_n) \geq a \rightarrow \alpha(x_1, \ldots, x_n) = 0).$$

By this, we obtain that

$$\forall \alpha^*(\mathrm{Bd}\,(\alpha^*, \aleph_\tau) \rightarrow \Phi^*(\alpha_1, \ldots, \alpha_m, \alpha^*, a_1, \ldots, a_n))$$
$$\equiv \forall \alpha(\mathrm{Bd}\,(\alpha, \aleph_\tau) \rightarrow \Phi^*(\alpha_1, \ldots, \alpha_m, \alpha, a_1, \ldots, a_n)),$$
$$\exists \alpha^*(\mathrm{Bd}\,(\alpha^*, \aleph_\tau) \wedge \Phi^*(\alpha_1, \ldots, \alpha_m, \alpha^*, a_1, \ldots, a_n))$$
$$\equiv \exists \alpha(\mathrm{Bd}\,(\alpha, \aleph_\tau) \wedge \Phi^*(\alpha_1, \ldots, \alpha_m, \alpha, a_1, \ldots, a_n)).$$

Now we consider the formula Card (a) which is expressed by the formula of the form

$$\mathrm{Card}\,(a) = \forall \alpha(\mathrm{Bd}\,(\alpha, a) \rightarrow A(\alpha, a))$$

where $A(\alpha, a)$ is an arithmetical formula. The formulas We (a), St (a), Hyp_1 (a), Hyp (a, a, St), ... are expressed by the formula of the same form as for the case of Card (a). Let Ram (a) be the formula expressing that a is a Ramsey cardinal. Then it is described by the formula of the form

$$\mathrm{Ram}\,(a) = \forall \beta(\mathrm{Bd}\,(\beta, a) \rightarrow \exists \alpha(\mathrm{Bd}\,(\alpha, a) \wedge A(\alpha, \beta, a)))$$

where $A(\alpha, \beta, a)$ is arithmetical in α, β.

Therefore, we have $\mathrm{Ram}^*\,(\aleph_\tau) = \mathrm{Ram}\,(\aleph_\tau)$. And of course $\mathrm{Ram}\,(\aleph_\tau)$. So we have

$$\{\nu : \mathrm{Ram}\,((\aleph_\tau)_\nu)\} \in F_{\aleph_\tau}.$$

Let Ant (a) be the formula expressing that a is an antihomogeneous cardinal (cf. [**10**]). Then it is described by the formula of the form

$$\mathrm{Ant}\,(a) = \exists \beta(\mathrm{Bd}\,(\beta, a) \wedge \forall \alpha(\mathrm{Bd}\,(\alpha, a) \rightarrow A(\alpha, \beta, a)))$$

where $A(\alpha, \beta, a)$ is arithmetical in α, β.

Therefore we have the following equivalence:

$$\mathrm{Ant}\,(\aleph_\tau) \leftrightarrow \{\nu : \mathrm{Ant}\,((\aleph_\tau)_\nu)\} \in F_{\aleph_\tau}.$$

Namely, $\aleph_\tau$ cannot be the smallest antihomogeneous cardinal.

Now we consider the formula $b < 2^{\overline{\overline{a}}}$ which is described by the form

$$b < 2^{\overline{\overline{a}}} \equiv \forall \beta \exists \alpha(\mathrm{Bd}\,(\alpha, a) \wedge A(\alpha, \beta, a))$$

where $A(\alpha, \beta, a)$ is arithmetical in α, β. So

$$(b < 2^{\aleph_\tau})^* \equiv \forall \beta^* \exists \alpha(\mathrm{Bd}\,(\alpha, \aleph_\tau) \wedge A(\alpha, \beta^*, \aleph_\tau)).$$

Hence we obtain

$$b < 2^{\aleph_\tau} \to \{\nu : (b)_\nu < 2^{(\aleph_\tau)_\nu}\} \in F_{\aleph_\tau}.$$

Now we consider the formula $\exists x(\bar{a} < \bar{\bar{x}} \wedge x < 2^{\bar{\bar{a}}})$. Then we have

$$\exists x(\aleph_\tau < \bar{\bar{x}} \wedge x < 2^{\aleph_\tau}) \to \{\nu : \exists x((\aleph_\tau)_\nu < \bar{\bar{x}} \wedge x < 2^{(\aleph_\tau)_\nu})\} \in F_{\aleph_\tau}.$$

Let Meas (a) be the formula expressing that a is a normally measurable cardinal. Then it is described by the formula of the form

$$\text{Meas } (a) = \exists\beta\forall\alpha(\text{Bd } (\alpha, a) \to A(\alpha, \beta, a))$$

where $A(\alpha, \beta, a)$ is arithmetical in α, β.

Let $P_1(a), P_2(a), \ldots, P_n(a), \ldots$ be an enumeration of all Π_1^1-formulas such that $P_n(\aleph_\tau)$. Then we have $\{\nu : P_n((\aleph_\tau)_\nu)\} \in F_{\aleph_\tau}$ for all $n < \omega$. Since $F_{\aleph_\tau}$ is $\aleph_0$-complete, we have

$$\{\nu : P_n((\aleph_\tau)_\nu) \quad \text{for all } n < \omega, (\aleph_\tau)_\nu < \aleph_\tau\} \in F_{\aleph_\tau}.$$

From this we have a cardinal number $\aleph_\alpha$ $(<\aleph_\tau)$ such that $\aleph_a$ has all Π_1^1-properties of $\aleph_\tau$.

Let $\aleph_\tau$ be a normally measurable cardinal number. We consider a nonprincipal ultrafilter $F_{\aleph_\tau}$ with the character $\aleph_\tau > \aleph_0$. And we consider $\text{On}^{\aleph_\tau}/F_{\aleph_\tau}$ as before. By $a^{(\tau)}$ we denote the representative element of ath equivalence class of $\text{On}^{\aleph_\tau}/F_{\aleph_\tau}$. And by $(a)_\nu^{(\tau)}$, we denote the νth component of $a^{(\tau)}$.

The function $p_\tau(a)$ is defined by

$$p_\tau(a) = b \quad \text{if and only if} \quad \{\nu : (b)_\nu^{(\tau)} = a\} \in F_{\aleph_\tau}.$$

Let $f(a_1, \ldots, a_n)$ be a function. For any fixed ordinal number $a_1, \ldots, a_n$ we consider a function defined on $\aleph_\tau$ by

$$u(\nu) = f((a_1)_\nu^{(\tau)}, \ldots, (a_n)_\nu^{(\tau)}).$$

Since $u \in \text{On}^{\aleph_\tau}$, there is unique b such that

$$\{\nu : (b)_\nu^{(\tau)} = u(\nu)\} \in F_{\aleph_\tau}.$$

Now the function $\sigma_\tau f(a_1, \ldots, a_n)$ is defined by

$$\sigma_\tau f(a_1, \ldots, a_n) = b.$$

Namely, it has the property that

$$\{\nu : (\sigma_\tau f(a_1, \ldots, a_n))_\nu^{(\tau)} = f((a_1)_\nu^{(\tau)}, \ldots, (a_n)_\nu^{(\tau)})\} \in F_{\aleph_\tau}.$$

Let $\mathfrak{A}$ be a set of functions and let D be a set of ordinal numbers. Then the set D is called $\mathfrak{A}$-closed if for every function $f(a_1, \ldots, a_n)$ in $\mathfrak{A}$ and any ordinal numbers $a_1, \ldots, a_n \in D$, we have $f(a_1, \ldots, a_n) \in D$. Let B be a set of ordinal numbers. Then by $[B]_{\mathfrak{A}}$ we denote the smallest set which is $\mathfrak{A}$-closed and includes B. And $[B]_{\mathfrak{A}}$ is called the $\mathfrak{A}$-closure of B.

LEMMA 2. *Let* $\aleph_\tau$ *be a normally measurable cardinal and let* $F_{\aleph_\tau}$ *be a non-principal ultrafilter with the character* $\aleph_\tau > \aleph_0$. *Let* $\mathfrak{A}$ *be a set of functions and let* B *be a set of ordinal numbers such that*

(1) $\overline{\overline{B}} < \aleph_\tau$, $\aleph_\tau \in B$, *B is* $\mathfrak{A}$*-closed*, $\overline{\overline{\mathfrak{A}}} < \aleph_\tau$.

(2) $\mathfrak{A}$ *includes the identity function and the function* p_τ, $f \in \mathfrak{A}$ *implies* $\sigma_\tau f \in \mathfrak{A}$, $\mathfrak{A}$ *is classed under substitution.*

Then there is an ordinal number a_0 *such that*

(1) $[B \cup \{a_0\}]_{\mathfrak{A}} \cap a_0 \subset B$.

(2) $\sup (B \cap \aleph_\tau) \leq a_0 < \aleph_\tau$ *and* $a_0 \notin B$.

PROOF. Let $f_1, \ldots, f_\delta, \ldots, (\delta < \overline{\overline{\mathfrak{A}}})$ be an enumeration of functions in $\mathfrak{A}$. Now we define a set $A_{\delta a_1 \cdots a_n b}$ where $\delta < \overline{\overline{\mathfrak{A}}}$ and $a_1, \ldots, a_{n,b} \in B$ as follows:

$$A_{\delta a_1 \cdots a_n b} = \begin{cases} \{\nu : f_\delta((a_1)_\nu^{(\tau)}, \ldots, (a_n)_\nu^{(\tau)}) < (b)_\nu^{(\tau)}\} & \text{if } \sigma_\tau f_\delta(a_1, \ldots, a_n) < b, \\ \{\nu : f((a_1)_\nu^{(\tau)}, \ldots, (a_n)_\nu^{(\tau)}) = (b)_\nu^{(\tau)}\} & \text{if } \sigma_\tau f_\delta(a_1, \ldots, a_n) = b, \\ \{\nu : f((a_1)_\nu^{(\tau)}, \ldots, (a_n)_\nu^{(\tau)}) > (b)_\nu^{(\tau)}\} & \text{if } \sigma_\tau f_\delta(a_1, \ldots, a_n) > b. \end{cases}$$

By the definition of $\sigma_\tau f_\tau(a_1, \ldots, a_n)$, we obtain that

$$A_{\delta a_1 \cdots a_n b} \in F_{\aleph_\tau} \quad \text{for every } \delta < \overline{\overline{\mathfrak{A}}} \quad \text{and } a_1, \ldots, a_n, b \in B.$$

Since the cardinality of B, $\mathfrak{A}$ is less than $\aleph_\tau$, the cardinality of the set

$$\{\langle \delta, a_1, \ldots, a_n, b \rangle : n < \omega, \delta < \overline{\overline{\mathfrak{A}}}, a_1, \ldots, a_n, b \in B\}$$

is less than $\aleph_\tau$.

Let C be a subset of $\aleph_\tau$ defined by

$$C = \bigcap_{\substack{\delta < \overline{\overline{\mathfrak{A}}} \\ a_1, \ldots, a_n, b \in B}} A_{\delta a_1 \cdots a_n b}.$$

Since the character of the ultrafilter $F_{\aleph_\tau}$ is $\aleph_\tau$, we have $C \in F_{\aleph_\tau}$.

Let ν_0 be an ordinal number such that $\nu_0 \in C$. Let $f_\delta(a)$ be the identity function, namely $f_\delta(a) = a$. Then we have

$$\{\nu : (\aleph_\tau)_\nu^{(\tau)} < \aleph_\tau\} \supset A_{\delta \aleph_\tau p_\tau(\aleph_\tau)} \supset C.$$

Namely $(\aleph_\tau)_{\nu_0}^{(\tau)} < \aleph_\tau$. Now we shall show that

$$[B \cup \{(\aleph_\tau)_{\nu_0}^{(\tau)}\}]_{\mathfrak{A}} \cap (\aleph_\tau)_{\nu_0}^{(\tau)} \subset B.$$

Assume that this inclusion is not true. Then there is a function f_β such that

$$f_\beta((\aleph_\tau)_{\nu_0}^{(\tau)}, b_1, \ldots, b_n) = c \notin B \quad \text{and} \quad c < (\aleph_\tau)_{\nu_0}^{(\tau)}$$

where $b_1, \ldots, b_n \in B$.

By the assumption that $p_\tau(a) \in \mathfrak{A}$ and B is $\mathfrak{A}$-closed, we have

$$p_\tau(b_1), \ldots, p_\tau(b_n) \in B.$$

By $f_\beta \in \mathfrak{A}$, we obtain that $\sigma_\tau f_\beta \in \mathfrak{A}$. Hence by $\aleph_\tau \in B$,

$$\sigma_\tau f_\beta(\aleph_\tau, p_\tau(b_1), \ldots, p_\tau(b_n)) \in B.$$

We assume that

$$\sigma_\tau f_\beta(\aleph_\tau, p_\tau(b_1), \ldots, p_\tau(b_n)) \geq \aleph_\tau.$$

Then we obtain that

$$\{\nu : f_\beta((\aleph_\tau)_\nu^{(\tau)}, b_1, \ldots, b_n) \geq (\aleph_\tau)_\nu^{(\tau)}\} \supset A_{\beta' \aleph_\tau p_\tau(b_1) \cdots p_\tau(b_n) \aleph_\tau} \supset C$$

where $\sigma_\tau f = f_{\beta'}$. By this we obtain that

$$f_\beta((\aleph_\tau)_{\nu_0}^{(\tau)}, b_1, \ldots, b_n) \geq (\aleph_\tau)_{\nu_0}^{(\tau)}.$$

This contradicts our assumption. Namely, we obtain

$$\sigma_\tau f_\beta(\aleph_\tau, p_\tau(b_1), \ldots, p_\tau(b_n)) = e < \aleph_\tau \quad \text{and} \quad e \subset B.$$

Therefore, we obtain that

$$\{\nu : f_\beta((\aleph_\tau)_\nu^{(\tau)}, b_1, \ldots, b_n) = e\} \supset A_{\beta' \aleph_\tau p_\tau(b_1) \cdots p_\tau(b_n) e} \supset C.$$

Namely, we have

$$f_\beta((\aleph_\tau)_\nu^{(\tau)}, b_1, \ldots, b_n) \in B.$$

From this we obtain that

$$[B \cup \{(\aleph_\tau)_{\nu_0}^{(\tau)}\}]_{\mathfrak{A}} \cap (\aleph_\tau)_{\nu_0}^{(\tau)} \subset B.$$

Next we shall show that $\sup (B \cap \aleph_\tau) \leq (\aleph_\tau)_{\nu_0}^{(\tau)}$ and $(\aleph_\tau)_{\nu_0}^{(\tau)} \notin B$. We consider that identity function f_β of $\mathfrak{A}$. And let d be any element of $B \cap \aleph_\tau$. We obtain that

$$\{\nu : f_\beta(d) < (\aleph_\tau)_\nu^{(\tau)}\} = \{\nu : d < (\aleph_\tau)_\nu^{(\tau)}\} \supset A_{\beta d \aleph_\tau} \supset C.$$

From this we obtain that $\sup (B \cap \aleph_\tau) \leq (\aleph_\tau)_{\nu_0}^{(\tau)}$ and $(\aleph_\tau)_{\nu_0}^{(\tau)} \notin B$.

Let $\mathfrak{A}$ be a set of functions and let D be a set of ordinal numbers. Then a function $f(a)$ is called a generative function with respect to $\mathfrak{A}$ and D if the following conditions are satisfied:

(1) $f(a)$ is strictly increasing.
(2) $f(a) \notin [C \cup F_a]_{\mathfrak{A}}$ where $F_a = \{f(b) : b < a\}$.
(3) For any subset B of F_a, we have $[D \cup B \cup \{f(a)\}]_{\mathfrak{A}} \cap f(a) \subset [D \cup B]_{\mathfrak{A}}$.

THEOREM B. *Let* $\aleph_{\tau_1} < \cdots < \aleph_{\tau_\beta} < \cdots (\beta < \delta_0 < \aleph_{\tau_1})$ *be a sequence of normally measurable cardinals. And let* $\mathfrak{A}$ *be a set of functions and let* B *be a set of ordinal numbers such that*

(1) $\overline{\overline{B}} < \aleph_{\tau_1}$, $\aleph_{\tau_\beta} \in B$ *for all* $\beta < \delta_0$, B *is* $\mathfrak{A}$*-closed*, $\overline{\overline{\mathfrak{A}}} < \aleph_{\tau_1}$.
(2) $\mathfrak{A}$ *includes the identity function and the functions* $p_{\tau_\beta} \in \mathfrak{A}$, $f \in \mathfrak{A}$ *implies* $\sigma_{\tau_\beta} f \in \mathfrak{A}$ *for all* $\beta < \delta_0$, $\mathfrak{A}$ *is closed under substitution.*

Then there is a generative function $f_0(a)$ *such that*

(1) $f_0(a)$ *is a generative function with respect to* $\mathfrak{A}$ *and* B.
(2) *For every* $\aleph_{\tau_\beta}$, *we have* $\forall x(x < \aleph_{\tau_\beta} \rightarrow f_0(x) < \aleph_{\tau_\beta})$.

PROOF. We shall define a generative function $f_0(a)$ defined on $\sup_{\beta<\delta_0} \aleph_{\tau_\beta}$. By Lemma 2 we have an ordinal number a_0 such that

(1) $[B \cup \{a_0\}]_{\mathfrak{A}} \cap a_0 \subset B$.

(2) $\sup (B \cap \aleph_{\tau_1}) \leq a_0 < \aleph_{\tau_1}$ and $a_0 \notin B$.

Now $f_0(0)$ is defined by $f_0(0) = a_0$. We assume that for an ordinal number b_0 ($< \sup_{\beta<\delta_0} \aleph_{\tau_\beta}$) the values of the function f_0 for ordinals less than b_0 are already defined so that

(1) $\forall a \forall b(a < b < b_0 \rightarrow f_0(a) < f_0(b))$.

(2) $f_0(a) \notin [B \cup F_a]_{\mathfrak{A}}$ where $F_a = \{f_0(b): b < a\}$ for $a < b_0$.

(3) For every subset B of F_a, we have

$$[B \cup B' \cup \{f_0(a)\}]_{\mathfrak{A}} \cap f_0(a) \subset [B \cup B']_{\mathfrak{A}}.$$

(4) If $a < \aleph_{\tau_\beta}$ and $a < b_0$, then $f_0(a) < \aleph_{\tau_\beta}$.

Since $b_0 < \sup_{\beta<\delta_0} \aleph_{\tau_\beta}$, there is the least β_0 such that $b_0 < \aleph_{\tau_{\beta_0}}$. Now we consider the set $D_1 = [B \cup \{f_0(b): b < b_0\}]_{\mathfrak{A}}$. Then we have

(1) $\overline{\overline{D}}_1 < \aleph_{\tau_{\beta_0}}$, $\aleph_{\tau_{\beta_0}} \in D_1$, D_1 is $\mathfrak{A}$ closed, $\mathfrak{A} < \aleph_{\tau_{\beta_0}}$.

(2) $\mathfrak{A}$ includes the identity function and the function $p_{\tau_{\beta_0}}$, $f \in \mathfrak{A}$ implies $\sigma_{\tau_{\beta_0}} f \in \mathfrak{A}$, $\mathfrak{A}$ is closed under substitution.

Therefore by Lemma 2, we have an ordinal number a_0 such that

(1) For every subset B' of $F_{b_0} = \{f_0(a): a < b_0\}$, we have

$$[B \cup B' \cup \{a_0\}]_{\mathfrak{A}} \cap a_0 \subset [B \cup B']_{\mathfrak{A}}.$$

(2) $\sup (D_1 \cap \aleph_{\beta_0}) \leq a_0 < \aleph_{\tau_{\beta_0}}$ and $a_0 \notin D_1$.

Now $f_0(b_0)$ is defined to be this a_0. By the definition of f_0, we have

(1) $f_0(a)$ is a generative function with respect to $\mathfrak{A}$ and B defined on $\sup_{\beta<\delta_0} \aleph_{\tau_\beta}$.

(2) For every $\aleph_{\tau_\beta}$, we have $\forall x(x < \aleph_{\tau_\beta} \rightarrow f_0(x) < \aleph_{\tau_\beta})$.

Thus we complete the proof of the theorem.

3. φ-commutative functions. A set of functions $\mathfrak{A}$ is called arithmetically (analytically) closed if every function which is arithmetical (analytic) in the functions of $\mathfrak{A}$ belongs to $\mathfrak{A}$.

Let B be a set of ordinal numbers. And let φ be an order-preserving onto mapping such that $\varphi: \mathrm{On} \rightarrow B$. Then a function $g(a_1, \ldots, a_n)$ and function $f(a_1, \ldots, a_n)$ are said to be φ-commutative if

$$\forall x_1 \cdots \forall x_n(\varphi(g(x_1, \ldots, x_n)) = f(\varphi(x_1), \ldots, \varphi(x_n))).$$

An ordinal number a is said to be φ-invariant if $\varphi(a) = a$. A function $f(a_1, \ldots, a_n)$ is said to be φ-invariant if

$$\forall x_1 \cdots \forall x_n(\varphi(f(x_1, \ldots, x_n)) = f(\varphi(x_1), \ldots, \varphi(x_n))).$$

Let $A(\alpha, h_1, \ldots, h_m, a_1, \ldots, a_n)$ be an analytic formula in $\alpha, h_1, \ldots, h_m$ where α is the only one free function variable. Then a function $g(a_1, \ldots, a_n, b_1, \ldots, b_s)$ is called an R-function for the formula if the following condition is satisfied:

$$\forall x_1 \cdots \forall x_n(\exists \alpha A(\alpha, h_1, \ldots, h_m, x_1, \ldots, x_n) \\ \leftrightarrow A(\lambda z_1 \cdots z_s g(x_1, \ldots, x_n, z_1, \ldots, z_s), h_1, \ldots, h_m, x_1, \ldots, x_n)).$$

Let $A(\alpha, h_1, \ldots, h_m, a_1, \ldots, a_n)$ be an analytic formula in $\alpha, h_1, \ldots, h_m$. In the case where $\exists \alpha A(\alpha, h_1, \ldots, h_m, a_1, \ldots, a_n)$ is true for $a_1, \ldots, a_n$, we consider a function $f_{a_1 \cdots a_n}(b_1, \ldots, b_s)$ such that

$$A(\lambda x_1 \cdots x_s f_{a_1 \cdots a_n}(x_1, \ldots, x_s), h_1, \ldots, h_m, a_1, \ldots, a_n).$$

In the case where $\exists \alpha A(\alpha, h_1, \ldots, h_m, a_1, \ldots, a_n)$ is not true, we put

$$f_{a_1 \cdots a_n}(b_1, \ldots, b_s) = 0 \quad \text{for every } b_1, \ldots, b_s.$$

Now an R-function $g(a_1, \ldots, a_n, b_1, \ldots, b_s)$ for the formula $A(\alpha, h_1, \ldots, h_m, a_1, \ldots, a_n)$ is defined by

$$g(a_1, \ldots, a_n, b_1, \ldots, b_s) = f_{a_1 \cdots a_n}(b_1, \ldots, b_s).$$

A set of functions $\mathfrak{A}$ is called an analytic basis if for every analytic formula $A(\alpha, h_1, \ldots, h_m, a_1, \ldots, a_n)$ in $\alpha, h_1, \ldots, h_m$ where $h_1, \ldots, h_m \in \mathfrak{A}$, there is an R-function $g(a_1, \ldots, a_n, b_1, \ldots, b_s)$ for the formula in $\mathfrak{A}$.

LEMMA 3. *Let $\mathfrak{B}$ be a set of functions. Then there is a set of functions $\mathfrak{A}$ such that the following conditions are satisfied:*

(1) $\mathfrak{A} \supset \mathfrak{B}$ *and* $\overline{\overline{\mathfrak{A}}} \leq \max(\aleph_0, \overline{\overline{\mathfrak{B}}})$.

(2) $\mathfrak{A}$ *is an analytic basis.*

PROOF. Let $\mathfrak{B}^*$ be the set of all analytic formulas of the form

$$A(\alpha, h_1, \ldots, h_m, a_1, \ldots, a_n)$$

where $h_1, \ldots, h_m \in \mathfrak{B}$. For every formula of $\mathfrak{B}^*$, we take a representative R-function for the formula and consider the set $\mathfrak{B}^\#$ of all such representative R-functions. Clearly $\mathfrak{B}^\# \supset \mathfrak{B}$. By $g(\mathfrak{B})$, we denote the set of all functions analytic in the functions in $\mathfrak{B}^\#$. Let $g^{(n)}(\mathfrak{B})$ be a set of functions defined by

$$g^{(n)}(\mathfrak{B}) = \mathfrak{B} \quad \text{if } n = 0,$$
$$= g(g^{(n-1)}(\mathfrak{B})) \quad \text{if } n > 0.$$

Let $\mathfrak{A} = \bigcup_{n<\omega} g^{(n)}(\mathfrak{B})$. Then $\mathfrak{A}$ satisfies the conditions (1) and (2).

LEMMA 4. *Let B be a class of ordinal numbers and let φ be an order-preserving onto mapping such that φ:On $\to B$. Let $\mathfrak{A}$ be a set of functions satisfying the following conditions:*

(1) $\mathfrak{A}$ *is arithmetically closed.*

(2) *Let $\lambda x_1 \cdots x_n z_1 \cdots z_s g(x_1, \ldots, x_n, z_1, \ldots, z_s)$ be a function in $\mathfrak{A}$ and let $a_1, \ldots, a_s \in B$. Then function the $\lambda x_1 \cdots x_n g(x_1, \ldots, x_n, a_1, \ldots, a_s)$ belongs to $\mathfrak{A}$.*

(3) *B is $\mathfrak{A}$-closed.*

(4) $h_1, \ldots, h_m \in \mathfrak{A}$.

(5) $\varphi(\tilde{h}_i(a_1, \ldots, a_s)) = h_i(\varphi(a_1), \ldots, \varphi(a_s)), \quad i = 1, \ldots, m.$

Then we have

$$\varphi(f(\tilde{h}_1, \ldots, \tilde{h}_m, a_1, \ldots, a_n)) = f(h_1, \ldots, h_m, \varphi(a_1), \ldots, \varphi(a_n)).$$

PROOF. In the case where the defining schema is one of (I) to (III), we have

$$\varphi(\mathrm{Iq}\ (a, b)) = \mathrm{Iq}\ (\varphi(a), \varphi(b)).$$
$$\varphi(j(a, b)) = j(\varphi(a), \varphi(b)).$$
$$\varphi(g^1(a)) = g^1(\varphi(a)).$$
$$\varphi(g^2(a)) = g^2(\varphi(a)).$$

In the case where the last schema is (IV), the function $f(\alpha_1, \ldots, \alpha_m, a_1, \ldots, a_n)$ is defined by $f(\alpha_1, \ldots, \alpha_m, a_1, \ldots, a_n) = \alpha_i(a_1, \ldots, a_n)$. By the assumption of the lemma, we have

$$\varphi(\tilde{h}_i(a_1, \ldots, a_n)) = h_i(\varphi(a_1), \ldots, \varphi(a_n)).$$

Namely, we have

$$\varphi(f(\tilde{h}_1, \ldots, \tilde{h}_m, a_1, \ldots, a_n)) = f(h_1, \ldots, h_m, \varphi(a_1), \ldots, \varphi(a_n)).$$

The case (V) is treated as usual. In the case where the last schema is (VI), the function is defined by

$$f(\alpha_1, \ldots, \alpha_m, a_1, \ldots, a_n) = \mu x g(\alpha_1, \ldots, \alpha_m, a_1, \ldots, a_n, x).$$

By the assumption of the induction, we have

$$\varphi(g(\tilde{h}_1, \ldots, \tilde{h}_m, a_1, \ldots, a_n, a)) = g(h_1, \ldots, h_m, \varphi(a_1), \ldots, \varphi(a_n), \varphi(a)).$$

And it is easily proved that $\exists x(f(\tilde{h}_1, \ldots, \tilde{h}_m, a_1, \ldots, a_n, x) = 0)$ if and only if

$$\exists x(g(h_1, \ldots, h_m, \varphi(a_1), \ldots, \varphi(a_n), x) = 0).$$

In the case where $\neg\exists x(g(\tilde{h}_1, \ldots, \tilde{h}_m, a_1, \ldots, a_n, x) = 0)$, we have

$$f(\tilde{h}_1, \ldots, \tilde{h}_m, a_1, \ldots, a_n) = 0, \qquad f(h_1, \ldots, h_m, \varphi(a_1), \ldots, \varphi(a_n)) = 0.$$

Namely, we have

$$\varphi(f(\tilde{h}_1, \ldots, \tilde{h}_m, a_1, \ldots, a_n)) = f(h_1, \ldots, h_m, \varphi(a_1), \ldots, \varphi(a_n)).$$

In the case where $\exists x(g(\tilde{h}_1, \ldots, \tilde{h}_m, a_1, \ldots, a_n, x) = 0)$, we have

$$g(\tilde{h}_1, \ldots, \tilde{h}_m, a_1, \ldots, a_n, f(\tilde{h}_1, \ldots, \tilde{h}_m, a_1, \ldots, a_n)) = 0.$$

Namely, we have

$$g(h_1, \ldots, h_m, \varphi(a_1), \ldots, \varphi(a_n), \varphi(f(\tilde{h}_1, \ldots, \tilde{h}_m, a_1, \ldots, a_n))) = 0.$$

Therefore we obtain that

$$f(h_1, \ldots, h_m, \varphi(a_1), \ldots, \varphi(a_n)) \leq \varphi(f(\tilde{h}_1, \ldots, \tilde{h}_m, a_1, \ldots, a_n)).$$

Since $f(h_1, \ldots, h_m, b_1, \ldots, b_n)$ is arithmetical in $h_1, \ldots, h_m$ where $h_1, \ldots, h_m \in \mathfrak{A}$ and $\varphi(a_1), \ldots, \varphi(a_n) \in B$, we have $f(h_1, \ldots, h_m, \varphi(a_1), \ldots, \varphi(a_n)) \in B$. Namely, there is an ordinal number c such that

$$\varphi(c) = f(h_1, \ldots, h_m, \varphi(a_1), \ldots, \varphi(a_n)), \qquad c \leq f(\tilde{h}_1, \ldots, \tilde{h}_m, a_1, \ldots, a_n).$$

Therefore we have

$$g(h_1, \ldots, h_m, \varphi(a_1), \ldots, \varphi(a_n), \varphi(c)) = g(\tilde{h}_1, \ldots, \tilde{h}_m, a_1, \ldots, a_n, c) = 0.$$

Hence we obtain that

$$f(\tilde{h}_1, \ldots, \tilde{h}_m, a_1, \ldots, a_n) \leq c.$$

Namely, we have

$$\varphi(f(\tilde{h}_1, \ldots, \tilde{h}_m, a_1, \ldots, a_n)) = f(h_1, \ldots, h_m, \varphi(a_1), \ldots, \varphi(a_n)).$$

In the case where the last schema is (VII), the function $f(\alpha_1, \ldots, \alpha_m, a_1, \ldots, a_n)$ is defined by

$$f(\alpha_1, \ldots, \alpha_m, a_1, \ldots, a_n, a) = g(\alpha_1, \ldots, \alpha_m, f^a, a_1, \ldots, a_n, a).$$

For fixed $a_1, \ldots, a_n$, we assume that there is an ordinal a such that

$$\varphi(f(\tilde{h}_1, \ldots, \tilde{h}_m, a_1, \ldots, a_n, a)) \neq f(h_1, \ldots, h_m, \varphi(a_1), \ldots, \varphi(a_n), \varphi(a)).$$

Let x_0 be the smallest ordinal number such that

$$\varphi(f(\tilde{h}_1, \ldots, \tilde{h}_m, a_1, \ldots, a_n, x_0)) \neq f(h_1, \ldots, h_m, \varphi(a_1), \ldots, \varphi(a_n), \varphi(x_0)).$$

Then $\lambda x f^{x_0}(\tilde{h}_1, \ldots, \tilde{h}_m, a_1, \ldots, a_n, x)$ and $\lambda x f^{\varphi(x_0)}(h_1, \ldots, h_m, \varphi(a_1), \ldots, \varphi(a_n), x)$ are φ-commutative and the $\lambda x f^{\varphi(x_0)}(h_1, \ldots, h_m, \varphi(a_1), \ldots, \varphi(a_n), x)$ belongs to $\mathfrak{A}$. By the assumption of the induction we obtain

$$\varphi(g(\tilde{h}_1, \ldots, \tilde{h}_m, f^{x_0}, a_1, \ldots, a_n, x_0)) = g(h_1, \ldots, h_m, f^{\varphi(x_0)}, \varphi(a_1), \ldots, \varphi(a_n), \varphi(x_0)).$$

This means that

$$\varphi(f(\tilde{h}_1, \ldots, \tilde{h}_m, a_1, \ldots, a_n, x_0)) = f(h_1, \ldots, h_m, \varphi(a_1), \ldots, \varphi(a_n), \varphi(x_0)).$$

This is a contradiction. Namely, for all $a_1, \ldots, a_n, a$, we have

$$\varphi(f(\tilde{h}_1, \ldots, \tilde{h}_m, a_1, \ldots, a_n, a)) = f(h_1, \ldots, h_m, \varphi(a_1), \ldots, \varphi(a_n), \varphi(a)).$$

Thus we complete the proof of the lemma.

THEOREM C. *Let B be a class of ordinal numbers and let φ be an order-preserving onto mapping such that φ:On $\to B$. Let $\mathfrak{A}$ be a set of functions satisfying the following conditions*:

(1) *$\mathfrak{A}$ is an analytically closed, analytic basis.*
(2) *Let $\lambda x_1 \cdots x_n z_1 \cdots z_s g(x_1, \ldots, x_n, z_1, \ldots, z_s)$ be a function in $\mathfrak{A}$ and let $a_1, \ldots, a_s \in B$. Then the function $\lambda x_1 \cdots x_n g(x_1 \cdots x_n, a_1, \ldots, a_s)$ belongs to $\mathfrak{A}$.*
(3) *B is $\mathfrak{A}$-closed.*
(4) *$h_1, \ldots, h_m \in \mathfrak{A}$.*
(5) *$\varphi(\tilde{h}_i(a_1, \ldots, a_s)) = h_i(\varphi(a_1), \ldots, \varphi(a_s)), \quad i = 1, \ldots, m.$*

Then we have

$$\tilde{\Phi}(\tilde{h}_1, \ldots, \tilde{h}_m, a_1, \ldots, a_n) \leftrightarrow \Phi(h_1, \ldots, h_m, \varphi(a_1), \ldots, \varphi(a_n))$$

where $\tilde{\Phi}(\alpha_1, \ldots, \alpha_m, a_1, \ldots, a_n)$ *is obtained from* $\Phi(\alpha_1, \ldots, \alpha_m, a_1, \ldots, a_n)$ *by restricting all function quantifiers to the functions* $\alpha(a_1, \ldots, a_s)$ *such that*

$$\forall x_1 \cdots \forall x_s(\varphi(\alpha(x_1, \ldots, x_s)) = h(\varphi(x_1), \ldots, \varphi(x_s)))$$

for some function $h(a_1, \ldots, a_s)$ *in* $\mathfrak{A}$.

PROOF. The case where $\Phi(\alpha_1, \ldots, \alpha_m, a_1, \ldots, a_n)$ is of the form

$$f(\alpha_1, \ldots, \alpha_m, a_1, \ldots, a_n) = g(\alpha_1, \ldots, \alpha_m, a_1, \ldots, a_n).$$

By Lemma 4 we have the following equivalence

$$\begin{aligned} & f(\tilde{h}_1, \ldots, \tilde{h}_m, a_1, \ldots, a_n) = g(\tilde{h}_1, \ldots, \tilde{h}_m, a_1, \ldots, a_n) \\ & \quad \leftrightarrow \varphi(f(\tilde{h}_1, \ldots, \tilde{h}_m, a_1, \ldots, a_n)) = \varphi(g(\tilde{h}_1, \ldots, \tilde{h}_m, a_1, \ldots, a_n)) \\ & \quad \leftrightarrow f(h_1, \ldots, h_m, \varphi(a_1), \ldots, \varphi(a_n)) = g(h_1, \ldots, h_m, \varphi(a_1), \ldots, \varphi(a_n)). \end{aligned}$$

Namely, we obtain that

$$\tilde{\Phi}(\tilde{h}_1, \ldots, \tilde{h}_m, a_1, \ldots, a_n) \leftrightarrow \Phi(h_1, \ldots, h_m, \varphi(a_1), \ldots, \varphi(a_n)).$$

The case where the outermost symbol is $<$, $\neg$, $\vee$ is treated as usual.

In the case where $\Phi(\alpha_1, \ldots, \alpha_m, a_1, \ldots, a_n)$ is of the form

$$\exists x \Psi(\alpha_1, \ldots, \alpha_m, a_1, \ldots, a_n, x),$$

assume that $\exists x \tilde{\Psi}(\tilde{h}_1, \ldots, \tilde{h}_m, a_1, \ldots, a_n, x)$. Then there is an ordinal number a such that $\tilde{\Psi}(\tilde{h}_1, \ldots, \tilde{h}_m, a_1, \ldots, a_n, a)$. By the assumption of induction, we have $\Psi(h_1, \ldots, h_m, \varphi(a_1), \ldots, \varphi(a_n), \varphi(a))$. Hence we obtain

$$\exists x \Psi(h_1, \ldots, h_m, \varphi(a_1), \ldots, \varphi(a_n), x).$$

Conversely, we assume the above formula. Now we consider a function defined by

$$f(h_1, \ldots, h_m, b_1, \ldots, b_n) = \mu x \Psi(h_1, \ldots, h_m, b_1, \ldots, b_n, x).$$

Then $f(h_1, \ldots, h_m, b_1, \ldots, b_n)$ is analytic in $h_1, \ldots, h_m$. And therefore $f(h_1, \ldots, h_m, b_1, \ldots, b_n)$ belongs to $\mathfrak{A}$. By $\varphi(a_1), \ldots, \varphi(a_n) \in B$, we obtain

$$f(h_1, \ldots, h_m, \varphi(a_1), \ldots, \varphi(a_n)) \in B.$$

So there is an ordinal number c such that $\Psi(h_1, \ldots, h_m, \varphi(a_1), \ldots, \varphi(a_n), \varphi(c))$. By the assumption of the induction, we have $\tilde{\Psi}(\tilde{h}_1, \ldots, \tilde{h}_m, a_1, \ldots, a_n, c)$. Namely we get $\exists x \tilde{\Psi}(\tilde{h}_1, \ldots, \tilde{h}_m, a_1, \ldots, a_n, x)$.

In the case where $\Phi(\alpha_1, \ldots, \alpha_m, a_1, \ldots, a_n)$ is of the form

$$\exists \alpha \Phi(\alpha_1, \ldots, \alpha_m, \alpha, a_1, \ldots, a_n),$$

we assume that $\exists \tilde{\alpha} \tilde{\Psi}(\tilde{h}_1, \ldots, \tilde{h}_m, \tilde{\alpha}, a_1, \ldots, a_n)$. Then by the definition of the formula, we have a function $\tilde{h}(b_1, \ldots, b_s)$ and a function $h(b_1, \ldots, b_s)$ in $\mathfrak{A}$ such that

$$\forall x_1 \cdots \forall x_s(\varphi(\tilde{h}(x_1, \ldots, x_s)) = h(\varphi(x_1), \ldots, \varphi(x_s))),$$
$$\tilde{\Psi}(\tilde{h}_1, \ldots, \tilde{h}_m, \tilde{h}, a_1, \ldots, a_n).$$

By the assumption of the induction, we have $\Psi(h_1, \ldots, h_m, h, \varphi(a_1), \ldots, \varphi(a_n))$. Namely, we obtain that $\exists\alpha\Psi(h_1, \ldots, h_m, \alpha, \varphi(a_1), \ldots, \varphi(a_n))$. Conversely we assume the above formula. Then by $h_1, \ldots, h_m \in \mathfrak{A}$ and by the fact that $\mathfrak{A}$ is an analytic basis, there is a function $h(b_1, \ldots, b_n, c_1, \ldots, c_s)$ in $\mathfrak{A}$ such that

$$\forall x_1 \cdots \forall x_n(\exists\alpha\Psi(h_1, \ldots, h_m, \alpha, x_1, \ldots, x_n)$$
$$\leftrightarrow \Psi(h_1, \ldots, h_m, \lambda z_1 \cdots z_s h(x_1, \ldots, x_n, z_1, \ldots, z_s), x_1, \ldots, x_n)).$$

Now we consider the function $\tilde{h}(b_1, \ldots, b_n, c_1, \ldots, c_s)$ defined by

$$\tilde{h}(b_1, \ldots, b_n, c_1, \ldots, c_s) = \varphi^{-1}(h(\varphi(b_1), \ldots, \varphi(b_n), \varphi(c_1), \ldots, \varphi(c_s))).$$

This function $\tilde{h}$ is totally defined because $h \in \mathfrak{A}$ and B is $\mathfrak{A}$ closed. Namely, $\lambda z_1 \cdots z_s\tilde{h}(a_1, \ldots, a_n, z_1, \ldots, z_s)$ and $\lambda z_1 \cdots z_s h(\varphi(a_1), \ldots, \varphi(a_n), z_1, \ldots, z_s)$ are φ-commutative and $\lambda z_1 \cdots z_s h(\varphi(a_1), \ldots, \varphi(a_n), z_1, \ldots, z_s)$ belongs to $\mathfrak{A}$. By the definition of $\lambda z_1 \cdots z_s h(\varphi(a_1), \ldots, \varphi(a_n), z_1, \ldots, z_s)$, we obtain

$$\Psi(h_1, \ldots, h_m, \lambda z_1 \cdots z_s h(\varphi(a_1), \ldots, \varphi(a_n), z_1, \ldots, z_s), \varphi(a_1), \ldots, \varphi(a_n)).$$

By the assumption of the induction, we have

$$\tilde{\Psi}(\tilde{h}_1, \ldots, \tilde{h}_m, \lambda z_1 \cdots z_s\tilde{h}(a_1, \ldots, a_n, z_1, \ldots, z_s), a_1, \ldots, a_n).$$

Namely we have $\exists\tilde{\alpha}\tilde{\Psi}(\tilde{h}_1, \ldots, \tilde{h}_m, \tilde{\alpha}, a_1, \ldots, a_n)$. Thus we complete the proof of the theorem.

As a simple application of this theorem, we obtain that if $\Phi(h_1, \ldots, h_m, a_1, \ldots, a_n)$ is a Π^1_1-formula in $h_1, \ldots, h_m$ where $h_1, \ldots, h_m$ are φ-invariant functions in $\mathfrak{A}$, then we have

$$\Phi(h_1, \ldots, h_m, a_1, \ldots, a_n) \rightarrow \Phi(h_1, \ldots, h_m, \varphi(a_1), \ldots, \varphi(a_n)).$$

If the formula is Σ^1_1 in $h_1, \ldots, h_m$, we have

$$\Phi(h_1, \ldots, h_m, \varphi(a_1), \ldots, \varphi(a_n)) \rightarrow \Phi(h_1, \ldots, h_m, a_1, \ldots, a_n).$$

If the formula is Δ^1_1 in $h_1, \ldots, h_m$, we have

$$\Phi(h_1, \ldots, h_m, a_1, \ldots, a_n) \leftrightarrow \Phi(h_1, \ldots, h_m, \varphi(a_1), \ldots, \varphi(a_n)).$$

If $f(h_1, \ldots, h_m, a_1, \ldots, a_n)$ is a Δ^1_1-function in φ-invariant functions in $\mathfrak{A}$, then

$$\varphi(f(h_1, \ldots, h_m, a_1, \ldots, a_n)) = f(h_1, \ldots, h_m, \varphi(a_1), \ldots, \varphi(a_n)).$$

Especially, every Δ^1_1-function is φ-invariant.

By the similar method as the proof of Theorem C, we obtain

THEOREM C′. *Let $B\ (\subset \aleph_a)$ be a set of ordinal numbers and let φ be an order-preserving onto mapping such that $\varphi: \aleph_b \to B$. Let $\mathfrak{A}$ be a set of functions considering $\aleph_a$ as universe satisfying the following conditions:*

(1) *$\mathfrak{A}$ is an analytically closed, analytic basis.*

(2) *Let $\lambda x_1 \cdots x_n z_1 \cdots z_s g(x_1, \ldots, x_n, z_1, \ldots, z_s)$ be a function in $\mathfrak{A}$ and let $a_1, \ldots, a_s \in B$. Then the function $\lambda x_1 \cdots x_n g(x_1, \ldots, x_n, a_1, \ldots, a_s)$ belongs to $\mathfrak{A}$.*

(3) *B is $\mathfrak{A}$-closed.*

(4) *$h_1, \ldots, h_m \in \mathfrak{A}$.*

(5) *$\varphi(\tilde{h}_i(a_1, \ldots, a_s)) = h_i(\varphi(a_1), \ldots, \varphi(a_s))$.*

Then we have

$$\tilde{\Phi}^{(\aleph_b)}(\tilde{h}_1, \ldots, \tilde{h}_m, a_1, \ldots, a_s) \leftrightarrow \Phi^{(\aleph_a)}(h_1, \ldots, h_m, \varphi(a_1), \ldots, \varphi(a_s))$$

where $\tilde{\Phi}^{(\aleph_b)}(\alpha_1, \ldots, \alpha_m, a_1, \ldots, a_s)$ is obtained from $\Phi^{(\aleph_b)}(\alpha_1, \ldots, \alpha_m, a_1, \ldots, a_s)$ by restricting all function quantifiers to the functions $\alpha(a_1, \ldots, a_s)$ such that

$$\forall x_1 \cdots \forall x_s(\varphi(\alpha(x_1, \ldots, x_s)) = h(\varphi(x_1), \ldots, \varphi(x_s)))$$

for some function $h(a_1, \ldots, a_s)$ in $\mathfrak{A}$.

THEOREM 1. *Let $\aleph_\tau$ be a normally measurable cardinal. And let*

$$f(g_1, \ldots, g_m, a_1, \ldots, a_n)$$

be a Δ_1^1-function in $g_1, \ldots, g_m$ considering $\aleph_\tau$ as universe such that

(1) *$g_i(a_1, \ldots, a_n) \leq c_0 < \aleph_\tau$ for every $a_1, \ldots, a_n < \aleph_\tau$.*

(2) *$g_i(a_1, \ldots, a_n) = 0$ if $\max(a_1, \ldots, a_n) \geq c_0$.*

Then we have

$$\overline{\overline{f(g_1, \ldots, g_m, a_1, \ldots, a_n)}} \leq \max(\aleph_0, c_0, a_1, \ldots, a_n).$$

PROOF. Let $\mathfrak{A}$ be a set of functions in $(\aleph_\tau)$ such that

(1) $\overline{\overline{\mathfrak{A}}} = \aleph_0$.

(2) $\mathfrak{A}$ is an analytically closed analytic basis in $(\aleph_\tau)$.

(3) $g_1, \ldots, g_m \in \mathfrak{A}$.

Let g be a function in $(\aleph_\tau)$. Then by g^* we denote the function defined by

$$g^*(a_1, \ldots, a_n) = g(a_1, \ldots, a_n) \quad \text{if } \max(a_1, \ldots, a_n) < \aleph_\tau,$$
$$= 0 \quad \text{otherwise}.$$

By $\mathfrak{A}^*$ we denote the smallest set of functions such that

(1) $g^* \in \mathfrak{A}^*$ for every $g \in \mathfrak{A}$.

(2) $\mathfrak{A}^*$ includes the identity function and p_τ, $\mathfrak{A}^*$ is closed under substitution.

From this definition, we have $\overline{\overline{\mathfrak{A}^*}} = \aleph_0$.

Now let $a_1, \ldots, a_n$ be fixed ordinals less than $\aleph_\tau$. And let

$$B = \{a : a \leq \max(c_0, a_1, \ldots, a_n) \vee a = \aleph_\tau\}.$$

By B^* we denote the set $[B]_{\mathfrak{A}^*}$. Then $\overline{\overline{B^*}} \leq \max(\aleph_0, c_0, a_1, \ldots, a_n)$.

By Theorem B we have a generative function with respect to $\mathfrak{A}^*$ and B^* such that $f_0(0) \geq \sup (B^* \cap \aleph_\tau)$.

Let $E = [B^* \cup \{f_0(a): a < \aleph_\tau\}]_{\mathfrak{A}^*} \cap \aleph_\tau$. Then we have

(1) E is $\mathfrak{A}$-closed.

(2) $\overline{\overline{E}} = \aleph_\tau$.

Let d be an ordinal number such that $d \in E \cap f_0(0)$. Then there are ordinals $c_1 < \cdots < c_s$ such that

$$d \in [B^* \cup \{f_0(c_1), \dots, f_0(c_s)\}]_{\mathfrak{A}^*} \cap f_0(0).$$

By the property on f_0 we have

$$[B^* \cup \{f_0(c_1), \dots, f_0(c_s)\}]_{\mathfrak{A}^*} \cap f_0(c_s) \subset [B^* \cup \{f_0(c_1), \dots, f_0(c_{s-1})\}]_{\mathfrak{A}^*}.$$

Therefore we obtain that

$$[B^* \cup \{f_0(c_1), \dots, f_0(c_s)\}]_{\mathfrak{A}^*} \cap f_0(0) = [B^* \cup \{f_0(c_1), \dots, f_0(c_{s-1})\}]_{\mathfrak{A}^*} \cap f_0(0).$$

By repeating this, we have

$$[B^* \cup \{f_0(c_1), \dots, f_0(c_s)\}]_{\mathfrak{A}^*} \cap f_0(0) \subset [B^*]_{\mathfrak{A}^*} \cap f_0(0) \subset B^*.$$

Namely, we have $E \cap f_0(0) \subset B^*$.

Let $f(g_1, \dots, g_m, a_1, \dots, a_n)$ be a Δ_1^1-function in $g_1, \dots, g_m$. Then by the definition of $\mathfrak{A}$, we have $f(g_1, \dots, g_m, a_1, \dots, a_n) \in \mathfrak{A}$. Let $\varphi: \aleph_\tau \to E$ be an order-preserving onto mapping. Then we have

$$f(g_1, \dots, g_m, a_1, \dots, a_n) \in E \cap f_0(0), \qquad \varphi(a) = f(g_1, \dots, g_m, a_1, \dots, a_n).$$

Namely, we have

$$\overline{\overline{a}} \leq \overline{\overline{E \cap f_0(0)}} \leq \overline{\overline{B^*}} \leq \max (\aleph_0, c_0, a_1, \dots, a_n).$$

By $c_0 \subset E$, we have

$$\forall x_1 \cdots \forall x_n (\varphi(g_i(x_1, \dots, x_n)) = g_i(\varphi(x_1), \dots, \varphi(x_n))) \quad \text{for } i = 1, \dots, m.$$

By Theorem C, we have

$$\begin{aligned} \varphi(f(g_1, \dots, g_m, a_1, \dots, a_n)) &= f(g_1, \dots, g_m, \varphi(a_1), \dots, \varphi(a_n)) \\ &= f(g_1, \dots, g_m, a_1, \dots, a_n). \end{aligned}$$

Namely, we obtain that

$$\overline{\overline{f(g_1, \dots, g_m, a_1, \dots, a_n)}} \leq \max (\aleph_0, c_0, a_1, \dots, a_n).$$

THEOREM 2. *Let $\aleph_\tau$ be a normally measurable cardinal. And let*

$$f(g_1, \dots, g_m, a_1, \dots, a_n)$$

be a Δ_1^1-function in $g_1, \dots, g_m$ defined by $\Phi^{(\aleph_\tau)}(g_1, \dots, g_m, a_1, \dots, a_n, a)$. And let $f'(g_1, \dots, g_m, a_1, \dots, a_n)$ be a Δ_1^1-function in $g_1, \dots, g_m$ defined by

$\Phi^{(\aleph_a)}(g_1, \ldots, g_m, a_1, \ldots, a_n, a)$ *where*

(1) $g_i(a_1, \ldots, a_n) \leq c_0 < \aleph_a < \aleph_\tau$ *for every* $a_1, \ldots, a_n < \aleph_\tau$.

(2) $g_i(a_1, \ldots, a_n) = 0$ *if* $\max(a_1, \ldots, a_n) \geq c_0$.

Then we have

$$\forall x_1 \cdots \forall x_n(x_1, \ldots, x_n < \aleph_a \to f'(g_1, \ldots, g_m, x_1, \ldots, x_n) = f(g_1, \ldots, g_m, x_1, \ldots, x_n)).$$

Proof. The sets $\mathfrak{A}^*$, B^*, E and the functions f_0, φ are used here without changing their meaning. Now let $\max(a_1, \ldots, a_n) < \aleph_a$. We define the set E' by

$$E' = [B^* \cup \{f_0(b): b < \aleph_a\}]_{\mathfrak{A}^*} \cap \aleph_\tau.$$

Then we have

(1) E' is $\mathfrak{A}$-closed.

(2) The order-type of E' is exactly same as $\aleph_a$.

(3) $E \cap f_0(\aleph_a) = E'$.

Let $\psi: \aleph_a \to E'$ be an order-preserving onto mapping. Then $\psi(a) = \varphi(b) \in E'$ implies $a = b$. By Theorem C′ we have

$$\psi(f'(g_1, \ldots, g_m, a_1, \ldots, a_n)) = f(g_1, \ldots, g_m, \psi(a_1), \ldots, \psi(a_n))$$
$$= f(g_1, \ldots, g_m, \varphi(a_1), \ldots, \varphi(a_n)) = \varphi(f(g_1, \ldots, g_m, a_1, \ldots, a_n)).$$

Therefore we have

$$f'(g_1, \ldots, g_m, a_1, \ldots, a_n) = f(g_1, \ldots, g_m, a_1, \ldots, a_n).$$

Thus we complete the proof of Theorem 2.

Theorem 3. *Let* $A(g_1, \ldots, g_m, a, a_1, \ldots, a_n)$ *be a* Δ_1^1*-formula in* $g_1, \ldots, g_m$ *in* $(\aleph_\tau)$. *And let* $g_1, \ldots, g_m$ *be functions such that*

(1) $g_i(a_1, \ldots, a_n) \leq c_0$ *for all* $a_1, \ldots, a_n < \aleph_\tau$.

(2) $g_i(a_1, \ldots, a_n) = 0$ *if* $\max(a_1, \ldots, a_n) \geq c_0$.

And let $\aleph_0, c_0 < \aleph_{c_1} < \cdots < \aleph_{c_n} < \aleph_\tau$ *and* $\aleph_0, c_0 < \aleph_{d_1} < \cdots < \aleph_{d_n} < \aleph_\tau$ *be two sequences of cardinals. Then we have*

$$\forall x(x < \min(\aleph_{c_1}, \aleph_{d_1}) \to (A(g_1, \ldots, g_m, x, \aleph_{c_1}, \ldots, \aleph_{c_n}) \leftrightarrow A(g_1, \ldots, g_m, x, \aleph_{d_1}, \ldots, \aleph_{d_n}))).$$

Proof. The sets and the functions in the proof of Theorem 1 are used here without changing their meaning. Let $\aleph_e = \max(\aleph_{c_n}, \aleph_{d_n}) < \aleph_\tau$. The sets D_1 and D_2 are defined by

$$D_1 = \{f_0(a): a < \aleph_{c_1} \vee \aleph_e \cdot 2 \leq a < \aleph_e \cdot 2 + \aleph_{c_2} \vee \cdots \vee \aleph_e \cdot 2(n-1) \leq a < \aleph_e \cdot 2(n-1) + \aleph_{c_n} \vee \aleph_e \cdot 2n \leq a < \aleph_\tau\},$$

$$D_2 = \{f_0(a): \aleph_e \leq a < \aleph_e + \aleph_{d_1} \vee a = \aleph_e \cdot 2 \vee \aleph_e \cdot 3 \leq a < \aleph_e \cdot 3 + \aleph_{d_2} \vee \cdots \vee a = \aleph_e \cdot 2(n-1) \vee \aleph_e \cdot (2n-1) \leq a < \aleph_e(2n-1) + \aleph_{d_n} \vee \aleph_e \cdot 2n \leq a < \aleph_\tau\},$$

where the sum and the product means that of ordertypes. Clearly we have

$$D_1 \cap D_2 = \{f_0(a): a = \aleph_e \cdot 2 \vee \cdots \vee a = \aleph_e \cdot 2(n-1) \vee \aleph_e \cdot 2n \leq a < \aleph_\tau\}.$$

Let ψ and φ be two order-preserving onto mappings such that

$$\psi: \aleph_\tau \to [B^* \cup D_1]_{\mathfrak{A}^*} \cap \aleph_\tau, \qquad \varphi: \aleph_\tau \to [B^* \cup D_2]_{\mathfrak{A}^*} \cap \aleph_\tau.$$

By the properties of f_0, D_1 and D_2, we obtain that

$$\psi(\aleph_{c_i}) = \varphi(\aleph_{d_i}) = \aleph_e \cdot 2i \qquad \text{for every } i = 1, \ldots, n.$$

Now we consider a Δ_1^1-function defined by

$$f(g_1, \ldots, g_m, a, a_1, \ldots, a_n) = 0 \quad \text{if } A(g_1, \ldots, g_m, a, a_1, \ldots, a_n),$$
$$= 1 \quad \text{otherwise.}$$

By Theorem C we have

$$\psi(f(g_1, \ldots, g_m, a_0, \aleph_{c_1}, \ldots, \aleph_{c_n})) = f(g_1, \ldots, g_m, a_0, \psi(\aleph_{c_1}), \ldots, \psi(\aleph_{c_n}))$$
$$= f(g_1, \ldots, g_m, a_0, \varphi(\aleph_{d_1}), \ldots, \varphi(\aleph_{d_n})) = \varphi(f(g_1, \ldots, g_m, a_0, \aleph_{d_1}, \ldots, \aleph_{d_n})).$$

But $f(g_1, \ldots, g_m, a_0, \aleph_{c_1}, \ldots, \aleph_{c_n})$, $f(g_1, \ldots, g_m, a_0, \aleph_{d_1}, \ldots, \aleph_{d_n}) < 2$. We have

$$f(g_1, \ldots, g_m, a_0, \aleph_{c_1}, \ldots, \aleph_{c_n}) = f(g_1, \ldots, g_m, a_0, \aleph_{d_1}, \ldots, \aleph_{d_n}).$$

Namely, we obtain that

$$A(g_1, \ldots, g_m, a_0, \aleph_{c_1}, \ldots, \aleph_{c_n}) \leftrightarrow A(g_1, \ldots, g_m, a_0, \aleph_{d_1}, \ldots, \aleph_{d_n}).$$

Thus we complete the proof of the theorem.

THEOREM 4 (Σ_1^1-TRANSCENDENCY OF CARDINALS). *Let $P(a, a_1, \ldots, a_n)$ be a Σ_1^1-formula considering $\aleph_\tau$ as universe. Then we have for any $\aleph_c$ $(\aleph_0 < \aleph_c < \aleph_\tau)$,*

$$\forall x_1 \cdots \forall x_n(x_1, \ldots, x_n < \aleph_c \wedge \exists x P(x, x_1, \ldots, x_n)$$
$$\to \exists x(x < \aleph_c \wedge P(x, x_1, \ldots, x_n))).$$

PROOF. Let $a_1, \ldots, a_n$ be ordinals less than $\aleph_c$. And assume that

$$\exists \alpha A(\alpha, a_0, a_1, \ldots, a_n).$$

Let $\mathfrak{A}$ be a set of functions defined on $\aleph_\tau$ and having values in it such that

(1) $\overline{\overline{\mathfrak{A}}} = \aleph_0$.

(2) $\mathfrak{A}$ is an analytically closed analytic basis.

By $\mathfrak{A}^*$ we denote the smallest set of functions such that

(1) $g^* \in \mathfrak{A}^*$ for every $g \in \mathfrak{A}$.

(2) $\mathfrak{A}^*$ includes the identity functions and p_τ, $\mathfrak{A}^*$ is closed under substitution.

From this definition, we have $\overline{\overline{\mathfrak{A}^*}} = \aleph_0$.

Now let $a_1, \ldots, a_n$ be fixed ordinals less than $\aleph_\tau$. And let

$$B = \{a: a \leq \max(a_1, \ldots, a_n) \vee a = a_0 \vee a = \aleph_\tau\}.$$

By B^* we denote the set $[B]_{\mathfrak{A}^*}$. Then $\overline{\overline{B^*}} \leq \max(\aleph_0, a_1, \ldots, a_n)$.

By Theorem B, there is a generative function with respect to $\mathfrak{A}^*$ and B^* such that $f_0(0) > \sup(B^* \cap \aleph_\tau)$.

Let E be the set $[B^* \cup \{f_0(a): a < \aleph_\tau\}]_{\mathfrak{A}^*} \cap \aleph_\tau$. And let $\varphi: \aleph_\tau \to E$ be the order-preserving onto mapping. As before, if $a_0 = \varphi(b)$ then

$$\overline{\overline{b}} \leq \max(\aleph_0, a_1, \ldots, a_n) < \aleph_c.$$

By the definition of E, we have E is $\mathfrak{A}$-closed. Let $\mathfrak{B}$ be the set of functions

$$\lambda x_1 \cdots x_n g(x_1, \ldots, x_n, e_1, \ldots, e_s)$$

where $e_1, \ldots, e_s \in E$ and $\lambda x_1 \cdots x_n z_1 \cdots z_s g(x_1, \ldots, x_n, z_1, \ldots, z_s) \in \mathfrak{A}$. Then we have

(1) $\mathfrak{B}$ is an analytically closed, analytic basis.

(2) Let $\lambda x_1 \cdots x_n z_1 \cdots z_s g(x_1, \ldots, x_n, z_1, \ldots, z_s)$ be a function in $\mathfrak{B}$ and. let $e_1, \ldots, e_s \in E$. Then the function $\lambda x_1 \cdots x_n g(x_1, \ldots, x_n, e_1, \ldots, e_s)$ belongs to $\mathfrak{B}$.

(3) E is $\mathfrak{B}$-closed.

Let $\varphi(b) = a_0$. Then by the property of E, we have $\varphi(a_1) = a_1, \ldots, \varphi(a_n) = a_n$. Therefore, by Theorem C, we have

$$\begin{aligned} \exists \alpha A(\alpha, a_0, a_1, \ldots, a_n) &\to \exists \alpha A(\alpha, \varphi(b), \varphi(a_1), \ldots, \varphi(a_n)) \\ &\to \exists \alpha A(\alpha,, b, a_1, \ldots, a_n). \end{aligned}$$

By the fact that $b < \aleph_c$, we obtain that

$$\exists x(x < \aleph_c \wedge \exists \alpha A(\alpha, x, a_1, \ldots, a_n)).$$

Thus we complete the proof of Theorem 4.

Let $P(a)$ be a Σ_1^1-formula in $(\aleph_\tau)$. Then by Theorem 4, we have

$$\exists x P(x) \to \exists x(x < \aleph_1 \wedge P(x)).$$

Since every ordinal number definable in L (cf. Gödel [3]) is arithmetically definable, we see that every ordinal number definable in L is a countable ordinal.

THEOREM 5. *Let $\aleph_{\tau_1} < \ldots < \aleph_{\tau_n} < \aleph_\sigma$ be a sequence of normally measurable cardinals and let $\aleph_0 < \aleph_{c_1} < \cdots < \aleph_{c_n}$ be a sequence of cardinals such that $\aleph_{c_i} \leq \aleph_{\tau_i}$ $(i = 1, \ldots, n)$. And let $P(a, a_1, \ldots, a_n)$ be a Σ_1^1-formula in $\aleph_\sigma$. Then we have*

$$\forall x(x < \aleph_{c_1} \wedge P(x, \aleph_{\tau_1}, \ldots, \aleph_{\tau_n}) \to P(x, \aleph_{c_1}, \ldots, \aleph_{c_n})).$$

PROOF. Let $\mathfrak{A}$ be a set of functions in $(\aleph_\sigma)$ such that

(1) $\overline{\overline{\mathfrak{A}}} = \aleph_0$.

(2) $\mathfrak{A}$ is an analytically closed analytic basis.

Let g be a function in $(\aleph_\sigma)$. Then by g^* we denote the function defined by

$$\begin{aligned} g^*(a_1, \ldots, a_n) &= g(a_1, \ldots, a_n) \quad \text{if } \max(a_1, \ldots, a_n) < \aleph_\sigma, \\ &= 0 \quad \text{otherwise.} \end{aligned}$$

By $\mathfrak{A}^*$ we denote the smallest set of functions such that

(1) $g^* \in \mathfrak{A}^*$ for every $g \in \mathfrak{A}$.

(2) $\mathfrak{A}^*$ includes the identity function, $p_{\tau_1}, \ldots, p_{\tau_n}$ and p_σ, $\mathfrak{A}^*$ is closed under substitution.

Let $B^* = [\{a : a < a_0 \vee a = \aleph_{\tau_1} \vee \cdots \vee a = \aleph_{\tau_n} \vee a = \aleph_\sigma\}]_{\mathfrak{A}^*}$ where a_0 is an arbitrary but fixed ordinal $< \aleph_{c_1}$. And we consider the ordinal number $d_0 = \sup (B^* \cap \aleph_{\tau_1}) < \aleph_{\tau_1}$. By Theorem B, there is a generative function f_0 such that

$$d_0 \leq f_0(0) \quad \text{and} \quad \forall x(x < \aleph_{\tau_i} \to f_0(x) < \aleph_{\tau_i}) \quad \text{for } i = 1, \ldots, n.$$

Now we consider the set defined by

$$E = [B^* \cup \{f_0(a) : a < \aleph_{c_1} \vee \aleph_{\tau_1} \leq a < \aleph_{\tau_1} + \aleph_{c_2} \vee \cdots \vee \aleph_{\tau_n} \leq a < \aleph_\sigma\}]_{\mathfrak{A}^*} \cap \aleph_\sigma.$$

By this definition, we have that E is $\mathfrak{A}$-closed. And there is an order-preserving onto mapping φ such that $\varphi : \aleph_\sigma \to E$. By the definition of E, we have

$$\varphi(\aleph_{c_1}) = \aleph_{\tau_1}, \ldots, \varphi(\aleph_{c_n}) = \aleph_{\tau_n}.$$

Since E is $\mathfrak{A}$-closed and by the method used in the proof of Theorem 4 and by Theorem C, we have, for the ordinal a_0 such that $a_0 < \aleph_{c_1}$,

$$P(a_0, \aleph_{\tau_1}, \ldots, \aleph_{\tau_n}) \to P(a_0, \varphi(\aleph_{c_1}), \ldots, \varphi(\aleph_{c_n})) \to P(a_0, \aleph_{c_1}, \ldots, \aleph_{c_n}).$$

Thus we complete the proof of the theorem.

Let $\aleph_\tau < \aleph_\sigma$ be two normally measurable cardinals. And let $P(a)$ be a Σ_1^1-formula in $(\aleph_\sigma)$. Then we have

$$P(\aleph_\tau) \to P(\aleph_c) \quad \text{for every } \aleph_c \text{ such that } \aleph_0 < \aleph_c \leq \aleph_\tau.$$

Let $P(a)$ be a Π_1^1-formula. Then we have

$$P(\aleph_c) \to P(\aleph_\tau) \quad \text{for every } \aleph_c \text{ such that } \aleph_0 < \aleph_c \leq \aleph_\tau.$$

Namely, a normally measurable cardinal number has Π_1^1-properties any of which is a property of smaller cardinals (except $\aleph_0$). Let $P(a)$ be a Σ_1^1-formula. Then by the fact that $\{\nu : \text{Card}((\aleph_\tau)_\nu)\} \in F_{\aleph_\tau}$, we have

$$P(\aleph_\tau) \to \{\nu : P((\aleph_\tau)_\nu)\} \in F_{\aleph_\tau}.$$

By the theorem proved before, for any Π_1^1-formula $P(a)$ we have

$$P(\aleph_\tau) \to \{\nu : P((\aleph_\tau)_\nu)\} \in F_{\aleph_\tau}.$$

Let $P(a)$ be a formula obtained by connecting Σ_1^1- and Π_1^1-formulas by logical connectives $\neg$, $\vee$. Then we have

$$P(\aleph_\tau) \leftrightarrow \{\nu : P((\aleph_\tau)_\nu)\} \in F_{\aleph_\tau}.$$

Namely, there is a cardinal number $\aleph_c$ $(< \aleph_\tau)$ such that there is neither a Σ_1^1- nor a Π_1^1-formula $P(a)$ which distinguishes these two cardinals, one of which is $\aleph_0$-complete and the other not $\aleph_0$-complete.

Let $P(a)$ be the Σ_1^1-formula which states that there is a standard model in which

a is the ath regular cardinal number in the model defined in $\aleph_\sigma$. Since $R(\aleph_\sigma)$ is a model for Σ^*, there is a one-one onto mapping $h: R(\aleph_\sigma) \to \aleph_\sigma$ such that

(1) $x \in y \to h(x) < h(y)$.

(2) $h(\aleph_\tau) = \aleph_\tau$.

Therefore we have $P(\aleph_\tau)$. By the previous theorem, we have $P(\aleph_a)$ for every cardinal $\aleph_a \leq \aleph_\tau$. Namely $\aleph_a$ is a regular cardinal in L. Hence, every infinite set of cardinals is not constructible, for example, the set $\{\aleph_0, \aleph_1, \ldots, \aleph_n, \ldots\} \notin L$.

4. Ultraproducts and direct-limits. The purpose of this section is to prove the Π_1^1-transcendency of normally measurable cardinals. In order to show this we shall begin with a fundamental lemma which is essentially due to Haim Gaifman. Throughout this section we use the notations defined in [**4**].

Let $A(a, b, c)$ and $B(a, b, c)$ be the formulas defined by

$$A(a, b, c) \equiv c \in a \wedge \mathfrak{D}(c) = \mathfrak{S}(b) \wedge \mathfrak{F}_n(c).$$

$$\begin{aligned} B(a, b, c) \equiv \exists\alpha\exists\beta(\alpha < \beta \wedge \mathfrak{F}_n(c) \wedge \mathfrak{D}(c) = \beta - \alpha \\ \wedge \forall\gamma(\alpha \leq \gamma < \beta \to c'\gamma \in a''\{\gamma\}) \\ \wedge \forall\gamma\forall\delta(\alpha \leq \gamma < \delta < \beta \to \langle c'\gamma\, c'\delta\rangle \in b''\{\langle\gamma\delta\rangle\})). \end{aligned}$$

LEMMA 5. *Let $\langle \bar{u};\ u, m_0, \in\rangle$ be a transitive $\in$ structure satisfying the following conditions:*

(1) *Let $\Phi^{(u)}(a_1, \ldots, a_n)$ be a formula with the parameters in $\bar{u}$ and u which is restricted to the set u. Then we have*

$$\{\langle x_1 \cdots x_n\rangle : x_1, \ldots, x_n \in u \wedge \Phi^{(u)}(x_1, \ldots, x_n)\} \in \bar{u},$$

$$t_1, \ldots, t_n \in u \to \{\langle x_1 \cdots x_n\rangle : x_1 \in t_1 \wedge \cdots \wedge x_n \in t_n \wedge \Phi^{(u)}(x_1, \ldots, x_n)\} \in u,$$

$$t \in u \wedge \forall x(x \in t \to \exists y \Phi^{(u)}(y, x)) \to \exists p(p \in u \wedge \forall x(x \in t \to \Phi^{(u)}(p'x, x))).$$

(2) *There is a mapping $\gamma \in \bar{u}$ and an ordinal number $\rho_0 \in \bar{u} - u$ such that*

$$\gamma : u \to \rho_0 \quad \textit{and} \quad \gamma'x = \sup_{y \in x} \gamma'y.$$

(3) *There is a mapping $g \in \bar{u}$ such that $\forall p(A(u, m_0, p) \to \exists x(\langle x\, p\rangle \in g))$ and*

$$\begin{aligned} \forall z\forall p(\langle z\, p\rangle \in g \leftrightarrow A(u, m_0, p) \wedge \forall t(t \in z \leftrightarrow \exists q(A(u, m_0, q) \\ \wedge \{v : q'v \in p'v\} \in m_0 \wedge \langle t\, q\rangle \in g))). \end{aligned}$$

And the sets s, $\bar{s}$, m, a, f, and $\bar{f}$ are defined by simultaneous induction as follows:

$$\begin{aligned} \langle x\, \alpha\rangle \in s \equiv \alpha < \rho_0 \wedge ((\alpha = 0 \wedge x \in u) \vee \exists\beta(\alpha = \beta + 1 \wedge \exists p(A(s''\{\beta\}, m''\{\beta\}, p) \\ \wedge \langle x\, p\rangle \in a''\{\alpha\})) \vee (\alpha > 0 \wedge \forall\beta(\alpha \neq \beta + 1) \\ \wedge \exists p(B(s \restriction \alpha, f \restriction \alpha^2, p) \wedge \langle x\, p\rangle \in a''\{\alpha\}))). \end{aligned}$$

$$\langle x\, \alpha\rangle \in m \equiv \alpha < \rho_0 \wedge ((\alpha = 0 \wedge x \in m_0) \vee (0 < \alpha \wedge \langle x\, m''\{0\}\rangle \in f''\{\langle\alpha\, 0\rangle\})).$$

$$\begin{aligned}\langle\langle x\,y\rangle\alpha\rangle \in a \equiv \alpha < \rho_0 \wedge \exists\beta(\alpha = \beta + 1 \wedge A(s''\{\beta\}, m''\{\beta\}, y)\\ \wedge \forall z(z \in x \leftrightarrow \exists p(A(s''\{\beta\}, m''\{\beta\}, p) \wedge \{v : p'v \in y'v\} \in m''\{\beta\}\\ \wedge \langle z\,p\rangle \in a''\{\alpha\}))) \vee (\alpha > 0 \wedge \forall\beta(\alpha \neq \beta + 1) \wedge B(s \upharpoonright \alpha, f \upharpoonright \alpha^2, y)\\ \wedge \forall z(z \in x \leftrightarrow \exists p(B(s \upharpoonright \alpha, f \upharpoonright \alpha^2, p) \wedge \exists\beta(\beta < \alpha \wedge p'\beta \in y'\beta)\\ \wedge \langle z\,p\rangle \in a''\{\alpha\}))).\end{aligned}$$

$$\begin{aligned}\langle\langle z\,x\rangle\langle\beta\,\alpha\rangle\rangle \in f \equiv \beta < \rho_0 \wedge ((\beta = \alpha \wedge x = z \wedge x \in s''\{\alpha\})\\ \vee \exists\gamma(\beta = \gamma + 1 \wedge ((\alpha < \gamma \wedge \exists y(\langle y\,x\rangle \in f''\{\langle\gamma\,\alpha\rangle\}\}\\ \wedge \langle z\,y\rangle \in f''\{\langle\beta\,\gamma\rangle\}) \vee (\alpha = \gamma \wedge \exists p(A(s''\{\alpha\}, m''\{\alpha\}, p)\\ \wedge \mathfrak{W}(p) = \{x\} \wedge \langle z\,p\rangle \in a''\{\beta\}))))) \vee (\beta > \alpha \wedge \forall\gamma(\beta \neq \gamma + 1)\\ \wedge \exists p(\mathfrak{D}(p) = \beta - \alpha \wedge B(s \upharpoonright \beta, f \upharpoonright \beta^2, p)\\ \wedge p'\alpha = x \wedge \langle z\,p\rangle \in a''\{\beta\}))).\end{aligned}$$

$$\begin{aligned}\langle\langle z\,x\rangle\langle\beta\,\alpha\rangle\rangle \in \bar{f} \equiv \beta < \rho_0 \wedge ((\beta = \alpha \wedge x = z \wedge z \in \bar{s}''\{\alpha\})\\ \vee \exists\gamma(\beta = \gamma + 1 \wedge ((\alpha < \gamma \wedge \exists y(\langle y\,x\rangle \in \bar{f}''\{\langle\gamma\,\alpha\rangle\}\\ \wedge \langle z\,y\rangle \in \bar{f}''\{\langle\beta\,\gamma\rangle\}) \vee (\alpha = \gamma \wedge x \in \bar{s}''\{\alpha\}\\ \wedge \forall t(t \in z \leftrightarrow \exists p(A(s''\{\alpha\}, m''\{\alpha\}, p) \wedge \{v : P'v \in x\} \in m''\{\alpha\}\\ \wedge \langle t\,p\rangle \in a''\{\beta\})))))) \vee (\beta > \alpha \wedge \forall\gamma(\beta \neq \gamma + 1)\\ \wedge \exists p(\mathfrak{D}(p) = \beta - \alpha \wedge B(\bar{s} \upharpoonright \beta, \bar{f} \upharpoonright \beta^2, p) \wedge p'\alpha = x\\ \wedge \forall t(t \in z \leftrightarrow \exists q(B(s \upharpoonright \beta, f \upharpoonright \beta^2, q) \wedge \exists\gamma(\gamma < \beta \wedge q'\gamma \in p'\gamma)\\ \wedge \langle t\,q\rangle \in a''\{\beta\}))))).\end{aligned}$$

$$\langle x\,\alpha\rangle \in \bar{s} \equiv \langle x\,\alpha\rangle \in s \vee \exists t(t \in \bar{u} \wedge \langle x\,t\rangle \in \bar{f}''\{\langle\alpha\,0\rangle\}).$$

Then we have the following properties for all $\alpha, \beta, \gamma < \rho_0$:

(1) $\alpha = \beta + 1 \wedge A(s''\{\beta\}, m''\{\beta\}, p) \to \exists x(\langle x\,p\rangle \in a''\{\alpha\})$.

(2) $\mathrm{Lim}\,(\alpha) \wedge B(s \upharpoonright \alpha, f \upharpoonright \alpha^2, p) \to \exists x(\langle x\,p\rangle \in a''\{\alpha\})$.

(3) $x \in \bar{s}''\{\alpha\} \wedge \alpha \leq \beta \to \exists!\, y(\langle y\,x\rangle \in \bar{f}''\{\langle\beta\,\alpha\rangle\})$.

(4) $\langle x\,y\rangle \in \bar{f}''\{\langle\alpha\,\beta\rangle\} \to x \in \bar{s}''\{\alpha\} \wedge y \in \bar{s}''\{\beta\}$.

(5) $\langle x\,y\rangle \in \bar{f}''\{\langle\alpha\,\beta\rangle\} \wedge \langle y\,z\rangle \in \bar{f}''\{\langle\beta\,\gamma\rangle\} \to \langle x\,z\rangle \in \bar{f}''\{\langle\alpha\,\gamma\rangle\}$.

By means of (3), *we define a function* $\bar{f}^{\alpha}_{\beta}$ *so that*

$$y = \bar{f}^{\alpha}_{\beta}(x) \quad \textit{if and only if } \langle y\,x\rangle \in \bar{f}''\{\langle\beta\,\alpha\rangle\}.$$

Then (5) *means that* $\bar{f}^{\beta}_{\gamma}\bar{f}^{\alpha}_{\beta}(x) = \bar{f}^{\alpha}_{\gamma}(x)$.

(6) $\bar{f}^{\alpha}_{\beta}(m''\{\alpha\}) = m''\{\beta\}, \bar{f}^{\alpha}_{\beta}(s''\{\alpha\}) = s''\{\beta\}$ *and* $\bar{f}^{\alpha}_{\beta}(\rho_0) = \rho_0$.

(7) $\bar{f}^{\alpha}_{\beta}$ *is an elementary mapping such that*

$$\bar{f}^{\alpha}_{\beta} : \langle \bar{s}''\{\alpha\};\ s''\{\alpha\}, m''\{\alpha\}, \in\rangle \to \langle \bar{s}''\{\beta\};\ s''\{\beta\}, m''\{\beta\}, \in\rangle.$$

PROOF. By the definition of $s''\{0\}$, $m''\{0\}$ and $s''\{0\}$ we have

$$m''\{0\} = m_0, \quad s''\{0\} = u \quad \text{and} \quad \bar{s}''\{0\} = \bar{u}.$$

Now we assume that (1) to (7) are true for all $\alpha, \beta, \gamma < \delta\ (<\rho_0)$. Let ρ be an ordinal number less than δ. Then by the fact that s, m, a, and f belongs to $\bar{s}''\{0\}$, we obtain the following.

$$\begin{aligned}\langle x\ \alpha\rangle \in \bar{f}^0_\rho(s) \equiv \alpha < \rho_0 \wedge ((\alpha = 0 \wedge x \in s''\{\rho\}) \\ \vee \exists\beta(\alpha = \beta + 1 \wedge \exists p(A(\bar{f}^0_\rho(s)''\{\beta\}, \bar{f}^0_\rho(m)''\{\beta\}, p) \\ \wedge \langle x\ p\rangle \in \bar{f}^0_\rho(a)''\{\alpha\})) \vee (\alpha > 0 \wedge \forall\beta(\alpha \neq \beta + 1) \\ \wedge \exists p(B(\bar{f}^0_\rho(s) \restriction \alpha, \bar{f}^0_\rho(f) \restriction \alpha^2, p) \wedge \langle x\ p\rangle \in \bar{f}^0_\rho(a)''\{\alpha\}))).\end{aligned}$$

$$\begin{aligned}\langle x\ \rho + \alpha\rangle \in s \equiv \alpha < \rho_0 \wedge ((\alpha = 0 \wedge x \in s''\{\rho\}) \\ \vee \exists\beta(\alpha = \beta + 1 \wedge \exists p(A(s''\{\rho + \beta\}, m''\{\rho + \beta\}, p) \\ \wedge \langle x\ p\rangle \in a''\{\rho + \alpha\})) \vee (\alpha > 0 \wedge \forall\beta(\alpha \neq \beta + 1) \\ \wedge \exists p(B(s \restriction \rho + \alpha, f \restriction (\rho + \alpha)^2, p) \wedge \langle x\ p\rangle \in a''\{\rho + \alpha\}))).\end{aligned}$$

$$\begin{aligned}\langle x\ \alpha\rangle \in \bar{f}^0_\rho(m) \equiv \alpha < \rho_0 \wedge ((\alpha = 0 \wedge x \in m''\{\rho\}) \\ \vee (0 < \alpha \wedge \langle x\ m''\{\rho\}\rangle \in \bar{f}^0_\rho(f)''\{\langle\alpha\ 0\rangle\})).\end{aligned}$$

$$\begin{aligned}\langle x\ \rho + \alpha\rangle \in m \equiv \alpha < \rho_0 \wedge ((\alpha = 0 \wedge x \in m''\{\rho\}) \\ \vee (0 < \alpha \wedge \langle x\ m''\{0\}\rangle \in f''\{\langle\rho + \alpha\ 0\rangle\})).\end{aligned}$$

$$\begin{aligned}\langle\langle x\ y\rangle\alpha\rangle \in \bar{f}^0_\rho(a) \equiv \alpha < \rho_0 \wedge \exists\beta(\alpha = \beta + 1 \wedge A(\bar{f}^0_\rho(s)''\{\beta\}, \bar{f}^0_\rho(m)''\{\beta\}, y) \\ \wedge \forall z(z \in x \leftrightarrow \exists p(A(\bar{f}^0_\rho(s)''\{\beta\}, \bar{f}^0_\rho(m)''\{\beta\}, p) \\ \wedge \{v : P'v \in y'v\} \in \bar{f}^0_\rho(m)''\{\beta\} \wedge \langle z\ p\rangle \in \bar{f}^0_\rho(a)''\{\alpha\}))) \\ \vee (\alpha > 0 \wedge \forall\beta(\alpha \neq \beta + 1) \wedge B(\bar{f}^0_\rho(s) \restriction \alpha, \bar{f}^0_\rho(f) \restriction \alpha^2, y) \\ \wedge \forall z(z \in x \leftrightarrow \exists p(A(\bar{f}^0_\rho(s) \restriction \alpha, \bar{f}^0_\rho(f) \restriction \alpha^2, p) \\ \wedge \exists\beta(\beta < \alpha \wedge p'\beta \in y'\beta) \wedge \langle z\ p\rangle \in \bar{f}^0_\rho(a)''\{\alpha\}))).\end{aligned}$$

$$\begin{aligned}\langle\langle x\ y\rangle\rho + \alpha\rangle \in a \equiv \alpha < \rho_0 \wedge \exists\beta(\alpha = \beta + 1 \wedge A(s''\{\rho + \beta\}, m''\{\rho + \beta\}, y) \\ \wedge \forall z(z \in x \leftrightarrow \exists p(A(s''\{\rho + \beta\}, m''\{\rho + \beta\}, p) \\ \wedge \{v : p'v \in y'v\} \in m''\{\rho + \beta\} \wedge \langle z\ p\rangle \in a''\{\rho + \alpha\}))) \\ \vee (\alpha > 0 \wedge \forall\beta(\alpha \neq \beta + 1) \wedge B(s \restriction \rho + \alpha, f \restriction (\rho + \alpha)^2, y) \\ \wedge \forall z(z \in x \leftrightarrow \exists p(B(s \restriction \rho + \alpha, f \restriction (\rho + \alpha)^2, p) \\ \wedge \exists\beta(\beta < \alpha \wedge p'\beta \in y'\beta) \wedge \langle z\ p\rangle \in a''\{\rho + \alpha\}))).\end{aligned}$$

$$\begin{aligned}\langle\langle z\ x\rangle\langle\beta\ \alpha\rangle\rangle \in \bar{f}^0_\rho(f) \equiv \beta < \rho_0 \wedge ((\beta = \alpha \wedge x = z \wedge x \in \bar{f}^0_\rho(s)''\{\alpha\}) \\ \vee \exists\gamma(\beta = \gamma + 1 \wedge ((\alpha < \gamma \wedge \exists y(\langle y\ x\rangle \in \bar{f}^0_\rho(f)''\{\langle\gamma\ \alpha\rangle\} \\ \wedge \langle z\ y\rangle \in \bar{f}^0_\rho(f)''\{\langle\beta\ \gamma\rangle\}) \\ \vee (\alpha = \gamma \wedge \exists p(A(\bar{f}^0_\rho(s)''\{\alpha\}, \bar{f}^0_\rho(m)''\{\alpha\}, p) \\ \wedge \mathfrak{W}(p) = \{x\} \wedge \langle z\ p\rangle \in \bar{f}^0_\rho(a)''\{\beta\}))))) \\ \vee (\beta > \alpha \wedge \forall\gamma(\beta \neq \gamma + 1) \wedge \exists p(\mathfrak{D}(p) = \beta - \alpha \\ \wedge B(\bar{f}^0_\rho(s) \restriction \beta, \bar{f}^0_\rho(f) \restriction \beta^2, p) \wedge p'\alpha = x \wedge \langle z\ p\rangle \in \bar{f}^0_\rho(a)''\{\beta\})).\end{aligned}$$

$$\begin{aligned}\langle\langle z\,x\rangle\langle\rho+\beta\;\rho+\alpha\rangle\rangle\in f\equiv \beta<\rho_0\wedge((\beta=\alpha\wedge x=z\wedge x\in s''\{\rho+\alpha\}))\\ \vee\,\exists\gamma(\beta=\gamma+1\wedge((\alpha<\gamma\\ \wedge\,\exists y(\langle y\,x\rangle\in f''\{\langle\rho+\gamma\;\rho+\alpha\rangle\}\\ \wedge\,\langle z\,y\rangle\in f''\{\langle\rho+\beta\;\rho+\gamma\rangle\})\vee(\alpha=\gamma\\ \wedge\,\exists p(A(s''\{\rho+\alpha\},m''\{\rho+\alpha\},p)\wedge\mathfrak{W}(p)=\{x\}\\ \wedge\,\langle z\,p\rangle\in a''\{\rho+\beta\})))))\vee(\beta>\alpha\wedge\forall\gamma(\beta\neq\gamma+1)\\ \wedge\,\exists p(\mathfrak{D}(p)=\beta-\alpha\wedge B(s\restriction\rho+\beta,f\restriction(\rho+\beta)^2,p)\\ \wedge\,p'\rho+\alpha=x\wedge\langle z\,p\rangle\in a''\{\rho+\beta\})).\end{aligned}$$

By these properties we have

$$\bar{f}^0_\rho(m)''\{0\}=m''\{\rho+0\},\bar{f}^0_\rho(s)''\{0\}=s''\{\rho+0\},\bar{f}^0_\rho(f)''\{\langle 0\;0\rangle\}=f''\{\langle\rho+0\;\rho+0\rangle\}.$$

Assume that for all $\beta,\gamma<\alpha\;(<\bar{f}^0_\rho(\rho_0))$ the following properties are satisfied;

(1) $\neg\operatorname{Lim}(\beta)\to(\langle x\,p\rangle\in\bar{f}^0_\rho(a)''\{\beta\}\leftrightarrow\langle x\,p\rangle\in a''\{\rho+\beta\})$.

(2) $\operatorname{Lim}(\beta)\to(\langle x\,p\rangle\in\bar{f}^0_\rho(a)''\{\beta\}\leftrightarrow\langle x\,p^\rho\rangle\in a''\{\rho+\beta\})$ where $y^{\rho}{}'\rho+\nu=y'\nu$ for $\nu\in D(y)$ and $D(y^\rho)=\{\rho+\nu:\nu\in D(y)\}$.

(3) $\bar{f}^0_\rho(s)''\{\beta\}=s''\{\rho+\beta\}$.

(4) $\bar{f}^0_\rho(m)''\{\beta\}=m''\{\rho+\beta\}$.

(5) $\bar{f}^0_\rho(f)''\{\langle\beta\,\gamma\rangle\}=f''\{\langle\rho+\beta\;\rho+\gamma\rangle\}$.

If α is not a limit ordinal, then we have

$$\langle\langle x\,y\rangle\alpha\rangle\in\bar{f}^0_\rho(a)\leftrightarrow\langle\langle x\,y\rangle\rho+\alpha\rangle\in a.$$

By this we obtain that $\bar{f}^0_\rho(s)''\{\alpha\}=s''\{\rho+\alpha\}$ and $\bar{f}^0_\rho(m)''\{\alpha\}=m''\{\rho+\alpha\}$. And we also have

$$\langle\langle z\,x\rangle\langle\alpha\;\alpha\rangle\rangle\in\bar{f}^0_\rho(f)\leftrightarrow\langle\langle z\,x\rangle\langle\rho+\alpha\;\rho+\alpha\rangle\rangle\in f,$$

$$\langle\langle z\,x\rangle\langle\alpha\;\alpha-1\rangle\rangle\in\bar{f}(f)\leftrightarrow\langle\langle z\,x\rangle\langle\rho+\alpha\;\rho+\alpha-1\rangle\rangle\in f.$$

Hence we have $\bar{f}^0_\rho(f)''\{\langle\alpha\,\beta\rangle\}=f''\{\langle\rho+\alpha\;\rho+\beta\rangle\}$.

If α is a limit ordinal, then we have by the induction on the rank of x,

$$\begin{aligned}\langle\langle x\,y\rangle\alpha\rangle\in\bar{f}^0_\rho(a)&\leftrightarrow B(\bar{f}^0_\rho(s)\restriction\alpha,\bar{f}^0_\rho(f)\restriction\alpha^2,y)\wedge\forall z(z\in x\leftrightarrow\exists p(B(\bar{f}^0_\rho(s),\bar{f}^0_\rho(f)\restriction\alpha^2,p)\\ &\quad\wedge\exists\beta(\beta<\alpha\wedge p'\beta\in y'\beta)\wedge\langle z\,p\rangle\in\bar{f}^0_\rho(a)''\{\alpha\}))\\ &\leftrightarrow B(s\restriction\rho+\alpha,f\restriction(\rho+\alpha)^2,y^\rho)\\ &\quad\wedge\forall z(z\in x\leftrightarrow\exists p^\rho(B(s\restriction p+\alpha,f\restriction(\rho+\alpha)^2,p^\rho)\\ &\quad\wedge\exists\beta(\beta<\alpha\wedge p^{\rho}{}'\rho+\beta\in y^{\rho}{}'\rho+\beta)\wedge\langle z\,p^\rho\rangle\in a''\{\rho+\alpha\}))\\ &\leftrightarrow\langle\langle x\,y^\rho\rangle\,\rho+\alpha\rangle\in a.\end{aligned}$$

By this we have

$$f^0_\rho(s)''\{\alpha\}=s''\{\rho+\alpha\},\quad\bar{f}^0_\rho(m)''\{\alpha\}=m''\{\rho+\alpha\}$$

and

$$\bar{f}^0_\rho(f)''\{\langle\alpha\,\beta\rangle\}=f''\{\langle\rho+\alpha\;\rho+\beta\rangle\}.$$

By these we obtain for all $\rho < \delta$ and $\alpha, \beta < \bar{f}^0_\rho(\delta)$,

(1) $\neg \operatorname{Lim}(\alpha) \to (\langle x\, p\rangle \in \bar{f}^0_\rho(a)''\{\alpha\} \leftrightarrow \langle x\, p\rangle \in a''\{\rho + \alpha\})$.

(2) $\operatorname{Lim}(\alpha) \to (\langle x\, p\rangle \in \bar{f}^0_\rho(a)''\{\alpha\} \leftrightarrow \langle x\, p^\rho\rangle \in a''\{\rho + \alpha\})$.

(3) $\bar{f}^0_\rho(s)''\{\alpha\} = s''\{\rho + \alpha\}$.

(4) $\bar{f}^0_\rho(m)''\{\alpha\} = m''\{\rho + \alpha\}$.

(5) $\bar{f}^0_\rho(f)''\{\langle \alpha\, \beta\rangle\} = f''\{\langle \rho + \alpha\; \rho + \beta\rangle\}$.

In the case where δ is not a limit ordinal, there is an ordinal number α such that $\delta = \alpha + 1$. Now by the assumption of the lemma there is a set $f \in \bar{s}''\{0\}$ such that

$$\langle z\, p\rangle \in g \leftrightarrow A(s''\{0\}, m''\{0\}, p) \wedge \forall t(t \in z \to \exists q(A(s''\{0\}, m''\{0\}, q) \wedge \{v : q'v \in p'v\} \in m''\{0\} \wedge \langle t\, q\rangle \in g)),$$

$$A(s''\{0\}, m''\{0\}, p) \to \exists x(\langle x\, p\rangle \in g).$$

By the induction hypothesis, we have

$$\langle z\, p\rangle \in \bar{f}^0_\rho(g) \leftrightarrow A(s''\{\alpha\}, m''\{\alpha\}, p) \wedge \forall t(t \in z \leftrightarrow \exists q(A(s''\{\alpha\}, m''\{\alpha\}, g) \wedge \{v : q'v \in p'v\} \in m''\{\alpha\} \wedge \langle t\, q\rangle \in \bar{f}^0_\alpha(g))),$$

$$A(s''\{\alpha\}, m''\{\alpha\}, p) \to \exists x(\langle x\, p\rangle \in \bar{f}^0_\alpha(g)).$$

By the induction on the rank of z, we have $\bar{f}^0_\alpha(g) = a''\{\delta\}$ and therefore

$$A(s''\{\alpha\}, m''\{\alpha\}, p) \to \exists x(\langle x\, p\rangle \in a''\{\delta\}).$$

In the case where δ is a limit ordinal and there is an ordinal ρ less than δ such that $\rho < \delta < \rho + \bar{f}(\delta)$, there is an ordinal number α such that $\delta = \rho + \alpha$ and $\alpha < \bar{f}^0_\rho(\delta)$. Hence we have

$$\langle\langle x\, p\rangle \alpha\rangle \in \bar{f}^0_\rho(a) \leftrightarrow \langle\langle x\, p^\rho\rangle \delta\rangle \in a.$$

By the assumption of induction, we obtain

$$B(\bar{f}^0_\rho(s) \restriction \alpha, \bar{f}^0_\rho(f) \restriction \alpha^2, p) \to \exists x(\langle x\, p\rangle \in \bar{f}^0_\rho(a)''\{\alpha\}).$$

And therefore we get

$$B(s \restriction \delta, f \restriction \delta^2, p^\rho) \to \exists x(\langle x\, p^\rho\rangle \in a''\{\delta\}).$$

Assume $B(s \restriction \delta, f \restriction \delta^2, q)$. Then there is a function p such that

$$B(s \restriction \delta, f \restriction \delta^2, p^\rho) \quad \text{and} \quad \exists \beta(\beta < \delta \wedge p^{\rho}{}'\beta = q'\beta).$$

By this we have $\exists x(\langle x\, q\rangle \in a''\{\delta\})$.

In the case where δ is a limit ordinal and for all $\rho < \delta$, $\delta = \rho + \bar{f}_\rho(\delta)$. In this case we have $\bar{f}_\rho(\delta) = \delta$ for all $\rho < \delta$. Let $x \in s''\{0\}$. And by x^* we denote the function defined by

$$x^*{}'v = \bar{f}^0_\rho(x) \quad \text{for} \quad v < \delta \quad \text{and} \quad D(x^*) = \delta.$$

Clearly we have $B(s \restriction \delta, f \restriction \delta^2, x^*)$. Let $v'x_0 = \eta$ and $x_0 \in s''\{0\}$. And we assume

$$\exists x(v'x < \eta \wedge B(s \restriction \delta, f \restriction \delta^2, p) \wedge \exists \rho(\rho < \delta \wedge p'\rho \in \bar{f}^0_\rho(x))) \to \exists t(\langle t\, x\rangle \in a''\{\delta\}).$$

Therefore by the properties on $\bar{f}$, we have

$$\exists x(x \in s''\{\rho\} \wedge \bar{f}(\nu)'x < \bar{f}(\eta) \wedge B(s \upharpoonright \delta, f \upharpoonright \delta^2, p) \wedge \exists \nu(\nu < \delta \wedge p' \rho + \nu \in \bar{f}^{\rho}_{\rho+\nu}(x))) \rightarrow \exists t(\langle t\, x\rangle \in \bar{f}^0_\rho(a)''\{\delta\}).$$

Let v be the set defined by

$$v = \{p : \exists x(x \in s''\{\rho\} \wedge \bar{f}^0_\rho(\nu)'x < \bar{f}(\eta) \wedge B(s \upharpoonright \delta, f \upharpoonright \delta^2, p)) \wedge \exists \nu(\nu < \delta \wedge p' \rho + \nu \in \bar{f}^{\rho}_{\rho+\nu}(x))\}.$$

Then v has the property that

$$p \in v \wedge B(s \upharpoonright \delta, f \upharpoonright \delta^2, q) \wedge \exists \nu(\nu < \delta \wedge q'\nu \in p'\nu) \rightarrow q \in v.$$

Therefore by the induction on $(\bar{f}^0_\rho(a)''\{\delta\})'p$, we have

$$p \in v \rightarrow (\langle x\, p\rangle \in \bar{f}^0_\rho(a)''\{\delta\} \leftrightarrow \langle x\, p\rangle \in a''\{\delta\}).$$

Now we consider a function q such that

$$B(s \upharpoonright \delta, f \upharpoonright \delta^2, q) \quad \text{and} \quad \exists \beta(\beta < \delta \wedge q'\beta \in x^{*}_0{}'\beta).$$

Then we have $\bar{f}^0_\rho(\nu)'(q'\rho) < \bar{f}^0_\rho(\eta)$ for some $\rho < \delta$. Hence we obtain $\exists t(\langle t\, q\rangle \in a''\{\delta\})$. Now we put

$$z = \{t : \langle t\, p\rangle \in a''\{\delta\} \wedge B(s \upharpoonright \delta, f \upharpoonright \delta^2, p) \wedge \exists \beta(\beta < \delta \wedge P'\beta \in x^{*}_0{}'\beta)\}.$$

Then by the definition of a, we have $\langle z\, x^{*}_0\rangle \in a''\{\delta\}$. Namely, we have

$$\forall p(B(s \upharpoonright \delta, f \upharpoonright \delta^2, p) \rightarrow \exists x(\langle x\, p\rangle \in a''\{\delta\})),$$

$$x \in \bar{s}''\{\alpha\} \wedge \alpha \leq \delta \rightarrow \exists!\, y(\langle y\, x\rangle \in \bar{f}''\{\langle \delta\, \alpha\rangle\}),$$

$$\langle x\, y\rangle \in \bar{f}''\langle \delta\, \alpha\rangle\} \wedge \alpha \leq \delta \rightarrow x \in \bar{s}''\{\delta\} \wedge y \in \bar{s}''\{\alpha\},$$

$$\langle x\, y\rangle \in \bar{f}''\{\langle \delta\, \alpha\rangle\} \wedge \langle y\, z\rangle \in \bar{f}''\{\langle \alpha\, \beta\rangle\} \wedge \beta \leq \alpha \leq \delta \rightarrow \langle x\, z\rangle \in \bar{f}''\{\langle \delta\, \beta\rangle\}$$

are proved by (1) and (2).

Now we consider $\bar{f}^{\alpha}_{\sigma}(m''\{\alpha\}) = \bar{f}^{\alpha}_{\delta}(\bar{f}^0_{\alpha}(m''\{0\})) = m''\{\delta\}$. We shall show

$$\bar{f}^{\alpha}_{\delta}(s''\{\alpha\}) = s''\{\delta\}.$$

In the case where δ is not a limit ordinal number, it is enough to show it in the case $\delta = \alpha + 1$. By the definition of $\bar{f}^{\alpha}_{\delta}(s''\{\alpha\})$, we have

$$\forall t(t \in \bar{f}^{\alpha}_{\delta}(s''\{\alpha\}) \leftrightarrow \exists p(A(s''\{\alpha\}, m''\{\alpha\}, p) \wedge \{\nu : p'\nu \in s''\{\alpha\}\} \in m''\{\alpha\} \wedge \langle t\, p\rangle \in a''\{\delta\})).$$

By $A(s''\{\alpha\}, m''\{\alpha\}, p) \rightarrow \{\nu : p'\nu \in s''\{\alpha\}\} \in m''\{\alpha\}$ and by the definition of $s''\{\delta\}$, we have $s''\{\delta\} = \bar{f}^{\alpha}_{\delta}(s''\{\alpha\})$. Now we treat the case where δ is a limit ordinal, we have

$$\forall t(t \in \bar{f}^{\alpha}_{\delta}(s''\{\alpha\}) \leftrightarrow \exists q(B(s \upharpoonright \delta, f \upharpoonright \delta^2, q) \wedge \exists \gamma(\gamma < \delta \wedge q'\gamma \in x^{*}{}'\gamma) \wedge \langle t\, q\rangle \in a''\langle \delta\rangle))$$

where $x^{*}{}'\nu = s''\{\nu\}$ for $\alpha \leq \nu < \delta$. By $B(s \upharpoonright \delta, f \upharpoonright \delta^2, q) \rightarrow q'\gamma \in s''\{\gamma\}$ for some $\gamma < \delta$. And by the definition of $s''\{\delta\}$, we have $s''\{\delta\} = \bar{f}^{\alpha}_{\delta}(s''\{\alpha\})$.

In order to show (7) it is enough to show that $\bar{f}^{\alpha}_{\delta}(\alpha < \delta)$ is an elementary mapping such that

$$\bar{f}^{\alpha}_{\delta}: \langle \bar{s}''\{\alpha\};\ s''\{\alpha\}, m''\{\alpha\}, \in\rangle \to \langle \bar{s}''\{\delta\};\ s''\{\delta\}, m''\{\delta\}, \in\rangle.$$

We shall first treat the case where δ is not a limit ordinal. In this case it is enough to treat the case where $\delta = \alpha + 1$. Now we shall show

$$\begin{aligned}&\langle a_1 p_1\rangle \in a''\{\delta\}, \dots, \langle a_n p_n\rangle \in a''\{\delta\}, t_1 \in \bar{s}''\{0\}, \dots, t_m \in \bar{s}''\{0\}\\ &\quad \to (\{\nu : \Phi^{(s''\{\alpha\})}(p_1'\nu, \dots, p_n'\nu, \bar{f}^0_\alpha(t_1), \dots, \bar{f}^0_\alpha(t_m))\} \in m''\{\alpha\}\\ &\qquad \leftrightarrow \Phi^{(s''\{\delta\})}(a_1, \dots, a_n, \bar{f}^0_\delta(t_1), \dots, \bar{f}^0_\delta(t_m))).\end{aligned}$$

In the case where Φ is a prime formula, we have

$$\begin{gathered}\langle a_1 p_1\rangle \in a''\{\delta\}, \langle a_2 p_2\rangle \in a''\{\delta\} \to (\{\nu : p_1'\nu \in p_2'\nu\} \in m''\{\alpha\} \leftrightarrow a_1 \in a_2),\\ \langle a_1 p_1\rangle \in a''\{\delta\}, t_1 \in \bar{s}''\{0\} \to (\{\nu : p_1'\nu \in \bar{f}^0_\alpha(t_1)\} \in m''\{\alpha\} \leftrightarrow a_1 \in \bar{f}^0_\delta(t_1)),\\ \langle a_1 p_1\rangle \in a''\{\delta\}, t_1 \in \bar{s}''\{0\} \to (\{\nu : \bar{f}^0_\alpha(t_1) \in p_1'\nu\} \in m''\{\alpha\} \leftrightarrow \bar{f}^0_\delta(t_1) \in a_1),\\ t_1 \in \bar{s}''\{0\}, t_2 \in \bar{s}''\{0\} \to (\bar{f}^0_\alpha(t_1) \in \bar{f}^0_\alpha(t_2) \leftrightarrow \bar{f}^0_\delta(t_1) \in \bar{f}^0_\delta(t_2)).\end{gathered}$$

In the last two cases, if $t_1 \notin s''\{0\}$, then both sides of the equivalences are not true and if $t \in s''\{0\}$, then $\langle \bar{f}^0_\delta(t_1)u^*\rangle \in a''\{\delta\}$ where $u^{*\prime}\nu = \bar{f}^0_\alpha(t_1)$ for $v \in \mathfrak{S}(m''\{\alpha\})$. The case where the outermost symbol of Φ is one of $\neg$, $\vee$ is treated as usual. In the case where the outermost symbol of Φ is $\exists$, we assume

$$\begin{gathered}\langle a_1 p_1\rangle \in a''\{\delta\}, \dots, \langle a_n p_n\rangle \in a''\{\delta\}, t_1 \in \bar{s}''\{0\}, \dots, t_m \in \bar{s}''\{0\},\\ \{\nu : \exists x(x \in s''\{\alpha\} \wedge \Phi^{(s''\{\alpha\})}(p_1'\nu, \dots, p_n'\nu, x, \bar{f}^0_\alpha(t_1), \dots, \bar{f}^0_\alpha(t_m)))\} \in m''\{\alpha\}.\end{gathered}$$

Now we consider the formula which is true in $\langle \bar{s}''\{0\};\ s''\{0\}, m''\{0\}, \in\rangle$.

$$\begin{aligned}&\forall p_1 \cdots \forall p_n(A(s''\{0\}, m''\{0\}, p_1) \wedge \cdots \wedge A(s''\{0\}, m''\{0\}, p_n)\\ &\quad \wedge \{\nu : \exists x(x \in s''\{0\} \wedge \Phi^{(s''\{0\})}(p_1'\nu, \dots, p_n^1\nu, x, t_1, \dots, t_m))\} \in m''\{0\}\\ &\quad \to \exists p(A(s''\{0\}, m''\{0\}, p)\\ &\qquad \wedge \{\nu : \Phi^{(s''\{0\})}(p_1'\nu, \dots, p_n'\nu, p'\nu, t_1, \dots, t_m)\} \in m''\{0\})).\end{aligned}$$

By the assumption we have

$$\begin{aligned}&\forall p_1 \cdots \forall p_n(A(s''\{\alpha\}, m''\{\alpha\}, p_1) \wedge \cdots \wedge A(s''\{\alpha\}, m''\{\alpha\}, p_n) \wedge \{\nu : \exists x(x \in s''\{\alpha\}\\ &\quad \wedge \Phi^{(s''\{\alpha\})}(p_1'\nu, \dots, p_n'\nu, x, \bar{f}^0_\alpha(t_1), \dots, \bar{f}^0_\alpha(t_m)))\} \in m''\{\alpha\}) \to \exists p(A(s''\{\alpha\}, m''\{\alpha\}, p)\\ &\qquad \wedge \{\nu : \Phi^{(s''\{\alpha\})}(p_1'\nu, \dots, p_n'\nu, p'\nu, \bar{f}^0_\alpha(t_1), \dots, \bar{f}^0_\alpha(t_m))\} \in m''\{\alpha\}).\end{aligned}$$

By this property we have a function p such that $A(s''\{\alpha\}, m''\{\alpha\}, p)$ and

$$\{\nu : \Phi^{(s''\{\alpha\})}(p_1'\nu, \dots, p_n'\nu, p'\nu, \bar{f}^0_\alpha(t_1), \dots, \bar{f}^0_\alpha(t_m))\} \in m''\{\alpha\}.$$

Since there is a set b such that $\langle b\, p\rangle \in a''\langle\delta\rangle$, we have

$$\Phi^{(s''\{\delta\})}(a_1, \dots, a_n, b, \bar{f}^0_\alpha(t_1), \dots, \bar{f}^0_\alpha(t_m)).$$

Namely, we have

$$\exists x(x \in s''\{\delta\} \wedge \Phi^{(s''\{\alpha\})}(a_1, \dots, a_n, x, \bar{f}^0_\alpha(t_1), \dots, \bar{f}^0_\alpha(t_m))).$$

Conversely, we assume the above formula, then there is a set b such that

$$b \in s''\{\delta\} \wedge \Phi^{(s''\{\delta\})}(a_1, \ldots, a_n, b, \bar{f}^0_\delta(t_1), \ldots, \bar{f}^0_\delta(t_m)).$$

Since $b \in s''\{\delta\}$, we have a function p such that $\langle b\, p\rangle \in a''\{\delta\}$. Namely, we have

$$\{\nu : \Phi^{(s''\{\alpha\})}(p_1'\nu, \ldots, p_n'\nu, p'\nu, \bar{f}^0_\alpha(t_1), \ldots, \bar{f}^0_\alpha(t_m))\} \in m''\{\alpha\}.$$

Hence we have

$$\{\nu : \exists x(x \in s''\{\alpha\} \wedge \Phi^{(s''\{\alpha\})}(p_1'\nu, \ldots, p_n'\nu, x, \bar{f}^0_\alpha(t_1), \ldots, \bar{f}^0_\alpha(t_m)))\} \in m''\{\alpha\}.$$

From this we obtain that for $a_1, \ldots, a_n \in s''\{\alpha\}$ and $t_1, \ldots, t_m \in \bar{s}''\{\alpha\}$,

$$\Phi^{(s''\{\alpha\})}(a_1, \ldots, a_n, t_1, \ldots, t_m) \leftrightarrow \Phi^{(s''\{\delta\})}(\bar{f}^\alpha_\delta(a_1), \ldots, \bar{f}^\alpha_\delta(a_n), \ldots, \bar{f}^\alpha_\delta(t_m)).$$

Now we treat the case where δ is a limit ordinal. We shall show for all $\alpha < \delta$,

$$a_1, \ldots, a_n \in s''\{\alpha\} \quad \text{and} \quad t_1, \ldots, t_m \in \bar{s}''\{\alpha\}$$

$$\Phi^{(s''\{\alpha\})}(a_1, \ldots, a_n, t_1, \ldots, t_m) \leftrightarrow \Phi^{(s''\{\delta\})}(\bar{f}^\alpha_\delta(a_1), \ldots, \bar{f}^\alpha_\delta(a_n), \bar{f}^\alpha_\delta(t_1), \ldots, \bar{f}^\alpha_\delta(t_m)).$$

In the case where Φ is a prime formula, we have

$$a_1, a_2 \in s''\{\alpha\} \to (a_1 \in a_2 \leftrightarrow \bar{f}^\alpha_\delta(a_1) \in \bar{f}^\alpha_\delta(a_2)),$$
$$a_1 \in s''\{\alpha\}, t_1 \in \bar{s}''\{\alpha\} \to (a_1 \in t_1 \leftrightarrow \bar{f}^\alpha_\delta(a_1) \in \bar{f}^\alpha_\delta(t_1)),$$
$$a_1 \in s''\{\alpha\}, t_1 \in \bar{s}''\{\alpha\} \to (t_1 \in a_1 \leftrightarrow \bar{f}^\alpha_\delta(t_1) \in \bar{f}(a_1)),$$
$$t_1, t_2 \in \bar{s}''\{\alpha\} \to (t_1 \in t_2 \leftrightarrow \bar{f}^\alpha_\delta(t_1) \in \bar{f}^\alpha_\delta(t_2)).$$

In the last two cases, if $t_1 \notin s''\{\alpha\}$, then both sides of the equivalences are not true. The case where the outermost symbol of Φ is one of $\neg$, $\vee$ is treated as usual. In the case where the outermost symbol of Φ is $\exists$, we assume

$$a_1, \ldots, a_n \in s''\{\alpha\} \quad \text{and} \quad t_1, \ldots, t_m \in \bar{s}''\{\alpha\},$$

$$\exists x(x \in s''\{\alpha\} \wedge \Psi^{(s''\{\alpha\})}(a_1, \ldots, a_n, x, t_1, \ldots, t_m)).$$

By this we have a set $d \in s''\{\alpha\}$ such that $\Psi^{(s''\{\alpha\})}(a_1, \ldots, a_n, d, t_1, \ldots, t_m)$. Therefore, we obtain that $\Psi^{(s''\{\delta\})}(\bar{f}^\alpha_\delta(a_1), \ldots, \bar{f}^\alpha_\delta(a_n), \bar{f}^\alpha_\delta(d), \bar{f}^\alpha_\delta(t_1), \ldots, \bar{f}^\alpha_\delta(t_m))$. Namely, we have

$$\exists x(x \in s''\{\delta\} \wedge \Psi^{(s''\{\delta\})}(\bar{f}^\alpha_\delta(a_1), \ldots, \bar{f}^\alpha_\delta(a_n), x, \bar{f}^\alpha_\delta(t_1), \ldots, \bar{f}^\alpha_\delta(t_m))).$$

Conversely we assume the above formula. Then we have a set $b \in s''\{\delta\}$ such that

$$\Psi^{(s''\{\delta\})}(\bar{f}^\alpha_\delta(a_1), \ldots, \bar{f}^\alpha_\delta(a_n), b, \bar{f}^\alpha_\delta(t_1), \ldots, \bar{f}^\alpha_\delta(t_m)).$$

By $b \in s''\{\delta\}$, we have a function p such that $\langle b\, p\rangle \in a''\{\delta\}$. Therefore, we have an ordinal β such that $\alpha \leq \beta < \delta$ and

$$\Psi^{(s''\{\beta\})}(\bar{f}^\alpha_\beta(a_1), \ldots, \bar{f}^\alpha_\beta(a_n), p'\beta, \bar{f}^\alpha_\beta(t_1), \ldots, \bar{f}^\alpha_\beta(t_m)).$$

Hence we have

$$\exists x(x \in s''\{\beta\} \wedge \Psi^{(s''\{\beta\})}(\bar{f}^\alpha_\beta(a_1), \ldots, \bar{f}^\alpha_\beta(a_n), x, \bar{f}^\alpha_\beta(t_1), \ldots, \bar{f}^\alpha_\beta(t_m))).$$

Namely, by the assumption of the induction, we have

$$\exists x(x \in s''\{\alpha\} \wedge \Psi^{(s''\{\alpha\})}(a_1, \ldots, a_n, x, t_1, \ldots, t_m)).$$

Thus we complete the proof of (7). Since ρ_0 is the set of all ordinal numbers in $s''\{0\}$, we have by (7),

$$f_\delta^\alpha(\rho_0) = \rho_0.$$

Thus we complete the proof of the lemma.

THEOREM D. *Let the following conditions be satisfied:*

(1) $\aleph_{\tau_1} < \aleph_{\tau_2} < \cdots < \aleph_{\tau_n}$ *is a sequence of normally measurable cardinals.*

(2) $\aleph_{c_1} < \aleph_{c_2} < \cdots < \aleph_{c_n}$ *is a sequence of cardinals such that*

$$\aleph_{\tau_i} = \aleph_{c_i} \vee (2^{\aleph_{\tau_i}} < \aleph_{c_i} \wedge \aleph_{\tau_i} < \mathrm{cf}(\aleph_{c_i})) \quad \textit{for } i = 1, \ldots, n.$$

(3) $\aleph_\sigma$ *is a regular cardinal number such that* $2^{\aleph_{\tau_n}} < \aleph_\sigma$ *and* $\aleph_{c_n} < \aleph_\sigma$. *Then there is an elementary onto mapping f such that*

(1) $f: \langle \mathfrak{P}(R(\aleph_\sigma));\ R(\aleph_\sigma), \in\rangle \to \langle B;\ C, \in\rangle$

where $\langle B;\ C, \in\rangle$ *is a transitive* $\in$ *structure such that*

$$\aleph_\sigma \subset C \subset R(\aleph_\sigma) \quad \textit{and} \quad B \subset \mathfrak{P}(C).$$

(2) $f(\aleph_{\tau_i}) = \aleph_{c_i}$ *for* $i = 1, \ldots, n.$

PROOF. Let $F_{\aleph_{\tau_1}}, \ldots, F_{\aleph_{\tau_n}}$ be normal ultrafilters on $\aleph_{\tau_1}, \ldots, \aleph_{\tau_n}$, and let

$$\bar{u}_1 = \mathfrak{P}(R(\aleph_\sigma)),\ u_1 = R(\aleph_\sigma),\ m_{1,1}^* = F_{\aleph_{\tau_1}}, \ldots, m_{1,n}^* = F_{\aleph_{\tau_n}}.$$

Now we assume that the sets $\bar{u}_k, u_k, m_{k,1}^*, \ldots, m_{k,n}^*$ $(k \leq t)$ and g_k $(k < t)$ are already defined so that for $k < t$,

(1) g_k is an elementary mapping such that

$$g_k: \langle \bar{u}_k;\ u_k, \in\rangle \to \langle \bar{u}_{k+1};\ u_{k+1}, \in\rangle,$$

where $\aleph_\sigma \subset u_{k+1} \subset u_k$.

(2) $g_k(m_{k,i}^*) = m_{k+1,i}^*$ for $i = 1, \ldots, n.$

(3) $g_k \cdots g_1(\aleph_{\tau_i}) = \aleph_{c_i}$ for $i = 1, \ldots, k.$

(4) $g_k(\aleph_{c_i}) = \aleph_{c_i}$ for $i = k + 1, \ldots, n.$

By Lemma 5, we define the sets $s_t, \bar{s}_t, m_t, a_t$, and f_t so that

(1) $\bar{s}_t''\{0\} = \bar{u}_t$, $s_t''\{0\} = u_t$ and $m_t''\{0\} = m_{t,t}^*$.

(2) $f_{t,\beta}^\alpha$ is an elementary mapping such that

$$f_{t,\beta}^\alpha: \langle \bar{s}_t''\{\alpha\};\ s_t''\{\alpha\}, \in\rangle \to \langle \bar{s}_t''\{\beta\};\ s_t''\{\beta\}, \in\rangle.$$

In the case where $\aleph_{\tau_t} = \aleph_{c_t}$, we have $g_{t-1} \cdots g_1(\aleph_{\tau_t}) = \aleph_{c_t}$. In this case we define the sets by

$$\bar{u}_{t+1} = \bar{u}_t,\ u_{t+1} = u_t,\ m_{t+1,1}^* = m_{t,1}^*, \ldots, m_{t+1,n}^* = m_{t,n}^*.$$

And the elementary mapping g_t is defined to be the identity mapping such that

$$g_t: \langle \bar{u}_t;\ u_t, \in\rangle \to \langle \bar{u}_{t+1};\ u_{t+1}, \in\rangle.$$

In the case where $\aleph_{\tau_t} < \aleph_{c_t}$, we have

$$g_{t-1} \cdots g_1(\aleph_{\tau_{t-1}}) = \aleph_{c_{t-1}} < g_{t-1} \cdots g_1(\aleph_{\tau_t}) < g_{t-1} \cdots g_1(\aleph_{c_t}) = \aleph_{c_t}.$$

In this case we define g_t by $g_t = f^0_{t,\aleph_{c_t}}$. Now we define

$$\bar{u}_{t+1} = \bar{s}_t''\{\aleph_{c_t}\}, \qquad u_{t+1} = s_t''\{\aleph_{c_t}\}, \qquad m^*_{t+1,1} = g_t(m^*_{t,1}), \ldots, m^*_{t+1,n} = g_t(m^*_{t,n})$$

Then we have the following properties:

(1) g_t is an elementary mapping such that

$$g_t: \langle \bar{u}_t;\ u_t, \in\rangle \to \langle \bar{u}_{t+1};\ u_{t+1}, \in\rangle$$

where $\aleph_\sigma \subset u_{t+1} \subset u_t$.

(2) $g_t(m^*_{t,i}) = m^*_{t+1,i}$ for $i = 1, \ldots, n$.

(3) $g_t \cdots g_1(\aleph_{\tau_i}) = \aleph_{c_i}$ for $i = 1, \ldots, t$.

(4) $g_t(\aleph_{c_i}) = \aleph_{c_i}$ for $i = t+1, \ldots, n$.

Now we consider the function f defined by $f = g_n g_{n-1} \cdots g_1$. Then we have

$$f: \langle P(R(\aleph_\sigma));\ R(\aleph_\sigma), \in\rangle \to \langle \bar{u}_{n+1};\ u_{n+1}, \in\rangle,$$

where $\aleph_\sigma \subset u_{n+1} \subset \cdots \subset u_1 = R(\aleph_\sigma)$ and $\bar{u}_{n+1} \subset \mathfrak{P}(u_{n+1})$. And moreover $f(\aleph_{\tau_i}) = \aleph_{c_i}$ for $i = 1, \ldots, n$. Thus we complete the proof of the theorem.

THEOREM 6 (Π^1_1-TRANSCENDENCY OF MEASURABLE CARDINAL NUMBERS). *Let $\aleph_\tau < \aleph_\sigma$ be two normally measurable cardinal numbers and let $P(a, a_1, \ldots, a_n)$ be a Π^1_1-formula considering $\aleph_\sigma$ as universe. Then we have*

$$\forall x_1 \cdots \forall x_n(x_1, \ldots, x_n < \aleph_\tau \wedge \exists x P(x, x_1, \ldots, x_n) \to \exists x(x < \aleph_\tau \wedge P(x, x_1, \ldots, x_n))).$$

PROOF. In order to show this it is enough to show that for every Σ^1_1-formula $Q(a, a_1, \ldots, a_n)$, the following formula

$$\forall x_1 \cdots \forall x_n(x_1, \ldots, x_n < \aleph_\tau \wedge \forall x(x < \aleph_\tau \to Q(x, x_1, \ldots, x_n)) \to \forall x(Q(x, x_1, \ldots, x_n)))$$

is true considering $\aleph_\sigma$ as universe. Let $a_1, \ldots, a_n$ be any but fixed ordinal numbers less than $\aleph_\tau$. And assume that

$$\forall x(x < \aleph_\tau \to Q(x, a_1, \ldots, a_n)).$$

Let b be any ordinal less than $\aleph_\tau$. Then we have $Q(a, a_1, \ldots, a_n)$ in $(\aleph_\sigma)$. Therefore, by Theorem C′, we have $Q^{(\aleph_\tau)}(a, a_1, \ldots, a_n)$. Namely we obtain that

$$\forall x(x < \aleph_\tau \to Q^{(\aleph_\tau)}(x, a_1, \ldots, a_n)).$$

This means that $\forall x Q(x, a_1, \ldots, a_n)^{(\aleph_\tau)}$ is true in the structure

$$\langle P(R(\aleph_{\sigma^+}));\ R(\aleph_{\sigma^+}), \in\rangle.$$

Therefore, by Theorem D, we have that $\forall x Q(x, a_1, \ldots, a_n)^{(\aleph_\sigma)}$ is true in the structure $\langle \bar{u};\ u, \aleph_\sigma, \in\rangle$ where $\aleph_{\sigma^+} \subset u \subset R(\aleph_{\sigma^+})$. Since $(Q(a, a_1, \ldots, a_n)$ is a Σ_1^1-formula, we have

$$\forall x(x < \aleph_\sigma \to Q^{(\aleph_\sigma)}(x, a_1, \ldots, a_n)).$$

Namely we obtain that

$$\forall x_1 \cdots \forall x_n(x_1, \ldots, x_n < \aleph_\tau \wedge \exists x P(x, x_1, \ldots, x_n) \to \exists x(x < \aleph_\tau \wedge P(x, x_1, \ldots, x_n)))$$

is true considering $\aleph_\sigma$ as universe. Thus we complete the proof of the theorem.

Now let $P(a)$ be a Π_1^1-formula considering $\aleph_\sigma$ as universe. Then, by the above theorem, we have

$$\exists x P(x) \to \exists x(x < \aleph_\tau \wedge P(x))$$

where $\aleph_\tau$ is the smallest 2-valued measurable cardinal number. Namely any Π_1^1-axiom of strong infinity of the form $\exists x P(x)$ does not imply the existence of normally measurable cardinal number.

THEOREM 7. *Let* $\aleph_{\tau_1} < \aleph_{\tau_2} < \cdots < \aleph_{\tau_n} < \aleph_\tau$ and $\aleph_{\sigma_1} < \aleph_{\sigma_2} < \cdots < \aleph_{\sigma_n} < \aleph_\tau$ *be two sequences of normally measurable cardinals and let* $P(a, a_1, \ldots, a_n)$ *be a* Π_1^1- $(\Sigma_1^1$-$)$ *formula considering* $\aleph_\tau$ *as universe. Then we have*

$$\forall x(x < \aleph_{\tau_1}, \aleph_{\sigma_1} \to (P(x, \aleph_{\tau_1}, \ldots, \aleph_{\tau_n}) \leftrightarrow P(x, \aleph_{\sigma_1}, \ldots, \aleph_{\sigma_n}))).$$

PROOF. It is enough to prove the theorem in the case where $P(a, a_1, \ldots, a_n)$ is a Σ_1^1-formula. Let E be the set $\{\aleph_{\tau_1}, \ldots, \aleph_{\tau_n}, \aleph_{\sigma_1}, \ldots, \aleph_{\sigma_n}\}$, and let

$$\aleph_{\rho_0} = \max(E), \qquad \aleph_{\rho_{i+1}} = \max(E - \{\aleph_{\rho_0}, \ldots, \aleph_{\rho_i}\}) \quad \text{for } i = 0, \ldots, n-1.$$

Then we obtain

$$\aleph_{\tau_i} = \aleph_{\sigma_i} \vee (2^{\aleph_{\tau_1}} < \aleph_{\rho_i} \wedge \aleph_{\rho_i} < \mathrm{cf}(\aleph_{\rho_i})),$$
$$\aleph_{\sigma_i} = \aleph_{\rho_i} \vee (2^{\aleph_{\sigma_i}} < \aleph_{\rho_i} \wedge \aleph_{\sigma_i} < \mathrm{cf}(\aleph_{\rho_i})).$$

By Theorem 5, we have

$$\forall x(x < \aleph_{\tau_1}, \aleph_{\sigma_1} \to (P(x, \aleph_{\rho_1}, \ldots, \aleph_{\rho_n}) \to P(x, \aleph_{\tau_1}, \ldots, \aleph_{\tau_n}))),$$
$$\forall x(x < \aleph_{\tau_1}, \aleph_{\sigma_1} \to (P(x, \aleph_{\rho_1}, \ldots, \aleph_{\rho_n}) \to P(x, \aleph_{\sigma_1}, \ldots, \aleph_{\sigma_n}))).$$

Now we assume that for any but fixed ordinal a $(<\aleph_\tau, \aleph_{\sigma_1})$, $P(a, \aleph_{\tau_1}, \ldots, \aleph_{\tau_n})$. Then, by Theorem D, there is an elementary mapping f such that

(1) $f\colon \langle P(R(\aleph_\tau));\ R(\aleph_\tau), \in\rangle \to \langle B;\ C, \in\rangle$
where $\aleph_\tau \subset C \subset R(\aleph_\tau)$.

(2) $f(\aleph_{\tau_i}) = \aleph_{\rho_i}$ for $i = 1, \ldots, n$.

Since $P(a, a_1, \ldots, a_n)$ is a Σ_1^1-formula in $(\aleph_\tau)$, we obtain $P(a, \aleph_{\rho_1}, \ldots, \aleph_{\rho_n})$. Namely we obtain

$$\forall x(x < \aleph_{\tau_1}, \aleph_{\sigma_1} \to (P(x, \aleph_{\tau_1}, \ldots, \aleph_{\tau_n}) \to P(x, \aleph_{\rho_1}, \ldots, \aleph_{\rho_n}))).$$

By the symmetry of the reason, we have

$$\forall x(x < \aleph_{\tau_1}, \aleph_{\sigma_1} \rightarrow (P(x, \aleph_{\sigma_1}, \ldots, \aleph_{\sigma_n}) \rightarrow P(x, \aleph_{\rho_1}, \ldots, \aleph_{\rho_n}))).$$

From these we have

$$\forall x(x < \aleph_{\tau_1}, \aleph_{\sigma_1} \rightarrow (P(x, \aleph_{\tau_1}, \ldots, \aleph_{\tau_n}) \leftrightarrow P(x, \aleph_{\sigma_1}, \ldots, \aleph_{\sigma_n}))).$$

Thus we complete the proof of the theorem.

REFERENCES

1. Paul J. Cohen, *The independence of the continuum hypothesis.* I, Proc. Nat. Acad. Sci. U.S.A. **50** (1963), 1143–1148; II, ibid. **51** (1964), 105–110.

2. Haim Gaifman, *Uniform extension operators for models and their applications; sets, models and recursion theory, studies in logic*, North-Holland, Amsterdam, 1967, pp. 122–155.

3. ———, *Further consequences of the existence of measurable cardinals* (abstract). Presented to the 1964 Internat. Congr. of Logic, Methodology and Philosophy of Sciences, Jerusalem, 1964.

4. Kurt Gödel, *The consistency of the axiom of choice and of the generalized continuum hypothesis with the axioms of set theory*, Princeton Univ. Press, Princeton, N.J., 1951.

5. Simon Kochen, *Ultraproducts in the theory of models*, Ann. of Math. **74** (1961), 221–261.

6. Azriel Lévy, *Axiom schemata of strong infinity in axiomatic set theory*, Pacific J. of Math. **10** (1960), 223–238.

7. Kanji Namba, *On $\aleph_0$-complete cardinals*, J. Math. Soc. Japan **19** (1967), 347–358.

8. ———, *$\aleph_0$-complete cardinals and transcendency of cardinals*, J. Symbolic Logic **32** (1967), 452–472.

9. ———, *Axiom of strong infinity and analytic hierarchy of ordinal numbers*, Comment. Math. Univ. St. Paul **16** (1967), 21–55.

10. J. Novák, *Problimes*, Colloq. Math. **3** (1954–55), 171.

11. Dana Scott, *Measurable cardinals and constructible sets*, Bull. Acad. Polon. Sci. Sér. Sci. Math. Astronom. Phys. **9** (1961), 521–524.

12. Jack Silver, *The consistency of the generalized continuum hypothesis with the existence of a measurable cardinal*, Notices Amer. Math. Soc. **13** (1966), 721.

13. ———, *Some applications of model theory in set theory*, Doctoral Dissertation, University of California, Berkeley, Calif., 1966.

14. Robert M. Solovay, *Real valued measurable cardinals*, Notices Amer. Math. Soc. **13** (1966), 721.

15. Gaisi Takeuti, *On the recursive functions of ordinal numbers*, J. Math. Soc. Japan **12** (1960), 119–127.

16. ———, *Transcendency of cardinals*, J. Symbolic Logic **30** (1965), 1–7.

TOKYO UNIVERSITY OF EDUCATION

LIBERAL INTUITIONISM AS A BASIS FOR SET THEORY

LAWRENCE POZSGAY

My object is to formulate philosophically a certain intuitive approach to sets which I think represents the basic insight underlying the Zermelo-Fraenkel axioms. Similar views have already been presented in the literature, as for example the restricted Platonism referred to by Paul Bernays and the type of conceptualism developed by Robert McNaughton.[1] But to my knowledge no one has yet presented the precise position which I have in mind, at least not with the thoroughness which I think it deserves.

A leading incentive is of course the hope that by a careful examination of the notion of a "set" we may come to discover some new set-theoretical axioms which will help to resolve such otherwise unsolvable problems as the Continuum Hypothesis.[2] I must admit that I have come up with nothing of the sort so far, and it may well be that such epistemological speculation is simply not calculated to produce any mathematical insights of the kind that are required.[3] But I for one am not convinced that all the possibilities along such lines have already been exhausted,

---

[1] See [3] and [15]. Similar views are also expressed or referred to in [9], [10], [13], [22], and [23, pp. 424–426]. We shall have occasion to refer to some of these later.

[2] Hope of finding such axioms has been expressed by Kurt Gödel in [13] and more recently—in spite of Cohen's independence result—by R. H. Bing in [4].

[3] See the related remarks by A. Fraenkel in [11, p. 66] and [12]. Andrzej Mostowski and others have proposed that we simply forget epistemological questions and develop two kinds of set theory, one accepting and the other rejecting the Continuum Hypothesis. See [16] and [17, pp. 83–84, 149].

and my purpose here is simply to pursue one line of investigation which I think still lies open and which might possibly bear some fruit.[4]

What I will do is first state some vague heuristic principles which more or less define the position which I propose and then attempt to make my meaning clearer by interpreting the paradoxes and the usual axioms from this viewpoint. Towards the end I will say something about the question of a suitable formalization for this particular approach, and also about Skolem's paradox and the Continuum Hypothesis.

I call the approach "liberal intuitionism" because it regards sets as mental fictions or constructs, but unlike ordinary intuitionism allows for the construction of actually infinite sets of arbitrarily high cardinality.[5] These two basic attitudes may be stated in two "principles" as follows:

PRINCIPLE A (THE INTUITIONISTIC ASPECT). A set is a mental construct which results from the mental operation of thinking of a collection of objects which have been previously discovered or constructed and then regarding the entire collection as a single new object called the "set" of the given objects.

PRINCIPLE B (THE LIBERAL ASPECT). Any well-defined mental process for constructing sets which has been clearly envisioned without ambiguities or contradictions may be regarded as already completed, regardless of any merely practical difficulties which may prevent one from actually carrying it out.

The first principle enunciates a definition of a "set" which seems quite close to Cantor's notion of a "collection into a whole of definite and separate objects of our intuition or our thought."[6] But here I have tried to emphasize that the objects to be collected into a whole must somehow be given prior to the construction of the new set. This may be implicit in Cantor's phrase but I think it deserves to be brought out explicitly, especially since it is crucial for handling the paradoxes.

Also note that I have placed no restriction on the objects which may be formed into a set, provided only that they be given previously. In a well-known letter to Dedekind, Cantor spoke of collections which could not be formed into a single totality and called them "absolutely infinite" or "inconsistent" collections. He may have been saying the same thing in different words, but I think it more appropriate to say that *any* collection of given objects may be formed into a set and put all the emphasis on the demand that they be previously given.[7]

---

[4] If nothing else, a deeper understanding of the notion of a "set" may reveal why it is that insights bearing on the Continuum Hypothesis are so hard to come by.

[5] Maybe "Liberal Conceptualism" or "Liberal Constructivism" would be preferable so as to dissociate the approach entirely from Brouwer's basic intuition regarding the natural numbers. The term used was chosen merely to locate the position relative to the three traditional categories of Platonism, Formalism, and Intuitionism.

[6] See [**5**, p. 481].

[7] See [**6**, p. 443]. Paul Finsler in [**9**] explicitly denies that any collection of given (vorgegebene) objects can be formed into a set (Menge). He seems to resort to this denial simply to avoid the paradoxes without giving any clear intuitive basis for it. As a result Skolem seems justified in calling his distinction between collections (Gesamtheiten, Systeme) and sets a "Wortspiel" (see [**21**] and [**10**, p. 200]). See my discussion of "proper classes" below for an attempt to formulate a more genuine conceptual distinction.

For future reference we might sum up these remarks in two further clarifications of Principle A:

A.1. Any collection whatever of objects previously given through discovery or construction may be regarded as forming a single totality or "set."

A.2. If S is any set and x is a member of S, then x must have been given prior to the construction of S.

The objects which are given through "discovery" rather than through "construction" are usually called "individuals" or "urelements." The development of set theory is thus a step-by-step mental process which begins with certain individuals and proceeds to form sets of these individuals, then other sets containing possibly individuals and sets, and so on indefinitely. If we say that an object x is "essentially in"[8] a set S if it is somehow involved in the total construction of S (i.e., by being in the transitive closure of S), then we can add the following implication of Principle A:

A.3. Any set S is distinct from all the objects x which are essentially in S.

By now, of course, we can easily dispense with the paradoxes. Within such an intuitive context Cantor's paradox cannot arise because in order to construct a set M of all sets you would have to presuppose that all sets including M had already been constructed, so there would be a contradiction in the very presuppositions required to construct M.[9] Also, since no set can be a member of itself, any attempt to construct a set of all sets which are not members of themselves must amount to an attempt to construct a set of all sets whatsoever, so Russell's paradox vanishes with Cantor's. And as for Burali-Forti, you cannot presuppose that all ordinals are "given" and at the same time pretend to construct a set which gives rise to an altogether new ordinal—which is precisely what would happen if you tried to construct a set of all ordinals.[10]

Now for the usual axioms, by which of course I mean the Zermelo-Fraenkel axioms including the Axiom of Choice. Since the Axiom of Extensionality says merely that a set is determined purely and simply by its members and by nothing else, we need only concern outselves with those axioms which assert the existence of sets. Among these the Axioms of Pairing and of Unions can be handled quite

[8] A term coined by Finsler in [**9**, p. 693].

[9] Hao Wang in [**23**, p. 425] draws the same conclusion from what he terms the "genetic" aspect of Cantor's notion of a set. As I see it, Cantor's paradox goes directly counter to any Platonistic attitude which would want to speak of *all* sets as "given" or "existing" all by themselves, just waiting for us to "investigate" them. Such an attitude seems to be shared by Finsler (see [**10**, p. 202]), a fact which goes a long way towards explaining the arbitrary nature of his distinction between collections and sets (see note 7 above). It is because I cannot share this attitude that I have avoided the use of Bernays' term "restricted Platonism" [**3**, p. 277] to describe the approach I am presenting here.

[10] A much more thorough discussion of the paradoxes in much the same spirit is given by Erik Stenius in [**22**], though I think he makes matters seem more complicated than they are. He also discusses at some length the semantical paradoxes, which do not concern us here.

easily. The first says that we can form sets containing one or two given elements, which is nothing but the simplest possible application of A.1. The Axiom of Unions follows from A.1. and A.2.: Given a set S of sets, by A.2. every element of the elements of S was constructed prior to the construction of S and so by A.1. we may think of all these elements as forming the desired set $U(S)$.

Most of the remaining existence axioms rely heavily on Principle B, so this is a good place to pause and take a good look at that principle. It gives us a free hand in devising constructions for sets, with only the minimal limitation that the constructions be clearly and unambigously defined. The basic idea is that all that matters is that the construction be theoretically sound, so that one knows exactly what is would mean to perform it and even to carry it to completion. If this is the case, one may immediately regard it as already completed in spite of any merely practical limitations on one's ability actually to carry it out.

Of course the main point of this principle is to justify the construction of actually infinite sets. There have always been those who had qualms about such sets, and when the paradoxes were encountered it was taken as a sort of confirmation of this viewpoint.[11] But as we have seen, the paradoxes can be handled quite adequately without any reference to infinity at all. The only real argument against infinite sets is that they seem to be so completely out of our reach: we can never actually produce any of them by an actual construction, so why even think about them? The answer, of course, is why *not*? Mathematicians have always felt free to conceive of ideal objects like perfect circles and perfectly straight lines of infinite length and zero thickness without regard for their inability ever to realize such notions in actual physical constructions or even in vivid mental images. To use the phrase of Bernays, mathematics in general and set theory in particular is "an ideal projection of a domain of thought."[12] All that matters is clarity of ideas, and merely practical considerations are out of place except insofar as they contribute to the ideas.[13]

Of course there will always be the question of how we can be sure that a given process for constructing sets is in fact clear and unambiguous and free from contradictions, and for this I can offer no sure-fire testing procedures. All I know is that the constructions which I shall propose seem to me to be clear enough. Since it seems that we have to do without combinatorial consistency proofs for set theory, the next best thing, as I see it, is to find some plausible intuitive basis for the axioms, and that is all I am attempting to provide here.

Before coming to the existence axioms, another remark is in order concerning the "proper classes" which are distinguished from "sets" in some formulations of set theory. For a liberal intuitionist the notion of a set of all sets is an absurdity, and Cantor's paradox is just a proof of this fact. If we accept Principle B, what

[11] Thus for example Henri Poincaré in [**18**, pp. 189–196].
[12] See [**3**, p. 277].
[13] On this point we are in full agreement with Paul Finsler (see [**10**, p. 182]). See also the references to F. P. Ramsey and Bertrand Russell in note 18 below. The opposite view would hold that what is not realizable in practice is simply meaningless. See for example [**1**, pp. 102–103].

this means is that there can never be a well-defined mental process for constructing all possible sets.[14] This provides an intuitive interpretation for the term "proper class." We may think of it as an accumulation of objects which must always remain in a state of development and which can never be fully envisioned within any well-defined mental process for constructing its elements. That is, the term "set" is so absolutely general that no well-defined process can ever exhaust all the possibilities to which it might apply. The totality of sets which have been constructed at any given stage in the development of set theory can always be extended to other sets which have not been previously envisioned. The possibilities are not merely infinite, in the numerical sense of this term, but are unlimited in the most absolute sense of being positively inconceivable. Maybe this is what Cantor meant by the term "absolutely infinite." In any case, what his paradox shows is that the class of all sets is a "proper class" in this sense.

Of course the Burali-Forti paradox shows the same thing about ordinals. This can come as a surprise, since we seem to have some pretty powerful methods for constructing ordinals. But closer inspection shows that either these methods do not give all possible ordinals, or else their definitions make use of notions like "set" which are at least equally as unlimited as that of "ordinal."

Now for the other existence axioms. Here we will take the usual tack that all we are interested in is sets, so we will have no use for any individuals besides the empty set. Principle A as it stands does not provide such a set, but the following quite natural assumption does the trick and may be regarded as one application of Principle B:

B.1. Given a set S, any well-defined process for either adding elements to S or taking elements out of S always produces another set S'.[15]

Much more important is the Axiom of Infinity, which provides the prime illustration of the chief import of Principle B. By repeated application of the Axiom of Pairing, starting with the empty set, we can construct successively the natural numbers $0, 1, 2, \ldots$. Principle B says that we may think of this whole process as completed and so may regard all the natural numbers as given once and for all and by A.1. may form the set ω containing all of them. The reason is that the process is perfectly well defined: it is perfectly clear what needs to be done, even though no normal human being could ever carry it out completely in practice.

Notice the sharp distinction between this situation and that which leads to Cantor's paradox (for example). In the case of the paradox there is no talk at all

[14] In this we seem to disagree with a remark made by Paul Cohen in [**8**, p. 85], where he proposes precisely such a process as a distinct possibility.

[15] Serious objections have been raised against the idea of a set containing only one object, or no objects at all (see for example [**22**, p. 65] and [**14**, pp. 67, 74–78]). Admittedly one has to stretch the definition of a "set" expressed in our Principle A to get such sets. But I see a definite intuitive distinction between thinking of an object x and thinking of that same object as the sole element of a set S. The latter is a mental construct, whereas x may be something as concrete as the moon. Accordingly it seems quite natural to remove x from S and still have something left, namely the "empty set" $\varnothing$. One may wish with Hao Wang in [**23**, p. 424] to classify such sets as "just a manner of speaking," but I would insist that there is a good intuitive basis for it.

about any process for constructing all sets: it is merely assumed without proof of any kind that all sets are somehow "given." And what the paradox shows is precisely how absurd it is to suppose that there could be such a process. In the case of the natural numbers, however, we have a definite, clearcut process for constructing them, even though it happens to be a process which we cannot fully execute in practice.

The Axiom of Separation states that if we are given a set S and a condition $A(x)$, then we can form the set T consisting precisely of those elements of S which satisfy $A(x)$. Actually the liberal intuitionist can generalize this considerably. For the role of the condition $A(x)$ is simply to ensure that those elements of S which are to belong to T have been carefully distinguished from those which are not. And there is no need for a condition $A(x)$ to make the decision as long as the decision is somehow made. Since sets are mental constructs, and free choices are perfectly good mental operations, it seems quite natural to allow for the possibility that the elements of T have been chosen quite arbitrarily from S. So it seems that we can generalize the Axiom of Separation as follows:

B.2. Given a set S, a new set T can be formed out of any of the elements of S whatsoever, no matter how arbitrarily they have been chosen, provided only that for each element x of S it has been definitely decided whether x is to belong to T or not.

From here it is just one small step to the Axiom of Choice, which not merely allows for the possibility of a mental operation composed of arbitrary choices as a means of constructing a set, but positively asserts that under certain circumstances one may simply go ahead and assume that such a process has already been completed. This is quite an assumption, especially if the number of choices to be made is, say, uncountably infinite. But even here it seems clear that the difficulties involved are merely practical ones: if one is given a set S of nonempty sets and is told to pick one element from each element of S, it seems quite clear what needs to be done, even though the size of S may prevent one from actually doing it.

The situation seems essentially the same, though tremendously more complicated, in the case of the Axiom of the Power Set. In a sense this axiom encompasses both the Axioms of Separation and of Choice combined by asserting that *all possible* free choices of elements from a given set S may be thought of as already completed. So here it is not a question of executing just one cluster of choices and producing a single new set, but rather of executing a whole cluster of clusters of choices—one cluster of choices for each subset of the given set S. In spite of the greater complication, though, I think it still remains clear precisely what needs to be done, and all the difficulties involved are merely practical ones arising from the vast number of operations which may need to be performed. Certainly for each cluster of choices—that is, for each possible subset T of S—it is quite clear what needs to be done: one has to decide, for each element x of S, whether x is to belong to T or not. So any particular subset T may be thought of as constructed, by Principle B, and this is just another way of saying that all possible subsets may be thought of as already constructed. Applying A.1., we may proceed to combine them all into the single set called the Power Set $P(S)$.[16]

Of course the fact that we are allowing for arbitrary choices is crucial here. If one insisted on some sort of condition, as in the usual formulation of the Axiom of Separation, one might well balk at the idea that all subsets can be given once and for all, since there could well be conditions which require the previous construction of certain subsets of S in order to be meaningful. That is, there would arise the whole problem of impredicative definitions, and the need to distinguish different "types" within $P(S)$.[17] But for the liberal intuitionist free choices are perfectly good operations for constructing sets and there is no need to worry about conditions at all.[18]

The Axiom of Replacement says that, given a set S such that with each element x of S there is associated a way of constructing a definite set $F(x)$, we may think of all the sets $F(x)$ as forming a new set T. The strongest applications of this axiom are had when the set S is well ordered and the construction of any particular $F(x)$ depends on the completion of the constructions of all the $F(y)$ for y preceding x in the ordering. So what the axiom provides is an extremely powerful tool for combining constructions and piling them up end-on-end to form composite constructions of possibly massive proportions. Nevertheless, in spite of the multiplication of the practical difficulties involved, the new constructions will always be well defined and theoretically sound provided only that the set S is well defined and the individual constructions for the $F(x)$ are themselves clearly mapped out—which of course is the basic hypothesis of the axiom. So by Principle B we may regard all the sets $F(x)$ as well constructed and may think of them as forming the required set T.

So much for the usual axioms except to note that the Axiom of Regularity follows from the fact that ultimately all sets are built up from the empty set so that infinite descending epsilon-chains are impossible.

Now for a word about formalization. Though I sympathize with Brouwer's view that formalization is something secondary and subordinate to the intuitive development of mathematics, nevertheless it does have a place, as we all know. So it is of some importance to ask what formal system best expresses the point of view of the liberal intuitionist.

---

[16] This sort of procedure for constructing power sets was called by Bernays "quasi-combinatorial" [**3**, pp. 275–276]. The same idea was taken up in more detail by McNaughton in [**15**].

[17] This concern for predicativity has been involved in varying degrees in some of the work of Bertrand Russell, Hermann Weyl, Hao Wang, and many others.

[18] F. P. Ramsey's simplification of Russell's theory of types [**19**, pp. 1–61] seems based on much the same sort of insight. Russell's remark in [**20**, p. 85] is very relevant here: "An infinite decimal may be recurring, or may follow some more complicated law by which its successive places can be calculated. But if there is no such law, there is no way of mentioning the decimal in question. This seems, however, to be due to what is, from a logical point of view, a mere accident, namely the fact that we do not live for ever. Pure logic must surely be independent of such trivial circumstances. And if the lawless decimal is admissible, so are Ramsey's functions in extension."

As far as set-theoretical axioms go, the best available seem to be the Zermelo-Fraenkel axioms, and the main question here is whether the underlying logic should be intuitionistic or classical. What we want is a logic which preserves the same attitude to existence which we adopted in regard to the axioms. That is, if A is an existential statement $(\exists x)B(x)$, A should be asserted only if there is at hand a definite construction for producing a set x with the property $B(x)$. Since this attitude is basically intuitionistic, it immediately raises some doubts about the Law of the Excluded Middle. Certainly we would not want to assert "not-A" unless it had been proved conclusively that sets with the property $B(x)$ are altogether impossible. And between the extremes of sets which have definitely been constructed and those whose construction is definitely impossible there is that nebulous "no man's land" of sets which are in some sense theoretically possible but which have not *yet* been constructed and hence do not yet "exist" in our sense.

It turns out that because of the Law of the Excluded Middle and the way the Axioms of Replacement and of Separation are usually formulated, the classical first-order formalization of Zermelo-Fraenkel set theory does assert the existence of sets for which no definite constructions are available at the present state of our knowledge. For example, using the Axiom of Replacement in conjunction with the Law of the Excluded Middle one can prove the existence of a set S which is the first inaccessible cardinal if such exists and otherwise is the empty set. Applying the liberal intuitionist notion of existence throughout, this says that S is the empty set until such time as someone manages to construct the first inaccessible cardinal, and then S will be equal to that cardinal. This is certainly a strange way to define a set. Certainly at the present stage of our knowledge there is no definite construction available for S, and there will not be one until someone either manages to construct an inaccessible cardinal or else proves that such a construction is impossible.

Similarly the Axiom of Separation permits one to prove the existence of set T which equals the natural number 1 if there exists an inaccessible cardinal and otherwise equals 0. Here the problem is not so much to provide a definite construction procedure for T as to decide which of two well-constructed sets the symbol "T" is supposed to represent. As long as inaccessible cardinals remain in that shady area between existence and nonexistence, the liberal intuitionist will be inclined to question whether T is a well-defined set.

Nevertheless he is not forced to abandon classical logic altogether. For he accepts a great many sets as well constructed and fully determined once and for all, and questions concerning the properties and relationships of sets within this range always have a definite answer—even though at a given stage of our knowledge we may not be able to determine what that answer is. In symbolic terms, the Law of the Excluded Middle does hold for sentences with quantifiers restricted to definite, well-constructed sets. For this reason the liberal intuitionist readily accepts the results of classical mathematics, since they are formulated and proved within some definite context such as the real numbers, a definite topological space, etc.

Actually to me it seems unlikely that the totally unrestricted Law of the Excluded

Middle will ever play an essential role in a proof except in connection with foundational investigations where formal sentences are considered in themselves and apart from any intuitive context. Such proofs do exist, and I think some questions should be asked about their significance.[19] But in any case I do not think it has much bearing on the vast bulk of mathematics.

So as regards the formalization appropriate to set theory as viewed by the liberal intuitionist, I suppose it might be something like Zermelo-Fraenkel set theory over the intuitionistic predicate calculus, with the added rule that the Law of the Excluded Middle should hold for statements with restricted quantifiers. Some special attention would have to be given to the Axioms of Replacement and Separation, but I have not yet worked out the details.

Now for Skolem's Paradox. Among the statements which the liberal intuitionist regards as absolutely decided once and for all is that which asserts the uncountability of $P(\omega)$. For him this set is a perfectly well-defined mathematical entity and the proof of Cantor's theorem is conclusive evidence that there cannot be a 1-1 correspondence between it and itself. To put it another way, the set S of all 1-1 functions from ω into $P(\omega)$ is a definite subset of $P(P(P(\omega \cup P(\omega))))$, and what Cantor's proof shows is that no function in S has range equal to $P(\omega)$. In fact, even if we did not have Cantor's proof, the question of the countability of $P(\omega)$ concerns the relationships of sets within the range of those already well constructed and so would have a definite answer even though we might not be able to find out what that answer is.

Of course the existence of countable models—even standard transitive ones—for any reasonable formalization of set theory offers no real difficulty, since any model-theoretic interpretation, in the case when the model is a set, involves a drastic change from its original interpretation as a theory about all possible sets. The fact that a model is countable simply shows how drastic a reinterpretation is involved, and has little to do with whether or not certain "real" sets are uncountable.

Another statement which for the liberal intuitionist has a well-determined truth value is the Continuum Hypothesis. This may be seen by noting that the sets S of all 1-1 functions from ω into $P(\omega)$ and T of all 1-1 functions from $P(\omega)$ into itself are well-defined totalities, and the question of whether or not the Continuum Hypothesis is true amounts to asking whether there is a subset of $P(\omega)$ which is not the range of any function in either S or T. Accordingly for the liberal intuitionist the Continuum Hypothesis is either true or false, even though at present we do not know how to find out which.

So the liberal intuitionist would have to differ with anyone who would suggest that because of Cohen's independence result we may now be free either to accept or to reject the Continuum Hypothesis as we please.[20] There is a close parallel between Cohen's result and Gödel's incompleteness theorems. Both show that

---

[19] See for example [**8**, p. 82].
[20] See footnote 3, p. 321.

certain formal statements are formally independent of the axioms of their respective theories but leave essentially untouched the question of whether the intuitive understanding of the axioms determines the truth or falsity of the statements in question under their natural interpretation. For the liberal intuitionist such a determination is had in the case of the Continuum Hypothesis, even though he is unable to find out what direction it takes.

Of course the Generalized Continuum Hypothesis represents a somewhat different situation. For any given, well-constructed ordinal α it is certainly either true or false that $\aleph_{\alpha+1} = 2^{\aleph_\alpha}$. But whether it is true or false for "all" ordinals may depend on a particular stage of development of the class of ordinals.

BIBLIOGRAPHY

1. S. Barker, *The philosophy of mathematics*, Prentice-Hall, Englewood Cliffs, N.J., 1964.

2. P. Benecerraf and H. Putnam, *Philosophy of mathematics: selected readings*, Prentice-Hall, Englewood Cliffs, N.J., 1964.

3. P. Bernays, *On Platonism in mathematics*, English translation of a 1934 lecture published in [**2**, pp. 274–286].

4. R. H. Bing, *Challenging conjectures*, Amer. Math. Monthly **74**(1967), 56–64.

5. G. Cantor, *Beiträge zur Begründung der transfiniten Mengenlehre*, Math. Ann. **46**(1895), 481–512, and **49**(1897), 207–246, (tr. by P. Jourdain: Open Court, Chicago, 1915).

6. ———, Letter to Dedekind of July 28, 1899, in [**7**, pp. 443–447].

7. ———, *Gesammelte Abhandlungen*, Editor E. Zermelo, Berlin, 1932; new printing: Georg Olms, Hildesheim, 1962.

8. P. Cohen, *Set theory and the continuum hypothesis*, W. A. Benjamin, New York, 1966.

9. P. Finsler, *Über die Grundlegung der Mengenlehre. Erster Teil. Die Mengen und ihre Axiome*, Math. Z. **25**(1926), 683–713.

10. ———, *Über die Grundlegung der Mengenlehre. Zweiter Teil. Verteidigung*, Comment. Math. Helv. **38**(1964), 172–218.

11. A. Fraenkel, *Mengenlehre und Logik*, Duncker und Humblot, Berlin, 1959.

12. ———, "Epistemology and logic," in *Logic and language: Studies dedicated to Prof. Rudolf Carnap*, D. Reidel, Dordrecht, Holland, 1962.

13. K. Gödel, *What is Cantor's continuum problem?*, in [**2**, pp. 258–273].

14. E. Luschei, *The logical systems of Lesniewski*, North-Holland, Amsterdam, 1962.

15. R. McNaughton, *Conceptual schemes in set theory*, Philos. Rev. **66**(1957), 66–80.

16. A. Mostowski, *Widerspruchsfreiheit und Unabhängigkeit der Kontinuumhypothese*, Elem. Math. **19**(1964), 121–125.

17. ———, *Thirty years of foundational studies*, Barnes and Noble, New York, 1966.

18. H. Poincaré, *Science and method*, English translation of the 1908 book by F. Maitland, Dover, New York.

19. F. P. Ramsey, *The foundations of mathematics*, Harcourt, New York, 1931.

20. B. Russell, Review of [**19**], in Philosophy **7**(1932), 84–86.

21. T. Skolem, Review of [**9**], in Jahrbuch über die Fortschritte der Mathematik **54**, Jahrgang 1928(1932), 90, (quoted fully in [**10**].)

22. E. Stenius, *Das Problem der logischen Antinomien*, Soc. Sci. Fenn. Comment, Phys.-Math., **14**(1949), no. 11.

23. H. Wang, *Survey of mathematical logic*, North-Holland, Amsterdam, 1963.

ST. LOUIS UNIVERSITY

FORCING WITH PERFECT CLOSED SETS

GERALD E. SACKS[1]

0. **Introduction.** Forcing with perfect closed sets has its origin in Cohen's invention of forcing with finite conditions [**1**] and in Spector's construction of a minimal Turing degree [**28**]. Cohen taught that truth can be approximated more easily and more completely in the intermediate stages of a construction than any recursion theorist would have believed possible, and Spector revealed that infinite approximation by means of perfect closed conditions is radically different from finite approximation. The difference, it seems to us, is manifested in the sequential lemma, Lemma 1.4 of § 1 (see Footnote 2, p. 335).

In § 1 we study perfect closed forcing in the context of set theory. We show: if ZF (Zermelo-Fraenkel axioms for set theory) is consistent, then so is ZF plus there exist precisely two degrees of nonconstructibility. Two sets of natural numbers have the same degree of nonconstructibility if each is constructible from the other. All the constructible subsets of ω constitute one degree of nonconstructibility denoted by 0. A degree c is said to be less than or equal to a degree d if some set of degree c is constructible from some set of degree d. A nonzero degree is called minimal if the only degree less than it is 0. We lift Spector's construction of a minimal Turing degree from recursion theory to set theory in order to force all the nonconstructible subsets of ω to occupy the same degree. We also show by means of a transformation lemma that sets which are generic with respect to perfect closed forcing have the following curious property: if T is generic, then every set which has the same degree of nonconstructibility as T is generic.

In § 2 we examine perfect closed forcing in the hyperarithmetic case. The

[1] The preparation of this paper was supported in part by U.S. Army Contract DAHCO-4-67-C-0052. The author wishes to thank Professor J. Barkley Rosser for introducing him to Spector's minimal degree argument.

principal result of [5] is reviewed in order to bring out certain technical difficulties which crop up in the hyperarithmetic case and which have no counterpart in the set-theoretic case. It has to be shown in the hyperarithmetic case that the forcing relation, restricted to sentences of recursive ordinal rank, is Π_1^1, and this is seen to be the case only after the forcing relation is defined with certain precautions; if it is defined by routine analogy with the set-theoretic case, then its classification becomes uncertain. The corresponding problem of definability of the forcing relation in the set-theoretic case presents no problem at all.

In § 3 we touch on the arithmetic case and introduce the notion of pointed, perfect closed set. Using that notion, we obtain an order-theoretic characterization of the Turing degree of the truth-set for arithmetic in terms of the degrees of the arithmetic sets and the jump operator. We also briefly describe Spector's minimal Turing degree construction in terms of perfect closed forcing as well as several results based on forcing with pointed perfect closed sets whose detailed proofs will appear elsewhere [23].

Let $2 = \{0, 1\}$ have the discrete topology and let 2^ω have the product topology [8]. Let $P, Q, R, \ldots$ be perfect closed subsets of 2^ω. We conceive of a typical P as a tree with binary branching; each path of the tree is an infinite sequence of 0's and 1's and is viewed as the characteristic function of a subset of ω. Every path of P branches infinitely often; this is equivalent to saying that P is perfect. Suppose we adopt some standard recursive one-to-one correspondence between all finite sequences of 0's and 1's and ω. Then P can be encoded as a set of sequence numbers corresponding to finite, initial segments of paths of P. We will use $P, Q, R, \ldots$ ambiguously to denote perfect closed subsets of 2^ω and their encodings as subsets of ω. Whenever we classify P, we refer to the encoding of P. Thus if we say P is arithmetic, we mean that the encoding of P as a subset of ω is arithmetic. Let T be an arbitrary subset of ω. If we say $T \in P$, then clearly we are thinking of P as a subset of 2^ω. If we say $P \subseteq Q$, then we mean $(T)(T \in P \rightarrow T \in Q)$.

Let $p, q, r, \ldots$ ambiguously denote sequence numbers and the finite, initial segments of characteristic functions they encode. We say $p \geq r$ if $p = r$ or if p is extended by r. Let P be a nonempty set of sequence numbers. Then P is an encoding of a perfect closed subset of 2^ω if and only if

$$(p)_{p\in P}(\exists q)_{q\in P}(\exists r)_{r\in P}[p \geq q \;\&\; p \geq r \;\&\; q \not\geq r \;\&\; r \not\geq q].$$

We say $T \in p$ (or T satisfies p) if p is an initial segment of the characteristic function of T.

1. **The set-theoretic case.** Let $\mathscr{M}$ be a countable, transitive set which is a model of ZF plus V = L (Gödel's axiom of constructibility [6]). Let $T \subseteq \omega$. $\mathscr{M}(T)$ is the set of all sets constructible from T via the ordinals of $\mathscr{M}$. In order to make the definition of $\mathscr{M}(T)$ more precise, we adopt the symbolism of Tharp [30]. The symbols of $\mathscr{L}^0$ are: $\in$ (membership); unranked set variables $x, y, z, \ldots$; ranked set variables $x^\alpha, y^\alpha, z^\alpha, \ldots$ for each ordinal $\alpha \in \mathscr{M}$; propositional connectives; and quantifiers for both ranked and unranked variables. The atoms of $\mathscr{L}^0$ are of the form $t_1 \in t_2$, where t_1 and t_2 are variables. The formulas of $\mathscr{L}^0$ are

constructed from the atoms and the logical symbols in the usual fashion. A class $\mathscr{C}$ of constants intended to name the members of $\mathscr{M}(T)$ is defined by induction on the ordinals of $\mathscr{M}$.

$\mathscr{C}(0) = \{\bar{n} \mid n < \omega\}$.

$\mathscr{C}(\alpha + 1)$: $\mathscr{T} \in \mathscr{C}(\alpha + 1)$; let $\phi(x^\alpha, y_1, \ldots, y_n)$ be a formula of $\mathscr{L}^0$ whose only free variables are $x^\alpha, y_1, \ldots, y_n$ $(n \geq 0)$ and whose quantified variables are all of the form x^β for β's $\leq \alpha$; then $\hat{x}^\alpha\phi(x^\alpha, c_1, \ldots, c_n)$, where each c_i $(1 \leq i \leq n)$ is either $\mathscr{T}$ or a member of $\bigcup\{\mathscr{C}(\beta) \mid \beta \leq \alpha\}$, is a typical member of $\mathscr{C}(\alpha + 1)$.

$\mathscr{C}(\lambda) = \bigcup\{\mathscr{C}(\beta) \mid \beta < \lambda\}$ for each limit ordinal λ.

Let $\mathscr{C} = \bigcup\{\mathscr{C}(\alpha) \mid \alpha \in \mathscr{M}\}$. The symbols of $\mathscr{L}(\mathscr{T})$ are those of $\mathscr{L}^0$ together with the members of $\mathscr{C}$. The atomic formulas of $\mathscr{L}(\mathscr{T})$ are of the form $t_1 \in t_2$, where t_1 and t_2 are variables or members of $\mathscr{C}$.

A sentence $\mathscr{F}$ of $\mathscr{L}(\mathscr{T})$ is said to be *ranked* if all its variables are ranked. The ordinal rank of a ranked sentence $\mathscr{F}$, denoted by $o(\mathscr{F})$, is the least upper bound of $\{\beta \mid (x^\beta) \text{ or } (\exists x^\beta) \text{ occurs in } \mathscr{F}\} \cup \{\beta + 1 \mid \hat{x}^\beta \text{ occurs in } \mathscr{F}\}$. If $t \in \mathscr{C}$, then $o(t) = o(\bar{0} \in t)$. For each $T \subseteq \omega$ it is a routine matter to simultaneously inductively define the following concepts: $c \in \mathscr{C}(\alpha)$ and $b \in \mathscr{M}_\alpha(T)$ and c denotes b; $o(\mathscr{F}) \leq \alpha$ and $\mathscr{F}$ is true in $\mathscr{M}_\alpha(T)$. Of course $\mathscr{T}$ is interpreted at T, x^α ranges over $\mathscr{M}_\alpha(T)$, and x ranges over $\mathscr{M}(T) = \bigcup\{\mathscr{M}_\alpha(T) \mid \alpha \in \mathscr{M}\}$. We write $\mathscr{M}(T) \models \mathscr{F}$ to mean $\mathscr{F}$ is true in $\mathscr{M}(T)$. Let 0 denote the empty set, and let $\mathscr{M}_\alpha$ be $\mathscr{M}_\alpha(0)$. Then $\mathscr{M} = \bigcup\{\mathscr{M}_\alpha \mid \alpha \in \mathscr{M}\}$.

Now we define the forcing relation $P \Vdash \mathscr{F}$, where $P \in \mathscr{M}$ and $\mathscr{F}$ is a sentence of $\mathscr{L}(\mathscr{T})$. Of course $P \in \mathscr{M}$ means P is a perfect closed subset of 2^ω whose standard encoding as a subset of ω is a member of $\mathscr{M}$. The definition is orthodox save for the interpretation of P.

(1) $P \Vdash (\exists x)\mathscr{F}(x)$ if $P \Vdash \mathscr{F}(c)$ for some $c \in \mathscr{C}$.

(2) $P \Vdash (\exists x^\alpha)\mathscr{F}(x^\alpha)$ if $P \Vdash \mathscr{F}(c)$ for some $c \in \mathscr{C}(\alpha)$.

(3) $P \Vdash \mathscr{F} \;\&\; \mathscr{G}$ if $P \Vdash \mathscr{F}$ and $P \Vdash \mathscr{G}$.

(4) $P \Vdash \sim\mathscr{F}$ if $(Q)[P \supseteq Q \to \sim(Q \Vdash \mathscr{F})]$.

(5) $P \Vdash \bar{n} \in \mathscr{T}$ if $(T)[T \in P \to \mathscr{M}(T) \models \bar{n} \in \mathscr{T}]$.

(6) $P \Vdash \bar{m} \in \bar{n}$ if $m < n$.

(7) $P \Vdash t_1 \in t_2$ if $o(t_1) < o(t_2)$, t_2 is $\hat{x}^\beta\phi(x^\beta)$ and $P \Vdash \phi(t_1)$.

(8) $P \Vdash t_1 \in t_2$ if $o(t_1) > o(t_2)$, t_1 is $\hat{x}^\alpha\phi(x^\alpha)$ and $P \Vdash (\exists y^\alpha)[(x^\alpha)(x^\alpha \in y^\alpha \leftrightarrow \phi(x^\alpha)) \;\&\; y^\alpha \in t_2]$.

(9) $P \Vdash t_1 \in t_2$ if $o(t_1) = o(t_2) > 0$. Similar to (8).

We say T is *generic* if for each sentence $\mathscr{F}$ of $\mathscr{L}(T)$ there is a perfect closed $P \in \mathscr{M}$ such that $T \in P$ and either $P \Vdash \mathscr{F}$ or $P \Vdash \sim\mathscr{F}$. Standard arguments [7] suffice to show: (i) $(\mathscr{F})(P) \sim [P \Vdash \mathscr{F}$ and $P \Vdash \sim\mathscr{F}]$; (ii) $(P)(\mathscr{F})(\exists Q)[P \supseteq Q \;\&\; Q \Vdash \mathscr{F}$ or $Q \Vdash \sim\mathscr{F}]$; (iii) $(P)(\mathscr{F})(Q)[P \supseteq Q \;\&\; P \Vdash \mathscr{F} \to Q \Vdash \mathscr{F}]$; (iv) generic T's exist. A delicate but not troublesome point arises in the course of the proof of (v): if T is generic, then $\mathscr{M}(T) \models \mathscr{F}$ if and only if $(\exists P)[T \in P \;\&\; P \Vdash \mathscr{F}]$. The difficulty can be put as follows. Suppose T is generic, $T \in P \cap Q$, $P \Vdash \mathscr{F}$, and

$Q \Vdash \mathscr{S}$; then there should be an $R \subseteq P \cap Q$ such that $T \in R$ and $R \Vdash \mathscr{F} \,\&\, \mathscr{S}$. In the case of Cohen-forcing with finite conditions, it was safe to let R be $P \cap Q$, but now a different argument is required, because the intersection of two overlapping, perfect closed sets need not be perfect.

It is not difficult to show by a Cantor-Bendixson analysis that if $T \notin \mathscr{M}$ and $T \in P \cap Q$, then $T \in R \subseteq P \cap Q$, where R is the perfect kernel of $P \cap Q$. In short, if a perfect closed set is encoded by a subset of ω belonging to $\mathscr{M}$, then so is its perfect kernel. But Kreisel **[13]** has shown that this type of argument will not work in the hyperarithmetic case of § 2. His result is: there is a hyperarithmetically encodable, closed set whose perfect kernel is not hyperarithmetically encodable. The proof of proposition 1.1 (ii) does work in the hyperarithmetic case.

PROPOSITION 1.1. (i) *If T is generic, then $T \notin \mathscr{M}$.* (ii) *If T is generic and $T \in P \cap Q$, then $T \in R$ for some $R \subseteq P \cap Q$.*

PROOF. Let $c \in \mathscr{C}$ denote a subset of ω, and let $\bar{c}(x^0)$ denote the sequence number which represents the characteristic function of c restricted to arguments $< x^0$.

To prove (i) let T be generic and suppose $T = d \in \mathscr{M}$. Then there is a P such that $T \in P$ and (a) $P \Vdash (\exists x^0)(\bar{\mathscr{T}}(x^0) \neq \bar{d}(x^0))$ or (b) $P \Vdash \sim(\exists x^0)(\bar{\mathscr{T}}(x^0) \neq \bar{d}(x^0))$. If (a) holds, then for some n, $P \Vdash \bar{\mathscr{T}}(\bar{n}) \neq \bar{d}(\bar{n})$, and consequently $\bar{T}(n) \neq \bar{d}(n)$. If (b) holds, then $(T)(T \in P \to T = d)$.

To prove (ii) let $\mathscr{F}$ be $(\exists x^0)(\bar{\mathscr{T}}(x^0) \notin P \cap Q)$. Then there is an R such that $T \in R$ and $R \Vdash \mathscr{F}$ or $R \Vdash \sim\mathscr{F}$. If $R \Vdash \mathscr{F}$, then $T \notin P \cap Q$. So $R \Vdash \sim\mathscr{F}$. This means that

$$(R_1)_{R \supseteq R_1}(n) \sim [R_1 \Vdash \mathscr{T}(n) \notin P \cap Q].$$

But then $R \subseteq P \cap Q$.

Let A be an arbitrary set. We say A is $\mathscr{M}$-definable if there exists a formula $\mathscr{F}(x)$ of ZF with constants denoting elements of $\mathscr{M}$ such that $A = \{b \mid b \in \mathscr{M} \,\&\, \mathscr{M} \models \mathscr{F}(b)\}$.

LEMMA 1.2. *For each $n \geq 0$: the relation $P \Vdash \mathscr{F}$, restricted to sentences $\mathscr{F}$ of $\mathscr{L}(\mathscr{T})$ having at most n unranked quantifiers, is $\mathscr{M}$-definable.*

The proof of 1.2 is standard [7]. It follows from 1.2 that if T is generic, then the replacement axiom holds in $\mathscr{M}$. It suffices to prove: let $\mathscr{F}(x, y)$ be a formula of $\mathscr{L}(\mathscr{T})$ whose only free variables are x and y: if $P \Vdash (x^\alpha)(\exists y)\mathscr{F}(x^\alpha, y)$, then $P \Vdash (x^\alpha)(\exists y^\beta)\mathscr{F}(x^\alpha, y^\beta)$ for some β. P has the property that

$$(c)_{c \in \mathscr{C}(\alpha)}(Q)_{P \supseteq Q}(\exists R)_{Q \supseteq R}(\exists d)_{d \in \mathscr{C}(\alpha)}[R \Vdash \mathscr{F}(c, d)].$$

By 1.2, d can be regarded as an $\mathscr{M}$-definable function of c and Q. Since $\mathscr{C}(\alpha)$ and $\{Q \mid P \supseteq Q\}$ are sets of $\mathscr{M}$, the range of d, restricted to $\mathscr{C}(\alpha) \times \{Q \mid P \supseteq Q\}$, is a set of $\mathscr{M}$ and thus a subset of $\mathscr{C}(\beta)$ for some sufficiently large β. But this means $P \Vdash (x^\alpha)(\exists y^\beta)\mathscr{F}(x^\alpha, y^\beta)$.

LEMMA 1.3. *For each* $n \geq 0$: *There exists an* $\mathscr{M}$*-definable function* $\lambda\mathscr{F} \mid \mathscr{F}^*$. *defined for all sentences* $\mathscr{F}$ *of* $\mathscr{L}(\mathscr{T})$ *having at most n unranked quantifiers, such that*

$$(P)[P \Vdash \mathscr{F} \leftrightarrow P \Vdash \mathscr{F}^*]$$

and the ordinal rank of $\mathscr{F}^*$ *is less than* ω_2 (*in the sense of* $\mathscr{M}$).

PROOF. Standard. Let $K = \{P \mid P \Vdash \mathscr{F}\}$. By 1.2, K is a subset of 2^ω in $\mathscr{M}$. By Gödel [6], $K \in \mathscr{M}_\alpha$ for some $\alpha < \omega_2$, where ω_2 is the ω_2 of $\mathscr{M}$. Let $\mathscr{F}^*$ be $(\exists P)(P \in K \ \& \ \mathscr{T} \in P)$. Clearly the ordinal rank of $\mathscr{F}^* < \omega_2$. Suppose $P \Vdash \mathscr{F}^*$. Then for some Q, $P \Vdash Q \in K \ \& \ \mathscr{T} \in Q$. It follows $Q \in K$, since Q and K are members of $\mathscr{M}$ and no falsehood about $\mathscr{M}$ can ever be forced. $P \subseteq Q$, because $P \Vdash \mathscr{T} \in Q$ implies $(p)(p \in P \rightarrow p \in Q)$. But then $P \Vdash \mathscr{F}$.

If one forces with finite conditions in the manner of Cohen, then the ω_2 in Lemma 1.3 is replaced by ω_1. On the other hand it can be shown that if one is forcing with perfect closed conditions, then the ω_2 of Lemma 1.3 cannot be reduced to ω_1.

We turn now to the problem of showing for generic T that the cardinals of $\mathscr{M}(T)$ are the same as those of $\mathscr{M}$. In the case of Cohen-forcing with finite conditions [1] or Solovay-forcing with closed sets of positive measure [25], [26], the preservation of cardinals in generic extensions is a consequence of the countable chain condition: if J is a set of pairwise incomparable conditions (i.e., $P, Q \in J \ \& \ P \neq Q \rightarrow P \nsubseteq Q \ \& \ Q \nsubseteq P$), then J is countable. It is not difficult to see that the countable chain condition fails for perfect closed conditions. Nonetheless, cardinals are preserved, thanks to Lemma 1.4, the sequential lemma.[2]

LEMMA 1.4. *Let* $\{\mathscr{F}_i \mid i \in \omega\}$ *be a countable* (*in the sense of* $\mathscr{M}$) *set of sentences of* $\mathscr{L}(\mathscr{T})$. *Let P have the property that*

$$(i)(Q)_{P \supseteq Q}(\exists R)_{Q \supseteq R}[R \Vdash \mathscr{F}_i].$$

Then there exists an $R \subseteq P$ *such that for each i*,

$$R = \bigcup_{j<2^i} R^i_j, \quad \text{and} \quad R^i_j \mid \mathscr{F}_i$$

for all $j < 2^i$. *In addition*, $R^i_j \cap R^i_k = 0$ *if* $j \neq k$.

PROOF. Suppose $(i)(Q)_{P \supseteq Q}(\exists R)_{Q \supseteq R}[R \Vdash \mathscr{F}_i]$. By 1.2, there is an $\mathscr{M}$-definable function $f(i, Q)$ such that if $P \supseteq Q$, then $Q \supseteq f(i, Q)$ and $f(i, Q) \Vdash \mathscr{F}_i$. Let us say Q_1 and Q_2 are basic, disjoint perfect closed subsets of Q if there exist sequence numbers q_1 and $q_2 \in Q$ such that $q_1 \ngeq q_2$, $q_2 \ngeq q_1$, $Q_1 = \{p \mid p \in Q \ \& \ q_1 \geq p\}$, and $Q_2 = \{p \mid p \in P \ \& \ q_2 \geq p\}$. By iterating $f(i, Q)$, it is not difficult to

[2] Lemma 1.4 is called the sequential lemma in order to be consistent with [5], where a hyperarithmetic version (Lemma 2.2 of the present paper) of 1.4 occurs. Mathias [15] formulates 1.4 more abstractly and calls the result the fusion lemma. Shoenfield [24] invokes the term "splitting" to describe the proof of 1.4.

develop an $\mathcal{M}$-definable function $\lambda ij \mid Q_j^i$ with the following properties: $Q_0^0 \subseteq P$ and $Q_0^0 \Vdash \mathcal{F}_0$; for each i and $j < 2^i$, Q_j^{i+1} and $Q_{j+2^i}^{i+1}$ are basic, disjoint perfect closed subsets of Q_j^i, $Q_j^{i+1} \Vdash \mathcal{F}_{i+1}$, and $Q_{j+2^i}^{i+1} \Vdash \mathcal{F}_{i+1}$. Let $R = \bigcap_i \bigcup_{j<2^i} Q_j^i$, and let $R_j^i = R \cap Q_j^i$.

We say P weakly forces $\mathcal{F}$ ($P \Vdash^* \mathcal{F}$) if $P \Vdash \sim\sim\mathcal{F}$. An immediate corollary of Lemma 1.4 is

$$(i)(Q)_{P \supseteq Q}(\exists R)_{Q \supseteq R}[R \Vdash \mathcal{F}_i] \rightarrow (\exists R)_{P \supseteq R}(i)[R \Vdash^* \mathcal{F}_i].$$

Another way of expressing the message of 1.4 is: for all P and all sequences $\{\mathcal{F}_i \mid i \in \omega\}$ (in the sense of $\mathcal{M}$) of sentences of $\mathcal{L}(\mathcal{T})$,

$$(\exists R)_{P \supseteq R}(i)[R \Vdash^* \mathcal{F}_i \vee R \Vdash^* \sim\mathcal{F}_i].$$

Thus perfect closed forcing gives us the power to "decide" every one of a countable set of sentences by means of a single forcing condition. If we concentrate on sentences of countable ordinal rank, then we can do even more; this additional power is important in the hyperarithmetic case (§ 2), where the sentences of recursive ordinal rank behave very much like the sentences of countable ordinal rank in the set-theoretic case.

LEMMA 1.5. *Let $\mathcal{F}$ be a sentence of $\mathcal{L}(\mathcal{T})$ of countable (in the sense of $\mathcal{M}$) ordinal rank. Then for each P there is a $Q \subseteq P$ such that either* (1) *or* (2) *holds:*

(1) $(T)(T \in Q \rightarrow \mathcal{M}(T) \models \mathcal{F})$;

(2) $(T)(T \in Q \rightarrow \mathcal{M}(T) \models \sim\mathcal{F})$.

PROOF. By transfinite induction on the rank of $\mathcal{F}$. We give only the principal inductive step. Let $\mathcal{F}$ be $(\exists x^\alpha)\mathcal{F}(x^\alpha)$, where α is countable in the sense of $\mathcal{M}$. Let $\{c_i \mid i \in \omega\}$ be an enumeration of $\mathcal{C}(\alpha)$; $\mathcal{C}(\alpha)$ is countable (in the sense of $\mathcal{M}$) because α is. For each i, $\mathcal{F}(c_i)$ has lower rank than $(\exists x^\alpha)\mathcal{F}(x^\alpha)$. It follows from the inductive hypothesis that for each i and each P, there is a $Q \subseteq P$ such that either (1) or (2) holds:

(1) $(T)(T \in Q \rightarrow \mathcal{M}(T) \models \mathcal{F}(c_i))$;

(2) $(T)(T \in Q \rightarrow \mathcal{M}(T) \models \sim\mathcal{F}(c_i))$.

Fix P. If there is a $Q \subseteq P$ such that (1) holds, then the argument is complete; suppose there is no such Q. Then it follows that

$$(i)(Q)_{P \supseteq Q}(\exists R)_{Q \supseteq R}(T)[T \in R \rightarrow \mathcal{M}(T) \models \sim\mathcal{F}(c_i)].$$

To obtain a $Q \subseteq P$ such that $(R)[T \in Q \rightarrow \mathcal{M}(T) \models (x^\alpha) \sim \mathcal{F}(x^\alpha)]$, one repeats the construction used to prove Lemma 1.4 with one small change. The $\mathcal{M}$-definable function $f(i, Q)$ now has the property that if $P \supseteq Q$, then $Q \supseteq f(i, Q)$ and

$$(T)(T \in f(i, Q) \rightarrow \mathcal{M}(T) \models \sim\mathcal{F}(c_i)).$$

LEMMA 1.6. *Let* $\{\mathscr{F}_i\}$ *be a countable (in the sense of* $\mathscr{M}$*) set of sentences of* $\mathscr{L}(\mathscr{T})$ *of countable ordinal rank. Let P have the property that*

$$(i)(Q)_{P \supseteq R}(\exists R)_{Q \supseteq R}[R \Vdash \mathscr{F}_i].$$

Then there exists a $Q \subseteq P$ *such that*

$$(i)(T)[T \in Q \rightarrow \mathscr{M}(T) \models \mathscr{F}_i].$$

PROOF. We claim that P has the property that

$$(i)(Q)_{P \supseteq Q}(\exists R)_{Q \supseteq R}(T)[T \in R \rightarrow \mathscr{M}(T) \models \mathscr{F}_i].$$

Fix i and Q. Then there is an $R \subseteq Q$ such that $R \Vdash \mathscr{F}_i$. By Lemma 1.5, there is an $R_1 \subseteq R$ such that either

(1) $(T)(T \in R_1 \rightarrow \mathscr{M}(T) \models \mathscr{F}_i)$ or

(2) $(T)(T \in R_1 \rightarrow \mathscr{M}(T) \models \sim\mathscr{F}_i)$.

Since there is a generic $T \in R_1 \subseteq R$ and $R \Vdash \mathscr{F}_i$, it is clear that (1) holds. But the existence of the desired Q follows from our claim and a repetition of the argument occurring at the end of the proof of Lemma 1.5.

LEMMA 1.7. *If T is generic, then* $\mathscr{M}$ *and* $\mathscr{M}(T)$ *have the same cardinals.*

PROOF. Let α be an uncountable cardinal in the sense of $\mathscr{M}$, and let $\beta < \alpha$. Let $f \in \mathscr{C}$ denote a function which maps β into α. For each $\gamma < \beta$, let

$$k(\gamma) = \{\tau \mid \tau < \alpha \;\&\; (\exists P)(P \Vdash f(\gamma) = \tau)\}.$$

By Lemma 1.2, $k(\gamma) \in \mathscr{M}$; and the cardinality of $k(\gamma)$ is at most ω_1, since the set of all P's has cardinality equal to ω_1. If $\alpha > \omega_1$, then $K = \bigcup\{k(\gamma) \mid \gamma < \beta\}$ has cardinality less than α. If T is generic, then in $\mathscr{M}(T)$ the range of f is a subset of K.

Suppose $\alpha = \omega_1$ and $\beta = \omega_0$. Now we need the sequential lemma, 1.4. Suppose for the sake of a contradiction $P \Vdash f$ is a 1-1 map of ω_0 onto ω_1. It suffices to find an $R \subseteq P$ and a $\gamma < \omega_1$ such that $R \Vdash^*$ range of f is bounded above by γ. P has the property that

$$(i)(Q)_{P \supseteq Q}(\exists R)_{Q \supseteq R}[R \Vdash (\exists \tau)(f(i) = \tau < \omega_1)].$$

By Lemma 1.4 there is an $R \subseteq P$ such that for each i,

$$R = \bigcup_{j < 2^i} R_j^i, \quad \text{and} \quad R_j^i \Vdash f(i) = \tau_j^i < \omega_1$$

for all $j < 2^i$, in addition, $R_j^i \cap R_k^i = 0$ if $j \neq k$. Let $\gamma = \text{lub}\,\{\tau_j^i \mid i \in \omega \;\&\; j < 2^i\}$. Then $\gamma < \omega_1$ since $\lambda ij \mid \tau_j^i$ is $\mathscr{M}$-definable; and $R \Vdash^*$ range of f is bounded above by γ.

The next lemma lifts the heart of Spector's minimal degree argument **[28]** from recursion theory to set theory. Let $c \in \mathscr{C}$ denote a subset of ω. For each T, generic or not generic, let

$$c(T) = \{n \mid \mathscr{M}(T) \models \bar{n} \in c\}.$$

LEMMA 1.8. *Let $c \in \mathscr{C}$ denote a subset of ω. For each P there exists a $Q \subseteq P$ such that either* (1) *or* (2) *holds:*

(1) $(T)[T \in Q \text{ \& } T \text{ is generic} \rightarrow T \in \mathscr{M}(c(T))]$

(2) $(\exists b)(T)[T \in Q \text{ \& } T \text{ is generic} \rightarrow c(T) = b \in \mathscr{M}]$.

If in addition c has countable (in the sense of $\mathscr{M}$) ordinal rank, then either (1′) *or* (2′) *holds:*

(1′) $(T)[T \in Q \rightarrow T \in \mathscr{M}_{\omega_1}(c(T))]$;

(2′) $(\exists b)(T)[T \in Q \rightarrow c(T) = b \in \mathscr{M}]$.

PROOF. We continue to exploit the construction occurring in the proof of the sequential lemma, 1.4.

Case 1. $(Q)_{P \supseteq Q}(\exists Q_1)_{Q \supseteq Q_1}(\exists Q_2)_{Q \supseteq Q_2}(\exists n)[Q_1 \Vdash \bar{n} \in c \text{ \& } Q_2 \Vdash \bar{n} \notin c]$. By Lemma 1.2, Q_1, Q_2 and n can be regarded as $\mathscr{M}$-definable functions of Q. Iteration of these functions in the style of the proof of Lemma 1.4 leads to an $\mathscr{M}$-definable function $\lambda im \mid Q_i^m$ with the following properties: $Q_0^0 \subseteq P$; for each $m \geq 0$ and $i < 2^m$, Q_i^{m+1} and $Q_{i+2^m}^{m+1}$ are basic, disjoint perfect closed subsets of Q_i^m, and

$$(\exists n)[Q_i^{m+1} \Vdash \bar{n} \in c \text{ \& } Q_{i+2^m}^{m+1} \Vdash \bar{n} \notin c].$$

We define Q by

$$T \in Q \leftrightarrow (m)(\exists_1 i)_{i<2^m}(T \in Q_i^m).$$

We say that t puts T in Q if $(m)(T \in Q_{t(m)}^m)$. Each $T \in Q$ is put in Q by a unique t and is arithmetic in that t, $\lambda im \mid Q_i^m$. We claim Q satisfies condition (1) of the lemma. Fix a generic $T \in Q$. To see that $T \in \mathscr{M}(c(T))$, it is enough to see that the unique t which puts T in Q is a member of $\mathscr{M}(c(T))$. Consider the following definition of t:

$$t(0) = 0 \text{ \& } [t(m+1) = t(m) \vee t(m+1) = t(m) + 2^m];$$
$$t(m+1) = t(m) \leftrightarrow (\exists n)[Q_{t(m)}^{m+1} \Vdash \bar{n} \in c \text{ \& } Q_{t(m)+2^m}^{m+1} \Vdash \bar{n} \notin c \text{ \& } n \in c(T)].$$

Clearly, t is arithmetic in $c(T)$, $\lambda im \mid Q_i^m$, and the forcing relation $R \Vdash \mathscr{F}$ restricted to R's of the form Q_i^m and $\mathscr{F}$'s of the form $\bar{n} \in c$ and $\bar{n} \notin c$. Since $\lambda im \mid Q_i^m$ and $R \Vdash \mathscr{F}$ restricted both belong to $\mathscr{M} \subseteq \mathscr{M}(c(T))$, it follows that $T \in \mathscr{M}(c(T))$. We had to assume that T was generic in order to be sure that

$$T \in Q_{t(m)}^{m+1} \text{ \& } Q_{t(m)}^{m+1} \Vdash \bar{n} \in c \rightarrow n \in c(T).$$

The assumption of genericity can be dropped if c has countable ordinal rank. Because then Lemma 1.5 makes it possible to choose Q_i^{m+1} and $Q_{i+2^m}^{m+1}$ so that there exists an n such that for all T,

$$T \in Q_i^{m+1} \rightarrow \mathscr{M}(T) \models \bar{n} \in c,$$
$$T \in Q_{i+2^m}^{m+1} \rightarrow \mathscr{M}(T) \models \bar{n} \notin c.$$

Note that if X, $Y \in Q$ and $X \neq Y$, then $c(X) \neq c(Y)$.

Case 2. Case 1 fails; then there is an $R \subseteq P$ such that

$$(Q_1)_{R \supseteq Q_1}(Q_2)_{R \supseteq Q_2}(n) \sim [Q_1 \Vdash \bar{n} \notin c \;\&\; Q_2 \Vdash \bar{n} \in c].$$

For each $n \in \omega$, let

$$\mathscr{F}_n \text{ be } \begin{cases} \bar{n} \in c & \text{if } (\exists Q)_{R \supseteq Q}[Q \Vdash n \in c] \\ \bar{n} \notin c & \text{if } (\exists Q)_{R \supseteq Q}[Q \Vdash n \notin c]. \end{cases}$$

The defining property of R guarantees that $\mathscr{F}_n$ is well defined and that

$$(n)(Q_1)_{R \supseteq Q_1}(\exists Q_2)_{Q_1 \supseteq Q_2}[Q_2 \Vdash \mathscr{F}_n].$$

By Lemma 1.4 there is a $Q \subseteq R \subseteq P$ such that $(n)[Q \Vdash^* \mathscr{F}_n]$. If $T \in Q$ and T is generic, then

$$\mathscr{M}(T) \models \bar{n} \in c \leftrightarrow Q \Vdash^* \bar{n} \in c.$$

So Q satisfies condition (2) of the lemma, since $Q \Vdash^* \bar{n} \in c$ is $\mathscr{M}$-definable. Suppose c has countable ordinal rank. By Lemma 1.6 there is a $Q \subseteq R$ such that

$$(n)(T)[T \in Q \rightarrow \mathscr{M}(T) \models \mathscr{F}_n].$$

It follows that

$$(n)(T)[T \in Q \rightarrow (n \in c(T) \leftrightarrow \mathscr{F}_n \text{ is } \bar{n} \in c)].$$

So Q satisfies condition (2′) of the lemma, since $\lambda n \mid \mathscr{F}_n$ is $\mathscr{M}$-definable.

THEOREM 1.9. *If T is generic, then $\mathscr{M}(T)$ is a model of* ZF *which satisfies the following sentences: every constructible cardinal is a cardinal; there exist exactly two degrees of nonconstructibility.*

PROOF. We noted in the remarks subsequent to Lemma 1.2 that the replacement axiom holds in $\mathscr{M}(T)$. Clearly, the following also holds in $\mathscr{M}(T)$: there is a set $T \subseteq \omega$ such that every set is constructible from T. By Lemma 1.7 the cardinals of $\mathscr{M}(T)$ are the same as the cardinals of $\mathscr{M}$. But then the argument of the last chapter of Gödel [6] shows that the generalized continuum hypothesis holds in $\mathscr{M}(T)$. So the power set axiom must hold in $\mathscr{M}(T)$. (Of course one can show the power set axiom holds in $\mathscr{M}(T)$ without using 1.7 or the sequential lemma.)

By Proposition 1.1 T is not constructible in the sense of $\mathscr{M}(T)$.

Suppose $K \subseteq \omega$ and $K \in \mathscr{M}(T)$. Let $c \in \mathscr{C}$ define K: $\bar{n} \in K \leftrightarrow \mathscr{M}(T) \models \bar{n} \in c$. Consider the sentence $\mathscr{F}$:

$$\mathscr{T} \in \mathscr{M}(c(\mathscr{T})) \vee c(\mathscr{T}) \in \mathscr{M}.$$

By Lemma 1.8 there is no P that forces the negation of $\mathscr{F}$. Since T is generic, it follows that

$$T \in \mathscr{M}(c(T)) \quad \text{or} \quad c(T) \in \mathscr{M}.$$

In other words, $T \in \mathscr{M}(K)$ or $K \in \mathscr{M}$.

THEOREM 1.10. *If constructible ω_2 is countable, then there exists a minimal degree of nonconstructibility.*

PROOF. By Lemma 1.3 T is generic if and only if T is generic with respect to all sentences of ordinal rank less than the ω_2 of $\mathscr{M}$. So if constructible ω_2 is countable, then generic T's actually exist and are of minimal degree by Theorem 1.9.

It would be interesting to find some axiom of infinity which, when added to ZF, decides every elementary question about the partial ordering of degrees of nonconstructibility. R. Solovay suggested that a measurable cardinal might suffice for this purpose.

If T is Cohen-generic (i.e., generic in the sense of forcing with finite conditions), then it is easy to see that T does not have minimal degree in $\mathscr{M}(T)$ by looking at the even and odd parts of T. R. Solovay has shown: if T is Cohen-generic and $T' \in \mathscr{M}(T)$, then there exists a Cohen-generic T'' such that $\mathscr{M}(T') = \mathscr{M}(T'')$. It follows that minimal degrees of nonconstructibility do not occur in Cohen-generic extensions of $\mathscr{M}$.

Thus far we have been a little vague about how to encode a perfect closed $P \subseteq 2^\omega$ as a set of sequence numbers. Now we choose a particular method of encoding for the sake of the proof of Lemma 1.11, a transformation lemma associated with perfect closed forcing. As in § 0, $p, q, r, \ldots$ ambiguously denote sequence numbers and the finite, initial segments of characteristic functions represented by those sequence numbers. We encode P in the form of a function $\lambda ji \mid p_i^j$ with the following properties: p_0^0 is the null initial segment; for each $j \geq 0$ and $i < 2^j$, $p_i^j \geq p_i^{j+1}$, $p_i^j \geq p_{i+2^j}^{j+1}$, $p_i^{j+1} \ngeq p_{i+2^j}^{j+1}$ and $p_{i+2^j}^{j+1} \ngeq p_i^{j+1}$. Then

$$(T)[T \in P \leftrightarrow (j)(\exists_1 i)(i < 2^j \;\&\; T \in p_i^j)].$$

As in § 0, $T \in p$ means p is an initial segment of the characteristic function of T.

Suppose Q is similarly encoded by $\lambda ji \mid q_i^j$. Then we define:

$$h_{P,Q}(p_i^j) = q_i^j.$$

For each $T \in P$, let $h_{P,Q}(T)$ be the unique member of $\bigcap\{h_{P,Q}(p_i^j) \mid i < 2^j \;\&\; T \in p_i^j\}$. Then $h_{P,Q}$ is a 1-1 map of P onto Q. We will call $h_{P,Q}$ a *canonical homeomorphism of P onto Q*. If $R \subseteq P$, then $h_{P,Q}(R) = \{h_{P,Q}(T) \mid T \in R\}$ is a perfect closed subset of Q. The map $h_{P,Q}$ induces a map $h_{P,Q}: \mathscr{L}(\mathscr{T}) \to \mathscr{L}(\mathscr{T})$ as follows. Let $\mathscr{T} \in p$ denote a finite conjunction of sentences of the form $\bar{n} \in \mathscr{T}$ and $\bar{m} \notin \mathscr{T}$ with the defining property that

$$(T)(T \in p \leftrightarrow \mathscr{M}(T) \models \mathscr{T} \in p).$$

For each n, let $\{r_i^n\}$ be the finite set

$$\{p_i^{n+1} \mid i < 2^{n+1} \;\&\; (T)(T \in p_i^{n+1} \to n \in T)\},$$

and similarly, let $\{s_i^n\}$ be

$$\{p_i^{n+1} \mid i < 2^{n+1} \;\&\; (T)(T \in p_i^{n+1} \to n \notin T)\}.$$

Then for every $T \in P$:

$$n \in T \leftrightarrow (\exists i)(T \in r_i^n) \;\&\; (i)(T \notin s_i^n).$$

Let c be a number-theoretic term of $\mathscr{L}(\mathscr{T})$. Then $h_{P,Q}(c \in \mathscr{T})$ is defined to be

$$(\exists i)[\mathscr{T} \in h_{P,Q}(r_i^c)] \;\&\; (i)[\mathscr{T} \notin h_{P,Q}(s_i^c)];$$

if $\mathscr{F}$ is a formula of $\mathscr{L}(\mathscr{F})$, then $h_{P,Q}(\mathscr{F})$ is the result of replacing each occurrence of $c \in \mathscr{T}$ in $\mathscr{F}$ by $h_{P,Q}(c \in \mathscr{T})$; $h(d)$, where $d \in \mathscr{C}$, is defined similarly. If $h_{Q,P}(\mathscr{T}) \in \mathscr{C}(1)$, then the following transformation lemma (cf. Feferman **[2]**) can be established by induction on the rank of $\mathscr{F}$: for all $T \in P$, $\mathscr{M}(T) \models \mathscr{F}$ if and only if $\mathscr{M}(h_{P,Q}(T)) \models h_{P,Q}(\mathscr{F})$.

LEMMA 1.11. *Let h be a canonical homeomorphism of P_1 onto Q_1. Then for each $T \in P_1$, T is generic if and only if $h(T)$ is generic.*

PROOF. Let K be an $\mathscr{M}$-definable set of P's. We say K is *dense* if

$$(R)(\exists P)[R \supseteq P \;\&\; P \in K].$$

It is standard to observe that T is generic if and only if $T \in \bigcup\{P \mid P \in K\}$ for every $\mathscr{M}$-definable, dense K. Clearly, if K is $\mathscr{M}$-definable and dense, then there cannot be an R that forces

$$\mathscr{T} \notin \bigcup\{P \mid P \in K\};$$

and for each sentence $\mathscr{F}$ of $\mathscr{L}(\mathscr{T})$, the set

$$\{P \mid P \Vdash \mathscr{F} \quad \text{or} \quad P \Vdash \sim\!\mathscr{F}\}$$

is $\mathscr{M}$-definable and dense.

Fix P_1. We say K_1 is *dense in* P_1 if $(P)[P \in K_1 \rightarrow P \subseteq P_1]$ and

$$(R)(\exists P)[R \subseteq P_1 \rightarrow R \supseteq P \;\&\; P \in K_1].$$

The above standard observation can be relativized to P_1: $T \in P_1$ is generic if and only if $T \in \bigcup\{P \mid P \in K_1\}$ for every $\mathscr{M}$-definable K_1 dense in P_1. Suppose that $T \in P_1$ is generic, and that K_1 is dense in P_1. There cannot be an R such that $T \in R$ and

$$R \Vdash \mathscr{T} \in P_1 \quad \& \quad \mathscr{T} \notin \bigcup\{P \mid P \in K_1\}.$$

If there were such an R, then by Proposition 1.1 there would be an $R_1 \subseteq R \cap P_1$. But then the density of K_1 in P_1 would provide a $P \subseteq R$, such that

$$P \Vdash \mathscr{T} \in \bigcup\{P \mid P \in K_1\}.$$

Since h is a canonical homeomorphism of P_1 onto Q_1, $\bigcup\{P \mid P \in K_1\}$ is dense in P_1 if and only if $\bigcup\{h(P) \mid P \in K_1\}$ is dense in Q_1; furthermore, if K_2 is dense in Q_1, then $K_2 = \bigcup\{h(P) \mid P \in K_1\}$ for some K_1 dense in P_1.

THEOREM 1.12. *If T_1 is generic and $T_2 \in \mathscr{M}(T_1) - \mathscr{M}$, then T_2 is generic.*

PROOF. Since $\mathcal{M}$ and $\mathcal{M}(T_1)$ have the same cardinals, $T_2 = c(T_1) = \{n \mid \mathcal{M}(T_1) \Vdash \bar{n} \in c\}$ for some $c \in \mathcal{C}(\omega_1)$. By Lemma 1.8 there is a Q such that $T_1 \in Q$ and

$$(T)(T \in Q \rightarrow T \in \mathcal{M}_{\omega_1}(c(T)).$$

Recall the end of Case 1 of the proof of Lemma 1.8. Q has the property that

$$(X)(Y)[X, Y \in Q \ \& \ X \neq Y \rightarrow c(X) \neq c(Y)].$$

Let $R = \{c(T) \mid T \in Q]$; then c is a 1-1 map of Q onto R. We claim a small addition to Case 1 of Lemma 1.8 guarantees that c can be construed as a canonical homeomorphism of Q onto R. The addition is: for all m and $i < 2^{m+1}$, require that (i) or (ii) hold:

(i) $(T)(T \in Q_i^{m+1} \rightarrow \mathcal{M}(T) \Vdash \bar{m} \in c)$;
(ii) $(T)(T \in Q_i^{m+1} \rightarrow \mathcal{M}(T) \Vdash \bar{m} \notin c)$.

Then Lemma 1.11 implies that $T_2 = c(T_1)$ is generic.

For more information concerning the properties of $\mathcal{M}(T)$ for generic T see Prikry [**17**] and Mathias [**15**].

The minimality of the degree of a generic T was established in the proof of Theorem 1.9 by clauses (1) and (2) of Lemma 1.8. If one wishes to build a more complicated model of ZF in which the degrees of subsets of ω are controlled by perfect closed forcing, one cannot in general prove Lemma 1.8 (1) or Lemma 1.8 (2). In that emergency one falls back on clauses (1′) and (2′) of Lemma 1.8, which suffice if cardinals are preserved.

Since the relation A is constructible from B is analytical (in fact, Σ_2^1) in the sense of Kleene [**9**], it can be used to study the independence of various questions concerning the analytical hierarchy. A. Lévy has shown by means of his cardinal-collapsing method that if ZF is consistent, then ZF plus the negation of the axiom of choice for some Π_2^1 predicate is consistent. The axiom of choice for a predicate $P(n, X)$ is:

$$(n)(\exists X)P(n, X) \rightarrow (\exists X)(n)P(n, (X)_n),$$

where $(X)_n = \{m \mid 2^m \cdot 3^n \in X\}$. In Lévy's model, constructible ω_n is countable for all finite n. His result limiting the provable versions of the axiom of choice can be obtained by extending the arguments of this section to build a model $\mathcal{N}$ in which the degrees of nonconstructibility are isomorphic to ω and in which every constructible cardinal is a cardinal. Thus it is true in $\mathcal{N}$ that: for each $n > 0$, there is an X such that $(X)_0, (X)_1, \ldots, (X)_n$ have distinct degrees, but there is no X such that for all $n > 0$, $(X)_0, (X)_1, \ldots, (X)_n$ have distinct degrees.

S. Kripke and D. A. Martin carried on in the above vein and built a model of ZF in which a Δ_4^1 subset of ω failed to be constructible.[3] Mathias [**15**] gives an intuitive account of how this is done. The general idea is as follows. Choose an appropriate sequence $\langle \mathcal{S}_n \rangle$ of logically independent sentences from the theory of partial order. Than a set $A \subseteq \omega$ can be encoded as an analytical set in a model $\mathcal{N}$

[3] The question: is every analytical subset of ω constructible? was first suggested to us by G. Kreisel.

by taking steps so that $n \in A$ if and only if $\mathscr{S}_n$ is true in the partial ordering of degrees of nonconstructibility of $\mathscr{N}$. Jensen succeeded in using these ideas to obtain the best possible result: he lowered Δ_4^1 to Δ_3^1 and arranged for the axiom of choice to be true in his model. Jensen's result was first obtained by Solovay, who managed it without any use of degrees.

Silver has found another way of building a model in which minimal degrees of nonconstructibility occur. He makes use of a special kind of perfect closed condition he calls coinfinite. A coinfinite condition $P = (A, B)$ is given by a pair $A, B \subseteq \omega$ such that $A \cap B = 0$ and $\omega - (A \cup B)$ is infinite; we say $T \in P$ if $A \subseteq T$ and $T \cap B = 0$. The sequential lemma (1.4), mildly altered, holds for coinfinite conditions. (For more information about coinfinite conditions, see Mathias [**15**].) Coinfinite conditions were discussed but not used in Spector's paper on minimal Turing degrees [**28**]. They were used to prove Theorem 2 of [**21**].

2. **The hyperarithmetic case.** In the present section our purpose is to expound forcing with perfect closed sets in a context where the definability of the forcing relation is a more delicate matter than it is in the set-theoretic context of § 1. With this end in mind we review the minimal hyperdegree construction of [**5**] and add some intuitive remarks that were unfortunately left out of [**5**].

The language $\mathscr{L}(\mathscr{T})$ is little different from Feferman's language $\mathscr{L}^*(\mathscr{S})$ [**2**, p. 335]. $\mathscr{L}(\mathscr{T})$ is first order number theory augmented by a constant set symbol $\mathscr{T}$, certain set variables, and the membership symbol $\in$. Let O_1 be a Π_1^1 set of unique notations [**3**] for recursive ordinals: if $b \in O_1$ is the unique notation for β, then we write $|b| = \beta$. In addition we suppose there exists a recursively enumerable relation $R(c, b)$ such that for all $b \in O_1$, $\{c \mid R(c, b)\} = \{c \mid |c| < |b|\}$. For each $b \in O_1$, $\mathscr{L}(\mathscr{T})$ has ranked set variables $X^b, X^b, Z^b, \ldots$; $\mathscr{L}(\mathscr{T})$ also has unranked set variables $X, Y, Z, \ldots$, number variables $x, y, z, \ldots$, a numeral $\bar{n}$ for each natural number n, and symbols for equality, successor, addition and multiplication.

A formula $\mathscr{F}$ of $\mathscr{L}(\mathscr{T})$ is said to be *ranked* if every set variable occurring in $\mathscr{F}$ is ranked. A formula of $\mathscr{L}(\mathscr{T})$ is said to be existential (or Σ_1^1) if it is ranked or it is of the form $(\exists X)\mathscr{F}$ with X the only unranked variable occurring in $\mathscr{F}$. The *ordinal rank* of a ranked formula $\mathscr{F}$ is least ordinal α such that $\alpha \geq |b|$ for every free variable X^b of $\mathscr{F}$ and such that $\alpha > |b|$ for every bound variable X^b of $\mathscr{F}$.

Let T be an arbitrary subset of ω. Following Feferman [**2**], for each $b \in O_1$ we inductively define a structure $\mathscr{M}_b(T)$ and truth in $\bigcup\{\mathscr{M}_a(T) \mid |a| < |b|\}$.

(i) A sentence $\mathscr{F}$ of ordinal rank $\leq |b|$ is true in $\bigcup\{\mathscr{M}_a(T) \mid |a| < |b|\}$ if it is true when $\mathscr{T}$ is interpreted as T, the number variables of $\mathscr{F}$ are restricted to ω, and each ranked variable X^a of $\mathscr{F}$ is restricted to $\mathscr{M}_a(T)$.

(ii) For each formula $\mathscr{G}(x)$ (with only x free) of ordinal rank $\leq |b|$, let $\hat{x}\mathscr{G}(x)$ be $\{n \mid \mathscr{G}(\bar{n})$ is true in $\bigcup\{\mathscr{M}_a(T) \mid |a| < |b|\}\}$; then $\mathscr{M}_b(T)$ consists of all such sets $\hat{x}\mathscr{G}(x)$.

Let $\mathscr{M}(T)$ be $\bigcup\{\mathscr{M}_b(T) \mid b \in O_1\}$. Then $\mathscr{F}$ is true in $\mathscr{M}(T)$ (written $\mathscr{M}(T) \models \mathscr{F}$) if it is true when each unranked variable of $\mathscr{F}$ is restricted to $\mathscr{M}(T)$ and the remaining symbols of $\mathscr{F}$ are interpreted as in (i) above. If T is hyperarithmetic, then $\mathscr{M}(T)$ is the set of all hyperarithmetic sets, which we denote by $\mathscr{M}$.

$\mathscr{F}(\hat{x}\mathscr{G}(x))$ denotes the result of replacing each occurrence of $t \in \mathscr{U}$ in $\mathscr{F}(\mathscr{U})$ by $\mathscr{G}(t)$, where t is a number-theoretic term and $\mathscr{U}$ is a set variable.

As in § 0 we use $P, Q, R, \ldots$ to ambiguously denote perfect closed subsets of 2^ω and their encodings as subsets of ω. But now we insist that $P, Q, R, \ldots$, be hyperarithmetic. Thus we can write, as we did in § 1, $P \in \mathscr{M}$. It is routine to assign indices to hyperarithmetic, perfect closed sets so that the following relation in x and Y is Π_1^1: x is the index of a hyperarithmetic, perfect closed set P and $Y \in P$. For this reason it makes sense to regard the set of all hyperarithmetic, perfect closed (h.p.c.) sets as a Π_1^1 subset of ω. Similarly, one regards the set of all formulas of $\mathscr{L}(\mathscr{T})$ as a Π_1^1 subset of ω.

It is instructive to see what happens when the analogy between the constructible hierarchy of § 1 and the hyperarithmetic hierarchy of the present section is pursued. In both cases a perfect closed set is acceptable as a forcing condition only if it can be encoded by a subset of ω belonging to $\mathscr{M}$. The analogy suggests that the forcing relation $P \Vdash \mathscr{F}$, where P is h.p.c. set and $\mathscr{F}$ is a sentence of $\mathscr{L}(\mathscr{T})$, should be defined by the following inductive clauses:

(1) $P \Vdash (\exists X)\mathscr{F}(X)$ if $(\exists b)_{b \in O_1}[P \Vdash (\exists X^b)\mathscr{F}(X^b)]$.

(2a) $P \Vdash (\exists X^b)\mathscr{F}(X^b)$ if $P \Vdash \mathscr{F}(\hat{x}\mathscr{G}(x))$ for some $\mathscr{G}(x)$ of ordinal rank $\leq |b|$.

(2b) $P \Vdash (\exists x)\mathscr{F}(x)$ if $P \Vdash \mathscr{F}(\bar{n})$ for some $n \in \omega$.

(3) $P \Vdash \mathscr{F} \vee \mathscr{G}$ if $P \Vdash \mathscr{F}$ or if $P \Vdash \mathscr{G}$.

(4) $P \Vdash \sim\mathscr{F}$ if $(Q)_{P \supseteq Q} \sim [Q \Vdash \mathscr{F}]$.

(5) $P \Vdash \bar{n} \in \mathscr{T}$ if $(T)[T \in P \rightarrow \mathscr{M}(T) \models \bar{n} \in \mathscr{T}]$.

(6) $P \Vdash \mathscr{F}$ if $\mathscr{F}$ is a true, quantifier-free sentence of arithmetic.

The key property of the forcing relation in § 1 is its $\mathscr{M}$-definability as expressed by Lemma 1.2. The analogous fact for the present section (Lemma 2.1) is: the forcing relation restricted to Σ_1^1 $\mathscr{F}$'s is Π_1^1. As we shall see, Lemma 2.1 suffices to establish the sequential lemma and the hyperarithmetic comprehension axiom in $\mathscr{M}(T)$ for all generic T. (To see the analogy between Lemmas 1.2 and 2.1, recall that a Π_1^1-relation becomes Σ_1^1 [**4**, 29] when the universe is restricted to $\mathscr{M}$ = the hyperarithmetic sets. Thus 2.1 says that the forcing relation restricted to Σ_1^1 $\mathscr{F}$'s is $\mathscr{M}$-definable by means of a Σ_1^1 formula.) But we are unable to show that the forcing relation $\Vdash$, defined by (1)—(6) above and restricted to Σ_1^1 sentences, is Π_1^1. The difficulty is in clause (4): even for a ranked $\mathscr{F}$, the definition of $P \Vdash \sim\mathscr{F}$ involves quantification over the set of all Q's $\subseteq P$, that is to say, quantification over a Π_1^1 set of indices. We would be very interested to know the precise classification of $\Vdash$ restricted to Σ_1^1 $\mathscr{F}$'s. A more intuitive way of stating the difficulty is: the collection of all P's is a class of $\mathscr{M}$ but not a set of $\mathscr{M}$. If we restricted P to be recursive in some fixed hyperarithmetic set H, then the difficulty would vanish, but a generic set would not in general be of minimal hyperdegree, and the sequential lemma would hold only for sentences below a certain ordinal rank determined by H.

So we drop the analogy and define $\Vdash_h$ in the manner of [**5**].

(i) $P \Vdash_h (\exists X)\mathscr{F}(X)$ if $(\exists b)_{b \in O_1}[P \Vdash_h (\exists X^b)\mathscr{F}(X^b)]$.

(iia) $P \Vdash_h (\exists X^b)\mathscr{F}(X^b)$ if $\mathscr{F}(X^b)$ is unranked and $P \Vdash_h \mathscr{F}(\hat{x}\mathscr{G}(x)$ for some $\mathscr{G}(x)$ of ordinal rank $\leq |b|$.

(iib) $P \Vdash_h (\exists x)\mathscr{F}(x)$ if $\mathscr{F}(x)$ is unranked and $P \Vdash_h \mathscr{F}(\bar{n})$ for some $n \in \omega$.

(iii) $P \Vdash_h \mathscr{F} \vee \mathscr{G}$ if $\mathscr{F} \vee \mathscr{G}$ is unranked and $P \Vdash_h \mathscr{F}$ or $P \Vdash_h \mathscr{G}$.

(iv) $P \Vdash_h \sim\mathscr{F}$ if $\mathscr{F}$ is unranked and $(Q)_{P \supseteq Q} \sim [Q \Vdash_h \mathscr{F}]$.

(v) $P \Vdash_h \mathscr{F}$ if $\mathscr{F}$ is ranked and $(T)[T \in P \rightarrow \mathscr{M}(T) \models \mathscr{F}]$.

LEMMA 2.1 [**5**]. *The relation* $P \Vdash_h \mathscr{F}$, *restricted to* Σ_1^1 $\mathscr{F}$*'s of* $\mathscr{L}(\mathscr{T})$, *is* Π_1^1.

The proof of Lemma 2.1 is nothing more than inspection of clauses (v) and (i) of the definition of $\Vdash_h$. But we must pay a price, fortunately within our means, for the use of clause (v). It is no longer obvious as it was in § 1 that

$$(\mathscr{F})(P)(\exists Q)_{P \supseteq Q}[Q \Vdash_h \mathscr{F} \text{ or } Q \Vdash_h \sim\mathscr{F}],$$

because clause (iv) does not apply to ranked sentences. But this difficulty can be overcome by a sequential lemma (2.2) argument.

LEMMA 2.2 [**5**]. *Let* $\{\mathscr{F}_i\}$ *be a hyperarithmetic sequence of* Σ_1^1 *sentences of* $\mathscr{L}(\mathscr{T})$. *Let P have the property that*

$$(i)(Q)_{P \supseteq Q}(\exists R)_{Q \supseteq R}[R \Vdash_h \mathscr{F}_i].$$

Then there exists a $Q \subseteq P$ *such that*

$$(i)[Q \Vdash_h \mathscr{F}_i].$$

Note that Lemma 2.2 is closer in flavor to Lemma 1.6 than to Lemma 1.4. But the proof of 2.2 is essentially the same as that of 1.4. The only real difference is that the $\mathscr{M}$-definable function $f(i, Q)$ of 1.4 becomes a partial Π_1^1 function defined for all i and all hyperarithmetic $Q \subseteq P$. (A function is called partial Π_1^1 if its graph is Π_1^1.) The existence of f follows from Lemma 2.1 and an important lemma of Kreisel [**14**] which says: if $P(x, y)$ is Π_1^1, then there exists a partial Π_1^1 function g such that

$$(x)[(\exists y)P(x, y) \rightarrow g(x) \text{ is defined } \& \ P(x, g(x))].^4$$

The desired $Q \subseteq P$ is built, as it was in Lemma 1.4, by iterating f. To see that Q is hyperarithmetic one has to use the fact that a partial Π_1^1 function, restricted to a hyperarithmetic subset of its domain, is hyperarithmetic.

LEMMA 2.3 [**5**]. $(\mathscr{F})(P)(\exists Q)_{P \supseteq Q}[Q \Vdash_h \mathscr{F} \vee Q \Vdash_h \sim\mathscr{F}]$.

The proof of Lemma 2.3 is very much like that of Lemma 1.5 and proceeds by transfinite induction on the rank, appropriately defined, of $\mathscr{F}$. The only interesting

---

[4] Kreisel's lemma generalizes a familiar lemma of recursion theory which says the same thing with Π_1^1 replaced by recursively enumerable.

inductive step occurs when $\mathscr{F}$ is a ranked sentence of the form $(\exists X^b)\mathscr{F}(X^b)$. If there is a $\mathscr{G}(x)$ of ordinal rank $\leq |b|$ and a $Q \subseteq P$ such that $Q \Vdash_h \mathscr{F}(\hat{x}\mathscr{G}(x))$, then $Q \Vdash_h (\exists X^b)\mathscr{F}(X^b)$. Suppose there is no such $\mathscr{G}(x)$ and Q. Let $\{\mathscr{G}_i(x)\}$ be a hyperarithmetic enumeration of all $\mathscr{G}(x)$'s of ordinal rank $\leq |b|$. Then the induction hypothesis implies

$$(i)(Q)_{P \supseteq Q}(\exists R)_{Q \supseteq R}[R \Vdash_h \sim\mathscr{F}(\hat{x}\mathscr{G}_i(x))].$$

Now the sequential lemma (2.2) can be applied, as it was in Lemma 1.5, to obtain a $Q \subseteq P$ such that $Q \Vdash (X^b) \sim \mathscr{F}(X^b)$.

T is said to be *generic* if for every sentence $\mathscr{F}$ of $\mathscr{L}(\mathscr{T})$ there is a P such that $T \in P$ and either $P \Vdash_h \mathscr{F}$ or $P \Vdash_h \sim\mathscr{F}$. The existence of generic T's follows from Lemma 2.3. The minimal degree argument of Lemma 1.8 is easily modified to prove the next lemma.

LEMMA 2.4 [**5**]. *Let $\mathscr{G}(x)$ be a ranked formula of $\mathscr{L}(\mathscr{T})$ with only x free. For each P there exists a $Q \subseteq P$ such that either* (1) *or* (2) *holds:*

(1) $(T)[T \in Q \rightarrow T$ *is hyperarithmetic in* $\{n \mid \mathscr{M}(T) \models \mathscr{G}(\bar{n})\}]$;

(2) $(\exists H)(T)[T \in Q \rightarrow \{n \mid \mathscr{M}(T) \models \mathscr{G}(\bar{n})\} = H$ *is hyperarithmetic*$]$.

LEMMA 2.5 [**5**]. *If T is generic, then every set hyperarithmetic in T belongs to $\mathscr{M}(T)$.*

Before we discuss the proof of Lemma 2.5, we make some related observations without proof. For every T it is the case that every member of $\mathscr{M}(T)$ is hyperarithmetic in T. If T is generic, then $\omega_1^T = \omega_1$, where ω_1^T is the least ordinal not recursive in T.

According to Kreisel [**13**] the conclusion of Lemma 2.5 is equivalent to the statement that the hyperarithmetic comprehension axiom (h.c.a.) holds in $\mathscr{M}(T)$. A typical instance of the h.c.a. is the universal closure of a formula of the form

$$(x)[(\exists Y)\mathscr{A}(x, Y) \leftrightarrow (Z)\mathscr{B}(x, Z)] \rightarrow (\exists X)(x)[x \in X \leftrightarrow (\exists Y)\mathscr{A}(x, Y)],$$

where $\mathscr{A}(x, Y)$ and $\mathscr{B}(x, Z)$ are arithmetic formulas which may contain free set variables other than Y or Z. By an argument of Feferman [**2**, p. 339] which holds for *all* T, the hyperarithmetic comprehension axiom is true in $\mathscr{M}(T)$ if the following Σ_1^1-bounding principle is true in $\mathscr{M}(T)$: let $\mathscr{F}(x, Y)$ be a formula of $\mathscr{L}(\mathscr{T})$ whose only free variables are x and Y and whose only unranked variable is Y; if $(x)(\exists Y)\mathscr{F}(x, Y)$, then $(x)(\exists Y^b)\mathscr{F}(x, Y^b)$ for some $b \in O_1$. The principle holds in $\mathscr{M}(T)$ for all generic T by Lemmas 2.6 and 2.7.

LEMMA 2.6. *Let $\mathscr{F}(x, Y)$ be a formula of $\mathscr{L}(\mathscr{T})$ whose only free variables are x and Y and whose only unranked variable is Y. If $P \Vdash_h (x)(\exists Y)\mathscr{F}(x, Y)$, then $Q \Vdash_h (x)(\exists Y^b)\mathscr{F}(x, Y^b)$ for some $Q \subseteq P$ and $b \in O_1$.*

The proof of Lemma 2.6 is another application of the sequential lemma (2.2). Suppose $P \Vdash_h (x)(\exists Y)\mathscr{F}(x, Y)$, where $\mathscr{F}(x, Y)$ is as in Lemma 2.6. Then

$$(n)(Q)_{P \supseteq Q}(\exists R)_{Q \supseteq R}[R \Vdash_h (\exists Y)\mathscr{F}(\bar{n}, Y)].$$

By Lemma 2.2 there is a $Q \subseteq P$ such that

$$(n)(\exists b)[b \in O_1 \ \& \ Q \Vdash_h (\exists Y^b)\mathscr{F}(\bar{n}, Y)].$$

By Lemma 2.1 and Kreisel's lemma (as stated just before 2.2), there is a hyperarithmetic function f such that

$$(n)[f(n) \in O_1 \ \& \ Q \Vdash_h (\exists Y^{f(n)})\mathscr{F}(n, Y^{f(n)})].$$

By Spector [27] there is a $b \in O_1$ such that $(n)(|f(n)| \leq |b|)$. So

$$Q \Vdash_h (x)(\exists Y^b)\mathscr{F}(x, Y^b).$$

Lemma 2.6 almost establishes that the Σ_1^1-bounding principle holds in $\mathscr{M}(T)$ for all generic T. What is lacking are some technical details concerning definability that are immediate in the set-theoretic context of § 1 but that require proof in the hyperarithmetic case.

Let A be a set of indices of hyperarithmetic perfect closed sets. We write $P \in A$ to mean that the index of the perfect closed set P, encoded as a hyperarithmetic set of sequence numbers, belongs to A. A is said to be *dense* if $(P)(\exists Q)[Q \in P \ \& \ Q \subseteq P]$.

A set A is said to be $\mathscr{M}$-definable if there is a formula $\mathscr{F}(x)$ of $\mathscr{L}(\mathscr{T})$ in which $\mathscr{T}$ does not occur such that $A = \{n \mid \mathscr{M} \models \mathscr{F}(\bar{n})\}$. For each T: a set A is said to be $\mathscr{M}(T)$-definable if there is a formula $\mathscr{F}(x)$ of $\mathscr{L}(\mathscr{T})$ such that

$$A = \{n \mid \mathscr{M}(T) \models \mathscr{F}(\bar{n})\}.$$

A is called *generically persistent*[5] if there is a formula $\mathscr{F}(x)$ of $\mathscr{L}(\mathscr{T})$ such that for all generic T, $A = \{n \mid \mathscr{M}(T) \quad \mathscr{F}(\bar{n})\}$.

LEMMA 2.7. *T is generic if and only if $T \in \bigcup\{P \mid P \in A\}$ for every dense A arithmetic in Kleene's O.*

LEMMA 2.8. *A is arithmetic in Kleene's O if and only if A is $\mathscr{M}$-definable if and only if A is generically persistent.*

Let T be generic; we use Lemmas 2.6 and 2.7 to show the Σ_1^1-bounding principle holds in $\mathscr{M}(T)$. Let $\mathscr{F}(x, Y)$ be as in the hypothesis of 2.6. Let A be the set of all Q's such that for some $b \in O_1$,

$$Q \Vdash_h (x)(\exists Y)\mathscr{F}(x, Y) \rightarrow (x)(\exists Y^b)\mathscr{F}(x, Y^b).$$

A is arithmetic in Kleene's O, since both O_1 and the indexing of all P's is Π_1^1. A is dense by 2.6. Then 2.7 provides a $Q \in A$ such that $T \in Q$.

To complete the proof of Lemma 2.5 we need only prove Lemmas 2.7 and 2.8.

[5] Persistence is a term used by H. Putnam [18] to describe a well-known property of the hyperarithmetic sets. Kreisel [13] formulates the persistence property as follows: a set is hyperarithmetic if and only if it belongs to every ω-model of the hyperarithmetic comprehension axiom. Another way of putting it is: a set A is hyperarithmetic if and only if there is a ranked formula $\mathscr{G}(x)$ of $\mathscr{L}(\mathscr{T})$ such that for all T, $A = \{n \mid \mathscr{M}(T) \models \mathscr{F}(\bar{n})\}$. This formulation of persistence is a consequence of a slight modification of Kleene [10, p. 35].

Suppose A is dense and arithmetic in Kleene's O. By 2.8 A is generically persistent. Let $\mathscr{A}(x)$ be a formula such that for *all* generic T, $A = \{n \mid \mathscr{M}(T) \models \mathscr{A}(\bar{n})\}$. We claim there is an $\mathscr{F}$ such that for all generic T,

$$\mathscr{M}(T) \models \mathscr{F} \leftrightarrow T \in \bigcup\{Q \mid Q \in A\}.$$

The desired $\mathscr{F}$ is $(\exists x)[\mathscr{A}(x) \; \& \; (Y)(\mathscr{B}(x, Y) \rightarrow \mathscr{D}(Y, \mathscr{T}))]$, where $\mathscr{B}(x, Y)$ and $\mathscr{D}(Y, Z)$ are arithmetic formulas with the following properties. Let I be the Π^1_1 set of indices of hyperarithmetic perfect closed sets. If $n \in I$, let P_n be the hyperarithmetic perfect closed set whose index is n. For each $n \in I$, $(\exists_1 Y)\mathscr{B}(n, Y)$ and $(Y)[\mathscr{B}(n, Y) \rightarrow (Y)_0 = P_n]$.[6] $\mathscr{D}(P, T)$ says $T \in P$. To see that every generic $T \in \bigcup\{Q \mid Q \in A\}$, it suffices to see there is no P such that $P \Vdash_h \sim\mathscr{F}$. If there were such a P, then by the density of A there would be a $Q \subseteq P$ such that $Q \in A$ and $Q \Vdash_h \sim\sim\mathscr{F}$.

Suppose T is arbitrary but that $T \in \bigcup\{Q \mid Q \in A\}$ for every dense A arithmetic in Kleene's O. T is generic if $T \in \bigcup\{Q \mid Q \Vdash_h \mathscr{F}$ or $Q \Vdash_h \sim\mathscr{F}\} = K_{\mathscr{F}}$ for every sentence $\mathscr{F}$ of $\mathscr{L}(\mathscr{T})$. But it is easy to check that $K_{\mathscr{F}}$ is dense and arithmetic in Kleene's O.

We turn to Lemma 2.8. It is enough to show: (i) O is $\mathscr{M}$-definable; (ii) O is generically persistent; (iii) if A is generically persistent, then A is arithmetic in O. We dispose immediately of (iii). Suppose $A = \{n \mid \mathscr{M}(T) \models \mathscr{A}(\bar{n})\}$ for all generic T. Then $n \in A \leftrightarrow (\exists P)(P \Vdash_h \mathscr{A}(\bar{n}))$. A is arithmetic in O since $P \Vdash_h \mathscr{F}$, restricted to $\mathscr{F}$'s of the form $\mathscr{A}(\bar{n})$, is arithmetic in O. Mostowski's conjecture [**4**] provides an arithmetic formula $C(X, n)$ such that for all n, $n \in O \leftrightarrow (\exists X)[X \in \mathrm{HYP} \; \& \; C(X, n)]$. Thus (i) O is $\mathscr{M}$-definable by a Σ^1_1 formula. O will then be generically persistent if every arithmetic formula $C(X)$ and every generic T are such that if $\mathscr{M}(T) \models (\exists X)C(X)$, then $(\exists X)[X \in \mathrm{HYP} \; \& \; C(X)]$. Suppose $P \Vdash_h (\exists X)C(X)$. Then $P \Vdash_h C(\hat{x}\mathscr{G}(x))$ for some ranked $\mathscr{G}(x)$. Let $H \in P$ be hyperarithmetic. Then $J = \{n \mid \mathscr{M}(H) \models \mathscr{G}(\bar{n})\}$ is hyperarithmetic and $C(J)$.

THEOREM 2.9 [**5**]. *If T is generic, then T has minimal hyperdegree.*

If T is generic, then T is not hyperarithmetic by the same mode of argument employed in Proposition 1.1 to show that a set generic in the sense of § 1 is not a constructible set of $\mathscr{M}$. The fact that T has minimal hyperdegree follows from Lemmas 2.4, 2.5 and 2.7.

THEOREM 2.10. *If T is generic, then every set of the same hyperdegree as T is generic.*

Let S have the same hyperdegree as some generic T. Then

$$S = \{n \mid \mathscr{M}(T) \models \mathscr{G}(\bar{n})\},$$

where $\mathscr{G}(x)$ is a ranked formula. By Lemmas 2.4 and 2.7 there is a Q such that $T \in Q$ and Q satisfies clause (1) of 2.4. Then the argument of Theorem 1.12,

[6] The existence of $\mathscr{B}$ follows from the uniform implicit arithmetic definability of hyperarithmetic sets of arbitrarily high Turing degree, cf. Kreisel [**7**, p. 307].

routinely modified, establishes that S is generic. The modification includes Lemmas 2.7 and 2.8.

3. **The arithmetic case.** In accord with § 0, P, Q, R, ... ambiguously denote arithmetic perfect closed subsets of 2^ω and their encodings as arithmetic subsets of ω. One of the principal objectives of the present section is to introduce pointed, perfect closed sets and to apply them to a problem concerning the Turing degrees of certain hyperarithmetic sets and in particular the truth set for arithmetic.

The arithmetic language $\mathscr{L}(\mathscr{T})$ is the language of first order number theory together with a constant set symbol $\mathscr{T}$ and a predicate letter $\mathscr{R}$ denoting a certain recursive predicate. $\mathscr{L}(\mathscr{T})$ has symbols for membership ($\in$), addition ($+$), multiplication ($\cdot$), and successor ($'$), and a numeral $\bar{n}$ for each natural number n. $\mathscr{L}(\mathscr{T})$ is not essentially different from Feferman's language $L^*(S)$ [**2**, p. 328]. The predicate letter $\mathscr{R}$ is associated with the jump operator [**19**] for Turing degrees. By the enumeration theorem there is a recursive predicate $\mathscr{R}_2(T, x_1, x_2, y)$ such that for each recursive predicate $\mathscr{P}(T, x_2, y)$ there is an e with the property that

$$(T)(x_2)[(\exists y)\mathscr{R}_2(T, e, x_2, y) \leftrightarrow (\exists y)\mathscr{P}(T, x_2, y)].$$

The *jump* of T, denoted by T', can be defined by

$$2^m \cdot 3^n \in T' \leftrightarrow (\exists y)\mathscr{R}_2(T, m, n, y)$$

with the understanding that every member of T' is of the form $2^m \cdot 3^n$. T' can be thought of as the disjoint recursive union of all sets recursively enumerable in T. $T^{(n)}$, the n-jump of T, is defined by: $T^{(0)} = T$; $T^{(n+1)} = (T^{(n)})'$. 0 is the empty set. A set is arithmetic if and only if it is recursive in $0^{(n)}$ for some n. $T^{(\omega)}$, the ω-jump of T, is the disjoint recursive union of all n-jumps of T:

$$2^m \cdot 3^n \in T^{(\omega)} \leftrightarrow m \in T^{(n)}.$$

We call $0^{(\omega)}$ the truth set for arithmetic.

It is essential to note the existence of a recursive predicate $\mathscr{R}(T, n, x, y)$ such that for all T and n:

$$n \in T^{(2)} \leftrightarrow (\exists x)(y)\mathscr{R}(T, n, x, y).$$

The language $\mathscr{L}(\mathscr{T})$ includes $\mathscr{R}$; thus the 2-jump of T is definable by a Σ_2 formula of $\mathscr{L}(\mathscr{T})$. It is convenient to assume that the truth or falsity of $\mathscr{R}(T,n,x,y)$ for each choice of T, n, x and y is determined by an initial segment of the characteristic function of T of length at most y.

Let $\mathscr{F}$ be a sentence of $\mathscr{L}(\mathscr{T})$, and let T be an arbitrary subset of ω. We say T makes $\mathscr{F}$ true (or more briefly, $\mathscr{F}$ is true) when $\mathscr{F}$ is true subject to the interpretation of $\mathscr{T}$ as T and to the standard interpretation of the arithmetic symbols and $\mathscr{R}$. The forcing relation, $P \Vdash \mathscr{F}$, where $\mathscr{F}$ is a sentence of $\mathscr{L}(\mathscr{T})$ and P is an arithmetic perfect closed set, is defined by: $P \Vdash \mathscr{F} \leftrightarrow (T)[T \in P \rightarrow \mathscr{F}]$. Since we require every $T \in P$ to make $\mathscr{F}$ true, it is perhaps not appropriate to say P forces $\mathscr{F}$. On the other hand, if forcing is thought of as an approximation of truth, it seems reasonable to expect perfect closed forcing to provide a perfect approximation of truth.

We say T is *generic* if for each $\mathscr{F}$ there is a P such that $T \in P$ and either $P \Vdash \mathscr{F}$ or $P \Vdash \sim\mathscr{F}$. The existence of generic T's is an immediate consequence of Lemma 3.1. We could prove 3.1 by the sequential lemma approach of § 1 and 2, but it is more instructive to invoke what might be termed the *local forcing* approach. Let $p, q, r, \ldots$ be finite, initial segments of characteristic functions of subsets of ω. ($p \geq q$ and $T \in p$ are defined in § 0.) The definition of $p \Vdash \mathscr{F}$, where $\mathscr{F}$ is a sentence of $\mathscr{L}(\mathscr{T})$, is given by four inductive clauses:

(i) $p \Vdash \mathscr{F}$ if $\mathscr{F}$ is a true equation of arithmetic, or if $\mathscr{F}$ is of the form $\bar{n} \in \mathscr{T}$ or $\mathscr{R}(\mathscr{T}, \bar{m}, \bar{n}, \bar{c})$ and every $T \in p$ makes $\mathscr{F}$ true.

(ii) $p \Vdash (\exists x)\mathscr{F}(x)$ if $(\exists n)[p \Vdash \mathscr{F}(\bar{n})]$.

(iii) $p \Vdash \sim\mathscr{G}$ if $(q)_{p \geq q} \sim [q \Vdash \mathscr{G}]$.

(iv) $p \Vdash \mathscr{F}_1 \,\&\, \mathscr{F}_2$ if $p \Vdash \mathscr{F}_1$ and $p \Vdash \mathscr{F}_2$.

In short, $p \Vdash \mathscr{F}$ follows the rules of Cohen's forcing with finite conditions as formulated by Feferman [**2**] in the arithmetic case. Let P be an arithmetic perfect closed set. The forcing relation $p \Vdash^P \mathscr{F}$, where $p \in P$ and $\mathscr{F}$ is a sentence of $\mathscr{L}(\mathscr{T})$, is defined exactly as $p \Vdash \mathscr{F}$ was defined above except that clause (iii) becomes:

$$(\text{iii})^P \; p \Vdash^P \sim\mathscr{G} \quad \text{if} \quad (q)_{p \geq q}[q \in P \rightarrow \sim q \Vdash^P \mathscr{G}].$$

We call $p \Vdash^P \mathscr{F}$ the *localization* of $p \Vdash \mathscr{F}$ to P. The idea of local forcing was used by Spector [**28**] in his construction of a minimal Turing degree. Spector's fundamental lemma [**28**, p. 588] was in essence the following: let $\mathscr{F}$ be a Σ_2 sentence of $\mathscr{L}(\mathscr{T})$, and let P be a recursive perfect closed set; then there exists a recursive perfect closed $Q \subseteq P$ such that either $(T)(T \in Q \rightarrow \mathscr{F})$ or $(T)(T \in Q \rightarrow \sim\mathscr{F})$. The details of Spector's proof coincide with the details of the local forcing argument given in Lemma 3.1 for the case $\mathscr{F} \in \Sigma_2$. Spector uses his lemma to construct a sequence $P_0 \supseteq P_1 \supseteq P_2 \supseteq \cdots$ of recursive perfect closed sets with the following properties: for all $T \in P_e$, if $\{e\}^T$ is total, then $\{e\}^T$ is recursive or T is recursive in $\{e\}^T$ ($\{e\}^T$ is the eth function partial recursive in T according to the standard enumeration); for all $T \in P_e$, T is not the eth partial recursive function. Thus if $T \in \bigcap\{P_e \mid e \geq 0\}$, it is clear that T has minimal Turing degree.

The properties of $\Vdash^P$ are virtually the same as those of $\Vdash$; intuitively, this is the case because P and 2^ω are homeomorphic by means of an arithmetic map. T is *generic for* P if for each sentence $\mathscr{F}$ of $\mathscr{L}(\mathscr{T})$ there is a $p \in P$ such that $T \in p$ and either $p \Vdash^P \mathscr{F}$ or $p \Vdash^P \sim\mathscr{F}$. If T is generic for P, then $T \in P$, and T makes $\mathscr{F}$ true if and only if there is a $p \in P$ such that $T \in p$ and $p \Vdash^P \mathscr{F}$.

LEMMA 3.1. *Let P be an arithmetic perfect closed set, and let $\mathscr{F}$ be a sentence of the arithmetic language $\mathscr{L}(\mathscr{T})$. Then there exists an arithmetic perfect closed $Q \subseteq P$ such that either* (a) *or* (b) *holds:*

(a) $(T)(T \in Q \rightarrow \mathscr{F})$;

(b) $(T)(T \in Q \rightarrow \sim\mathscr{T})$.

Furthermore, if $\mathscr{F}$ is Σ_2, then Q is recursive in P.

PROOF. There must be a $p \in P$ such that either (a) $p \Vdash^P \mathscr{F}$ or (b) $p \Vdash^P \sim\mathscr{F}$. Suppose (a). We develop a $Q \subseteq P$ satisfying clause (a) of the lemma. Let $\{\mathscr{F}_m \mid m \geq 0\}$ be an enumeration of all sentences of $\mathscr{L}(\mathscr{T})$ having no more quantifiers than $\mathscr{F}$ does. The relation $q \Vdash^P \mathscr{F}_m$, viewed as a relation on q and m, is arithmetic, because a finite bound has been imposed on the number of quantifiers that can occur in $\mathscr{F}_m$. It follows there must exist a partial arithmetic function $f(q, m)$ such that if $q \in P$, then $f(q, m) \in P$, $q \geq f(q, m)$, and either $f(q, m) \Vdash^P \mathscr{F}_m$ or $f(q, m) \Vdash^P \sim\mathscr{F}_m$. By iteration of f, it is possible to define a partial arithmetic function $\lambda im \mid q_i^m$ with the following properties: $q_0^0 = p$; for each $m \geq 0$ and $i < 2^m$, $q_i^m \geq q_i^{m+1}$, $q_i^m \geq q_{i+2^m}^{m+1}$, $q_i^{m+1} \ngeq q_{i+2^m}^{m+1}$, $q_{i+2^m}^{m+1} \ngeq q_i^{m+1}$, $q_i^{m+1} \Vdash^P \mathscr{F}_m$ or $q_i^{m+1} \Vdash^P \sim\mathscr{F}_m$, and $q_{i+2^m}^{m+1} \Vdash^P \mathscr{F}_m$ or $q_{i+2^m}^{m+1} \Vdash^P \sim\mathscr{F}_m$. Then $Q = \{q_i^m \mid m \geq 0 \;\&\; i < 2^m\}$ is an arithmetic perfect closed set. If $T \in Q$, then T is generic for P with respect to all sentences in $\{\mathscr{F}_m\}$, and consequently, T makes $\mathscr{F}$ true since $T \in p = q_0^0$ and $p \Vdash^P \mathscr{F}$.

Now suppose $\mathscr{F}$ is Σ_2. Let $\mathscr{F}$ be $(\exists x)(y)\mathscr{S}(x, y)$, where $\mathscr{S}(x, y)$ is a quantifier-free formula of $\mathscr{L}(\mathscr{T})$. Suppose again that (a) $p \Vdash^P \mathscr{F}$. This means $p \Vdash^P (y)\mathscr{S}(\bar{n}, y)$ for some n, and so

$$(m)(q)_{q\in P}(\exists r)_{r\in P}[p \geq q \rightarrow q \geq r \;\&\; r \Vdash^P \mathscr{S}(\bar{n}, \bar{m})].$$

The relation $r \Vdash^P \mathscr{S}(\bar{n}, \bar{m})$, viewed as a relation on r, n and m, is recursive in P, because $\mathscr{S}$ has no quantifiers. It follows there exists a function $f(m, q)$ partial recursive in P such that if $q \in P$ and $p \geq q$, then $f(m, q) \in P$, $q \geq f(m, q)$, and $f(m, q) \Vdash^P \mathscr{S}(\bar{n}, \bar{m})$. Let $\mathscr{F}_m$ be $\mathscr{S}(\bar{n}, \bar{m})$, and define $\lambda im \mid q_i^m$ by iterating $f(m, q)$ as in the previous paragraph. Then $Q = \{q_i^m \mid m \geq 0 \;\&\; i < 2^m\}$ is recursive in P, and every $T \in Q$ makes $(\exists x)(y)\mathscr{S}(x, y)$ true.

Finally, suppose (b) $p \Vdash^P (x)(\exists y) \sim\mathscr{S}(x, y)$. Then

$$(n)(q)_{q\in P}(\exists m)(\exists r)_{r\in P}[p \geq q \rightarrow q \geq r \;\&\; r \Vdash^P \sim\mathscr{S}(\bar{n}, \bar{m})].$$

As above, we can make m and r into functions of n and q partial recursive in P, functions which are defined whenever $q \in P$ and $p \geq q$. Then as above r is iterated to obtain a Q recursive in P such that every $T \in Q$ makes $(x)(\exists y) \sim \mathscr{S}(x, y)$ true.

One of the advantages of using local forcing instead of an appropriate sequential lemma to prove Lemma 3.1 is that local forcing makes it easier to see Q is arithmetic.

We say P is a *pointed*, arithmetic perfect closed set if

$$(T)(T \in P \rightarrow P \text{ is recursive in } T).$$

Every perfect closed P has T's such that T is recursive in P; let T_P be the canonical one obtained by traveling up P viewed as a tree and turning left at every branch point. If P is pointed, then P and T_P have the same Turing degree, and that degree is less than or equal to the degree of every $T \in P$. Pointed, perfect closed sets are useful for forcing the degree of a generic T to be an upper bound for some given countable set of degrees. The notion of pointedness is used in **[23]** to show among other things: (1) every countable admissible **[12]**, **[16]** ordinal $> \omega$ is of the form ω_1^T for some $T \subseteq \omega$; (2) if α is a countable admissible ordinal $>\omega$, then $\omega_1^T = \alpha$ for some T with the minimal property that for every T' of lower hyperdegree than T, $\omega_1^{T'} < \alpha$. Recently Friedman and Jensen found a very interesting proof of (1) which replaces forcing by a compactness theorem of Barwise for infinitary languages. Their compactness approach does not seem to yield the minimal property of (2). A typical consequence of (2) is: the set of hyperdegrees of the Δ_2^1 sets has a minimal upper bound. The idea of pointedness goes back to the Kleene-Post proof [**11**] that there exists a set of lower degree than that of $0^{(\omega)}$ but of higher degree than that of $0^{(m)}$ for all m, i.e., the degree of the truth set for arithmetic is not a minimal upper bound for the degrees of the arithmetic sets. We say T is *pointedly generic* if for every sentence $\mathscr{F}$ of $\mathscr{L}(\mathscr{T})$ there is a pointed, arithmetic perfect closed P such that $T \in P$ and either $P \Vdash \mathscr{F}$ or $P \Vdash \sim\mathscr{F}$. The existence of a pointedly generic T follows from 3.1 and 3.2 and the mode of argument occurring in Lemma 3.3. We state without proof the fact that the Turing degree of any pointedly generic T is a minimal upper bound for the Turing degrees of the arithmetic sets, since it was shown in [**20**] that every countable set of Turing degrees has a minimal upper bound.

PROPOSITION 3.2. *If P is a pointed, arithmetic perfect closed set recursive in the arithmetic set A, then there exists a pointed, arithmetic perfect closed set $Q \subseteq P$ such that A and Q have the same Turing degree.*

PROOF. Although P may be encoded in a rather arbitrary way as an arithmetic set of sequence numbers, we can always re-encode P as $\{p_i^m \mid m \geq 0 \ \& \ i < 2^m\}$, where $\lambda im \mid p_i^m$ is recursive in the given encoding of P and is such that for all m and $i < 2^m$: $p_i^m \geq p_i^{m+1}$, $p_i^m \geq p_{i+2^m}^{m+1}$, $p_i^{m+1} \not\geq p_{i+2^m}^{m+1}$ and $p_{i+2^m}^{m+1} \not\geq p_i^{m+1}$. Since P is pointed, the re-encoding of P has the same Turing degree as the given encoding of P; i.e., the given encoding is recursive in the canonical path T_P, and T_P is recursive in $\lambda im \mid p_i^m$.

We define Q inductively: $p_0^0 \in Q$; if $m+1$ is odd and $p_i^m \in Q$, then p_i^{m+1}, $p_{i+2^m}^{m+1} \in Q$; if $m+1$ is even, $p_i^m \in Q$ and $(m+1)/2 \in A$, then $p_i^{m+1} \in Q$; if $m+1$ is even, $p_i^m \in Q$ and $(m+1)/2 \notin A$, then $p_{i+2^m}^{m+1} \in Q$. Clearly, $Q \subseteq P$, and Q is recursive in P, A. But P is recursive in A, so Q is recursive in A. Consider the canonical path $T_Q \in Q$. T_Q is recursive in Q, and $T_Q \in P$; so P by virtue of its pointedness is recursive in T_Q and hence in Q. But then A is recursive in Q, since A is recursive in P, Q as a consequence of

$$(m+1)/2 \in A \leftrightarrow (\exists i)(i < 2^m \ \& \ p_i^{m+1} \in Q).$$

To see that Q is pointed, fix $T \in Q$. A is recursive in T, P as a consequence of

$$(m+1)/2 \in A \leftrightarrow (\exists i)(i < 2^m \ \& \ T \in p_i^{m+1}).$$

Then Q is recursive in T, P, since Q is recursive in A. But P is recursive in T because $T \in P$; so Q is recursive in T.

THEOREM 3.3. *There exists a T such that $T^{(2)}$ has the same Turing degree as $0^{(\omega)}$, and such that $0^{(n)}$ is recursive in T for all n.*

PROOF. Let $\{\mathscr{F}_m\}$ be an enumeration of all Σ_2 sentences of $\mathscr{L}(\mathscr{T})$. We intend to define two functions, $\lambda m \mid P_m$ and $\lambda m \mid t(m)$, each recursive in $0^{(\omega)}$ and with the following properties: $P_0 = 2^\omega$; $P_m \supseteq P_{m+1}$; P_m has the same Turing degree as $0^{(m)}$; $P_{m+1} \Vdash \mathscr{F}_m$ or $P_{m+1} \Vdash \sim\mathscr{F}_m$; $t(m) = 0 \leftrightarrow P_{m+1} \Vdash \mathscr{F}_m$.

Assume that P_m has been defined and that P_m is a pointed, perfect closed set which has the same Turing degree as $0^{(m)}$. By Proposition 3.2, there is a pointed $R_m \subseteq P_m$ whose Turing degree is $0^{(m+1)}$. By Lemma 3.1, there is a $P_{m+1} \subseteq R_m$ such that P_{m+1} is recursive in R_m and either $P_{m+1} \Vdash \mathscr{F}_m$ or $P_{m+1} \Vdash \sim\mathscr{F}_m$. Since R_m is pointed, it follows that every $P \subseteq R_m$ and recursive in R_m is pointed and recursive in R_m. (This fact was demonstrated in 3.2.) So P_{m+1} is pointed and has the same Turing degree as $0^{(m+1)}$. If $P_{m+1} \Vdash \mathscr{F}_m$, put $t(m) = 0$; otherwise, put $t(m) = 1$.

We obtain $\lambda m \mid P_m$ and $\lambda m \mid t(m)$ recursive in $0^{(\omega)}$ by appealing to various uniformities. Examination of the proof of Proposition 3.2 reveals that R_m is recursive uniformly in P'_m; i.e., R_m is the result of applying a fixed recursive reduction procedure to P'_m, a procedure independent of the value of P_m. To develop P_{m+1} and $t(m)$ from Lemma 3.1, we must first decide if there is a $p \in R_m$ such that $p \Vdash^{R_m} \mathscr{F}_m$. Since $\mathscr{F}_m$ is Σ_2, the answer as a function of m is uniformly recursive in $R_m^{(2)}$. Thus P_{m+1} is uniformly recursive in $P_m^{(3)}$, and $\lambda m \mid t(m)$ is recursive in $\lambda m \mid R_m^{(2)}$. Since P_0 is recursive and $\lambda m \mid 0^{(m)}$ is recursive in $0^{(\omega)}$, it follows that $\lambda m \mid P_m$ and $\lambda m \mid t(m)$ are too.

Since $\{\mathscr{F}_m\}$ includes all sentences of the form $\bar{n} \in \mathscr{T}$, it must be that

$$\bigcap\{P_m \mid m \geq 0\}$$

has a unique member T recursive in $0^{(\omega)}$. And for each m, $0^{(m)}$ is recursive in T since $T \in P_m$ and P_m is pointed and of the same degree as $0^{(m)}$. Since the recursive predicate $\mathscr{R}$ is part of $\mathscr{L}(\mathscr{T})$, $\bar{m} \in \mathscr{T}^{(2)}$ is a Σ_2 formula of $\mathscr{L}(\mathscr{T})$. Assume $\mathscr{F}_{2m}$ is $\bar{m} \in \mathscr{T}^{(2)}$ for all m. Then $m \in T^{(2)} \leftrightarrow t(2m) = 0$, and consequently, $T^{(2)}$ is recursive in $0^{(\omega)}$. Putnam **[18]** has shown: for all T, if $(m)(0^{(m)}$ is recursive in $T)$, then $0^{(\omega)}$ is recursive in $T^{(2)}$.

For each set A, let $\mathbf{A}$ be the Turing degree of A. After Kleene and Post **[11]** showed that $\mathbf{0}^{(\omega)}$ is not a minimal upper bound for $\{\mathbf{0}^{(m)} \mid m \geq 0\}$, Spector **[28]**, proved that $\{\mathbf{0}^{(m)} \mid m \geq 0\}$ has no least upper bound. Putnam **[18]** calls $\mathbf{A}$ the *n-least upper bound* for $\{\mathbf{B}_i \mid i \geq 0\}$ if $\mathbf{A}$ is the least member of $\{\mathbf{C}^{(n)} \mid (i)[\mathbf{B}_i \leq \mathbf{C}]\}$. Thus Spector ruled out a 0-least upper bound for $\{\mathbf{0}^{(m)} \mid m \geq 0\}$. It is not difficult to show in the style of Theorem 3.3 that $\{\mathbf{0}^{(m)} \mid m \geq 0\}$ lacks a 1-least upper bound. And of course 3.3 says that $\mathbf{0}^{(\omega)}$ is the 2-least upper bound of $\{\mathbf{0}^{(m)} \mid m \geq 0\}$. By Spector's uniqueness theorem **[27]**, it is acceptable to define $\mathbf{0}^{(\alpha)}$ for each recursive ordinal α to be the Turing degree of the hyperarithmetic set H_b, where b is any notation for α. Then the argument of Theorem 3.3 is readily extended to give the following result.

THEOREM 3.4. *If α is a recursive limit ordinal, then $\mathbf{0}^{(\alpha)}$ is the 2-least upper bound of* $\{\mathbf{0}^{(\beta)} \mid \beta < \alpha\}$.

It can be shown that the set of all Turing degrees of hyperarithmetic sets has no 2-least upper bound.[7] This fact together with Theorem 3.4 leads to a tidy characterization of the hyperarithmetic hierarchy in terms of the jump operator and the notion of 2-least upper bound. H. Putnam **[18]** has extended Theorem 3.4 in a nontrivial fashion to the entire constructible hierarchy.

A set T is said to be *implicitly arithmetically definable* if there exists an arithmetic predicate $\mathscr{A}(X)$ such that $\mathscr{A}(T)$ and $(\exists_1 X)\mathscr{A}(X)$. Every implicitly arithmetically definable (i.a.d.) set is hyperarithmetic, and every hyperarithmetic set is recursive in some i.a.d. set. Feferman [2] found a hyperarithmetic set (in fact recursive in $0^{(\omega)}$) that was not i.a.d. We say two sets have the same arithmetic degree if each is arithmetic in the other.

THEOREM 3.5. *There exists a nonarithmetic, hyperarithmetic set T such that every nonarithmetic set arithmetic in T fails to be implicitly arithmetically definable.*

PROOF. Let $\{\mathscr{F}_n\}$ be an enumeration of the sentences of the arithmetic language $\mathscr{L}(\mathscr{T})$. We claim there is a function $\lambda n \mid P_n$ recursive in $0^{(\omega)}$ such that for each n, $P_n \supseteq P_{n+1}$, P_n is an arithmetic perfect closed set, and either $P_n \Vdash \mathscr{F}_n$ or $P_n \Vdash \sim\mathscr{F}_n$. The claim follows from some uniformities implicit in the proof of Theorem 3.1. Assume $\mathscr{F}$ is in prenex normal form and let $r(\mathscr{F}) = 1$ plus the number of alternations of quantifiers in the prenex of $\mathscr{F}$. Then the set $\{p \mid p \Vdash^P \mathscr{F}\}$ is recursive (uniformly) in the set $P^{(r(\mathscr{F}))}$. It follows that the first conclusion of Lemma 3.1 can be made to read: there exists a $Q \subseteq P$ such that either $P \Vdash \mathscr{F}$ or $P \Vdash \sim\mathscr{F}$ and such that Q is recursive (uniformly) in $P^{(1+r(\mathscr{F}))}$. But then the sequence $P_0 \supseteq P_1 \supseteq P_2 \supseteq \cdots$ can be constructed recursively from $0^{(\omega)}$. Let T be the unique member of $\bigcap\{P_n \mid n \geq 0\}$. T is generic, and the argument of Theorem 1.12, routinely modified, establishes that every set of the same arithmetic degree is generic. (T is of course of minimal arithmetic degree.) So it remains only to see that T is not implicitly arithmetically definable. This last is clear since if $\mathscr{A}(T)$ holds, then there is an arithmetic, perfect closed P such that $T \in P$ and

$$(T_1)(T_1 \in P \rightarrow \mathscr{A}(T_1)).$$

REFERENCES

1. P. J. Cohen, *The independence of the continuum hypothesis*, Proc. Nat. Acad. Sci. U.S.A. **50** (1963), 1143–1148 and **51** (1964), 105–110.

2. S. Feferman, *Some applications of the notion of forcing and generic sets*, Fund. Math. **56** (1965), 325–345.

3. S. Feferman and C. Spector, *Incompleteness along paths in progressions of theories*, J. Symbolic Logic **27** (1962), 383–390.

4. R. Gandy, *Proof of Mostowski's conjecture*, Bull. Acad. Polon. Sci. Sér. Math. **8** (1960), 571–575.

5. R. Gandy and G. E. Sacks, *A minimal hyperdegree*, Fund. Math. **61** (1967), 215–223.

[7] In fact it has no β-least upper bound for any recursive β.

6. K. Gödel, *The consistency of the axiom of choice and of the generalized continuum hypothesis*, Princeton Univ. Press, Princeton, N.J., 1966.

7. R. B. Jensen, "Concrete models of set theory" in *Sets, models and recursion theory*, Amsterdam, 1967, 44–74.

8. J. L. Kelley, *General topology*, New York, 1955.

9. S. C. Kleene, *Hierarchies of number-theoretic predicates*, Bull. Amer. Math. Soc. **61** (1955), 193–213.

10. ——, *Quantification of number-theoretic predicates*, Compositio Math. **15** (1959), 23–40.

11. S. C. Kleene and E. L. Post, *The upper semi-lattice of degrees of recursive unsolvability*, Ann. of Math. **59** (1954), 379–407.

12. S. Kripke, *Transfinite recursions on admissible ordinals*, I and II (abstracts), J. Symbolic Logic **29** (1964), 161–162.

13. G. Kreisel, *Set theoretic problems suggested by the notion of potential totality*, Proc. Sympos. Infinitistic Methods, Warsaw, 1961, 103–140.

14. ——, *The axiom of choice and the class of hyperarithmetic functions*, Indag. Math. **24** (1962), 307–319.

15. A. R. D. Mathias, *A survey of recent results in set theory*, these Proceedings, part II.

16. R. A. Platek, Ph.D. Thesis, Stanford University, 1965.

17. K. Prikry, *Models constructed using perfect sets*, Lectures Notes Axiomatic Set Theory (University of California, Los Angeles, 1967) pp. IV-K-1—IV-K-3.

18. H. Putnam, *Collected papers on hierarchy theory*, to appear.

19. H. Rogers, Jr., *Theory of recursive functions and effective computability*, McGraw-Hill, New York, 1967.

20. G. E. Sacks, *Degrees of unsolvability*, 2nd ed., Princeton Univ. Press, Princeton, N.J., 1966.

21. ——, *Post's problem, admissible ordinals, and regularity*, Trans. Amer. Math. Soc. **124** (1966), 1–23.

22. ——, *Measure-theoretic uniformity in recursion theory and set theory*, Trans. Amer. Math. Soc., **142**(1969), 381–420.

23. ——, *Countable admissible ordinals and hyperdegrees*, to appear.

24. J. R. Shoenfield, 1967 UCLA Recursion Theory Seminar Notes, Appendix B.

25. R. Solovay, *The measure problem*, abstract 65T-62, Notices Amer. Math. Soc. **12** (1965), 217.

26. ——, *The measure problem*, to appear.

27. C. Spector, *Recursive well-orderings*, J. Symbolic Logic, **20** (1955), 151–163.

28. ——, *On degrees of recursive unsolvability*, Ann. of Math. **64** (1956), 581–592.

29. ——, *Hyperarithmetical quantifiers*, Fund. Math., **48** (1959), 313–320.

30. L. Tharp, Set theory lecture notes, Massachusetts Institute of Technology, 1965.

MASSACHUSETTS INSTITUTE OF TECHNOLOGY

UNRAMIFIED FORCING[1]

J. R. SHOENFIELD

1. **Introduction.** The method of forcing was invented by Cohen in order to solve some classical independence problems. As it became apparent that the method was applicable in much more general circumstances, it was simplified and generalized by set theorists. One result of these efforts is the theory of Boolean models, discussed in the article by Scott and Solovay, part II of these Proceedings.

A feature of the Boolean approach is that the use of the ramified hierarchy of constructible sets has disappeared. Since forcing models can be obtained from Boolean models, it is apparent that this hierarchy is not needed for forcing models either.

One purpose of the present article is to give a direct construction of forcing models which does not use the ramified hierarchy. In addition, I have tried to present a summary of some of the simplifications and generalizations of forcing theory mentioned above. Many of these are only contained in the folk literature at present.

It would be an impossible task to list all the people who have contributed to each advance in the subject. The historical notes should at least show who the main contributors have been. The overall exposition owes much to notes by Scott and by Silver for this institute and to lectures by Rowbottom at UCLA in the fall of 1967. Conversations with Chang and Rowbottom have also been very helpful.

2. **Background.** We review here the principal facts about set theory which we need.

We use ZF for the Zermelo-Fraenkel axiom system (extensionality, regularity, infinity, union, replacement, and power set) and ZFC for ZF plus the axiom of

[1] The preparation of this paper was sponsored in part by NSF Grant GP-6726 and GP-5222.

choice. In considering a function as a set of ordered pairs, we put the value first and the argument second in the ordered pair. Thus the range (domain) of f is the set of first (second) elements of ordered pairs in f. Notation: $U(x)$ for the union of the sets in x, $S(x)$ for the power set of x, $\langle x, y\rangle$ for an ordered pair, $\mathrm{Ra}(x)$ for the range of x, $\mathrm{Do}(x)$ for the domain of x, ${}^x y$ for the set of mappings from x to y.

We identify an ordinal with the set of previous ordinals and a cardinal with the smallest ordinal having that cardinal (so that $\aleph_\alpha = \omega_\alpha$). We use Greek letters for ordinals and German letters for infinite cardinals (or infinite cardinals of a model). We write $|x|$ for the cardinal of x and $\mathfrak{m}^+$ for the next cardinal after $\mathfrak{m}$. Recall that $\mathrm{cf}(\alpha)$ is the smallest β such that there is a mapping of β onto a cofinal subset of α. An infinite cardinal $\mathfrak{m}$ is *regular* if $\mathrm{cf}(\mathfrak{m}) = \mathfrak{m}$ and *singular* if $\mathrm{cf}(\mathfrak{m}) < \mathfrak{m}$. Then $\mathrm{cf}(\alpha)$ is always 1 or a regular cardinal. If $\mathfrak{m}$ is regular, then the union of $<\mathfrak{m}$ sets each of cardinal $<\mathfrak{m}$ has cardinal $<\mathfrak{m}$. We use $S_\mathfrak{m}(x)$ for $[y : y \subset x \ \&\ |y| < \mathfrak{m}]$.

The $V(\alpha)$ (sometimes written $R(\alpha)$) are defined by

$$V(\alpha) = \bigcup_{\beta<\alpha} S(V(\beta)),$$

using transfinite induction. The *rank* of x, designated by $\mathrm{rk}(x)$, is the smallest α such that $x \in V(\alpha + 1)$. This is well defined, and $x \in y$ implies $\mathrm{rk}(x) < \mathrm{rk}(y)$. Moreover, $\mathrm{rk}(\alpha) = \alpha$.

We assume some elementary cardinal arithmetic. Recall that

$$|{}^x y| = |y|^{|x|}, \qquad |S(x)| = 2^{|x|}.$$

König's theorem says that $\mathrm{cf}(2^\mathfrak{m}) > \mathfrak{m}$. We write GCH for the generalized continuum hypothesis: $\forall \mathfrak{n}(2^\mathfrak{n} = \mathfrak{n}^+)$. *Weak powers* are defined by

$$\mathfrak{m}^{\underset{\smile}{\mathfrak{n}}} = \sum_{\mathfrak{p}<\mathfrak{n}} \mathfrak{m}^\mathfrak{p}.$$

(Here we allow $\mathfrak{p}$ to be finite; but these terms can be omitted if $\mathfrak{n} > \aleph_0$.) Then $\mathfrak{m}^{\underset{\smile}{\mathfrak{n}}^+} = \mathfrak{m}^\mathfrak{n}$. Since every subset of x of cardinal $\mathfrak{p}$ is the range of an element of ${}^\mathfrak{p}x$, we have $|S_\mathfrak{n}(x)| \leqq |x|^{\underset{\smile}{\mathfrak{n}}}$.

We have

$$\text{(3.1)} \qquad \mathfrak{n} < \mathrm{cf}(\mathfrak{m}) \ \&\ (\forall \mathfrak{p} < \mathfrak{m})(2^\mathfrak{p} \leqq \mathfrak{m}) \to \mathfrak{m}^\mathfrak{n} = \mathfrak{m}.$$

For $\mathfrak{n} < \mathrm{cf}(\mathfrak{m})$ implies ${}^\mathfrak{n}\mathfrak{m} = \bigcup_{\alpha<\mathfrak{m}} {}^\mathfrak{n}\alpha$. But $|{}^\mathfrak{n}\alpha| = |\alpha|^\mathfrak{n} \leqq 2^{|\alpha|\cdot\mathfrak{n}} \leqq \mathfrak{m}$ by hypothesis; so $\mathfrak{m}^\mathfrak{n} \leqq \mathfrak{m} \cdot \mathfrak{m} = \mathfrak{m}$. If the GCH holds, this reduces to $\mathfrak{n} < \mathrm{cf}(\mathfrak{m}) \to \mathfrak{m}^\mathfrak{n} = \mathfrak{m}$. It follows from (3.1) that

$$\mathfrak{m} \text{ regular } \&\ (\forall \mathfrak{p} < \mathfrak{m})(2^\mathfrak{p} \leqq \mathfrak{m}) \to \mathfrak{m}^{\underset{\smile}{\mathfrak{m}}} = \mathfrak{m}.$$

Thus $\mathfrak{m}^{\underset{\smile}{\mathfrak{m}}} = \mathfrak{m}$ if $\mathfrak{m}$ is $\aleph_0$ is strongly inacessible; and $\mathfrak{m}^{\underset{\smile}{\mathfrak{m}}} = \mathfrak{m}$ for all regular $\mathfrak{m}$ if the GCH holds.

We now turn to models of set theory. A model is *transitive* if its universe is a transitive set and its membership relation is the usual membership relation restricted to its universe. The collapsing technique of Mostowski [**8**] shows that every well-founded model of the extensionality axiom is isomorphic to a transitive model.

Henceforth, we shall take *model* to mean *transitive model*, and identify a model with its universe.

Let M be a model. To indicate that something is being considered in the model M, we append the phrase 'in M' or a superscript M. Thus if Φ is a sentence (possibly containing names of elements in M), we abbreviate 'Φ is true when interpreted in M' to 'Φ holds in M' or simply Φ^M. A *cardinal in M* is an element a of M such that 'a is a cardinal' holds in M. We designate by ω_1^M the element a of M such that $\Phi(a)^M$, where $\Phi(x)$ is the sentence of set theory saying that x is ω_1. Other examples are interpreted similarly.

A *class in M* is a set $[a : a \in M \,\&\, \Phi(a)^M]$, where $\Phi(x)$ contains only symbols from the language of ZFC and symbols for sets in M. Every set in M is a class in M. If M is a model of ZFC, then every class in M which is included in a set in M is itself a set in M. A *functional in M* is a function F such that for some formula $\Phi(x, y)$ of the type described above, $F(a) = b$ if and only if $a, b \in M$ and $\Phi(a, b)^M$. If M is a model of ZFC, then the image of a set in M under a functional in M is a set in M.

LEMMA 2.1. *Every transitive set M satisfying the following four conditions is a model of* ZF.

(a) $\omega \in M$.

(b) *Every class in M which is included in a set in M is itself a set in M.*

(c) *For every functional F in M and every set a in M included in the domain of F,* $U([F(b) : b \in a])$ *is included in a set in M.*

(d) *For every set a in M,* $S(a) \cap M$ *is included in a set in M.*

A relation symbol P defined in ZFC is *absolute* if for every model M of ZFC, P^M coincides with P on arguments in M. A similar definition holds for operation symbols (including constants, which are operation symbols with zero arguments). Most of the symbols introduced through the development of ordinals are absolute; S is an exception. For details, see [9] or [10]. If F is absolute, then F takes elements of M into elements of M (since F^M does). The absoluteness of 'is an ordinal' implies that the ordinals in M are the real ordinals which belong to M.

The *axiom of constructibility* [5] states that every set is constructible. It implies that there is a definable well-ordering of the universe and that the GCH holds. If M is a model of ZFC, the constructible sets in M form a model of ZFC plus the axiom of constructibility. The canonical function mapping the ordinals onto the constructible sets is absolute; so the constructible sets in a model M of ZFC are the images under this function of the ordinals in M. Thus if M and N have the same ordinals, then they have the same constructible sets.

3. **Notions of forcing.** Suppose that M is a model of ZFC, and that $a, b \in M$. We wish to extend M to a model N in which there is a mapping F of a onto b. To avoid obvious difficulties, we suppose that a is infinite and $b \neq 0$.

Let C be the set of all mappings form a finite subset of a into b. Then $C \in M$ by absoluteness. The set G of all finite subsets of F will be a subset of C; but it will not necessarily be in M. Our idea is first to select G, and then use G to build N.

Each $p \in C$ gives a condition which F must satisfy if p is to be in G; namely, we must have $p \subset F$. If $p \subset q$, the condition q gives more information than the condition p. We then say that q is an extension of p, and write $q \leqq p$. (This notation is to suggest that q allows fewer models N than p.) Then C is a partially ordered set with 0 as the largest element.

There are three obvious conditions which G must satisfy: (a) $0 \in G$; (b) if $p \in G$ and $p \leqq q$, then $q \in G$; (c) any two elements in G have a common extension in G. If G satisfies these conditions, $F = U(G)$ will be a mapping from a subset of a onto a subset of b.

In order to have $\mathrm{Do}(F) = a$ and $\mathrm{Ra}(F) = b$, G must also intersect certain sets, viz., the sets $[p : x \in \mathrm{Do}(p)]$ for $x \in a$ and the sets $[p : y \in \mathrm{Ra}(p)]$ for $y \in b$. These sets are in M and have the following property: every condition in C has an extension in the set. We shall therefore require that G intersect every set in M having this property.

We now generalize these notions. A *notion of forcing* is a partially ordered set C having a largest element. We write $\leqq_C$ or $\leqq$ for the ordering and 1_C or 1 for the largest element. The elements of C are called *conditions*. Conditions are generally designated by p, q, and r. If $p \leqq q$, we say p is an *extension* of q. A subset D of C is *C-dense* (or simply *dense*) if every condition in C has an extension in D.

Let C be a notion of forcing and let M be a set. A subset G of C is *C-generic* over M (or simply *generic*) if the following conditions hold.

(G1) $1 \in G$.

(G2) For all $p \in G$ and $q \geqq p$, $q \in G$.

(G3) For all $p, q \in G$, p and q have a common extension in G.

(G4) For all dense sets D in M, $G \cap D \neq 0$.

EXISTENCE THEOREM. *Let C be a notion of forcing; M a countable set; $p \in C$. Then there is a set G which is C-generic over M and contains p.*

PROOF. Let $a_0, a_1, \ldots$ be the elements of M. Choose p_n inductively as follows: $p_0 = p$; p_{n+1} is an extension of p_n in a_n if such an extension exists; $p_{n+1} = p_n$ otherwise. Then $G = [q : \exists n(p_n \leqq q)]$ has the required properties.

REMARK. This is the only place in which we make direct use of the countability of M. It would obviously suffice to assume that $M \cap S(C)$ is countable.

We let $H_{\mathfrak{m}}(A, B)$ be the set of all mappings p from an element of $S_{\mathfrak{m}}(A)$ to B. For $p, q \in H_{\mathfrak{m}}(A, B)$, we write $p \leqq q$ for $q \subset p$. Then $H_{\mathfrak{m}}(A, B)$ is a notion of forcing with largest element 0. We write $H(A, B)$ for $H_{\aleph_0}(A, B)$. Note that $H(A, B)$ is absolute.

Now suppose that M is a model of ZFC; $a, b \in M$; $b \neq 0$; $\mathfrak{m}$ is an infinite cardinal in M such that $\mathfrak{m} \leqq |a|$ in M; $C = H^M_{\mathfrak{m}}(a, b)$; G is C-generic over M; and $F = U(G)$. Then the remarks at the beginning of the section show that F is a mapping of a onto b. (We need $\mathfrak{m} \leqq |a|$ in M to guarantee that $[p : y \in \mathrm{Ra}(p)]$ is dense for $y \in b$.) Thus if we can construct an extension N of M containing G, we will obtain an extension in which there is a mapping of a onto b.

REMARK. This shows that if a is countably infinite and b is uncountable, then no C-generic set over M exists. Thus the requirement of countability in the existence theorem cannot be dropped.

HISTORICAL NOTE. The basic ideas of this section are due to Cohen **[1]**, **[2]**. The notion of a dense set is due to Solovay.

4. **The model.** Now suppose that C is a notion of forcing in[2] a countable model M of ZFC and that G is C-generic over M. We are going to construct an extension of M containing G.

We shall first define a structure which has universe M but has a new membership relation $\in_G$ defined by

$$a \in_G b \leftrightarrow (\exists p \in G)(\langle a, p\rangle \in b).$$

(Here and in what follows, a, b, c, and d represent elements of M.) We then use the collapsing technique to convert $\langle M, \in_G\rangle$ into a transitive model. We first note that

$$a \in_G b \rightarrow a \in \mathrm{Ra}(b) \tag{4.1}$$

and hence

$$a \in_G b \rightarrow \mathrm{rk}(a) < \mathrm{rk}(b). \tag{4.2}$$

We then define

$$K_G(b) = [K_G(a) : a \in_G b].$$

By (4.2), this is a legitimate definition by induction on $\mathrm{rk}(b)$. Finally, we define

$$M[G] = [K_G(a) : a \in M].$$

PRINCIPAL THEOREM. *Let M be a countable model of* ZFC; *C a notion of forcing in M; G a set which is C-generic over M. Then $M[G]$ is a countable model of* ZFC *which includes M and contains G; and it is the smallest such model.*

The proof of the fundamental theorem will be given in this and the next two sections. Throughout the rest of the paper, M, C, and G are assumed to satisfy the hypotheses of the fundamental theorem unless otherwise indicated. We shall write $\bar{a}$ for $K_G(a)$.

We begin with some simple observations. From the definition of $M[G]$ and the countability of M, we see that $M[G]$ is countable and transitive. Now define

$$\hat{b} = [\langle \hat{a}, 1\rangle : a \in b]$$

by induction on $\mathrm{rk}(b)$. This definition can be given in M; so the mapping from a to $\hat{a}$ is a functional in M. In particular, $\hat{a} \in M$. An easy induction (using (G1)) shows that $K_G(\hat{b}) = b$; so $M \subset M[G]$. Now set (with an abuse of notation)

$$\hat{G} = [\langle \hat{p}, p\rangle : p \in C].$$

Then $\hat{G} \in M$ and $K_G(\hat{G}) = G$. Hence $G \in M[G]$. In summary, $M[G]$ is a countable

[2] A notion of forcing C is in a model M if both the set C and the relation $\leqq$ are in M.

transitive set including M and containing G. The verification that it is a model of ZFC requires a new notion, which we now turn to.

5. **Forcing.** We introduce a language, called the *forcing language*, which is suitable for discussing $M[G]$. The symbols of the forcing language are the symbols of ZFC and the elements[3] of M. Each element a of M is regarded as a constant which designates the element $\bar{a}$ of $M[G]$; we say a is a *name* of $\bar{a}$. If Φ is a sentence of the forcing language, $\vdash_G \Phi$ means that Φ is true in $M[G]$.

Let $p \in C$, and let Φ be a sentence of the forcing language. We say that p *forces* Φ, and write $p \Vdash \Phi$, if $\vdash_G \Phi$ for every set G which is C-generic over M and contains p.

Our immediate object is to prove three lemmas about forcing.

Definability Lemma. *If $\Phi(x_1, \ldots, x_n)$ is a formula of* ZFC *containing only the free variables shown, then*

$$[\langle p, a_1, \ldots, a_n\rangle : p \Vdash \Phi(a_1, \ldots, a_n)]$$

is a class in M.

Extension Lemma. *If $p \Vdash \Phi$ and $q \leqq p$, then $q \Vdash \Phi$.*

Truth Lemma. *If G is generic, then $\vdash_G \Phi$ if and only if $(\exists p \in G)(p \Vdash \Phi)$.*

Our procedure is as follows. We define a modified notion of forcing, designated by $p \Vdash^* \Phi$. We then prove that the three lemmas hold when $\Vdash$ is replaced by $\Vdash^*$. From this, we show that

$$p \Vdash \Phi \leftrightarrow p \Vdash^* \neg\neg \Phi. \tag{5.1}$$

Then by taking the three lemmas for $\Vdash^*$ and replacing Φ by $\neg\neg\Phi$, we obtain the lemmas for $\Vdash$.

We first specify the undefined symbols of the forcing language. We let $\neg$ and $\vee$ be the undefined propositional connectives and let $\exists$ be the undefined quantifier. As undefined relation symbols, we take $\in$ and $\neq$. Of course, $x \notin y$ is defined as $\neg(x \in y)$ and $x = y$ is defined as $\neg(x \neq y)$.

We now define $p \Vdash^* \Phi$ by the following five clauses.

(a) $p \Vdash^* a \in b$ if $\exists c(\exists q \geqq p)(\langle c, q\rangle \in b \;\&\; p \Vdash^* a = c)$.

(b) $p \Vdash^* a \neq b$ if $\exists c(\exists q \geqq p)(\langle c, q\rangle \in a \;\&\; p \Vdash^* c \notin b)$ or
$\exists c(\exists q \geqq p)(\langle c, q\rangle \in b \;\&\; p \Vdash^* c \notin a)$.

(c) $p \Vdash^* \neg\Phi$ if $(\forall q \leqq p) \neg (q \Vdash^* \Phi)$.

(d) $p \Vdash^* \Phi \vee \Psi$ if $p \Vdash^* \Phi$ or $p \Vdash^* \Psi$.

(e) $p \Vdash^* \exists x\Phi(x)$ if $\exists b(p \Vdash^* \Phi(b))$.

We must first straighten out the circularities in the definition. If we trace back the definition of $p \Vdash^* a \neq b$, we find that it is defined in terms of certain $p' \Vdash^* a' \neq b'$ and $p' \Vdash^* b' \neq a'$ where $\text{rk}(a') < \text{rk}(a)$ and $\text{rk}(b') < \text{rk}(b)$. Thus we may define

[3] Those who do not like sets to be used as symbols can introduce a set of symbols which is in one-one correspondence with M.

$p \Vdash^* a \neq b$ by induction on max $(\text{rk}(a), \text{rk}(b))$. We may then use (c) to define $p \Vdash^* a = b$, (a) to define $p \Vdash^* a \in b$, and (c) to define $p \Vdash^* a \notin b$. We can then prove (b). This defines $p \Vdash^* \Phi$ for atomic Φ. We then use (c), (d) and (e) to define $p \Vdash^* \Phi$ for all Φ by induction on the length of Φ.

It is trivial to prove the definability lemma for $\Vdash^*$ by induction on the length of $\Phi(x_1, \ldots, x_n)$ (noting that the quantifiers in (a)–(e) vary through M or through the set C in M). Of course the definition of the class $[\langle p, a, b\rangle : p \Vdash^* a \neq b]$ will be by transfinite induction in M as described above.

We prove the other two lemmas by proving them for the sentences on the left of (a)–(e) under the assumption that they are true for the sentences on the right; this is seen to be a valid method of proof as above. The extension lemma is quite trivial; so we consider only the truth lemma.

We first show that if the truth lemma holds for Φ, then

(5.2) $$a \in_G b \;\&\; \vdash_G \Phi \leftrightarrow (\exists p \in G)(\exists q \geqq p)(\langle a, q\rangle \in b \;\&\; p \Vdash^* \Phi).$$

For if the left side holds, there are $q, r \in G$ such that $\langle a, q\rangle \in b$ and $r \Vdash^* \Phi$. Choosing a common extension p of q and r in G by (G3), we have $p \Vdash^* \Phi$ by the extension lemma. Conversely, let the right side of (5.2) hold for p and q. Then $\vdash_G \Phi$ by the truth lemma. Also $q \in G$ by (G2); so $a \in_G b$.

Now we turn to the cases of the truth lemma.

(a) By the definition of $\bar{b}$, $\vdash_G a \in b$ is equivalent to $\exists c(c \in_G b \;\&\; \vdash_G a = c)$. By the hypothesis and (5.2), this is equivalent to

$$\exists c(\exists p \in G)(\exists q \geqq p)(\langle c, q\rangle \in b \;\&\; p \Vdash^* a = c)$$

and hence to $(\exists p \in G)(p \Vdash^* a \in b)$.

(b) Clearly $\vdash_G a \neq b$ if and only if either $\exists c(c \in_G a \;\&\; \vdash_G c \notin b)$ or $\exists c(c \in_G b \;\&\; \vdash_G c \notin a)$. By the hypothesis and (5.2), this is equivalent to

$$\exists c(\exists p \in G)(\exists q \geqq p)(\langle c, q\rangle \in a \;\&\; p \Vdash^* c \notin b)$$
$$\vee\; \exists c(\exists p \in G)(\exists q \geqq p)(\langle c, q\rangle \in b \;\&\; p \Vdash^* c \notin a)$$

and hence to $(\exists p \in G)(p \Vdash^* a \neq b)$.

(c) By hypothesis, $\vdash_G \neg \Phi$ if and only if $\neg(\exists p \in G)(p \Vdash^* \Phi)$. Hence we must prove that exactly one of $(\exists p \in G)(p \Vdash^* \Phi)$ and $(\exists p \in G)(p \Vdash^* \neg \Phi)$ holds. To show that at least one holds, it will suffice by (G4) to show that $D = [p : p \Vdash^* \Phi$ or $p \Vdash^* \neg \Phi]$ is a dense set in M. It is in M by the definability lemma; so we must show every p has an extension in D. But either p has an extension q such that $q \Vdash^* \Phi$ and hence $q \in D$, or $p \Vdash^* \neg \Phi$ and hence p itself is in D.

Now suppose there are $p, q \in G$ such that $p \Vdash^* \Phi$ and $q \Vdash^* \neg \Phi$. By (G2), p and q have a common extension r; and by the extension lemma, $r \Vdash^* \Phi$. This contradicts $q \Vdash^* \neg \Phi$.

We leave the easy proofs of (d) and (e) to the reader.

Now we prove (5.1). Suppose $p \Vdash^* \neg\neg \Phi$. If G is generic and $p \in G$, then $\vdash_G \neg\neg \Phi$ by the truth lemma; so $\vdash_G \Phi$. Thus $p \Vdash \Phi$. Now suppose $\neg(p \Vdash^* \neg\neg \Phi)$. Then $q \Vdash^* \neg \Phi$ for some $q \leqq p$. Choose a generic G such that $q \in G$. By the truth lemma, $\vdash_G \neg \Phi$; and by (G2), $p \in G$. Hence $\neg(p \Vdash \Phi)$.

We have thus proved our lemmas. Substituting $\neg \neg \Phi$ for Φ in (c), we get

(5.3) $$p \Vdash \neg \Phi \text{ if and only if } (\forall q \leqq p) \neg (q \Vdash \Phi).$$

Replacing Φ by $\neg \Phi$ and noting that $p \Vdash \neg \neg \Phi$ is equivalent to $p \Vdash \Phi$,

(5.4) $$p \Vdash \Phi \text{ if and only if } (\forall q \leqq p) \neg (p \Vdash \neg \Phi).$$

HISTORICAL NOTE. Cohen's concept of forcing is essentially our $\Vdash^*$. Part (c) of the definition, which simplified Cohen's definition, is due to Scott. Our concept of forcing was introduced by Feferman **[4]**, who called it *weak forcing*. The three fundamental lemmas are due to Cohen.

6. **Completion of the proof.** We show that $M[G]$ is a model of ZF by showing that it satisfies the conditions of Lemma 2.1. Since $\omega \in M \subset M[G]$, condition (a) holds.

LEMMA 6.1. *Let A be a class in $M(G)$ such that $A \subset \bar{a}$. Then $A \in M[G]$, and A has a name c such that $c \subset \mathrm{Ra}(a) \times C$.*

PROOF. There is a formula $\Phi(x)$ of the forcing language such that

(6.1) $$\bar{b} \in A \leftrightarrow \vdash_G \Phi(b)$$

for all b. Set

$$c = [\langle b, p\rangle : b \in \mathrm{Ra}(a) \;\&\; p \Vdash \Phi(b)].$$

By the definability lemma, c is a class in M. Since $c \subset \mathrm{Ra}(a) \times C$, $c \in M$.

We must prove

$$\bar{b} \in \bar{c} \leftrightarrow \bar{b} \in A.$$

If $\bar{b} \in \bar{c}$, we may suppose, by changing b without changing $\bar{b}$, that $b \in_G c$. Then for some $p \in G$, $\langle b, p\rangle \in c$ and hence $p \Vdash \Phi(b)$. Hence $\vdash_G \Phi(b)$; so $\bar{b} \in A$ by (6.1). Now let $\bar{b} \in A$. Then $\bar{b} \in \bar{a}$; so we may suppose that $b \in_G a$. By (4.1), $b \in \mathrm{Ra}(a)$. By (6.1), $\vdash_G \Phi(b)$; so by the truth lemma, some $p \in G$ forces $\Phi(b)$. Then $\langle b, p\rangle \in c$; so $b \in_G c$; so $\bar{b} \in \bar{c}$.

It follows from Lemma 6.1 that (b) of Lemma 2.1 holds.

LEMMA 6.2. *If $x \subset M[G]$, and every element in x has a name in a, then x is included in a set in $M[G]$.*

PROOF. Let $b = a \times \{1\}$. Any element of x is $\bar{c}$ for some $c \in a$. Then $c \in_G b$ by (G1); so $\bar{c} \in \bar{b}$. Thus $x \subset \bar{b}$.

Now we prove that (c) of Lemma 2.1 holds. Let F be a functional in $M[G]$, and let $\bar{a}$ be a set in $M[G]$ included in the domain of F. There is a formula $\Phi(x, y)$ of the forcing language such that

$$F(\bar{b}) = \bar{c} \leftrightarrow \vdash_G \Phi(b, c)$$

for all b, c. Choose a set d in M with the following property: for each $p \in C$ and each $b \in \mathrm{Ra}(a)$, if there is a c such that $p \Vdash \Phi(b, c)$, then there is such a set in d. By replacing d by its transitive closure, we may suppose that d is transitive.

We now show that every element x of $U([F(\bar{b}):\bar{b} \in \bar{a}])$ has a name in d; in view of Lemma 6.2, this will complete our proof. We have $x \in F(\bar{b})$ with $\bar{b} \in \bar{a}$; and we may suppose that $b \in_G a$. Letting c be a name of $F(\bar{b})$, we have $\vdash_G \Phi(b, c)$; so some $p \in G$ forces $\Phi(b, c)$. Hence for some $c' \in d$, $p \Vdash \Phi(b, c')$. Then $\vdash_G \Phi(b, c')$; so $F(\bar{b}) = \bar{c}'$. Thus $x \in \bar{c}'$; whence $x = \bar{a}'$ with $a' \in_G c'$ and hence $a' \in \mathrm{Ra}(c')$. Since $c' \in d$ and d is transitive, $a' \in d$ as required.

Finally, we prove that (d) holds. Let $\bar{a} \in M[G]$ and let $\bar{b} \in S(\bar{a}) \cap M[G]$. By Lemma 6.1, $\bar{b}$ has a name c such that $c \subset \mathrm{Ra}(a) \times C$ and hence $c \in S^M(\mathrm{Ra}(a) \times C)$. The desired result now follows from Lemma 6.2. Thus $M[G]$ is a model of ZF.

LEMMA 6.3. *If N is a model of* ZF *which includes M and contains G, then there is a functional in N whose restriction to M is K_G.*

PROOF. We define (in N)

$$x \in^* y \leftrightarrow (\exists p \in G)(\langle x, p \rangle \in y), \qquad K^*(y) = [K^*(x): x \in^* y]$$

(using induction on $\mathrm{rk}(y)$). It is easy to see (using the transitivity of M) that $\in^*$ and K^* agree with $\in_G$ and K_G for arguments in M.

It follows from the lemma that $K_G(a) \in N$ for all $a \in M$, so that $M[G] \subset N$. Thus $M[G]$ is even the smallest model of ZF including M and containing G. Moreover, we can apply Lemma 6.3 to $M[G]$. We obtain a function K in $M[G]$ whose restriction to M is K_G.

Now let $\bar{a} \in M[G]$. Then there is a mapping of an ordinal onto $\mathrm{Ra}(a)$ which is in M and hence in $M[G]$. Composing K with this mapping, we get in $M[G]$ a mapping from an ordinal onto

$$[K(x): x \in \mathrm{Ra}(a)] = [\bar{b}: b \in \mathrm{Ra}(a)].$$

But by (4.1), this set includes $\bar{a}$. Thus in $M[G]$, the following holds: for every x, there is a mapping from an ordinal onto a set including x. This implies that the axiom of choice holds in $M[G]$. We have thus completed the proof of the fundamental theorem.

HISTORICAL NOTE. Most of the ideas of this section are due to Cohen. The proof that the power set axiom holds in $M[G]$ is essentially due to Solovay; it is simpler than Cohen's proof.

7. **The axiom of constructibility.** To make the best use of the fundamental theorem, we need information about the relation between M and $M[G]$. The following simple result is often useful.

LEMMA 7.1. *M and $M[G]$ have the same ordinals.*

PROOF. Since $M \subset M[G]$, we need only show that every ordinal α in $M[G]$ is in M. A simple induction shows that $\mathrm{rk}(\bar{a}) \leqq \mathrm{rk}(a)$ for all a. Taking a to be a name of α, $\alpha = \mathrm{rk}(\alpha) \leqq \mathrm{rk}(a)$. Since rk is absolute, $\mathrm{rk}(a) \in M$; so $\alpha \in M$ by the transitivity of M.

COROLLARY. *M and M[G] have the same constructible sets.*

Now let M be a countable model of ZFC, and let $C = H(\omega, 2)$. By absoluteness, C is in M. Take G generic over M, and set $F = U(G)$. As seen in §3, F is a mapping from ω to 2. Obviously $F \in M[G]$; we claim $F \notin M$. To see this, let f be a mapping from ω to 2 which is in M. Then $[p:(\exists n \in \omega)(n \in \mathrm{Do}(p) \;\&\; p(n) \neq f(n))]$ is a set in M which is easily seen to be dense. Hence it contains a $p \in G$; and this implies that $F \neq f$.

The set A having F as its characteristic function is thus in $M[G] - M$. By the corollary, A is not constructible in $M[G]$. Hence $M[G]$ is a model of ZFC′, where ZFC′ is ZFC with the additional axiom: there is a noncontructible subset of ω.

We have shown how to construct a model of ZFC′ from a countable model M of ZFC. The existence of M can be seen as follows. We start from any model N of ZFC. Using the Löwenheim-Skolem theorem to extract a countable submodel and then applying the collapsing technique, we obtain a countable model M of ZFC.

Unfortunately, the existence of a (transitive) model N of ZFC cannot be proved in ZFC, even from the assumption that ZFC is consistent. Hence if we wish a finitary proof of the relative consistency of ZFC′ to ZFC, we must proceed a little differently. We add to ZFC a constant N and axioms which say that N is transitive and nonempty and that each axiom of ZFC holds in N. The reflection principle **[6]** shows that this is a conservative extension of ZFC. We then define M and $M[G]$ as above. Our proof will then show that each axiom of ZFC′ holds in $M[G]$. For a slightly different technique, see **[10]**.

We can get a stronger result by "collapsing" a cardinal in M. Suppose that M satisfies the axiom of constructibility. (We can obtain such a model from a countable model N of ZFC by taking the sets constructible in N.) Take $C = H(\aleph_0, \aleph_1^M)$. Then in $M[G]$, there is a mapping of $\aleph_0$ onto $\aleph_1^M$; so $\aleph_1^M$ is countable in $M[G]$. Now there is a one-one correspondence between $S^M(\omega)$ and $\aleph_1^M$ which is in M and hence in $M[G]$; so $S^M(\omega)$ is countable in $M[G]$. But by the corollary, every subset of ω which is constructible in $M[G]$ is in M and hence in $S^M(\omega)$. Thus in $M[G]$, there are only countably many constructible subsets of ω.

HISTORICAL NOTE. The independence of the axiom of constructibility was proved by Cohen. Cardinal collapsing was introduced by Lévy.

8. **Products.** Let C_1 and C_2 be two notions of forcing in M. We shall put subscripts 1 or 2 on our previous notation to indicate that we are considering C_1 or C_2; e.g. p_1 for an element of C_1, G_2 for a C_2-generic set. We write $M[G_1, G_2]$ for $(M[G_1])[G_2]$.

We define a partial ordering on $C_1 \times C_2$ by

$$\langle p_1, p_2\rangle \leqq \langle q_1, q_2\rangle \leftrightarrow p_1 \leqq q_1 \;\&\; p_2 \leqq q_2.$$

Clearly $C_1 \times C_2$ is a notion of forcing in M with largest element $\langle 1_1, 1_2\rangle$.

Product Theorem. *Let C_1 and C_2 be notions of forcing in M. If G_1 is C_1-generic over M and G_2 is C_2-generic over $M[G_1]$, then $G_1 \times G_2$ is $(C_1 \times C_2)$-generic over M, and $M[G_1 \times G_2] = M[G_1, G_2]$. Every set which is $(C_1 \times C_2)$-generic over M is obtained in this way.*

Proof. It is easy to verify (G1), (G2) and (G3) for $G_1 \times G_2$. Let D be a $(C_1 \times C_2)$-dense set in M. We must show that $(G_1 \times G_2) \cap D \neq 0$, i.e. that $G_2 \cap D_2 \neq 0$, where

$$D_2 = [p_2 : (\exists p_1 \in G_1)(\langle p_1, p_2\rangle \in D)].$$

Since $D_2 \in M[G_1]$, it will suffice to show that D_2 is C_2-dense.

Let $q_2 \in C_2$; we must find $p_2 \leqq q_2$ and $p_1 \in G_1$ such that $\langle p_1, p_2\rangle \in D$. In other words, we must show $G_1 \cap D_1 \neq 0$, where

$$D_1 = [p_1 : (\exists p_2 \leqq q_2)(\langle p_1, p_2\rangle \in D)].$$

Since $D_1 \in M$, it suffices to show D_1 is C_1-dense. Let $q_1 \in C_1$, and choose $\langle p_1, p_2\rangle \leqq \langle q_1, q_2\rangle$ such that $\langle p_1, p_2\rangle \in D$. Then $p_1 \leqq q_1$ and $p_1 \in D_1$.

The equality $M[G_1 \times G_2] = M[G_1, G_2]$ holds because both are the smallest model of ZFC including M and containing G_1 and G_2.

Now let G be $(C_1 \times C_2)$-generic over M, and let G_1 and G_2 be the projections of G on C_1 and C_2 respectively. Clearly $G \subset G_1 \times G_2$. To prove $G = G_1 \times G_2$, let $\langle p_1, p_2\rangle \in G_1 \times G_2$. For some q_1 and q_2, $\langle p_1, q_2\rangle, \langle q_1, p_2\rangle \in G$. Hence they have a common extension $\langle r_1, r_2\rangle$ in G. Since $\langle r_1, r_2\rangle \leqq \langle p_1, p_2\rangle$, we have $\langle p_1, p_2\rangle \in G$.

The verification of (G1), (G2), and (G3) for G_1 and G_2 is easy. To verify (G4) for G_1, let D_1 be a C_1-dense set in M. Then $D_1 \times C_2$ is a $(C_1 \times C_2)$-dense set in M. Hence $G \cap (D_1 \times C_2) \neq 0$; so $G_1 \cap D_1 \neq 0$.

To verify (G4) for G_2, let D_2 be a C_2-dense set in $M[G_1]$. Let a be a name of D_2, and let Φ be the sentence of the forcing language which says that $\bar{a}$ is C_2-dense.

We show that

$$D = [\langle p_1, p_2\rangle : p_1 \Vdash \Phi \rightarrow \hat{p}_2 \in a]$$

is dense. Let $\langle q_1, q_2\rangle$ be given, and choose G_1' C_1-generic over M so that $q_1 \in G_1'$. If $K_{G_1'}(a)$ is C_2-dense, choose $p_2 \leqq q_2$ so that $p_2 \in K_{G_1'}(a)$; otherwise, let $p_2 = q_2$. In either case, $\vdash_{G_1'} \Phi \rightarrow \hat{p}_2 \in a$. Hence some $p_1 \in G_1'$ forces $\Phi \rightarrow \hat{p}_2 \in a$; and by (G3) and the extension lemma, we may suppose $p_1 \leqq q_1$. Then $\langle p_1, p_2\rangle \leqq \langle q_1, q_2\rangle$ and $\langle p_1, p_2\rangle \in D$.

Since $D \in M$, it follows that $G \cap D \neq 0$. Let $\langle p_1, p_2\rangle \in G \cap D$. Since $p_1 \in G_1$ and $p_1 \Vdash \Phi \rightarrow \hat{p}_2 \in a$, we have $\vdash_{G_1} \Phi \rightarrow \hat{p}_2 \in a$. But $\vdash_{G_1} \Phi$; so $p_2 \in \bar{a} = D_2$. But also $p_2 \in G_2$; so $G_2 \cap D_2 \neq 0$.

Corollary. *Let C_1 and C_2 be notions of forcing in M. Let G_1 be C_1-generic over M and let G_2 be C_2-generic over $M[G_1]$. Then G_1 is C_1-generic over $M[G_2]$, and $M[G_2, G_1] = M[G_1, G_2]$.*

Proof. By the theorem, $G_1 \times G_2$ is $(C_1 \times C_2)$-generic over M. Applying the obvious isomorphism of $C_1 \times C_2$ and $C_2 \times C_1$, $G_2 \times G_1$ is $(C_2 \times C_1)$-generic over M. Applying the theorem again, G_1 is C_1-generic over $M[G_2]$. We have $M[G_2, G_1] = M[G_1, G_2]$ because both are the smallest model of ZFC including M and containing G_1 and G_2.

We shall only consider one example of infinite products, the *weak power*. If C is a notion of forcing and I is any set, C^I is the set of all mappings p from I to C such that $[i \mid i \in I \;\&\; p(i) \neq 1]$ is finite. We then define $p \leqq q$ to mean $(\forall i \in I)(p(i) \leqq q(i))$. Clearly C^I is a notion of forcing whose largest element is the function constantly equal to 1_C. If C and I belong to a model M of ZFC, so does C^I; this is the reason for the finiteness restriction.

If J and K are disjoint subsets of I such that $J \cup K = I$, then C^I is isomorphic to $C^J \times C^K$ in a natural way. In particular, if $i \in I$ and $J = I - \{i\}$, then C^I is naturally isomorphic to $C^J \times C$. In either case, we can apply the product theorem.

9. **The axiom of choice.** By an *automorphism* of a notion of forcing C, we mean an automorphism of the partially ordered set C.

LEMMA 9.1. *Let π be an automorphism of C which is in M. Then $\pi(G)$ is C-generic over M, and $M[G] = M[\pi(G)]$.*

PROOF. Since π maps sets in M into sets in M, the first conclusion is trivial. Since $M[G]$ contains G and π, it contains $\pi(G)$; so $M[\pi(G)] \subset M[G]$ by the fundamental theorem. Substituting $\pi(G)$ for G and π^{-1} for π, we get $M[G] \subset M[\pi(G)]$; so $M[G] = M[\pi(G)]$.

Let $\mathfrak{A}$ be a set of automorphisms of C such that $\mathfrak{A} \in M$. An element a of M is *$\mathfrak{A}$-invariant* if $K_G(a) = K_{\pi(G)}(a)$ for every $\pi \in \mathfrak{A}$. For example, each $\hat{a}$ is invariant, since $K_G(\hat{a}) = K_{\pi(G)}(\hat{a}) = a$. A sentence Φ of the forcing language is *$\mathfrak{A}$-invariant* if every constant in Φ is $\mathfrak{A}$-invariant.

Let $\mathfrak{A}$ be a set of automorphisms of C. We say that C is *$\mathfrak{A}$-homogeneous* if for every $p, q \in C$, there is a $\pi \in \mathfrak{A}$ such that $\pi^{-1}(p)$ and q have a common extension. If $\mathfrak{A}$ is the set of all automorphisms of C, we say *homogeneous* for $\mathfrak{A}$-homogeneous.

If A is infinite, then $H(A, B)$ is homogeneous. For given $p, q \in H(A, B)$, we choose a permutation σ of A such that $\sigma(\mathrm{Do}(p)) \cap \mathrm{Do}(q) = 0$, and define the automorphism π by $\pi(r) = r \circ \sigma$. Then $\pi^{-1}(p) \cup q$ is a common extension of $\pi^{-1}(p)$ and q.

LEMMA 9.2. *Let $\mathfrak{A}$ be a set of automorphisms of C such that $\mathfrak{A} \in M$, and suppose that C is $\mathfrak{A}$-homogeneous in M. Let Φ be a sentence of the forcing language which is $\mathfrak{A}$-invariant. Then $\vdash_G \Phi$ if and only if $1 \Vdash \Phi$.*

PROOF. Since $1 \in G$, $1 \Vdash \Phi$ implies $\vdash_G \Phi$. Now suppose that $\vdash_G \Phi$ but that $\neg(1 \Vdash \Phi)$. By the former, some p forces Φ; and by the latter and (5.4), some q forces $\neg\, \Phi$. Choose $\pi \in \mathfrak{A}$ so that $\pi^{-1}(p)$ and q have a common extension. By the existence theorem and (G2), there is a generic G' containing $\pi^{-1}(p)$ and q. By Lemma 9.1, $\pi(G')$ is generic. Since $p \in \pi(G')$ and $q \in G'$, we have $\vdash_{\pi(G')} \Phi$ and $\vdash_{G'} \neg\, \Phi$. But $M[\pi(G')] = M[G']$ by Lemma 9.1, and the constants in Φ represent the same sets in these two models. Hence $\vdash_{\pi(G')} \Phi$ if and only if $\vdash_{G'} \Phi$, a contradiction.

We assume that the reader is familiar with OD (ordinal-definable) and HOD (hereditarily ordinal-definable) sets.[4] We need a slight generalization.

We say that u is OD *from* $v_1, \dots, v_k$ if there is an α such that $v_1, \dots, v_k \in V(\alpha)$ and u is definable in $\langle V(\alpha), \in, v_1, \dots, v_k \rangle$. We say that u is OD *over* w if for some $v_1, \dots, v_k \in w$, u is OD from $v_1, \dots, v_k, w$. We say that u is HOD *over* w if u is OD over w and every member of u is HOD over w; this is a definition by induction on $\mathrm{rk}(u)$.

The basic results about OD and HOD sets carry over to this situation. Thus every ordinal is OD from any $v_1, \dots, v_k$ and hence OD over any w. If $u_1, \dots, u_m$ are OD from $v_1, \dots, v_k$ (or over w), then $\mu(v_1, \dots, v_k)$ is also (where μ is a term defined in ZFC). The class of sets HOD over w is a model of ZF (but not necessarily of the axiom of choice). It is also clear that every member of w is OD over w.

LEMMA 9.3. *Let $\mathfrak{A}$ be a set of automorphisms of C such that $\mathfrak{A} \in M$, and suppose that C is $\mathfrak{A}$-homogeneous. Let u be* OD *from $v_1, \dots, v_k$ in $M[G]$, and suppose that $v_1, \dots, v_k$ have $\mathfrak{A}$-invariant names. Then $u \cap M \in M$.*

PROOF. There is a formula $\Phi(x)$ of the forcing language, containing names only for $v_1, \dots, v_k$ and some $\alpha \in M[G]$, such that

$$\bar{a} \in u \leftrightarrow \vdash_G \Phi(a) \tag{9.1}$$

for all a. Since $\alpha \in M$ by Lemma 7.1, $\hat{\alpha}$ is $\mathfrak{A}$-invariant; so we may suppose that every name in $\Phi(x)$ is $\mathfrak{A}$-invariant. Then putting $\hat{a}$ for a in (9.1) and using Lemma 9.2,

$$a \in u \leftrightarrow \vdash_G \Phi(\hat{a}) \leftrightarrow 1 \Vdash \Phi(\hat{a}).$$

Thus $u \cap M = [a : 1 \Vdash \Phi(\hat{a})]$ is a class in M. But if $\beta = \mathrm{rk}(u)$, then $u \cap M \subset V(\beta) \cap M = V^M(\beta)$; so $u \cap M$ is a set in M.

THEOREM 9.1. *Let C be homogeneous in M. If u is* OD *in $M[G]$, then $u \cap M \in M$; if u is* HOD *in $M[G]$, then $u \in M$.*

PROOF. The first conclusion is a special case of Lemma 9.3. The second conclusion follows easily from the first by induction on $\mathrm{rk}(u)$.

Now let M satisfy the axiom of constructibility, and let $C = H(\omega, 2)$. Using Theorem 9.1, the corollary to Lemma 7.1, and the fact that every constructible set is HOD, we see that in $M[G]$ the constructible sets coincide with the HOD sets. As seen in §7, there is a subset of ω which is nonconstructible in $M[G]$. Since every member of ω is HOD, it follows that this set is not OD in $M[G]$. From this, it follows that there is no OD mapping from an ordinal to $S(\omega)$ in $M[G]$; for if F is OD, then so is every $F(\alpha)$.

Next let $C = H(\omega, 2)^\omega$. Let G_i be the set of ith coordinates of elements in G, and let $H = [G_i : i \in \omega]$. Let N be the set of all sets which are HOD over H in $M[G]$. Then N is a model of ZF. We shall show that N is not a model of the axiom of choice; in fact, that there is no mapping from an ordinal onto $S(\omega)$ in N.

---

[4] See the article by Myhill and Scott, these Proceedings.

Suppose that there is such a mapping F. Then for some n, F is OD from $G_0, G_1, \ldots, G_{n-1}, H$ (in $M[G]$). Every element of $S(\omega) \cap N$ is $F(\alpha)$ for some α in $M[G]$ and hence is OD from $G_0, G_1, \ldots, G_{n-1}, H$.

We regard C as the product $C' \times C''$, where $C' = H(\omega, 2)^n$ and $C'' = H(\omega, 2)^{\omega-n}$. Let G' and G'' be the projections of G on C' and C'' respectively. By the product theorem, G' is C'-generic over M; G'' is C''-generic over $M' = M[G']$; and $M'[G''] = M[G]$. Applying the product theorem again, G_n is $H(\omega, 2)$-generic over M'. Hence the set A having $U(G_n)$ as characteristic function is a subset of ω not in M'. Since $A \subset \omega \subset M'$, $A \cap M' = A$; so $A \cap M'$ is not in M'.

We complete the proof by using Lemma 9.3 to show that $A \cap M' \in M'$. For each permutation π of $\omega - n$, we define an automorphism π^* of C'' by $(\pi^*(p))_i = p_{\pi(i)}$. Let $\mathfrak{A}$ be the set of all π^* for π in M'. Clearly $\mathfrak{A} \in M'$. Noting that every permutation of $\omega - n$ which moves only finitely many numbers is in M', we see that C'' is $\mathfrak{A}$-homogeneous. Now A is OD from G_n and hence is in N; so A is OD from $G_0, G_1, \ldots, G_{n-1}, H$ in $M[G] = M'[G'']$. Hence we need only show that $G_0, G_1, \ldots, G_{n-1}, H$ have $\mathfrak{A}$-invariant names in M'.

Since $G_0, G_1, \ldots, G_{n-1}$ are in M', they have the $\mathfrak{A}$-invariant names $\hat{G}_0, \hat{G}_1, \ldots, \hat{G}_{n-1}$. Now set

$$a_i = \hat{G}_i \quad \text{if } i < n,$$
$$a_i = [\langle \hat{p}_i, p\rangle : p \in C''] \quad \text{if } i \geqq n,$$
$$a = [a_i : i \in \omega] \times \{1\}.$$

It is easy to see that a_i is a name of G_i and that a is a name of H. In $M'[\pi^*(G'')]$, a_i is a name of G_i if $i < n$ and a name of $\pi^*(G'')_i = G_{\pi^{-1}(i)}$ if $i \geqq n$; so a is a name of H. This shows that a is $\mathfrak{A}$-invariant.

Historical Note. The independence of the axiom of choice is due to Cohen. Models of ZFC in which there is no OD mapping from an ordinal to $S(\omega)$ were first constructed by Feferman [**4**]. Theorem 9.1 is due to Lévy [**7**].

10. **Preserving cardinals.** We now turn to the relation between cardinals in M and cardinals in $M[G]$.

Lemma 10.1 *Every cardinal in $M[G]$ is a cardinal in M.*

Proof. Let $\mathfrak{m}$ be a cardinal in $M[G]$. By Lemma 7.1, $\mathfrak{m}$ is an ordinal in M. If $\mathfrak{m}$ is not a cardinal in M, there is a mapping from an ordinal smaller than $\mathfrak{m}$ onto $\mathfrak{m}$ which is in M and hence in $M[G]$. This is impossible.

Since $0, 1, \ldots, \omega$ are absolute, they are cardinals in both M and $M[G]$. On the other hand, uncountable cardinals in M need not be cardinals in $M[G]$, as we saw in the last section.

If α is an ordinal in M, then, since $M \subset M[G]$, we have

$$\text{(10.1)} \qquad \mathrm{cf}^{M[G]}(\alpha) \leqq \mathrm{cf}^M(\alpha).$$

The converse holds when $\mathrm{cf}^M(\alpha) \leqq \omega$, since being a limit ordinal is absolute. Again, the last section shows that the converse may fail when $\mathrm{cf}(\alpha)$ is uncountable in M.

We shall now obtain some sufficient conditions for the converses to hold.

Let C be a notion of forcing, and let $p, q \in C$. We say p and q are *compatible* if they have a common extension; otherwise, we say p and q are *incompatible*. We say *C satisfies the* $\mathfrak{m}$*-chain condition* if every set of pairwise incompatible elements in C has cardinal $<\mathfrak{m}$.

LEMMA 10.2 *Let* $\mathfrak{m}$ *be a regular cardinal in M such that C satisfies the* $\mathfrak{m}$*-chain condition in M. Then*: (a) *if* $\alpha \in M$ *and* $\mathfrak{m} \leqq \mathrm{cf}^M(\alpha)$, *then* $\mathrm{cf}^M(\alpha) = \mathrm{cf}^{M[G]}(\alpha)$; *every cardinal in M which is* $\geqq\mathfrak{m}$ *is a cardinal in* $M[G]$.

PROOF. Let $x(y) = z$ be the formula of set theory which says that x is a function whose value at y is z. We say that γ is a *possible value* of a at β if some p forces $a(\hat{\beta}) = \hat{\gamma}$. We claim that the set of possible values of a at β has cardinal $<\mathfrak{m}$. For each such possible value γ, let p_γ force $a(\hat{\beta}) = \hat{\gamma}$. It will clearly suffice to show that if $\gamma \neq \delta$, then p_γ and p_δ are incompatible. Suppose they had a common extension q. Choose a generic G' such that $q \in G'$. By the extension lemma, q forces $a(\hat{\beta}) = \hat{\gamma}$ and $a(\hat{\beta}) = \hat{\delta}$; so by the truth lemma, $\vdash_{G'} a(\hat{\beta}) = \hat{\gamma}$ and $\vdash_{G'} a(\hat{\beta}) = \hat{\delta}$. Hence $\gamma = \bar{a}(\beta) = \delta$.

Now let α be as in (a). Let $\mathfrak{n} = \mathrm{cf}^{M[G]}(\alpha)$. Then $\mathfrak{n}$ is a cardinal in $M[G]$ and hence in M. Let $\bar{a}$ be a mapping from $\mathfrak{n}$ onto a cofinal subset of α. Let b be the set of possible values of a at ordinals $<\mathfrak{n}$. Clearly $b \in M$. If $\sigma < \mathfrak{n}$ and $\bar{a}(\sigma) = \tau$, then some $p \in G$ forces $a(\hat{\sigma}) = \hat{\tau}$; so τ is a possible value of a at σ. Thus $\mathrm{Ra}(\bar{a}) \subset b$; so b is cofinal in α. Hence $\mathrm{cf}(\alpha) \leqq |b|$ in M.

By the result above, b is, in M, the union of $\mathfrak{n}$ sets, each having cardinal $<\mathfrak{m}$. If $\mathfrak{n} < \mathfrak{m}$, then $|b| < \mathfrak{m}$ in M (since $\mathfrak{m}$ is regular in M). This is impossible since $\mathfrak{m} \leqq \mathrm{cf}(\alpha) \leqq |b|$ in M. Thus $\mathfrak{m} \leqq \mathfrak{n}$; so $|b| \leqq \mathfrak{m} \cdot \mathfrak{n} = \mathfrak{n}$ in M. Thus $\mathrm{cf}(\alpha) \leqq |b| \leqq \mathfrak{n}$ in M, i.e. $\mathrm{cf}^M(\alpha) \leqq \mathrm{cf}^{M[G]}(\alpha)$. Using (10.1), we get equality.

Now let $\mathfrak{n}$ be a cardinal in M such that $\mathfrak{m} \leqq \mathfrak{n}$; we show by induction on $\mathfrak{n}$ that $\mathfrak{n}$ is a cardinal in $M[G]$. If $\mathfrak{n}$ is regular in M, then $\mathrm{cf}^M(\mathfrak{n}) = \mathfrak{n} \geqq \mathfrak{m}$; so $\mathrm{cf}^{M[G]}(\mathfrak{n}) = \mathrm{cf}^M(\mathfrak{n}) = \mathfrak{n}$ by (a); so $\mathfrak{n}$ is a cardinal in $M[G]$. If $\mathfrak{n}$ is singular in M, then $\mathfrak{n}$ is the supremum of the set of cardinals $\mathfrak{p}$ in M such that $\mathfrak{m} \leqq \mathfrak{p} < \mathfrak{n}$. Since these are all cardinals in $M[G]$ and the supremum of a set of cardinals is a cardinal, $\mathfrak{n}$ is a cardinal in $M[G]$.

COROLLARY. *If C satisfies the* $\aleph_1$*-chain condition in M, then* $\mathrm{cf}^M = \mathrm{cf}^{M[G]}$, *and M and* $M[G]$ *have the same cardinals.*

We wish to apply these results to $H_\mathfrak{m}(A, B)$. We note first that for each $\mathfrak{p} < \mathfrak{m}$ (including finite $\mathfrak{p}$), there are at most $|A|^\mathfrak{p}$ subsets D of A such that $|D| = \mathfrak{p}$, and that for each such D, there are $|B|^\mathfrak{p}$ mappings from D to B. Thus

$$|H_\mathfrak{m}(A, B)| \leqq \sum_{\mathfrak{p}<\mathfrak{m}} |A|^\mathfrak{p} \cdot |B|^\mathfrak{p};$$

so

$$|H_\mathfrak{m}(A, B)| \leqq (|A| \cdot |B|)^{\underline{\mathfrak{m}}}. \tag{10.2}$$

Lemma 10.3 *If* $\mathfrak{m}^{\underset{\smile}{\mathfrak{m}}} = \mathfrak{m}$ *and* $|B| \leqq \mathfrak{m}$, *then* $H_{\mathfrak{m}}(A, B)$ *satisfies the* $\mathfrak{m}^+$*-chain condition.*

Proof Let I be a set of pairwise incompatible elements of $H_{\mathfrak{m}}(A, B)$. We define a subset A_α of A for each α by induction. Let $A_0 = 0$; and for α a limit number, let $A_\alpha = \bigcup_{\beta<\alpha} A_\beta$. Now let A_α be chosen. For each $p \in H_{\mathfrak{m}}(A_\alpha, B)$, choose a $q \in I$ whose restriction to A_α is p, provided that such a q exists. Let $A_{\alpha+1}$ be the union of the domains of these q's and A_α.

We prove by induction that $|A_\alpha| \leqq \mathfrak{m}$ for $\alpha \leqq \mathfrak{m}$. This is trivial if $\alpha = 0$ or α is a limit number. Suppose $|A_\alpha| \leqq \mathfrak{m}$. By (10.2),

$$|H_{\mathfrak{m}}(A_\alpha, B)| \leqq (\mathfrak{m} \cdot |B|)^{\underset{\smile}{\mathfrak{m}}} = \mathfrak{m}^{\underset{\smile}{\mathfrak{m}}} = \mathfrak{m}. \tag{10.3}$$

Then clearly

$$|A_{\alpha+1}| \leqq |A_\alpha| + |H_{\mathfrak{m}}(A_\alpha, B)| \cdot \mathfrak{m} \leqq \mathfrak{m} + \mathfrak{m} \cdot \mathfrak{m} = \mathfrak{m}.$$

In particular, $|A_{\mathfrak{m}}| \leqq \mathfrak{m}$; so $|H_{\mathfrak{m}}(A_{\mathfrak{m}}, B)| \leqq \mathfrak{m}$ by (10.3).

We complete the proof by showing that $I \subset H_{\mathfrak{m}}(A_{\mathfrak{m}}, B)$. Let $p \in I$. Since $|\mathrm{Do}(p)| < \mathfrak{m}$, there is an $\alpha < \mathfrak{m}$ such that $\mathrm{Do}(p) \cap A_\alpha = \mathrm{Do}(p) \cap A_{\alpha+1}$. Choose $q \in I$ so that p and q have the same restriction to A_α and $\mathrm{Do}(q) \subset A_{\alpha+1}$. If $x \in \mathrm{Do}(p) \cap \mathrm{Do}(q)$, then $x \in A_{\alpha+1}$; so $x \in \mathrm{Do}(p) \cap A_{\alpha+1} \subset A_\alpha$; so $p(x) = q(x)$. It follows that p and q are compatible. Since $p, q \in I$, this implies $p = q$; so $\mathrm{Do}(p) \subset A_{\alpha+1} \subset A_{\mathfrak{m}}$ and hence $p \in H_{\mathfrak{m}}(A_{\mathfrak{m}}, B)$.

For conditions under which $\mathfrak{m}^{\underset{\smile}{\mathfrak{m}}} = \mathfrak{m}$ holds, see §2.

A subset D of C is a *section* if every extension of a condition in D is in D.

Lemma 10.4. *Let D be a section in M. If $D \cap G' \neq 0$ for every G' which is C-generic over M, then D is dense.*

Proof. Given p, choose G' generic such that $p \in G'$. Choose $q \in D \cap G'$, and let r be a common extension of p and q. Then $r \leqq p$ and $r \in D$.

A notion of forcing C is $\mathfrak{m}$-*closed* if for each $\alpha < \mathfrak{m}$ and each decreasing sequence $\{p_\beta\}_{\beta<\alpha}$ of conditions in C, there is a $p \in C$ such that $p \leqq p_\beta$ for all $\beta < \alpha$. For example, if $\mathfrak{m}$ is regular, then $H_{\mathfrak{m}}(A; B)$ is $\mathfrak{m}$-closed; for we may take $p = \bigcup_{\beta<\alpha} p_\beta$.

Lemma 10.5. *If C is $\mathfrak{m}$-closed in M, $\alpha < \mathfrak{m}$, and $\{D_\beta\}_{\beta<\alpha}$ is a sequence in M of dense sections, then $\bigcap_{\beta<\alpha} D_\beta$ is dense.*

Proof. Given p, we may define inductively in M a decreasing sequence $\{p_\beta\}_{\beta<\alpha}$ such that $p_\beta \in D_\beta$ and $p_\beta \leqq p$. (We must use the hypothesis that C is $\mathfrak{m}$-closed when β is a limit number.) Choosing q so that $q \leqq p_\beta$ for all $\beta < \alpha$, we have $q \leqq p$ and $q \in \bigcap_{\beta<\alpha} D_\beta$.

Lemma 10.6. *Let C be $\mathfrak{m}$-closed in M, and let $\alpha < \mathfrak{m}$. Then $({}^\alpha a)^M = ({}^\alpha a)^{M[G]}$ for all $a \in M$, and $S^M(\alpha) = S^{M[G]}(\alpha)$.*

PROOF. Clearly $({}^{\alpha}a)^{M} \subset ({}^{\alpha}a)^{M[G]}$. Now let $\bar{b} \in ({}^{\alpha}a)^{M[G]}$. Let $\Phi(x, y)$ be

$$\exists z(z \in \hat{a} \;\&\; \langle z, y\rangle \in b) \rightarrow \langle x, y\rangle \in b;$$

and for each $\beta < \alpha$, let

$$D_\beta = [p : \exists c(p \Vdash \Phi(\hat{c}, \hat{\beta}))].$$

Then D_β is a section in M. We use Lemma 10.4 to see that it is dense. Let G' be generic. Then for some $c \in a$, $\vdash_{G'} \Phi(\hat{c}, \hat{\beta})$; so by the truth lemma, $D_\beta \cap G' \neq 0$.

By Lemma 10.5, $D = \bigcap_{\beta<\alpha} D_\beta$ is dense. Hence there is a $q \in G \cap D$. If $q \Vdash \Phi(\hat{c}, \hat{\beta})$,then $c = \bar{b}(\beta)$. Hence $\bar{b}(\beta)$ is the unique c such that $q \Vdash \Phi(\hat{c}, \hat{\beta})$. It readily follows that $\bar{b} \in M$.

The last conclusion follows by taking $a = 2$ and using the correspondence between a set and its characteristic function.

COROLLARY. *Let C be $\mathfrak{m}$-closed in M. Then:* (a) *if $\alpha \in M$ and* $\mathrm{cf}^M(\alpha) \leqq \mathfrak{m}$, *then* $\mathrm{cf}^M(\alpha) = \mathrm{cf}^{M[G]}(\alpha)$; (b) *every cardinal $\mathfrak{n}$ in M which is $\leqq \mathfrak{m}$ is a cardinal in $M[G]$.*

PROOF. (a) If not, then $\mathrm{cf}^{M[G]}(\alpha) < \mathrm{cf}^M(\alpha) \leqq \mathfrak{m}$. Let f be a mapping from $\mathrm{cf}^{M[G]}(\alpha)$ onto a cofinal subset of α in $M[G]$. By the theorem, $f \in M$; so $\mathrm{cf}^{M[G]}(\alpha) \geqq \mathrm{cf}^M(\alpha)$, a contradiction. (b) If not, there is a mapping f from an ordinal $<\mathfrak{n}$ onto $\mathfrak{n}$ in $M[G]$. By the theorem, $f \in M$; and this is impossible.

We note that if we prove that M and $M[G]$ have the same cardinals by means of the theorems of this section, then $\mathrm{cf}^M = \mathrm{cf}^{M[G]}$. This suggests a problem: can we choose C so that M and $M[G]$ have the same cardinals, but $\mathrm{cf}^M \neq \mathrm{cf}^{M[G]}$? Prikry has shown that it is possible if there is a measurable cardinal in M.

HISTORICAL NOTE. The results on $\mathfrak{m}$-closed notions are due to Solovay; the remaining results are due to Cohen.

11. **The continuum hypothesis.** We first investigate the size of power sets in $M[G]$.

LEMMA 11.1. *Let C satisfy the $\mathfrak{m}$-chain condition in M. Then for every infinite cardinal $\mathfrak{n}$ in M,*

$$|S(\mathfrak{n})|^{M[G]} \leqq ((|C|^{\mathfrak{m}}_{\smile})^{\mathfrak{n}})^M.$$

PROOF. For $a \in M$ and $\alpha < \mathfrak{n}$, let $\phi_a(\alpha) = [p : p \Vdash \hat{\alpha} \in a]$. Then

$$\bar{a} \subset \mathfrak{n} \;\&\; \bar{b} \subset \mathfrak{n} \;\&\; \phi_a = \phi_b \rightarrow \bar{a} = \bar{b}.$$

By symmetry, it suffices to show that $\bar{a} \subset \bar{b}$. Let $\alpha \in \bar{a}$. Then some $p \in G$ is in $\phi_a(\alpha)$ and hence in $\phi_a(\beta)$; so $\vdash_G \hat{\alpha} \in b$; so $\alpha \in \bar{b}$.

It follows that

$$|S(\mathfrak{n})|^{M[G]} \leqq |[\phi_a : a \in M]|^M.$$

Now ϕ_a is a mapping from $\mathfrak{n}$ to Q, where Q is the set of all sets $[p : p \Vdash \Phi]$. It will thus suffice to prove $|Q| \leqq |C|^{\mathfrak{m}}_{\smile}$ in M.

Let $a \in Q$. Using Zorn's lemma, choose a maximal pairwise incompatible subset b of a. Then a can be recovered from b by the equivalence

$$p \in a \leftrightarrow (\forall q \leqq p)(\exists r \in b)(q \text{ and } r \text{ are compatible}).$$

The implication from left to right holds by the extension lemma and the maximality of b. Suppose $p \notin a$. If $a = [p : p \Vdash \Phi]$, then by (5.4) there is a $q \leqq p$ such that $q \Vdash \neg\Phi$. By (5.3) and the extension lemma, no r compatible with q can force Φ; so no such r can be in b.

It follows that $|Q|$ is at most equal to the number of pairwise incompatible subsets b of C. Since $|b| < \mathfrak{m}$ for all such b,

$$|Q| \leqq |S_{\mathfrak{m}}(C)| \leqq |C|^{\underset{\smile}{\mathfrak{m}}}.$$

Now let $\mathfrak{m}$ and $\mathfrak{n}$ be cardinals in M. Can we choose C so that M and $M[G]$ have the same cardinals and $2^{\mathfrak{m}} = \mathfrak{n}$ in $M[G]$? If we can, then

$$\mathfrak{n}^{\mathfrak{m}} = (2^{\mathfrak{m}})^{\mathfrak{m}} = 2^{\mathfrak{m}} = \mathfrak{n}$$

in $M[G]$. Since $({}^{\mathfrak{m}}\mathfrak{n})^{M} \subset ({}^{\mathfrak{m}}\mathfrak{n})^{M[G]}$, and M and $M[G]$ have the same cardinals, we have $(\mathfrak{n}^{\mathfrak{m}})^{M} \leq (\mathfrak{n}^{\mathfrak{m}})^{M[G]}$. Hence $\mathfrak{n}^{\mathfrak{m}} = \mathfrak{n}$ in M.

Let us therefore assume that $\mathfrak{n}^{\mathfrak{m}} = \mathfrak{n}$ in M. To get $2^{\mathfrak{m}} = \mathfrak{n}$, we should introduce $\mathfrak{n}$ subsets of $\mathfrak{m}$. We actually introduce a mapping F from $\mathfrak{n} \times \mathfrak{m}$ to 2, and take $[\beta : F(\alpha, \beta) = 0]$ to be the αth subset of $\mathfrak{m}$.

Let $C = H(\mathfrak{n} \times \mathfrak{m}; 2)$. Setting $F = \bigcup_{p \in G} p$, F is a mapping from $\mathfrak{m} \times \mathfrak{n}$ to 2 in $M[G]$. Set $A_{\alpha} = [\beta : \beta < \mathfrak{m} \;\&\; F(\alpha, \beta) = 0]$ for $\alpha < \mathfrak{n}$. Then A_{α} is a subset of $\mathfrak{m}$ in $M[G]$. Moreover, $\alpha \neq \alpha' \rightarrow A_{\alpha} \neq A_{\alpha'}$; this follows from the fact that

$$[p : \exists\beta(p(\alpha, \beta) \;\&\; p(\alpha', \beta) \text{ are defined and unequal})]$$

is a dense set in M.

By Lemma 10.3 and the Corollary to Lemma 10.2, $\mathrm{cf}^{M} = \mathrm{cf}^{M[G]}$ and M and $M[G]$ have the same cardinals. In particular, $\mathfrak{m}$ and $\mathfrak{n}$ are cardinals in $M[G]$. Since we have exhibited $\mathfrak{n}$ distinct subset of $\mathfrak{m}$ in $M[G]$, $2^{\mathfrak{m}} \geqq \mathfrak{n}$ in $M[G]$. By Lemma 11.1,

$$(2^{\mathfrak{m}})^{M[G]} \leqq ((|C|^{\aleph_0})^{\mathfrak{m}})^{M}.$$

Calculating in M, we have by (10.2)

$$|C| \leqq (\mathfrak{m} \cdot \mathfrak{n})^{\underset{\smile}{\aleph_0}} = \mathfrak{n}$$

(since $\mathfrak{n}^{\mathfrak{m}} = \mathfrak{n}$ implies $\mathfrak{m} < \mathfrak{n}$); so

$$(|C|^{\aleph_0})^{\mathfrak{m}} = |C|^{\mathfrak{m}} \leqq \mathfrak{n}^{\mathfrak{m}} = \mathfrak{n}.$$

Hence $2^{\mathfrak{m}} = \mathfrak{n}$ in $M[G]$. This proves the following theorem.

THEOREM 11.1. *Let $\mathfrak{m}$ and $\mathfrak{n}$ be infinite cardinals in M such that $\mathfrak{n}^{\mathfrak{m}} = \mathfrak{n}$ in M. For a suitable choice of C, $\mathrm{cf}^{M} = \mathrm{cf}^{M[G]}$; M and $M[G]$ have the same cardinals; and $2^{\mathfrak{m}} = \mathfrak{n}$ in $M[G]$.*

Now suppose that we have defined a constant Γ in ZFC and proved in ZFC that Γ is a cardinal. We would like to show that $2^{\aleph_0} = \Gamma$ is consistent with ZFC. In view of König's theorem, we require that $\mathrm{cf}(\Gamma) > \omega$ is provable in ZFC. Assume this, and take M to satisfy the axiom of constructibility and hence the GCH. From the GCH and $\mathrm{cf}(\Gamma) > \omega$ we can prove $\Gamma^{\aleph_0} = \Gamma$; so $(\Gamma^{\aleph_0})^M = \Gamma^M$. Choosing C as in Theorem 11.1, we have $2^{\aleph_0} = \Gamma^M$ in $M[G]$. We then have the desired result if we can show $\Gamma^M = \Gamma^{M[G]}$. Recalling that M and $M[G]$ have the same cardinals and the same cf function, this holds if Γ is, say, $\aleph_2$ or $\aleph_{\omega_1}$ or the first weakly inaccessible cardinal (provided that there is a weakly inaccessible cardinal in M).

Suppose now that the GCH holds in M. If we make $2^{\mathfrak{m}} = \mathfrak{n}$ in $M[G]$, then $2^{\mathfrak{p}} \geqq \mathfrak{n}$ for $\mathfrak{p} \geqq \mathfrak{m}$; so the GCH may fail in $M[G]$ above $\mathfrak{m}$. We show that we can keep the GCH below $\mathfrak{m}$ if $\mathfrak{m}$ is regular in M.

THEOREM 11.2. *Let the* GCH *hold in M. Let $\mathfrak{m}$ and $\mathfrak{n}$ be infinite cardinals of M such that $\mathfrak{m}$ is regular in M and* $\mathrm{cf}(\mathfrak{n}) > \mathfrak{m}$ *in M. For a suitable choice of C,* $\mathrm{cf}^M = \mathrm{cf}^{M[G]}$*; M and $M[G]$ have the same cardinals;* $2^{\mathfrak{m}} = \mathfrak{n}$ *in $M[G]$; and* $\forall \mathfrak{p}(\mathfrak{p} < \mathfrak{m} \rightarrow 2^{\mathfrak{p}} = \mathfrak{p}^+)$ *holds in $M[G]$.*

PROOF. We take $C = H^M_{\mathfrak{m}}(\mathfrak{m} \times \mathfrak{n}, 2)$. The hypotheses show that $\mathfrak{m}^{\mathfrak{m}}_{\smile} = \mathfrak{m}$ in M. Hence Lemmas 10.2 and 10.3 and the corollary to Lemma 10.6 show that $\mathrm{cf}^M = \mathrm{cf}^{M[G]}$ and that M and $M[G]$ have the same cardinals. The proof that $2^{\mathfrak{m}} = \mathfrak{n}$ in $M[G]$ is essentially as before (noting that $\mathrm{cf}(\mathfrak{n}) > \mathfrak{m}$ and the GCH imply $\mathfrak{n}^{\mathfrak{m}} = \mathfrak{n}$). If $\mathfrak{p} < \mathfrak{m}$, $S^M(\mathfrak{p}) = S^{M[G]}(\mathfrak{p})$ by Lemma 10.6. Since M and $M[G]$ have the same cardinals and $2^{\mathfrak{p}} = \mathfrak{p}^+$ in M, we see that $2^{\mathfrak{p}} = \mathfrak{p}^+$ in $M[G]$.

It is not known if Theorem 11.2 holds when $\mathfrak{m}$ is singular in M. The simplest unsolved problem is: is it consistent with ZFC to assume that $\forall n(n < \omega \rightarrow 2^{\aleph_n} = \aleph_{n+1})$ and $2^{\aleph_\omega} \neq \aleph_{\omega+1}$?

HISTORICAL NOTE. Theorem 11.1 is due to Cohen; Theorem 11.2 is due to Solovay.

12. **Forcing with classes.** So far we have assumed that C is a set in M. Sometimes we can construct forcing models when C is merely a class in M. Since the general situation has not been investigated very thoroughly, we shall consider only a specific problem.

This problem is a generalization of that in the last section. Suppose that H is a mapping from the set of infinite cardinals of M to itself which is a functional in M. We want to choose C so that M and $M[G]$ have the same cardinals, and so that for every infinite cardinal $\mathfrak{m}$ in M, $2^{\mathfrak{m}} = H(\mathfrak{m})$ in $M[G]$.

We must clearly have

(12.1) $$\mathfrak{m} \leqq \mathfrak{n} \rightarrow H(\mathfrak{m}) \leqq H(\mathfrak{n}).$$

Moreover

(12.2) $$\mathfrak{m} < \mathrm{cf}(H(\mathfrak{m}))$$

must hold in M. For it will hold in $M[G]$ by König's theorem and hence in M by (10.1).

We shall assume in addition that M satisfies the axiom of constructibility.[5] We will then show that for a suitable C, M and $M[G]$ have the same cardinals, $\mathrm{cf}^M = \mathrm{cf}^{M[G]}$, and $2^{\mathfrak{m}} = H(\mathfrak{m})$ in $M[G]$ for every regular cardinal $\mathfrak{m}$ in M.

REMARK. Again not much is known about singular cardinals. Certainly further hypotheses on H must be added to cover this case. For example,

$$(\forall n \in \omega)(2^{\aleph_n} = \aleph_{\omega+1}) \rightarrow 2^{\aleph_\omega} = \aleph_{\omega+1}.$$

For

$$2^{\aleph_\omega} = 2^{\Sigma \aleph_n} = \Pi\, 2^{\aleph_n} = \aleph_{\omega+1}^{\aleph_0} = (2^{\aleph_0})^{\aleph_0} = 2^{\aleph_0} = \aleph_{\omega+1}.$$

We now describe C. We let $\mathfrak{m}$ and $\mathfrak{n}$ vary through *regular* cardinals of M. We set

$$Q_{\mathfrak{m}} = [\langle \mathfrak{n}, \alpha, \beta\rangle : \mathfrak{n} \leqq \mathfrak{m} \;\&\; \alpha < H(\mathfrak{n}) \;\&\; \beta < \mathfrak{n}].$$

We let C be the set of all functions p in M such that $\mathrm{Ra}(p) \subset 2$; $\mathrm{Do}(p) \subset \bigcup_{\mathfrak{m}} Q_{\mathfrak{m}}$; and $|\mathrm{Do}(p) \cap Q_{\mathfrak{m}}| < \mathfrak{m}$ in M for all $\mathfrak{m}$. For $p, q \in C$, $p \leqq q$ means $q \subset p$. Obviously, C is a notion of forcing which is a class in M.

It is natural to require our generic set to meet the dense classes in M. We therefore choose G to be C-generic over M', where M' is the set of all classes in M. Since M' is countable, this can be done by the existence theorem.

We now define $M[G]$ as before. Just as before, we prove that $M[G]$ is countable and transitive and that $M \subset M[G]$. In trying to extend §5, there is a difficulty: the inductive definition of $p \Vdash^* a \neq b$ can no longer be given in M. We therefore proceed differently.

Let

$$C_{\mathfrak{m}} = [p : p \in C \;\&\; \mathrm{Do}(p) \subset Q_{\mathfrak{m}}],$$
$$C^{\mathfrak{m}} = [p : p \in C \;\&\; \mathrm{Do}(p) \cap Q_{\mathfrak{m}} = 0].$$

Both $C_{\mathfrak{m}}$ and $C^{\mathfrak{m}}$ contain 1 and hence are notions of forcing. Moreover $C_{\mathfrak{m}}$ is a set in M; the function which maps $\mathfrak{m}$ into $C_{\mathfrak{m}}$ is a functional in M; and $C^{\mathfrak{m}}$ is a class in M.

For $p \in C$, let $p_{\mathfrak{m}}$ be the restriction of p to $Q_{\mathfrak{m}}$, and let $p^{\mathfrak{m}} = p - p_{\mathfrak{m}}$. It is easy to check that $p \rightarrow \langle p_{\mathfrak{m}}, p^{\mathfrak{m}}\rangle$ is an isomorphism of C and $C_{\mathfrak{m}} \times C^{\mathfrak{m}}$ whose inverse takes $\langle q, r\rangle$ into $q \cup r$. Let $G_{\mathfrak{m}} = [p_{\mathfrak{m}} : p \in G]$, $G^{\mathfrak{m}} = [p^{\mathfrak{m}} : p \in G]$. Then by the product theorem,[6] $G_{\mathfrak{m}}$ is $C_{\mathfrak{m}}$-generic over M and $G^{\mathfrak{m}}$ is $C^{\mathfrak{m}}$-generic over $M[G^{\mathfrak{m}}]$. Moreover, G corresponds to $G_{\mathfrak{m}} \times G^{\mathfrak{m}}$ under the above isomorphism, i.e.

(12.3) $$G = [p \cup q : p \in G_{\mathfrak{m}} \;\&\; q \in G^{\mathfrak{m}}].$$

[5] Only the GCH and the existence of a definable well-ordering of the universe are used, and the latter can be dispensed with. However, this is immaterial for consistency purposes.

[6] Strictly speaking, the product theorem is not applicable here; but the part of the proof needed carries over without difficulty.

It follows that

$$G \cap C_{\mathfrak{m}} = G_{\mathfrak{m}}. \tag{12.4}$$

Noting that C is the union of the $C_{\mathfrak{m}}$, we may define $\Delta(b)$ in M by induction on $\mathrm{rk}(b)$ as follows: $\Delta(b)$ is the smallest $\mathfrak{m}$ such that $\Delta(a) \leqq \mathfrak{m}$ for all $a \in \mathrm{Ra}(b)$ and $p \in C_{\mathfrak{m}}$ for all $p \in \mathrm{Do}(b)$. We prove by induction on $\mathrm{rk}(b)$ that

$$\Delta(b) \leqq \mathfrak{m} \to K_G(b) = K_{G_{\mathfrak{m}}}(b). \tag{12.5}$$

If $\langle a, p\rangle \in b$, then $\Delta(a) \leqq \Delta(b) < \mathfrak{m}$, so $K_G(a) = K_{G_{\mathfrak{m}}}(a)$ by induction hypothesis. Also $p \in C_{\mathfrak{m}}$; so $p \in G \leftrightarrow p \in G_{\mathfrak{m}}$ by (12.4). Hence $K_G(b) = K_{G_{\mathfrak{m}}}(b)$.

Now define

$$a^{(\mathfrak{m})} = [\langle b^{(\mathfrak{m})}, p\rangle : \langle b, p\rangle \in a \,\&\, p \in C_{\mathfrak{m}}].$$

Again this is a functional in M. We prove

$$K_G(a^{(\mathfrak{m})}) = K_{G_{\mathfrak{m}}}(a) \tag{12.6}$$

by induction on $\mathrm{rk}(a)$. By the induction hypothesis and (12.4),

$$\begin{aligned} K_G(a^{(\mathfrak{m})}) &= [K_G(b^{(\mathfrak{m})}) : (\exists p \in G)(\langle b, p\rangle \in a \,\&\, p \in C_{\mathfrak{m}})] \\ &= [K_{G_{\mathfrak{m}}}(b) : (\exists p \in G_{\mathfrak{m}})(\langle b, p\rangle \in a)] \\ &= K_{G_{\mathfrak{m}}}(a). \end{aligned}$$

An easy induction shows that

$$\Delta(a^{(\mathfrak{m})}) \leqq \mathfrak{m}. \tag{12.7}$$

From (12.5) and (12.6) we obtain

$$M[G] = \bigcup_{\mathfrak{m}} M[G_{\mathfrak{m}}]. \tag{12.8}$$

Moreover,

$$\mathfrak{m} \leqq \mathfrak{n} \to M[G_{\mathfrak{m}}] \subset M[G_{\mathfrak{n}}]. \tag{12.9}$$

For $K_{G_{\mathfrak{m}}}(a) = K_G(a^{(\mathfrak{m})}) = K_{G_{\mathfrak{n}}}(a^{(\mathfrak{m})})$ by (12.6), (12.7), and (12.5).

All the $C_{\mathfrak{m}}$ have the same forcing language as C. We define $p \Vdash^*_{\mathfrak{m}} \Phi$ for $C_{\mathfrak{m}}$ as before. We then set $\Delta(a, b) = \max(\Delta(a), \Delta(b))$, and define

$$\begin{aligned} p \Vdash^* a \in b \quad &\text{if} \quad p_{\Delta(a,b)} \Vdash^*_{\Delta(a,b)} a \in b, \\ p \Vdash^* a \neq b \quad &\text{if} \quad p_{\Delta(a,b)} \Vdash^*_{\Delta(a,b)} a \neq b. \end{aligned}$$

The definability and extension lemmas are trivial. To prove the truth lemma for, say, $a \in b$, we have for $\mathfrak{m} = \Delta(a, b)$:

$$\begin{aligned} \vdash_G a \in b &\leftrightarrow \vdash_{G_{\mathfrak{m}}} a \in b \quad \text{by (12.5)} \\ &\leftrightarrow (\exists p \in G_{\mathfrak{m}})(p \Vdash^*_{\mathfrak{m}} a \in b) \\ &\leftrightarrow (\exists p \in G)(p_{\mathfrak{m}} \Vdash^*_{\mathfrak{m}} a \in b) \\ &\leftrightarrow (\exists p \in G)(p \Vdash^* a \in b). \end{aligned}$$

We can now define $p \Vdash^* \Phi$ for nonatomic Φ and prove the three lemmas on forcing as before.

We next observe that $C_{\mathfrak{m}}$ satisfies the $\mathfrak{m}^+$-chain condition (in M). For by Lemma 10.3, $H_{\mathfrak{m}}(Q_{\mathfrak{m}}, 2)$ satisfies the $\mathfrak{m}^+$-chain condition. But $C_{\mathfrak{m}} \subset H_{\mathfrak{m}}(Q_{\mathfrak{m}}, 2)$; and compatible elements in $H_{\mathfrak{m}}(Q_{\mathfrak{m}}, 2)$ are compatible in $C_{\mathfrak{m}}$, since their union is in $C_{\mathfrak{m}}$. This gives the desired result. We also note that $C^{\mathfrak{m}}$ is $\mathfrak{m}^+$-closed in M.

We say that $\{D_\beta\}_{\beta<\mathfrak{m}}$ is a *sequence of classes in M* if the set of pairs $\langle p, \beta\rangle$ such that $p \in D_\beta$ is a class in M. We note that Lemma 10.4 extends to classes in M and that Lemma 10.5, extends to sequences of classes in M.

LEMMA 12.1. *Let $\{D_\beta\}_{\beta<\mathfrak{m}}$ be a sequence of classes in M such that each D_β is a dense section in C. Then there is a $q \in G^{\mathfrak{m}}$ such that*

$$(\forall\alpha < \mathfrak{m})(\exists p \in G_{\mathfrak{m}})(p \cup q \in D_\alpha).$$

PROOF. For $q \in C^{\mathfrak{m}}$, let

$$D_\alpha^q = [p : p \in C_{\mathfrak{m}} \,\&\, p \cup q \in D_\alpha].$$

It will suffice to choose $q \in G^{\mathfrak{m}}$ such that for each $\alpha < \mathfrak{m}$, D_α^q is $C_{\mathfrak{m}}$-dense. Hence setting

$$D'_\alpha = [q : q \in C^{\mathfrak{m}} \;\&\; D_\alpha^q \text{ is } C_{\mathfrak{m}}\text{-dense}],$$

it will suffice to show that $\bigcap_{\alpha<\mathfrak{m}} D'_\alpha$ is dense. In view of the above remarks, we need only show that D'_α is dense.

Let $q' \in C^{\mathfrak{m}}$. Choose r_β inductively in M so that $r_\beta \leqq q'$; $r_\beta \in D_\alpha$; and for all $\gamma < \beta$, $(r_\beta)^{\mathfrak{m}} \leqq (r_\gamma)^{\mathfrak{m}}$ and $(r_\beta)_{\mathfrak{m}}$ and $(r_\gamma)_{\mathfrak{m}}$ are incompatible. If there are many such r_β, we use the definable well-ordering of D_α given by the axiom of constructibility to choose one; if there is no such r_β, then r_β is undefined.

Since $C_{\mathfrak{m}}$ satisfies the $\mathfrak{m}^+$-chain condition in M, we see that the smallest β such that r_β is undefined exists and satisfies $|\beta| \leqq \mathfrak{m}$ in M. Hence there is a $q \in C^{\mathfrak{m}}$ such that $q \leqq (r_\gamma)^{\mathfrak{m}}$ for $\gamma < \beta$ and $q \leqq q'$. We must show that $q \in D'_\alpha$, i.e. that D_α^q is dense.

Let $p \in C_{\mathfrak{m}}$. Then $p \cup q$ has an extension r in D_α. Since r is not a possible value for r_β, $r_{\mathfrak{m}}$ is compatible with some $(r_\gamma)_{\mathfrak{m}}$. Let p' be a common extension of $r_{\mathfrak{m}}$ and $(r_\gamma)_{\mathfrak{m}}$. Then $p' \leqq r_{\mathfrak{m}} \leqq p$; and $p' \cup q \leqq (r_\gamma)_{\mathfrak{m}} \cup (r_\gamma)^{\mathfrak{m}} = r_\gamma$, so $p' \cup q \in D_\alpha$ and hence $p' \in D_\alpha^q$.

LEMMA 12.2. *Let F be a functional in $M[G]$ such that $\mathfrak{m}$ is included in the domain of F. Then the restriction f of F to $\mathfrak{m}$ is in $M[G]$. If $[F(\alpha) : \alpha < \mathfrak{m}]$ is included in a set in $M[G_{\mathfrak{m}}]$, then f is in $M[G_{\mathfrak{m}}]$.*

PROOF. Let $\Phi(x, y)$ be a formula of the forcing language such that

$$F(\bar{a}) = \bar{b} \leftrightarrow \vdash_G \Phi(a, b). \tag{12.10}$$

For each $\alpha < \mathfrak{m}$, let

$$D_\alpha = [p : \exists b(p \Vdash \exists x\Phi(\hat{\alpha}, x) \rightarrow \Phi(\hat{\alpha}, b))].$$

Then $\{D_\alpha\}$ is a sequence of classes in M. Using Lemma 10.4, we see that D_α is a dense section. Hence by Lemma 12.1, there is a $q \in G^{\mathfrak{m}}$ such that

$$(12.11) \qquad (\forall \alpha < \mathfrak{m})(\exists p \in G_{\mathfrak{m}})(p \cup q \in D_\alpha).$$

For $\alpha < \mathfrak{m}$ and $p \in C_{\mathfrak{m}}$, let $A_{\alpha,p}$ be the set of b such that

$$(12.12) \qquad p \cup q \Vdash \exists x\Phi(\hat{\alpha}, x) \rightarrow \Phi(\hat{\alpha}, b).$$

Let g be a function in M whose domain is the set of $\langle \alpha, p\rangle$ in $\mathfrak{m} \times C_{\mathfrak{m}}$ such that $A_{\alpha,p} \neq 0$, and such that $g(\langle \alpha, p\rangle) \in A_{\alpha,p}$ for each such $\langle \alpha, p\rangle$. We show that

$$(12.13) \qquad f = [\langle K_G(b), \alpha\rangle : (\exists p \in G_{\mathfrak{m}})(b = g(\langle \alpha, p\rangle))].$$

By (12.11), there is for each $\alpha < \mathfrak{m}$ a $p \in G_{\mathfrak{m}}$ such that $A_{\alpha,p} \neq 0$, and hence such that $g(\langle \alpha, p\rangle)$ is defined. Hence we must show that if $p \in G_{\mathfrak{m}}$ and $b = g(\langle \alpha, p\rangle)$, then $K_G(b) = f(\alpha)$. Since (12.12) holds and since $p \cup q \in G$ by (12.3), $\vdash_G \exists x\Phi(\hat{\alpha}, x) \rightarrow \Phi(\hat{\alpha}, b)$. From this and (12.10), $F(\alpha) = K_G(b)$; so $f(\alpha) = K_G(b)$.

Choose $\mathfrak{n}$ so that $\mathfrak{m} \leqq \mathfrak{n}$ and $\Delta(b) \leqq \mathfrak{n}$ for $b \in \mathrm{Ra}(g)$. By (12.4) and (12.5), (12.13) becomes

$$f = [\langle K_{G_{\mathfrak{n}}}(b), \alpha\rangle : (\exists p \in G_{\mathfrak{n}})(b = g(\langle \alpha, p\rangle))].$$

Since $G_{\mathfrak{n}} \in M[G_{\mathfrak{n}}]$ and there is a functional in $M[G_{\mathfrak{n}}]$ coinciding with $K_{G_{\mathfrak{n}}}$ on arguments in M, it follows that $f \in M[G_{\mathfrak{n}}]$. Hence $f \in M[G]$.

Now suppose that $[F(\alpha) : \alpha < \mathfrak{m}] \subset K_{G_{\mathfrak{m}}}(a)$; we want to show that we may take $\mathfrak{n} = \mathfrak{m}$. We modify D_α to be the set of p such that

$$(\exists b \in \mathrm{Ra}(a^{(\mathfrak{m})}))[p \Vdash \exists x(x \in a^{(\mathfrak{m})} \;\&\; \Phi(\hat{\alpha}, x)) \rightarrow \Phi(\hat{\alpha}, b)].$$

In proving this is dense, we have to note that every member of $K_G(a^{(\mathfrak{m})})$ has a name in $\mathrm{Ra}(a^{(\mathfrak{m})})$ by (4.1). The fact that $\vdash_G \exists x(x \in a^{(\mathfrak{m})} \;\&\; \Phi(\alpha, x)) \rightarrow \Phi(\alpha, b)$ yields $F(\alpha) = K_G(b)$ now uses (12.6). We can now suppose that every b in $\mathrm{Ra}(g)$ is in $\mathrm{Ra}(a^{(\mathfrak{m})})$ and hence is of the form $b'^{(\mathfrak{m})}$. But $\Delta(b'^{(\mathfrak{m})}) \leqq \mathfrak{m}$; so we may indeed take $\mathfrak{n} = \mathfrak{m}$.

LEMMA 12.3. *If $\bar{a} \in M[G]$, and $\bar{a} \neq 0$, then there is an $\mathfrak{m}$ such that $\bar{a} \in M[G_{\mathfrak{m}}]$ and a mapping f of $\mathfrak{m}$ onto $\bar{a}$ which is in $M[G_{\mathfrak{m}}]$.*

PROOF. Choose $\mathfrak{n}$ by 12.8 so that $\bar{a} \in M[G_{\mathfrak{n}}]$. Choose $\mathfrak{m}$ so that $\mathfrak{n} \leqq \mathfrak{m}$ and $|\bar{a}| \leqq \mathfrak{m}$ in $M[G_{\mathfrak{n}}]$. Then there is an f in $M[G_{\mathfrak{n}}]$ mapping $\mathfrak{m}$ onto $\bar{a}$; and $\bar{a}, f \in M[G_{\mathfrak{m}}]$ by (12.9).

Now we verify that $M[G]$ satisfies the conditions of Lemma 2.1. Since $\omega \in M \subset M[G]$, (a) holds. Now let $\bar{a} \in M[G]$, $\bar{a} \neq 0$; and choose $\mathfrak{m}$ and f as in Lemma 12.3. Suppose A is a class in $M[G]$ such that $A \subset \bar{a}$ and $A \neq 0$. Since $f \in M[G]$, it is easy to define a functional F in $M[G]$ with domain $\mathfrak{m}$ and range A. By Lemma 12.2, $F \in M[G_{\mathfrak{m}}]$; so $A \in M[G_{\mathfrak{m}}] \subset M[G]$. This proves (b). It also shows that every subset of $\bar{a}$ in $M[G]$ is in $M[G_{\mathfrak{m}}]$ and hence in the power set of $\bar{a}$ in $M[G_{\mathfrak{m}}]$. This proves (d).

Now let F be a functional in $M[G]$, and let $\bar{a}$ be a nonempty subset of the domain of F. Again let $\mathfrak{m}$ and f be as in Lemma 12.3, and let F_1 be the functional defined by $F_1(\alpha) = F(f(\alpha))$. Then $F_1 \in M[G]$ by Lemma 12.2. Hence for some $\mathfrak{n}$, $F_1 \in M[G_{\mathfrak{n}}]$; so $U(\mathrm{Ra}(F_1)) \in M[G_{\mathfrak{n}}] \subset M[G]$. But $U(\mathrm{Ra}(F_1)) = U([F(x):x \in \bar{a}])$; so we have proved (c).

By Lemma 2.1, $M[G]$ is a model of ZF. Using Lemma 12.3, we see that $M[G]$ is a model of ZFC.

Now we show that $\mathrm{cf}^M = \mathrm{cf}^{M[G]}$. If not, we would have $\aleph_0 \leqq {}^{M[G]}(\alpha) < \mathrm{cf}^M(\alpha)$ for some α. Set $\mathfrak{m} = \mathrm{cf}^{M[G]}(\alpha)$. This is an infinite cardinal in M; and since $\mathfrak{m} = \mathrm{cf}^{M[G]}(\mathfrak{m}) \leqq \mathrm{cf}^M(\mathfrak{m}) \leqq \mathfrak{m}$, it is regular in M. Let f be a mapping of $\mathfrak{m}$ onto a cofinal subset of α such that $f \in M[G]$. By Lemma 12.2, $f \in M[G_{\mathfrak{m}}]$. Hence $\mathrm{cf}^{M[G_{\mathfrak{m}}]}(\alpha) \leqq \mathfrak{m}$. On the other hand, $\mathfrak{m} < \mathrm{cf}^M(\alpha)$ implies $\mathfrak{m}^+ \leqq \mathrm{cf}^M(\alpha)$; so $\mathrm{cf}^M(\alpha) = \mathrm{cf}^{M[G_{\mathfrak{m}}]}(\alpha)$ by Lemma 10.2. Thus $\mathfrak{m} < \mathrm{cf}^M(\alpha) = \mathrm{cf}^{M[G_{\mathfrak{m}}]}(\alpha) \leqq \mathfrak{m}$, a contradiction.

From $\mathrm{cf}^M = \mathrm{cf}^{M[G]}$ it follows that M and $M[G]$ have the same cardinals; the proof is like the proof of Lemma 10.2(b).

Any nonempty subset of $\mathfrak{m}$ in $M[G]$ is the image of $\mathfrak{m}$ under a functional in $M[G]$ and hence is in $M[G_{\mathfrak{m}}]$ by Lemma 12.2. Thus $S^{M[G]}(\mathfrak{m}) = S^{M[G_{\mathfrak{m}}]}(\mathfrak{m})$; so to prove $2^{\mathfrak{m}} = H(\mathfrak{m})$ in $M[G]$, it suffices to prove it in $M[G_{\mathfrak{m}}]$. Setting $F = U(G_{\mathfrak{m}})$ and

$$A_\alpha = [\beta:\beta < \mathfrak{m} \;\&\; F(\mathfrak{m}, \alpha, \beta) = 0]$$

for $\alpha < H(\mathfrak{m})$, we prove as before that the A_α are $H(\mathfrak{m})$ distinct subsets of $\mathfrak{m}$. Making use of (12.1), we have in M,

$$|Q_{\mathfrak{m}}| \leqq \mathfrak{m} \cdot H(\mathfrak{m}) \cdot \mathfrak{m} = H(\mathfrak{m})$$

(since (12.2) gives $\mathfrak{m} < H(\mathfrak{m})$). Since $C_{\mathfrak{m}} \subset H_{\mathfrak{m}}(Q_{\mathfrak{m}}, 2)$, we get

$$|C_{\mathfrak{m}}| \leqq H(\mathfrak{m})^{\mathfrak{m}}_{\smile} = H(\mathfrak{m})$$

by (10.2) and (12.2). Then by Lemma 11.1,

$$|S(\mathfrak{m})|^{M[G_{\mathfrak{m}}]} \leqq (H(\mathfrak{m})^{\mathfrak{m}\cdot\mathfrak{m}})^M = H(\mathfrak{m}).$$

Thus $2^{\mathfrak{m}} = H(\mathfrak{m})$ in $M[G_{\mathfrak{m}}]$.

Historical Note. The results of this section are due to Easton [**3**].

Bibliography

1. P. J. Cohen, *The independence of the continuum hypothesis, Parts* I, II, Proc. Nat. Acad. Sci. U.S.A. **50** (1963), 1143–1148; **51** (1964), 105–110.

2. ———, *Set theory and the continuum hypothesis*, Benjamin, New York, 1966.

3. W. Easton, *Powers of regular cardinals*, Thesis, Princeton University, 1964.

4. S. Feferman, *Some applications of the notion of forcing and generic sets*, Fund. Math. **56** (1965), 325–345.

5. K. Gödel, *The consistency of the axiom of choice and of the generalized continuum hypothesis with the axioms of set theory*, Ann. of Math. Studies No. 3, Princeton University Press, Princeton, N.J., 1940.

6. A. Lévy, *Axiom schemata of strong infinity in axiomatic set theory*, Pacific J. Math. **10** (1960), 223–238.

7. ———, Definability in axiomatic set theory. I, (Proc. 1964 Internat. Congr.) *Logic, Methodology, and Philosophy of Science*, North-Holland, Amsterdam, 1966.

8. A. Mostowski, *An undecidable arithmetical statement*, Fund. Math. **36** (1949), 143–164.

9. J. C. Shepherdson, *Inner models for set theory, Part* I, J. Symbolic Logic **16** (1951), 161–190.

10. J. R. Shoenfield, *Mathematical logic*, Addison-Wesley, Reading, Mass., 1967.

DUKE UNIVERSITY AND
UNIVERSITY OF CALIFORNIA, LOS ANGELES

THE INDEPENDENCE OF KUREPA'S CONJECTURE AND TWO-CARDINAL CONJECTURES IN MODEL THEORY

JACK SILVER[1]

1. We begin with a word on terminology. Suppose $\mathfrak{A} = \langle A, \in \rangle$ is a structure where $0 \in A$ and $\in$ is the $\in$ relation on A. For each $a \in A$, a is an individual constant denoting a in $\mathfrak{A}$. The symbols $\in$, $=$, and $\{:\}$ have the customary (ZF) meanings. If σ is a sentence and τ a closed term for a first-order language involving only these symbols, then $\models_{\mathfrak{A}} \sigma$ means that σ is true in $\mathfrak{A}$ and $\tau^{\mathfrak{A}}$ is the denotation of τ in $\mathfrak{A}$. A and $\mathfrak{A}$ are used interchangeably.

By way of set-theoretic notation, $\bar{\bar{x}}$ is the cardinal number of x and Sx is the power set of x. κ, λ are always infinite cardinals. $f \upharpoonright X$ is the restriction of the function f to domain X.

In the remainder of this section, we sketch a convenient general framework for applying Cohen's methods, most of the basic ideas of which are due to Cohen and Solovay.

Suppose $\mathcal{M}$ is a countable transitive model of ZF + AC, and γ is the first ordinal not in $\mathcal{M}$. If S is a subset of $\mathcal{M}$, $\mathcal{M}[S]$ is defined to be the $\in$-structure whose universe is

$$\{X : (\exists \alpha \in \gamma, y \in \mathcal{M})(X \text{ is constructible from } y \text{ and } S \text{ at stage } \alpha)\}.$$

Suppose further that $\mathcal{P} = \langle P, \leq \rangle$ is a nonempty partially ordered structure such that $\mathcal{P} \in \mathcal{M}$. (Intuitively, $\mathcal{P}$ is the set of forcing conditions, and $p \leq q$

[1] This research was supported in part by an NSF Grant (Kleene, 1966–1967) at the University of Wisconsin and NSF GP-5632 (1967–1968) at the University of California.

means that p is stronger than q.) $S \subseteq \mathscr{P}$ is said to be *dense* in $\mathscr{P}$ if

$$(\forall p \in \mathscr{P})(\exists q \in S)(q \leq p).$$

p, q are said to be compatible in $\mathscr{P}$ if $(\exists r \in \mathscr{P})(r \leq p$ and $r \leq q)$. $G \subseteq \mathscr{P}$ is said to be $\mathscr{P}$-generic over $\mathscr{M}$ if the following conditions hold: (1) if $p, q \in G$, then there exists $r \in G$ such that $r \leq p$ and $r \leq q$; (2) if $p \geq q \in G$, then $p \in G$; (3) if $S \in \mathscr{M}$ and S is dense in $\mathscr{P}$, then $S \cap G \neq 0$.

Let $\mathscr{L}(\mathscr{M})$ be a first-order language with $=, \in, \{:\}$ having a name $\underset{\sim}{x}$ for each $x \in \mathscr{M}$ and an individual constant symbol $\mathscr{G}$. Define $\mathscr{M}[G]*$ to be $(\mathscr{M}[G], G)$, and for each $x \in \mathscr{M}$ set $\underset{\sim}{x}^{\mathscr{M}[G]*} = x$ and $\mathscr{G}^{\mathscr{M}[G]*} = G$.

We proceed now to list some basic facts concerning these notions.

I. (a) If $p \in \mathscr{P}$, then there exists a G containing p which is $\mathscr{P}$-generic over $\mathscr{M}$.

(b) If G is $\mathscr{P}$-generic over $\mathscr{M}$, then $\mathscr{M}[G]$ is a transitive model of ZF + AC containing the same ordinals as $\mathscr{M}$.

(c) Every element of $\mathscr{M}[G]$ is definable in $\mathscr{M}[G]*$ by a formula of $\mathscr{L}(\mathscr{M})$.

II. There is a relation $\Vdash_{\mathscr{P}}$ (called forcing) between members of $\mathscr{P}$ and sentences of $\mathscr{L}(\mathscr{M})$ such that:

(a) for any $p \in \mathscr{P}$ and sentence σ in $\mathscr{L}(\mathscr{M})$, $p \Vdash \sigma$ if and only if σ holds in every $\mathscr{M}[G]*$ where G is $\mathscr{P}$-generic over $\mathscr{M}$ and $p \in G$.

(b) If G is $\mathscr{P}$-generic over $\mathscr{M}$, then σ holds in $\mathscr{M}[G]*$ if and only if there is some $p \in G$ such that $p \Vdash \sigma$.

(c) For every formula θ in $\mathscr{L}(\mathscr{M})$ $\{\langle p, x_1, \ldots, x_m\rangle : p \Vdash \theta(x_1, \ldots, x_m)\}$ is definable in $\mathscr{M}$ from parameters in $\mathscr{M}$.

The above statements can be proved using either the ramified language techniques of Cohen [**2**] and Lévy [**5**], or Boolean methods of Scott-Solovay [**10**].

From facts I–II, one can now deduce III–VII.

III. If $\models_{\mathscr{M}} \underset{\sim}{\lambda}$ is a regular cardinal, and $\models_{\mathscr{M}}$ every set of pairwise incompatible members of $\underset{\sim}{\mathscr{P}}$ has cardinality $< \underset{\sim}{\lambda}$, then all cardinals of $\mathscr{M} \geq \lambda$ remain cardinals in $\mathscr{M}[G]$, provided G is $\mathscr{P}$-generic over $\mathscr{M}$.

III. is a generalization of Theorem 2, Cohen [**2**, p. 132]. See also Scott-Solovay [**10**].

IV. If $\{p : p \Vdash \sigma\}$ is dense in $\mathscr{M}$, then $\Vdash \sigma$ (i.e. σ holds in every $\mathscr{M}[G]*$ where G is $\mathscr{P}$-generic over $\mathscr{M}$).

This is clear from IIb, c and the definition of genericity.

V. If $\models_{\mathscr{M}}$(every countable compatible subset of $\mathscr{P}$ has a lower bound in $\mathscr{P}$), and G is $\mathscr{P}$-generic over $\mathscr{M}$, then any ω-sequence of ordinals which is a member of $\mathscr{M}[G]$ is already a member of $\mathscr{M}$.

By a principle similar to IV, V will be established if we can show that whenever $p \Vdash \tau$ is an ω-sequence of ordinals, then there is some $q \leq p$ and some $s \in \mathscr{M}$ such that $q \Vdash \tau = \underset{\sim}{s}$. To show the existence of such a q, build (in $\mathscr{M}$) a descending sequence $p_i, p_0 = p$ and an ω-sequence s satisfying: for each $i \in \omega$, $p_{i+1} \leq p_i$ and $p_{i+1} \Vdash \tau(\underset{\sim}{i}) = \underset{\sim}{s_i}$. Take q to be a lower bound for the p_i. (V is a well-known fact which is stated in Lévy-Solovay [**6**].)

VI. Suppose $\mathscr{P}_1$ and $\mathscr{P}_2$ are partially ordered structures in $\mathscr{M}$, each with a greatest element, and $\mathscr{P} = \mathscr{P}_1 \times \mathscr{P}_2$ (where $\langle p_1, p_2\rangle \leq \langle q_1, q_2\rangle$ iff $p_1 \leq q_1$ and $p_2 \leq q_2$).

(a) If G_1 is $\mathscr{P}_1$-generic over $\mathscr{M}$ and G_2 is $\mathscr{P}_2$-generic over $\mathscr{M}[G_1]$, then $G_1 \times G_2$ is $\mathscr{P}$-generic over $\mathscr{M}$.

(b) If G is $\mathscr{P}$-generic over $\mathscr{M}$, then there exists G_1 $\mathscr{P}_1$-generic over $\mathscr{M}$ and G_2 $\mathscr{P}_2$-generic over $\mathscr{M}[G_1]$ such that $G = G_1 \times G_2$.

Suppose that G is $\mathscr{P}$-generic over $\mathscr{M}$. Let 1 be the greatest element of both $\mathscr{P}_1$ and $\mathscr{P}_2$. If $G_1 = \{p_1 : \langle p_1, 1\rangle \in G\}$ and $G_2 = \{p_2 : \langle 1, p_2\rangle \in G\}$, then certainly $G = G_1 \times G_2$.

To see that G_1 is $\mathscr{P}$-generic over $\mathscr{M}$, notice that if D is a dense subset of $\mathscr{P}_1$ and $D \in \mathscr{M}$, then $D \times \mathscr{P}_2 \in \mathscr{M}$ is a dense subset of $\mathscr{P}$ and hence meets G.

To show that G_2 is $\mathscr{P}_2$-generic over $\mathscr{M}[G_1]$, suppose $D \in \mathscr{M}[G_1]$, say D is denoted by term τ of $\mathscr{L}(\mathscr{M})$ in $\mathscr{M}[G_1]*$, and suppose D is dense in $\mathscr{P}_2$. We proceed to show that G_2 meets D.

Take $p_1 \in G_1$ so that $p_1 \Vdash_1 \tau$ is dense in $\underline{\mathscr{P}}_2$ ($\Vdash_1$ is forcing appropriate for $\mathscr{P}_1$ over $\mathscr{M}$). Set

$$E = \{\langle q_1, q_2\rangle : q_1 \leq p_1 \quad \text{and} \quad q_1 \Vdash_1 \underline{q}_2 \in \tau\}.$$

Claim. E is dense beneath $p = \langle p_1, 1\rangle$, i.e., $(\forall r \leq p)(\exists q \leq r)(q \in E)$.

If $r = \langle r_1, r_2\rangle \leq p$, take an H_1 which is $\mathscr{P}_1$-generic over $\mathscr{M}$ and such that $r_1 \in H_1$. τ denotes a dense subset of $\mathscr{P}_2$ in $\mathscr{M}[H_1]*$ (since $p_1 \Vdash_1 \tau$ is dense in $\underline{\mathscr{P}}_2$), so $\exists q_2 \leq r_2$ such that $q_2 \in \tau^{\mathscr{M}[H_1]*}$. Hence $\exists q_1 \leq r_1$ such that $q_1 \Vdash_1 \underline{q}_2 \in \tau$. Thus $\langle q_1, q_2\rangle \leq \langle r_1, r_2\rangle$ and $\langle q_1, q_2\rangle \in E$.

E is dense beneath p and $E \in \mathscr{M}$, $p \in G$, so one has easily that $G \cap E \neq 0$. If $\langle s_1, s_2\rangle \in G \cap E$, then $s_2 \in G_2 \cap D$. Q.E.D.

VIa is not needed, so we omit its proof.

VII. Suppose $\models_{\mathscr{M}}$(every set of mutually incompatible elements of $\underline{\mathscr{P}}$ has cardinality $< \underline{\lambda}$), G is $\mathscr{P}$-generic over $\mathscr{M}$, κ is a cardinal in $\mathscr{M}[G]$, and $\mu = ((\sum_{\nu<\underline{\lambda}} \overline{\overline{\underline{\mathscr{P}}^\nu}})^\kappa)^{\mathscr{M}}$. Then $(2^{\underline{\kappa}})^{\mathscr{M}[G]} \leq \mu$.

This is best proved by associating (in $\mathscr{M}$) with $\mathscr{P}$ the Boolean algebra $\mathscr{B}$ of regular open subsets of $\mathscr{P}$ (under the topology generated by sets $T_p = \{q \in \mathscr{P} : q \leq p\}$), showing the cardinality of $\mathscr{B}$ to be $\leq \sum_{\nu<\lambda} \mathscr{P}^\nu$ (in $\mathscr{M}$) under our assumptions, and seeing that each subset of κ in $\mathscr{M}[G]$ comes from a map of κ into $\mathscr{B}$ in $\mathscr{M}$. $\mathscr{B}$ is the algebra to be used in the Scott-Solovay method to construct our model.

2. $K(\aleph_\alpha)$–Kurepa's conjecture [8] for $\aleph_\alpha$—is the statement that there exists a family $\mathscr{W}$ of $\aleph_{\alpha+2}$ subsets of $\omega_{\alpha+1}$ such that, for all $\beta \in \omega_{\alpha+1}$, the cardinality of $\{S \cap \beta : S \in \mathscr{W}\}$ is at most $\aleph_\alpha$.

We intend to construct a model of ZF + GCH + $\sim K(\aleph_0) + K(\aleph_1)$. The new result here is the independence of $K(\aleph_0)$. Lévy and Rowbottom [9] independently showed its consistency in 1963.

We start with a countable transitive model $\mathscr{M}$ of ZF + V = L + $\exists$ inaccessible cardinal. (V = L is only a matter of convenience. The GCH would suffice.) Let θ be an inaccessible cardinal in the sense of $\mathcal{M}$. Our model is obtained from $\mathcal{M}$ by adjoining a "generic sequence" $\langle f_\alpha : \alpha < \theta \rangle$ of collapsing functions, $f_\alpha : \omega_1^{\mathcal{M}} \xrightarrow{\text{onto}} \omega_{\underline{\alpha}}^{\mathcal{M}}$.

Define

$$\mathcal{P} = \{f : \mathcal{D}f \subseteq (\theta \times \omega_1^{\mathcal{M}}) \text{ and } f \text{ is a function and } (\forall \alpha \in \theta, \beta \in \omega_1)(f(\alpha, \beta) \in \omega_{\underline{\alpha}}^{\mathcal{M}}) \text{ and } f \in \mathcal{M} \text{ and } \models_{\mathcal{M}} \mathcal{D}f \text{ is countable}\},$$

ordered by: $f \leq g$ iff $f \supseteq g$. Clearly $\mathcal{P} \in \mathcal{M}$.

If G is $\mathcal{P}$-generic over $\mathcal{M}$, then, for $\alpha < \theta$, if $f_\alpha = \{\langle \beta, \gamma \rangle : \langle\langle \alpha, \beta \rangle, \gamma \rangle \in \bigcup G\}$, it is easy to see that $f_\alpha : \omega_1^{\mathcal{M}} \xrightarrow{\text{onto}} \omega_{\underline{\alpha}}^{\mathcal{M}}$. (So, if f is a condition such that $f(\alpha, \beta) = \gamma$, then f says that $f_\alpha(\beta) = \gamma$ for the resulting f_α of the model.) Clearly

$$\mathcal{M}[G] = \mathcal{M}[\langle f_\alpha : \alpha < \theta \rangle].$$

Let $\mathcal{N} = \mathcal{M}[G] = \mathcal{M}[\langle f_\alpha : \alpha < \theta \rangle]$, where G is an arbitrary set $\mathcal{P}$-generic over $\mathcal{M}$. Trivially $\omega^{\mathcal{M}} = \omega^{\mathcal{N}}$. Also $\models_{\mathcal{M}}$ every countable compatible subset of $\mathcal{P}$ has a lower bound in $\mathcal{P}$, so V tell us that no new countable sequences of ordinals appear in $\mathcal{N}$. Thus $(S\omega)^{\mathcal{M}} = (S\omega)^{\mathcal{N}}$ and $\omega_1^{\mathcal{M}} = \omega_1^{\mathcal{N}}$. Since $f_\alpha : \omega_1^{\mathcal{M}} \xrightarrow{\text{onto}} \omega_{\underline{\alpha}}^{\mathcal{M}}$, all $\alpha < \theta$, we have $\omega_{\underline{\alpha}}^{\mathcal{M}}$ equipollent to $\omega_{\underline{1}}^{\mathcal{M}}$ in $\mathcal{N}$. But the following lemma tells us that $\models_{\mathcal{M}}$ every mutually incompatible subset of $\mathcal{P}$ has cardinality $< \underline{\theta}$, so, by III, every cardinal in $\mathcal{M} \geq \theta$ remains a cardinal. Thus $\theta = \omega_2^{\mathcal{N}}$, $\omega_{\underline{\theta+1}}^{\mathcal{M}} = \omega_3^{\mathcal{N}}$, etc.

Let us make the following definitions:

DEFINITIONS. If $p \in \mathcal{P}$, $[p]_\nu = p \upharpoonright (\nu \times \omega_1)$, $[p]^\nu = p_\nu \upharpoonright (\theta \sim \nu) \times \omega_1)$. (So $[p]_\nu$ talks about f_α, $\alpha < \nu$, and $[p]^\nu$ about f_α, $\alpha \geq \nu$.)

The following lemma is stated and proved in the sense of $\mathcal{M}$.

LEMMA. *If $\mathcal{F}$ is a family of mutually incompatible elements of $\mathcal{P}$, then $\overline{\overline{\mathcal{F}}} < \theta$.*

PROOF. We can easily define a sequence $\langle(\mathcal{F}_\alpha, \nu_\alpha) : \alpha < \omega_1\rangle$ so that $\mathcal{F}_\alpha \subseteq \mathcal{F}$, $\mathcal{F}_\alpha$ is increasing, $\nu_\alpha < \theta$, ν_α strictly increasing, $\mathcal{F}_0$ has exactly one element,

$$(\forall p \in \mathcal{F}_\alpha)([p]_\nu = p), \ (\forall p \in \mathcal{F})(\exists q \in \mathcal{F}_{\alpha+1})([q]_{\nu_\alpha} = [p]_{\nu_\alpha}),$$

and $\overline{\overline{\mathcal{F}}}_\alpha < \theta$. *Claim.* $\mathcal{F} = \bigcup_{\alpha<\omega_1} \mathcal{F}_\alpha$ (which will show $\mathcal{F} < \theta$ since θ is regular, and each $\overline{\overline{\mathcal{F}}}_\alpha < \theta$). Let $\nu = \bigcup_{\alpha<\omega_1} \nu_\alpha$. If $p \in \mathcal{F}$, $p \notin \bigcup_{\alpha<\omega_1} \mathcal{F}_\alpha$, then for some α $[p]_\nu = [p]_{\nu_\alpha}$. Take $q \in \mathcal{F}_{\alpha+1}$ so that $[p]_{\nu_\alpha} = [q]_{\nu_\alpha}$. p and q can't be incompatible, which is a contradiction. (This is a well-known argument, the prototype for which can be found in Cohen [**2**, p. 131].)

We can now check the GCH in $\mathcal{N}$. $(S\omega)^{\mathcal{M}} = (S\omega)^{\mathcal{N}}$ gives $2^{\aleph_0} = \aleph_1$ in $\mathcal{N}$. Since $(\sum_{\alpha<\underline{\theta}} \overline{\overline{\mathcal{P}^\alpha}})^{\mathcal{M}} = \theta$, III gives us the GCH for cardinals in $\mathcal{N} \geq \omega_1$.

THEOREM. *$K(\aleph_0)$ fails in $\mathcal{N}$.*

PROOF. Suppose $K(\aleph_0)$ holds in $\mathscr{N}$, i.e. there is a family $\mathscr{W}$ of $\aleph_2$ subsets of ω_1 such that $\{S \cap \alpha : S \in \mathscr{W}\}$ is countable for each $\alpha < \omega_1$, all in the sense of $\mathscr{N}$. $\langle\{S \cap \alpha : S \in \mathscr{W}\} : \alpha < \omega_1\rangle$ can be coded up as a subset of ω_1. Since it is constructible in $\mathscr{N}$ from $\langle f_\alpha : \alpha < \theta\rangle$, $\theta = \omega_2^{\mathscr{N}}$, a theorem of Hajnal [3] shows that it is constructible from $\langle f_\alpha : \alpha < \delta\rangle$, some $\delta < \theta$. Assume $\omega_1 \leq \delta$. (Remark: The theorem of Hajnal says that any $A \subseteq \omega_1$ constructible from $B \subseteq \omega_2$ is already constructible from $B \cap \delta$, some $\delta \leq \omega_2$. Here it was convenient to use the fact that $\mathscr{M}$ is a model of V = L, but, so long as the GCH holds in $\mathscr{M}$, a chain condition argument can be substituted for constructibility considerations at this point in the proof of the theorem.)

Let $\mathscr{P}_1 = \{p \in \mathscr{P} : [p]_\delta = p\}$, $\mathscr{P}_2 = \{p \in \mathscr{P} : [p]^\delta = p\}$, $G_1 = G \cap \mathscr{P}_1$, $G_2 = G \cap \mathscr{P}_2$. We have canonically $\mathscr{P} \cong \mathscr{P}_1 \times \mathscr{P}_2$, $G \cong G_1 \times G_2$ under some isomorphism (see VI), and by VIb, G_2 is $\mathscr{P}_2$-generic over $\mathscr{M}[G_1]$ and G_1 is $\mathscr{P}_1$-generic over $\mathscr{M}$. So $\mathscr{M}' = \mathscr{M}[G_1] = \mathscr{M}[\langle f_\alpha : \alpha < \delta\rangle]$ is a model of ZF + AC. Clearly $\omega_1^{\mathscr{M}} = \omega_1^{\mathscr{M}'} = \omega_1^{\mathscr{N}}$; and, by VII, $(S\omega_1)^{\mathscr{M}'}$ has cardinality at most that of $(\delta^{\underline{\delta}})^{\mathscr{M}}$; hence $< \theta$. So, in $\mathscr{N}$, since everything $< \theta$ has been collapsed onto ω_1, $(S\omega_1)^{\mathscr{M}'}$ has cardinality $\aleph_1$.

From now on in this proof we regard $\mathscr{N}$ as a Cohen extension of $\mathscr{M}'$, $\mathscr{N} = \mathscr{M}'[G_2]$, where G_2 is $\mathscr{P}_2$-generic over $\mathscr{M}'$. Thus $\Vdash$ will always be forcing appropriate for $\mathscr{P}_2$ over $\mathscr{M}'$. Since $(S\omega_1)^{\mathscr{M}'}$ has cardinality $\aleph_1$ in N, there is $S \in \mathscr{W}$ such that $S \notin \mathscr{M}'$. Let τ be a term of $\mathscr{L}(\mathscr{M}')$ denoting S in $\mathscr{M}'[G_2]*$ and take $p \in G_2$ so that (setting $u = (S\omega_1)^{\mathscr{M}'}$)

$$p \Vdash \tau \notin \underline{u} \wedge \tau \subseteq \omega_1 \wedge (\forall \alpha \in \omega_1)(\tau \cap \alpha \in \underline{K_\alpha})$$

where $K = \langle\{S \cap \alpha : S \in \mathscr{W}\} : \alpha < \omega_1\rangle$. (Note that $K \in \mathscr{M}[\langle f_\alpha : \alpha < \delta\rangle] = \mathscr{M}'$.)

Claims. (a) $(\forall q \leq p)(\exists \beta < \omega_1)(\text{neither } q \Vdash \underline{\beta} \in \tau \text{ nor } q \Vdash \underline{\beta} \notin \tau)$.

(b) $(\forall q \leq p)(\exists \beta < \omega_1)(\exists r_1, r_2 \leq q)(r_1 \Vdash \underline{\beta} \in \tau \text{ and } r_2 \Vdash \underline{\beta} \notin \tau)$.

(b) is an immediate consequence of (a). If (a) fails for some q, then, taking a $\mathscr{P}_2$-generic H over $\mathscr{M}'$ containing q, we see that $\tau^{\mathscr{M}'[H]*} = \{\beta : q \Vdash \underline{\beta} \in \tau\} \in (S\omega_1)^{\mathscr{M}'}$ by IIc, which contradicts $q \leq p$ and $p \Vdash \tau \notin \underline{u}$.

In $\mathscr{M}'$, let us form a tree indexed by finite sequences of 0's and 1's as follows. We define $p_s \in \mathscr{P}_2$ and $\alpha_s \in \omega_1$. (*Notation*: if $s = \langle i_0, \ldots, i_{k-1}\rangle$ and $i = 0$ or 1, $s^\frown\langle i\rangle = \langle i_0, \ldots, i_{k-1}, i\rangle$.) Let $p_0 = p$. Given p_s, s a $\{0, 1\}$ sequence of length n, use (b) to get $p_{s^\frown\langle 0\rangle}, p_{s^\frown\langle 1\rangle} \leq p_s$ and α_s so that $p_{s^\frown\langle 1\rangle} \Vdash \underline{\alpha_s} \in \tau$, $p_{s^\frown\langle 0\rangle} \Vdash \underline{\alpha_s} \notin \tau$.

Still working in $\mathscr{M}'$, if f is a function from ω into $\{0, 1\}$ (i.e. $f \in {}^\omega 2$), $f \restriction n$ is the sequence of length n agreeing with the first n values of f. If $f \in {}^\omega 2$, let $p_f = \bigcup_{n \in \omega} p_{f \restriction n}$. ($\mathscr{P}_2$ is still closed under countable monotone unions in $\mathscr{M}'$, essentially by V, which guarantees that $\mathscr{M}'$ contains no countable sequences of ordinals not in $\mathscr{M}$.)

Let $\eta = \bigcup_{s \in {}^n 2} (\alpha_s + 1)$. Working in $\mathscr{M}'$, for each $f \in {}^\omega 2$, get $q_f \leq p_f$ and $T_f \subseteq \eta$ so that $q_f \Vdash \tau \cap \underline{\eta} = \underline{T_f}$. (This can be done by forming a countable montone sequence of conditions below p_f to decide successively each $\underline{\beta} \in \tau \cap \underline{\eta}$ and taking the union. Trivially, $\eta < \omega_1$.) But $f \neq g$ implies $T_f \neq T_g$, for, if n is the least number at which f and g differ, then $p_{f\upharpoonright n}$ and $p_{g\upharpoonright n}$ decide $\alpha_{f\upharpoonright n} \in \tau$ in different ways, so $\alpha_{f\upharpoonright n} \in T_f$ iff $\alpha_{f\upharpoonright n} \notin T_g$.

Finally, one shows that, for all $f \in {}^{\omega}2$ in $\mathscr{M}'$, $T_f \in K_\eta$: if H is a set $\mathscr{P}_2$-generic over $\mathscr{M}'$ such that $q_f \in H$, then $T_f = \tau^{\mathscr{M}[H]*} \cap \eta \in K_\eta$, since $q_f \Vdash \tau \cap \underline{\eta} = \underline{T}_f$ and $p \Vdash (\forall\alpha < \omega_1)(\tau \cap \alpha \in K_\alpha)$. (Notice that we have made strong use of the fact that K is in the ground model $\mathscr{M}'$, and hence has constant denotation.) But K_η is countable and there are $2^{\aleph_0}$ different T_f's (in the sense of $\mathscr{M}'$). Contradiction.

THEOREM. *$K(\aleph_1)$ holds in $\mathscr{N}$.*

PROOF. Let $\mathscr{W} = (S\underline{\theta})^{\mathscr{M}}$. $\overline{\overline{\underline{\mathscr{W}}}}^{\mathscr{M}} = \aleph_{\underline{\theta}+1}^{\mathscr{M}}$, so $\overline{\overline{\underline{\mathscr{W}}}}^{\mathscr{N}} = \aleph_3^{\mathscr{N}}$. But, if $\alpha < \theta = \omega_2^{\mathscr{N}}$, then $\{S \cap \alpha : S \in \mathscr{W}\}$ has cardinality $< \theta$ in $\mathscr{M}$, since θ is inaccessible in $\mathscr{M}$, and hence has cardinality $\leq \aleph_1$ in $\mathscr{N}$ since everything $< \theta$ has been collapsed onto ω_1. (This proof is essentially due to Lévy and Rowbottom [**8**].)

3. These facts have an immediate application in model theory. Suppose we have a first-order language L with equality and predicate symbols $\underline{U}$, $\underline{R}_1$, $\underline{R}_2$, . . . , where $\underline{U}$ is unary. A structure $\langle A, U, R_1, \ldots\rangle$ is said to be of type $\langle\kappa, \lambda\rangle$ if $\overline{\overline{A}} = \kappa$ and $\overline{\overline{U}} = \lambda$. We write $\langle\kappa, \lambda\rangle \Rightarrow \langle\kappa', \lambda'\rangle$ if any sentence σ of L having a model of type $\langle\kappa, \lambda\rangle$ also has a model of type $\langle\kappa', \lambda'\rangle$. Vaught [**7**] has shown $(\forall\alpha)[\langle\aleph_{\alpha+1}, \aleph_\alpha\rangle \Rightarrow \langle\aleph_1, \aleph_0\rangle]$ and Chang [**1**], using the GCH, has strengthened that to

$$(\forall\alpha)(\forall \text{ regular } \aleph_\beta)[\langle\aleph_{\alpha+1}, \aleph_\alpha\rangle \Rightarrow \langle\aleph_{\beta+1}, \aleph_\beta\rangle].$$

(For a fuller discussion, see Vaught [**12**].)

The gap-two two-cardinal conjecture states:

$$(\forall\alpha, \beta)[\langle\aleph_{\alpha+2}, \aleph_\alpha\rangle \Rightarrow \langle\aleph_{\beta+2}, \aleph_\beta\rangle].$$

The independence of this conjecture (modulo an inaccessible cardinal) from the GCH is shown here. Vaught [**12**] had already observed that there is a sentence σ_0 of L such that

$$(\forall\alpha)(\sigma_0 \text{ has a model of type } \langle\aleph_{\alpha+2}, \aleph_\alpha\rangle \text{ iff } K(\aleph_\alpha)).$$

In $\mathscr{N}$, $K(\aleph_0)$ fails while $K(\aleph_1)$ holds. So σ_0 has a model of type $\langle\aleph_3, \aleph_1\rangle$ but no model of type $\langle\aleph_2, \aleph_0\rangle$. In other words, $\langle\aleph_3, \aleph_1\rangle \not\Rightarrow \langle\aleph_2, \aleph_0\rangle$ in $\mathscr{N}$. Recall that the GCH holds in $\mathscr{N}$. (With a little more work, one can get a model $\mathscr{N}'$ of the GCH in which $\langle\aleph_2, \aleph_0\rangle \not\Rightarrow \langle\aleph_3, \aleph_1\rangle$.)

One can also get a GCH model in which $\langle\aleph_7, \aleph_5\rangle \not\Rightarrow \langle\aleph_3, \aleph_1\rangle$ and a GCH model which $\langle\aleph_3, \aleph_1\rangle \not\Rightarrow \langle\aleph_7, \aleph_5\rangle$ (though I don't see how to get the $\not\Rightarrow$ both ways simultaneously). All of the results mentioned depend upon having a model of ZF + AC + $\exists$ inaccessible cardinal.[2]

[2] This sharp form depends upon a method of Stewart (unpublished) for proving the consistency of Kurepa's conjecture relative to ZF (without an inaccessible cardinal).

A consistency result here is possible. Assuming only the existence of a model of ZF, I can construct a GCH model in which

$$\langle \aleph_6, \aleph_4 \rangle \Leftrightarrow \langle \aleph_3, \aleph_1 \rangle.$$

But I don't know how to get the consistency of the full gap-two two-cardinal conjecture. (A more general theorem reads: if m, n, k are natural numbers such that $m \geq 1$, $n \geq m + k + 1$, then one can get a GCH model in which:

$$\langle \aleph_{n+k}, \aleph_n \rangle \Leftrightarrow \langle \aleph_{m+k}, \aleph_m \rangle.)$$

4. The above results can be generalized from $\aleph_0$ to arbitrary $\aleph_\alpha$. Thus, if $\aleph_\alpha$ is a cardinal in some countable transitive model $\mathscr{M}$ of ZF + AC < some $\underline{\theta}$ inaccessible in $\mathscr{M}$, then we can find a Cohen extension $\mathscr{N}$ of $\mathscr{M}$ in which $\aleph_{\underline{\alpha}}^{\mathscr{N}} = \aleph_{\underline{\alpha}}^{\mathscr{M}}$, $\aleph_{\underline{\alpha+1}}^{\mathscr{N}} = \aleph_{\underline{\alpha+1}}^{\mathscr{M}}$ while $\sim K(\aleph_{\underline{\alpha}}) \wedge K(\aleph_{\underline{\alpha+1}})$ holds in $\mathscr{N}$.

In the customary fashion, all of these results can be recast as relative consistency results. Thus, we have shown:

ZF + AC + '$\exists$ inaccessible cardinal' is consistent

$\rightarrow$ ZF + GCH + $K(\aleph_1)$ + $\sim K(\aleph_0)$ is consistent.[3]

5. Recently, I have gotten a further consistency result in the direction of this paper. Chang's conjecture (CC) is the statement:

Any relational structure $\langle \omega_2, \omega_1, R \rangle$ has an elementary substructure $\langle A, A \cap \omega_1, {}^2A \cap R \rangle$ where $\bar{\bar{A}} = \aleph_1$ and $A \cap \omega_1$ is countable.

The following is an equivalent formulation in algebraic terms:

Any algebra $\langle \omega_2, f \rangle$ where f is a binary operation on ω_2 has a subalgebra $\langle A, f \rangle$ such that $\bar{\bar{A}} = \aleph_1$ and $A \cap \omega_1$ is countable.

Rowbottom [**8**] has shown that CC contradicts the axiom of constructibility. Indeed, Keisler-Rowbottom [**4**] show that CC implies the existence of weakly compact cardinals in the constructible universe.

My result states: If ZF + AC + '$\exists$ measurable cardinal' is consistent, then ZF + GCH + CC is consistent. '$\exists$ measurable cardinal' can be weakened to: '$\exists$ a cardinal κ for which $\kappa \rightarrow (\aleph_1)^{<\aleph_0}$' (see Silver [**11**] for notation).

Chang's conjecture bears on the present paper because it is easily seen to imply the negation of Kurepa's conjecture. Indeed, it implies the negation of an ostensibly weaker statement which has been considered by Erdös and myself (independently):

There is a family $\mathscr{F}$ of $\aleph_2$ functions from ω_1 into ω such that any two members of $\mathscr{F}$ eventually differ (i.e., if $f, g \in \mathscr{F}$ and $f \neq g$, there is $\alpha \in \omega_1$ such that, for all $\beta > \alpha$, $f(\beta) \neq g(\beta)$).

Whether the independence of this statement can be proved without assuming the consistency of some strong axioms of infinity remains open.

---

[3] Using a result of Solovay, I have observed that the converse is also true. More precisely: if $K(\aleph_0)$ fails, then ω_2 is inaccessible in L, the constructible universe.

BIBLIOGRAPHY

1. C. C. Chang, *A note on the two cardinal problem*, Proc. Amer. Math. Soc. **16**(1965), 1148–1155.

2. P. J. Cohen, *Set theory and the continuum hypothesis*, Benjamin, New York, 1966.

3. A. Hajnal, *On a consistency theorem connected with the generalized continuum hypothesis*, Acta. Math. Acad. Sci. Hungar., **12**(1961), 321–374.

4. H. J. Keisler and F. Rowbottom, *Constructible sets and weakly compact cardinals*, Notices Amer. Math. Soc., **12**(1965), p. 373.

5. A. Lévy, Definability in axiomatic set theory, *Logic, methodology, and the philosophy of science*, (Proc. 1964 Internat. Congr.) North-Holland, Amsterdam, 1965, pp. 127–151.

6. A. Lévy and R. M. Solovay, *Measurable cardinals and the continuum hypothesis*, Israel J. Math., **5**(1967), 234–248.

7. M. Morley and R. L. Vaught, *Homogeneous universal models*, Math. Scand., **11**(1962), 37–57.

8. R. A. Ricabarra, *Conjuntos ordinados y ramificados*, Bahia Blanca, 1958.

9. F. Rowbottom, *Doctoral dissertation*, University of Wisconsin, Madison, 1964.

10. D. Scott and R. M. Solovay, *Boolean valued models for set theory*, these Proceedings, part II.

11. J. Silver, *Some applications of model theory in set theory*, Doctoral dissertation, University of California, Berkeley, 1966.

12. R. L. Vaught, The Löwenheim-Skolem theorem, *Logic, methodology, and the philosophy of science*, (Proc. 1964 Internat. Congr.), North-Holland, Amsterdam, 1965, pp. 81–89.

UNIVERSITY OF WISCONSIN, MADISON AND
UNIVERSITY OF CALIFORNIA, BERKELEY

THE CONSISTENCY OF THE GCH WITH THE EXISTENCE OF A MEASURABLE CARDINAL

JACK SILVER[1]

1. We begin with a word on terminology. Consider $\mathfrak{A} = \langle A, \in \rangle$ where $\in$ is the $\in$-relation on A. $\mathscr{L}_{\mathfrak{A}}$ is a first-order language with equality which has a $\{:\}$ symbol for the formation of terms (to be interpreted after the manner of ZF), a predicate symbol $\in$ to denote $\in$, and for each $a \in A$, an individual constant $\underline{a}$ which denotes a in $\mathfrak{A}$. If σ is a sentence of $\mathscr{L}_{\mathfrak{A}}$, $\mathfrak{A} \models \sigma$ means that σ is true in $\mathfrak{A}$. (If A is a proper class, this can be formalized in Gödel-Bernays set theory for a single standard sentence at a time, but not in full generality.) If τ is a closed term of $\mathscr{L}_{\mathfrak{A}}$, $\tau^{\mathfrak{A}}$ is the denotation of τ in $\mathfrak{A}$. Finally, we often use A and $\mathfrak{A}$ interchangeably.

We pass on to some preliminary remarks on the notion of relative constructibility, which was discovered independently by Lévy [3] and others.

If A is a set, we define inductively: $M_0(A) = 0$, $M_{\alpha+1}(A)$ is the collection of sets $X \subset M_\alpha(A)$ such that, for some unary formula φ in $\mathscr{L}_{\mathscr{M}_\alpha(A)}$ (where $\mathscr{M}_\alpha(A) = \langle M_\alpha(A), \in, A \cap M_\alpha(A) \rangle$), $X = \{y : \mathscr{M}_\alpha(A) \models \varphi(\underline{y})\}$, and if α is a limit ordinal, $M_\alpha(A) = \bigcup_{\beta<\alpha} M_\beta(A)$. Finally, L_A, the class of sets relatively constructible from A, is defined to be $\bigcup \{M_\alpha(A) : \alpha \text{ is an ordinal}\}$.

It is a standard fact [3] that L_A is a model of ZF + AC. Moreover, it is easy to see that $A \cap L_A \in L_A$. Indeed, L_A can be characterized as (in a sufficiently strong set theory) the smallest transitive model $\mathscr{M}$ of ZF containing all ordinals such that $A \cap \mathscr{M} \in \mathscr{M}$. It is also easy to see that

$$L_A = L_{A \cap L(A)} = L^{L(A)}_{\underline{A \cap L(A)}},$$

[1] This work was partially supported by an NSF Cooperative Graduate Fellowship (1965–66) and by NSF GP-5632 (1967–68), both at the University of California, Berkeley.

i.e. $L_A \models$ every set is relatively constructible from $\underline{A \cap L_A}$. It should also be noted that each $M_\alpha(A)$ is transitive, and, if $\alpha < \beta$ are ordinals, then $M_\alpha(A) \subset M_\beta(A)$.

2. Henceforward, we assume the axiom of choice. Suppose κ is a measurable cardinal (i.e. there is a nonprincipal, κ-complete ultrafilter on κ and $\kappa > \aleph_0$—in the language of Keisler-Tarski [**2**], $\kappa \notin C_1$ and $\kappa > \aleph_0$). By definition, a *normal ultrafilter on* κ is a nonprincipal, κ-complete ultrafilter E on κ with the following additional property: if f is any function from κ into κ such that $f(\alpha) \in \alpha$ whenever $\alpha \neq 0$, then f is constant on some set S where $S \in E$. (The notion is due to Scott.) Keisler-Tarski show that there exists a normal ultrafilter on κ (they use the term: strongly κ-complete).

Fix a normal ultrafilter D' on κ for the remainder of this paper. We intend to show that $L_{D'}$, the class of sets relatively constructible from D', is a model of ZF + AC + GCH + $\exists$ measurable cardinal. ($L_{D'}$ was proposed as a candidate for these properties by Solovay, who had seen that κ remains measurable in $L_{D'}$ and that the GCH holds from κ upwards in $L_{D'}$.) This will establish the relative consistency of the GCH with ZF + AC + $\exists$ measurable cardinal. (Jensen [**1**] independently obtained this result by a different method. For earlier work in this area, see Lévy-Solovay [**4**].)

It is easy to see that $D = D' \cap L_{D'}$ is a normal ultrafilter on κ in the sense of $L_{D'}$, so $L_{D'} \models \kappa$ is measurable. Only the GCH remains to be verified. By the remark at the end of §1, $L_{D'} \models$ every set is relatively constructible from $\underline{D}$, a normal ultrafilter on κ. Therefore, the proof will be complete if we can show that this assumption, that the universe $V = L_D$ where D is a normal ultrafilter on κ, yields the GCH.

3. Throughout §§ 3–5, assume that $V = L_D$ where D is a normal ultrafilter on κ. As a model for doing the hard case, the GCH beneath κ, we do the case $2^\kappa = \kappa^+$ (κ^+ is the successor cardinal of κ) in this section. The argument for cardinals $> \kappa$ is almost identical.

But first it is convenient to state a technical lemma, whose proof we omit. In the statement of this lemma, we ignore the distinction between a sentence (or term) of the language of ZF and a term formalizing that sentence in ZF.

LEMMA (*). *There is a sentence σ_0 provable in* ZF *such that* (A) *and* (B) *are provable in* ZF:

(A). *If $\mathcal{N} = \langle N, \in \rangle$ is a transitive model of σ_0, $A \in N$, α is an ordinal in N, then $M_\alpha(\underline{A})^{\mathcal{N}} = M_\alpha(A)$.*

(B). *If $\mathcal{N} = \langle N, \in \rangle$ is a transitive model of σ_0, $E \cap N \in N$, α is an ordinal in N, then*

$$M_\alpha(E) = M_\alpha(E \cap N) = M_\alpha(\underline{E \cap N})^{\mathcal{N}}.$$

(A) is a generalization of the well-known fact that "being constructible at stage α" is absolute with respect to appropriate transitive structures. σ_0 should contain enough information to make the M construction go satisfactorily in $\mathcal{N}$.

(B) is an easy consequence of (A), the second equality being an instance of (A) and the first equality following thereupon by induction on α.

To prove $2^\kappa = \kappa^+$, it suffices to show that $S\kappa$, the power set of κ, is included in $M_\kappa + (D)$, since one can easily show by induction that $\overline{\overline{M_\alpha(D)}} \leq \overline{\overline{\alpha}} + \aleph_0$. (Here $\overline{\overline{X}}$ is the cardinal number of X.) The rest of this section is devoted to the proof of this claim. (Notice that this proof is a straightforward generalization of one proof that the GCH follows from the axiom of constructibility.)

Suppose $B \subset \kappa$. We seek to show that $B \in M_\kappa + (D)$. There is some λ for which $B \in M_\lambda(D)$. Using the reflection theorem for ZF, there is a model $\langle R_\beta, \in\rangle$ of σ_0 such that $D \in R_\beta$, $B \in R_\beta$, $\lambda < \beta$. (Here R_β is the collection of sets having rank less than β.)

By the Löwenheim-Skolem theorem, the structure $\mathfrak{A} = \langle R_\beta, \in\rangle$ has an elementary substructure $\mathscr{X} = \langle X, \in\rangle$ where B, $D \in X$, $\lambda \in X$, $\kappa + 1 \subset X$, while $\overline{\overline{X}} = \kappa$. There exists an isomorphism f of $\mathscr{X}$ with some $\mathscr{Y} = \langle Y, \in\rangle$ where Y is transitive. Since f is the identity on κ, it is clear that $f(B) = B$ and $f(D) = Y \cap D$. Since $\mathfrak{A}$ is a transitive model of σ_0, we have $\mathfrak{A} \models \underline{B} \in M_{\underline{\lambda}}(\underline{D})$, whence $\mathscr{X} \Vdash \underline{B} \in M_{\underline{\lambda}}(\underline{D})$, whence, by the isomorphism, $\mathscr{Y} \Vdash \underline{B} \in M_{\underline{f(\lambda)}}(\underline{Y \cap D})$.

Since $\mathscr{Y}$ is a transitive model of σ_0, part (B) of Lemma (*) tells us that $B \in M_{f(\lambda)}(D)$. Since $\overline{\overline{Y}} = \kappa$ and Y is transitive, $f(\lambda) < \kappa^+$, so $B \in M_\kappa + (D)$, as desired.

In this section, we have not yet used the fact that D is an ultrafilter, but only $D \subset S\kappa$.

4. We turn now to the GCH beneath κ. The proof rests on the following theorem of Rowbottom [**4**] (see also Silver [**5**]):

THEOREM. *Suppose D is a normal ultrafilter on κ, $A \geq \kappa$, $W \subset A$, $\overline{\overline{W}} = \lambda \geq \aleph_0$. If $Z \in A$ and* $\overline{\overline{Z}} < \kappa$, *then $\langle A, \in\rangle$ has an elementary substructure $\langle X, \in\rangle$ where $X \cap \kappa \in D$, $W \subset X$, $Z \in X$, and $\overline{\overline{Z \cap X}} \leq \lambda$.*

Suppose $\lambda < \kappa$. We present the proof of $2^\lambda = \lambda^+$ in some detail. Though new subsets of λ are constructed after stage κ, the following will be shown: if $T \subset \lambda$ and $T \in M_{\alpha+1}(D) - M_\alpha(D)$, then $M_{\alpha+1}(D) \cap S\lambda$ has cardinality $\leq \lambda$. From this, $2^\lambda = \lambda^+$ is immediate.

So suppose $T \subset \lambda$ and $T \in M_{\alpha+1}(D) - M_\alpha(D)$. Choose an ordinal β so that $D \in R_\beta$, $\alpha + 1 < \beta$, and $\langle R_\beta, \in\rangle$ is a model of σ_0 (see Lemma (*)). In Rowbottom's theorem, take $W = (\lambda + 1) \cup \{T, D, \alpha\}$ and $Z = M_{\alpha+1}(D) \cap S\lambda$. Thus, $\mathfrak{A} = \langle R_\beta, \in\rangle$ has an elementary substructure $\mathscr{X} = \langle X, \in\rangle$ such that $\lambda \subset X$, Z, T, D, $\alpha \in X$, $X \cap \kappa \in D$, and $\overline{\overline{Z \cap X}} \leq \lambda$. Let f be an isomorphism of $\mathscr{X}$ with a structure $\mathscr{Y} = \langle Y, \in\rangle$ where Y is transitive. Since $\lambda \subset X$, $f(T) = T$.

Now we make use of the normality of D to see that $F = \{\beta \in \kappa \cap X : f(\beta) = \beta\}$ is a member of D. Clearly $f(\beta) \leq \beta$ for all $\beta \in \kappa \cap X$. So, if our conclusion fails, using $\kappa \cap X \in D$, we would have

$$\{\beta \in \kappa \cap X : f(\beta) < \beta\} \in D,$$

whence f is constant on a set of "measure 1" (by the normality of D), contradicting the fact that f is 1-1.

But, for any $A \in S\kappa \cap X$, the symmetric difference of A and $f(A)$ is included in $\kappa - F$. Thus, $A \in D$ if and only if $f(A) \in D$. Hence $f(D) = Y \cap D$.

Suppose $f(\alpha) = \eta$. Since $\mathfrak{A}$ is a model of σ_0, $\mathfrak{A} \models \underline{T} \in M_{\underline{\alpha+1}}(\underline{D}) - M_{\underline{\alpha}}(\underline{D})$. Since f^{-1} is an elementary embedding of $\mathcal{Y}$ into $\mathfrak{A}$,

$$\mathcal{Y} \models \underline{T} \in M_{\underline{\eta+1}}(\underline{Y \cap D}) - M_{\underline{\eta}}(\underline{Y \cap D}).$$

Recalling that $\mathcal{Y}$ is a transitive model of σ_0, part (B) of Lemma (*) gives $T \in M_{\eta+1}(D) - M_\eta(D)$. Since $M_\beta(D)$ is increasing in β and $T \in M_{\alpha+1}(D) - M_\alpha(D)$, we conclude $\eta = \alpha$, i.e. $f(\alpha) = \alpha$.

From $\models \mathfrak{A}$ $\underline{Z} = M_{\underline{\alpha+1}}(\underline{D}) \cap S\underline{\lambda}$, we have $\mathcal{Y} \models \underline{f(Z)} = M_{\underline{\alpha+1}}(\underline{Y \cap D}) \cap S\underline{\lambda}$. Applying Lemma (*) again, $f(Z) = M_{\alpha+1}(D) \cap S\lambda = Z$. But $f(Z) = \{f(x) : x \in Z \cap X\}$ has cardinality at most λ, since $\overline{\overline{Z \cap X}} \leq \lambda$. Thus the cardinality of $M_{\alpha+1}(D) \cap S\lambda$ is at most λ, as was to be proved.

5. By the method of indiscernibles (see Silver [6] and Solovay [7]) it is possible to sharpen Rowbottom's theorem somewhat so as to have the substructure $\langle X, \in \rangle$ generated by some collection of Skolem functions from a set of indiscernibles which is a member of D. The normality of D then implies that the set of indiscernibles is closed relative to X. Analyzing the argument of the preceding section in this light, we obtain:

THEOREM. *If* $T \subset \lambda < \kappa$, $T \in M_{\alpha+1}(D) - M_\alpha(D)$, *then, for every* $W \in S\kappa \cap M_{\alpha+1}(D)$, *there exists a closed cofinal subset* C *of* κ *such that* $C \subset W$ *or* $C \cap W = 0$.

One can take for C an appropriate terminal segment of f^*S where S is the set of indiscernibles and f is the function of §4.

The above theorem shows that, in some sense, the $M_\alpha(D)$ construction is very absolute through all stages at which we are getting new subsets of κ.

A further application of the method of indiscernibles enables us to deduce from this theorem (still under the hypothesis $V = L_D$):

THEOREM. *There is a* Δ^1_3 *well-ordering of the set of reals* (*i.e. of* $S\omega$).

Similar results hold for all cardinals $\leq \kappa$.

6. The methods of this paper have been applied by Solovay [7] to prove the following theorem: if ZF + AC + '∃ real-valued measurable cardinal ' is consistent, then ZF + AC + '∃ a (2-valued) measurable cardinal ' is consistent.

It is also possible to generalize the arguments here to countably many measurable cardinals. Jensen [1] is able to handle more measurable cardinals, but doesn't get such nice structure in his model (e.g. the Δ^1_3 well-orderings).

BIBLIOGRAPHY

1. R. Jensen, *Measurable cardinals and the* GCH, Univ. of California, Los Angeles, 1967 (mimeographed notes).

2. H. J. Keisler and A. Tarski, *From accessible to inaccessible cardinals*, Fund. Math. **53** (1964), 225–308.

3. A. Lévy, *A generalization of Gödel's notion of constructibility*, J. Symbolic Logic **25** (1960), 147–155.

4. A. Lévy and R. Solovay, *Measurable cardinals and the continuum hypothesis*, Israel J. Math. **5** (1967), 234–248.

5. F. Rowbottom, Doctoral dissertation, Univ. of Wisconsin, Madison, 1964.

6. J. Silver, *Some applications of model theory in set theory*, Doctoral dissertation, Univ. of California, Berkeley, 1966.

7. R. Solovay, *A nonconstructible* Δ_3^1 *set of integers*, Trans. Amer. Math. Soc. **127** (1967), 58–75.

8. ———, *Real-valued measurable cardinals*, these Proceedings.

UNIVERSITY OF CALIFORNIA, BERKELEY

REAL-VALUED MEASURABLE CARDINALS[1]

ROBERT M. SOLOVAY[2]

Our starting point is the following question: Is there an extension of Lebesgue measure to a countably additive measure μ defined on every set of reals?

A classical result of Vitali (cf. [**4**, p. 67]) states that such a μ cannot be translation invariant.

Ulam [**21**] showed that the existence of μ implied that the continuum hypothesis is "badly violated". Precisely, let κ be the least cardinal such that a set of positive μ-measure is the union of κ sets of μ-measure zero. Clearly,

$$\aleph_0 < \kappa \leqq 2^{\aleph_0}.$$

Ulam showed that κ is weakly inaccessible. Using ideas of Keisler and Tarski [**6**], we prove

THEOREM 1.[3] (a) *κ is the κth weakly inaccessible cardinal.*

(b) *Let $K \subseteq \kappa$ be relatively closed and unbounded. Then K contains a weakly inaccessible cardinal.* (*I.e. κ is weakly Mahlo.*)

The notion of a (two-valued) measurable cardinal will be recalled in §1 below. The following theorem shows that Lebesgue measure does have extensions to everywhere defined countably additive measures in certain models of set-theory.

---

[1] The main results of this paper were proved in the spring of 1966.

[2] The author is a Sloan Foundation Fellow. The writing of this paper was supported in part by NSF grant GP-8746.

[3] Theorem 1 was proved independently by R. Jensen. Part (a) of Theorem 1 has also been proved by D. H. Fremlin.

THEOREM 2. *The following two theories are equiconsistent.*

(1) **ZFC**[4] + "*There is a measurable cardinal*".

(2) **ZFC** + "*Lebesgue measure has a countably additive extension* μ *defined on every set of reals*".

This paper is organized as follows: §1 relates the questions considered in the introduction to the notion of a λ-saturated ideal. We collect various known results, in and out of the literature concerning this notion. §2 is devoted to the proof of Theorem 1. We prove this theorem in the more general context of a κ-complete κ-saturated nonprincipal ideal (showing that κ is large). The proof of Theorem 2 splits into two parts, which are given in §§3-4. §5 is devoted to the solution of a problem of Födor and Hajnal [**1**]. We show that if κ is regular, every stationary subset of κ can be decomposed into κ disjoint stationary subsets. §6 is devoted to the proof of a theorem of Kunen. (Kunen remarked, in his thesis, that this theorem could be proved by the methods of the present paper.) Finally, in §7, we announce some results to be published elsewhere related to the present work.

1. Ideals in fields of sets.

1.1. If X is a set, let $P(X)$ be the family of all subsets of X. A *nontrivial measure* on X is a map

$$\mu : P(X) \to [0, 1]^5$$

such that μ is a countably additive measure vanishing on points with $\mu(X) = 1$.

If μ_1 is a countably additive extension of Lebesgue measure to $P(\mathbf{R})$, then μ_1 restricted to $P([0, 1])$ is a nontrivial measure on $[0, 1]$.

Let μ be a nontrivial measure on X. We say that μ is *two-valued* if $\mu : P(X) \to \{0, 1\}$.

Now let μ, X be as above. Let κ be a cardinal. We say μ is κ-additive if whenever $\{A_\xi, \xi < \lambda\}$ is a family of sets of measure zero, and $\lambda < \kappa$, then $\bigcup_{\xi<\lambda} A_\xi$ has measure zero. It is easily seen that there is a largest κ such that μ is κ-additive; we call this κ add (μ). (add (μ) is the least κ such that the union of κ sets of measure zero has positive measure. Clearly $\aleph_0 <$ add (μ) (since μ is countably additive) and add $(\mu) <$ card (X) (since X is the union of its one-element subsets).)

We are going to construct a real-valued measure on add (μ). Let $\kappa =$ add (μ). Let A be a set of positive μ-measure which is the disjoint union of κ sets of measure zero:

$$A = \bigcup_{\xi<\kappa} A_\xi.$$

Define a map $f : A \to \kappa$ by putting $f(x) = \xi$ if and only if $x \in A_\xi$. Define a real-valued countably additive measure γ on κ by the formula

$$\gamma(B) = \mu(f^{-1}(B))/\mu(A).$$

One checks easily that γ is a nontrivial κ-additive measure on κ.

---

[4] **ZFC** is Zermelo-Frankel set-theory with the axiom of choice.

[5] $[0, 1] = \{x \in \mathbf{R} : 0 \leqq x \leqq 1\}$.

DEFINITION 1. κ is a real-valued measurable cardinal if and only if κ carries a nontrivial κ-additive measure.

Our discussion yields the following:

PROPOSITION 1. *Suppose that Lebesgue measure has a countably additive extension defined on all sets of reals. Then there is a real-valued measurable cardinal, κ, with $\kappa \leqq 2^{\aleph_0}$.*

(This proposition is essentially due to Ulam [**21**].)

DEFINITION 2. κ is (two-valued) measurable if and only if κ carries a nontrivial two-valued κ-additive measure.

PROPOSITION 2 (ULAM [**21**]). *Let κ be real-valued measurable. Then either κ is 2-valued measurable or $\kappa \leqq 2^{\aleph_0}$. If κ is 2-valued measurable, then κ is strongly inaccessible. If $\kappa \leqq 2^{\aleph_0}$, then there is an extension γ of Lebesgue measure, defined on all subsets of* **R** *with* add $(\gamma) = \kappa$.

(For the last sentence, cf. also [**4**, §41].)

1.2. Let κ be an uncountable cardinal. Let $P(\kappa)$ be the power set of κ. A subset $I \subseteq P(\kappa)$ is an *ideal* if

(1) $A \in I,\ B \subseteq A \rightarrow B \in I$;

(2) $A_1 \in I$ and $A_2 \in I \rightarrow A_1 \cup A_2 \in I$.

I is *κ-complete* if whenever $\xi < \kappa$, and $\{A_\alpha, \alpha < \xi\}$ is a collection of members of I, then $\bigcup_{\alpha<\xi} A_\alpha \in I$.

I is nontrivial if I is κ-complete, $\kappa \notin I$, and $\{\lambda\} \in I$ for each $\lambda < \kappa$.

The following proposition relates the concept of ideal to that of real-valued measure.

PROPOSITION 3. *Let κ be a real-valued measurable cardinal. Let μ be a nontrivial κ-additive real-valued measure on κ. Put $I = \{A \subseteq \kappa : \mu(A) = 0\}$. Then I is a nontrivial ideal in $P(\kappa)$.*

PROOF. Clear.

We say that I is a measure ideal if I arises from a nontrivial κ-additive measure μ as in Proposition 3.

Inspired by this example, we introduce the following terminlogy. A set $A \subseteq \kappa$ has *measure zero* (rel. I) if and only if $A \in I$; A has *positive measure* (rel. I) if and only if $A \notin I$; A has *measure* 1 (rel. I) if and only if $(\kappa - A) \in I$. (We frequently drop "(rel. I)" if I is clear from the context.)

1.3. Let $\mathscr{B}$ be a Boolean algebra. We say that a sequence of elements of $\mathscr{B}$, $\{b_\xi : \xi < \lambda\}$ is *pairwise disjoint* if $\xi < \xi' < \lambda$ implies

$$b_\xi \cdot b_{\xi'} = \mathbf{0}.$$

Let λ be a cardinal. $\mathscr{B}$ is *λ-saturated* if whenever $\{b_\xi : \xi < \lambda\}$ is a pairwise disjoint family of elements from $\mathscr{B}$, then $b_{\xi_0} = \mathbf{0}$ for some $\xi_0 < \lambda$. If λ is greater than card $(|\mathscr{B}|)$, $\mathscr{B}$ is clearly λ-saturated. Let sat $(\mathscr{B})$ be the least λ for which $\mathscr{B}$ is λ-saturated. Clearly $\mathscr{B}$ is λ'-saturated for any $\lambda' \geqq$ sat $(\mathscr{B})$.

Now let I be a nontrivial ideal in $P(\kappa)$. We form the quotient algebra

$$\mathscr{B} = P(\kappa)/I.$$

We say that I is *λ-saturated* if and only if $\mathscr{B}$ is λ-saturated. We put sat $(I) =$ sat $(\mathscr{B})$. Note that I is 2-saturated if and only if I is prime if and only if $\mathscr{B} = \mathbf{2}$.

We can reformulate this as follows: Let $\{A_\xi : \xi < \lambda\}$ be a sequence of subsets of κ. We say that this sequence is *almost disjoint* if for $\xi < \xi' < \lambda$, $A_\xi \cap A_{\xi'}$ has measure zero. Then I is λ-saturated if and only if whenever $\{A_\xi, \xi < \lambda\}$ is an almost-disjoint family, then $A_{\xi_0} \in I$ for some $\xi_0 < \lambda$.

PROPOSITION 4. *The following conditions on a nontrivial ideal I are equivalent (for any $\lambda \leqq \kappa$).*

(a) *I is λ-saturated.*

(b) *If $\{A_\xi, \xi < \lambda\}$ is pairwise disjoint, then $A_\xi \in I$ for some $\xi < \lambda$.*

PROOF. Obviously (a) implies (b). Suppose I satisfies (b), and that $\{A_\xi, \xi < \lambda\}$ is almost disjoint.

Put $B_\alpha = A_\alpha - \bigcup_{\beta<\alpha} A_\beta$. Then clearly $\{B_\alpha, \alpha < \lambda\}$ is a disjoint family. By (b), B_ξ has measure zero for some $\xi < \lambda$. But then

$$A_\xi \subseteq B_\xi \cup \bigcup_{\beta<\xi} (A_\beta \cap A_\xi)$$

is the union of less than κ sets of measure zero. Since I is κ-complete, A_ξ has measure zero. Thus I satisfies (a).

The following proposition is due to A. Tarski [**20**, Theorem 4.5]. We feel it should be mentioned here, although we do not in fact use it.

PROPOSITION 5. *Let $\mathscr{B}$ be a Boolean algebra. Let* sat $(\mathscr{B}) \geqq \aleph_0$. *Then* sat $(\mathscr{B}) > \aleph_0$ *and* sat $(\mathscr{B})$ *is regular.*

PROPOSITION 6 (ULAM). *Let κ be a real-valued measurable cardinal. Let μ be a nontrivial measure on κ. Let I be the ideal of sets of measure zero of μ. Then I is $\aleph_1$-saturated.*

PROOF. Suppose that $\{A_\xi, \xi < \aleph_1\}$ is a collection of almost disjoint subsets of κ. Then for each n, there are at most n ξ's for which $\mu(A_\xi) \geqq 1/n$. So there are at most countably many ξ's for which $\mu(A_\xi) > 0$.

PROPOSITION 7 ([**20**, THEOREM 4.12]). *Let I be a nontrivial ideal in $P(\kappa)$. Suppose I is λ-saturated for some $\lambda < \kappa$. Suppose further that $\aleph < \lambda \rightarrow 2^\aleph < \kappa$. Then κ is measurable, and the quotient algebra $\mathscr{B} = P(\kappa)/I$ is totally atomic. (I.e., every nonzero element of $\mathscr{B}$ is the* sup *of the atoms it contains.)*

We remark that Lévy and Silver have independently used the idea of Tarski's proof to show the following:

PROPOSITION 8. *Let κ be strongly inaccessible. Let I be a nontrivial ideal in $P(\kappa)$ with* sat $(I) = \kappa$. *Then there is a well-founded κ Souslin tree of power κ (and so κ is not weakly compact).*

For the following proposition, see [**16,** 21.3, p. 65]. If κ is a cardinal, let κ^+ be the least cardinal greater than κ.

PROPOSITION 9. *Let κ be an uncountable cardinal. Let I be a nontrivial κ^+-saturated ideal in $P(\kappa)$. Then the quotient algebra, $P(\kappa)/I$, is complete.*

2. **Proof of Theorem 1.**

2.1. We fix once and for all an uncountable cardinal κ carrying a nontrivial κ-complete ideal, I.

LEMMA 1. *Let $\xi < \kappa$. Then $\{\alpha \mid \alpha < \xi\}$ has measure zero.*

PROOF. Clear since I is κ-complete and points have measure zero.

LEMMA 2. *κ is regular.*

PROOF. Otherwise, there exists $\alpha < \kappa$ and $\lambda_\xi < \kappa$ defined for $\xi < \alpha$ such that $\kappa = \sup\{\lambda_\xi : \xi < \alpha\}$. But then $\kappa \subseteq \bigcup_{\xi<\alpha} \lambda_\xi$ is the union of α sets of measure zero (by Lemma 1). Thus κ has measure zero, which is absurd.

2.2. Let κ, I be as above. We say that I is *normal* if whenever $B \subseteq \kappa$ has positive measure, and $f: B \to \kappa$ is such that $f(\xi) < \xi$, $\xi \in B$, then there is $B' \subseteq B$ of positive measure, and a $\lambda < \kappa$ such that

$$f(\xi) = \lambda, \qquad \xi \in B'.$$

THEOREM 3. *Let κ be an uncountable cardinal, I a κ-saturated nontrivial ideal on $P(\kappa)$. Then there is a normal nontrivial ideal, J on $P(\kappa)$, with*

$$\text{sat}\,(J) \leqq \text{sat}\,(I).$$

If I is a measure ideal, so is J.

PROOF.[6] We begin with some definitions. Let $f: A \to \kappa$.

We say that f is *almost bounded* if there is a $\lambda < \kappa$ such that $\{\xi \in A \mid f(\xi) > \lambda\}$ has measure zero. We say that f is *nowhere bounded* if for each $\lambda < \kappa$, $\{\xi \in A \mid F(\xi) \leqq \lambda\}$ has measure zero. Finally, f is incompressible if

(1) f is nowhere bounded.

(2) If $B \subseteq A$ has positive measure, $g: B \to \kappa$ and $g(\xi) < f(\xi)$ for all $\xi \in B$, then g is almost bounded.

These concepts are only important if A has positive measure. If A has measure zero, every map $f: A \to \kappa$ is simultaneously almost bounded, nowhere bounded, and incompressible.

LEMMA 3. *Let $f: A \to \kappa$ be nowhere bounded. Then we can write A as the disjoint union of sets B and C such that*

(1) *$f \mid B$ is incompressible.*

(2) *There is a $g: C \to \kappa$ with $g(\xi) < f(\xi)$, $\xi \in C$, such that g is nowhere bounded.*

[6] C. C. Chang has helped simplify our proof.

PROOF. We use Zorn's lemma to get a maximal family

$$\mathscr{F} = \{\langle C_i, g_i\rangle : i \in K\}$$

such that

(1) $C_i \subseteq A$ has positive measure.

(2) $g_i : C_i \to \kappa$ is nowhere bounded.

(3) $\xi \in C_i \to g_i(\xi) < f(\xi)$.

(4) If i, j are distinct elements of K, then $C_i \cap C_j = \varnothing$.

Using (1) and (4), we see that since I is κ-saturated, card $(K) < \kappa$. We put

$$C = \bigcup_{i \in K} C_i; \qquad g = \bigcup_{i \in K} g_i.$$

Since I is κ-complete and card $(K) < \kappa$, g is nowhere bounded. Clearly $\xi \in C \to g(\xi) < f(\xi)$.

Put $B = A - C$. If B has measure zero, $f \mid B$ is trivially incompressible. If B has positive measure, the maximality of $\mathscr{F}$ implies that $f \mid B$ is incompressible.

LEMMA 4. *There exists an incompressible function $f : \kappa \to \kappa$.*

PROOF. We define a sequence of sets $\{A_n\}$ and functions $h_n : A_n \to \kappa$ by induction on n.

(1) $n = 0$. Put $A_0 = \kappa$. Let $h_0(\alpha) = \alpha$ for $\alpha < \kappa$. (By Lemma 1, h_0 is nowhere bounded.)

(2) $n = k + 1$. Say $h_k : A_k \to \kappa$ is nowhere bounded. We apply Lemma 3 to A_k, h_k and get the following:

(1) $A_{k+1} \subseteq A_k$; (2) a map $h_{k+1} : A_{k+1} \to \kappa$ which is nowhere constant; moreover, (3) for $\xi \in A_{k+1}$, $h_{k+1}(\xi) < h_k(\xi)$; and (4) $h_k \mid A_k - A_{k+1}$ is incompressible.

I claim next that $\bigcap_n A_n = \varnothing$. For if ξ lies in this intersection, the sequence $h_0(\xi) > h_1(\xi) > \cdots$ is a descending sequence of ordinals.

We thus have a disjoint union

$$\kappa = \bigcup_n (A_n - A_{n+1})$$

and $h_n \mid A_n - A_{n+1}$ is incompressible. We define $h : \kappa \to \kappa$ by

$$h(\xi) = h_n(\xi) \quad \text{if } \xi \in A_n - A_{n-1}.$$

Since κ is uncountable, and I is κ-complete, one checks without difficulty, that h is incompressible.

We can now prove Theorem 3. Let $h : \kappa \to \kappa$ be incompressible. Define a subset $J \subseteq P(\kappa)$ by

$$J = \{A \subseteq \kappa \mid h^{-1}(A) \in I\}.$$

It is straightforward to check that J is a κ-complete ideal, and that $\kappa \notin J$. Since h is nowhere bounded, $h^{-1}(\{\lambda\})$ has I-measure zero. Hence $\{\lambda\}$ has J-measure zero. An easy check shows that sat $(J) \leqq$ sat (I).

We check that J is normal. Suppose A has positive J-measure, and $g : A \to \kappa$ is such that $g(\xi) < \xi$ for $\xi \in A$. Put $B = h^{-1}(A)$. Then B has positive I-measure (since A has positive J-measure). Let $f: B \to \kappa$ be the composition

$$B \xrightarrow{h} A \xrightarrow{g} \kappa.$$

Then if $\gamma \in B, f(\gamma) = g(h(\gamma)) < h(\gamma)$. But h is incompressible. It follows that there is a $\lambda < \kappa$ such that $\{\gamma \in B \mid f(\gamma) \leqq \lambda\}$ has positive I-measure. Since I is κ-complete and $\lambda < \kappa$, there is a $\lambda' \leqq \lambda$ such that $D = \{\gamma \in B \mid f(\gamma) = \lambda'\}$ has positive I-measure. Let $E = \{\gamma \in A \mid g(\gamma) = \lambda'\}$. Then $\gamma \in D \leftrightarrow g(h(\gamma)) = \lambda' \leftrightarrow h(\gamma) \in E$. Thus $D = h^{-1}(E)$. It follows that E has positive J-measure. I.e. given g, A as above, there is an $E \subseteq A$ of positive J-measure on which g is constant. Thus J is normal.

Finally, let κ be a real-valued measurable cardinal. Let μ be a nontrivial κ-additive real-valued measure on κ. Let I be the ideal of sets of measure zero of μ. Let $h: \kappa \to \kappa$ be I-incompressible. Define $\nu: P(\kappa) \to [0, 1]$ by

$$\nu(A) = \mu(h^{-1}(A)).$$

Let J be as above. Then ν is easily checked to be κ-additive and nontrivial, and J is the ideal of sets of ν-measure zero. Thus J is a measure ideal. The proof of Theorem 3 is complete.

In the course of §6, we reprove Theorem 3 under the weaker hypothesis that I is κ^+-saturated. One could also get this result by modifying the proof given here.

2.3. We now assume that κ is as above, but that J is a nontrivial κ-saturated normal ideal.

LEMMA 5. *Let A have positive measure. Let $h: A \to \kappa$ be such that $h(\xi) < \xi$, for all $\xi \in A$. Then h is almost bounded.*

PROOF. Let

$$E = \{\lambda < \kappa \mid h^{-1}(\{\lambda\}) \text{ has positive measure}\}.$$

Since J is κ-saturated, card $(E) < \kappa$. Let

$$B = \{\gamma \in A \mid h(\gamma) \notin E\}.$$

I say B has measure zero. Otherwise, by normality, there is a $\lambda < \kappa$ and a $B' \subseteq B$ of positive measure such that $h(\gamma) = \lambda$ for $\gamma \in B'$. But then $\lambda \in E$, contradicting the definition of B.

Since κ is regular, sup $(E) = \lambda_0$, say, is $<\kappa$. But

$$\{\gamma \in A \mid h(\gamma) > \lambda_0\} \subseteq B.$$

Thus h is almost bounded.

We say that a property P holds for almost all α if $\{\alpha < \kappa \mid P(\alpha)\}$ has measure one.

LEMMA 6. *For almost all α, α is a regular cardinal.*

PROOF. Suppose not. We shall get a contradiction. Let $E = \{\alpha \mid \text{cf}(\alpha) < \alpha\}$. Our assumption is that E has positive measure. Hence, by normality, there is a

$\lambda < \kappa$ such that

$$E_1 = \{\alpha \mid \lambda = \text{cf}(\alpha)\}$$

has positive measure. For each $\alpha \in E_1$, we select $h_\alpha : \lambda \to \alpha$ such that $\sup(h_\alpha[\lambda]) = \alpha$. Define for each $\xi < \lambda$ a map $g_\xi : E_1 \to \kappa$ by $g_\xi(\alpha) = h_\alpha(\xi)$. Then for all $\alpha \in E_1$, $g_\xi(\alpha) = h_\alpha(\xi) < \alpha$.

We apply Lemma 5 to g_ξ. It follows that there is a set of measure zero, N_ξ, and an ordinal $\gamma_\xi < \kappa$ such that

$$g_\xi(\alpha) \leqq \gamma_\xi \quad \text{if } \alpha \in E_1 - N_\alpha.$$

Put $\gamma = \sup\{\gamma_\xi \mid \xi < \lambda\}$. Since κ is regular, $\gamma < \kappa$. Put $E_2 = E_1 - \bigcup_{\alpha<\lambda} N_\alpha$. Since J is κ-additive, E_2 has positive measure. For $\alpha \in E_2$,

$$\alpha = \sup\{g_\xi(\alpha) \mid \xi < \lambda\} \leqq \sup\{\gamma_\xi \mid \xi < \lambda\} \leqq \gamma.$$

Thus $E_2 \subseteq \{\alpha \mid \alpha < \gamma\}$. But this is absurd since E_2 has positive measure, while by Lemma 1, $\{\alpha \mid \alpha < \gamma\}$ has measure zero.

COROLLARY 7. *κ is weakly inaccessible.*

PROOF. We write β^+ for the least cardinal greater than β. We know that κ is regular. (Lemma 2.) If κ is not weakly inaccessible, $\kappa = \beta^+$, for some $\beta < \kappa$. But then

$$\{\alpha < \kappa \mid \alpha \text{ is regular}\} \subseteq \{\alpha < \kappa \mid \alpha \leqq \beta\}. \tag{1}$$

By Lemma 6, the left-hand side of (6) has measure 1. The right-hand side has measure zero by Lemma 1. This contradiction shows that κ is weakly inaccessible.

COROLLARY 8. *Almost all α are weakly inaccessible.*

PROOF. Suppose not. Then by Lemma 6,

$$E = \{\alpha^+ \mid \alpha \text{ a cardinal less than } \kappa\}$$

has positive measure. Define $h: E \to \kappa$ by $h(\alpha^+) = \alpha$ for α a cardinal $<\kappa$. By normality there is $E_1 \subseteq E$ of positive measure and a $\lambda < \kappa$ such that $h(\gamma) = \lambda$ for $\gamma \in E_1$. But then $E_1 = \{\lambda^+\}$ which is absurd since singletons have measure zero.

COROLLARY 9. *κ is the κth weakly inaccessible cardinal.*

PROOF. $D = \{\alpha < \kappa \mid \alpha \text{ is weakly inaccessible}\}$ has measure 1. Since J is κ-complete, D has power κ. Since $D \subseteq \kappa$, the order type of D must be exactly κ.

2.4. Let α be a regular cardinal. A set $K \subseteq \alpha$ is closed unbounded if (1) $\text{lub}\, K = \alpha$; (2) if $S \subseteq K$, then $\text{lub}\, S \in K \cup \{\alpha\}$.

Let S be a class of ordinals. A cardinal κ is S-Mahlo if (1) κ is weakly inaccessible; (2) if $K \subseteq \kappa$ is closed unbounded, then $K \cap S \neq \varnothing$.

THEOREM 4. *Let κ, J be as above. Let $S \subseteq \kappa$ have positive measure. Then κ is S-Mahlo. Moreover, almost all $\alpha < \kappa$ are S-Mahlo.*

PROOF. We begin with the following lemma:

LEMMA 10. *Suppose $K \subseteq \kappa$ is closed unbounded. Then K has measure one.*

PROOF. Suppose not. Then $\kappa - K$ has positive measure. Define a map $g: \kappa - K \to \kappa$ by

$$g(\alpha) = \sup (\alpha \cap K).$$

Since K is closed unbounded, $g(\alpha) \in K$. Thus $g(\alpha) < \alpha$, for $\alpha \in \kappa - K$. We now invoke normality. There is a set D of positive measure and an ordinal $\lambda < \kappa$ such that $D \subseteq \kappa - K$, and $g(\alpha) = \lambda$ for $\alpha \in D$.

But this is absurd: Since lub $K = \kappa$, we can pick $\lambda_1 \in K$ with $\lambda_1 > \lambda$. Since D has positive measure, there is a λ_2 in D with $\lambda_2 > \lambda_1$. But then

$$g(\lambda_2) = \sup (K \cap \lambda_2) \geqq \lambda_1 > \lambda.$$

This contradicts our choice of D and λ.

COROLLARY 11. *κ is S-Mahlo.*

PROOF. Let $K \subseteq \kappa$ be closed unbounded. If $S \cap K = \varnothing$, S has measure zero, by Lemma 10. Thus $S \cap K \neq \varnothing$, since S has positive measure.

We now assume that there is a set $B \subseteq \kappa$ of positive measure such that if $\gamma \in B$, γ is not S-Mahlo. By Corollary 8, we may assume that each $\gamma \in B$ is weakly inaccessible. Since γ is not S-Mahlo, there is a closed unbounded set K_γ of γ with $K_\gamma \cap S = \varnothing$. We let $f_\gamma: \gamma \to \gamma$ enumerate K_γ in increasing order. Note that since K_γ is closed,

$$f_\gamma(\lambda) = \sup \{f_\gamma(\xi) \mid \xi < \lambda\},$$

if λ is a limit ordinal less than γ.

Define $g_\xi: B \to \kappa$ as follows:

$$g_\xi(\gamma) = f_\gamma(\xi) \quad \text{if } \xi < \gamma;$$
$$g_\xi(\gamma) = 0 \qquad \text{if } \xi \geqq \gamma.$$

Then $g_\xi(\gamma) < \gamma$ for all γ in B. Let $E_\xi = \{\lambda < \kappa : g_\xi^{-1}(\{\lambda\})$ has positive measure. Let $N_\xi = \{\gamma \in B \mid g_\xi(\gamma) \notin E_\xi\}$. Then E_ξ has power less than κ and N_ξ has measure zero (cf. the proof of Lemma 5).

LEMMA 12. *For almost all γ in B, $f_\gamma(\xi) \in E_\xi$ for all $\xi < \gamma$.*

PROOF. Otherwise, there is a $D \subseteq B$ of positive measure such that for each $\gamma \in D$, there is a $\xi_\gamma < \gamma$ such that $f_\gamma(\xi_\gamma) \notin E_{\xi_\gamma}$. We invoke normality and get a $D_1 \subseteq D$ of positive measure and a $\xi < \kappa$ such that

$$f_\gamma(\xi) \notin E_\xi \quad \text{for } \gamma \in D_1.$$

I.e., $D_1 \subseteq N_\xi$. But this is absurd since D_1 has positive measure and N_ξ has measure zero.

We now throw away a set of measure zero from B, if necessary, and assume that for all $\gamma \in B$, and all $\xi < \gamma, f_\gamma(\xi) \in E_\xi$.

Put $\gamma_\xi = \sup(E_\xi)$. Since κ is regular, and card $(E_\xi) < \kappa$, we have $\gamma_\xi < \kappa$.

The following lemma is an easy consequence of normality:

LEMMA 13. *For almost all* α, $\alpha \geqq \sup\{\gamma_\xi : \xi < \alpha\}$.

We are now ready to complete the proof of Theorem 4. Since S has positive measure, there is a $\lambda_0 \in S$ such that (1) λ_0 is weakly inaccessible, and (2) $\gamma_\xi < \lambda_0$ for all $\xi < \lambda_0$.

Since B is unbounded, there is a $\lambda_1 \in B$ with $\lambda_1 > \lambda_0$. I shall show that $\lambda_0 \in K_{\lambda_1} \cap S$. This will contradict our choice of K_{λ_1}.

It suffices to show $f_{\lambda_1}(\lambda_0) = \lambda_0$. Since the inequality $f_{\lambda_1}(\lambda_0) \geqq \lambda_0$ is trivial, we must show $f_{\lambda_1}(\lambda_0) \leqq \lambda_0$.

Since λ_0 is a limit ordinal,

$$f_{\lambda_1}(\lambda_0) = \sup\{f_{\lambda_1}(\xi) : \xi < \lambda_0\}. \tag{2}$$

But the right-hand side of (2) is at most

$$\sup\{\gamma_\xi : \xi < \lambda_0\} \leqq \lambda_0.$$

The proof of Theorem 4 is complete.

2.5. One can now extend the remainder of the Keisler-Tarski formalism in a purely formal way. For example, if R is a well-ordering of κ and D has measure zero, $M^R(D)$ has measure zero (cf. Hanf **[5]** for this notion).

Theorem 1 is now clear from the results of this section together with Propositions 1, 3, and 6 of §1.

2.6. We now assume that κ, J are as above, and that κ is λ-saturated for some uncountable regular cardinal $\lambda < \kappa$. This assumption is harmless for applications, since if κ is λ-saturated, it is a fortiori λ^+ saturated and $\lambda^+ < \kappa$ by Corollary 7. Alternatively, J is sat (J)-saturated, and sat (J) is regular by Tarski's Proposition 5.

We are going to generalize a theorem of Rowbottom **[13]** to the present context. The generalization will be used in the proof of Theorem 2.

THEOREM 5. *Let* $\mathscr{A} = \langle X : U, R, \ldots\rangle$ *be a relational system with less than* λ *predicates and constants. Let* U *be a one place predicate such that* $\{x \in X : \mathscr{A} \models Ux\}$ *has power* $<\kappa$. *Suppose finally that* $\kappa \subseteq X$. *Then there is an elementary submodel* $\mathscr{A}'$ *of* $\mathscr{A}$: $\mathscr{A}' = \langle X' : U', R', \ldots\rangle$ *with* $X' \cap \kappa$ *of measure* 1, *and* $\{x \in X : \mathscr{A}' \models Ux\}$ *has power* $<\lambda$.

For an exposition of Rowbottom's proof we refer the reader to **[9]**. It suffices to prove the following lemma. (One applies Lemma 14 to a set of Skolem functions for $\mathscr{A}$.)

LEMMA 14. Let $D^{[n]}$ *be the set of subsets of* D *of cardinality* n. *Let* $\gamma < \kappa$. *Let* $f : \kappa^{[n]} \to \gamma$. *Then there is a* $D \subseteq \kappa$ *of measure one, with* $f[D^{[n]}]$ *of cardinality less than* λ.

PROOF. The proof proceeds by induction on n.

Case 1. $n = 1$. Let $E = \{\xi \mid f^{-1}(\{\xi\})$ has positive measure$\}$. Since J is λ-saturated, card $(E) < \lambda$. Let $N = \{\gamma \mid f(\gamma) \notin E\}$. Since f is bounded below κ, normality implies that N has measure zero (cf. the proof of Lemma 5). Thus it suffices to take $D = \kappa - N$.

Case 2. $n = k + 1$. For each α we define a map $h_\alpha : \kappa^{[k]} \to \gamma$. $h_\alpha(x) = f(x \cup \{\alpha\})$ if $\alpha \notin x$; otherwise $h_\alpha(x) = 0$.

We apply our inductive assumption to h_α and get the following:

(1) A set D_α of measure 1 with $D_\alpha \subseteq \{\gamma \mid \alpha < \gamma < \kappa\}$.

(2) A nonvoid set S_α of power less than λ of γ such that $h_\alpha[D_\alpha^{[k]}] \subseteq S_\alpha$.

Our next task is to construct a set D' of measure one such that $\bigcup \{S_\alpha : \alpha \in D'\}$ has power less than λ. In the first place, consider the map $g : \kappa \to \lambda$ defined by $g(\alpha) = \text{card}\,(S_\alpha)$.

Since g is bounded below κ, there is a set E' of cardinality less than λ and a set D'' of measure 1 with $g[D''] \subseteq E'$. Since λ is regular, $\sup (E') = \lambda' < \lambda$.

For each $\alpha \in D''$, let $s_\alpha : \lambda' \to S_\alpha$ by a surjection. Define $g_\xi : D'' \to \gamma$ for $\xi < \lambda'$ by

$$g_\xi(\alpha) = s_\alpha(\xi).$$

Since g_ξ is bounded below κ, there is a set E_ξ of power less than λ and a set N'_ξ of measure zero such that $g_\xi(\alpha) \in E_\xi$ if $\alpha \notin N'_\xi$. Put

$$D' = D'' - \bigcup \{N'_\xi : \xi < \lambda'\}; \quad E = \bigcup \{E_\xi : \xi < \lambda'\}.$$

Since λ is regular, card $(E) < \lambda$. Since $\lambda' < \kappa$, D' has measure 1. Finally if $\alpha \in D'$

$$S_\alpha = \{g_\xi(\alpha) \mid \xi < \lambda'\} \subseteq E.$$

Now we take D a set of measure 1 with the following properties: (1) $D \subseteq D'$; (2) if $\gamma \in D$, then $\gamma \in D_\alpha$ for each $\alpha < \gamma$. (The existence of D follows easily from normality.) To complete the proof it suffices to show

$$f[D^{[n]}] \subseteq E$$

(since E has cardinal $< \lambda$).

Let $x \in D^{[n]}$. Let α be the least element of x. Let $y \in D^{[k]}$ be such that $x = \{\alpha\} \cup y$. By our choice of D, $y \in D_\alpha^{[k]}$. Thus $f(x) = h_\alpha(y) \in S_\alpha$. Since $\alpha \in D \subseteq D'$, $S_\alpha \subseteq E$. Thus $f(x) \in E$. This completes the proof.

3. **Getting measurable cardinals.**

3.1. In this section, we show how to prove the consistency of the theory "**ZFC** + 'There is a measurable cardinal' " from the consistency of the theory "**ZFC** + 'There is a real-valued measurable cardinal' ". Our method is the method of inner models of Gödel [**2**]. We use in an essential way some recent work of Silver [**17**].

3.2. We begin by recalling the inner model construction in the form in which we will use it. Let A be a set. We are going to construct a class, $L[A]$, with the following properties:

(1) $L[A]$ is an $\in$-model of **ZFC**. (This is a theorem scheme.)

(2) $L[A]$ is transitive and contains all ordinals.

(3) $A \cap L[A]$ is in $L[A]$.

(4) $L[A]$ is the smallest transitive class satisfying (1)-(3).

We begin by defining a preliminary function $D(x; A)$. Suppose that x is a set. Let $\in_x$ be the relation $\{\langle y, z\rangle : y \in x \wedge z \in x \wedge y \in z\}$. Let A_x be $A \cap x$. Let $\mathscr{L}$ be a first order language with one two place predicate, $\in$, and one one place predicate **A**. We interpret $\mathscr{L}$ in the relational system $\mathscr{A}_x = \langle x;\ \in_x, A_x\rangle$ in the obvious way. A set $B \subseteq x$ is $\mathscr{A}_x$-definable if there are elements $\mathbf{y}_1, \ldots, \mathbf{y}_n$ of x (possibly $n = 0$) and a formula $\psi(y, y_1, \ldots, y_n)$ of $\mathscr{L}$ such that

$$B = \{y \in x \mid \mathscr{A}_x \models \psi(y, \mathbf{y}_1, \ldots, \mathbf{y}_n)\}.$$

We put $D(x, A) = \{B \subseteq x \mid B \text{ is } \mathscr{A}_x\text{-definable}\}$.

We now define $L_\alpha[A]$ for $\alpha \in \mathrm{OR}$[7] by induction on α:

$\alpha = 0 : L_0[A] = \varnothing$,

$\alpha = \beta + 1 : L_\alpha[A] = D(L_\beta[A], A)$,

α is a limit ordinal: $L_\alpha[A] = \bigcup_{\beta<\alpha} L_\beta[A]$.

3.3. There is a canonical bijection

$$F_A : \mathrm{OR} \to L[A].$$

F_A is definable in set-theory. I.e., there is a set-theoretical formula $\psi(x, y, z)$ which expresses

$$\text{“}y \text{ is an ordinal and } F_x(y) = z\text{”}.$$

We shall not recall the explicit construction of F_A, but shall recall the properties we use.

(1) F_A induces a well-ordering of $L[A]$. If α is an ordinal, then $L_\alpha[A]$ is an initial segment of $L[A]$ with respect to this well-ordering.

3.4. Let σ be the conjunction of a large finite number of axioms of **ZFC**. (It would be possible though tedious to explicitly exhibit σ.) We take σ “large enough” so that the discussions of subsections 3.2–3.3 can be formalized in the theory “σ”. Moreover, if σ is “large enough” we have the following absoluteness result:

Let N be a transitive model of σ. Let A be a set. Suppose that $A \cap N \in N$. Put $\bar{A} = A \cap N$.

Then for α an ordinal of N, $L_\alpha[A] = L_\alpha[\bar{A}] = L_\alpha^N[\bar{A}]$ and $F_A(\alpha) = F_{\bar{A}}(\alpha) = F_{\bar{A}}^N(\alpha)$. (Here the superscipt N means “computed in N”.)

Let α_0 be the least ordinal not in N. Then $\alpha_0 = \mathrm{OR}^N$, and $L[\bar{A}]^N = L_{\alpha_0}[\bar{A}] = L_{\alpha_0}[A]$.

[7] OR is the class of all ordinals.

3.5. Now let κ be an uncountable cardinal. Let J be a normal nontrivial λ-saturated ideal in $P(\kappa)$. Let $L[J]$ be defined as in subsection 3.2. Put $\bar{J} = L[J]$. It is straightforward to check that in $L[J]$, $\bar{J}$ is a normal nontrivial λ-saturated ideal in $P(\kappa)$.

We are going to prove in **ZFC** the following theorem.

THEOREM 6. *Suppose that J, κ, λ are as above and that $\lambda < \kappa$. Then in $L[J]$, κ is measurable.*

Theorem 6 readily implies one half of Theorem 2. It follows from some unpublished work of the author that one can prove the following in **ZFC** + "There are two measurable cardinals":

THEOREM A. *Let J, κ, λ be as above. Assume that $\lambda \leqq \kappa$. Then in $L[\bar{J}]$, $\bar{J}$ is prime (and so κ is measurable).*

Of course, a proof of Theorem A from two measurable cardinals is irrelevant to the question of whether "**ZFC** + 'There is one measurable cardinal'" is consistent. Recently, Kunen has announced a proof of Theorem A in **ZFC**.

3.6. Let κ be a measurable cardinal. Let μ be a normal measure on κ and J the ideal of sets of measure zero. Then Silver [**18**] has shown that the generalized continuum hypothesis holds in $L[J]$.

Our plan is to adapt Silver's argument to the context of Theorem 6. Replacing λ by λ^+ if necessary, we may as well assume that λ is regular and greater than $\aleph_0$. We are unable to prove the full GCH (though this has now been proved (using our Theorem 6) in work of Kunen [**8**, Theorem 7.8]). However, we can show that "$\aleph < \lambda \rightarrow 2^{\aleph} \leqq \lambda$" is valid in $L[J]$. By Tarski's result (Proposition 7 of §1), this will imply that κ is measurable in $L[J]$.

3.7. We now place ourselves inside $L[J]$. We write J instead of $\bar{J}$. Our assumption is that J is λ-saturated for some regular λ with $\aleph_0 < \lambda < \kappa$.

We assume that for some $\aleph < \lambda$, $2^{\aleph} > \lambda$. We let x be the λth subset of $\aleph$ in the canonical well-ordering of $L[J]$. (The set x exists since $2^{\aleph} > \lambda$.)

Let δ be an ordinal such that

(1) κ, J, and x appear in $L_\delta[J]$.

(2) σ is valid in $L_\delta[J]$.

We define a relational system $\mathscr{A}$ as follows: The underlying set of $\mathscr{A}$ is $L_\delta[J]$; the restriction of $\in$ to $L_\delta[J]$ is a relation of $\mathscr{A}$. Moreover, $\mathscr{A}$ has constants for x, κ, J, λ, and for each $\xi \leqq \aleph$.

We apply Theorem 5 to $\mathscr{A}$, getting an elementary submodel $\mathscr{A}'$ of $\mathscr{A}$ with the following properties: Let M be the underlying set of $\mathscr{A}'$. Then

(P1) $M \cap \kappa$ has J-measure 1.

(P2) $M \cap \lambda$ has cardinality less than λ.

We now let ψ be the unique $\in$-isomorphism of M with a transitive set [**10**, Theorem 3]. Say $\psi: M \simeq N$.

Since $\xi \in M$ for $\xi \leqq \aleph$, and $x \subseteq \aleph$, we have
(Q1) $\psi(\xi) = \xi$ for $\xi \leqq \aleph$.
(Q2) $\psi(x) = x$.
(P1) and (P2) have the following consequences:
(Q3) $\psi(\kappa) = \kappa$.
(Q4) $\psi(\lambda) < \lambda$.
The next two lemmas are due to Silver (in a slightly different context):

LEMMA 1. *There is a set $D \subseteq M$ of measure 1 such that $\psi(\xi) = \xi$ for $\xi \in D$.*

PROOF. Since $\kappa \cap M$ has measure 1, it suffices to show that the set

$$B = \{\alpha \in M \mid \alpha < \kappa \text{ and } \psi(\alpha) < \alpha\}$$

has measure zero. But if B had positive measure, we could find, by normality, a $B' \subseteq B$ of positive measure, and a $\gamma < \kappa$ such that $\psi(\alpha) = \gamma$ for all $\alpha \in B'$. This is absurd, since ψ is one-to-one.

LEMMA 2. *Let $A \in M$. Then $A \in J$ if and only if $\psi(A) \in J$.*

PROOF. Clearly $A \subseteq \kappa$ if and only if $\psi(A) \subseteq \kappa$ by (Q3). Say then that $A \subseteq \kappa$, $A \in M$. Let

$$D = \{\alpha \in M \mid \alpha < \kappa \text{ and } \psi(\alpha) = \alpha\}.$$

Since $\psi(A) = \{\psi(\alpha) \mid \alpha \in A\}$ and ψ is one-to-one,

$$A \cap D = \psi(A) \cap D.$$

Since D has measure 1, $A \in J \leftrightarrow A \cap D \in J \leftrightarrow \psi(A) \cap D \in J \leftrightarrow \psi(A) \in J$. This proves the lemma.

Lemma 2 has the corollary
(Q5) $\psi(J) = J \cap N$.
We now invoke the absoluteness results of subsection 3.4. They imply that if α_0 is the least ordinal not in N, then $N = L_{\alpha_0}[J]$ and that the canonical well-ordering on N given by $F_{\psi(J)}$ is an initial segment of the canonical well-ordering of $L[J]$. Thus:

(R1) In N, x is the λth subset of $\aleph$ in the canonical well-ordering of $L[\psi(J)]$.

On the other hand, the statement

"x is the λth subset of $\aleph$ in the canonical well-ordering of $L[J]$"

holds in $L_\delta[J]$ and hence in the elementary submodel M. Applying ψ to this statement and using (Q1) and (Q2) we see

(R2) In N, x is the $\psi(\lambda)$th subset of $\aleph$ in the canonical well-ordering of $L[\psi(J)]$.

Since, by (Q4), $\psi(\lambda) < \lambda$, (R1) contradicts (R2). This contradiction shows that "$\aleph < \lambda \rightarrow 2^\aleph \leqq \lambda$" is valid in $L[J]$. Thus (cf. subsection 3.6), κ is measurable in $L[J]$.

4. Creating real-valued measurable cardinals.

4.1. We are going to use the Boolean-valued version of Cohen's forcing method [15]. Our main result will be the following:

THEOREM 7.[8] *Let κ be a real-valued measurable cardinal. Let X be a finite measure space with measure algebra $\mathscr{B}$. Then in $V^{(\mathscr{B})}$, κ is real-valued measurable.*

Theorem 7 readily yields a proof of the remainder of Theorem 2. Thus suppose that κ is measurable in V. We let X_λ be the cartesian product of λ copies of $[0, 1]$ with the product measure. Let $\mathscr{B}_\lambda$ be the measure algebra of X_λ. Then in **[15]** it is shown that if $\aleph = \text{card}(\lambda)^{\aleph_0}$,

$$\|2^{\aleph_0} = \aleph\|^{(\mathscr{B})} = \mathbf{1}.$$

Thus if we take $\lambda = \kappa$, then in $V^{(\mathscr{B})}$, $\kappa = 2^{\aleph_0}$, and by Theorem 7, κ is real-valued measurable. Proposition 2 now completes the proof of Theorem 2.

Theorem 7 was inspired by a theorem of Prikry **[11]** which is part (a) of Theorem 8:

THEOREM 8. *Let κ be an uncountable cardinal. Let I be a nontrivial ideal in $P(\kappa)$. Let λ be an uncountable regular cardinal less than κ, and $\mathscr{B}$ a complete λ-saturated Boolean algebra. Let $J \in V^{(\mathscr{B})}$ be the ideal generated by $I^\vee$ in $P(\kappa)^{(\mathscr{B})}$. Thus*

$$\|A \in J\| = \sum \{\|A \subseteq B^\vee\| : B \in I\}.$$

Then J is nontrivial. Moreover,

(a) I λ-saturated $\rightarrow J$ $\lambda^\vee$-saturated.

(b) I κ-saturated $\rightarrow J$ $\kappa^\vee$-saturated.

We shall give an example later where κ is measurable, $\mathscr{B}$ is a complete κ-saturated Boolean algebra and in $V^{(\mathscr{B})}$, $\kappa^\vee = \aleph_1$. This example is due to A. Lévy.

4.2. We turn to the proof of Theorem 7. We are given the following data:

(1) κ is a real-valued measurable cardinal. We fix a nontrivial κ-additive real-valued measure μ_1 on $P(\kappa)$. We use differential notation for integrals. Thus

$$\int f(\alpha)\, d\mu_1(\alpha).$$

(2) X is a set, $\mathscr{S}$ a σ-algebra of subsets of X, with $X \in \mathscr{S}$, and $\mu_2 : \mathscr{S} \rightarrow [0, 1]$ is a countably additive measure with $\mu_2(X) = 1$. Let $\mathscr{I} = \{A \in \mathscr{S} \mid \mu_2(A) = 0\}$. Let $\mathscr{B} = \mathscr{S}/\mathscr{I}$ be the quotient σ-algebra. $\mathscr{B}$ is complete. (Cf. Halmos [**3**, §15].) If $A \in \mathscr{S}$, we denote by $[A]$ the class of A in $\mathscr{B}$.

4.3. We shall need a characterization of the reals of $V^{(\mathscr{B})}$ due to Dana Scott.

Let $\mathbf{Q}$ be the set of rationals. To each real x, we associate the set

$$A_x = \{r \in \mathbf{Q} : r < x\}.$$

We identify x with A_x, which is permissible since A_x determines x.

Now look at

$$\mathbf{R}^{(\mathscr{B})} = \{\mathbf{x} \in V^{(\mathscr{B})} : \|\mathbf{x} \text{ is a real}\| = \mathbf{1}\}.$$

---

[8] Originally, we only proved Theorem 7 for measurable κ. Ken Kunen observed that our proof gives the following more general result.

Each $\mathbf{x} \in \mathbf{R}^{(\mathscr{B})}$ determines a $\mathscr{B}$-valued subset of $\mathbf{Q}^{\vee}$. Thus $\mathbf{x}$ determines a function $\{b_r : r \in \mathbf{Q}\}$ such that

(a) $\prod_{s \in \mathbf{Q}} b_s = \mathbf{0}$;

(b) $\sum_{s \in \mathbf{Q}} b_s = \mathbf{1}$;

(c) $b_s = \sum_{t > s} b_t \ (s \in \mathbf{Q})$.

Conversely, it is easy to check that if $\{b_s : s \in \mathbf{Q}\}$ satisfies (a)-(c), then for a unique[9] real $\mathbf{x} \in \mathbf{R}^{(\mathscr{B})}$,

$$b_s = \|s < x\|.$$

($\{b_s\}$ determines a subset A of $\mathbf{Q}^{\vee}$ in $V^{(\mathscr{B})}$. (b) says that A is nonempty with probability $\mathbf{1}$. (a) and (c) imply that A is bounded above with probability $\mathbf{1}$. Take $\mathbf{x} = \text{lub } A$. Use (c) to check $A = A_x$.)

Now let $f : X \to \mathbf{R}$. Using the remark just made, one sees there is a unique real $\mathbf{x}_f \in \mathbf{R}^{(\mathscr{B})}$ such that

$$\|r < x_f\| = [\{x \in X : f(x) < r\}].$$

It is easy to see that $x_f = x_g$ if and only if $f(x) = g(x)$ almost everywhere.

Conversely, let $x \in \mathbf{R}^{(\mathscr{B})}$. Let

$$b_s = \|s < x\|, \qquad s \in Q.$$

Let $B_s \in \mathscr{S}$ be a representative of b_s. Since (a), (b), and (c) represent only countably many equations, we can find a single set $N \in \mathscr{S}$ of measure zero such that

(a′) $\bigcap_{s \in \mathbf{Q}} B_s \subseteq N$;

(b′) $(\bigcup_{s \in \mathbf{Q}} B_s) - N = \mathbf{R} - N$;

(c′) For $s \in \mathbf{Q}$, $B_s - N = \bigcup_{t > s} (B_t - N)$.

Define $g : X \to \mathbf{R}$ by

$$g(x) = 0, \qquad x \in N,$$
$$g(x) = \text{lub } \{r \mid x \in B_r\} \quad \text{for } r \notin N.$$

Using (a′)-(c′), one shows that g is well defined, and that $B_s - N = \{x \mid g(x) > r\} - N$. It follows that $\mathbf{x} = \mathbf{x}_g$. Thus we have proved the following lemma.

LEMMA 1 (SCOTT). *Let X, $\mathscr{B}$ be as above. Then there is a natural one-one correspondence between*

(a) $\mathbf{R}^{(\mathscr{B})}$.

(b) *Equivalence classes of measurable functions $f : X \to \mathbf{R}$, where the equivalence relation is "equal almost everywhere".*

The Scott models for the reals of $V^{(\mathscr{B})}$ has a transparent interpretation in the case of countable transitive models. We describe this interpretation without proofs. (The reader familiar with **[15]** and **[19]** can easily work out the proofs if he desires.)

Let $\mathscr{M}$ be a countable transitive model of **ZFC**. Let α be an ordinal of $\mathscr{M}$.

---

[9] We identify elements of $V^{(\mathscr{B})}$ which are equal with probability $\mathbf{1}$.

In $\mathcal{M}$, we form the measure space I^α, with measure algebra $\mathcal{B}$. We also form the measure space I^α in the real world. Let $X =$ "real world's" I^α; $X_{\mathcal{M}} =$ "$\mathcal{M}$'s" I_α.

Certain of the points of X determine $\mathcal{M}$-completely additive homomorphisms of $\mathcal{B}$. The precise statement is that there is a set N of measure zero of X such that $X - N$ is in natural one-one correspondence with the set of $\mathcal{M}$-completely additive homomorphisms of $\mathcal{B}$. Thus each point $x \in X - N$ determines a Cohen extension $\mathcal{M}[x]$ of type $\mathcal{B}$.

Next, given a Borel function $f: X_{\mathcal{M}} \to R_{\mathcal{M}}$ in $\mathcal{M}$, there is a canonical way to prolong it to a Borel function $\tilde{f}: X \to R$ which preserves all reasonable properties.

We now let $a \in \mathcal{M}^{(\mathcal{B})}$ be such that $\|a \text{ is a real}\| = \mathbf{1}$. In each model $\mathcal{M}[x]$, a determines an ordinary real, a_x. Then one has the following interpretation of the "Scott model": Let f be, in $\mathcal{M}$, a Borel function mapping $X_{\mathcal{M}}$ into $R_{\mathcal{M}}$. Let f correspond to the real a of $\mathcal{M}^{(\mathcal{B})}$. Then for all $x \in X - N$, $\tilde{f}(x) = a_x$.

(We remark that from this standpoint, Lemmas 2 and 3, below, are obvious.)

4.4. LEMMA 2. *Let $f, g: X \to \mathbf{R}$ be measurable. Let f, g correspond to reals $\mathbf{x}$, $\mathbf{y}$ in $\mathbf{R}^{(\mathcal{B})}$. Then*

(1) *$f + g$ corresponds to $\mathbf{x} + \mathbf{y}$;*

(2) *$\{x \mid f(x) < g(x)\}$ represents $\|\mathbf{x} < \mathbf{y}\|$;*

(3) *$\{x \mid f(x) = g(x)\}$ represents $\|\mathbf{x} = \mathbf{y}\|$.*

PROOF. We prove (1). The proofs of (2) and (3) are similar and we leave them to the reader.

We first compute $\{x \mid (f + g)(x) > r\}$. This is $\bigcup \{A_{s,t}: s + t > r, s, t \in \mathbf{Q}\}$, where $A_{s,t} = \{x \mid f(x) > s \text{ and } g(x) > t\}$.

Similarly $\|\mathbf{x} + \mathbf{y} > r\|$ is $\sum \{a_{s,t}: s + t > r, s, t \in \mathbf{Q}\}$, where $a_{s,t} = \|\mathbf{x} > s\| \cdot \|\mathbf{y} > t\|$. (1) is now clear.

In a similar way, we can prove

LEMMA 3. *Let $\{f_n, n \in \omega\}$ be a sequence of measurable functions, $f_n: X \to \mathbf{R}$. We suppose that $f_{n+1}(x) > f_n(x)$ almost everywhere, for $n \in \omega$. Suppose that $f_n \to f$ almost everywhere. Let $\mathbf{x}_n$ correspond to f_n, and $\mathbf{x}$ to f. Then*

$$\lim_n \mathbf{x}_n = \mathbf{x}$$

with probability **1**. (*Much more than this is true, but this special case will suffice for our needs.*)

4.5. We shall also need the "model" for $P(\kappa)^{(\mathcal{B})}$. Let

$$E = \{A \in V^{(\mathcal{B})} \mid \|A \subseteq \kappa\| = \mathbf{1}\}.$$

We define a map $\psi: E \to \mathcal{B}^\kappa$ as follows: $\psi(A)(\alpha) = \|\alpha^\vee \in A\|$.

The following facts are proved in **[15]**:

(1) ψ is a bijection.

(2) Let $A, B \in E$. Put $\psi(A) = f_A$; $\psi(B) = f_B$. Let $b \in \mathcal{B}$. Then $b \leqq \|A = B\|$ if and only if

$$b \cdot f_A(\alpha) = b \cdot f_B(\alpha), \quad \text{for all } \alpha \in \kappa.$$

4.6. We are going to define a $\mu \in V^{(\mathscr{B})}$ which is a real-valued measure on $P(\kappa)^{(\mathscr{B})}$. We will need the following lemma from [**15**]:

LEMMA 4. *Let $A, B \in V^{(\mathscr{B})}$. Suppose $\|A = \varnothing\| = \|B = \varnothing\| = \mathbf{0}$. Put*

$$A\hat{} = \{x \in V^{(\mathscr{B})} \mid \|x \in A\| = \mathbf{1}\};$$
$$B\hat{} = \{x \in V^{(\mathscr{B})} \mid \|x \in B\| = \mathbf{1}\}.$$

Let $h: A\hat{} \to B\hat{}$ be a map which satisfies

$$\|h(x) = h(y)\| \geqq \|x = y\|,$$

for any $x, y \in A\hat{}$. Then there is a unique $f \in V^{(\mathscr{B})}$ such that:
(1) $\|f: A \to B\| = \mathbf{1}$.
(2) *Let $x \in A\hat{}$, $y \in B\hat{}$ and suppose $y = h(x)$. Then $\|y = f(x)\| = \mathbf{1}$.*

4.7. We now start to describe μ. To each $A \in E$, we will associate a real $\mathbf{x}_A \in \mathbf{R}^{(\mathscr{B})}$.

Let $A \in E$. Let $f_A: \kappa \to \mathscr{B}$ represent A. So $\|\alpha^{\vee} \in A\| = f_A(\alpha)$.

We are going to define a map $\gamma_A: \mathscr{B} \to [0, 1]$. Let $b \in \mathscr{B}$. Let $\mu_3: \mathscr{B} \to \mathbf{R}$ be induced by μ_2. Put

$$\gamma_A(b) = \int \mu_3(b \cdot f_A(\alpha))\, d\mu_1(\alpha).$$

LEMMA 5. (1) $\gamma_A(b) \leqq \mu_3(b)$.
(2) *γ_A is a countably additive measure on $\mathscr{B}$.*

PROOF. (1) comes from integrating over κ the inequality

$$\mu_3(b \cdot f_A(\alpha)) \leqq \mu_3(b).$$

To prove (2), let $b = \sum_{i \in \omega} b_i$ (disjoint sum). Then

$$1 \geqq \mu_3(b \cdot f_A(\alpha)) = \sum_{i \in \omega} \mu_3(b_i \cdot f_A(\alpha)).$$

By the Lebesgue dominated convergence theorem (cf. [**4**, p. 110]), we can integrate this equation to get

$$\gamma(b) = \sum_{i \in \omega} \gamma(b_i).$$

This proves Lemma 5.

By Lemma 5 and the Radon-Nikodym theorem [**4**, p. 128], there is a function $g_A: X \to \mathbf{R}$ such that

$$\gamma([C]) = \int_C g_A(x)\, d\mu_2(x)$$

for all C in $\mathscr{S}$. Moreover, the equivalence class of g_A is uniquely determined by γ_A in this way. Thus by Lemma 1, g_A determines a unique real $\mathbf{x}_A \in \mathbf{R}^{(\mathscr{B})}$.

LEMMA 6. *Let $A, A' \in E$. Let $b = \|A = A'\|$. Then $b \leqq \|\mathbf{x}_A = \mathbf{x}_{A'}\|$.*

PROOF. If $b' \leqq b$, then

$$b' \cdot f_A(\alpha) = b' \cdot f_{A'}(\alpha), \qquad \alpha \in \kappa,$$

by subsection 4.5(2). Integrating with respect to α, we get $\gamma_A(b') = \gamma_{A'}(b')$.

Pick representatives g_A, $g_{A'}$ as above. Let B represent b. Then for any set $B' \subseteq B$,

$$\int_{B'} g_A(x)\, d\mu_2(x) = \int_{B'} g_{A'}(x)\, d\mu_2(x).$$

It follows that $g_A(x) = g_{A'}(x)$ almost everywhere in B. By Lemma 2, $b \leqq \|\mathbf{x}_A = \mathbf{x}_{A'}\|$.

LEMMA 7. *For all* $A \in E$, $\|0 \leqq \mathbf{x}_A \leqq 1\| = \mathbf{1}$.

PROOF. We know by Lemma 5, that

$$0 \leqq \gamma_A(b) \leqq \mu_3(b), \qquad b \in \mathscr{B}.$$

Hence $0 \leqq g_A(x) \leqq 1$ a.e. Lemma 2 completes the proof.

4.8. By Lemmas 6 and 4, there is a $\mu \in V^{(\mathscr{B})}$ such that

(1) $\|\mu$ maps $P(\kappa)$ into $[0, 1]\| = \mathbf{1}$.

(2) For $A \in E$, $\|\mu(A) = \mathbf{x}_A\| = \mathbf{1}$.

The following lemma will complete the proof of Theorem 7.

LEMMA 8. μ *is a nontrivial* κ*-additive measure on* κ.

PROOF. That $\mu(\kappa) = 1$, and that μ vanishes on singletons is easy to check. We show that μ is κ-additive. Let $A \in E$. Then we put

$$\gamma(A) = \int \mu_3(f_A(\alpha))\, d\mu_1(\alpha).$$

It is easy to see that $\gamma(A) = 0 \leftrightarrow \mu_3(f_A(\alpha)) = 0$, for almost all $\alpha \leftrightarrow f_A(\alpha) = \mathbf{0}$ for almost all α. Finally note that if $\gamma(A) = 0$, $\gamma_A \equiv 0$ so $\mathbf{x}_A = 0$.

Suppose now that $A_1, \ldots, A_n \in E$ and $\|A_i \cap A_j = \varnothing\| = \mathbf{1}$ for $1 \leqq i < j \leqq n$. Then the elements $\{f_{A_1}(\alpha), \ldots, f_{A_n}(\alpha)\}$ are pairwise disjoint. It follows that $\gamma(A_1) + \cdots + \gamma(A_n) \leqq 1$.

Using the finite additivity of γ, it is easy to see that $\gamma(A_\xi) = 0$ for all ξ outside some countable set S. (Cf. the proof of Proposition 6.) Thus by an earlier remark, there is for $\xi \in \lambda - S$ a set $N_\xi \subseteq \kappa$ of measure zero such that $f_{A_\xi}(\alpha) = \mathbf{0}$ for $\alpha \notin N_\xi$, $\xi \notin S$.

Since $\lambda < \kappa$, we can amalgamate the N_ξ's into a single set N of measure zero such that $f_{A_\xi}(\alpha) = \mathbf{0}$ if $\alpha \notin N$, $\xi \notin S$.

It follows that for $b \in \mathscr{B}$, $\alpha \notin N$,

$$\mu_3(b \cdot f_A(\alpha)) = \sum_{\xi \in S} \mu_3(b \cdot f_{A_\xi}(\alpha)).$$

Integrating, we get $\gamma_A = \sum_{\xi \in S} \gamma_{A_\xi}$, and therefore $g_A = \sum_{\xi \in S} g_{A_\xi}$. By Lemma 3 (applied to the sequence of partial sums), $\mathbf{x}_A = \sum_{\xi \in S} \mathbf{x}_{A_\xi}$. But if $\xi \in \lambda - S$, $\gamma(A_\xi) = 0$ so $\mathbf{x}_{A_\xi} = 0$. Thus $\mathbf{x}_A = \sum_{\xi < \lambda} \mathbf{x}_{A_\xi}$, and the lemma is proved.

(Our means of computing the measure of a subset A of $\kappa^\vee$ no doubt looks a little ad hoc. We give a heuristic interpretation of our formula as a "Fubini" formula in the countable model case.

So let $\mathscr{M}$ be a countable transitive model of **ZFC**, κ a real-valued measurable cardinal in $\mathscr{M}$ with measure μ; α an ordinal of $\mathscr{M}$, $\mathscr{B}$ the measure algebra of I^α in $\mathscr{M}$, and X the real world version of I^α, with measure μ.

Let $A \in \mathscr{M}^{\mathscr{B}}$ be such that $\|A \subseteq \kappa\| = \mathbf{1}$. For almost all $x \in X$, the model $\mathscr{M}[x]$ is defined and A determines a subset A_x of κ in $\mathscr{M}[x]$. Let us suppose, heuristically, that γ extends in a "nice" way to a measure γ_x in $\mathscr{M}[x]$. We seek to determine the function $x \to \gamma_x(A_x)$. For this we must determine

$$\int_B \gamma_x(A_x)\, dx$$

for B a Borel set describable in $\mathscr{M}$.

We write this as

$$\int_B \int_\kappa \chi_{A_x}(\alpha)\, d\alpha\, dx.$$

If we reverse the order of integration, this becomes

$$\int_\kappa \int_B \chi_{A_x}(\alpha)\, dx\, d\alpha.$$

Now let B_α be a Borel set of X representing $\|\alpha \in A\|$. Then $\alpha \in A_x$ if and only if $x \in B_\alpha$. Thus our formula becomes

$$\int_\kappa \int_B \chi_{B_\alpha}(x)\, d\alpha = \int_\kappa \mu(B \cap B_\alpha)\, d\alpha.$$

This is exactly the formula implicit in our recipe for extending γ.)

4.9. We turn now to the proof of Theorem 8. We first show that if $A \in J$ with probability $\mathbf{1}$, then there is a single $B \in I$ with $\|A \subseteq B^\vee\| = \mathbf{1}$.

Let $\{\langle b_i, B_i\rangle : i \in K\}$ be maximal subject to the following:

(1) $B_i \in I$.

(2) $\mathbf{0} < b_i \leqq \|A \subseteq B_i^\vee\|$.

(3) $\{b_i, i \in K\}$ is pairwise disjoint.

Since $\mathscr{B}$ is κ-saturated, card $(K) < \kappa$. Since $\|A \in J\| = \mathbf{1}$, $\sum_{i \in K} b_i = \mathbf{1}$. Put

$$B = \bigcup_{i \in K} B_i.$$

Since I is κ-complete, $B \in I$. Clearly

$$b_i \leqq \|A \subseteq B_i^\vee\| \leqq \|A \subseteq B^\vee\|.$$

Since $\sum_{i \in K} b_i = \mathbf{1}$, $\|A \subseteq B^\vee\| = \mathbf{1}$, so our claim is proved.

It is now easy to check that J is κ-complete. Let $\{A_\xi, \xi < \alpha\}$ be a sequence of elements of $V^{(\mathscr{B})}$, with $\|A_\xi \in J\| = \mathbf{1}$, all $\xi < \alpha$ and $\alpha < \kappa$. We must show $\|\bigcup_{\xi<\alpha} A_\xi \in J\| = \mathbf{1}$. By our preceding remark, we can select $B_\xi \in I$, $\xi < \alpha$, so that $\|A_\xi \subseteq B_\xi^\vee\| = \mathbf{1}$. Let $B = \bigcup_{\xi<\alpha} B_\xi$. Then $B \in I$, and $\|\bigcup_{\xi<\alpha} A_\xi \subseteq B^\vee\| = \mathbf{1}$. Thus $\|\bigcup_{\xi<\alpha} A_\xi \in J\| = \mathbf{1}$. We have shown that J is κ-complete. We leave the remainder of the proof that J is nontrivial to the reader.

We treat cases (a) and (b) in parallel as much as possible. Let $\gamma = \lambda$ in case (a) and κ in case (b). Suppose $\{A_\xi; \xi < \gamma\}$ is a sequence of elements of $V^{(\mathscr{B})}$ with $\|A_\xi \subseteq \kappa\| = \mathbf{1}$ and for $\xi < \xi' < \gamma$, $\|A_\xi \cap A_{\xi'} = \varnothing\| = \mathbf{1}$. We show that for some $\xi < \gamma$, $\|A_\xi \in J\| = \mathbf{1}$.

Let for $\alpha < \kappa, \beta < \gamma, b(\alpha, \beta) = \|\alpha \in A_\beta\|$. Since for $\beta < \beta' < \gamma, A_\beta \cap A_{\beta'} = \varnothing$ with probability $\mathbf{1}$, $\{b(\alpha, \beta) \mid \beta < \gamma\}$ is pairwise disjoint. Since $\mathscr{B}$ is λ-saturated, the set

$$S_\alpha = \{\beta \mid b(\alpha, \beta) \neq \mathbf{0}\}$$

has power less than λ.

In case (a), we can improve our estimate on the size of S_α. Let $f: \kappa \to \lambda$ send α into the order type of S_α. Since I is λ-saturated, there is a set E of power less than λ and a set N of measure zero (rel. I) such that $f(\alpha) \in E$ for $\alpha \notin N$. Since λ is regular, $\sup(E) < \lambda$.

Put $\gamma' = \sup(E)$ in case (a); $\gamma' = \lambda$ in case (b). Then $\gamma' < \gamma$ in either case, and for $\alpha \notin N$, S_α has order type $\leqq \gamma'$.

Let $A_{\xi,\eta}$ be defined for $\xi < \gamma$, $\eta < \gamma'$ as follows:

$$A_{\xi,\eta} = \{\alpha \in \kappa \mid \xi \text{ is the } \eta\text{th member of } S_\alpha\}.$$

Clearly, $\xi < \xi' \to A_{\xi,\eta} \cap A_{\xi',\eta} = \varnothing$. Let $T_\eta = \{\xi \mid A_{\xi,\eta}$ has positive measure$\}$. Since I is γ-saturated, T_η has power less than γ. Since γ is regular, $\bigcup_{\eta<\gamma'} T_\eta$ has power less than γ. Pick $\xi_0 < \gamma$, with $\xi_0 \notin T_\eta$ for any $\eta < \gamma'$. But we have

$$A_{\xi_0} \subseteq \left[\bigcup_{\eta<\gamma'} A_{\xi_0,\eta}\right]^\vee \cup N^\vee$$

with probability $\mathbf{1}$. Since $\xi_0 \notin T_\eta$, any $\eta < \gamma'$, each $A_{\xi_0,\eta}$ lies in I. Since $\gamma' < \gamma \leqq \kappa$,

$$\bigcup_{\eta<\gamma'} A_{\xi_0,\eta}$$

lies in I. Thus $A_{\xi_0} \in J$, and the theorem is proved.

4.10. We now briefly describe Lévy's counterexample. For details, see [**19**, I§3]. Let κ be a measurable cardinal. For each ordinal ξ with $0 < \xi < \kappa$, one adjoins a generic collapsing map $f_\xi: \omega \to \xi$. The conditions give only finitely much information about the f_ξ's. Let $\mathscr{B}$ be the Boolean algebra associated (in the sense of [**15**]) to this notion of forcing. Lévy proves that $\mathscr{B}$ is κ-saturated. By construction, $\kappa \leqq \aleph_1^{(\mathscr{B})}$. Thus $\kappa = \aleph_1^{(\mathscr{B})}$. But $P(\aleph_1)$ has no $\aleph_1$-saturated nontrivial ideal by a theorem of Ulam (reproved in §2).

4.11. We close this section with some questions.

(1) Let I be a nontrivial ideal in $P(\kappa)$ which is κ-saturated. We know that

$P(\kappa)/I$ is complete. Is it ever countably generated (qua complete Boolean algebra)? We are especially interested in the case when I arises from a real-valued measure. To avoid trivialities, assume that κ is not measurable.

(2) Is it consistent for an I as in question (1) to have sat $(I) = \kappa$? Can sat (I) be a weakly inaccessible cardinal less than κ? If so, which ones are possible? Proposition 8 of §1 is relevant here. Note also that if it is possible to have a strongly inaccessible κ carrying an I with sat $(I) = \kappa$, one can also have $\kappa = 2^{\aleph_0}$ and κ carries an I with sat $(I) = \kappa$ (by (b) of Theorem 8). Is the existence of an I with sat $(I) = \kappa$ compatible with GCH?

(3) It would also be interesting to have examples of I with sat $(I) = \kappa^+$. For example, can $\aleph_1$ carry a nontrivial $\aleph_2$-saturated ideal? Again, can it if GCH holds?

5. **Solution of a problem of Födor and Hajnal.**

5.1. In this section we use the results of §2 to settle a problem of Födor and Hajnal [**1**].

We begin by stating the problem we shall solve. Let κ be a regular cardinal greater than $\aleph_0$. A set $K \subseteq \kappa$ is *closed* (in κ) if $\varnothing \neq S \subseteq K \rightarrow \text{lub}\, S \in K \cup \{\kappa\}$; K is *unbounded* (in κ) if lub $K = \kappa$. A set $A \subseteq \kappa$ is stationary if $A \cap K \neq \varnothing$ for any $K \subseteq \kappa$ which is closed unbounded.

Our result is as follows:

THEOREM 9. *Let $A \subseteq \kappa$ be stationary. Then A is the disjoint union of κ stationary subsets.*

5.2. The concept "stationary" is closely connected with the notions of §2. In fact, define a set I as follows:

$$I = \{A \subseteq \kappa \mid A \cap K = \varnothing \quad \text{for some closed unbounded } K \subseteq \kappa\}.$$

The following lemma is due to Födor and Hajnal [**1**]:

LEMMA 1. *I is a nontrivial ideal in κ. A is stationary if and only if A has positive I-measure. I is normal. That is, if $A \subseteq \kappa$ has positive I-measure, $f: A \rightarrow \kappa$, and $f(\alpha) < \alpha$ for all $\alpha \in A$, then there is a $B \subseteq A$ of positive I-measure and a $\lambda < \kappa$ such that $f(\alpha) = \lambda$ for all $\alpha \in B$.*

We remark that Födor-Hajnal give the following alternative characterization of stationary sets. A is stationary if and only if for every $f: A \rightarrow \kappa$, with $f(\alpha) < \alpha$, for all $\alpha \in A$, there is a set $B \subseteq A$ of power κ and a $\lambda < \kappa$, such that $f(\alpha) = \lambda$ for all $\alpha \in B$.

5.3. We collect some lemmas about closed unbounded sets which we will need in the proof of Theorem 9. Recall that α is a limit point of a set of ordinals A if $\alpha = \sup S$ for some nonempty $S \subseteq A \cap \alpha$.

LEMMA 2. *Let κ be a regular cardinal $> \aleph_0$. Let K be a closed unbounded subset of κ. Then if we put $K' = \{\beta < \kappa \mid \beta \text{ is a limit point of } K\}$ then K' is closed unbounded. Also $K' \subseteq K$.*

PROOF. Let $\beta \in K'$. Then $\beta = \sup S$ for some $S \subseteq K$. Since K is closed, $\beta \in K$. Thus $K' \subseteq K$. Let $S \subseteq K'$, with $\beta = \text{lub } S < \kappa$. If $\beta \in S$, $\beta \in K'$, otherwise $S \subseteq \beta \cap K' \subseteq \beta \cap K$. So $\beta = \text{lub } S \in K'$. Finally, let $\beta < \kappa$. We show K' contains elements greater than β. Since K is unbounded we can define α_n inductively in K with $\alpha_0 > \beta$, and $\alpha_{n+1} > \alpha_n$, for all n. Let $\alpha = \sup \{\alpha_n \mid n \in \omega\}$. Then $\alpha < \kappa$, since κ is not cofinal with ω. Clearly $\alpha \in K'$.

LEMMA 3. *Let κ be a regular cardinal greater than $\aleph_0$. Let K be a closed unbounded subset of κ. Let α be a regular cardinal greater than $\aleph_0$, with $\alpha \in K'$. Then $K \cap \alpha$ is closed unbounded in α.*

PROOF. Clearly $K \cap \alpha$ is closed in α. Since $\alpha \in K'$, $\alpha = \text{lub } (K \cap \alpha)$, so $K \cap \alpha$ is unbounded in α.

5.4. Suppose that Theorem 9 is false. Let A be a stationary set which is not the union of κ stationary subsets. We define a set J as follows:

$$J = \{B \subseteq \kappa \mid B \cap A \in I\}.$$

LEMMA 4. *J is a normal nontrivial κ-saturated ideal in $P(\kappa)$.*

PROOF. That J is a normal nontrivial ideal in $P(\kappa)$ follows readily from Lemma 1. We show that J is κ-saturated. Suppose not. By Proposition 4 of §1, there is a sequence $\{B_\xi \mid \xi < \kappa\}$ which is pairwise disjoint and such that B_ξ has J positive measure for all $\xi < \kappa$. Thus $\{A \cap B_\xi \mid \xi < \kappa\}$ is a pairwise disjoint family of sets of positive I-measure. For $\xi \neq 0$, put $A_\xi = A \cap B_\xi$. Let $A_0 = \{\alpha \mid \alpha \notin A_\xi$ for any $\xi > 0\}$. Then $A = \bigcup_{\xi<\kappa} A_\xi$ and the union is disjoint. If $\xi > 0$, A_ξ is stationary by our previous remarks. Since $A \cap B_0 \subseteq A_0$, A_0 is also stationary. But this contradicts our assumption on A.

LEMMA 5. *There is a closed unbounded set $K \subseteq \kappa$ such that $K \cap A \subseteq \{\alpha \mid \alpha$ is regular, $>\aleph_0$, and A-Mahlo$\}$.*

PROOF. We apply the results of §2 to J. (This is justified by Lemma 4.) Since A has J measure 1, the set $E = \{\alpha \mid \alpha$ is singular or $\alpha \leqq \aleph_0$ or α is not A-Mahlo$\}$, has J-measure zero. Hence $E \cap A \in I$. I.e., there is a closed unbounded set K such that $E \cap A \cap K = \varnothing$. This proves the lemma.

We now let K be as in Lemma 5. Let K' be the set of limit points of K. By Lemma 2, K' is closed unbounded. Hence $K' \cap A \neq \varnothing$. Let α be the least member of $K' \cap A$. By Lemma 5, α is regular, $>\aleph_0$, and A-Mahlo. By Lemma 3, $K \cap \alpha$ is closed unbounded in α. Hence, by Lemma 2, $K' \cap \alpha$ is closed unbounded in α. Since α is A-Mahlo, there is a $\beta < \alpha$ with $\beta \in K' \cap A$. This contradicts our choice of α as the least member of $K' \cap A$. This contradiction establishes Theorem 9.

6. **Proof of a theorem of Kunen.**

6.1. Before stating Kunen's theorem, let me describe my original motivation that led me to the proofs of Theorem 1. Let κ be a real-valued measurable cardinal.

One wants to imitate, for κ, the Keisler-Tarski results for two-valued measurable cardinals. I tried to extend the ultraproduct version of their proof.

When one tries to extend this version, one runs into the following problem. For ultraproduct methods, the important fact about $\{0, 1\}$ is that it is a Boolean algebra. But alas, $[0, 1]$ is not a Boolean algebra, so the methods do not extend.

Here is the way out of the impasse: Let I be the ideal of sets of measure zero relative to μ, and let $\mathscr{B} = P(\kappa)/I$. We get a $\mathscr{B}$-valued measure γ in the following trivial way:

$$\gamma(A) \text{ is the image of } A \text{ in } \mathscr{B}.$$

Using γ one can form ultraproducts in pretty much the usual way. The ultraproducts are now $\mathscr{B}$-valued relational systems. In particular, taking the ultrapower of the universe of sets, V, yields a structure isomorphic to a transitive subclass, $V_*^{(\mathscr{B})}$ of $V^{(\mathscr{B})}$. One can now extend one of the usual proofs of the existence of a normal measure to prove Theorem 3. (To get the "incompressible function" f, take the function which represents κ in the ultraproduct.)

6.2. The usual ultraproduct method yields two types of results about a measurable cardinal κ:

(1) κ is very badly Mahlo.

(2) κ is Π_1^2 indescribable. (This notion of indescribability is stated in **[8]**.)

We have shown in §2 that results of type (1) generalize readily to real-valued measurable cardinals. However, I tried without success to generalize (2) to the case of real-valued measurable cardinals.

Kunen, in **[8,** Theorem 16.8] showed (almost) that there was no such generalization. He produced a model in which there was an uncountable cardinal κ, with $P(\kappa)$ carrying an $\aleph_1$-saturated nontrivial ideal, such that $\langle \kappa; < \rangle$ had a Π_1^1-characterization. Since the methods I was using applied equally well to κ's carrying an $\aleph_1$-saturated nontrivial ideal, they could not prove that a real-valued measurable cardinal is not Π_1^1-characterizable.

6.3. Kunen also discovered that there was a useful indescribability theorem true for real-valued measurable cardinals. In order to explain it, I shall briefly outline the usual proof that measurable cardinals are Π_n^1-indescribable.

Let κ be a measurable cardinal. Let μ be a normal measure on κ. It turns out that if V^* is the transitive class isomorphic to V^κ/μ, then the identity function $i: \kappa \to \kappa$ represents κ in the ultraproduct. (Precisely, the equivalence class of i in V^κ/μ corresponds to κ in V^*.) Let φ be a Π_n^1 formula true of α for almost all α. Then φ is true of κ in V^* by the fundamental theorem on ultraproducts. It turns out however that all subsets of κ appear in V^*. Thus φ holds of κ in V^* if and only if it holds of κ in V.

One can push through a similar argument if μ is a normal nontrivial real-valued measure on κ. What one gets is the following: Let φ be Π_n^1. Say φ holds of α for almost all $\alpha < \kappa$. Then φ holds of κ in $V^{(\mathscr{B})}$.

Alas, $V^{(\mathscr{B})} \neq V$. This is the obstruction to proving indescribability results about κ.

Let me now state a version of Kunen's theorem, and see how it falls out of the situation we've just described: Suppose $\mathbf{ZFC} \vdash \kappa$ real-valued measurable $\to \varphi(\kappa)$, where φ is Π^1_n. Then $\mathbf{ZFC} \vdash \kappa$ real-valued measurable, μ a normal real-valued measure on $\kappa \to \mu(\{\alpha \mid \varphi(\alpha)\}) = 1$.

SKETCH OF PROOF. Otherwise, we would have $\|\varphi(\kappa)\| < \mathbf{1}$, in $V^{(\mathscr{B})}$. But $\|\kappa$ is real-valued measurable$\| = \mathbf{1}$ in $V^{(\mathscr{B})}$. Since $V^{(\mathscr{B})}$ is a model of **ZFC**, $\|\varphi(\kappa)\| = \mathbf{1}$.

In the remainder of this section, we develop the details that turn this sketch into a proof, and state Kunen's theorem (as he did) with the appropriate level of generality.

6.4. Recall that κ^+ is the least cardinal greater than κ. We fix, until further notice, an uncountable cardinal κ, and a nontrivial κ^+-saturated ideal I in $P(\kappa)$. Let $\mathscr{B} = P(\kappa)/I$. By Proposition 9 of §1, $\mathscr{B}$ is a complete Boolean algebra. If $A \subseteq \kappa$, $[A]$ will denote its image in $\mathscr{B}$.

6.5. We shall assume, throughout §6, that the reader is familiar with the theory of Boolean-valued models for set-theory. (Cf. **[15]**.) We make a few remarks on the conventions followed in the present paper.

(1) We always assume that $V^{(\mathscr{B})}$ is separated. That is, if $x, y \in V^{(\mathscr{B})}$ and $\|x = y\| = \mathbf{1}$, then $x = y$.

(2) In $V^{(\mathscr{B})}$ there is a unique element h with the following properties.

(a) $h: \mathscr{B}^{\vee} \to \mathbf{2}$.

(b) h is a homomorphism.

(c) $\|h(b^{\vee}) = \mathbf{1}\| = b$.

We fix the letter h as a notation for this map.

(3) We frequently identify x in V with $x^{\vee}$ in $V^{(\mathscr{B})}$, if it seems harmless.

6.6. If A and B are classes, $A^B = \{F: F$ is a function with domain B and range $(F) \subseteq A\}$. Let V be the class of all sets. We make V^κ into a $\mathscr{B}$-valued relational system, by defining relations $\simeq$ $\tilde{\in}$ as follows:

Let $f, g \in V^\kappa$. Then

$$\|f \simeq g\| = [\{\alpha \mid f(\alpha) = g(\alpha)\}]$$
$$\|f \tilde{\in} g\| = [\{\alpha \mid f(\alpha) \in g(\alpha)\}].$$

Let $\tilde{\varphi}(x_1, \ldots, x_n)$ be a first order formula in the predicates $\tilde{\in}$ and $\simeq$. Then, in the usual way, we define $\|\tilde{\varphi}(f_1, \ldots, f_n)\|$. (Since V^κ is a proper class, this is a scheme of definitions.)

The following lemma is proved in the usual way. (Cf. **[7,** Proof of Theorem 5.1.].)

LEMMA 2. *Let $\varphi(x_1, \ldots, x_n)$ be a first-order formula in the predicates $\in$ and $=$; let $\tilde{\varphi}(x_1, \ldots, x_n)$ be the corresponding formula in the predicates $\tilde{\in}$ and $\simeq$. Let $f_1, \ldots, f_n$ be elements of V^κ. Then*

$$\|\tilde{\varphi}(f_1, \ldots, f_n)\| = [\{\alpha \mid \varphi(f_1(\alpha), \ldots, f_n(\alpha))\}].$$

(This is a scheme of theorems.)

COROLLARY 3. *The axioms for equality (stating that $\simeq$ is an equivalence relation substitutive with respect to $\tilde{\in}$) are $\mathscr{B}$-valid in W.*

COROLLARY 4. *W is a $\mathscr{B}$-model for* **ZFC**.

(Again, Corollary 4 is a scheme of theorems.)

6.7. Consider for a moment the classical case when κ is measurable, and $\mathscr{B} = \mathbf{2}$. In this case the **2**-valued relational system, V^κ, is essentially the ultrapower V^κ/I. Moreover, there is a canonical embedding of V^κ/I into V whose image is a transitive class.

We are going to show, in a similar way, that V^κ maps onto a transitive subclass of $V^{(\mathscr{B})}$ in the general case. Our proof will be the usual proof in the classical case carried out inside $V^{(\mathscr{B})}$.

6.8. We work inside $V^{(\mathscr{B})}$. V^κ determines a class of $V^{(\mathscr{B})}$, which we again denote by V^κ, by

$$\|x \in V^\kappa\| = \sum \{\|x = y^\vee\| : y \in V^\kappa\}.$$

We define an equivalence relation on V^κ, $\approx$, as follows:

$$x \approx y \quad \text{iff} \quad h(\|x \simeq y\|) = \mathbf{1}.$$

We wish to pass to equivalence classes, relative to $\approx$. There is a standard difficulty since the equivalence classes will be proper classes, and a standard way around the difficulty (due to Scott [**14**]). The upshot is that, in an explicit way, one can associate to each $x \in V^\kappa$ a set $[x]$ such that $[x] = [y]$ if and only if $x \approx y$.

6.9. We return to V. Then if $f \in V^\kappa$, the discussion of §6.8 assigns a $[f] \in V^{(\mathscr{B})}$ which is the "Scott equivalence class" of f relative to $\approx$ with probability **1**.

LEMMA 5. $\|[f] = [g]\| = \|f \simeq g\|$. *In particular,* $[f] = [g]$ *if and only if* $\|f \simeq g\| = \mathbf{1}$ *if and only if f and g are equal almost everywhere.*

PROOF. $\|[f] = [g]\| = \|f \approx g\| = \|h(\|f \simeq g\|) = \mathbf{1}\| = \|f \simeq g\|$. The second sentence follows from (1) of §6.5.

We let $W = \{[f] : f \in V^\kappa\}$ viewed as a class in $V^{(\mathscr{B})}$. I.e.,

$$\|x \in W\| = \sum \{\|x = [f]\| : f \in V^\kappa\}.$$

LEMMA 6. *Let* $x \in V^{(\mathscr{B})}$. *Suppose* $\|x \in W\| = \mathbf{1}$. *Then there is an* $f \in V^\kappa$ *such that* $[f] = x$.

CAUTION. This lemma is stated in V, so f is an element of the V-class V^κ (and not just the $V^{(\mathscr{B})}$-class V^κ defined in §6.8).

PROOF. By Zorn, we pick a maximal family $\{\langle b_i, f_i\rangle : i \in K\}$ such that

(1) For all $i \in K$, $\mathbf{0} < b_i \leqq \|x = [f_i]\|$ (so $b_i \in \mathscr{B}, f_i \in V^\kappa$).

(2) $\{b_i :\ i \in K\}$ is pairwise disjoint.

Since $\|x \in W\| = \mathbf{1}$, maximality implies that $\sum_{i \in K} b_i = \mathbf{1}$.

Since $\mathscr{B}$ is κ^+-saturated, card $(K) \leqq \kappa$. Thus we may as well assume that K is an ordinal $\lambda \leqq \kappa$. For each $\xi < \lambda$, select A'_ξ which represents b_ξ. Since for $\xi < \xi'$, $b_\xi \cdot b_{\xi'} = \mathbf{0}$, $A'_\xi \cdot A'_{\xi'}$ has measure zero. Since I is κ-complete, A'_ξ differs from

$$A_\xi = A'_\xi - \bigcup \{A'_\alpha \mid \alpha < \xi\}$$

on a set of measure zero. So $[A_\xi] = b_\xi$. By construction $\{A_\xi \mid \xi < \lambda\}$ is pairwise disjoint.

Let $f : \kappa \to V$ be such that

$$f(x) = f_\alpha(x) \quad \text{for } x \in A_\alpha.$$

Then $[A_\alpha] \leqq \|[f] = [f_\alpha]\|$. Since $b_\alpha = [A_\alpha] \leqq \|[f_\alpha] = x\|$, we have

$$\mathbf{1} = \sum_{\alpha<\lambda} b_\alpha \leqq \|[f] = x\|.$$

Thus $x = [f]$ by (1) of §6.5.

6.10. We return to $V^{(\mathscr{B})}$. We make W into a relational system as follows: Let x, y in W. Then $[f] \mathrel{\tilde{\in}} [g]$ if and only if $h(\|f \mathrel{\tilde{\in}} g\|) = \mathbf{1}$. (Using Corollary 3, one checks that this is well defined, and does not depend on the choice of f and g.) We put $[f] \simeq [g]$ just in case $[f] = [g]$.

6.11. Thus W is a relational system inside $V^{(\mathscr{B})}$. Let $\tilde{\varphi}(x_1, \ldots, x_n)$ be a first order formula in $\tilde{\in}$ and $\simeq$. Let $f_1, \ldots, f_n \in V^\kappa$. The following scheme of lemmas is easy to check.

LEMMA 7. $\|\tilde{\varphi}(f_1, \ldots, f_n)\|_{V^\kappa} = \|W \models \tilde{\varphi}([f_1], \ldots, [f_n])\|$.

PROOF. Let $\tilde{\varphi}(x_1, \ldots, x_n)$ be $(\exists x)\tilde{\psi}(x, x_1, \ldots, x_n)$. As a sample, we prove the lemma for $\tilde{\varphi}$, assuming that it has been proved for $\tilde{\psi}$.

Fix $f_1, \ldots, f_n \in V^\kappa$. Let

$$b_0 = \|\tilde{\varphi}(f_1, \ldots, f_n)\|;$$
$$b_1 = \|W \models \tilde{\varphi}[f_1], \ldots, [f_n]\|.$$

We show $b_0 = b_1$. Let $f \in V^\kappa$. Then by induction hypothesis,

$$\|\tilde{\psi}(f, f_1, \ldots, f_n)\| = \|W \models \tilde{\psi}([f], [f_1], \ldots, [f_n])\| \leqq b_1.$$

Hence, $b_0 \leqq b_1$. To prove the reverse inequality, we exploit the fact that $V^{(\mathscr{B})}$ is full. It follows that there is an $x_0 \in V^{(\mathscr{B})}$ such that the following are $\mathscr{B}$-valid: (a) $x_0 \in W$; (b) in W, $\tilde{\varphi}([f_1], \ldots, [f_n]) \leftrightarrow \tilde{\psi}(x_0, [f_1], \ldots, [f_n])$. Hence

$$b_1 = \|W \models \tilde{\psi}(x_0, [f_1], \ldots, [f_n])\|.$$

By Lemma 6, we can pick $f_0 \in V^\kappa$ with $x = [f_0]$. By induction hypothesis,

$$b_1 = \|\tilde{\psi}(f_0, f_1, \ldots, f_n)\| \leqq b_0.$$

COROLLARY 8. *Let φ be an axiom of* **ZFC**. *Then* $\|W \models \tilde{\varphi}\| = \mathbf{1}$.

PROOF. Clear from Lemma 7 and Corollary 4.

6.12. We wish to show that, in $V^{(\mathscr{B})}$, $\langle W; \simeq, \tilde{\in}\rangle$ is isomorphic to $\langle A; =, \in\rangle$, where A is a transitive class. By a known theorem [**10,** Theorem 3] we have to verify that the following are $\mathscr{B}$-valid:

(1) W is well founded.

(2) If $x \in W$, $\{y \in W \mid y \tilde{\in} x\}$ is a set.

This will be done in the next two lemmas.

LEMMA 9. *(2) is $\mathscr{B}$-valid.*

PROOF. Let $x \in V^{(\mathscr{B})}$ such that $\|x \in W\| = \mathbf{1}$. By Lemma 6, $x = [f]$ for some $f: \kappa \to V$. Let y be a transitive set such that range $(f) \subseteq y$. (One can take y to be the transitive hull of f, for example.) Then,

$$B = \{[g]: g: \kappa \to y\}$$

is a set in $V^{(\mathscr{B})}$. We show that if $y \in V^{(\mathscr{B})}$, $\|y \in W\| = \mathbf{1}$, then the statement

$$\text{“} y \tilde{\in} x \to y \in B \text{”}$$

is $\mathscr{B}$-valid.

In fact, by Lemma 6, $y = [f_1]$ for some $f_1: \kappa \to V$. Let

$$A = \{\alpha \mid f_1(\alpha) \in f(\alpha)\}.$$

Then $[A] = \|y \tilde{\in} x\|$ (by Lemma 7). Let $f_2(x) = f_1(x)$, if $x \in A$; $f_2(x) = f(x)$, otherwise. Then since y is transitive, $f_2 \in y^\kappa$. Thus

$$\|y \tilde{\in} x\| = [A] \leqq \|y = [f_2]\| \leqq \|y \in B\|.$$

This proves the lemma.

LEMMA 10. *With probability* **1**, *W is well founded.*

PROOF. Let b be the truth value:

$$\|W \text{ is not well founded}\|.$$

Since $V^{(\mathscr{B})}$ is full there is an $s \in V^{(\mathscr{B})}$ such that (1) $\|s$ maps $\omega^\vee$ into $W\| = \mathbf{1}$; (2) $\|s$ is a descending $\in$-sequence $\| = b$. Let $x_n \in V^{(\mathscr{B})}$ be $s(n^\vee)$. (I.e., $\|s(n^\vee) = x_n\| = \mathbf{1}$.) Then $\|x_n \in W\| = \mathbf{1}$, $\|x_{n+1} \in x_n\| \geqq b$.

By Lemma 6, we can find $f_n \in V^\kappa$ with $x_n = [f_n]$. Let $B \subseteq \kappa$ represent $b \in \mathscr{B}$. Then for almost all $\alpha \in B$, $f_{n+1}(\alpha) \in f_n(\alpha)$. Since $\kappa > \aleph_0$ and I is κ-complete, we have $\{f_n(\alpha): n \in \omega\}$ is a descending $\in$-sequence for almost all α in B. But there are no descending $\in$-sequences. Hence, B has measure zero. Thus $b = \mathbf{0}$, and the lemma is proved.

6.13. By the remark at the beginning of 6.12 there is a transitive class of $V^{(\mathscr{B})}$, A, and an isomorphism $\psi: W \simeq A$ in $V^{(\mathscr{B})}$. By abuse of notation, if $f \in V^{(\mathscr{B})}$ we denote by $\psi(f)$ the unique z in $V^{(\mathscr{B})}$ such that

$$\|\psi([f]) = \check{z}\| = \mathbf{1}.$$

The following lemma is now easy to prove.

LEMMA 11. *Let A, ψ be as above.*

(1) $\|A$ *is a transitive class*$\| = \mathbf{1}$.

(2) *Let* $f_1, \ldots, f_n \in V^\kappa$. *Let* $\varphi(v_1, \ldots, v_n)$ *be a set-theoretical formula. Then if*

$$b = [\{\alpha \mid \varphi(f_1(\alpha), \ldots, f_n(\alpha))\}],$$

then

$$b = \|\langle A; \in\rangle \models \varphi(\psi(f_1), \ldots, \psi(f_n))\|.$$

(3) *Let* $\|x \in A\| = \mathbf{1}$. *Then* $x = \psi(f)$ *for some* $f \in V^\kappa$.

PROOF. (1) is clear. (2) follows from Lemmas 2, 7, and the fact that ψ is an isomorphism of W with A. (3) follows readily from Lemma 6.

6.14. LEMMA 12. *With probability* $\mathbf{1}$, *A contains all ordinals.*

PROOF. Let x be a set. Define $c_x \in V^\kappa$ by $c_x(\alpha) = x$, for all $\alpha \in \kappa$. By Lemma 11, we have

(1) If $\alpha \in \mathrm{OR}$, $\psi(c_\alpha)$ is an ordinal, with probability $\mathbf{1}$.

(2) If $\alpha, \beta \in \mathrm{OR}$ with $\alpha < \beta$ then

$$\|\psi(c_\alpha) < \psi(c_\beta)\| = \mathbf{1}.$$

Using (1) and (2), one checks easily by induction that (3) $\|\psi(c_\alpha) \geqq \alpha\| = \mathbf{1}$, if $\alpha \in \mathrm{OR}$.

Since A is transitive, it follows that $\alpha \in A$ with probability $\mathbf{1}$.

LEMMA 13. (1) *If* $\alpha < \kappa$, $\psi(c_\alpha) = \alpha$.

(2) $\|\psi(c_\kappa) > \kappa\| = \mathbf{1}$.

(3) $\psi(f) = \kappa$ *if and only if f is incompressible.*

(4) *Let* $1_\kappa \in V^\kappa$ *be defined by* $1_\kappa(\alpha) = \alpha$, *for* $\alpha < \kappa$.

Then I is normal if and only if $\psi(1_\kappa) = \kappa$.

PROOF. We prove (1) by induction on α. Let $f_\alpha \in V^\kappa$ be such that $\psi(f_\alpha) = \alpha^\vee$. Then $f_0(\gamma) \geqq c_0(\gamma)$, for almost all γ. Thus $\|\psi(c_0)$ is an ordinal $\leqq 0\| = \mathbf{1}$. Thus $\|\psi(c_0) = 0\| = 1$.

Say now that by induction hypothesis, $\psi(c_\beta) = \beta$ for $\beta < \alpha$. Then since f_α represents α, we have $f_\alpha(\gamma) > c_\beta(\gamma) = \beta$ for almost all γ. Since I is κ-complete, and $\alpha < \kappa$, we have $f_\alpha(\gamma) \geqq c_\alpha(\gamma)$ a.e. On the other hand, since $c_\alpha(\gamma) > c_\beta(\gamma)$ a.e., for all $\beta < \gamma$, we have $\|\psi(c_\alpha) > \beta\| = \mathbf{1}$ so $\|\psi(c_\alpha) \geqq \alpha^\vee\| = \mathbf{1}$, so $c_\alpha(\gamma) \geqq f_\alpha(\gamma)$ a.e. Thus $f_\alpha(\gamma) = c_\alpha(\gamma)$, a.e., and (1) is proved. (2) is clear since $1_\kappa > c_\beta$ a.e., so $\psi(c_\kappa) > \psi(1_\kappa) \geqq \kappa$.

Let f be such that $\psi(f) = \kappa$. Then we have already seen that $f(\gamma) < \kappa$, almost everywhere. If f is constant and equal to λ on a set B of positive measure, then by (1),

$$\|\psi(f) = \lambda\| \geqq [B].$$

Thus f is nowhere constant. If $g \in \kappa^\kappa$ is less than f on the set B of positive measure, then

$$\|\psi(g) < \kappa\| = [B] > 0.$$

Hence for some $\lambda < \kappa$, $\|\psi(g) = \lambda\| > 0$. So $\{\alpha \mid g(\alpha) = \lambda\}$ has positive measure by (1). Thus f is incompressible. If g is incompressible, one sees easily that $g = f$, a.e. For example if $g < f$ on a set B of positive measure, then g is somewhere constant, since f is incompressible, contradicting the fact that g is nowhere constant. This proves (3).

Clearly I is normal if and only if 1_κ is incompressible so (4) is clear.

REMARK. Thus the techniques of this section supply a new proof of Theorem 3. This, in fact, was the author's first proof of Theorem 3.

6.15. We now assume that I is normal. As stated in Theorem 3, if κ is real-valued measurable, we can choose I so that $\mathscr{B}$ is a measure algebra.

LEMMA 14. *Let $D \in V^{(\mathscr{B})}$. Suppose $\|D \subseteq \kappa\| = \mathbf{1}$. Then $D \in A$.*

PROOF. Let $b_\xi = \|\xi \in D\|$. Let $A_\xi \subseteq \kappa$ represent b_ξ. Let

$$D_\xi = \{\gamma < \xi \mid \xi \in A_\gamma\}.$$

Let $f \in V^\kappa$ be defined by $f(\xi) = D_\xi$.

Since I is normal, and for all $\alpha < \kappa$, $D_\alpha \subseteq \alpha$, we have $\|\psi(f) \subseteq \kappa\| = \mathbf{1}$. We compute

$$\|\xi \in \psi(f)\| = [\{\alpha \mid \xi \in D_\alpha\}] = [\{\alpha \mid \alpha \in A_\xi\}] = b_\xi.$$

Thus $\psi(f) = D$.

6.16. We are now in a position to expound Kunen's work. Let α be an uncountable cardinal. Let L_α be the set of sets "constructible before stage α" [2]. Let $P(L_\alpha)$ be the power set of L_α. Since L_α is transitive, $L_\alpha \subseteq P(L_\alpha)$. Let X be a variable ranging over subsets of $P(L_\alpha)$. Then if φ is a first order sentence in the predicates $\in$, $=$, and a one place predicate, U, we say $\langle P(L_\alpha);\ X\rangle \models \varphi$ if φ becomes true in $P(L_\alpha)$ when we interpret U to be X.

A Σ_1^2 sentence about α has the form

$$\exists X \in P(P(L_\alpha)) : \langle P(L_\alpha);\ X\rangle \models \varphi.$$

The following lemma is obvious from Lemma 14.

LEMMA 15. *Let $\varphi(\alpha)$ be a Σ_1^2 sentence about α. Then $\|A \models \varphi(\kappa)\| \leqq \|\varphi(\kappa)\|$.*

LEMMA 16. *Suppose that $\varphi(\alpha)$ is a Σ_1^2 sentence about α. Then if $\{\alpha \mid \varphi(\alpha)\}$ has positive measure, then $\|\varphi(\kappa)\| > \mathbf{0}$.*

PROOF. This is clear from Lemmas 11 and 15, and (4) of Lemma 13.

6.17. A Π_1^2 statement about α is simply the negation of a Σ_1^2 statement about κ. We fix a Π_1^2 statement, $\varphi(\alpha)$, about α.

We have three situations to consider simultaneously. Let $H_1(\kappa)$ express: κ is a real-valued measurable cardinal. Let $H_2(\kappa)$ express: κ is uncountable, and $P(\kappa)$ carries a nontrivial $\aleph_1$-saturated ideal. Let $H_3(\kappa)$ express: κ is uncountable, and $P(\kappa)$ carries a nontrivial λ-saturated ideal for some $\lambda < \kappa$.

For $i < 3$, we let $J_i(I)$ express: I is a normal nontrivial λ-saturated ideal and: ($i = 1$), $\lambda = \aleph_1$, and $P(\kappa)/I$ is a measure algebra; ($i = 2$), $\lambda = \aleph_1$; ($i = 3$), $\lambda < \kappa$.

The following is a metatheorem, provable in Peano arithmetic.

THEOREM 10 (KUNEN). *Suppose that $\varphi(\alpha)$ is a Π^2_1 statement. Suppose*

$$\mathbf{ZFC} \vdash H_i(\kappa) \rightarrow \varphi(\kappa).$$

Then the following is a theorem of **ZFC**: *Let κ be a cardinal, I an ideal of subsets of κ. Suppose $H_i(\kappa)$ and $J_i(I)$. Then $\{\alpha \mid \varphi(\alpha)\}$ has I-measure* 1.

PROOF. Let $\mathscr{B} = P(\kappa)/I$. Using Theorem 7 if $i = 1$, and Theorem 8(a) for $i = 2, 3$, we see that $\|H_i(\kappa)\| = \mathbf{1}$ in $V^{(\mathscr{B})}$. Hence by our assumption on φ, $\|\varphi(\kappa)\| = \mathbf{1}$ in $V^{(\mathscr{B})}$. Applying Lemma 16 to the negation of φ, we see that $\{\alpha \mid \varphi(\alpha)\}$ has I-measure 1.

7. A generalization of the work of §6.

7.1. We consider the following situation:

(1) I is a κ^+-saturated nontrivial ideal in $P(\kappa)$; κ is an uncountable cardinal.

Thus the quotient algebra $\mathscr{B} = P(\kappa)/I$ is complete. (By Proposition 9 of §1.)

7.2. Our hypotheses on κ, I imply that κ is regular. So far as I know, they do not exclude the possibility that $\kappa = \aleph_1$.

Using a version of Gaifman's iterated ultraproduct technique, I have shown the following. (The details will be published elsewhere.)

Let κ, I satisfy the hypotheses of §7.1; let A be a set of ordinals with lub $A < \kappa$. Then κ is strongly inaccessible in $L[A]$. Moreover, the relational system

$$\langle L_\kappa[A];\ \in, A\rangle$$

has a set of κ indiscernibles. (It follows, by work of Silver [17], that $V \neq L[A]$.)

7.3. It follows from the result mentioned in §2 that the following proposition holds in L: Let κ be an uncountable cardinal. Let I be a nontrivial ideal in $P(\kappa)$. Then $P(\kappa)/I$ is incomplete.

In fact, if $P(\kappa)/I$ is complete and I is κ^+-saturated, §7.2 implies that $V \neq L$. To verify that I is κ^+-saturated, we use the following lemma.

LEMMA. *Suppose that κ is an uncountable cardinal, I is a nontrivial ideal in $P(\kappa)$, $\mathscr{B} = P(\kappa)/I$ is complete, and $2^\kappa < 2^{\kappa^+}$. Then I is κ^+-saturated.*

PROOF. Clearly card $(\mathscr{B}) \leqq 2^\kappa$. Suppose that $\mathscr{B}$ is not κ^+-saturated. Let $\{b_\xi \mid \xi < \kappa^+\}$ be a pairwise disjoint family of nonzero elements of $\mathscr{B}$. For $S \subseteq \kappa^+$, let $c_S = \sup \{b_\xi \mid \xi \in S\}$. Then $b_\alpha \cap c_S \neq \mathbf{0}$ if and only if $\alpha \in S$. Thus the map $S \rightarrow c_S$ is one-one, and

$$2^{\kappa^+} \leqq \text{card}\,(\mathscr{B}) \leqq 2^\kappa.$$

This contradicts our assumption that $2^\kappa < 2^{\kappa^+}$.

ADDED IN PROOF. Kunen's argument, alluded to earlier, establishes the following. Let $\mathscr{I}$ be a normal κ^+-saturated nontrivial ideal in $P(\kappa)$. Let $A \subseteq \kappa$ with lub $A < \kappa$. Then in $L[\mathscr{I}, A]$, κ is measurable and $\mathscr{I} \cap L[\mathscr{I}, A]$ is a normal prime ideal. This gives a much simpler proof of my result on indiscernibles for $L[A]$, and I do not plan to publish my proof.

Kunen also has a proof of the following. Let **ZFC** + "There is a measurable cardinal" be consistent. Then so is **ZFC** + "$2^{\aleph_0}$ carries an exactly $2^{\aleph_0}$-saturated nontrivial ideal". This settles one of the questions mentioned earlier in the paper.

References

1. G. Födor, *On stationary sets and regressive functions*, Acta Sci. Math. (Szeged) **27** (1966), 105–110.

2. K. Gödel, *The consistency of the axiom of choice and of the generalized continuum hypothesis with the axioms of set theory*, Ann. of Math. Studies No. 3, Princeton Univ. Press, Princeton, N.J., 1940.

3. P. R. Halmos, *Lectures on Boolean algebras*, Van Nostrand, Princeton, N.J., 1963.

4. ———, *Measure theory*, Van Nostrand, Princeton, N.J., 1963.

5. W. Hanf, *Incompactness in languages with infinitely long expressions*, Fund. Math. **53** (1964), 309–324.

6. H. J. Keisler and A. Tarski, *From accessible to inaccessible cardinals*, Fund. Math. **53** (1964), 225–308.

7. S. Kochen, *Ultraproducts in the theory of models*, Ann. of Math. **74** (1961), 221–261.

8. K. Kunen, *Inaccessibility properties of cardinals*, Doctoral dissertation, Stanford University, 1968.

9. D. A. Martin and R. M. Solovay, *A basis theorem for Σ_3^1 sets of reals*, Ann. of Math. **89** (1969), 138–159.

10. A. Mostowski, *An undecidable arithmetical statement*, Fund. Math. **36** (1949), 143–164.

11. K. Prikry, *Changing measurable into accessible cardinals*, Doctoral dissertation, University of California, Berkeley, 1968.

12. ———, *Measurable cardinals and saturated ideals*, Notices Amer. Math. Soc. **13** (1966), 720.

13. F. Rowbottom, Doctoral dissertation, University of Wisconsin, 1964.

14. D. Scott, *The notion of rank in set theory*, Summaries of talks presented at the Summer Institute of Symbolic Logic in 1957 at Cornell University, vol. II, 267.

15. D. Scott and R. M. Solovay, *Boolean valued models of set theory*, these Proceedings, part II.

16. R. Sikorski, *Boolean algebras*, Springer-Verlag, Berlin, 1960.

17. J. Silver, *Some applications of model theory in set theory*, Doctoral dissertation, University of California, Berkeley, 1966.

18. ———, *The consistency of the* GCH *with the existence of a measurable cardinal*, these Proceedings.

19. R. Solovay, *A model of set theory in which every set of reals is Lebesgue measurable*, Ann. of Math. (to appear).

20. A. Tarski, *Ideale in vollständige Mengenkörpen*. II, Fund. Math. **33** (1945), 51–65.

21. S. Ulam, *Zur Masstheorie in der allgemeinen Mengenlehre*, Fund. Math. **16** (1930), 140–150.

UNIVERSITY OF CALIFORNIA, BERKELEY

TRANSFINITE SEQUENCES OF AXIOM SYSTEMS FOR SET THEORY

G. L. SWARD

1. **Introduction.** Just as in the case of arithmetic, it is impossible to give a recursively enumerable axiom system for set theory from which all of the true statements of set theory are provable. More generally, this incompleteness result of Gödel holds for axiom systems S for which the formalization $\mathrm{Pr}_S(\ulcorner\psi\urcorner)$ of the provability-from-S relation satisfies the usual condition: if $S \vdash \psi$, then $S \vdash \mathrm{Pr}_S(\ulcorner\psi\urcorner)$. It is appropriate, therefore, to strengthen set theory by adding to the axiom system some sentences which are intuitively true or by adding some rules of inference which intuitively preserve validity. We shall do both. We shall add the ω-rule (from $\psi(\bar{0}), \psi(\bar{1}), \psi(\bar{2}), \ldots$, infer $\forall x \in \omega\psi(x)$) at the outset, and then we shall add some intuitively true sentences, repeating the addition of true sentences transfinitely many times. Thus our program is to study certain transfinite sequences of axiom systems for set theory where the proof relation includes the ω-rule.

Throughout the paper the notion of a true sentence or valid formula of set theory is to be understood informally with respect to the standard interpretation of set theory when formalized in the first-order predicate calculus. We shall frequently say that a term or formula of set theory expresses a function or relation. This is also to be understood informally and by "express" we mean express extensionally. A function or relation may be expressed extensionally in many ways, and occasionally we shall make the vague distinction between "express in a natural way" and "express in an unnatural way". The difference is that the natural way formally copies as closely as possible in set theory the informal notion. For a discussion of the importance of natural definitions see Feferman [1].

Since the discovery of forcing by Cohen many sentences of set theory have been proved to be independent of Zermelo-Fraenkel set theory; for example, the

Continuum Hypothesis. We do not propose to add sentences of this kind for which there is no general agreement as to which alternative is true. Rather we shall make a transfinite sequence of safe additions each of which merely expresses, via an assumed Gödel numbering, the soundness of the preceding axiom system.

Let S be an axiom system for set theory. The sentences which collectively express the soundness of S are given by the following reflection principle for S:

$$\mathrm{Pr}_S(\ulcorner\psi\urcorner) \to \psi, \text{ for all sentences } \psi$$

where $\ulcorner\psi\urcorner$ is the g.n. (Gödel number) of the sentence ψ and $\mathrm{Pr}_S(\ulcorner\psi\urcorner)$ is a formula of set theory which expresses the provability from S of ψ. For each sentence ψ, this is a formalization of the statement: If ψ is provable from S, then ψ is true. Therefore, if S is sound, the sentences given by the reflection principle are true. The reflection principle, as formulated above, however, is ambiguous; a precise formulation depends on a particular formalization of both S and the proof relation.

These ideas lead to the consideration of transfinite sequences S_α of axiom systems which satisfy the following *fundamental conditions:*

$$S_0 \text{ is a sound and sufficiently strong axiom system,}$$
$$S_{\alpha+1} = S_\alpha \cup \{\mathrm{Pr}_{S_\alpha}(\ulcorner\psi\urcorner) \to \psi \mid \text{for all sentences } \psi\},$$
$$S_\alpha = \bigcup_{\beta<\alpha} S_\beta, \text{ for limit ordinals } \alpha.$$

These conditions, for a fixed S_0, do not uniquely determine the sequence S_α because the second condition concerning $S_{\alpha+1}$ incorporates the ambiguity of the reflection principle just mentioned. Similar sequences of axiom systems, but without the ω-rule, were first studied by Turing [**5**] and more recently by Feferman [**2**]. Turing and Feferman dealt with arithmetic instead of set theory, and Feferman obtained some interesting completeness results. Sequences of axiom systems for set theory with the ω-rule which are based on a much stronger reflection principle are studied by Takeuti [**4**] and the formalization of proof relations which include the ω-rule was studied by Rosser [**3**].

The writer wishes to thank Professor Gaisi Takeuti for suggesting this topic and for his interest and advice.

2. **Preliminary remarks and the main results.** Let us assume that we have a fixed system of Gödel numbering of the symbols and expressions of set theory. We say that a wff $\phi(x)$ with one free variable x *defines an axiom system* S if $\phi(x)$ expresses the relation that x is the g.n. of a member of S. The wff $\mathrm{Pr}_{S_\alpha}(\ulcorner\psi\urcorner)$ and therefore the definition of $S_{\alpha+1}$ depend on a formula $\phi_\alpha(x)$ which defines S_α. Hence in order to define the transfinite sequence S_α, it is necessary to define a transfinite sequence $\phi_\alpha(x)$ of wffs such that $\phi_\alpha(x)$ defines S_α. But more than this is required because we want the $\phi_\alpha(x)$ to possess certain formal properties which make them useful definitions of the S_α. This is achieved by formally copying as closely as possible the informal definition; that is, by constructing definitions in as natural a way as possible. It is easy to formally copy the informal proof relation and obtain, for each α uniformly, a wff $\mathrm{Pr}_{\phi_\alpha}(x)$ which expresses the relation that x

is the g.n. of a wff which is provable from S_α. Then, in terms of $\mathrm{Pr}_{\phi\alpha}(x)$, we immediately obtain a wff $\phi_{\alpha+1}(x)$ which defines $S_{\alpha+1}$ and which is natural to the extent that $\phi_\alpha(x)$ is a natural definition of S_α.

But even within a fixed formalization of syntax including the formalization of S_0 and passage from S_α to $S_{\alpha+1}$, the S_α still depend on the definition $\phi_\alpha(x)$ of S_α for limit ordinals α, for which no natural definition is immediately evident. Indeed, there is a natural barrier obstructing the naive attempt; for example, we would like to define $\phi_\omega(x)$ as $\phi_0(x) \vee \phi_1(x) \vee \phi_2(x) \vee \cdots$, but, of course, we cannot express an infinite disjunction in set theory. Turing and Feferman overcame this barrier by making use of the set $\mathcal{O}$ of notations for the constructive ordinals; $\mathcal{O}$ is a partially ordered subset of ω such that each path is well ordered and all paths have the same initial element. We shall overcome this barrier by making use of wffs $A(i,j)$ which well-order ω. The entire sequence $\phi_\alpha(x)$ will then depend on $A(i,j)$ and the order-type of $A(i,j)$ will be the upper bound for α.

Thus we shall assume a fixed formalization of the syntax. Then, in terms of a wff $A(i,j)$ which well-orders ω with order-type τ, we define a transfinite sequence $\phi_\alpha(x)$ of formulas for $\alpha < \tau$ such that $\phi_\alpha(x)$ defines S_α and the sequence S_α satisfies the fundamental conditions. Clearly this will be a definition schema; the S_α will depend on $A(i,j)$. However, the theorems of S_α will be independent of $A(i,j)$ as follows. Let S_α and S'_α be the axiom systems determined by formulas $A(i,j)$ and $A'(i,j)$ with order-types τ and τ', and let T_α and T'_α be the theorems of S_α and S'_α. Then $T_\alpha = T'_\alpha$ as long as S_α and S'_α are both defined; that is, if $\alpha < \tau$ and $\alpha < \tau'$. This is the first main result.

The second main result is that the entire sequence S_α is meaningful in the sense that we get new theorems at each step: $T_{\alpha+1} \neq T_\alpha$. This is an immediate consequence of the extension of the first part of Gödel's incompleteness theorem to each S_α and the assumption that each S_α is consistent. This assumption of consistency is implied by the original motivating assumptions that (1) the standard interpretation of set theory is meaningful, (2) all of the sentences of S_0 are true, (3) the rules of inference preserve validity, and (4) the formalization $\mathrm{Pr}_S(x)$ of the proof relation is correct. Then, following Gödel, for each α, there is a sentence ψ such that ψ is $\neg\mathrm{Pr}_{S_\alpha}(\ulcorner\psi\urcorner)$, and, since we are assuming S_α is consistent, not $S_\alpha \vdash \neg\mathrm{Pr}_{S_\alpha}(\ulcorner\psi\urcorner)$. Therefore, for the same ψ, not $S_\alpha \vdash \mathrm{Pr}_{S_\alpha}(\ulcorner\psi\urcorner) \rightarrow \psi$, and hence $T_{\alpha+1} \neq T_\alpha$. Moreover, since $\mathrm{Pr}_{S_\alpha}(\ulcorner 0 = 1\urcorner) \rightarrow 0 = 1 \in S_{\alpha+1}$, $S_{\alpha+1} \vdash \neg\mathrm{Pr}_{S_\alpha}(\ulcorner 0 = 1\urcorner)$ and therefore $S_{\alpha+1}$ is at least as strong as $S_\alpha \cup \{\mathrm{Con}_{S_\alpha}\}$ where Con_{S_α} is a wff which expresses in a natural way the consistency of S_α.

So far we have mentioned only one assumption concerning $A(i,j)$; namely, $A(i,j)$ well-orders ω, and this is sufficient in order for the wffs $\phi_\alpha(x)$ to define S_α. However, both main results depend on two other assumptions concerning $A(i,j)$, the second of which implies that $A(i,j)$ well-orders ω. Let $\bar{n}$ be the numeral for $n \in \omega$, and let (τ^ω) be the set of 1-1, onto functions from ω to τ. Our two assumptions concerning $A(i,j)$ are (1) if $A(i,j)$, then $S_0 \vdash A(\bar{i},\bar{j})$, which we call *local provability* for $A(i,j)$, and (2) $S_0 \vdash \exists\tau\exists f\{f \in (\tau^\omega) \wedge \forall i \in \omega\, \forall j \in \omega[f(i) < f(j) \leftrightarrow A(i,j)]\}$, which we call *global provability* for $A(i,j)$.

The first main result also depends on transfinitely many special applications of the ω-rule, and more will be said about this at that point in the paper. However, although a restricted ω-rule is sufficient in order to obtain the main results, within the context of set theory as opposed to arithmetic, it is more natural to assume the full ω-rule as we have done.

3. **Set-theoretic preliminaries.** The basic language for the set theories is the first order predicate calculus with identity and the relational constant $\in$. This language is then extended by introducing by definition the usual individual, functional, and relational constants. The logical axioms of the set theories are any primitive recursive set of formulas which is logically complete for the extended language when modus ponens and generalization are the rules of inference.

The initial axiom system S_0 may be any sound, primitive recursive extension of the axioms for Zermelo-Fraenkel set theory. However, roughly speaking, all that is required for the following definitions and theorems are the countable ordinals. In particular, $A(i,j)$ with order-type τ is used to define a 1-1, onto function from ω to τ, which then provides a notation for each $\alpha < \tau$.

It is convenient to use the natural numbers in the usual way for the representation in set theory of the symbols and expressions of set theory. Then, as is well known, we can form in the natural way the following primitive recursive terms and formulas which express, via Gödel numbers, the usual syntactical functions and relations: $\mathrm{Var}(a)$, a is a variable; $\mathrm{Wff}(a)$, a is a well-formed formula; $\mathrm{Clf}(a)$, a is a closed wff; $\mathrm{Lax}(a)$, a is a logical axiom; $\mathrm{num}(a)$, the numeral for a; $\mathrm{Sub}^b_c a$, the term or wff obtained from a by substituting the term c for all free occurrences of variable b; $a \ulcorner\!\rightarrow\!\urcorner b$, the implication with a as antecedent and b as consequent; $\ulcorner\forall\urcorner ab$, the generalization with respect to a of b; $\ulcorner\forall x \in \omega\urcorner a$, the generalization with respect to x and relativized to ω of a.

We extend the notion of proof just enough to accommodate the ω-rule and call the new notion an ω-proof. An *ω-proof* from an axiom system S is a countable transfinite sequence of wffs with a last member such that each wff is either (1) a logical axiom, (2) a member of S, (3) inferred from two earlier wffs by modus ponens, (4) inferred from one earlier wff by generalization, or (5) of the form $\forall x \in \omega\psi(x)$ and inferred from ω earlier wffs by the ω-rule.

Let ω_1 be the first uncountable ordinal, and let us assume that $\phi(x)$ defines the axiom system S. We shall now give the formal definition $\mathrm{Prf}(\phi; x, y)$ in terms of $\phi(x)$ of the relation that y is an ω-proof from S of x. In order to be definite we shall assume throughout the paper that the individual variables x, y, z are assigned the g.n.s 1, 3, 5. Then $\mathrm{Prf}(\phi; x, y)$ is

$$\begin{aligned}&\exists\alpha < \omega_1(y \in \mathrm{Wff}^{\alpha+1} \wedge x = y(\alpha) \wedge \forall\beta \leq \alpha\{\mathrm{Lax}(y(\beta)) \vee \phi(y(\beta))\\ &\quad \vee \exists\gamma < \beta\exists\delta < \beta[y(\gamma) = y(\delta) \ulcorner\!\rightarrow\!\urcorner y(\beta)] \vee \exists\gamma < \beta\exists u \in \omega[y(\beta) = \ulcorner\forall\urcorner u\, y(\gamma)]\\ &\quad \vee \exists u \in \omega\exists f \in \beta^\omega[y(\beta) = \ulcorner\forall x \in \omega\urcorner u \wedge \forall i \in \omega\exists j \in \omega[y(f(j)) = \mathrm{Sub}^1_{\mathrm{num}(i)}u]]\}).\end{aligned}$$

$\mathrm{Prf}(\phi; x, y)$ is a natural definition of the ω-proof-from-S relation to the extent that $\phi(x)$ is a natural definition of S. This is illustrated by the following theorems, which are provable without the use of the ω-rule.

THEOREM 1. $S_0 \vdash \mathrm{Lax}(x) \rightarrow \exists y \mathrm{Prf}(\phi; x, y)$.

THEOREM 2. $S_0 \vdash \phi(x) \wedge \mathrm{Wff}(x) \rightarrow \exists y \mathrm{Prf}(\phi; x, y)$.

THEOREM 3. $S_0 \vdash \exists y \mathrm{Prf}(\phi; x_1 \ulcorner\rightarrow\urcorner x_2, y) \wedge \exists y \mathrm{Prf}(\phi; x_1, y) \rightarrow \exists y \mathrm{Prf}(\phi; x_2, y)$.

THEOREM 4. $S_0 \vdash \exists y \mathrm{Prf}(\phi; x_2, y) \wedge \mathrm{Var}(x_1) \rightarrow \exists y \mathrm{Prf}(\phi; \ulcorner\forall\urcorner x_1 x_2, y)$.

THEOREM 5. $S_0 \vdash \forall i \in \omega \exists y \mathrm{Prf}(\phi; \mathrm{Sub}^1_{\mathrm{num}(i)} x, y) \rightarrow \exists y \mathrm{Prf}(\phi; \ulcorner\forall x \in \omega\urcorner x, y)$.

Let $\phi(x)$ and $\psi(x)$ be wffs with one free variable x.

THEOREM 6. $S_0 \vdash \forall x \in \omega[\phi(x) \rightarrow \psi(x)] \wedge \exists y \mathrm{Prf}(\phi; x, y) \rightarrow \exists y \mathrm{Prf}(\psi; x, y)$.

THEOREM 7.

$$
\begin{aligned}
S_0 \vdash \{&\forall x[\mathrm{Lax}(x) \rightarrow \psi(x)] \wedge \forall x[\phi(x) \wedge \mathrm{Wff}(x) \rightarrow \psi(x)] \\
&\wedge \forall x_1 \forall x_2[\psi(x_1 \ulcorner\rightarrow\urcorner x_2) \wedge \psi(x_1) \wedge \mathrm{Wff}(x_1 \ulcorner\rightarrow\urcorner x_2) \rightarrow \psi(x_2)] \\
&\wedge \forall x_1 \forall x_2[\psi(x_1) \wedge \mathrm{Wff}(\ulcorner\forall\urcorner x_2 x_1) \rightarrow \psi(\ulcorner\forall\urcorner x_2 x_1)] \\
&\wedge \forall x[\forall i \in \omega \psi(\mathrm{Sub}^1_{\mathrm{num}(i)} x) \wedge \mathrm{Wff}(x) \rightarrow \psi(\ulcorner\forall x \in \omega\urcorner x)]\} \\
&\qquad\qquad\qquad\qquad \rightarrow \forall x[\exists y \mathrm{Prf}(\phi; x, y) \rightarrow \psi(x)].
\end{aligned}
$$

Furthermore, let $\mathrm{Prf}^*(\phi; x, y)$ be the natural definition of the usual proof-from-S relation.

THEOREM 8. $S_0 \vdash \exists y \mathrm{Prf}^*(\phi; x, y) \rightarrow \exists y \mathrm{Prf}(\phi; x, y)$.

4. **The formal definition of the sequence S_α.** In order to formally define the S_α we must first formalize the reflection principle. It is clear that the g.n.

$$\ulcorner\exists y \mathrm{Prf}(\phi; x, y)\urcorner$$

is a primitive recursive function of the g.n. $\ulcorner\phi(x)\urcorner$ when $\ulcorner\phi(x)\urcorner$ is regarded as a parameter. That is, in analogy to $\mathrm{Sub}^b_c a$, let $\mathrm{Sub}^{*\ulcorner\theta\urcorner}_{\ulcorner\phi\urcorner} \ulcorner\exists y \mathrm{Prf}(\theta; x, y)\urcorner = \ulcorner\exists y \mathrm{Prf}(\phi; x, y)\urcorner$, where θ is any fixed wff with one free variable; then we are interested in the primitive recursive function $\mathrm{Sub}^{*\ulcorner\theta\urcorner}_{y} \ulcorner\exists y \mathrm{Prf}(\theta; x, y)\urcorner$ as a function of y. We may now form the following primitive recursive formula $B(x, y)$:

$$\exists u < x[\mathrm{Clf}(u) \wedge x = \mathrm{Sub}^1_{\mathrm{num}(u)} \mathrm{Sub}^{*\ulcorner\theta\urcorner}_{y} \ulcorner\exists y \mathrm{Prf}(\theta; x, y)\urcorner \ulcorner\rightarrow\urcorner u]$$

which, when y is $\ulcorner\phi(x)\urcorner$, defines the set $\{\exists y \mathrm{Prf}(\phi; \ulcorner\psi\urcorner, y) \rightarrow \psi \mid \text{for all sentences } \psi\}$ of wffs. Therefore, if $\phi(x)$ defines S, then $B(x, \ulcorner\phi\urcorner)$ formalizes the particular instance of the reflection principle for S given by $\phi(x)$.

We shall first define the transfinite sequence $\phi_\alpha(x)$ of formulas by defining a single formula $\Phi(x, \alpha)$ such that $\Phi(x, \alpha)$ is $\phi_\alpha(x)$. In an effort to provide some intuitive understanding of the meaning of the formula and why it works we shall arrive at it by successive approximations. To begin with, we have already observed that, for limit ordinals α, we would like to define $\phi_\alpha(x)$ as $\phi_0(x) \vee \phi_1(x) \vee \cdots \vee \phi_\beta(x) \vee \cdots$ for $\beta < \alpha$. Since $S_\alpha \subset S_{\alpha+1}$, for all ordinals α, $S_\alpha = S_0 \cup \bigcup_{<\alpha\,\beta}(S_{\beta+1} - S_\beta)$. Then, if $S_{\beta+1} - S_\beta$ is taken to be the set $\{\exists y \mathrm{Prf}(\phi_\beta; \ulcorner\psi\urcorner, y) \rightarrow \psi\}$, which is defined by $B(x, \ulcorner\phi_\beta\urcorner)$, we may replace the above improper expression by another improper expression

$$\phi_0(x) \vee B(x, \ulcorner\phi_0\urcorner) \vee \cdots \vee B(x, \ulcorner\phi_\beta\urcorner) \vee \cdots \quad \text{for } \beta < \alpha.$$

This works for all ordinals α and can be written as

$$\Phi_0(x) \vee \exists\beta[\beta < \alpha \wedge B(x, \ulcorner\phi_\beta\urcorner)]$$

where $\Phi_0(x)$ is a fixed primitive recursive formula defining S_0.

The "occurrence" of the g.n. $\ulcorner\phi_\beta\urcorner$ in the above expressions reduces the problem to that of defining a function G such that

$$G(\alpha) = \ulcorner\phi_\alpha(x)\urcorner = \ulcorner\Phi_0(x) \vee \exists\beta[\beta < \alpha \wedge B(x, G(\beta))]\urcorner.$$

This equation, however, is not a formula of set theory and cannot be used to define G. This suggests that the situation may be circular; in order to define $\phi_\alpha(x)$ we must define G, and in order to define G we must define $\phi_\alpha(x)$.

This circularity can be avoided by exploiting the ambiguity of the reflection principle. Instead of using the g.n. $\ulcorner\phi_\beta(x)\urcorner$ of $\phi_\beta(x)$, which is taken as the definition of S_β, in constructing the definition $\phi_{\beta+1}(x)$ of $S_{\beta+1}$, we may use the g.n. $\ulcorner\phi_\beta^*(x)\urcorner$ of some other formula $\phi_\beta^*(x)$ which also defines S_β. This amounts to taking $S_{\beta+1} - S_\beta$ to be the set $\{\exists y \mathrm{Prf}(\phi_\beta^*; \ulcorner\psi\urcorner, y) \rightarrow \psi\}$, although S_β is defined by $\phi_\beta(x)$. Then the function G such that $G(\alpha) = \ulcorner\phi_\alpha^*(x)\urcorner$ will work just as well as the one such that $G(\alpha) = \ulcorner\phi_\alpha(x)\urcorner$, and this reduces the problem to that of defining G where no circularity is implied.

Let F be the 1-1, onto function from ω to the order-type of $A(i,j)$ such that $F(i) < F(j)$ if and only if $A(i,j)$. By using the fixed-point theorem for formulas, we form the following formula $C(x, y)$, in which occurs its own g.n.

$$\begin{aligned} &\exists g \exists f \exists \tau\{f \in (\tau^\omega) \wedge \forall i \in \omega \forall j \in \omega[f(i) < f(j) \leftrightarrow A(i,j)] \wedge g \in \omega^\tau \\ &\quad \wedge \forall\alpha \in \tau \forall i \in \omega[\alpha = f(i) \rightarrow g(\alpha) = \mathrm{Sub}^3_{\mathrm{num}(i)}\ulcorner C(x, y)\urcorner] \\ &\qquad\qquad \wedge (\Phi_0(x) \vee \exists\beta[\beta < f(y) \wedge B(x, g(\beta))])\}. \end{aligned}$$

Then G is defined for $\alpha <$ the order-type of $A(i,j)$ by

$$G(\alpha) = \mathrm{Sub}^3_{\mathrm{num}(i)}\ulcorner C(x, y)\urcorner$$

where $\alpha = F(i)$.

Inside the curly brackets of $C(x, y)$, τ and f are uniquely satisfied by the order-type of $A(i,j)$ and F, and g is uniquely satisfied by G. The earlier circularity now shows up in $C(x, y)$ and is rendered noncircular by the occurrence in $C(x, y)$ of its own g.n. For $\alpha = F(y)$, the last conjunct expresses the same relation as $\phi_\alpha(x)$ and hence $C(x, y)$ also expresses this same relation. If $\beta = F(i)$, $C(x, \bar{i})$ plays the role of a formula $\phi_\beta^*(x)$ which also defines S_β, and its g.n. $\ulcorner C(x, \bar{i})\urcorner$ may be used in the definition $\phi_\alpha(x)$ of S_α in accordance with the ambiguity of the reflection principle.

Therefore $\phi_\alpha(x)$ defines S_α, and, for $\alpha = F(y)$, so does $C(x, y)$. With respect to $C(x, y)$ no appeal to the ambiguity of the reflection principle is necessary; however, we prefer to think in terms of $\phi_\alpha(x)$ because it leads to the following simplification.

In $\phi_\alpha(x)$ we replace $\beta < \alpha$ by $A(i, n)$ and then obtain a formula $\phi_n(x)$ as follows:

$$\Phi_0(x) \vee \exists i < x[A(i, n) \wedge B(x, \text{Sub}^3_{\text{num}(i)} \ulcorner C(x, y)\urcorner)].$$

$\phi_n(x)$ is primitive recursive in $A(i, j)$; therefore if $A(i, j)$ is general recursive, so is $\phi_n(x)$. Since $F(i) < F(n)$ if and only if $A(i, n)$ and since $G(\beta) = \text{Sub}^3_{\text{num}(i)} \ulcorner C(x, y)\urcorner$ if $\beta = F(i)$, $\phi_\alpha(x)$ and $\phi_n(x)$ express the same relation for $\alpha = F(n)$. Hence, for $\alpha = F(n)$, $\phi_n(x)$ also defines S_α, and we shall take $\phi_n(x)$ for our definition of S_α.

The obvious advantage of $\phi_n(x)$ is that it yields the following theorem.

THEOREM 9. *Let* $\alpha = F(n)$. *For each wff* ψ,
(1) $\psi \in S_\alpha \Rightarrow S_0 \vdash \phi_{\bar{n}}(\ulcorner\psi\urcorner)$,
(2) $\psi \notin S_\alpha \Rightarrow S_0 \vdash \neg\phi_{\bar{n}}(\ulcorner\psi\urcorner)$.

Let us write $C_n(x)$ for $C(x, n)$. The obvious disadvantage of $\phi_n(x)$ is that it takes $S_{\beta+1} - S_\beta$ to be the set $\{\exists y\text{Prf}(C_{\bar{i}}; \ulcorner\psi\urcorner, y) \rightarrow \psi\}$ instead of the set

$$\{\exists y\text{Prf}(\phi_{\bar{i}}; \ulcorner\omega\urcorner, y) \rightarrow \psi\},$$

where $F(i) = \beta$. This disadvantage is overcome in our proofs by use of global provability for $A(i, j)$, which yields the following theorems.

THEOREM 10. $S_0 \vdash \phi_n(x) \leftrightarrow C_n(x)$.

THEOREM 11. $S_0 \vdash \exists y\text{Prf}(\phi_n; x, y) \leftrightarrow \exists y\text{Prf}(C_n; x, y)$.

THEOREM 12. *For each wff* ψ,

$$S_0 \vdash [\exists y\text{Prf}(\phi_n; \ulcorner\psi\urcorner, y) \rightarrow \psi] \leftrightarrow [\exists y\text{Prf}(C_n; \ulcorner\psi\urcorner, y) \rightarrow \psi].$$

THEOREM 13. *Let* $\alpha = F(n)$. *For each wff* ψ,
(1) $\psi \in S_\alpha \Rightarrow S_0 \vdash C_{\bar{n}}(\ulcorner\psi\urcorner)$,
(2) $\psi \notin S_\alpha \Rightarrow S_0 \vdash \neg C_{\bar{n}}(\ulcorner\psi\urcorner)$.

If $C_n(x)$ were taken as the definition of S_α, global provability for $A(i, j)$ would still be necessary for the proof of Theorem 13. By the use of local provability for $A(i, j)$ we may prove the following theorem.

THEOREM 14. *For each* $m, n \in \omega$,
(1) $F(m) < F(n) \Rightarrow S_0 \vdash \phi_{\bar{m}}(x) \rightarrow \phi_{\bar{n}}(x)$,
(2) $F(m) < F(n) \Rightarrow S_0 \vdash C_{\bar{m}}(x) \rightarrow C_{\bar{n}}(x)$,
(3) $F(n) = F(m) + 1 \Rightarrow S_0 \vdash \phi_{\bar{n}}(x) \leftrightarrow \phi_{\bar{m}}(x) \vee B(x, \text{Sub}^3_{\text{num}(\bar{m})} \ulcorner C(x, y)\urcorner)$.

We shall now prove the main lemma; here it is convenient to extend the designation of local provability in the obvious way.

THEOREM 15. *Let* $\alpha = F(n)$. *For each wff* ψ,

(1) $S_\alpha \vdash \psi \Rightarrow S_0 \vdash \exists y \mathrm{Prf}(\phi_{\bar{n}}; \ulcorner\psi\urcorner, y)$,

(2) $S_\alpha \vdash \psi \Rightarrow S_0 \vdash \exists y \mathrm{Prf}(C_{\bar{n}}; \ulcorner\psi\urcorner, y)$.

PROOF. The proof is by transfinite induction on the length of the ω-proof from S_α of ψ.

Basis step: ψ is a logical axiom or a member of S_α. In the first case use Theorem 1 and local provability for $\mathrm{Lax}(x)$; in the second case use Theorem 2 and local provability for $\phi_n(x)$ or $C_n(x)$ and $\mathrm{Wff}(x)$.

Induction step: We may assume that ψ is neither a logical axiom nor a member of S_α. Then there are three cases corresponding to the three rules of inference. In the case of modus ponens use Theorem 3, the inductive hypothesis, and local provability for $x = x_1 \ulcorner\rightarrow\urcorner x_2$; in the case of generalization use Theorem 4, the inductive hypothesis and local provability for $\mathrm{Var}(x_1)$ and $x = \ulcorner\forall\urcorner x_1 x_2$. In the case of the ω-rule use the inductive hypothesis and the ω-rule to obtain

$$S_0 \vdash \forall i \in \omega \exists y \mathrm{Prf}(\phi_{\bar{n}}; \mathrm{Sub}^1_{\mathrm{num}(i)} \ulcorner\psi_1(x)\urcorner, y)$$

where ψ is $\forall x \in \omega \psi_1(x)$. Then use Theorem 5 and local provability for $x = \ulcorner\forall x \in \omega\urcorner x_1$.

5. **Proofs of the main results.** Let wffs $A(i, j)$ and $A'(i, j)$ be well-orderings of ω which satisfy local and global provability, and let F and F' be the 1-1, onto functions from ω to their respective order-types. Let $C(x, y)$ and $C'(x, y)$ be the associated wffs, and let $\phi_n(x)$ and $\phi'_m(x)$ be the definitions of the associated axiom systems S_α and S'_α where $\alpha = F(n) = F'(m)$. That is, $\phi_n(x)$ and $\phi'_m(x)$ are, respectively,

$$\Phi_0(x) \vee \exists i < x[A(i, n) \wedge B(x, \mathrm{Sub}^3_{\mathrm{num}(i)} \ulcorner C(x, y)\urcorner)]$$

$$\Phi_0(x) \vee \exists i < x[A'(i, m) \wedge B(x, \mathrm{Sub}^3_{\mathrm{num}(i)} \ulcorner C'(x, y)\urcorner)];$$

and S_α and S'_α are, respectively,

$$S_0 \cup \bigcup_{F(i) < \alpha} \{\exists y \mathrm{Prf}(C_{\bar{i}}; \ulcorner\psi\urcorner, y) \rightarrow \psi\} \qquad S_0 \cup \bigcup_{F'(j) < \alpha} \{\exists y \mathrm{Prf}(C'_{\bar{j}}; \ulcorner\psi\urcorner, y) \rightarrow \psi\}.$$

In order to prove that the theorems of S_α and S'_α are the same, it is sufficient to prove

$$F(i) = F'(j) \Rightarrow S_0 \vdash \exists y \mathrm{Prf}(C_{\bar{i}}; \ulcorner\psi\urcorner, y) \leftrightarrow \exists y \mathrm{Prf}(C'_{\bar{j}}; \ulcorner\psi\urcorner, y).$$

Since we make the same assumptions on the primed and unprimed formulas, it is sufficient to consider explicitly only one direction of the equivalence, and by Theorem 10, it is sufficient to prove

$$F(i) = F'(j) \Rightarrow S_0 \vdash \exists y \mathrm{Prf}(\phi_{\bar{i}}; \ulcorner\psi\urcorner, y) \rightarrow \exists y \mathrm{Prf}(C'_{\bar{j}}; \ulcorner\psi\urcorner, y).$$

This will be proved by proving by transfinite induction on $F(i) = F'(j)$ the stronger result

$$F(i) = F'(j) \Rightarrow S_0 \vdash \forall x \in \omega[\phi_{\bar{\imath}}(x) \to \exists y \operatorname{Prf}(C'_{\bar{\jmath}}; x, y)],$$
$$\Rightarrow S_0 \vdash \exists y \operatorname{Prf}(\phi_{\bar{\imath}}; \ulcorner\psi\urcorner, y) \to \exists y \operatorname{Prf}(C'_{\bar{\jmath}}; \ulcorner\psi\urcorner, y).$$

We shall first prove two lemmas and then the above result.

THEOREM 15. *For each $i, j \in \omega$ such that $F(i) = F'(j)$,*

$$S_0 \vdash \forall x \in \omega[\phi_{\bar{\imath}}(x) \to \exists y \operatorname{Prf}(C'_{\bar{\jmath}}; x, y)] \Rightarrow$$
$$\textit{for each wff } \psi, S_0 \vdash \exists y \operatorname{Prf}(\phi_{\bar{\imath}}; \ulcorner\psi\urcorner, y) \to \exists y \operatorname{Prf}(C'_{\bar{\jmath}}; \ulcorner\psi\urcorner; y).$$

The proof is a straightforward formalization of the informal idea of placing at the beginning of an ω-proof from S_α of an ω-proof from S'_α of all of the members of S_α. The hypothesis is needed in order to prove formally from S_0 that the resulting formalized ω-proof from S'_α is a formalized ω-proof of S'_α.

THEOREM 16. *Let $j, k, l \in \omega$ be such that $F'(j) = F(k) + 1 = F'(l) + 1$. If, for each wff ψ, $S_0 \vdash \exists y \operatorname{Prf}(\phi_{\bar{k}}; \ulcorner\psi\urcorner, y) \to \exists y \operatorname{Prf}(C'_{\bar{l}}; \ulcorner\psi\urcorner, y)$, then, for each wff ψ, $B(\ulcorner\psi\urcorner, \operatorname{Sub}^3_{\operatorname{num}(k)} \ulcorner C(x, y)\urcorner) \Rightarrow S_0 \vdash \exists y \operatorname{Prf}(C'_{\bar{\jmath}}; \ulcorner\psi\urcorner, y)$.*

PROOF. Suppose $B(\ulcorner\psi\urcorner, \operatorname{Sub}^3_{\operatorname{num}(k)} \ulcorner C(x, y)\urcorner)$. Then ψ is $\exists y \operatorname{Prf}(C_{\bar{k}}; \ulcorner\psi'\urcorner, y) \to \psi'$ for some wff ψ', and, by the hypothesis and Theorem 11,

$$S_0 \vdash [\exists y \operatorname{Prf}(C'_{\bar{l}}; \ulcorner\psi'\urcorner, y) \to \psi'] \to [\exists y \operatorname{Prf}(C_{\bar{k}}; \ulcorner\psi'\urcorner, y) \to \psi'].$$

Let us write $\psi'' \to \psi$ for the above wff; then by Theorems 6, 14, and 15, $S_0 \vdash \exists y \operatorname{Prf}(C'_{\bar{\jmath}}; \ulcorner\psi'' \to \psi\urcorner, y)$. Since $\psi'' \in S'_{F'(j)}$, by Theorem 15, $S_0 \vdash \exists y \operatorname{Prf}(C'_{\bar{\jmath}}; \ulcorner\psi''\urcorner, y)$ and hence, by Theorem 3, $S_0 \vdash \exists y \operatorname{Prf}(C'_{\bar{\jmath}}; \ulcorner\psi\urcorner, y)$.

THEOREM 17. *For each $i, j \in \omega$ such that $F(i) = F'(j)$ and each wff ψ,*

$$S_0 \vdash \forall x \in \omega[\phi_{\bar{\imath}}(x) \to \exists y \operatorname{Prf}(C'_{\bar{\jmath}}; x, y)],$$

$$S_0 \vdash \exists y \operatorname{Prf}(\phi_{\bar{\imath}}; \ulcorner\psi\urcorner, y) \to \exists y \operatorname{Prf}(C'_{\bar{\jmath}}; \ulcorner\psi\urcorner, y).$$

PROOF. The proof is by transfinite induction on $F(i) = F'(j)$, and by Theorem 15 it is sufficient to prove the first result. Furthermore in each of the cases below it is sufficient to prove, for each $x \in \omega$, $\phi_i(x) \Rightarrow S_0 \vdash \exists y \operatorname{Prf}(C'_{\bar{\jmath}}; \bar{x}, y)$. For, by Theorem 9, this yields, for each $x \in \omega$, $S_0 \vdash \phi_{\bar{\imath}}(\bar{x}) \to \exists y \operatorname{Prf}(C'_{\bar{\jmath}}; \bar{x}, y)$ to which we apply the ω-rule to obtain the first result.

Case I. $F(i) = F'(j) = 0$ and $\phi_i(x)$. Then $x = \ulcorner\psi\urcorner$ for some $\psi \in S_0$. Hence by Theorem 15, $S_0 \vdash \exists y \operatorname{Prf}(C'_{\bar{\jmath}}; \ulcorner\psi\urcorner, y)$.

Case II. $F(i) = F'(j) = F(k) + 1 = F'(l) + 1$ and $\phi_i(x)$. Then $x = \ulcorner\psi\urcorner$ for some $\psi \in S_{F(i)}$, and therefore $\phi_k(\ulcorner\psi\urcorner)$ or $B(\ulcorner\psi\urcorner, \operatorname{Sub}^3_{\operatorname{num}(k)} \ulcorner C(x, y)\urcorner)$. By the first inductive hypothesis, Theorem 14, and Theorem 6, $S_0 \vdash \phi_{\bar{k}}(\ulcorner\psi\urcorner) \to \exists y \operatorname{Prf}(C'_{\bar{\jmath}}; \ulcorner\psi\urcorner, y)$. Then by Theorem 9, $\phi_k(\ulcorner\psi\urcorner) \Rightarrow S_0 \vdash \exists y \operatorname{Prf}(C'_{\bar{\jmath}}; \ulcorner\psi\urcorner, y)$. By the second inductive hypothesis and Theorem 16, $B(\ulcorner\psi\urcorner, \operatorname{Sub}^3_{\operatorname{num}(k)} \ulcorner C(x, y)\urcorner) \Rightarrow S_0 \vdash \exists y \operatorname{Prf}(C'_{\bar{\jmath}}; \ulcorner\psi\urcorner, y)$.

Case III. $F(i) = F'(j)$ is a limit ordinal. Then $x = \ulcorner\psi\urcorner$ for some ψ and some $S_{F(k)}$ such that $\psi \in S_{F(k)}$ and $F(k) = F'(l) < F(i) = F'(j)$. Again by the first inductive hypothesis, Theorem 14, Theorem 6, and Theorem 9, $S_0 \vdash \exists y \operatorname{Prf}(C'_{\bar{\jmath}}; \ulcorner\psi\urcorner, y)$.

The use of the ω-rule in each case of the last proof is our second and final use of the ω-rule. The first use is in the proof of the main lemma, Theorem 15, but there the ω-rule was needed merely to accommodate the inclusion of the ω-rule in our notion of proof. In contrast to this the use of the ω-rule appears to be essential in order to obtain the result of Theorem 17 because it concerns the unprimed wff $\phi_i(x)$ and the primed wff $C_j'(x)$ without making any assumptions except local and global provability for $A(i,j)$ and $A'(i,j)$.

We now obtain our first main result.

THEOREM 18. *Suppose* $\alpha = F(n) = F'(m)$. *Then, for each wff* Ψ,

$$S_\alpha \vdash \Psi \Leftrightarrow S_\alpha' \vdash \Psi.$$

PROOF. It is sufficient to prove, for each wff Ψ, (1) $\Psi \in S_\alpha \Rightarrow S_\alpha' \vdash \Psi$ and (2) $\Psi \in S_\alpha' \Rightarrow S_\alpha \vdash \Psi$; and moreover it is sufficient to consider explicitly only (1). If $\Psi \in S_0$, the result is immediate. Suppose $\Psi \in S_\alpha$, but $\Psi \notin S_0$. Then Ψ is $\exists y \mathrm{Prf}(C_{\bar{i}}; \ulcorner\psi\urcorner, y) \rightarrow \psi$ for some wff ψ and some $i \in \omega$ such that $F(i) < \alpha$. The companion wff Ψ': $\exists y \mathrm{Prf}(C_{\bar{j}}'; \ulcorner\psi\urcorner, y) \rightarrow \psi$, where $F(i) = F'(j)$, is a member of S_α'. By Theorems 10 and 17, $S_0 \vdash \exists y \mathrm{Prf}(C_{\bar{i}}; \ulcorner\psi\urcorner, y) \rightarrow \exists y \mathrm{Prf}(C_{\bar{j}}'; \ulcorner\psi\urcorner, y)$. Therefore $S_0 \vdash \Psi' \rightarrow \Psi$; and since $S_\alpha' \vdash \Psi'$, $S_\alpha' \vdash \Psi$.

In order to show that if S_α is consistent then $T_{\alpha+1} \neq T_\alpha$, we merely use Theorem 15 to extend the first part of Gödel's incompleteness theorem.

THEOREM 19. *Let* $\alpha = F(n)$ *and let* ψ *be the wff such that* ψ *is* $\neg\exists y \mathrm{Prf}(C_{\bar{n}}; \ulcorner\psi\urcorner, y)$. *Then, if* S_α *is consistent, not* $S_\alpha \vdash \psi$.

PROOF. Suppose $S_\alpha \vdash \psi$; i.e. $S_\alpha \vdash \neg\exists y \mathrm{Prf}(C_{\bar{n}}; \ulcorner\psi\urcorner, y)$. By Theorem 15, $S_0 \vdash \exists y \mathrm{Prf}(C_{\bar{n}}; \ulcorner\psi\urcorner, y)$, and therefore $S_\alpha \vdash \exists y \mathrm{Prf}(C_{\bar{n}}; \ulcorner\psi\urcorner, y)$ which contradicts the assumed consistency of S_α.

THEOREM 20. *If* S_α *is consistent, then* $T_{\alpha+1} \neq T_\alpha$.

PROOF. Let $\alpha = F(n)$ and let ψ be the wff such that ψ is $\neg\exists y \mathrm{Prf}(C_{\bar{n}}; \ulcorner\psi\urcorner, y)$. By Theorem 19, not $S_\alpha \vdash \psi$. Since $\vdash [\exists y \mathrm{Prf}(C_{\bar{n}}; \ulcorner\psi\urcorner, y) \rightarrow \psi] \leftrightarrow \psi$, not $S_\alpha \vdash \exists y \mathrm{Prf}(C_{\bar{n}}; \ulcorner\psi\urcorner, y) \rightarrow \psi$, but, of course, $S_{\alpha+1} \vdash \exists y \mathrm{Prf}(C_{\bar{n}}; \ulcorner\psi\urcorner, y) \rightarrow \psi$.

REFERENCES

1. S. Feferman, *Arithmetization of metamathematics in a general setting*, Fund. Math. **49** (1960), 35–92.

2. ——— *Transfinite recursive progressions of axiomatic theories*, J. Symbolic Logic **27** (1962), 259–316.

3. J. B. Rosser, *Gödel theorems for non-constructive logics*, J. Symbolic Logic **2** (1937), 129–137.

4. G. Takeuti, "Formalization principle" in *Logic, methodology and philosophy of science. III*, Editors van Rooselaar and Stall, North-Holland, Amsterdam, 1968, pp. 105–108.

5. A. M. Turing, *Systems of logic based on ordinals*, Proc. London Math. Soc. **45** (1939), 161–228.

THE CATHOLIC UNIVERSITY OF AMERICA

HYPOTHESES ON POWER SET

GAISI TAKEUTI[1]

Though stronger and stronger axioms of infinity have been developed, the continuum hypothesis seems to be independent from them. We believe that we should develop axioms on power set beside axioms of strong infinity. As for axioms of strong infinity, we have the following leading principle.

Reflection principle. There exists an arbitrarily large good approximation of the universe.

Now, is there any good leading principle on axioms on power set? It seems to us that somehow many logicians are getting a feeling that a power set of an infinite set is very rich. (For one such opinion, see the last section of Cohen [**1**].) If this feeling is really the case, this should be an excellent leading principle on axioms on power sets. The easy way to express such a feeling is to say that the cardinality of a power set of an infinite set is very big with respect to the notion of cardinal. For this purpose, the hierarchy of predicates on ordinal numbers is very useful. It is understood in this paper that $\alpha, \beta, \gamma, \ldots$ stand for ordinals. Predicates of the form $\forall f A(f, \alpha)$ or $\exists f A(f, \alpha)$ is said to be in $\Pi_1^{1,\mathrm{Ord}}$ or $\Sigma_1^{1,\mathrm{Ord}}$ respectively, where $\forall f$ or $\exists f$ is a quantifier over functions of ordinals and $A(f, \alpha)$ is an arithmetical predicate on ordinal numbers (cf. [**12**]). A predicate is said to be in $\Delta_1^{1,\mathrm{Ord}}$ if it is in both $\Pi_1^{1,\mathrm{Ord}}$ and $\Sigma_1^{1,\mathrm{Ord}}$. The following was considered at the end of §2 in Chapter II of [**13**].

Hypothesis 1. Let $\alpha \geq \omega$ and g be in $\Delta_1^{1,\mathrm{Ord}}$ in Reg and f_0, where Reg is the class of all regular cardinals and f_0 is any function from an ordinal $\alpha_0 < \overline{\overline{P(\alpha)}}$ into $\overline{\overline{P(\alpha)}}$. Then $g(\beta) < \overline{\overline{P(\alpha)}}$ for every $\beta < \overline{\overline{P(\alpha)}}$. As is shown in [**13**], this certainly implies that $\overline{\overline{P(\alpha)}}$ is very big.

---

[1] Work partially supported by National Science Foundation grant GP-6132.

Another idea is that the degree of $\aleph$'s increasing is negligible compared with the magnitude of $\overline{\overline{P(\alpha)}}$ if α is infinite We have an analogous relation between recursive function of ordinal numbers and $N(\alpha)$, where $N(\alpha)$ is the least cardinal greater than α. Namely we have the following theorem in [**11**]. If $\alpha \geq \omega$ and f is recursive, then $f(\alpha) < N(\alpha)$, whence follows $\forall\beta < N(\alpha)(N(f(\beta)) \leq N(\alpha))$. Replacing α, f, and N by $\aleph_\alpha$, $\aleph$, and $\overline{\overline{P(\ \ )}}$, we present the following hypothesis.

Hypothesis 2. $\forall\beta < \overline{\overline{P(\aleph_\alpha)}}(\overline{\overline{P(\aleph_\beta)}} \leq \overline{\overline{P(\aleph_\alpha)}})$.

This hypothesis also implies that $2^{\aleph_\alpha}$ is weakly inaccessible. For if $2^{\aleph_\alpha} = \aleph_{\beta+1}$, then $2^{\aleph_{\beta+1}} \leq 2^{\aleph_\alpha}$ which is a contradiction and if $\beta_0 = \mathrm{cf}(2^{\aleph_\alpha}) < 2^{\aleph_\alpha}$, then by virtue of König's lemma, $\mathrm{cf}(2^{\aleph_\alpha}) = \mathrm{cf}(2^{\beta_0}) > \beta_0$, which is a contradiction. Especially this implies that $\aleph_\beta < 2^{\aleph_0} \rightarrow 2^{\aleph_\beta} = 2^{\aleph_0}$. As a special case, we have Lusin's hypothesis $2^{\aleph_0} = 2^{\aleph_1}$, which also follows from some hypothesis in Hausdorff [**5**].

These two hypotheses simply express that the cardinality of power set is very big in comparison with the notion of $\aleph$. However, our feeling about the richness of power set is rather vague at this moment (cf. Gödel [**4**]) and we are not sure that the cardinality of power set is very big. It is perfectly possible that our feeling of richness of power set may simply mean the richness of the structure of power set and the cardinality of power set might be rather small. Therefore it seems much more interesting to develop hypotheses on structural richness of power set. Our leading principle is the following.

Reflection principle on power set. If m is transitive and infinite, then there exists an arbitrarily large good approximation M of the universe V such that $P(m) \notin M$.

This hypothesis means that $P(m)$ is so rich that there exists a good approximation of the universe which excludes $P(m)$. If we delete the condition $P(m) \notin M$, then this principle is a usual reflection principle. We shall consider several forms of this principle and their implication.

Hypothesis 3. If m is transitive and infinite, then for every $r \in P(m)$ there exists a transitive class M satisfying the following conditions.

(1) $\mathrm{On} \subseteq M \wedge r \in M \wedge P(m) \notin M \wedge m \in M$.

(2) $\langle M, m, r; k; \alpha\rangle_{k\in m,\alpha\in\mathrm{On}}$ is elementarily equivalent to $\langle V, m, r; k; \alpha\rangle_{k\in m,\alpha\in\mathrm{On}}$.

PROPOSITION. *Hypothesis* 3 *implies the following. There is no well-ordering of $P(m)$, which is definable by using m, a member of $P(m)$, members of m, and ordinal numbers.*

PROOF. Suppose not, i.e. suppose that $\psi(m, k_1, \ldots, k_n, r, \alpha, a, b)$, which will be denoted $a \overset{\psi}{<} b$, is a definable well-ordering of $P(m)$. Then using ψ, we can define $\psi_1(m, k_1, \ldots, k_n, r, \alpha, a, \beta)$ (abbreviated to $\psi_1(a, \beta)$) and β_0 satisfying the following conditions.

(1) $\forall\beta < \beta_0 \exists!\, a \in P(m)\psi_1(a, \beta)$.

(2) $\forall a \in P(m)\exists!\, \beta < \beta_0\psi_1(a, \beta)$.

Now take M in Hypothesis 3 for this r. There must exist $r_0 \in P(m)$ and a unique $\beta_1 < \beta_0$ such that $r_0 \notin M$ and the following hold on V.

(3) $\exists x \subseteq m\psi_1(x, \beta_1) \wedge \forall x \forall y(\psi_1(x, \beta_1) \wedge \psi_1(y, \beta_1) \rightarrow x = y)$.

(4) $\exists x(k \in x \wedge \psi_1(x, \beta_1))$ for every $k \in r_0$ and $\exists x(k \notin x \wedge \psi_1(x, \beta_1))$ for every $k \notin r_0$.

Therefore M satisfies (3) and (4), which implies $r_0 \in M$, i.e. a contradiction.

In order to state another application of Hypothesis 3, we give the following definition.

DEFINITION. Let a be a transitive set. $L(a)$ is defined to be the intersection of all transitive classes which have a and all ordinals and are models of ZF.

$L(a)$ is a generalization of L in Gödel [**3**]. Following Easton [**2**] and Lévy [**7**], we shall give another definition of $L(a)$ by using ramified hierarchy.

Ramified hierarchy. Our ramified hierarchy has a predicate constant $A(\ )$ and individual constants $\underline{k}$'s for all k in a. For every ordinal α, we introduce ranked variables $x_0^\alpha, x_1^\alpha, x_2^\alpha, \ldots$. The limited formula and the abstraction term are defined as follows.

(1) Let b and c be either individual constants, ranked variables, or abstraction terms. Then $A(b)$ and $b \in c$ are limited formulas.

(2) Let ψ_1 and ψ_2 be limited formulas. Then $\neg \psi_1$, $\psi_1 \wedge \psi_2$ and $\forall x_i^\alpha \psi_1$ are limited formulas.

(3) Let $\psi(x^\alpha)$ be a limited formula such that the following conditions hold:

(a) $\psi(x^\alpha)$ contains no free variables other than x^α,

(b) if $\underline{k}$ is an individual constant that appears in $\psi(x^\alpha)$, then the rank of k is less than α,

(c) if the abstraction term $\hat{y}^\beta\psi_1(y^\beta)$ appears in $\psi(x^\alpha)$ then $\beta < \alpha$,

(d) if the quantifier $\forall y^\beta$ appears in $\psi(x^\alpha)$ then $\beta \leq \alpha$.

Then $\hat{x}^\alpha\psi(x^\alpha)$ is an abstraction term.

A constant term is either an individual constant or an abstraction term. The rank ρ of a constant term is defined as $\rho(\underline{k}) = \text{rank}\ (k)$ and $\rho(\hat{x}^\alpha\psi(x^\alpha)) = \alpha$.

We define a denotation operator D as follows, which will be applied to a limited formula or a constant term. First we define $T_\alpha = \{t \mid t \text{ is a constant term with a rank less than } \alpha\}$.

(1) $D(\underline{k}) = k$.

(2) $D(A(t))$ is $D(t) \in a$.

(3) $D(t_1 \in t_2)$ is $D(t_1) \in D(t_2)$.

(4) $D(\neg \psi)$ is $\neg D(\psi)$.

(5) $D(\psi_1 \wedge \psi_2)$ is $D(\psi_1) \wedge D(\psi_2)$.

(6) $D(\forall x^\alpha\psi(x^\alpha))$ is $\forall t \in T_\alpha D(\psi(t))$.

(7) $D(\hat{x}^\alpha\psi(x^\alpha)) = \{D(t) \mid t \in T_\alpha \wedge D(\psi(t))\}$.

It is easy to show that $L(a) = \{D(t) \mid t \text{ is a constant term}\}$ and $L(a)$ is a model of ZF and $L(a) = L_a$ if $a \subseteq L$, where L_a is the class defined in Lévy [**6**] and Shoenfield [**10**].

PROPOSITION. *Hypothesis* 3 *implies that* $L(P(m))$ *does not satisfy* AC (*axiom of choice*), *where* m *is transitive and infinite.*

PROOF. Suppose $L(P(m))$ satisfies AC. Then there exists a constant term t such that $D(t)$ well-orders $P(m)$. Let $\underline{k}_1, \ldots, \underline{k}_n$ and $\alpha_1, \ldots, \alpha_m$ be all individual constants and ordinals used to define t. Then $D(t)$ is definable by using $k_1, \ldots, k_n, \alpha_1, \ldots, a_m$, and m. This contradicts the previous proposition.

However it has been known that $L(P(\aleph_\alpha))$ satisfies several weaker forms of axioms of choice. We shall give definitions of them.

DEFINITION. $\mathrm{AC}_{\aleph_\alpha}$ is defined to be the following: for every family a of disjoint sets, such that $0 \notin a$ and $\overline{\overline{a}} \leq \aleph_\alpha$, there exists a choice set.

DEFINITION. DC (axiom of depending choice) is defined to be the following: for every set X and R $\forall x \in X \, \exists y \in X(\langle xy \rangle \in R) \rightarrow$

$$\forall x \in X \, \exists f \in {}^\omega X (f'0 = x \wedge \forall n(\langle f'n, f'n + 1 \rangle \in R)).$$

DEFINITION. Let j be Gödel's pairing function of ordinals, i.e. j be a 1-1 map from On × On onto On. If $a \subseteq$ On, then $(a)_\alpha$ is defined to be $\{\beta \mid j(\alpha, \beta) \in a\}$.

PROPOSITION. *$L(P(\aleph_\alpha))$ satisfies* $\mathrm{AC}_{\aleph_\alpha}$.

PROOF. Let a be a member of $L(P(\aleph_\alpha))$ satisfying conditions in the definition of $\mathrm{AC}_{\aleph_\alpha}$. Let $a_0, a_1, \ldots, a_\beta, \ldots$ $(\beta < \aleph_{\beta_0})$ be all members of a, where $\beta_0 \leq \alpha$. Choose sequences $b_0, \ldots, b_\beta, \ldots$ and $t_0, \ldots, t_\beta, \ldots$ in V (outside $L(P(\aleph_\alpha))$) such that $b_\beta \in a_\beta$ and $D(t_\beta) = b_\beta$ for every $\beta < \aleph_{\beta_0}$. Let $\underline{k}_0, \ldots, \underline{k}_\beta, \ldots$ $(\beta < \aleph_{\beta_0})$ be all individual constants in $t_0, \ldots, t_\beta, \ldots$. There exists k in $P(\aleph_\alpha)$ such that $(k)_\beta = k_\beta$ for every $\beta < \aleph_{\beta_0}$. Let $\aleph_\gamma$ be a regular cardinal which is greater than any of $\rho(t_\beta)(\beta < \aleph_{\beta_0})$ and $\aleph_\alpha$. Define S' to be $\{t \mid \rho(t) < \aleph_\gamma$ and t does not contain any individual constants other than $k\}$ and S to be $\{D(t) \mid t \in S'\}$. Obviously, S' and S are well-ordered in $L(P(\aleph_\alpha))$ and $b_0, b_1, \ldots, b_\beta, \ldots$ are members of S. Therefore there exists a choice set of a in $L(P(\aleph_\alpha))$.

PROPOSITION. *$L(P(\aleph_\alpha))$ satisfies* DC.

PROOF. Let X and R be sets in $L(P(\aleph_\alpha))$ satisfying $\forall x \in X \exists y \in X(\langle xy \rangle \in R)$ and x be a member of X. There exists a set S of constant terms such that every member t in S has only one individual constant and $\forall y \in X \, \exists t \in S(D(t) = y)$ Well-order S in V (outside $L(P(\aleph_\alpha))$). Let $\beta_0 = \sup_{t \in S} \rho(t)$. $T_{\beta_0}(k)$ is defined to be $\{t \mid t \in T_{\beta_0}$ and t has only one individual constant $\underline{k}\}$. There exists a set $\{\langle <_k, k \rangle \mid k \in P(\aleph_\alpha)\}$ in $L(P(\aleph_\alpha))$ such that $<_k$ is a well-ordering of $T_{\beta_0}(k)$. Take the first t_1 in S such that $\langle x, D(t_1) \rangle \in R$. Let $\underline{k}_1$ be the individual constant in t_1 and t^1 be the first member in $S \cap T_{\beta_0}(k_1)$ w.r.t. $<_{k_1}$ such that $\langle x, D(t^1) \rangle \in R$. Now assume that $t^1, \ldots, t^n$ and $k_1, \ldots, k_n$ have been chosen. Take t_{n+1} in S such that $\langle D(t_{n+1}), D(t^n) \rangle \in R$. Let k_{n+1} be the only individual constant in t_{n+1} and t^{n+1} be the first member in $S \cap T_{\beta_0}(k_{n+1})$ w.r.t. $<_{k_{n+1}}$ such that $\langle D(t^n), D(t^{n+1}) \rangle \in R$. Now take k in $P(\aleph_\alpha)$ such that $(k)_n = k_n$ for $n \in \omega$. Sequences $x, t^1, t^2, \ldots$ and $x, D(t^1), D(t^2), \ldots$ are uniquely determined from k, β_0, x and S and so are members of $L(P(\aleph_\alpha))$.

Hypothesis 4. Let m be transitive and infinite. Then for any $a \subseteq P(m)$ and any ordinal α_0 there exists a transitive set M satisfying the following conditions.

(1) $m \in M \wedge \alpha_0 \in M \wedge P(m) \notin M$.

(2) $\langle M, a \cap M, \alpha_0, m; k\rangle_{k\in m}$ is elementarily equivalent to $\langle V, a, \alpha_0, m; k\rangle_{k\in m}$.

PROPOSITION. *Hypothesis* 4 *implies* $2^{\aleph_\alpha} > \aleph_{\alpha+1}$.

PROOF. Take m and α_0 to be $\aleph_\alpha$ and $\aleph_{\alpha+1}$ respectively. Let R be a subclass of On. We define $a \overset{R}{<} b$ to be $j(a, b) \in R$. We(R, d) is defined to be "$\overset{R}{<}$ is a well-ordering of d". $|R| = \beta$ is defined to be We$(R, \aleph_\alpha) \wedge$ "the order type of R is β". There exists a subset a_1 of $\aleph_\alpha$ satisfying

$$a_1 \subseteq P(\aleph_\alpha) \wedge \forall x \in a_1(\aleph_\alpha \leq |x| < \aleph_{\alpha+1}) \wedge \forall \beta < \aleph_{\alpha+1}(\aleph_\alpha \leq \beta \rightarrow \exists!\, x \in a_1(|x| = \beta)).$$

Suppose $2^{\aleph_\alpha} = \aleph_{\alpha+1}$. Then there exists a subset a of $P(\aleph_\alpha)$ such that the following conditions are satisfied.

(1) $\forall b \in a\, \forall \alpha \in b\, \exists \beta(\alpha = j(0, \beta) \vee \alpha = j(1, \beta))$.

(2) $\forall b \in a((b)_0 \in P(\aleph_\alpha) \wedge (b)_1 \in a_1)$.

(3) $\forall x \in P(\aleph_\alpha)\, \exists!\, b \in a(x = (b)_0)$.

(4) $\forall x \in a_1\, \exists!\, b \in a(x = (b)_1)$.

Then the following conditions hold.

(2)′ $\forall b \in a((b)_0 \in P(\aleph_\alpha) \wedge \exists \beta(\aleph_\alpha \leq \beta < \aleph_{\alpha+1} \wedge |(b)_1| = \beta)$.

(4)′ $\forall \beta(\aleph_\alpha \leq \beta < \aleph_{\alpha+1} \rightarrow \exists!\, b \in a(|(b)_1| = \beta))$.

(1), (2)′, (3), and (4)′ are expressible in our language with constants m, α_0, and a. Therefore Hypothesis 4 implies that for these m, α_0, and a there exists a transitive set M satisfying the conditions in Hypothesis 4. Hence there exists a set $r \in P(\aleph_\alpha)$ such that $r \notin M$. However there exists a set $b \in a$ such that $(b)_0 = r$. Define $\beta = |(b)_1|$. If we replace a by $a \cap M$ in (1), (2)′, (3), and (4)′, then M satisfies (1), (2)′, (3), and (4), whence follows $a \cap M = a$ and so $r \in M$, i.e. a contradiction.

REMARK 1. It is easily checked that only the following weaker axiom schemata is necessary to prove $2^{\aleph_\alpha} > \aleph_{\alpha+1}$.

$$\forall \beta\, \forall a \subseteq P(\aleph_\alpha)(\psi(a, \beta) \rightarrow \exists m(\text{"}m \text{ is transitive"} \wedge \beta \in m \wedge P(\aleph_\alpha) \notin m \wedge \psi^m(a \cap m, \beta)),$$

where a and b are only free variables in $\psi(a, b)$ and ψ^m is obtained from ψ by replacing all quantifiers $\forall x, \exists y, \ldots$ by $\forall x \in m, \exists y \in m, \ldots$ respectively.

REMARK 2. The similar method proves the following proposition: $\forall a \exists b \subseteq$ On ("*a is constructible from b*").

PROOF. This is proved by transfinite induction on rank(a). Let all the members of a be $a_0, a_1, \ldots, a_\beta, \ldots$ $(\beta < \alpha)$. By the inductive hypothesis there exist $b_0, b_1, \ldots, b_\beta, \ldots$ $(\beta < \alpha)$ such that every $a_\beta(\beta < \alpha)$ is constructible from b_β and $b_\beta \subseteq$ On. Let $a_\beta = F_{b_\beta}{}^{\text{`}}\gamma_\beta$ (cf. [**6**]). Define $b = \{j(j(\delta, \gamma_\beta), \beta) \mid \delta \in b_\beta\}$. Obviously $b \subseteq$ On. We can easily define $a_\beta$ and $\gamma_\beta$ from $b$. $a$ is expressed to be the set of $F_{b_\beta}{}^{\text{`}}\gamma_\beta$, which is certainly a member of $L(b)$.

In order to give a hypothesis which is a generalization of Hypothesis 4, we first give the following definition.

DEFINITION. A transitive class M is said to be semiuniversal for an infinite set m, an ordinal α, and $a \subseteq P(m)$, if and only if M satisfies the following conditions.

(1) $\mathrm{On} \subseteq M \wedge m \in M \wedge P(m) \notin M$.

(2) $\langle M, a \cap M, m; k; \alpha\rangle_{k\in m}$ is elementarily equivalent to $\langle V, a, m; k; \alpha\rangle_{k\in m}$.

Hypothesis 5. Let m be transitive and infinite. Then for any set $a \subseteq P(m)$, there exists a transitive class M which is semiuniversal for m and a.

PROPOSITION. *Hypothesis 5 implies the following. If $2^{\aleph_\beta} \leq \aleph_{\alpha+1}$, then there exists a semiuniversal class M for $\aleph_\alpha$ (and 0) such that $P(\aleph_\beta) \in M$.*

PROOF. There exists a subset a_1 of $P(\aleph_\alpha)$ satisfying the following conditions.

(1) $\forall x \in a_1(\aleph_\alpha \leq |x| < \aleph_\alpha + 2^{\aleph_\beta})$.

(2) $\forall \beta' < \aleph_\alpha + 2^{\aleph_\beta}(\aleph_\alpha \leq \beta' \rightarrow \exists!\, x \in a_1(|x| = \beta'))$.

Define $a \subseteq P(\aleph_\alpha)$ to satisfy the following.

(3) $\forall x \in a((x)_0 \in a_1 \wedge (x)_1 \subseteq \aleph_\beta)$.

(4) $\forall y \subseteq \aleph_\beta\, \exists!\, x \in a((x)_1 = y)$.

(5) $\forall y \in a_1\, \exists!\, x \in a((x)_0 = y)$.

Any M which is semiuniversal for $\aleph_\alpha$, constant ordinals used in (1)–(5), and a satisfies the proposition.

Hypothesis 6. Let κ be an infinite cardinal, α be an ordinal and a be a subset of $P(\kappa)$. Then there exists a transitive class M satisfying the following conditions.

(1) $\mathrm{On} \subseteq M \wedge P(\kappa) \notin M$.

(2) $\langle M, a \cap M; \alpha\rangle$ is $L_{\kappa,\kappa}$-equivalent to $\langle V, a, \alpha\rangle$.

We feel that if a statement Q expresses a richness of $P(m)$, and even if Q is too strong to be true on V, it is very likely that Q is true on $L(P(m))$. For $P(m)$ is relatively very rich in $L(P(m))$, though $P(m)$ is relatively very small in $P(P(m))$ and so in V. In this sense, the following questions seem very interesting. Is the axiom of determinateness true on $L(P(\omega))$? Is there any good reflection principle on power set which implies the axiom of determinateness on $L(P(\omega))$?

REMARK 3. The following special case of Hypothesis 4 is equivalent to $2^{\aleph_\alpha} > \aleph_{\alpha+1}$.

Hypothesis. For any $a \subseteq P(\aleph_\alpha)$ there exists a transitive set M satisfying the following conditions.

(1) $\aleph_{\alpha+1} \in M \wedge P(\aleph_\alpha) \notin M$.

(2) $\langle M, a \cap M, \aleph_{\alpha+1}, \aleph_\alpha; k\rangle_{k\in\aleph_\alpha}$ is elementarily equivalent to

$$\langle V, a, \aleph_{\alpha+1}, \aleph_\alpha; k\rangle_{k\in\aleph_\alpha}.$$

We have proved that this hypothesis implies $2^{\aleph_\alpha} > \aleph_{\alpha+1}$. Now assume $2^{\aleph_\alpha} > \aleph_{\alpha+1}$. Take the first order language of the set theory whose individual constants are a, $\aleph_{\alpha+1}$ and all k's $< \aleph_{\alpha+1}$. Let M' be a Skolem cover of $\aleph_{\alpha+1} \cup \{a\}$ and M be a transitive realization of M'. $\aleph_\alpha \subseteq M'$ implies $P(\aleph_\alpha) \cap M = P(\aleph_\alpha) \cap M'$. Since $a \subseteq P(\aleph_\alpha)$, a in M is $a \cap M$. $P(\aleph_\alpha) \notin M$ because $\overline{\overline{M}} = \aleph_{\alpha+1}$.

REMARK 4. The following axiom is a generalization of the existence of Ramsey cardinals.

Axiom R. If f is an increasing continuous function from On into On, then there exists an ordinal κ such that $\kappa = f(\kappa) \wedge \kappa \to (\kappa)^{<\omega}$.

This axiom implies that for any set $a \subseteq P(\aleph_\alpha)$ and for any ordinal $\beta < 2^{\aleph_\alpha}$ there exists a transitive class M satisfying the following conditions.

(1) $\mathrm{On} \subseteq M \wedge P(\aleph_\alpha) \notin M$.

(2) $\langle M, a \cap M; k\rangle_{k\in\beta}$ is elementarily equivalent to $\langle V, a; k\rangle_{k\in\beta}$.

PROOF. We may assume $\aleph_\alpha < \beta$. Take the first order language of the set theory whose individual constants are a and all k's ($<\beta$). Axiom R implies the existence of κ satisfying the following conditions.

(3) $\beta < \kappa \wedge \aleph_\alpha < \kappa$.

(4) $\langle R(\kappa), a; k\rangle_{k\in\beta}$ is an elementary substructure of $\langle V, a; k\rangle_{k\in\beta}$.

(5) $\kappa \to (\kappa)^{<\omega}$.

From a theorem of **[14]**, there exists a remarkable set $X \subseteq \kappa$ whose order type is κ. Using X and introducing an indiscernible class Y whose order type is On, we get a model $\mathcal{M} = \langle M, a^{\mathcal{M}}; k\rangle_{k\in\beta}$ satisfying the following conditions.

(6) $\mathrm{On} \subseteq M$.

(7) $\langle M, a^{\mathcal{M}}; k\rangle_{k\in\beta}$ is elementarily equivalent to $\langle V, a; k\rangle_{k\in\beta}$.

$a^{\mathcal{M}} = a \cap M$. Since the cardinality of $P(\aleph_\alpha)$ in M is $\overline{\overline{\beta}}$, $P(\aleph_\alpha) \notin M$.

REMARK 5. The following generalization of both Hypotheses 3 and 4 is inconsistent. (This may be still true in $L(P(m))$.)

Let m be transitive and infinite. Then for any set $a \subseteq P(m)$, there exists a transitive class M satisfying the following conditions.

(1) $\mathrm{On} \subseteq M \wedge m \in M \wedge P(m) \notin M$.

(2) $\langle M, a \cap M, m; k; \alpha\rangle_{k\in m, \alpha\in \mathrm{On}}$ is elementarily equivalent to

$$\langle V, a, m; k; \alpha\rangle_{k\in m, \alpha\in\mathrm{On}}.$$

PROOF. Let m be $\aleph_{\alpha_0}$ and $\lessdot$ be a well-ordering of $P(\aleph_{\alpha_0})$. Then there exists a set $a \subseteq P(\aleph_{\alpha_0})$ satisfying the following conditions.

$$\forall x_0 \in P(\aleph_{\alpha_0}) \forall x_1 \in P(\aleph_{\alpha_0})(x_0 \lessdot x_1 \leftrightarrow \exists x \in a((x)_0 = x_0 \wedge (x)_1 = x_1)).$$

Then every element of $P(\aleph_{\alpha_0})$ is definable by using an ordinal and a which contradicts the hypothesis.

Hypothesis 7. Let f_β ($\beta < \aleph_\alpha$) be functions from $\aleph_{\alpha+1} \times \aleph_{\alpha+1}$ into $\aleph_{\alpha+1}$. Then there exists a subset $X \subsetneqq \aleph_{\alpha+1}$ such that $\overline{\overline{X}} = \aleph_{\alpha+1}$ and X is closed with respect to all f_β's.

PROPOSITION. *Hypothesis* 7 *implies the hypothesis in Remark* 3 (*and* $2^{\aleph_\alpha} > \aleph_{\alpha+1}$).

PROOF. Take the first order language of the set theory whose individual constants are a and all k's $\leq \aleph_\alpha$. Let $f_0, f_1, \ldots, f_\beta, \ldots$ ($\beta < \aleph_\alpha$) be all Skolem functions having constants in the language. For each (n-nary) function f_β, we define g_β by the following.

For every $\alpha_1, \ldots, \alpha_n < \aleph_{\alpha+1}$

$$\begin{aligned} g_\beta(\alpha_1, \ldots, \alpha_n) &= f_\beta(\alpha_1, \ldots, \alpha_n) \quad \text{if} \quad f_\beta(\alpha_1, \ldots, \alpha_n) < \aleph_{\alpha+1} \\ &= 0 \qquad\qquad\qquad\qquad \text{otherwise.} \end{aligned}$$

By Hypothesis 7, there exists $X \subsetneqq \aleph_{\alpha+1}$ such that X is closed with respect to all g_β's. Let M' be a Skolem cover of $X \cup \{a\} \cup \aleph_\alpha$ and M be a transitive realization of M'. Then $M' \cap \aleph_{\alpha+1} = X$ and $\aleph_{\alpha+1}$ in M is $\aleph_{\alpha+1}$ itself. Since $a \subseteq P(\aleph_\alpha)$, a in M is $a \cap M$ and $P(\aleph_\alpha) \notin M$ because $\aleph_{\alpha+1} - M \neq 0$.

Hypothesis 8. In the same assumption of Hypothesis 7, there exists a subset $X \subseteq \aleph_{\alpha+1}$ such that $\overline{\overline{X}} = \overline{\overline{\aleph_{\alpha+1} - X}} = \aleph_{\alpha+1}$ and X is closed with respect to all f_β's.

REFERENCES

1. P. J. Cohen, *Set theory and the continuum hypothesis*, Benjamin, New York, 1966.
2. W. B. Easton, *Powers of regular cardinals*, Ph.D. Thesis, Princeton University, 1964.
3. K. Gödel, *The consistency of the axiom of choice and of the generalized continuum hypothesis with the axioms of set theory*, Princeton Univ. Press, Princeton, N.J., 1951.
4. ———, "What is Cantor's continuum problem?" in *Philosophy of mathematics*, edited by P. Benacerraf and H. Putnam, Englewood Cliffs, N.J., 1964, 258–273.
5. F. Hausdorff, *Untersuchungen über Ordnungstypen*, Berichte der Sachsische Acad. Wissenschaften, **59** (1907), 84–159.
6. A. Lévy, *A generalization of Gödel's notion of constructibility*, J. Symbolic Logic, **25** (1960), 147–155.
7. ———, *Definability in axiomatic set theory*. I. Logic, Methodology and Philos. Sci. (Proc. 1964 Internat. Congr.), North-Holland, Amsterdam, pp. 127–151, 1965.
8. J. Mycielski, *On the axiom of determinateness*, Fund. Math. **53** (1964), 205–224.
9. J. Mycielski and H. Steinhaus, *A mathematical axiom contradicting the axiom of choice*, Bull. Polon. Acad. Sci. Sér. Sci. Math. Astronom. Phys. **10** (1962), p. 1.
10. J. R. Shoenfield, *On the independence of the axiom of constructibility*, Amer. J. Math. **81** (1959), 537–540.
11. G. Takeuti, *On the recursive functions of ordinal numbers*, J. Math. Soc. Japan **12** (1960), 119–128.
12. ———, *Recursive functions and arithmetical functions of ordinal numbers*, Logic, Methodology and Philos. Sci. (Proc. 1964 Internat. Congr.), North-Holland, Amsterdam, 1965, pp. 179–196.
13. ———, *The universe of set theory*, Foundations of Mathematics, edited by J. Bulloff, T. Holyoke and S. Hahn, Springer, Berlin, 1969, pp. 74–128.
14. J. H. Silver, *Some applications of model theory in set theory*, Doctoral dissertation, University of California, Berkeley, 1961.

THE INSTITUTE FOR ADVANCED STUDY

MULTIPLE CHOICE AXIOMS

MARTIN M. ZUCKERMAN[1]

1. **Introduction.** A multiple choice axiom differs from the axiom of choice in that the former asserts the existence of a function which is defined on an arbitrary set X of sets, and which chooses subsets, rather than individual elements, of the elements of X. Various multiple choice axioms have been proposed ([**2**], [**3**], [**12**], [**23**]) in which the sets $f(y)$, $y \in x$, have been finite subsets of y, the number of whose elements satisfies some arithmetical formula. (In [**23**] there is also a restriction on the sets X considered.) In this article we discuss a fairly general class of multiple choice axioms; we characterize these axioms in terms of functions defined on ordinals and we consider relative consistency statements related to these axioms and the usual axioms of set theory. We also introduce new versions of multiple choice axioms in which the number of elements of each set $f(y)$, $y \in x$, is to be a "linear combination" of the prime factors or of powers of the prime factors of a fixed positive integer n. The interdependence of all of these axioms together with Mostowski's axioms of choice for finite sets [**15**] is discussed; some of the independence results are obtained by constructing Fraenkel-Mostowski-type models for set theory.

2. **Preliminaries.** Let σ be the set theory of [**14**]; this is a set theory of the Gödel-Bernays type. It permits the existence of *urelemente* (objects, other than the empty set, which are in the domain but not the range of the $\in$-relation) and it does not include the axiom of choice among its axioms. The first-order predicate calculus with identity serves as the logical framework for σ. In our metamathematical

[1] This research formed part of the author's Ph.D. thesis (Yeshiva University, 1967) under the supervision of Professor Martin Davis of New York University. The work was supported in part by the U.S. Air Force and in part by a National Science Foundation Science Faculty Fellowship.

investigations we shall assume the consistency of σ—or equivalently, the consistency of Gödel's system A, B, C, of [**9**].

Let 0 be the empty set, let $1 = \{0\}$, let $2 = 1 \cup \{1\}$, let $3 = 2 \cup \{2\}$, etc. By an *ordinal number* is meant a transitive set which is ordered by the $\in$-relation. (Regularity is included among the axioms of σ.) A set X is *finite* iff every nonempty set of subsets of X has a maximal element with respect to set inclusion; otherwise X is *infinite*. The finite ordinal numbers are thus the nonnegative integers. A set X is *denumerable* if there exists a function (which is itself a set) which maps X one-one onto the set of nonnegative integers. If there exists a function which maps the set X one-one onto the positive integer n, then X is called an n-element set; in this case we let $\mathbf{n}(X)$ denote the unique integer n for which such a mapping exists. For arbitrary sets X, Y, if there exists a one-one map of X into Y, we write $X \preceq Y$; if $X \preceq Y$ & $Y \npreceq X$, we write $X \prec Y$.

For each nonnegative integer n, I_n is the set of integers $\geq n$, J_n is the relative complement of I_{n+1} in I_1, $I_1 \setminus I_{n+1}$, and $K_n = J_n \setminus \{1\}$. Π represents the set of prime numbers and $\Pi_n = \Pi \cap I_n$.

For any set X, $\mathscr{P}(X)$ designates the power set of X, $\mathscr{P}^*(X) = \mathscr{P}(X) \setminus 1$, $\mathscr{P}^\#(X)$ is the set of finite subsets of X, $\mathscr{P}^{\#*}(X) = \mathscr{P}^\#(X) \setminus 1$, $\mathscr{P}^\infty(X) = \mathscr{P}(X) \setminus \mathscr{P}^\#(X)$, and $\mathscr{P}_k(X)$ denotes the set of all k-element subsets of X, $k \in I_1$. Furthermore, we let $X_{(r)} = \bigcup \{\mathscr{P}_r(Y) : Y \in X\}$, $r \in I_1$. Then $X_{(r)} \subseteq \mathscr{P}_r(\bigcup X)$; if X contains at least two nonempty sets and if $r \preceq \bigcup X$, then the inclusion is proper.

DEFINITION 1. A set X is *nontrivial* if X, as well as each element of X, is nonempty.

DEFINITION 2. By the *axiom of choice* (AC) we mean the following statement of σ: "For every nontrivial set X there is a function f with domain X such that $f(Y) \in Y$ for each $Y \in X$."

DEFINITION 3. By a *weak form of the axiom of choice* (WAC) we mean an axiom, Ax, satisfying

(i) (AC) $\vdash_\sigma$ Ax. (To be read: "The axioms of σ together with (AC) yield Ax.")

(ii) Ax $\nvdash_\sigma$ (AC).

(iii) Ax is independent of σ.

It follows from (i) and from Gödel's results ([**8**] and [**9**]) that any (WAC) is relatively consistent with σ. Of course, a version of the axiom of choice which is obtained from (AC) by some restriction on the sets X, Y, or $f(Y)$ of Definition 2 can be sufficiently weak so that (iii) fails to hold; this is the case, for example, if the sets X of Definition 2 are required to be finite or effectively denumerable. We shall be concerned with determining which of the axioms we consider are (WAC)'s.

In general, if Ax stands for the axiom of choice or for some variation thereof, then it will be understood that the choice functions f, under consideration, are to be defined for each nontrivial set of the universe. Ax$^\#$ will mean that the choice functions f are to be defined only for sets X consisting of finite, nonempty sets. On the other hand, if X is any set, then Ax(X) will denote the existence of a choice function defined on X; it will be convenient to agree that if $X = 0$, then

(1) 0 itself is a choice function which realizes Ax(0).

Henceforth, we shall write the implication "$\Gamma \to \Delta$" when we mean "$\Gamma \vdash_{\sigma} \Delta$". Moreover, "$\Gamma_1 \to \Gamma_2 \to \cdots \to \Gamma_n$" will stand for "$(\Gamma_1 \to \Gamma_2)$ & $(\Gamma_2 \to \Gamma_3)$ & $\cdots$ & $(\Gamma_{n-1} \to \Gamma_n)$". If Ax_1 and Ax_2 are any axioms considered in the preceding paragraph, then "$\mathrm{Ax}_1 \overset{*}{\to} \mathrm{Ax}_2$" will mean the following: "For any set X for which Ax_1 guarantees the existence of a choice function and for any such choice function f, f also realizes Ax_2 for X." Similar remarks apply with "$\leftrightarrow$" replacing "$\to$" and "$\overset{*}{\leftrightarrow}$" replacing "$\overset{*}{\to}$".

3. **On the interdependence of the axioms** FS_m **and** $C(m)$. We first define the multiple choice axioms FS_m introduced by M. N. Bleicher in **[2]** in connection with properties of vector spaces, and further studied in **[3]**.

DEFINITION 4. If $m \in I_0$, FS_m will denote the following axiom: "For every nontrivial set X there exists a function f such that for every $A \in X, f(A) \in \mathscr{P}^{\#*}(A)$ and $(\mathbf{n}(f(A)), m) = 1$." (Here, (r, s) is to be 0 if $s = 0$.)

In particular, FS_0 is equivalent to the full axiom of choice (AC) and FS_1 is the statement which asserts that for each $A \in X, f(A) \in \mathscr{P}^{\#*}(A)$.

If $Z = \{m_1, m_2, \cdots, m_k\} \in \mathscr{P}^{\#*}(I_1)$, FS_Z will denote FS_{m_1} & FS_{m_2} & $\cdots$ & FS_{m_k}.

Any function f which is a choice function for a set X, obviously, realizes $FS_m(X)$ for each $m \in I_1$; in view of Theorems 10 and 13 of **[3]**, each FS_m is a (WAC), as is $(\forall m)FS_m$.

In **[15]**, Mostowski introduced the *axioms of choice for finite sets*; these have also been studied in **[3]**, **[12]**, **[17]**, **[18]**, **[19]**, **[20]**, **[21]**, **[22]**, and **[23]**.

DEFINITION 5. Let $n \in I_1$ and let $C(n)$ be the following statement of set theory. "For every set X of n-element sets there is a function f defined on X such that for each $Y \in X, f(Y) \in Y$." Let $C(0) =$ "truth," and for $Z \in \mathscr{P}^{\#}(I_1)$ let $C(Z)$ be the conjunction of the statements $C(z), z \in Z$.

(Since positive integers are not subsets of I_1, no confusion will result from our writing $C(\{n\})$ as $C(n), n \in I_1$.)

Mostowski **[15**, Theorem IV**]** shows the following condition, (M), to be necessary for an implication of the form $C(Z) \to C(n)$, $Z \in \mathscr{P}^{\#}(I_1)$, $n \in I_1$.

DEFINITION 6. $Z(\in \mathscr{P}^{\#}(I_1))$ and $n(\in I_1)$ satisfy *Condition* (M) if for any decomposition of n into a sum of (not necessarily distinct) primes

$$n = p_1 + p_2 + \cdots + p_s,$$

there are nonnegative integers $r_1, r_2, \ldots, r_s$ such that

$$r_1p_1 + r_2p_2 + \cdots + r_sp_s \in Z.$$

The necessity of (M) for an implication $C(Z) \to C(n)$ has as a consequence that

(2) for each $n \in I_2$, $C(n)$ is a (WAC).

For letting Z be the empty set, it follows that Z and $n(\geq 2)$ fail to satisfy Condition (M). Thus $C(0) \nrightarrow C(n)$, and, consequently, Condition (iii) of Definition 3 holds for each such $C(n)$. (Of course, for $n = 1$ this is not the case, since $C(1)$ is, trivially, a theorem of σ.) Finally, for $n \in I_2$, let $p(n)$ be the smallest prime which is bigger than n. Then n and $p(n)$ fail to satisfy Condition (M); consequently $C(n) \nrightarrow C(p(n))$. Using Condition (i) of Definition 3, $C(n) \nrightarrow$ (AC); hence Condition (ii) of this definition holds.

Furthermore, using [**12**, Theorem 6], it follows that $(\forall n)C(n)$ is a (WAC).

In the theorem just cited, Lévy exhibited a model for $\sigma \cup \{FS_1;\ (\forall n \in I_2)C(n)\}$; a modification of Lévy's method will yield not only the independence of $(\forall n \in I_2)C(n)$ from $\sigma \cup \{FS_1\}$, but, in fact, the independence of $(\exists n \in I_2)C(n)$ from $\sigma \cup \{FS_1\}$, as well.

LEMMA 1. $(\forall n, k \in I_1)(C(nk) \to C(k))$.

(The proof of this lemma, which is attributed to A. Tarski, is given in [**17**, p. 99, Theorem 2].)

THEOREM 1. *If σ (or Gödel's system A, B, C, of* [9]) *is consistent, then $FS_1 \to (\exists n \in I_2)C(n)$ is unprovable in σ.*[2]

PROOF. Let σ^* be σ together with (AC) and an axiom which asserts the existence of denumerably many urelemente. Then σ^* is relatively consistent with σ;[3] we shall work within σ^*.

Let $\mathscr{M}$ be a denumerable set of urelemente. Define $\mathscr{M}_0$ to be $\mathscr{M}$ and for each ordinal number $\eta > 0$, let $\mathscr{M}_\eta = \mathscr{M} \cup \bigcup \{\mathscr{P}(\mathscr{M}_\mu) : \mu \in \eta\}$.

Let $\mathfrak{G}_0$ be the group of all one-one transformations of $\mathscr{M}$ into itself. By transfinite induction, if $x \in \mathscr{M}_\eta \setminus \bigcup \mathscr{M}_\xi$ $(\xi \in \eta)$ for some $\eta > 0$ and if $\phi \in \mathfrak{G}_0$, we "extend" ϕ by letting $\phi(x) = \{\phi(y) : y \in x\}$.

Let T be the subset of $I_0 \times I_0$ consisting of all ordered pairs $\langle i, t\rangle$, where i ranges over I_0 and where $t \in 2$ if i is even and $t \in 3$ if i is odd. Since T is denumerable, there is a one-one correspondence between T and $\mathscr{M}$. Choose any such one-one correspondence, and let $m_{i,t}$ be the member of $\mathscr{M}$ which corresponds to $\langle i, t\rangle$ under this mapping. For each $i_0 \in I_0$, let $\mathscr{M}^{(i_0)} = \{m_{i,t} : i = i_0\}$. Then $\mathscr{M}$ is the pairwise disjoint union of the $\mathscr{M}^{(i)}$, $i \in I_0$.

For $i \in I_0$, let $p_i = 2$ if i is even, and $p_i = 3$ if i is odd. Let χ_i be the element of $\mathfrak{G}_0$ which maps $m_{j,t}$ into itself for $j \neq i$, and which maps $m_{i,t}$ into $m_{i,s}$ for $s \in p_i$ and $s \equiv t + 1 (\operatorname{mod} p_i)$. Let $\mathfrak{G}_1$ be the subgroup of $\mathfrak{G}_0$ generated by $\{\chi_i : i \in I_0\}$. If $Z \in \mathscr{P}^{\#}(I_0)$ and if $\phi \in \mathfrak{G}_1$, then ϕ is said to be *Z-identical* if $\phi(m_{i,t}) = m_{i,t}$ for every pair $\langle i, t\rangle$ corresponding to $i \in Z$. If $x \in \mathscr{M}_\eta$ for some ordinal number η, then x is said to be *Z-symmetric* if $\phi(x) = x$ for every Z-identical $\phi \in \mathfrak{G}_1$.

For each ordinal number η, we define $\mathscr{K}_\eta$ by transfinite induction:

$$\mathscr{K}_0 = \mathscr{M} \cup \{0\}, \text{ for each ordinal number } \eta > 0, x \in \mathscr{K}_\eta \leftrightarrow$$
$$((\forall y \in x)(\exists \xi \in \eta)(y \in \mathscr{K}_\xi) \ \&\ (\exists Z \in \mathscr{P}^{\#}(I_0))\ (x \text{ is } Z\text{-symmetric})).$$

x is said to be an *$\mathscr{M}$-element* if there exists an ordinal number η such that $x \in \mathscr{K}_\eta$. A class X is called an *$\mathscr{M}$-class* if every element of X is an $\mathscr{M}$-element and if

---

[2] As is remarked in [**12**, p. 478], using Mendelson's method of [**13**], one can prove this theorem with the conclusion replaced by "$FS_1 \to (\exists n \in I_2)C(n)$ is unprovable in Gödel's system A, B, C".

[3] See the discussion in [**12**, pp. 478–479].

there is some $Z \in \mathscr{P}^{\#}(I_0)$ with the property that for every Z-identical $\phi \in \mathfrak{G}_1$, $\phi(y) \in X$ for every $y \in X$. If X and Y are classes, define $X \in_{\mathscr{M}} Y$ to be true iff X is an $\mathscr{M}$-element, Y is an $\mathscr{M}$-class and $X \in Y$. Clearly, if Y is an $\mathscr{M}$-element, then for any set X, $X \in Y$ iff (X is an $\mathscr{M}$-element and) $X \in_{\mathscr{M}} Y$. Then if we interpret σ in σ^* by replacing the primitive notions "element","class", "$\in$", and "0" by the notions "$\mathscr{M}$-element", "$\mathscr{M}$-class", "$\in_{\mathscr{M}}$", and "0", respectively, all of the axioms and theorems of σ will become theorems of σ^*.

In **[14]** and **[15]** one can find a discussion of various notions as they apply to Fraenkel-Mostowski models, as well as the verification that certain axioms are true in these models. (In the proof of Lemma 12 of **[15]**, we now make use of the fact that if $m_{i,t} \in \mathscr{M}_0$ and if $\phi \in \mathfrak{G}_1$, then $\phi(m_{i,t}) = m_{i,s}$ for some $s \in p_i$. Thus, it can be shown that if x is an $\mathscr{M}$-set which is Z-symmetric, then for any $\phi \in \mathfrak{G}_1$, $\phi(x)$ is also Z-symmetric.) The notion of finiteness is absolute with respect to our interpretation [**11**, Lemma 1]. If Π is the set of primes of σ^*, then Π is absolute because $\Pi(\subset \mathscr{K}_\omega)$ is 0-symmetric.

Under this interpretation, for each prime p, $C(p)$ is transformed into the statement: "For every $\mathscr{M}$-element x which is a set of p-element sets, there is an $\mathscr{M}$-element, f, which is a function having the property that for every $y \in x, f(y) \in x$." We must show that this sentence is refutable in the interpretation.

For each prime p, we consider the $\mathscr{K}_2$-element x_p defined as follows:

$$x_2 = \{\mathscr{M}^{(2i)} : i \in I_0\},$$
$$x_3 = \{\mathscr{M}^{(2i+1)} : i \in I_0\}.$$

For $p \in \Pi_5$, p can be written uniquely as a sum

(3) $$3 + 2k, \quad k \in I_1;$$

let $x_p = \{\mathscr{M}^{(2kj+1)} \cup \bigcup \mathscr{M}^{(2(kj+l))}(l \in k) : j \in I_0\}$.

With respect to any prime p, let $x = x_p$, let $k = (k(p))$ be as in (3), and suppose that f is a choice function for x_p. Then f must be Z-symmetric for some $Z \in \mathscr{P}^{\#}(I_0)$.

For $j \in I_0$ let R_j be the set

$$\{2(kj + l) : l \in k\} \cup \{2kj + 1\}.$$

Choose j_0 to be the smallest integer j for which $Z \cap R_j = 0$. Now for each $r \in R_{j_0}$, χ_r is Z-identical, and hence

$$\begin{aligned}\chi_{2kj_0}(f) &= \chi_{2kj_0+1}(f) = \chi_{2kj_0+2}(f) = \chi_{2kj_0+4}(f)\\ &= \chi_{2kj_0+6}(f) = \cdots = \chi_{2(kj_0+k-1)}(f)\\ &= f.\end{aligned}$$

Let $(x_p)_{j_0} = \mathscr{M}^{(2kj_0+1)} \cup \bigcup \mathscr{M}^{(2(kj_0+l))}(l \in k)$. Since $f((x_p)_{j_0}) \in (x_p)_{j_0}$ and the $\mathscr{M}^{(i)}$ are pairwise disjoint, it follows that $f((x_p)_{j_0}) \in \mathscr{M}^{(r)}$ for some unique $r \in R_{j_0}$, and, hence, that $f((x_p)_{j_0}) = \mathscr{M}_{r,t}$ for some $t \in p_r$. Thus

(4) $$\langle (x_p)_{j_0}, \mathscr{M}_{r,t} \rangle \in f$$

and, hence, $\langle (x_p)_{j_0}, \chi_r(\mathcal{M}_{r,t})\rangle = \chi_r(\langle (x_p)_{j_0}, \mathcal{M}_{r,t}\rangle) \in \chi_r(f) = f$. Now $\chi_r(\mathcal{M}_{r,t}) \neq \mathcal{M}_{r,t}$; therefore

$$\text{(5)} \qquad \langle (x_p)_{j_0}, \chi_r(\mathcal{M}_{r,t})\rangle \in f.$$

(4) and (5) together contradict the assumption that f is a function. Thus for each prime p, $C(p)$ is false in this model. Finally, by Lemma 1, for each $n \in I_0$, $C(n)$ is false in the model; hence $(\exists n \in I_0)(C(n))$ is false in the model.

The proof that FS_1 is true in the model is the same as in [**12**, pp. 480–482].

We now consider the relationship between $C(n)$ and $FS_m^{\#}$[4] for $m, n \in I_2$. Bleicher shows in [**3**] that if m and n have the same prime factors then $C(n) \to FS_m^{\#}$; he also shows that in this case $FS_m \overset{*}{\leftrightarrow} FS_n$, $FS_m^{\#} \overset{*}{\leftrightarrow} FS_n^{\#}$, and for any set X, $FS_m(X) \overset{*}{\leftrightarrow} FS_n(X)$. Finally, Theorem 13 of [**3**] states that if σ is consistent, then FS_m is not a theorem of σ for any $m \in I_2$. The question naturally arises as to the independence of the weaker axioms $FS_m^{\#}$ from the axioms of σ. For multiples of 2 or of 3 we obtain this independence as a corollary of the following:

THEOREM 2. (a) $C(2) \leftrightarrow C(4) \leftrightarrow (\forall k \in I_1)FS_{2^k}^{\#}$.
(a′) *For any set* X, $C(2)(\mathcal{P}_2(X)) \leftrightarrow (\forall k \in I_1)FS_{2^k}(\mathcal{P}^{\#*}(X))$.
(b) $C(3) \leftrightarrow (\forall k \in I_1)FS_{3^k}^{\#}$.
(b′) *For any set* X, $C(3)(\mathcal{P}_3(X)) \leftrightarrow (\forall k \in I_1)FS_{3^k}(\mathcal{P}^{\#*}(X))$.
(c) $C(6) \leftrightarrow (\forall k \in I_1)FS_{6^k}^{\#}$.

PROOF. In cases (a), (b), and (c), by the remarks preceding the theorem, we need only show that $FS_n^{\#} \to C(n)$ for $n = 2$, 3, and 6, and that for any set X and for $n = 2$ and 3, $FS_n(\mathcal{P}^{\#*}(X)) \to C(n)(\mathcal{P}_n(X))$.

The implication $FS_2^{\#} \to C(2)$ is immediate, since if X is any nonempty set of 2-element sets and if f realizes $FS_2(X)$, then f, obviously, generates a choice function F on X; $F(y)$ is defined to be the unique element of $f(y)$ for each $y \in X$.

We now assume $FS_3^{\#}$ is logically true and let X be any nonempty set of 3-element sets. If f realizes $FS_3^{\#}(X)$, then for $y \in X$, $f(y)$ must have either 1 or 2 elements. In the former case we let $g(y)$ be the unique element of $f(y)$; in the latter case we let $g(y)$ be the unique element of $y \setminus f(y)$. Then g is a choice function for X, and we have proved the implication $FS_3^{\#} \to C(3)$.

In order to obtain (a′) from (a) and (b′) from (b) we let $Y = \mathcal{P}^{\#*}(X)$, and proceed as in the proof of (a) (respectively (b)). For example, for $n = 2$, if f realizes $FS_2(Y)$, we define $F(y)$, for $y \in \mathcal{P}_2(X)$, to be the unique element of $f(y)$. The proof in the other direction uses the fact that n is prime, and follows from Theorem 3 of [**3**].

Finally, by the above remarks together with [**3**, Lemma 1] and [**17**, p. 101, Theorem 4], we have

$$FS_6^{\#} \overset{*}{\to} (FS_2^{\#} \ \& \ FS_3^{\#}) \to (C(2) \ \& \ C(3)) \to C(6).$$

COROLLARY 1. $FS_{kn}^{\#} \to C(n)$ *for* $n = 2, 3, 4$, *and* 6 *and* $k \in I_1$.

[4] $FS_m^{\#}$ is denoted by FS_m^{*} in [**3**].

COROLLARY 2. *If σ is consistent, then $FS^{\#}_{kn}$ is not a theorem of σ for $n = 2$ and 3 and for $k \in I_1$.*

PROOF. For each $n \in I_2$, $C(n)$ is independent of σ (by (2)).

Now for $n = 2$ and 3 and $k \in I_1$, $FS^{\#}_{kn} \overset{*}{\to} FS^{\#}_{n} \to C(n)$; hence if $FS^{\#}_{kn}$ were a theorem of σ, then $C(n)$ would be one, as well.

COROLLARY 3. $FS^{\#}_{6} \to C(\{2, 3, 4, 6, 8, 9\})$.

COROLLARY 4. *$FS^{\#}_{6}$ is independent of $C(2)$; $FS^{\#}_{6}$ is independent of $C(3)$.*

PROOF. Otherwise, $C(2) \to C(6)$ ($C(3) \to C(6)$); but $\{2\}$ and 6 ($\{3\}$ and 6) fail to satisfy Condition (M).

COROLLARY 5. (a) *$(\exists k \in I_1)FS^{\#}_{2^k}$ and $(\exists l \in I_1)FS^{\#}_{3^l}$ are independent of each other.*

(b) *$(\exists k \in I_1)FS^{\#}_{6^k}$ is independent of $(\exists l \in I_1)FS^{\#}_{2^l}$.*

(c) *$(\exists k \in I_1)FS^{\#}_{6^k}$ is independent of $(\exists l \in I_1)FS^{\#}_{3^l}$.*

PROOF. $(\exists k \in I_1)FS^{\#}_{2^k} \overset{*}{\to} FS^{\#}_{2m}$, some $m \in I_1$; moreover, $FS^{\#}_{2m} \overset{*}{\to} FS^{\#}_{2} \to C(2)$. Now $C(3)$ is independent of $C(2)$ (by [**15**, Theorem VII]), and $C(3) \leftrightarrow (\exists l \in I_1)FS^{\#}_{3^l}$. Consequently, $(\exists l \in I_1)FS^{\#}_{3^l}$ is independent of $(\exists k \in I_1)FS^{\#}_{2^k}$.

The other independence arguments proceed along similar lines.

We have shown that $FS^{\#}_{n} \to C(n)$ for $n = 4$ and 6. We shall now show that this is not the case for composites bigger than 6.

LEMMA 2. *For any $Z \in \mathscr{P}^{\#*}(I_1)$ and for any $n \in I_1$, Condition (M) is necessary for the implication $FS^{\#}_{Z} \to C(n)$.*

PROOF. (By Theorem 1, we can confine our attention to $Z \in \mathscr{P}^{\#*}(I_2)$.)

(6) Suppose Z and n fail to satisfy (M);

by the necessity of (M) for an implication $C(Z) \to C(n)$, there is a model $\mathscr{M}$ for $\sigma \cup \{C(Z), \sim C(n)\}$. Since $(\forall m \in I_2)(C(m) \to FS^{\#}_{m})$ is a theorem of set theory, $\mathscr{M}$ is also a model for $\sigma \cup \{FS^{\#}_{Z}, \sim C(n)\}$. Thus the Supposition (6) assures that $C(n)$ cannot be proved from $\sigma \cup \{FS^{\#}_{Z}\}$.

For $n \in I_2$, we shall let $\mathbf{P}(n)$ denote the set of prime factors of n.

THEOREM 3. *Let n be a composite bigger than 6. Then $C(n)$ is independent of $FS^{\#}_{n}$.*

PROOF. Let n be a composite >6. We have to show that $C(n)$ is independent of $FS^{\#}_{\mathbf{P}(n)}$.

Theorem 5 of [**20**] asserts that for such n, $C(n)$ is independent of $C(\mathbf{P}(n))$. Since $C(\mathbf{P}(n)) \to FS^{\#}_{\mathbf{P}(n)}$, $FS^{\#}_{\mathbf{P}(n)}$ cannot imply $C(n)$. (In fact, Theorem 5 of [**20**] is proved by showing that for such n, $\mathbf{P}(n)$ together with n fail to satisfy Condition (M). Thus our theorem follows directly from Lemma 2.)

Theorem 3 proves a conjecture of Bleicher [**3**, p. 251] that $FS^{\#}_{Z}$ is, in general, weaker than $C(Z)$.

We have already utilized the implication

$$(7) \qquad FS^{\#}_{kn} \to FS^{\#}_{n}, \qquad k, n \in I_0,$$

which follows directly from the definition of $FS^{\#}_{n}$; however, the converse of (7) is false.

COROLLARY 1. *For $n = 2$ and for $k \in I_4$, or for $n = 3$ and for $k \in I_3$, or for $n \in I_4$ and for $k \in I_2$, the following two statements are valid:*
(i) *$FS^{\#}_{kn}$ is independent of $C(n)$.*
(ii) *$FS^{\#}_{kn}$ is independent of $FS^{\#}_{n}$.*

PROOF. (i) for k and n as indicated, if $C(n) \to FS^{\#}_{kn}$, then since $C(kn) \to C(n)$, we would have $C(kn) \to FS^{\#}_{kn}$, contradicting the theorem.
(ii) Similarly, if $FS^{\#}_{n} \to FS^{\#}_{kn}$, then, since $C(n) \to FS^{\#}_{n}$, we would have a contradiction of (i).

COROLLARY 2. *If n is divisible by 2 or 3, then $FS^{\#}_{n}$ is a* (WAC).

PROOF. For such n, surely, (AC) $\to FS^{\#}_{n}$; moreover, if $FS^{\#}_{n} \to$ (AC), then $FS^{\#}_{n} \to FS^{\#}_{kn}$ for all $k \in I_1$, contradicting the preceding corollary. Finally, Corollary 2 of Theorem 2 guarantees the independence of $FS^{\#}_{n}$ from σ.

COROLLARY 3. $(\forall n \in I_1) FS^{\#}_{n}$ *is a* (WAC).

PROOF. This follows from Corollary 2 together with [3, Theorem 10].
Theorem 2, Corollary 4 of Theorem 2, and Corollary 1 of Theorem 3 completely settle the status of the implications

$$(8) \qquad C(2) \to FS^{\#}_{2k}$$

and

$$(9) \qquad C(3) \to FS^{\#}_{3k}$$

for $k \in I_1$. Moreover, the proof of Theorem 6 of [12] also shows that if σ (or Gödel's system A, B, C) is consistent, then $(FS_1 \;\&\; (\forall n)C(n)) \to FS^{\#}_{0}$ is unprovable in σ (or in Gödel's system A, B, C); the counterexample given in this proof is, in fact, a set of finite sets. It follows that $FS^{\#}_{0}$ is independent of $C(2)$ and of $C(3)$ and, thus, the status of the implications (8) and (9) are determined for all $k \in I_0$.
REMARK. Our methods do not enable us to ascertain the status of the implications $FS^{\#}_{p} \to C(p)$ for primes $p \geq 5$. However, certain weaker implications follow from a theorem of Bleicher.
DEFINITION 7. $Z(\in \mathscr{P}^{\#}(I_1)$ and $n(\in I_1)$ satisfy *Condition* (S) if for any decomposition of n into a sum of (not necessarily distinct) primes,

$$n = p_1 + p_2 + \cdots + p_s,$$

there is an $r \in I_1$ such that for some $i \in J_s$, $rp_i \in Z$.
Theorem 7* of [3] asserts that Condition (S) is sufficient for the implication

$FS_Z \to C(n)$; in fact, Theorem 7 (*ibid.*) clearly indicates that

(10) (S) is sufficient for the (stronger) implication $FS^{\#}_Z \to C(n)$.

It follows from (10) that the following implications are valid:

(a) $(FS^{\#}_{\{2,5\}} \vee FS^{\#}_{\{3,5\}}) \to C(5)$.

(a′) For any set X, $(FS_{\{2,5\}}(\mathscr{P}^{\#*}(X)) \vee FS_{\{3,5\}}(\mathscr{P}^{\#*}(X))) \to C(5)(\mathscr{P}_5(X))$.

(b) $FS^{\#}_{\{2,7\}} \to C(7)$.

(b′) For any set X, $FS_{\{2,7\}}(\mathscr{P}^{\#*}(X)) \to C(7)(\mathscr{P}_7(X))$.

(c) For any $p \in \Pi_{11}$ if p' is the largest prime $\leq (p-2)/3$, and if $P_p = \{p\} \cup (\Pi \cap p'+1)$, then $FS^{\#}_{P_p} \to C(p)$.

(c′) Under the hypothesis of (c), for any set X,

$$FS_p(\mathscr{P}^{\#*}(X)) \to C(p)(\mathscr{P}_p(X)).$$

4. **The axioms** LC_n. Henceforth, unless otherwise specified, for each $n \in I_2$, under consideration, we shall let

$$n = p_1^{e_1} p_2^{e_2} \cdots p_s^{e_s}, \qquad p_1 < p_2 < \cdots < p_s,$$

be the unique prime factorization of n. Thus $\mathbf{P}(n) = \{p_1, p_2, \ldots, p_s\}$.

DEFINITION 8. Let $n \in I_1$ and let $m_1, m_2, \ldots, m_n$ be distinct integers of I_2. Define *Lin Comb*$(m_1, m_2, \ldots, m_n)$ to be the set of all integers of the form $z_1 m_1 + z_2 m_2 + \cdots + z_n m_n$, where each $z_i \in I_1$, for $i \in J_n$.

We shall write Lin Comb(M) for Lin Comb $(m_1, m_2, \ldots, m_n)$ when $M = \{m_1, m_2, \ldots, m_n\} \in \mathscr{P}^{\#*}(I_2)$.

DEFINITION 9. For $n \in I_2$, let LC_n be the statement: "For every nontrivial set X there is a function f defined on X with the property that if $A \in X$ and if $p_1 + p_2 + \cdots + p_s \leq A$, then $f(A)$ is a finite subset of A such that $\mathbf{n}(f(A)) \in$ Lin Comb$(\mathbf{P}(n))$; if $A \in X$ and if $A \prec p_1 + p_2 + \cdots + p_s$, then $f(A) = A$".

Let $LC_1 = FS_1$.

Let X be any set. It is immediate that

(11) if m and n have the same prime factors, then
$$LC_m \overset{*}{\leftrightarrow} LC_n, \qquad LC^{\#}_m \overset{*}{\leftrightarrow} LC^{\#}_n, \quad \text{and} \quad LC_m(X) \overset{*}{\leftrightarrow} LC_n(X).$$

In particular, any statement about LC_p (respectively $LC^{\#}_p$, $LC_p(X)$), $p \in \Pi$, also applies to LC_m (respectively, $LC^{\#}_m$, $LC_m(X)$) if m is a power of p. For the most part, we shall omit this generalization in stating our results. Furthermore,

(12) $$(\exists n \in I_1) LC_n \overset{*}{\to} LC_1; \qquad (\exists n \in I_1) LC_n(X) \overset{*}{\to} LC_1(X)$$

are, obviously, valid.

THEOREM 4. *If σ is consistent, then for each $n \in I_1$, LC_n is independent of σ.*

PROOF. By (12), $LC_n \overset{*}{\to} LC_1$ for each $n \in I_1$. In [**3**, Theorem 13], Bleicher shows that $FS_1 (= LC_1)$ is independent of σ.

THEOREM 5. *Let $n \in I_2$. If $q \in \Pi \cap$ Lin Comb$(\mathbf{P}(n))$, then $LC_q(X) \overset{*}{\rightarrow} LC_n(X)$ for any nontrivial set X, of sets A none of which satisfies $p_1 + p_2 + \cdots + p_s = m \prec A \prec q$. In particular, if $m \in \Pi$, then $LC_m \overset{*}{\rightarrow} LC_n$ and $LC_m^{\#} \overset{*}{\rightarrow} LC_n^{\#}$.*

THEOREM 6. *Let $n \in I_2$ and let X be any infinite set. Then $LC_{\mathbf{P}(n)}(\mathscr{P}^{\infty}(X)) \rightarrow LC_n(\mathscr{P}^{\infty}(X))$.*

PROOF. Let f_1 realize $LC_{p_i}(\mathscr{P}^{\infty}(X))$ for $i \in J_s$ and let A_1 be any infinite subset of X. For $j \in K_s$ let $A_j = A_{j-1} \setminus f_{j-1}(A_{j-1})$. Then $B = \bigcup f_j(A_j)$ $(j \in J_s)$ is a finite subset of A_1 with the property that $\mathbf{n}(B) \in$ Lin Comb(P).

DEFINITION 10. Q is a *prime set for* n $(\in I_2)$ if $Q \in \mathscr{P}^{\#*}(\Pi)$ and $n \in$ Lin Comb(Q).

THEOREM 7. *Let $p \in \Pi$. Let X be a nonempty set of sets A satisfying $p \preceq A$, and suppose that $LC_p(X)$ is provable. Then there is a partition $\{X_z\}$ $(z \in Z)$ of X for some $Z \in \mathscr{P}^*(I_1)$ such that for each $z \in Z \cap I_2$, if Q is any prime set for z and if m is any integer whose set of prime factors equals Q, then $LC_m(X_z)$ is provable.*

PROOF. By $LC_p(X)$ and the restriction on the elements A of X, there is a function f defined on X such that for each $A \in X$, $\mathbf{n}(f(A)) = zp$ for some $z \in I_1$. Let Z be the set of integers z for which there is an $A \in X$ satisfying $\mathbf{n}(f(A)) = zp$; for $z \in Z$ let X_z be the subset of X consisting of those sets A such that $\mathbf{n}(f(A)) = zp$. We claim that $\{X_z\}$ $(z \in Z)$ is the required partition of X.

Let $z \in I_2$, let $\{q_1, q_2, \ldots, q_r\}$ be any prime set for z, and let $m = q_1^{e_1} q_2^{e_2} \cdots q_r^{e_r}$, $e_i \in I_1$, $i \in J_r$. Then for each $A \in X_z$, $\mathbf{n}(f(A)) = pz = pl_1q_1 + pl_2q_2 + \cdots + pl_rq_r$, where each $l_i \in I_1$. Thus f realizes $LC_m(X_z)$.

COROLLARY. *If $p = 2$, X can be any nontrivial set in the above theorem.*

There has been considerable interest in the investigation of the sets Lin Comb(M) corresponding to arbitrary $M \in \mathscr{P}^{\#*}(I_2)$. In particular, for sets $M = \{m_1, m_2, \ldots, m_s\}$ containing a pair of relatively prime integers let $B(m_1, m_2, \ldots, m_s)$ be the largest integer which is not in Lin Comb(M). Then $B(m_1, m_2, \ldots, m_s)$ has been determined for such sets M in which s is two or three as well as for various other sets M (see **[1]**, **[4]**, **[5]**, **[6]**, **[7]**, **[10]**, and **[16]**). For example, it is well known that if $(m_1, m_2) = 1$, then $B(m_1, m_2) = m_1m_2 - m_1 - m_2$. If $(m_1, m_2, \ldots, m_s) = 1$ and if $m_j = m_1 + (j-1)d$, $j \in J_s$, then

$$B(m_1, m_2, \ldots, m_s) = \left(\left[\frac{m_1 - 2}{s+1}\right] + 1\right)m_1 + (d-1)(m_1 - 1) - 1,$$

where $[x]$ denotes the greatest integer $\leq x$.

THEOREM 8. *If $m \in I_2$ is such that the sum of its prime factors is greater than $B(p_1, p_2, \ldots, p_s)$, then for any set X of sets A none of which satisfies $p_1 + p_2 + \cdots + p_s \prec A \preceq B(p_1, p_2, \ldots, p_s)$, we have $LC_m(X) \overset{*}{\rightarrow} LC_n(X)$.*

5. **The relationship of the LC_n to the previous axioms.** We first note that if, instead of the axioms $C(n)$, we consider the axioms $C(n)(X)$ for a particular nonempty set X of n-element sets, then the various implicational results for the $C(n)$,

mentioned in the literature, do not necessarily hold. For example, Lemma 1 asserts that $C(nk) \to C(k)$, $n, k \in I_1$, but it is by no means obvious whether or not $C(nk)(X) \to C(k)(X)$ for any arbitrary set X. Recall that $X_{(r)}$ was defined to be $\bigcup\{\mathscr{P}_r(Y): Y \in X\}$, $r \in I_1$. Then Tarski's proof of Lemma 1 actually shows the following to be true:

If X is a nonempty set of n-element sets and if $X' = \{x \times k : x \in X, k \in I_1\}$, then

$$(\forall k \in I_1)(C(kn)(X'_{(kn)}) \to C(n)(X)).$$

DEFINITION 11. Let (DAC) be the following statement: "For every denumerable set X of nonempty sets there is a function f on X such that for each $A \in X$, $f(A) \in A$."

THEOREM 9. *Let $p \in \Pi$ and let X be any set which consists of sets A satisfying $p \preceq A$.*[5] *Then*

$$((\text{DAC}) \,\&\, LC_p(X) \,\&\, (\forall j \in I_1)C(jp)(X_{(jp)})) \to FS_0(X).$$

PROOF. The conclusion is trivial if X is finite; assume that X is infinite.

Assume the hypothesis is valid and let f realize $LC_p(X)$. For $j \in I_1$, let Y_j be the set of all functions which realize $C(jp)(X_{(jp)})$, and let $Y = \{Y_j : j \in I_1\}$. Then Y is a denumerable set of nonempty sets; using (DAC), let $\{f_j : j \in I_1\}$ be a set consisting of a single element from each Y_j.

Now, as in Theorem 7, f induces a partition $\{X_j\}_{j \in J}$, $J \subseteq I_1$, on X; in fact, for $j \in J$, X_j is the subset of X consisting of those sets A such that $\mathbf{n}(f(A)) = jp$. Define the choice function g on X as follows: for $A \in X$, $g(A)$ is the unique element of $f_j(f(A))$ if $A \in X_j$.

THEOREM 10. *Let p and q be primes with $p < q$.*

(a) *Let X be any set consisting of nonempty finite sets A for which $\mathbf{n}(A) < pq$. Then $LC_p(X) \overset{*}{\to} FS_q(X)$.*

(b) *Suppose $p \not| \, n$. Then for any set Y, $LC_p(\mathscr{P}^{\#*}(Y)) \to FS_p(\mathscr{P}_n(Y))$.*

PROOF. (a) If $X = 0$, the conclusion follows from (1). Otherwise, let f realize $LC_p(X)$, and let $A \in X$. By the hypothesis, A is finite and nonempty. If $\mathbf{n}(A) \leq p$, then $f(A) = A$ and $(\mathbf{n}(f(A)), q) = 1$. If $\mathbf{n}(A) > p$, then $f(A)$ is a subset of A and $\mathbf{n}(f(A)) = kp$, where $k \in J_q$. Again, $(\mathbf{n}(f(A)), q) = 1$.

(b) If $n < p$, then we use the identity function on $\mathscr{P}_n(Y)$; thus we may assume $p < n$. If $Y \prec n$, we, again, use (1). If $n \prec Y$, let $A \in \mathscr{P}_n(Y)$, and let f realize $LC_p(\mathscr{P}^{\#*}(Y))$. We consider the sets A, $A_1 = A \setminus f(A)$, $A_2 = A_1 \setminus f(A_1)$, $A_3 = A_2 \setminus f(A_2)$, etc. In a finite number of steps we obtain a nonempty subset B of A for which $\mathbf{n}(B) < p$.

We now utilize the following axioms which were introduced by A. Lévy in [**12**].

DEFINITION 12. Let $n \in I_1$ and let $Z(n)$ be the statement: "On every nontrivial set X there exists a function f such that for every $A \in X$, $f(A) \in \mathscr{P}_j(A)$ for some $j \in J_n$."

[5] The corollary of Theorem 7, again, applies.

THEOREM 11. (a) $(\exists p \in \Pi)(FS_p^{\#} \ \& \ LC_p) \leftrightarrow (\exists p \in \Pi)(FS_p \ \& \ LC_p^{\#}) \leftrightarrow FS_0$. *Any of these implies* $(\forall p \in \Pi)(FS_p \ \& \ LC_p)$.

(b) *For any set* X,

$$(\exists p \in \Pi)(FS_p(\mathscr{P}^{\#*}(\bigcup X)) \ \& \ LC_p(\mathscr{P}^*(\bigcup X))) \leftrightarrow$$
$$(\exists p \in \Pi)(FS_p(\mathscr{P}^*(\bigcup X)) \ \& \ LC_p(\mathscr{P}^{\#*}(\bigcup X))) \leftrightarrow FS_0(X).$$

(c) $(\exists p \in \Pi)(FS_p^{\#} \ \& \ LC_p^{\#}) \leftrightarrow FS_0^{\#}$.

PROOF (of (a) and (*b*)).

(1) $(\exists p \in \Pi)(FS_p^{\#} \ \& \ LC_p) \rightarrow FS_0$:

Let X be a nontrivial set;

$$\text{let } f_1 \text{ realize } LC_p(\mathscr{P}^*(\bigcup X)). \tag{13}$$

Then for each $A \in X$, $f_1(A)$ is finite and nonempty; moreover,

$$\begin{aligned} \mathbf{n}(f_1(A)) &= k_1 p \text{ for some } k_1 \in I_1, && \text{if } p \preceq A, \\ &= \mathbf{n}(A), && \text{if } A \prec p. \end{aligned}$$

$$\text{Let } f_2 \text{ realize } FS_p(\mathscr{P}^{\#*}(\bigcup X)); \tag{14}$$

then for each $A \in X$,

$$\mathbf{n}(f_2 f_1(A)) = k_2 p + q, \qquad k_2 \in k_1, \ q \in J_{p-1}.$$

Similarly, for each $A \in X$,

$$\begin{aligned} \mathbf{n}(f_1 f_2 f_1(A)) &= k_3 p, \quad k_3 \in J_{k_2}, && \text{if } \mathbf{n}(f_2 f_1(A)) > p, \\ &= \mathbf{n}(f_2 f_1(A)), && \text{if } \mathbf{n}(f_2 f_1(A)) < p. \end{aligned}$$

Continue in this manner, and for each $k \in I_1$, let $g_{2k-1} = f_1 f_2 f_1 \cdots f_1 f_2 f_1$ and $g_{2k} = f_2 f_1 f_2 f_1 \cdots f_1 f_2 f_1$, where in each case the composite function is to contain k factors f_1.

Now, for each $A \in X$ there is a smallest integer n_A such that $g_n(A) = g_{n_A}(A)$ for all $n \in I_{n_A}$. (Actually, the integers n_A also depend on the particular functions f_1 and f_2 employed in (13) and (14).)

Let g be the function with domain X defined by

$$g(A) = g_{n_A}(A), \qquad A \in X.$$

Clearly, $\mathbf{n}(g(A)) \in J_{p-1}$ for each $A \in X$; since X is an arbitrary nontrivial set, $Z(p-1)$ is provable. Moreover, Theorem 2 of [**12**] asserts that $(\exists n)Z(n) \rightarrow FS_0$; hence

$$(\exists p \in \Pi)(FS_p^{\#} \ \& \ LC_p) \rightarrow Z(p-1) \rightarrow FS_0.$$

(2) $(\exists p \in \Pi)(FS_p \ \& \ LC_p^{\#}) \rightarrow FS_0$:

Here we begin with h_1 which realizes $FS_p(\mathscr{P}^*(\bigcup X))$, and compose it with h_2

which realizes $LC_p(\mathscr{P}^{\#*}(\bigcup X))$. Then for each $A \in X$, $h_1(A)$ is finite and nonempty; moreover,

$$\begin{aligned} \mathbf{n}(h_2h_1(A)) &= l_2p,\ l_2 \in I_1, \quad \text{if } \mathbf{n}(h_1(A)) > p, \\ &= \mathbf{n}(h_1(A)), \quad \text{if } \mathbf{n}(h_1(A)) < p. \end{aligned}$$

We continue along the same lines as in part (1).

(3) $FS_0 \to FS_p$ for all $p \in \Pi$, is immediate.

(4) $FS_0 \to LC_p$ for all $p \in \Pi$:

Let $p \in \Pi$, let X be an arbitrary nontrivial set, let f be a choice function on $\mathscr{P}^*(\bigcup X)$, and let $A \in X$. Define F on X as follows:

If A is finite and if either $\mathbf{n}(A) < p$ or $\mathbf{n}(A) = kp$, $k \in I_1$, let $F(A) = A$. Otherwise, let $A_1 = A \setminus \{f(A)\}$ and let $A_i = A_{i-1} \setminus \{f(A_{i-1})\}$ for $i \in K_{p-1}$. Let $F(A) = \{f(A), f(A_1), \ldots, f(A_{p-1})\}$.

The remaining implications result from the implications $FS_p \to FS_p^{\#}$ and $LC_p \to LC_p^{\#}$, $p \in \Pi$, together with elementary properties of the predicate calculus.

COROLLARY. *$(\exists p \in \Pi)(\exists k, l \in I_1)(FS^{\#}_{p^k}$ & $LC_{p^l}) \leftrightarrow (\exists p \in \Pi)(\exists k, l \in I_1)(FS_{p^k}$ & $LC^{\#}_{p^l}) \leftrightarrow FS_0$. Any of these implies $(\forall p \in \Pi)(\forall k, l \in I_1)(FS_{p^k}$ & $LC_{p^l})$. Moreover, generalizations of this sort can also be made corresponding to* (b) *and* (c) *of the theorem.*

THEOREM 15. (a) *Let $p \in \Pi$. Then*

$$LC_p^{\#} \to (\forall k \in I_0)C(kp + 1).$$

(b) *Let $p \in \Pi$ and let $r \in J_{p-1}$. Then*

$$(LC_p^{\#} \ \&\ C(r)) \to (\forall k \in I_0)C(kp + r).$$

(c) *Let $n \in I_2$. Then*

$$LC_n^{\#} \to C(p_1 + p_2 + \cdots + p_s + 1).$$

(Recall that $\{p_1, p_2, \ldots, p_s\}$ is the set of prime factors of n.)

(d) *If $n \in I_2$ and if $r \in J_{p_1-1}$, then*

$$(LC_n^{\#} \ \&\ C(r)) \to C(p_1 + p_2 + \cdots + p_s + r).$$

(e) *If $m \in I_2$ and if $q_1, q_2, \ldots, q_m$ are primes such that for each $k \in K_m$, $q_1 + q_2 + \cdots + q_{k-1} < q_k$, then*

$$LC^{\#}_{\{q_1, q_2, \ldots, q_m\}} \to C(q_1 + q_2 + \cdots + q_m + 1).$$

(f) *Under the assumptions of part* (e), *if $r \in J_{q_1-1}$, then*

$$(LC^{\#}_{\{q_1, q_2, \ldots, q_m\}} \ \&\ C(r)) \to C(q_1 + q_2 + \cdots + q_m + r).$$

PROOF. (a) If $k_0 \in I_1$ and if X_0 is a nonempty set of $(k_0p + 1)$-element sets, let f_0 realize $LC_p^{\#}(X_0)$. Let X_1 consist of all sets A_1 of the form $A_0 \setminus f(A_0)$, $A_0 \in X_0$. Then for each $A_1 \in X_1$, $\mathbf{n}(A_1) = k_1p + 1$, $k_1 \in k_0$. If $k_0 = 1$, then $k_1 = 0$, and we are finished; otherwise, let f_1 realize $LC_p^{\#}(X_1)$.

Similarly, for $i \in K_{k_0}$, let X_i consist of all sets A_i of the form $A_{i-1} \setminus f_{i-1}(A_{i-1})$, $A_{i-1} \in X_{i-1}$. Then for each $A_i \in X_i$, $\mathbf{n}(A_i) = k_i p + 1$, $k_i \in k_0 - i + 1$. Let f_i realize $LC_p^{\#}(X_i)$.

It follows that if $g(A_0)$ is the unique element of $f_{k_0} \cdot f_{k_0-1} \cdots f_2 \cdot f_1(A_0)$ for $A_0 \in X_0$, then g is a choice function for X_0.

(b) The proof here is similar to that of part (a) since X_{k_0} will be a set of r-element sets.

(c) Suppose $LC_n^{\#}$ and let X be a nonempty set consisting of $(p_1 + p_2 + \cdots + p_s + 1)$-element sets. If f realizes $LC_n^{\#}(X)$ for each $A \in X$, then $\mathbf{n}(f(A)) = p_1 + p_2 + \cdots + p_s$, and we let $g(A)$ be the unique element of $A \setminus f(A)$.

(e) If X_m ($\neq 0$) consists of $(q_1 + q_2 + \cdots + q_m + 1)$-element sets, let f_m realize $LC_{q_m}(X_m)$. Then for $A_m \in X_m$, $\mathbf{n}(A_m \setminus f_m(A_m)) = q_1 + q_2 + \cdots + q_{m-1} + 1$, because $\mathbf{n}(A_m) < 2q_m$.

Let X_{m-1} consist of sets $A_m \setminus f_m(A_m)$, $A_m \in X_m$, and let f_{m-1} realize $LC_{q_{m-1}}^{\#}(X_{m-1})$. Then for $A_{m-1} \in X_{m-1}$, $\mathbf{n}(A_{m-1} \setminus f_{m-1}(A_{m-1})) = q_1 + q_2 + \cdots + q_{m-2} + 1$.

Continuing in this fashion, we obtain, after m steps, a set X_1 and a function f_1 on X_1 such that $\mathbf{n}(f(A_1)) = 1$ for all $A_1 \in X_1$.

Setting $g(A_m)$ equal to the unique element of $f_1 \cdot f_2 \cdots f_m(A_m)$, $A_m \in X_m$, we obtain a choice function for X_m.

(d) follows from (c) and (f) from (e) as did (b) from (a).

COROLLARY. *For each $n \in I_2$, $LC_n^{\#}$ is independent of σ.*

PROOF. If for some $n \in I_2$, $LC_n^{\#}$ were a theorem of σ, then by part (c) of Theorem 12, $C(p_1 + p_2 + \cdots + p_s + 1)$ would, likewise, be a theorem of σ. This, clearly, contradicts (2).

THEOREM 12*. (a) *Let $p \in \Pi$, let $k \in I_0$, and let X be any set of $(kp + 1)$-element sets. Then*

$$(\forall j \in J_k) LC_p(X_{(jp+1)}) \to C(kp + 1)(X).$$

(b) *Let $p \in \Pi$, let $r \in J_{p-1}$, let $k \in I_0$, and let X be any set of $(kp + r)$-element sets. Then*

$$((\forall j \in J_k) LC_p(X_{(jp+1)}) \;\&\; C(r)(X_{(r)})) \to C(kp + r)(X).$$

(c) *Let $n \in I_2$ and let X be any set of $(p_1 + p_2 + \cdots + p_s + 1)$-element sets. Then*

$$LC_n(X) \to C(p_1 + p_2 + \cdots + p_s + 1)(X).$$

(d) *Let $n \in I_2$, let $r \in J_{p_1-1}$, and let X be any set of $(p_1 + p_2 + \cdots + p_s + r)$-element sets. Then*

$$(LC_n(X) \;\&\; C(r)(X_{(r)})) \to C(p_1 + p_2 + \cdots + p_s + 1)(X).$$

(e) *Let $m \in I_2$. Let $q_1, q_2, \ldots, q_m$ be primes such that for each $k \in K_m$, $q_1 + q_2 + \cdots + q_{k-1} < q_k$. For $j \in J_m$ let $s_j = q_1 + q_2 + \cdots + q_j + 1$. Let X be any* *set of* $(q_1 + q_2 + \cdots + q_m + 1)$*-element sets. Then*

$$(\forall j \in J_m) LC_{q_j}(X_{(s_j)}) \to C(s_m)(X).$$

(f) *Under the assumptions of* (e), *if* $r \in J_{q_1-1}$ *then*

$$((\forall j \in J_m) LC_{q_j}(X_{(s_j)}) \;\&\; C(r)(X_{(r)})) \to C(s_m)(X).$$

REMARK. (a) implies the following:

(a') Let $p \in \Pi$, let $k \in I_0$, and let Y be any set. Then

$$(\forall j \in J_k) LC_p(\mathscr{P}_{jp+1}(Y)) \to C(kp + 1)(\mathscr{P}_{kp+1}(Y)).$$

Similar implications follow from the other parts of the theorem.

LEMMA 3. *For each* $p \in \Pi$ *and* $k \in I_1$, $C(p + 1)$ *is independent of* $C(p^k)$.

PROOF. This follows from the necessity of Condition (M) for the implication $C(Z) \to C(n)$; in fact, no prime factor of $p + 1$ divides p^k.

THEOREM 13. (a) *Let* $m, n \in I_2$. *A sufficient condition for the independence of* $LC_n^{\#}$ *from* $C(m)$ *is that there exists an* $r \in J_{p_1-1}$ *for which* $C(p_1 + p_2 + \cdots + p_s + r)$ *is independent of* $C(r)$ & $C(m)$.

(b) *Any of the following conditions is sufficient for the independence of* $LC_n^{\#}$ *from* $C(n)$:

(i) n *is a prime power;*

(ii) *there exist* $q \in \Pi$ *and* $k \in I_1$ *such that*

$$p_1 + p_2 + \cdots + p_s < q^k < 2p_1 + p_2 + \cdots + p_s$$

and q *does not divide* $p_1 + p_2 + \cdots + p_s$;

(iii) (*special case of* (ii)) $p_1 + p_2 + \cdots + p_s + 1 \in \Pi$.

PROOF. (a) Suppose the stated condition is logically true and suppose that $C(m) \to LC_n^{\#}$. Now, by Theorem 12(d),

$$(LC_n^{\#} \;\&\; C(n)) \to C(p_1 + p_2 + \cdots + p_s + r);$$

hence $(C(m) \;\&\; C(r)) \to C(p_1 + p_2 + \cdots + p_s + r)$. Contradiction!

(b) follows from (a) together with Lemma 3. In (ii) we let $Z = \{p_1 + p_2 + \cdots + p_s, q^k - (p_1 + p_2 + \cdots + p_s)\}$; then Z does not contain a multiple of q.

6. The axioms PP_n. We would like to comment, briefly, on a variant of the axioms LC_n.

DEFINITION 13. For $n \in I_2$ let PP_n be the statement: "If X is a nontrivial set, then there is a function f defined on X such that if $A \in X$ and if $p_1 + p_2 + \cdots + p_s \preceq A$, then for positive integers $k_1, k_2, \ldots, k_s$, $f(A)$ is a $(p_1^{k_1} + p_2^{k_2} + \cdots + p_s^{k_s})$-element subset of A; if $A \in X$ and if $A \prec p_1 + p_2 + \cdots + p_s$, then $f(A) = A$.

Let $PP_1 = FS_1$.

Obviously,

(15) for all $p \in \Pi$ and for all sets X—

$$PP_p \overset{*}{\to} LC_p,\; PP_p^{\#} \overset{*}{\to} LC_p^{\#},\; PP_p(X) \overset{*}{\to} LC_p(X)—$$

because for $j \geq 1$, $p^j = p^{j-1}p$. Also, Statements (11) and (12) and the remarks in between, as well as Theorems 4, 6, 12 and its Corollary, 12*, and 13 apply with the symbol "PP" replacing "LC". Thus (15) can be extended to say that whenever m and n are each powers of p, then

$$PP_m \overset{*}{\to} LC_n,\ PP^{\#}_m \overset{*}{\to} LC^{\#}_n,\ PP_m(X) \overset{*}{\to} LC_n(X), \quad \text{for any set } X.$$

Theorem 8 is valid with the implication replaced by "$PP_m(X) \overset{*}{\to} LC_n(X)$". Theorem 9 goes over into the following:

THEOREM 9*. *Let $p \in \Pi$ and let X be any set which consists of sets A satisfying $p \preceq A$. Then*

$$((\mathrm{DAC})\ \&\ PP_p(X)\ \&\ (\forall j \in I_1)C(p^j)(X_{(p^j)})) \to FS_0(X).$$

Theorem 10 gets replaced by the following theorem, in which part (a) is considerably more general.

THEOREM 10*. *Let p and q be primes with $p < q$.*
(a) $PP_p \overset{*}{\to} FS_q$; $PP^{\#}_p \overset{*}{\to} FS^{\#}_q$; *for any set X,* $PP_p(X) \overset{*}{\to} FS_q(X)$.
(b) *Suppose $p \not| \, n$. Then for any set Y,* $PP_p(\mathscr{P}^{\#*}(Y)) \to FS_p(\mathscr{P}_n(Y))$.

Theorem 11 remains valid if "PP" is substituted for "LC"; in the new version we use the original version together with (15) to obtain those implications in which the conclusion is FS_0. In the proof of "$FS_0 \to PP_p$" there is a slight modification of the corresponding part of the proof of "$FS_0 \to LC_p$".

7. **Ordinal characterizations of multiple choice axioms.** We now formulate a precise definition of a multiple choice axiom of the type with which we have been concerned.

DEFINITION 14. Let $\Phi(X)$ be a formula of σ and let MC_Φ be the statement: "For every nontrivial set X there is a function f on X such that for each $A \in X$, $f(A) \in \mathscr{P}^{\#*}(A)$ and $\Phi(\mathbf{n}(f(A)))$". We call MC_Φ a *multiple choice axiom.*

In [**12**, Lemma 1], A. Lévy characterizes the multiple choice axioms $Z(n)$ (Definition 12, above) in terms of functions defined on ordinals; following Lévy's method, we now introduce some new axioms V_Φ and W_Φ which will similarly characterize the generalized multiple choice axioms MC_Φ.

DEFINITION 15. For any formula $\Phi(x)$ of σ, let V_Φ be the following statement: "For every nonempty set x there exists an ordinal number α and a function h on α such that for all $\beta \in \alpha$, $h(\beta) \in \mathscr{P}^{\#*}(x)$, $\bigcup h(\beta)(\beta \in \alpha) = x$, and $\Phi(\mathbf{n}(h(\beta)))$".

THEOREM 14. *For any formula $\Phi(x)$ of σ, $MC_\Phi \to V_\Phi$.*

PROOF. Let $\Phi(x)$ be a formula of σ and assume that MC_Φ is valid. Then if x is any nonempty set, there is a function f on $\mathscr{P}^*(x)$ such that for all $y \in \mathscr{P}^*(x)$, $f(y) \in \mathscr{P}^{\#*}(x)$ and $\Phi(\mathbf{n}(f(y)))$.

Define subsets h_β of x, by transfinite recursion, as follows:

$$\begin{aligned} h_\beta &= f(x \setminus \bigcup h_\gamma(\gamma \in \beta)) \quad \text{if } x \setminus \bigcup h_\gamma(\gamma \in \beta) \neq 0, \\ &= 0 \quad\quad\quad\quad\quad\quad\quad\quad\; \text{if } x = \bigcup h_\gamma(\gamma \in \beta). \end{aligned}$$

If $\beta_1 \neq \beta_2$, then $h_{\beta_1} \cap h_{\beta_2} = 0$. Further, if for every ordinal number β, $h_\beta \neq 0$, then the h_β would all be distinct subsets of x, and there would be a one-one mapping of the class of ordinal numbers into $\mathscr{P}^*(x)$. Contradiction!

Let α_0 be the smallest ordinal number α such that $h_\alpha = 0$. Clearly, $x = \bigcup h_\beta(\beta \in \alpha_0)$. Thus

$$h = \{\langle \beta, h_\beta \rangle : \beta \in \alpha_0\}$$

is a function on α_0; h together with α_0 satisfy the requirements of V_Φ.

DEFINITION 16. For any formula $\Phi(x)$ of σ, let W_Φ be the following statement: "Let x be any nontrivial set. Then there is a function H on x such that for each $y \in x$, $H(y) = \langle \alpha^y, h^y \rangle$, where α^y is an ordinal number and h^y is a function defined on α^y with the property that for all $\beta \in \alpha^y$, $h^y(\beta) \in \mathscr{P}^{\#*}(y)$, $\bigcup h^y(\beta)(\beta \in \alpha^y) = y$, and $\Phi(\mathbf{n}(h^y(\beta)))$".

THEOREM 15. *For any formula $\Phi(x)$ of σ, $MC_\Phi \leftrightarrow W_\Phi$.*

PROOF. Let $\Phi(x)$ be a formula of σ. Assume that MC_Φ is valid and let x be any nontrivial set. Let f be a function on x such that for all $y \in x$, $f(y) \in \mathscr{P}^{\#*}(x)$ and $\Phi(\mathbf{n}(f(y)))$.

For all $y \in x$, define the subsets $h_\beta(y)$ of y, by transfinite recursion, as follows:

$$\begin{aligned} h_\beta(y) &= f(y \setminus \bigcup h_\gamma(y))(\gamma \in \beta) \quad \text{if } y \setminus \bigcup h_\gamma(y)(\gamma \in \beta) \neq 0. \\ &= 0 \quad\quad\quad\quad\quad\quad\quad\quad\quad\quad \text{if } y = \bigcup h_\gamma(y)(\gamma \in \beta). \end{aligned}$$

If $\beta_1 \neq \beta_2$, then $h_{\beta_1}(y) \cap h_{\beta_2}(y) = 0$; if for every ordinal number β, $h_\beta(y) \neq 0$, then the $h_\beta(y)$ would all be distinct subsets of y, and there would be a one-one mapping of the class of ordinal numbers into $\mathscr{P}^{\#*}(y)$. Contradiction!

Let $\alpha_0(y)$ be the smallest ordinal number $\alpha(y)$ such that $h_{\alpha(y)}(y) = 0$. Then $y = \bigcup h_\beta(y)(\beta \in \alpha_0(y))$. $h^y = \{\langle \beta, h_\beta(y)\rangle : \beta \in \alpha_0(y)\}$ is a function on $\alpha_0(y)$.

Define g on x by $g(y) = \langle \alpha_0(y), h^y \rangle$, $y \in x$.

Now assume that W_Φ is valid, and let H realize W_Φ with respect to x. Then for each $y \in x$ we have an ordinal number a^y and a function h^y defined on α^y, such that for each $\beta \in \alpha^y$, $H(y) = \langle \alpha^y, h^y \rangle$, $h^y(\beta) \in \mathscr{P}^{\#*}(y)$, $\bigcup h^y(\beta)(\beta \in \alpha^y) = y$, and $\Phi(\mathbf{n}(h^y(\beta)))$. Since each $y \in x$ is nonempty, the corresponding ordinal number $\alpha^y > 0$.

Define f on x by $f(y) = h^y(0)$, $y \in x$.

8. **On generalized multiple choice axioms.** Let $\Pi(x)$ be a formula defining the set of primes. Let $\Lambda_p(x)$ be the formula,

$$\text{“}\Pi(p)\ \&\ (\exists y \in I_1)(x = yp)\text{”},$$

and let $\Xi_p(x)$ be the formula,

$$\text{“}\Pi(p)\ \&\ (\exists y \in I_1)(x = p^y)\text{”}.$$

Define $UF(n;\ p_1, p_2, \ldots, p_s)$ to be the $s + 1$-placed relation which is formally provable in σ iff $n \in I_2$ and $p_1, p_2, \ldots, p_s$ are all the distinct prime factors of n.

For $n \in I_2$, let $\Gamma_n(x)$ be the formula,

$$\text{“}UF(n;\ p_1, p_2, \ldots, p_s)\ \&\ x \in I_1\ \&\ \sim \Lambda_{p_1}(x)\ \&\ \sim \Lambda_{p_2}(x)\ \&\ \cdots\ \&\ \sim \Lambda_{p_s}(x)\text{”}.$$

Let $\Gamma_1(x)$ be the formula, "$x \in I_1$". Then for each $n \in I_1$, $FS_n \leftrightarrow MC_{\Gamma_n}$. It follows that for all $n \in I_1$, the axioms FS_n are characterizable in terms of functions defined on ordinal numbers (Theorems 14 and 15).

In order to characterize the axioms LC_n and PP_n in similar fashion, we must extend the notions of § 7 to cover the case of formulas involving two free variables. This is due to the fact that for each of the chosen sets $f(A)$, $\mathbf{n}(f(A))$ depends not only upon n but also upon A.

DEFINITION 14*. Let $\Phi(x, y)$ be a formula of σ and let MC_Φ be the statement: "For every nontrivial set X there is a function f defined on X such that for each $A \in X$, $f(A) \in \mathscr{P}^{\#*}(A)$ and $\Phi(A, \mathbf{n}(f(A)))$".

DEFINITION 15*. For any formula $\Phi(x, y)$ of σ, let V_Φ^* be the following statement: "For every nonempty set x there exists an ordinal number α and a function h on α such that for all $\beta \in \alpha$, $h(\beta) \in \mathscr{P}^{\#*}(x)$, $\bigcup h(\beta)\ (\beta \in \alpha) = x$, and $\Phi(x \backslash \bigcup h_\gamma\ (\gamma < \beta), \mathbf{n}(h(\beta)))$".

DEFINITION 16*. For any formula $\Phi(x, y)$ of σ, let W_Φ^* be the following statement: "Let x be any nontrivial set. Then there is a function H on x such that for each $y \in x$, $H(y) = \langle \alpha^y, h^y \rangle$, where α^y is an ordinal number and h^y is a function defined on α^y with the property that for all $\beta \in \alpha^y$, $h^y(\beta) \in \mathscr{P}^{\#*}(y)$, $\bigcup h^y(\beta)\ (\beta \in \alpha^y) = y$, and $\Phi(y \setminus \bigcup h^y(\gamma)\ (\gamma \in \beta), \mathbf{n}(h^y(\beta)))$".

THEOREM 14*. *For any formula $\Phi(x, y)$ of σ, $MC_\Phi \to V_\Phi^*$.*

THEOREM 15*. *For any formula $\Phi(x, y)$ of σ, $MC_\Phi \leftrightarrow W_\Phi^*$.*

The proofs of these two theorems are essentially the same as their counterparts for formulas with one free variable.

Now let $\Upsilon(x, y)$ be a formula expressing the existence of a one-one mapping of x into y. For $n \in I_2$, let $\Delta_n(x, y)$ be the formula,

$$\begin{aligned}&\text{“}UF(n;\ p_1, p_2, \ldots, p_s)\\ &\quad \&\ (\Upsilon(n, x) \to (y \in I_n\ \&\ (\exists y_1, y_2, \ldots, y_s \in I_2)(\Lambda_{p_1}(y_1)\ \&\ \Lambda_{p_2}(y_2)\ \&\ \cdots\ \&\ \Lambda_{p_s}(y_s)\\ &\qquad \&\ y = y_1 + y_2 + \cdots + y_s)))\ \&\ (\Upsilon(x, n) \to (\Upsilon(x, y)\ \&\ \Upsilon(y, x)))\text{”}.\end{aligned}$$

For $n \in I_2$, let $\Theta_n(x)$ be the formula,

$$\begin{aligned}&\text{“}UF(n;\ p_1, p_2, \ldots, p_s)\\ &\quad \&\ (\Upsilon(n, x) \to (y \in I_n\ \&\ (\exists y_1, y_2, \ldots, y_s \in I_2)(\Xi_{p_1}(y_1)\ \&\ \Xi_{p_2}(y_2)\ \&\ \cdots\ \&\ \Xi_{p_s}(y_s)\\ &\qquad \&\ y = y_1 + y_2 + \cdots + y_s)))\ \&\ (\Upsilon(x, n) \to (\Upsilon(x, y)\ \&\ \Upsilon(y, x)))\text{”}.\end{aligned}$$

Let $\Delta_1(x, y)$ and $\Theta_1(x, y)$ each be the formula, "$y \in I_1$". Then for each $n \in I_1$, $LC_n \leftrightarrow MC_{\Delta n}$ and $PP_n \leftrightarrow MC_{\Theta_n}$; thus, the axioms LC_n and PP_n, $n \in I_1$, are also characterizable in terms of functions defined on ordinal numbers.

We now generalize Theorem 6 of [**12**] as follows:

THEOREM 16. *If σ is consistent and if $\Psi(x)$ is a formula of σ with the property that there are infinitely many $n \in I_2$ for which the formula $\Pi(n)$ & $\sim\Psi(n)$ is provable in σ, then*

$$(FS_1 \ \& \ (\forall n \in I_1)\ C(n)) \rightarrow MC_{\Psi}$$

is unprovable in σ.[6]

PROOF. σ^* (see the proof of Theorem 1, above) is relatively consistent with σ; we shall work within σ^*.

The definitions of the $\mathscr{K}_\eta$, of $\mathfrak{G}_0$, and of $\phi(x)$ for $x \in \mathscr{K}_\eta \setminus \mathscr{K}_0$ are the same as in the proof of Theorem 1.

For $i \in I_0$ let p_i be the ith smallest prime. There is a one-one correspondence between $\mathscr{K}_0$ and the collection of ordered pairs $\langle p_i, l\rangle$ such that $i \in I_0$ and $l \in p_i$; let $k_{i,a}$ be the member of $\mathscr{K}_0$ corresponding to $\langle p_i, l\rangle$ under any such (fixed) mapping. Let $\mathscr{K}^{(i)} = \{k_{i,l} : l \in p_i\}$.

The remaining notions are the same as those in the proof of Theorem 1; the definition of the model $\mathscr{M}$ for σ is also the same. The comments concerning absoluteness carry over to the present case. The proof that FS_0 and $(\forall n \in I_1)\ C(n)$ are true in our interpretation is found in [**12**, Theorem 6].

We now show that the relativization of MC_Ψ is refutable in σ^*. MC_Ψ is replaced by a sentence equivalent to "For every nonempty $\mathscr{M}$-element which is a set of nonempty sets, there exists an $\mathscr{M}$-element f which is a function such that for each $y \in x$, $f(y)$ is a nonempty, finite subset of y and $\Psi(\mathbf{n}(f(y)))$."

Let x be the $\mathscr{M}$-element $\{\mathscr{K}^{(i)} : i \in I_0\}$. Suppose that f is an $\mathscr{M}$-element which is a function such that for each $y \in x$, $f(y)$ is a nonempty finite subset of y and $\Psi(\mathbf{n}(f(y)))$.

Let Z be a finite subset of I_0 with the property that f is Z-symmetric. Choose j such that $j \notin Z$ and $\Psi(p_j)$ is false. Since χ_j is Z-identical, $\chi_j(f) = f$. We have $\langle \mathscr{K}^{(j)}, f(\mathscr{K}^{(j)})\rangle \in f$; hence

$$\langle \mathscr{K}^{(j)}, \chi_j(f(\mathscr{K}^{(j)}))\rangle = \chi_j(\langle \mathscr{K}^{(j)}, f(\mathscr{K}^{(j)})\rangle) \in \chi_j(f) = f.$$

Now, $\mathbf{n}(\mathscr{K}^{(j)}) = p_j$ and $\Psi(p_j)$ is false; hence $f(\mathscr{K}^{(j)}) \neq \mathscr{K}^{(j)}$. Thus $\chi_j(f(\mathscr{K}^{(j)})) \neq f(\mathscr{K}^{(j)})$, and we have the distinct $\mathscr{M}$-elements

$$\langle \mathscr{K}^{(j)}, f(\mathscr{K}^{(j)})\rangle,\ \langle \mathscr{K}^{(j)}, \chi_j(f(\mathscr{K}^{(j)}))\rangle \in f.$$

This contradicts the assumption that f is a function. It follows that the relativiza- of MC_Ψ is refutable in σ^*.

---

[6] Again, σ can be replaced by the system A, B, C of [**9**] in either the hypothesis or the conclusion; see Footnote 2.

BIBLIOGRAPHY

1. P. T. Bateman, *Remark on a recent note on linear forms*, Amer. Math. Monthly **65** (1958), 517–518.

2. M. N. Bleicher, *Some theorems on vector spaces and the axiom of choice*, Fund. Math. **54** (1964), 95–107.

3. ———, *Multiple choice axioms and the axiom of choice for finite sets*, Fund. Math. **57** (1965), 247–252.

4. A. T. Brauer, *On a problem of partitions*. I, Amer. J. Math. **64** (1954), 298–312.

5. A. T. Brauer and B. M. Seelbinder, *On a problem of partitions*. II, Amer. J. Math. **76** (1954), 343–346.

6. A. T. Brauer and J. E. Shockley, *On a problem of Frobenius*, J. Reine Angew. Math. **211** (1962), 215–220.

7. A. L. Dulmage and N. S. Mendelsohn, *Gaps in the exponential set of primitive matrices*, Illinois J. Math. **8** (1964), 642–656.

8. K. Gödel, *The consistency of the axiom of choice and of the generalized continuum-hypothesis*, Proc. Nat. Acad. Sci. U.S.A. **24** (1938), 556–557.

9. ———, *The consistency of the axiom of choice and of the generalized continuum-hypothesis with the axioms of set theory*. 6th ed., Ann. of Math Studies No. 3, Princeton Univ. Press, Princeton, N.J., 1964.

10. S. M. Johnson, *A linear diophantine problem*, Canad. J. Math. **12** (1960), 390–398.

11. A. Lévy, *The independence of various definitions of finiteness*, Fund. Math. **46** (1958), 1–13.

12. ———, *Axioms of multiple choice*, Fund. Math. **50** (1962), 475–483.

13. E. Mendelson, *The independence of a weak axiom of choice*, J. Symbolic Logic **21** (1956), 350–366.

14. A. Mostowski, *Über die Unabhängigkeit des Wohlordnungssatzes vom Ordnungsprinzip*, Fund. Math. **32** (1939), 201–252.

15. ———, *Axiom of choice for finite sets*, Fund. Math. **33** (1945), 137–168.

16. J. B. Roberts, *Note on linear forms*, Proc. Amer. Math. Soc. **7** (1956), 465–469.

17. W. Sierpiński, *Cardinal and ordinal numbers*, 1st ed., Monografe Matematyczne, 34, PWN, Warsaw, 1958.

18. W. Szmielew, *On choices from finite sets*, Fund. Math. **34** (1947), 75–80.

19. K. Wiśniewski, *Remark on the axiom of choice for finite sets*, Bull. Acad. Polon. Sci. Sér. Sci. Math. Astronom. Phys. **15** (1967), 373–375.

20. M. M. Zuckerman, *Bertrand's postulate and conditions* (*M*) *and* (*S*) *for the axioms of choice for finite sets*, Illinois J. Math. (to appear).

21. ———, *A unifying condition for implications among the axioms of choice for finite sets*, Pacific J. Math. **28** (1969), 233–242.

22. ———, *Some theorems on the axioms of choice for finite sets* (to appear in Z. Math. Logik Grundlagen Math.)

23. ———, *On choosing subsets of n-element sets* (to appear in Fund. Math.)

NEW YORK UNIVERSITY AND
THE CITY COLLEGE OF THE CITY UNIVERSITY OF NEW YORK

Author Index

Roman numbers refer to pages on which a reference is made to an author or a work of an author.

Italic numbers refer to pages on which a complete reference to a work by the author is given.

Boldface numbers indicate the first page of the articles in the book.

Subject Index